Quantum Paths

Festschrift in Honor of

Berge Englert

on His 60th Birthday

QUANTUM PATHS

Festschrift in Honor of
Berge Englert
on His 60th Birthday

Edited by

Rui Han
Centre for Quantum Technologies,
National University of Singapore

Hui Khoon Ng
Yale-NUS College and Centre for Quantum Technologies,
National University of Singapore

World Scientific

NEW JERSEY • LONDON • SINGAPORE • BEIJING • SHANGHAI • HONG KONG • TAIPEI • CHENNAI

Published by

World Scientific Publishing Co. Pte. Ltd.

5 Toh Tuck Link, Singapore 596224

USA office: 27 Warren Street, Suite 401-402, Hackensack, NJ 07601

UK office: 57 Shelton Street, Covent Garden, London WC2H 9HE

Library of Congress Cataloging-in-Publication Data

Quantum paths : festschrift in honor of Berge Englert on his 60th birthday / [edited by] Hui Khoon Ng (Yale-NUS College, Singapore), Rui Han (Centre for Quantum Technologies, Singapore).

pages cm

ISBN 978-9814651837 (hard cover : alk. paper)

1. Quantum theory. 2. Englert, Berthold-Georg, 1953– honouree. I. Ng, Hui Khoon, editor. II. Han, Rui, 1985 or 1986– editor. III. Englert, Berthold-Georg, 1953– honouree.

QC173.97.Q36 2015

530.12--dc23

2014049413

British Library Cataloguing-in-Publication Data

A catalogue record for this book is available from the British Library.

Printed in Singapore

Preface

Age is strictly a case of mind over matter. If you don't mind, it doesn't matter.
American comedian, Jack Benny

And in the end, it is not the years in your life that count, it's the life in your years.
Abraham Lincoln

Sixty years: Many years of joy, tears, love, hate, wisdom, and folly, and many more to go. The life in Berge's years is certainly a rich one, but an inkling of which can be glimpsed from this collection of contributions. While old friends and colleagues from all over the world gathered here in Singapore for his 60th birthday conference (Berge Fest, 22–26 April 2014, NUS UTown), we collected contributions in celebration of Berge as a scientist, a teacher, and a friend. The contributions gathered here give a sense of how many of our lives Berge has touched, and in what significant ways.

The collection comprises scientific articles — either written by Berge, or inspired by his work — of particular impact suggested by contributors, as well as personal essays recounting past interactions with Berge. As far as possible, we have sought to include every journal article or book chapter in its entirety, but for some of the longer articles, this desire has proven impossible to reconcile with the goal of having a reasonably sized volume. In such cases, we have included only the first few pages of the article, in the hope that the reader will download his/her own copy directly from the journal or book website. We thank all contributors for thoughtful recollections and recommendations, giving us all insights into the many wondrous facets of Berge's life in these sixty years past. We are also grateful to the journal and book publishers for granting us permission to reprint the articles and book chapters included in this volume.

To Berge: A simple Happy Birthday would hardly suffice to express all our well-wishes; yet simplicity might serve well to convey our uncomplicated sense of utmost gratitude and admiration for you. So, Happy Birthday, Berge!

Rui Han
Hui Khoon Ng

Centre for Quantum Technologies
National University of Singapore
April 2014

Berge Fest, 26 April 2014, University Town, National University of Singapore

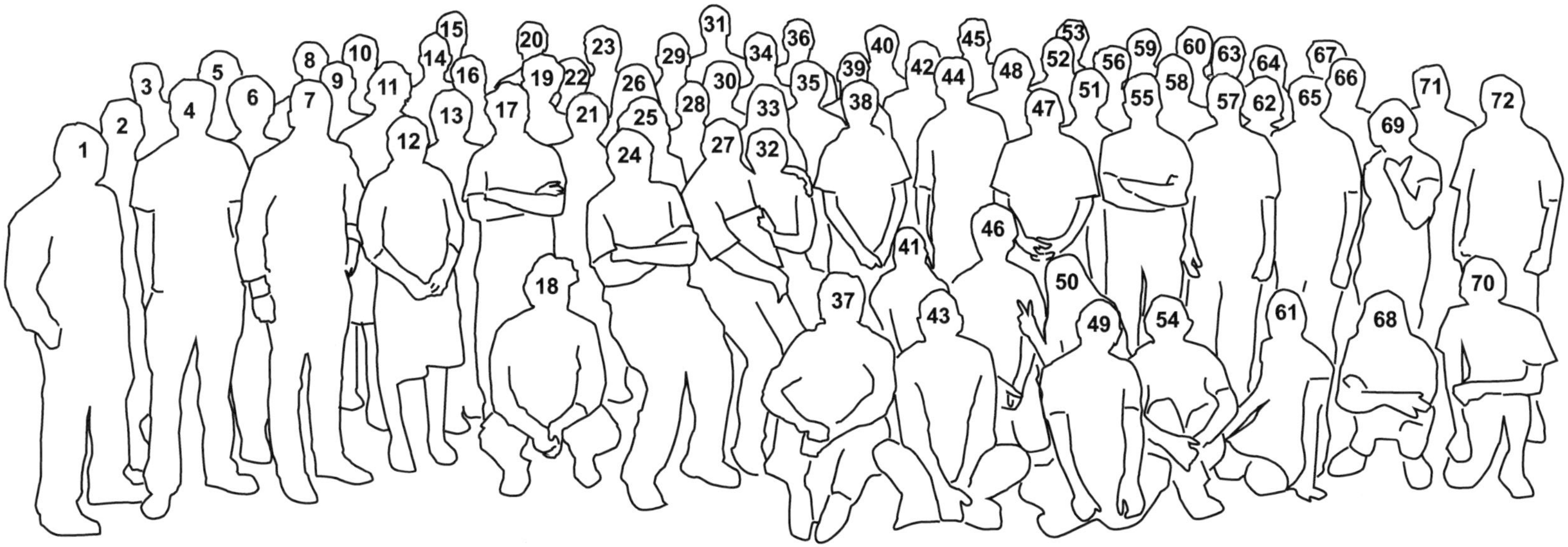

[1] Leong Chuan Kwek
[2] Yon Shin Teo
[3] Tomasz Karpiuk
[4] Amit Rai
[5] Helmut Rauch
[6] Tomotaka Kuwahara
[7] Hans Briegel
[8] Michael Fleischhauer
[9] Marek Wajs
[10] Georg Meyer
[11] Ewan Munro
[12] Bess Fang
[13] Jun Suzuki
[14] Nana Siddarth
[15] Heinz Siedentop
[16] Dario Poletti
[17] Benoît Grémaud
[18] Mathias Albert
[19] Wei Hou Tan
[20] Andreas Buchleitner
[21] Yink Loong Len
[22] Max Seah
[23] Thomas Durt
[24] Roy Glauber
[25] Shabnam Safaei
[26] Hui Khoon Ng
[27] Berge Englert
[28] Yun Li
[29] Luyao Ye
[30] Jiangwei Shang
[31] Rüdiger Schack
[32] Ola Englert
[33] Rui Han
[34] Sanjib Ghosh
[35] Jibo Dai
[36] Philippe Raynal
[37] Enrique Solano
[38] Mile Gu
[39] Martin Trappe
[40] Gerd Leuchs
[41] Jayne Thompson
[42] Kazik Rzążewski
[43] Christian Miniatura
[44] Yong Siah Teo
[45] David Hutchinson
[46] Sadegh Raeisi
[47] Lev Vaidman
[48] Reinhard Werner
[49] Patrick Coles
[50] Evon Tan
[51] Huangjun Zhu
[52] Wilhelm Becker
[53] Christophe Couteau
[54] Chang Jian Kwong
[55] Saverio Pascazio
[56] Luigi Amico
[57] Ady Mann
[58] Miguel Orszag
[59] Jook Walraven
[60] Zdének Hradil
[61] Si-Hui Tan
[62] Ulrike Herzog
[63] Sai Vinjanampathy
[64] Marek Kuś
[65] Ulrich Mohrhoff
[66] Karol Życzkowski
[67] *Nomen Nescimus*
[68] Ling Ming Zhao
[69] Danuta Rzążewska
[70] Wei Huang
[71] Jarda Řeháček
[72] Chu Guo

Berthold-Georg (Berge) Englert
Centre for Quantum Technologies and Department of Physics,
National University of Singapore

Contents

Acknowledgements

Various journal articles and book chapters contained in this volume are reprinted with permission of the following publishers:

- The American Physical Society

  Physical Review Letters **81**, 5705–5709 (1998); **91**, 037901 (2003); **107**, 020404 (2011); **110**, 260502 (2013); **113**, 170402 (2014).
  Physical Review A **47**, 3311–3329 (1993); **50**, 783–786 (1994); **65**, 040102(R) (2002); **70**, 032306 (2004); **70**, 052321 (2004); **74**, 022309 (2006); **80**, 043411 (2009); **81**, 052339 (2010); **85**, 052116 (2012).
- Springer (with kind permission of Springer Science+Business Media)

  Letters in Mathematical Physics **37**, 293–308 (1996) (Kluwer Academic Publishers).
  Foundations of Physics **28**, 549–559 (1998) (Plenum Publishing Corporation).
  The European Physical Journal D **67**, 238 (2013) (EDP Sciences, Società Italiana di Fisica, Springer-Verlag).
- European Physical Society (EDP Sciences, IOP Publishing and the Società Italiana di Fisica)

  EPL (Europhysics Letters) **104**, 50008 (2013).
- Taylor & Francis Ltd (`www.tandfonline.com`)

  Journal of Modern Optics **45**, 701–711 (1998); **60**, 255 (2013).
- Optical Society of America

  Optics Letters **39**, 2668–2671 (2014).
- Elsevier Science B. V.

  Physics Letters A **128**, 391–394 (1988); **284**, 1–5 (2001) (with permission from Elsevier).
  Nuclear Instruments & Methods in Physics Research A **440**, 562–567 (2000).
  Optics Communications **208**, 139–144 (2002).

- Cambridge University Press

  Science and Ultimate Reality: quantum theory, cosmology, and complexity, eds. J. D. Barrow, P. C. W. Davies and C. L. Harper, Jr., Chapter 15, 306–329 (2004).

- Nature Publishing Group (Macmillan Publishers Ltd)

  Nature **395**, 33–37 (1998).

- World Scientific Publishing Company

  International Journal of Quantum Information **8**, 535–537 (2010).

- Verlag der Zeitschrift für Naturforschung

  Zeitschrift für Naturforschung **47a**, 1175–1186 (1992); **56a**, 16–19 (2001).

- Taylor and Francis Group LLC Books (Chapman & Hall/CRC)

  Mathematics of Quantum Computation and Quantum Technology, eds. G. Chen, L. Kauffman and S. J. Lomonaco, Chapter 11, 309–310 (2007) (permission conveyed through Copyright Clearance Center, Inc.).

- Wiley-VCH Verlag GmbH & Co. KGaA

  Fortschritte der Physik **61**, 225–237 (2013).

- IOP Publishing Ltd and Deutsche Physikalische Gesellschaft

  New Journal of Physics **15**, 123026 (2013).

Chapter 1

Quantum Optics and Open Quantum Systems

Hans Briegel

Institut für Quantenoptik und Quanteninformation der Österreichischen Akademie der Wissenschaften, Innsbruck, Austria
Institut für Theoretische Physik, Universität Innsbruck, Innsbruck, Austria

Paper Title: **Quantum optical master equations: The use of damping bases**

Reference: Physical Review A **47**, 3311–3329 (1993)

Personal message:

This damping-bases paper was one of my first papers with Berge, who was my PhD supervisor and who introduced me to the field of quantum optics.

The paper introduced a new method for solving master equations that appear in quantum optics. It contains the full analytical solution of the "Jaynes-Cummings model with damping", which describes the interaction of a single atom with a quantized mode of the cavity field, when the field and the atom are subjected to damping and spontaneous decay, respectively. This model played an important role in cavity QED experiments and in the micromaser, which were studied at that time in the groups of Herbert Walther at MPQ in Garching and Serge Haroche at ENS in Paris. The paper became a starting point for a series of theoretical investigations of the micromaser. It allowed us, for example, to exactly compute the quantum state of the laser field under realistic experimental conditions, and to analyze the effect of non-Poissonian pumping on the laser field and its spectrum. It also enabled us to predict the properties of a single-atom laser and to explain the effect of dynamic noise reduction in lasers with multi-level pumping. Beyond laser physics and quantum optics, the damping bases method has seen many other applications, most recently in quantum information and in the analysis of quantum channels with memory.

Since there is only limited space in this Festschrift, I leave it to the editors to deal with a problem that Berge and I never addressed (and, fortunately, did not even think of), namely, how to shorten or compress this paper to the usual four pages ... (Editors' note: Since there is no need to subscribe to the illustrious 4-page limit here, we have included the 19-page article in its entirety, trusting readers to appreciate the deeper understanding one gains from the additional details, and not mind the extra weight of those 15 pages.)

Dear Berge,
You were a marvelous teacher and supervisor. I felt very proud and I was very lucky that you accepted me as your first PhD student. Working with you was a unique experience and rewarding in every respect. I don't remember any meeting with you which would not result in a new insight, a helpful calculation, or a useful advice that guided me in the right direction. Also from a personal perspective, the PhD was a very special time of my life. Working at the Coulombwall in Garching, with my desk facing towards the Isar meadows, I could experience theoretical research as it ought to be, with enough time to think and deeply concentrate on a problem. I was happy to have you next door, not only for advice, but also to bring me occasionally back down to Earth and go for a Leberkäs at the Cafeteria in the TU Mensa.

Thank you for this wonderful time, for our collaborations, and for your friendship. Happy Birthday!

With best wishes,
Hans

PHYSICAL REVIEW A VOLUME 47, NUMBER 4 APRIL 1993

Quantum optical master equations: The use of damping bases

Hans-Jürgen Briegel and Berthold-Georg Englert*
Sektion Physik, Universität München, Am Coulombwall 1, D-8046 Garching, Germany
(Received 21 July 1992)

We present the complete analytical solution of the cavity QED of a two-level atom and a field mode at zero temperature. It includes both dissipation of the field due to a finite Q value of the cavity and incoherent decay mechanisms for the atom. This analytical solution is provided by a powerful method for treating general master equations that appear in quantum optics. As distinct from the usual approaches we first deal with that part of the master equation which describes the dissipative coupling of the field and the atom to their reservoirs. Rather than using number-state or dressed-state bases we expand the density operator into the eigenstates of the nonunitary parts of the master equation which model the dissipative part of the dynamics. The set of these eigenstates is the damping basis. With the aid of this expansion we find the eigenvalues and eigenstates of the total Liouville operator. The evolution of an arbitrary initial state is then known. We employ these results to give an exact solution of the dynamics of the photon field in realistic experiments with one-atom masers at very low temperatures. It includes detuning, cavity leakage effects, spontaneous decay mechanisms for the atoms, a Fizeau-type velocity distribution for the atomic beam, and a statistical parameter for the probability of the excitation of incoming atoms, covering the limits of Poissonian pumping and of regular pumping. On the same grounds one can treat the one-atom laser, consisting of a single atom which stays in permanent interaction with the field mode and which is continuously pumped by external heat baths.

PACS number(s): 42.50.−p, 42.55.−f, 32.80.−t

I. INTRODUCTION

A. The problem

The dynamical evolution of a two-level atom interacting with one mode of the quantized photon field in a cavity or in a laser is well described by a master equation of the structure

$$\frac{\partial}{\partial t}P=\frac{1}{i\hbar}[H,P]+L_aP+L_\sigma P \tag{1.1}$$

in which the symbol P denotes the density operator that specifies the state of the system. The right-hand side consists of two parts, a unitary one and a nonunitary one. The unitary part involves the commutator with the Hermitian Hamilton operator

$$H=\hbar\omega a^\dagger a+\tfrac{1}{2}\hbar\Omega\sigma_z-\tfrac{1}{2}\hbar g(a^\dagger\sigma_-+a\sigma_+) \tag{1.2}$$

for the interaction between the photon mode with frequency ω and the atom with level spacing $\hbar\Omega$. The dynamical variables are the ladder operators $a^\dagger, a$ for the field and

$$\sigma_\pm=\sigma_x\pm i\sigma_y=(\sigma_\mp)^\dagger \tag{1.3}$$

for the atom, where σ_x, σ_y, and also σ_z are Pauli's spin operators. The constant g is the Rabi frequency that measures the coupling strength of this interaction.

The nonunitary part

$$LP=L_aP+L_\sigma P \tag{1.4}$$

accounts for losses to and gains from reservoirs. The first term

$$L_aP=-\frac{A}{2}(\nu+1)[a^\dagger aP+Pa^\dagger a-2aPa^\dagger]-\frac{A}{2}\nu[aa^\dagger P+Paa^\dagger-2a^\dagger Pa]\,, \tag{1.5}$$

with constants $A,\nu\geq 0$, describes the coupling of the field to a thermal reservoir at a temperature which corresponds to a mean number ν of thermal photons in the cavity or the laser. The second term reads

$$L_\sigma P=-\frac{B}{8}(1-s)[\sigma_+\sigma_-P+P\sigma_+\sigma_--2\sigma_-P\sigma_+]-\frac{B}{8}s[\sigma_-\sigma_+P+P\sigma_-\sigma_+-2\sigma_+P\sigma_-]-\frac{2C-B}{4}[P-\sigma_zP\sigma_z]\,, \tag{1.6}$$

with constants $2C\geq B\geq 0$ and $0\leq s\leq 1$; it models incoherent pumping and decay processes of the atom.

When $A,B,C=0$, (1.1) is the master equation of the familiar Jaynes-Cummings model [1] whose eigenvalues are well known.

When $g=0$, on the other hand, (1.1) describes pure relaxation of the oscillating mode and the atomic system, which then are decoupled. In this situation (and only then) the constant A gives the decay rate of the mean number $\langle a^\dagger a\rangle$ of photons in the cavity or in the laser towards the thermal-state value $\langle a^\dagger a\rangle_\infty=\nu$. The constants B and C are the corresponding atomic decay rates for the inversion measure $\langle\sigma_z\rangle$ and the polarization $\langle\sigma_\pm\rangle$ towards their equilibrium values $\langle\sigma_z\rangle_\infty=2s-1$ and $\langle\sigma_\pm\rangle_\infty=0$. Note that the third term in (1.6), involving the constant C, accounts for processes, which are sometimes termed *virtual*, such as atomic collisions that lead to a loss of atomic coherence without changing the level population. The parameter s, ranging from 0 to 1, further characterizes the atomic reservoir. If $s=0$, (1.6) only effects transitions from the upper to the lower level, modeling processes such as incoherent transitions due to spontaneous emission of photons into modes other than the privileged cavity (or laser) mode, or transitions induced by atomic collisions and electric scattering fields in a cavity. When $s>0$, there are also transitions from the lower to the upper level. The resulting equilibrium value of the inversion measure, $\langle\sigma_z\rangle_\infty=2s-1$, grows with increasing s. In particular, when $s>\frac{1}{2}$, the atomic bath effectively acts like a (incoherent) pumping reservoir under whose action the atomic population becomes inverted (i.e., the inversion becomes positive). So s can be called a pumping parameter, which is to be explained in more detail later on.

So far the total system under consideration consists of a single atom (Ω) interacting with a single field mode (ω), both being coupled to their specific reservoir. The characterizing constants are the system parameters ω,Ω,ν,s and the coupling constants g,A,B,C. To model a real laser, whose active medium consists of N two-level atoms, one could think of employing a well-established method given by Weidlich, Risken, and Haken [2] and by Risken [3] who, roughly speaking, replace the operators of the microscopic inversion, σ_z, and polarization, $\sigma_\pm$, by the macroscopic sums

$$S_z=\sum_{j=1}^{N}\sigma_z^j,\quad S_\pm=\sum_{j=1}^{N}\sigma_\pm^j \tag{1.7}$$

and discard any correlations $\langle\sigma^j\sigma^k\sigma^l\cdots\rangle$, $j\neq k\neq l\neq\cdots$, between the observables of different atoms. We shall return to this in a forthcoming paper with Ginzel and Schenzle [4].

As it stands, the master equation (1.1), together with (1.2), (1.5), and (1.6), is the basic equation of the quantum theory of the interaction between atoms and electromagnetic fields which are part of a thermodynamically open system. For $s=0$ it describes the cavity QED of an atom in a cavity with finite Q value and provides, in particular, the exact description of the dynamics in a one-atom maser. There, alternating periods of atomic interaction with the cavity field and of the pure decay of the latter obey Eq. (1.1) with $g\neq 0$ and $g=0$, respectively.

When $s>0$, an additional continuously pumping reservoir for the atom is included which leads to a positive atomic inversion when $s>\frac{1}{2}$ and thus describes a one-atom laser, which could be generalized to the many-atom laser by the standard method described in [2] and [3].

In the present paper we present a powerful approach for solving the master equation (1.1). In contrast to the standard methods which resort to approximate Fokker-Planck-type equations we employ an algebraic method that transforms (1.1) into coupled systems of 4×4-matrix equations. In the important case when $\nu=0$ and $s=0$, which is realized in the ongoing experiments with the one-atom maser [5], these can be solved *analytically*. In the most general situation, one is eventually led to 4×4-matrix continued fractions which can be solved numerically (with little effort) for any given set of parameters $\omega,\Omega,\nu,s,g,A,B,C$ without the need of additional structural approximations.

B. The method

In order to obtain all dynamical information about the atom-and-field coupled system described by the master equation (1.1) it is expedient to solve the corresponding eigenvalue equation

$$\mathcal{L}P\equiv\frac{1}{i\hbar}[H,P]+LP=\Gamma P\ , \tag{1.8}$$

where $\mathcal{L}$ denotes the total Liouville operator, L is the nonunitary part (1.4), and Γ is a, possibly complex, eigenvalue of $\mathcal{L}$. A note on the terminology is in order. In this paper we will consistently call P the *state* of the system, and L and the commutator with H are linear *operators* acting on this state. Of course, P itself is a linear operator acting on the Hilbert space *vectors*, and L or $\mathcal{L}$ are therefore called superoperators by some authors. We hope there will be no confusion.

When the system is closed, $L=0$, (1.8) refers to the Jaynes-Cummings model whose solutions are well known. The eigenstates of the commutator in (1.8) are then given by dyadic products of eigenvectors of H, popularly called "dressed" vectors.

When the system is open, $L\neq 0$, the nonunitary part that models the system's coupling to external reservoirs makes things more difficult. Although the master equation (1.1) does not describe a unitary evolution of the state P, the solution of the eigenvalue equation (1.8) nevertheless provides us with the solution of the time-dependent problem (1.1).

Suppose we have solved (1.8) and found all eigenvalues Γ and the corresponding eigenstates P_Γ. Then, if we can write the expansion

$$P(0)=\sum_\Gamma \check{c}_\Gamma P_\Gamma \tag{1.9}$$

for the initial state given at time $t=0$, we know the state of the system at any later time, namely,

$$P(t)=\sum_\Gamma \check{c}_\Gamma e^{\Gamma t}P_\Gamma\ . \tag{1.10}$$

Here and in the sequel the symbolic sum over Γ also properly takes account of the different eigenstates belonging to a degenerate eigenvalue. Thus, once (1.8) is solved, the remaining problem is, how does one find the expansion coefficients $\check{c}_\Gamma$ appearing in (1.9)? The answer is

$$\check{c}_\Gamma=\mathrm{Tr}\{\check{P}_\Gamma P(0)\}\ , \tag{1.11}$$

where Tr, the "total trace," symbolizes the injunction to trace over both the photon and the atom variables, and the states $\check{P}_\Gamma$ satisfy the duality relation

$$\mathrm{Tr}\{\check{P}_\Gamma P_{\Gamma'}\}=\delta_{\Gamma\Gamma'}\ . \tag{1.12}$$

Accordingly, we call the states $\check{P}_\Gamma$ *dual* to the eigenstates P_Γ.

In order to find these dual states, consider the time evolution of the expectation value of an observable $\mathcal{O}$.

An immediate consequence of the master equation (1.1) is

$$\frac{d}{dt}\langle \mathcal{O}\rangle = \frac{d}{dt}\mathrm{Tr}\{\mathcal{O}P\} = \mathrm{Tr}\left[\mathcal{O}\frac{\partial}{\partial t}P\right] = \mathrm{Tr}\{\mathcal{O}\mathcal{L}P\} \ , \quad (1.13)$$

provided that $\mathcal{O}$ is not explicitly time dependent itself. The dual conjugate $\check{\mathcal{L}}$ to $\mathcal{L}$ is the operator that obeys

$$\mathrm{Tr}\{\mathcal{O}\mathcal{L}P\} = \mathrm{Tr}\{(\check{\mathcal{L}}\mathcal{O})P\} = \mathrm{Tr}\{P\check{\mathcal{L}}\mathcal{O}\} \quad (1.14)$$

for all states P and all observables $\mathcal{O}$ [6]. Now, $\check{\mathcal{L}}$ has eigenstates, too, and these are just the states which we have denoted by $\check{P}_\Gamma$,

$$\check{\mathcal{L}}\check{P}_\Gamma = \Gamma\check{P}_\Gamma \ , \quad (1.15)$$

where the Γ are identical with the eigenvalues of $\mathcal{L}$. This can be seen immediately by inserting $P_{\Gamma'}$ and $\check{P}_\Gamma$, instead of P and $\mathcal{O}$, into (1.14), thereby confirming the duality relation (1.12).

One can read (1.14) as if $\mathcal{L}$ were acting to the left on $\mathcal{O}$ according to

$$\mathcal{O}\mathcal{L} = \check{\mathcal{L}}\mathcal{O} \ . \quad (1.16)$$

Then $\check{P}_\Gamma$ is simply the left eigenstate (or, more frivolously, "eigen-bra") to $\mathcal{L}$ with eigenvalue Γ,

$$\check{P}_\Gamma\mathcal{L} = \Gamma\check{P}_\Gamma \ , \quad (1.17)$$

whereas P_Γ is the corresponding right eigenstate (or "eigen-ket"),

$$\mathcal{L}P_\Gamma = P_\Gamma\Gamma \ . \quad (1.18)$$

In this context, the notion of duality is quite familiar.

The right eigenstates will be employed to expand arbitrary states P with the aid of the completeness relation

$$P = \sum_\Gamma P_\Gamma \mathrm{Tr}\{\check{P}_\Gamma P\} \ . \quad (1.19)$$

Likewise, the left eigenstates are used to expand arbitrary observables $\mathcal{O}$,

$$\mathcal{O} = \sum_\Gamma \check{P}_\Gamma \mathrm{Tr}\{\mathcal{O}P_\Gamma\} \ . \quad (1.20)$$

The duality relation (1.12) then supplies the "coordinate sum"

$$\langle \mathcal{O}\rangle = \mathrm{Tr}\{\mathcal{O}P\} = \sum_\Gamma \mathrm{Tr}\{\mathcal{O}P_\Gamma\}\mathrm{Tr}\{\check{P}_\Gamma P\} \quad (1.21)$$

for the expectation value $\langle \mathcal{O}\rangle$. Of course, for a given $\mathcal{L}$ the completeness of either set of eigenstates has to be established.

In summary, the master equation (1.1) is solved in general, if together with their eigenvalues all the eigenstates P_Γ and $\check{P}_\Gamma$ of the Liouville operator $\mathcal{L}$ and its dual $\check{\mathcal{L}}$, respectively, are found. Then, for a given initial state $P(0)$ the state at a later time t reads

$$P(t) = e^{\mathcal{L}t}P(0) \equiv \sum_\Gamma \mathrm{Tr}\{\check{P}_\Gamma P(0)\}e^{\Gamma t}P_\Gamma = \sum_\Gamma \check{c}_\Gamma e^{\Gamma t}P_\Gamma \ . \quad (1.22)$$

The temporal evolution of the expectation value of an observable

$$\mathcal{O} = \sum_\Gamma c_\Gamma \check{P}_\Gamma \quad (1.23)$$

is correspondingly computed according to

$$\langle \mathcal{O}\rangle_t = \mathrm{Tr}\{\mathcal{O}e^{\mathcal{L}t}P(0)\} \equiv \sum_\Gamma c_\Gamma e^{\Gamma t}\check{c}_\Gamma \ . \quad (1.24)$$

Now let us return to the question of how one, actually, solves the eigenvalue problem (1.8). As a first step one would expand the equation into eigenstates of an operator which is only a part of the total operator $\mathcal{L}$ and whose eigenstates are known. Conventionally this is the commutator with the free part of H, i.e., the atomic inversion σ_z and the photon number $a^\dagger a$, which would lead to a number-state representation of (1.8), or one could use the total Hamiltonian H, leading to an expansion of (1.8) into a dressed-state basis. In any case, the resulting systems of coupled differential equations for the expansion coefficients become rather involved.

In this paper, we will pursue a different approach, which consists in systematically using bases provided by the eigenstates of the *nonunitary* parts (summarized in L) of the total Liouville operator $\mathcal{L}$. This is a straightforward and powerful ansatz. It eventually supplies the complete analytic solution of the master equation in the important special case when $\nu, s = 0$, and substantially simplifies the numerical treatment of the master equation (1.1) in its most general form by reducing the problem to the calculation of 4×4-matrix continued fractions without any further structural approximations.

So the strategy is the following. First we investigate the eigenvalue problem posed by the nonunitary part (1.4)–(1.6) of the master equation:

$$LP_\lambda = \lambda P_\lambda \ , \quad (1.25)$$

where the symbols λ for the eigenvalues and P_λ for the eigenstates are used to distinguish them from the solutions of the total Liouville operator. Similar to (1.12) there are left eigenstates $\check{P}_\lambda$ of L, too, the right eigenstates of the dual conjugate $\check{L}$. We find that both the eigenstates P_λ and their duals $\check{P}_\lambda$ form complete sets into which any observable of the system, and in particular its state $P(t)$, can be expanded. It turns out that such an expansion is a very natural one, since all essential statistical information about the field is comprised in its first few coefficients only. In the language of phase-space functions this method removes, in a certain sense, the terms involving second-order derivatives in the corresponding Fokker-Planck-type equations from the very beginning and leaves one with differential equations of first order only.

C. Outline

In Sec. II we investigate the nonunitary parts L_a and L_σ of the master equation in detail. We find that L_a and L_σ together possess a complete set of eigenstates into which any function of the dynamical variables $a^\dagger, a, \sigma_+, \sigma_-$ can be expanded. These eigenstates and their eigenvalues are derived explicitly and their properties discussed. In Sec. III these results are used to attack the actual problem posed by Eq. (1.1). Upon expanding the state

$$P=P(a^\dagger,a,\sigma_+,\sigma_-,t) \tag{1.26}$$

into said eigenstates of $L=L_a+L_\sigma$ the master equation transforms into coupled systems of ordinary differential equations of first order for the expansion coefficients. Conveniently arranged, these equations form groups that do not mix. Finally one is left with separate systems of 4×4-matrix recurrence relations which can be solved by continued fractions or even analytically. In the following sections this general frame is further developed and employed for a rigorous treatment of cavity quantum electrodynamics and laser physics.

First, in Sec. IV the analytic solutions are presented. They supply a complete and exact description of the interaction of a two-level atom and a photon mode in a cavity at zero temperature, including leakage effects for the photons out of the cavity and incoherent transitions for the atom.

This is of particular interest for an exact description of the dynamics in the one-atom maser which is the subject of Sec. V. As the damping constants enter the system's eigenfrequencies, which are generalized Rabi frequencies, the trapping-state condition is modified, too. In this approach both parts of a one-atom-maser cycle are treated on almost equal footing: the interaction period and the period of pure decay of the cavity field are both described by the same equations with values $g\neq0$ and $g=0$, respectively, the latter being elementary. In particular, future experiments with highly regular pumping, where a second atom enters the cavity immediately after the preceding one has left it, cannot be described by current one-atom-maser theories which ignore cavity damping while the atom is present.

Finally, Sec. VI gives an outlook to the situation at finite temperature and when the atom is coupled to a heat bath which serves as a pumping reservoir. This is to be called the "theory of a one-atom *laser*." Even in this situation, when pumping of the atom is included, the eigenvalues determining the intensity dynamics are found to be mere solutions of an ordinary continued fraction equation. The whole spectrum of eigenvalues and the general dynamical behavior of the system can be found by solving 4×4-matrix continued fractions numerically, the results of which are reported in a paper with Ginzel, Martini, and Schenzle [7].

II. DAMPING BASES

In the following we will call L_a and L_σ as in (1.5) and (1.6), respectively, *damping operators.* Note, however, that L_σ can also describe incoherent pumping, which is a mechanism structurally identical with damping. The sets of eigenstates of these damping operators will consequently be referred to as *damping bases.*

A. Atomic damping

The eigenstates $P_\lambda(\sigma_+,\sigma_-)$ of L_σ depend on the dynamical variables σ_+,σ_- of the atom and satisfy the equation

$$L_\sigma P_\lambda=\lambda P_\lambda\ , \tag{2.1}$$

with L_σ defined in (1.6). Apart from the state corresponding to $\lambda=0$ in (2.1) these are simply the spin variables themselves. One immediately finds

$$\begin{aligned} &L_\sigma\sigma_0=0\ ,\\ &L_\sigma\sigma_z=-B\sigma_z\ ,\\ &L_\sigma\sigma_\pm=-C\sigma_\pm\ , \end{aligned} \tag{2.2}$$

where

$$\sigma_0=\tfrac{1}{2}[1+(2s-1)\sigma_z)]\ . \tag{2.3}$$

So the eigenvalues are zero and the negatives of the coupling constants B and C for the atom and its reservoir. Note that they are independent of s, which is just a parameter characterizing the equilibrium state of the atomic reservoir. This will be similar for the photon damping whose eigenvalues are also independent of the mean number ν of thermal photons in the cavity, i.e., of the reservoir temperature.

The dual operator $\check{L}_\sigma$ corresponding to (1.6) reads

$$\begin{aligned}\check{L}_\sigma\check{P}=\check{P}L_\sigma=&-\frac{B}{8}(1-s)\{\sigma_+\sigma_-\check{P}+\check{P}\sigma_+\sigma_--2\sigma_+\check{P}\sigma_-\}\\ &-\frac{B}{8}s\{\sigma_-\sigma_+\check{P}+\check{P}\sigma_-\sigma_+-2\sigma_-\check{P}\sigma_+\}\\ &-\frac{2C-B}{4}\{\check{P}-\sigma_z\check{P}\sigma_z\}\ . \end{aligned} \tag{2.4}$$

As discussed in Sec. I A it has the same eigenvalues as L_σ and its eigenstates are dual to those of L_σ:

$$\begin{aligned} &\check{\sigma}_0L_\sigma=\check{L}_\sigma\check{\sigma}_0=0\ ,\\ &\check{\sigma}_zL_\sigma=\check{L}_\sigma\check{\sigma}_z=-B\check{\sigma}_z\ ,\\ &\check{\sigma}_\pm L_\sigma=\check{L}_\sigma\check{\sigma}_\pm=-C\check{\sigma}_\pm\ , \end{aligned} \tag{2.5}$$

where

$$\check{\sigma}_0=1,\quad \check{\sigma}_z=\tfrac{1}{2}[\sigma_z-(2s-1)],\quad \check{\sigma}_\pm=\tfrac{1}{4}\sigma_\mp\ . \tag{2.6}$$

One immediately verifies the duality relations (1.12), here

$$\mathrm{tr}\{\check{\sigma}_j\sigma_k\}=\delta_{jk},\quad j,k\in\{0,z,\pm\}\ . \tag{2.7}$$

If we write symbolically

$$\left.\frac{\partial P}{\partial t}\right|_\sigma=L_\sigma P \tag{2.8}$$

for that part of the evolution of the state which originates in the coupling to the atomic reservoir, the eigenstates of the dual $\check{L}_\sigma$ obey

$$\left.\frac{d}{dt}\right|_\sigma\langle\check{\sigma}_0\rangle=0\ , \tag{2.9a}$$

$$\left.\frac{d}{dt}\right|_\sigma\langle\check{\sigma}_z\rangle=-B\langle\check{\sigma}_z\rangle\ , \tag{2.9b}$$

$$\left.\frac{d}{dt}\right|_\sigma \langle \check{\sigma}_\pm \rangle = -C\langle \check{\sigma}_\pm \rangle \ . \tag{2.9c}$$

Equation (2.9a) simply states that the atomic damping keeps the normalization of the state constant. In view of (2.6), (2.9b) and (2.9c) say that the atomic inversion and polarization decay with mean inverse lifetimes B and C, respectively, which motivates the parametrization of L_σ in (1.6). The expectation values in the steady state σ_0, reached in the limit $t \to \infty$, are

$$\langle \sigma_z \rangle_\infty = 2s-1 \quad \text{and} \quad \langle \sigma_\pm \rangle_\infty = 0 \ . \tag{2.10}$$

This means that under the influence of the atomic reservoir the atomic coherence gets lost, independent of the bath parameter $s = \langle (1+\sigma_z)/2 \rangle_\infty$, as it should be. The inversion approaches a steady-state value which is between -1 and 1, depending on the value of s which ranges from 0 to 1. If $s > \frac{1}{2}$, then $\langle \sigma_z \rangle_\infty > 0$ and the atomic population becomes inverted. For these reasons we call s the *pumping parameter* for the atom.

Naturally, any state or observable $\mathcal{O} = \mathcal{O}(\sigma_+, \sigma_-)$ depending solely on the dynamical variables of the atom can uniquely be expanded into the eigenstates of both L_σ and $\check{L}_\sigma$:

$$\mathcal{O}(\sigma_+,\sigma_-) = \check{c}_0\sigma_0 + \check{c}_z\sigma_z + \check{c}_-\sigma_- + \check{c}_+\sigma_+ \ , \tag{2.11}$$

where

$$\check{c}_0 = \mathrm{tr}\{\check{\sigma}_0 \mathcal{O}\}, \quad \check{c}_z = \mathrm{tr}\{\check{\sigma}_z \mathcal{O}\}, \quad \check{c}_\pm = \mathrm{tr}\{\check{\sigma}_\pm \mathcal{O}\} \ . \tag{2.12}$$

Similarly,

$$\mathcal{O}(\sigma_+,\sigma_-) = c_0\check{\sigma}_0 + c_z\check{\sigma}_z + c_-\check{\sigma}_- + c_+\check{\sigma}_+ \ , \tag{2.13}$$

with

$$c_0 = \mathrm{tr}\{\mathcal{O}\sigma_0\}, \quad c_z = \mathrm{tr}\{\mathcal{O}\sigma_z\}, \quad c_\pm = \mathrm{tr}\{\mathcal{O}\sigma_\pm\} \ . \tag{2.14}$$

When $\mathcal{O} = \mathcal{O}(a^\dagger, a, \sigma_+, \sigma_-)$ depends on the dynamical variables of the photon, too, then the expansions (2.11) and (2.13) hold similarly if in (2.12) and (2.14) the trace tr is taken only over the atomic variables and then the expansion coefficients c_j and $\check{c}_j$ are still functions of $a^\dagger$ and a.

B. Photon damping

The solution of the eigenvalue problem for the photon damping operator L_a in (1.5) is somewhat more involved than for L_σ but still a straightforward business. The eigenstates ρ_λ of L_a and their eigenvalues are determined by

$$L_a\rho_\lambda = \lambda\rho_\lambda \tag{2.15}$$

subject to the constraint that

$$\mathrm{tr}\{B(a^\dagger,a)\rho_\lambda(a^\dagger,a)\} < \infty \tag{2.16}$$

holds for any bound observable $B(a^\dagger, a)$. Pure damping does not couple different diagonals of a state in the number-vector representation, so that the ansatz

$$\rho_\lambda(a^\dagger,a) = a^{\dagger\kappa} f(a^\dagger a) \tag{2.17a}$$

or

$$\rho_\lambda(a^\dagger,a) = f(a^\dagger a)a^k \tag{2.17b}$$

for $k = 0, 1, 2, \ldots$, is invited, where f is a function of the photon number $a^\dagger a$. Inserted into (2.15), this ansatz yields a recursion for $f(a^\dagger a)$. Considerably more convenient is the use of a normally ordered function $:f(a^\dagger a):$ rather than $f(a^\dagger a)$ itself. Then (2.15) implies an ordinary second-order differential equation for f, viz.,

$$(\nu+1)zf''(z) + [z+(\nu+1)(k+1)]f'(z) + \left[\frac{k}{2}+1-\lambda/A\right]f(z) = 0 \ , \tag{2.18}$$

where $z = a^\dagger a$. Its derivation uses the identities

$$\begin{aligned} &:f(a^\dagger a):a^\dagger = a^\dagger :f(a^\dagger a) + f'(a^\dagger a): \ , \\ &a^\dagger :f(a^\dagger a):a = :a^\dagger a f(a^\dagger a): \ . \end{aligned} \tag{2.19}$$

Equation (2.18) turns into Kummer's differential equation if $-z/(\nu+1)$ is regarded as the variable. Its analytical solutions obey the constraint (2.16) only if

$$-\frac{\lambda}{A} - \frac{k}{2} = n = 0, 1, 2, \ldots \tag{2.20}$$

for any given k. For these values of λ the solutions of (2.18) are

$$f(z) = \exp\left[-\frac{z}{1+\nu}\right] L_n^{(k)}\left[\frac{z}{1+\nu}\right] , \tag{2.21}$$

with the generalized Laguerre polynomials $L_n^{(k)}$. So the result is

$$L_a\rho_n^{(k)} = -A\left[n+\frac{|k|}{2}\right]\rho_n^{(k)} , \quad n = 0, 1, 2, \ldots, \quad k = 0, \pm 1, \pm 2, \ldots , \tag{2.22}$$

with

$$\rho_n^{(k)} = \begin{cases} a^{\dagger k}\dfrac{(-1)^n}{(1+\nu)^{k+1}} :L_n^{(k)}\left[\dfrac{a^\dagger a}{1+\nu}\right]\exp\left[-\dfrac{a^\dagger a}{1+\nu}\right]: & \text{for } k \ge 0 \ , \\[2ex] \dfrac{(-1)^n}{(1+\nu)^{|k|+1}} :L_n^{(|k|)}\left[\dfrac{a^\dagger a}{1+\nu}\right]\exp\left[-\dfrac{a^\dagger a}{1+\nu}\right]:a^{|k|} & \text{for } k \le 0 \ . \end{cases} \tag{2.23}$$

The two possibilities of (2.17) are now distinguished by the sign of k. Reversing the sign of k is equivalent to taking the adjoint; the eigenvalues $-A(n+|k|/2)$ involve the magnitude of k only. The $a^\dagger a$-independent factors in (2.23) provide us with a convenient normalization.

The eigenstates $\rho_n^{(k)}$ are linearly independent and complete, as for any given k the $L_n^{(k)}$ are polynomials of order n and thus independent for different values of n; for different values of k the corresponding $\rho_n^{(k)}$ involve different powers of the variables $a^\dagger$ or a, so they are independent in the first place.

Note that, although the eigenstates of L_a depend on ν, its eigenvalues are independent of this parameter which is related to the bath temperature. This is similar to the situation of atomic damping, where the pumping parameter s only enters the eigenstates of L_σ, but not its eigenvalues.

For the dual damping operator $\check{L}_a$, given by

$$\check{\rho}L_a=\check{L}_a\check{\rho}=-\frac{A}{2}(\nu+1)\{a^\dagger a\check{\rho}+\check{\rho}a^\dagger a-2a^\dagger\check{\rho}a\}$$
$$-\frac{A}{2}\nu\{aa^\dagger\check{\rho}+\check{\rho}aa^\dagger-2a\check{\rho}a^\dagger\}\ , \tag{2.24}$$

which is to be compared with (1.5), the derivation of its eigenstates is done analogously. The result is

$$\check{\rho}_n^{(k)}L_a=\check{L}_a\check{\rho}_n^{(k)}=-A\left[n+\frac{|k|}{2}\right]\check{\rho}_n^{(k)}\ ,$$
$$n=0,1,2,\ldots,\quad k=0,\pm1,\pm2,\ldots, \tag{2.25}$$

with

$$\check{\rho}_n^{(k)}=\begin{cases}\left[\dfrac{-\nu}{1+\nu}\right]^n\dfrac{n!}{(n+k)!}:L_n^{(k)}\left[\dfrac{a^\dagger a}{\nu}\right]:a^k & \text{for } k\ge0\\[2ex] \left[\dfrac{-\nu}{1+\nu}\right]^n\dfrac{n!}{(n+|k|)!}a^{\dagger|k|}:L_n^{(|k|)}\left[\dfrac{a^\dagger a}{\nu}\right]: & \text{for } k\le0.\end{cases} \tag{2.26}$$

After some algebra, which is reported in the Appendix, one can convince oneself that these $\check{\rho}_n^{(k)}$ are indeed dual to the $\rho_n^{(k)}$ of (2.23),

$$\mathrm{tr}\{\check{\rho}_n^{(k)}\rho_{n'}^{(k')}\}=\delta_{nn'}\delta_{kk'}$$
$$\text{for } n,n'=0,1,2,\ldots,\ k,k'=0,\pm1,\pm2,\ldots\ . \tag{2.27}$$

The states $\rho_n^{(k)}$ are all traceless, except for $n=0$ and $k=0$:

$$\mathrm{tr}\{\rho_n^{(k)}\}=\delta_{n0}\delta_{k0}\ . \tag{2.28}$$

Note that the ground state $\rho_0^{(0)}$, which is the steady state of L_a, is the thermal state:

$$\rho_0^{(0)}=\frac{1}{1+\nu}:\exp\left[-\frac{a^\dagger a}{1+\nu}\right]:=\frac{1}{1+\nu}\left[\frac{\nu}{1+\nu}\right]^{a^\dagger a}\ , \tag{2.29}$$

if we recognize that the number of thermal photons, ν, is related to the temperature T by $\nu/(1+\nu)=\exp(-\hbar\omega/k_BT)$.

Now, any state or observable $\mathcal{O}=\mathcal{O}(a^\dagger,a)$ depending only on the photon variables can be expanded uniquely into the eigenstates of both L_a and $\check{L}_a$:

$$\mathcal{O}(a^\dagger,a)=\sum_{n,k}\check{\alpha}_{nk}\rho_n^{(k)}\ , \tag{2.30}$$

with

$$\check{\alpha}_{nk}=\mathrm{tr}\{\check{\rho}_n^{(k)}\mathcal{O}\}\ , \tag{2.31}$$

and similarly

$$\mathcal{O}(a^\dagger,a)=\sum_{n,k}\alpha_{nk}\check{\rho}_n^{(k)}\ , \tag{2.32}$$

where

$$\alpha_{nk}=\mathrm{tr}\{\mathcal{O}\rho_n^{(k)}\}\ . \tag{2.33}$$

In particular, $\mathcal{O}$ can be the reduced state of the joint atom-photon system, the state of the electromagnetic field alone:

$$\rho(a^\dagger,a,t)=\mathrm{tr}_\sigma P(a^\dagger,a,\sigma_+,\sigma_-,t)\ , \tag{2.34}$$

where tr_σ traces over the atom variables only. Equation (2.30) then reads

$$\rho(t)=\sum_{n,k}\check{\alpha}_{nk}(t)\rho_n^{(k)}\ , \tag{2.35}$$

with the expansion coefficients

$$\check{\alpha}_{nk}(t)=\langle\check{\rho}_n^{(k)}\rangle_t=\left[\frac{-\nu}{1+\nu}\right]^n\frac{n!}{(n+k)!}$$
$$\times\left\langle:L_n^{(k)}\left[\frac{a^\dagger a}{\nu}\right]:a^k\right\rangle_t \tag{2.36}$$

for $k\ge0$, and the complex conjugates for $k\le0$.

Note that, although the expectation values in (2.36) might look rather involved on first sight, the expansion (2.35) is actually a very natural one, as the first few coefficients already contain the basic statistical information about the field. For example,

$$\langle a^\dagger a\rangle=\nu+(1+\nu)\check{\alpha}_{10} \tag{2.37a}$$

is the mean number of photons, and

$$\langle a\rangle=\check{\alpha}_{01} \tag{2.37b}$$

is the mean electric field, and similarly for higher powers of $a,a^\dagger$ and $a^\dagger a$. In contrast to expansions into a number-vector basis where one needs *all* diagonal expansion coefficients to calculate just the mean photon number $\langle a^\dagger a\rangle$, the most essential statistical quantities of the field are provided by a few coefficients only in (2.35).

We close this section with a comment on the special case $\nu=0$. For $\nu=0$, the right eigenstates of L_a acquire the form

$$\rho_n^{(k)}=\begin{cases}a^{\dagger k}(-1)^{a^\dagger a+n}\begin{bmatrix}n+k\\ a^\dagger a+k\end{bmatrix}, & k\ge0\\[2ex] (-1)^{a^\dagger a+n}\begin{bmatrix}n+|k|\\ a^\dagger a+|k|\end{bmatrix}a^{|k|}, & k\le0\ ,\end{cases} \tag{2.38}$$

and the dual states are

$$\check{\rho}_n^{(k)}=\begin{cases}\frac{n!}{(n+k)!}\begin{bmatrix}a^\dagger a\\ n\end{bmatrix}a^k=a^{\dagger n}a^{n+k}/(n+k)! & \text{for } k\geq 0\\ a^{\dagger|k|}\begin{bmatrix}a^\dagger a\\ n\end{bmatrix}\frac{n!}{(n+|k|)!}=a^{\dagger n+|k|}a^n/(n+|k|)! & \text{for } k\leq 0\,.\end{cases} \tag{2.39}$$

The statements (2.22), (2.25) and (2.27) can be verified easily. The corresponding eigenvalues are, of course, still the same, as they are independent of v.

In this situation, the coefficients of a corresponding expansion (2.35) of the state ρ of the field have an even more immediate significance, for instance,

$$\begin{aligned}\langle a^\dagger a\rangle&=\check{\alpha}_{10}\,,\\ \langle a\rangle&=\check{\alpha}_{01}\end{aligned} \tag{2.40}$$

for the mean number of photons and the mean electric field, and

$$\begin{aligned}&\langle (a^\dagger a)^2\rangle-\langle a^\dagger a\rangle^2=2\check{\alpha}_{20}+\check{\alpha}_{10}-\check{\alpha}_{10}^2\,,\\ &\langle a^\dagger a\rangle-\langle a^\dagger\rangle\langle a\rangle-|\langle a^2\rangle-\langle a\rangle^2|\\ &\qquad=\check{\alpha}_{10}-|\check{\alpha}_{01}|^2-|2\check{\alpha}_{02}-\check{\alpha}_{01}^2|\end{aligned} \tag{2.41}$$

for the variance of $\langle a^\dagger a\rangle$ and the squeezing measure of the field.

III. EXPANSIONS OF THE MASTER EQUATION

In the following we will apply the results of Sec. II and successively expand the state P into eigenstates of L_σ and L_a. The master equation then transforms into groups of matrix equations for the expansion coefficients.

A. Expansion into eigenstates of L_σ

The state P in Eq. (1.1) is a function of the dynamical variables $a^\dagger, a$ and, in particular, σ_+,σ_-. We can therefore employ the expansion (2.11) and write

$$P=\rho\sigma_0+\rho_z\sigma_z+\rho_-\sigma_-+\rho_+\sigma_+\,. \tag{3.1}$$

The coefficients appearing in (3.1) are functions of the photon variables:

$$\begin{aligned}\rho&=\mathrm{tr}_\sigma\{P\}\,,\\ \rho_z&=\mathrm{tr}_\sigma\{\check{\sigma}_zP\}\,,\\ \rho_\pm&=\mathrm{tr}_\sigma\{\check{\sigma}_\pm P\}\,.\end{aligned} \tag{3.2}$$

When inserted into the master equation (1.1), (3.1) yields a coupled system of differential equations for the photon functions $\rho,\rho_z,\rho_\pm$:

$$\begin{aligned}\left[\frac{\partial}{\partial t}-L_a\right]\rho&=-i\omega[a^\dagger a,\rho]\\ &\quad+2ig([a,\rho_-]+[a^\dagger,\rho_+])\,,\\ \left[\frac{\partial}{\partial t}-L_a+B\right]\rho_z&=-i\omega[a^\dagger a,\rho_z]\\ &\quad+ig(\{a,\rho_-\}-\{a^\dagger,\rho_+\}\\ &\qquad-(2s-1)[a,\rho_-]\\ &\qquad-(2s-1)[a^\dagger,\rho_+])\,,\\ \left[\frac{\partial}{\partial t}-L_a+C\right]\rho_-&=-i\omega[a^\dagger a,\rho_-]+i\Omega\rho_-\\ &\quad+i\frac{g}{4}([a^\dagger,\rho]+2\{a^\dagger,\rho_z\}\\ &\qquad+(2s-1)\{a^\dagger,\rho\})\,,\\ \left[\frac{\partial}{\partial t}-L_a+C\right]\rho_+&=-i\omega[a^\dagger a,\rho_+]-i\Omega\rho_+\\ &\quad+i\frac{g}{4}([a,\rho]-2\{a,\rho_z\}\\ &\qquad-(2s-1)\{a,\rho\})\,,\end{aligned} \tag{3.3}$$

where $\{\,,\,\}$ denotes the anticommutator and L_a is the operator of (1.5). If one is only interested in the reduced state of the field then ρ_z and $\rho_\pm$ in (3.3) are merely auxiliary functions for the calculation of ρ. This system of equations could now be treated further by using "quasiprobability functions" [3] to get a c-number representation of (3.3).

In this paper we will, however, stick to our strategy and first take care of the nonunitary damping operators.

B. Expansions into eigenstates of L_a

Since the objects of interest ρ, ρ_z, and $\rho_\pm$ are all functions of the photon variables $a^\dagger$ and a we can expand them into eigenstates $\rho_n^{(k)}$ of the photon damping operator L_a. Employing (2.35) for $\rho(t)$ and similar expansions for ρ_z and $\rho_\pm$ we write

$$\begin{aligned}\rho(t)&=\sum_{n,k}\check{\alpha}_{nk}(t)\rho_n^{(k)},\quad \check{\alpha}_{n,-k}=\check{\alpha}_{n,k}^*\,,\\ \rho_z(t)&=\sum_{n,k}\check{\beta}_{nk}(t)\rho_n^{(k)},\quad \check{\beta}_{n,-k}=\check{\beta}_{n,k}^*\,,\\ \rho_-(t)&=\sum_{n,k}\check{\gamma}_{n,k-1}(t)\rho_n^{(k)},\quad \check{\gamma}_{n,-k}=\check{\eta}_{n,k}^*\,,\\ \rho_+(t)&=\sum_{n,k}\check{\eta}_{n,k+1}(t)\rho_n^{(k)},\quad \check{\eta}_{n,-k}=\check{\gamma}_{n,k}^*\,,\end{aligned} \tag{3.4}$$

where the restrictions on the coefficients are due to the fact that ρ and ρ_z are Hermitian whereas ρ_+ and ρ_- are Hermitian conjugates of each other.

Now we use the following commutation and anticommutation relations:

$$\begin{aligned}&[a^\dagger,\rho_n^{(k)}]=\rho_n^{(k+1)},\quad [a,\rho_n^{(k+1)}]=-\frac{n+1}{\nu+1}\rho_{n+1}^{(k)}\,,\\ &\{a^\dagger,\rho_n^{(k)}\}=2(\nu+\tfrac{1}{2})\rho_n^{(k+1)}+2(\nu+1)\rho_{n-1}^{(k+1)},\quad \{a,\rho_n^{(k+1)}\}=2(n+k+1)\rho_n^{(k)}+2(\nu+\tfrac{1}{2})\frac{n+1}{\nu+1}\rho_{n+1}^{(k)}\end{aligned}\tag{3.5}$$

for $n,k=0,1,2,\ldots$, and the adjoint relations for $k=-1,-2,\ldots$, and insert the expansions (3.4) into (3.3). After conveniently arranging the coefficients we eventually arrive at a system of differential equations which is equivalent to the master equation (1.1):

$$\frac{d}{dt}\begin{bmatrix}\check\alpha_{n+1,k}\\ \check\beta_{n,k}\\ \frac{1}{i}\check\gamma_{n,k}\\ \frac{1}{i}\check\eta_{n+1,k}\end{bmatrix}$$

$$=\begin{bmatrix}-A\left[n+\frac{k}{2}+1\right]-i\omega k & 0 & 2g\frac{n+1}{\nu+1} & -2g\\ 0 & -A\left[n+\frac{k}{2}\right]-B-i\omega k & -2g(n+k+1) & 2g(\nu+1)\\ g(\nu+1)\left[s-\frac{1}{2}\right] & g\left[\nu+\frac{1}{2}\right] & -A\left[n+\frac{k}{2}+\frac{1}{2}\right]-C-i\omega k-i\Delta & 0\\ -g(n+k+1)\left[s-\frac{1}{2}\right] & -g\left[\nu+\frac{1}{2}\right]\frac{n+1}{\nu+1} & 0 & -A\left[n+\frac{k}{2}+\frac{1}{2}\right]-C-i\omega k+i\Delta\end{bmatrix}\begin{bmatrix}\check\alpha_{n+1,k}\\ \check\beta_{n,k}\\ \frac{1}{i}\check\gamma_{n,k}\\ \frac{1}{i}\check\eta_{n+1,k}\end{bmatrix}$$

$$+\begin{bmatrix}0&0&0&0\\0&0&0&0\\0&g(\nu+1)&0&0\\0&-g(n+k+1)&0&0\end{bmatrix}\begin{bmatrix}\check\alpha_{n+2,k}\\ \check\beta_{n+1,k}\\ \frac{1}{i}\check\gamma_{n+1,k}\\ \frac{1}{i}\check\eta_{n+2,k}\end{bmatrix}+\begin{bmatrix}0&0&0&0\\ 0&0&-2g(\nu+s)\frac{n}{\nu+1}&2g(\nu+s)\\ \frac{g}{4}[1+(2\nu+1)(2s-1)]&0&0&0\\ -\frac{g}{4}[1+(2\nu+1)(2s-1)]\frac{n+1}{\nu+1}&0&0&0\end{bmatrix}\begin{bmatrix}\check\alpha_{n,k}\\ \check\beta_{n-1,k}\\ \frac{1}{i}\check\gamma_{n-1,k}\\ \frac{1}{i}\check\eta_{n,k}\end{bmatrix}$$

$$\text{for } n=0,1,2,3,\ldots,\ k=1,2,3,\ldots\,;\tag{3.6a}$$

$$\frac{d}{dt}\begin{bmatrix}\check\alpha_{n+1,0}\\ \check\beta_{n,0}\\ \frac{1}{i}\check\gamma_{n,0}\\ \frac{1}{i}\check\eta_{n,0}\end{bmatrix}=\begin{bmatrix}-A(n+1)&0&2g\frac{n+1}{\nu+1}&-2g\frac{n+1}{\nu+1}\\ 0&-An-B&-2g(n+1)&2g(n+1)\\ g(\nu+1)(s-\frac{1}{2})&g(\nu+\frac{1}{2})&-A(n+\frac{1}{2})-C-i\Delta&0\\ -g(\nu+1)(s-\frac{1}{2})&-g(\nu+\frac{1}{2})&0&-A(n+\frac{1}{2})-C+i\Delta\end{bmatrix}\begin{bmatrix}\alpha_{n+1,0}\\ \check\beta_{n,0}\\ \frac{1}{i}\check\gamma_{n,0}\\ \frac{1}{i}\check\eta_{n,0}\end{bmatrix}$$

$$+\begin{bmatrix}0&0&0&0\\0&0&0&0\\0&g(\nu+1)&0&0\\0&-g(\nu+1)&0&0\end{bmatrix}\begin{bmatrix}\check\alpha_{n+2,0}\\ \check\beta_{n+1,0}\\ \frac{1}{i}\check\gamma_{n+1,0}\\ \frac{1}{i}\check\eta_{n+1,0}\end{bmatrix}+\begin{bmatrix}0&0&0&0\\ 0&0&-2g(\nu+s)\frac{n}{\nu+1}&2g(\nu+s)\frac{n}{\nu+1}\\ \frac{g}{4}[1+(2\nu+1)(2s-1)]&0&0&0\\ -\frac{g}{4}[1+(2\nu+1)(2s-1)]&0&0&0\end{bmatrix}\begin{bmatrix}\check\alpha_{n,0}\\ \check\beta_{n-1,0}\\ \frac{1}{i}\check\gamma_{n-1,0}\\ \frac{1}{i}\check\eta_{n-1,0}\end{bmatrix}$$

$$\text{for } n=0,1,2,\ldots,\ k=0\,;\tag{3.6b}$$

$$\frac{d}{dt}\begin{bmatrix}\check\alpha_{0,k}\\ \frac{1}{i}\check\eta_{0,k}\end{bmatrix}=\begin{bmatrix}-A\frac{k}{2}-i\omega k & -2g\\ -gk\left[s-\frac{1}{2}\right] & -A\left[\frac{k}{2}-\frac{1}{2}\right]-C-i\omega k+i\Delta\end{bmatrix}\begin{bmatrix}\check\alpha_{0,k}\\ \frac{1}{i}\check\eta_{0,k}\end{bmatrix}-gk\begin{bmatrix}0\\ \check\beta_{0,k}\end{bmatrix}\quad\text{for } n=-1,\ k=1,2,3,\ldots\,;\tag{3.6c}$$

$$\frac{d}{dt}\check\alpha_{00}=0\quad\text{for } n=-1,\ k=0\,.\tag{3.6d}$$

This is supplemented by the complex conjugate equations. Note that in the system (3.6) coefficients with different indices k do not couple. Therefore we obtain closed groups of equations with respect to the index k. With respect to the index n, every group has the form of a three-term recurrence relation.

For the further discussion we write (3.6) compactly in the form

$$\frac{d}{dt}\check{X}_{nk}=M_n^{(k)}\check{X}_{nk}+G_n^{(k)}\check{X}_{n+1,k}+F_n\check{X}_{n-1,k}\ , \tag{3.7}$$

with $n=-1,0,1,2,\ldots$ and $k=0,\pm 1,\pm 2,\ldots$, where $\check{X}_{nk}$ denotes a column of four coefficients and $M_n^{(k)}$, $G_n^{(k)}$, and F_n are the corresponding 4×4 matrices involved [8]. The system of differential equations is the "coordinate representation" of the master equation in the damping bases. An eigenvalue of the master equation is then a simultaneous eigenvalue of a group of equations with common index k,

$$\mathcal{L}P=\Gamma P \Longleftrightarrow \Gamma\check{X}_{nk}=M_n^{(k)}\check{X}_{nk}+G_n^{(k)}\check{X}_{n+1,k}+F_n\check{X}_{n-1,k}\ . \tag{3.8}$$

Note that the matrices $G_n^{(k)}$ and F_n couple the dynamics of the nth column $\check{X}_{nk}$ to the columns with neighboring indices n, $\check{X}_{n+1,k}$ and $\check{X}_{n-1,k}$, respectively. Most generally, the recursion in (3.8) can be solved by matrix continued fractions, which is further discussed in [7].

For a complete solution of (1.1) we also have to calculate the left eigenstates of $\mathcal{L}$ which are the right-eigenstates of its dual $\check{\mathcal{L}}$ according to (1.15) as was pointed out in Sec. I B. To do this, we proceed as above, except that we now employ the expansions into the eigenstates of the dual damping operators. That is,

$$\check{P}=\check{\rho}\check{\sigma}_0+\check{\rho}_z\check{\sigma}_z+\check{\rho}_-\check{\sigma}_-+\check{\rho}_+\check{\sigma}_+\ , \tag{3.9}$$

where

$$\check{\rho}=\mathrm{tr}_\sigma\{\check{P}\sigma_0\}\ ,\quad \check{\rho}_z=\mathrm{tr}_\sigma\{\check{P}\sigma_z\}\ ,$$
$$\check{\rho}_\pm=\mathrm{tr}_\sigma\{\check{P}\sigma_\pm\}\ , \tag{3.10}$$

in analogy to (2.13). Similarly, we apply (2.32) to further expand the photon functions (3.10):

$$\check{\rho}=\sum_{n,k}\alpha_{nk}\check{\rho}_n^{(k)}\ ,\quad \check{\rho}_z=\sum_{n,k}\beta_{nk}\check{\rho}_n^{(k)}\ ,$$
$$\check{\rho}_-=\sum_{n,k}\gamma_{n,k-1}\check{\rho}_n^{(k)}\ ,\quad \check{\rho}_+=\sum_{n,k}\eta_{n,k+1}\check{\rho}_n^{(k)}\ , \tag{3.11}$$

as in (3.4). Using the following commutation relations for the dual eigenstates $\check{\rho}_n^{(k)}$:

$$\begin{aligned}
&[a^\dagger,\check{\rho}_n^{(k+1)}]=-\check{\rho}_n^{(k)}\ ,\\
&[a,\check{\rho}_n^{(k)}]=\frac{n}{1+\nu}\check{\rho}_{n-1}^{(k+1)}\ ,\\
&\{a^\dagger,\check{\rho}_n^{(k+1)}\}=2(\nu+\tfrac{1}{2})\check{\rho}_n^{(k)}+2(\nu+1)\check{\rho}_{n+1}^{(k)}\ ,\\
&\{a,\check{\rho}_n^{(k)}\}=2(n+k+1)\check{\rho}_n^{(k+1)}+2(\nu+\tfrac{1}{2})\frac{n}{1+\nu}\check{\rho}_{n-1}^{(k+1)}
\end{aligned} \tag{3.12}$$

for $n,k=0,1,2,\ldots$, and the adjoint relations for $k=-1,-2,\ldots$, we finally arrive at a system of equations which is equivalent to the eigenvalue equation (1.15) for the dual Liouville operator:

$$\check{\mathcal{L}}\check{P}=\Gamma\check{P}\Longleftrightarrow \Gamma X_{nk}=X_{nk}M_n^{(k)}+X_{n-1,k}G_{n-1}^{(k)}+X_{n+1,k}F_{n+1}\ , \tag{3.13}$$

with $n=-1,0,1,2,\ldots$ and $k=0,\pm 1,\pm 2,\ldots$. This is to be compared with the corresponding equation (3.8). Now, the X_{nk} denote *rows*, containing the coefficients appearing in (3.11), namely,

$$\begin{aligned}
&X_{nk}=(\alpha_{n+1,k},\beta_{nk},i\gamma_{nk},i\eta_{n+1,k}) && \text{for } n=0,1,2,\ldots,\quad k=1,2,\ldots,\\
&X_{n0}=(\alpha_{n+1,0},\beta_{n0},i\gamma_{n0},i\eta_{n0}) && \text{for } n=0,1,2,\ldots,\quad k=0\ ,\\
&X_{-1,k}=(\alpha_{0k},0,0,i\eta_{0k}) && \text{for } n=-1,\quad k=1,2,\ldots,\\
&X_{-1,0}=(\alpha_{00},0,0,0) && \text{for } n=-1,\quad k=0\ ,
\end{aligned} \tag{3.14}$$

in contrast to the $\check{X}_{nk}$ in (3.8) which denote *columns*, comprising the coefficients of the expansion (3.4). Note that the matrices M,G,F appearing in (3.6) or (3.7) are the *same* as for the dual coefficients in (3.13).

We can summarize these results compactly by writing

$$X_k=(X_{-1,k},X_{0,k},X_{1,k},\ldots),\quad \check{X}_k=\begin{bmatrix}\check{X}_{-1,k}\\ \check{X}_{0,k}\\ \check{X}_{1,k}\\ \vdots\end{bmatrix}\ ,$$

and

$$\Lambda_k=\begin{bmatrix} M_{-1}^{(k)} & G_{-1}^{(k)} & & \\ F_0 & M_0^{(k)} & G_0^{(k)} & 0\\ & F_1 & M_1^{(k)} & G_1^{(k)}\\ 0 & \ddots & \ddots & \ddots\end{bmatrix}\ , \tag{3.15}$$

which comprises all the coefficients of (3.4) and (3.11) for a fixed k in a (super) column $\check{X}_k$ and in a row X_k and all the matrices in (3.6) or (3.7) in the single tridiagonal (super) matrix Λ_k.

In this notation the master equation (1.1) and the corresponding eigenvalue equations have a clear coordinate counterpart

$$\frac{\partial}{\partial t}P=\mathcal{L}P \Longleftrightarrow \frac{d}{dt}\check{X}_k=\Lambda_k\check{X}_k,\quad k=0,\pm 1,\pm 2,\ldots,$$
$$\left.\begin{array}{l}\mathcal{L}P_\Gamma=\Gamma P_\Gamma\\ \check{\mathcal{L}}\check{P}_\Gamma=\Gamma\check{P}_\Gamma\end{array}\right\}\Longleftrightarrow\left\{\begin{array}{l}\Gamma\check{X}_k^{(\Gamma)}=\Lambda_k\check{X}_k^{(\Gamma)}\\ \Gamma X_k^{(\Gamma)}=X_k^{(\Gamma)}\Lambda_k,\quad k=0,\pm 1,\pm 2,\ldots\ .\end{array}\right. \tag{3.16}$$

So the eigenvalues and (right) eigenstates of the Liouville operator $\mathcal{L}$ and its dual conjugate $\check{\mathcal{L}}$ are given by the eigenvalues and eigencolumns and eigenrows, respectively, of the matrices Λ_k as in (3.15). Please note how naturally the right eigenstates of $\check{\mathcal{L}}$ appear as left eigenstates of $\mathcal{L}$ in (3.16).

The subsequent sections are devoted to the solutions of (3.16).

IV. ANALYTICAL SOLUTIONS

In this section we concentrate on the special situation in which $\nu=0$ and $s=0$. The master equation (1.1) then reads

$$\frac{\partial}{\partial t}P=\frac{1}{i\hbar}[H,P]-\frac{A}{2}(a^\dagger aP+Pa^\dagger a-2aPa^\dagger)$$
$$-\frac{B}{8}\{\sigma_+\sigma_-P+P\sigma_+\sigma_--2\sigma_-P\sigma_+\}$$
$$-\frac{2C-B}{4}\{P-\sigma_zP\sigma_z\}\ , \qquad (4.1)$$

with H as in (1.2). It describes a two-level atom with level spacing $\hbar\Omega$ interacting with a photon mode of frequency ω in a cavity with a vanishing number of thermal photons; the nonunitary terms account for photon leakage out of the cavity [9] and incoherent transitions of the atoms. These can be either real transitions from the upper to the lower level or virtual transitions [the last term in (4.1)] that only destroy atomic coherence without actually changing the inversion [cf. Eq. (2.9c)].

As we have seen in Sec. III, the master equation (4.1) is equivalent to the system (3.7), or (3.6) for $\nu=0=s$. For these parameters, however, the matrices F_n that couple the coefficients $\check{X}_{nk}(t)$ in (3.7) to the coefficients $\check{X}_{n-1,k}(t)$ *vanish.*

Let us first consider initial states $P(t=0)$ whose coefficients $\check{X}_{nk}(0)$ vanish for all k except for $k=0$; then they will do so for all times,

$$\check{X}_{nk}(0)=0,\quad k=1,2,3,\ldots \Longrightarrow \check{X}_{nk}(t)\equiv 0\ ,$$
$$k=1,2,3,\ldots \qquad (4.2)$$

because in (3.7) coefficients with different indices k are decoupled. If additionally $\check{X}_{n0}(t=0)=0$ for $n>N$, which implies that at time $t=0$ there are not more than $N+1$ photons in the cavity, these coefficients remain zero as well, as their dynamics only depends on coefficients with higher n, which all vanish initially:

$$\check{X}_{n0}(0)\equiv\check{X}_n(0)=0,\quad n=N+1,N+2,\ldots \Longrightarrow \check{X}_n(t)\equiv 0\ ,$$
$$n=N+1,N+2,\ldots . \qquad (4.3)$$

In this situation the master equation corresponds to a finite number of coupled differential equations in the form of two-term recursions:

$$\frac{\partial}{\partial t}P=\mathcal{L}P \Longleftrightarrow \begin{cases} \dfrac{d}{dt}\check{X}_n=M_n\check{X}_n+G\check{X}_{n+1}, \\ \qquad n=0,1,2,\ldots,\quad N-1, \\ \dfrac{d}{dt}\check{X}_N=M_N\check{X}_N\ , \end{cases} \qquad (4.4)$$

where we have temporarily omitted the subscript $k=0$, for notational simplicity. Accordingly, the corresponding eigenvalue problem in this finite-dimensional situation reads

$$\Gamma P=\mathcal{L}P \Longleftrightarrow \begin{cases} \Gamma\check{X}_n=M_n\check{X}_n+G\check{X}_{n+1}, \\ \qquad n=0,1,2,\ldots,\quad N-1, \\ \Gamma\check{X}_N=M_N\check{X}_N\ . \end{cases} \qquad (4.5)$$

Therefore the eigenvalues of the matrix M_N are also eigenvalues of the Liouville operator $\mathcal{L}$. The same reasoning applies to the groups of equations with $k\neq 0$, and for arbitrary N.

So we have the result that the eigenvalues of the master equation (4.1) are identical with the eigenvalues of the matrices $M_n^{(k)}$ appearing in (3.6) with $\nu=0,s=0$:

$$\det(M_n^{(k)}-\Gamma)=0\ . \qquad (4.6)$$

In general, the solutions of (4.6) are the roots of a fourth-order polynomial in Γ, namely,

$$\Gamma'^4+(2C-B)\Gamma'^3+\left[4g^2\left[n+\frac{k}{2}+1\right]-\left[\frac{A-B}{2}\right]^2+\left[\frac{2C-B}{2}\right]^2+\Delta^2\right]\Gamma'^2$$
$$+(2C-B)\left[4g^2\left[n+\frac{k}{2}+1\right]-\left[\frac{A-B}{2}\right]^2\right]\Gamma'+\left[kg^2-\frac{A-B}{2}i\Delta\right]^2-\left[\frac{2C-B}{2}\right]^2\left[\frac{A-B}{2}\right]^2=0 \qquad (4.7a)$$

for $n=0,1,2,\ldots$, $k=0,1,2,\ldots$, with $\Gamma'=\Gamma+A(n+k/2+\frac{1}{2})+B/2+i\omega k$ and, for $n=-1$, the residual equations

$$\left[\Gamma+i\omega k+A\frac{k}{2}\right][\Gamma+i\omega k+A(k-1)/2+C-i\Delta]+kg^2=0 \quad \text{for } n=-1,\ k=1,2,\ldots \qquad (4.7b)$$

and

$$\Gamma=0 \quad \text{for } n=-1,\ k=0\ . \qquad (4.7c)$$

The eigenstate to $\Gamma=0$ is, of course, the steady-state solution of (4.1). Equations (4.7b) similarly give the "ground-state values" for the groups of equations with different indices k. For instance, (4.7b) with $k=1$ determines the long-time decay constant for an initial perturbation $\delta\langle a\rangle$ of the electric field in the cavity [10]:

$$\Gamma^{\pm}_{-1,1} = -\frac{A}{4} - \frac{C}{2} - i\omega + i\frac{\Delta}{2} \pm i\left[g^2 - \left(\frac{A}{4} - \frac{C}{2} + i\frac{\Delta}{2}\right)^2\right]^{1/2} . \tag{4.8}$$

The quartic equation (4.7a) cannot be brought into a simpler form and we will not bore the reader by giving solutions explicitly. For $C = B/2$, which means a vanishing last term in the master equation (i.e., no dephasing collisions for the atom), (4.7a), however, simplifies to a biquadratic equation. The roots of (4.6) are then explicitly given by the following expressions [11]:

$$\Gamma^{\pm\pm}_{nk} = -A\left[n + \frac{k}{2} + \frac{1}{2}\right] - \frac{B}{2} - i\omega k \pm 2i\Omega^{\pm}_{nk} ,$$

where

$$\Omega^{\pm}_{nk} = \frac{1}{\sqrt{2}}\left\{\left[g^2\left[n + \frac{k}{2} + 1\right] - \frac{(A-B)^2}{16} + \frac{\Delta^2}{4}\right] \pm \left\{\left[g^2\left[n + \frac{k}{2} + 1\right] - \frac{(A-B)^2}{16} + \frac{\Delta^2}{4}\right]^2 - \left[\frac{k}{2}g^2 - i\Delta\frac{A-B}{4}\right]^2\right\}^{1/2}\right\}^{1/2}$$

$$\text{for } n = 0,1,2,\dots,\ k = 0,1,2,\dots, \tag{4.9a}$$

$$\Gamma^{\pm}_{-1,k} = -A\left[\frac{k}{2} - \frac{1}{4}\right] - \frac{B}{4} - i\omega k + i\frac{\Delta}{2} \pm i\left[kg^2 - \left(\frac{A-B}{4} + i\frac{\Delta}{2}\right)^2\right]^{1/2} \quad \text{for } n = -1,\ k = 1,2,3,\dots, \tag{4.9b}$$

and

$$\Gamma_{-1,0} = 0 \quad \text{for } n = -1,\ k = 0 . \tag{4.9c}$$

Here the signs in the labeling of $\Gamma^{\pm\pm}_{nk}$ refer to their order of appearance on the right-hand side and $\Delta = \omega - \Omega$ denotes the detuning. Together with the complex conjugate values, Eqs. (4.9) give the complete set of eigenvalues for the master equation (4.1) if $C = B/2$.

To get familiar with these expressions let us first look at the special case of vanishing damping constants and detuning, $A, B, \Delta = 0$. Then, on the main diagonal ($k = 0$) we have

$$\Gamma^{\pm\pm}_{n0} = \pm(1\pm1)i\Omega_n = \begin{cases} 0, & \text{twofold} \\ \pm 2i\Omega_n, & \Omega_n = g\sqrt{n+1} , \end{cases} \tag{4.10}$$

which are the eigenvalues of the Jaynes-Cummings commutator $(1/i\hbar)[H,\cdot]$ associated with the projector eigenstates $|\gamma^{\pm}_n\rangle\langle\gamma^{\pm}_n|$, where the Hilbert space vectors $|\gamma^{\pm}_n\rangle$ obey the eigenvalue equation

$$H|\gamma^{\pm}_n\rangle = |\gamma^{\pm}_n\rangle\{\hbar\omega(n+\tfrac{1}{2}) \pm \hbar\Omega_n\} ,$$
$$\text{with } n = 0,1,2,\dots, \tag{4.11}$$

and Ω_n is the Rabi frequency. Similarly, for off-diagonal eigenvalues with $k > 0$

$$\Gamma^{\pm\pm}_{nk} = -i\omega k \pm ig(\sqrt{n+k+1} \pm \sqrt{n+1})$$
$$= -i\omega k \pm i(\Omega_{n+k} \pm \Omega_n) , \tag{4.12}$$

which are eigenvalues corresponding to dyadic products $|\gamma^{\pm}_{n+k}\rangle\langle\gamma^{\pm}_n|$. The "ground-state" values in (4.9) reduce to

$$\Gamma^{\pm}_{-1,k} = -i\omega k \pm ig\sqrt{k} \quad \text{for } k = 1,2,3,\dots \tag{4.13}$$

and $\Gamma_{-1,0} = 0$, of course, corresponding to products $|\gamma^{\pm}_{k-1}\rangle\langle\gamma_{-1}|$ and $|\gamma_{-1}\rangle\langle\gamma_{-1}|$, respectively, where $|\gamma_{-1}\rangle = |(a^\dagger a)' = 0, \sigma'_z = -1\rangle$ obeys

$$H|\gamma_{-1}\rangle = |\gamma_{-1}\rangle(-\tfrac{1}{2}\hbar\omega) . \tag{4.14}$$

When the damping A, B is switched on, (4.10), for example, changes into

$$\Gamma^{\pm\pm}_{n0} = -A\left[n + \frac{1}{2}\right] - \frac{B}{2} \pm 2i\Omega^{\pm}_n ,$$
$$\Omega^{\pm}_n = \frac{1\pm1}{2}\left[g^2(n+1) - \frac{(A-B)^2}{16}\right]^{1/2} , \tag{4.15}$$

where Ω^{+}_n is the modified Rabi frequency. More generally, for $k = 0$, (4.9) involves two frequencies Ω^{+}_n and Ω^{-}_n, one of which is purely imaginary, thus attenuating or enhancing the (real) damping part $-A(n+\frac{1}{2}) - B/2$. The second one is positive and real and defines the generalized Rabi frequencies (which are the imaginary parts of the eigenvalues).

As an immediate application consider the time dependence of the mean number of photons $\langle a^\dagger a\rangle$ in the cavity, when an atom is present. The general form of that expectation value reads (for simplicity we choose $\omega = \Omega$, $B = 0$)

$$\langle a^\dagger a\rangle_t = \sum_{n=0}^{\infty} e^{-A(n+1/2)t}\left\{a_n + b_n\cos\left[2\left[g^2(n+1) - \frac{A^2}{16}\right]^{1/2} t\right] + c_n\sin\left[2\left[g^2(n+1) - \frac{A^2}{16}\right]^{1/2} t\right]\right\} , \tag{4.16}$$

where the constants a_n, b_n, c_n are determined by the initial state of the field. For a large initial photon number $\langle a^\dagger a\rangle_0$, the terms oscillating at different frequencies, each being damped differently, add up to produce oscillations around the classical exponential decay $\langle a^\dagger a\rangle_t \sim e^{-At}$, since the atom will now and then absorb and emit one of the photons. As long as their mean number is large the modifications are mere fluctuations of the relative order $1/\langle a^\dagger a\rangle$. After the elapse of a sufficiently long time, however, the only surviving term in (4.16) is the one damped by the factor $e^{-At/2}$:

$$\langle a^\dagger a\rangle_t \cong e^{-At/2}\left\{a_0 + b_0\cos\left[2\left[g^2-\frac{A^2}{16}\right]^{1/2} t\right] + c_0\sin\left[2\left[g^2-\frac{A^2}{16}\right]^{1/2} t\right]\right\} \tag{4.17}$$

as $t\to\infty$ [13]. Now the presence of the atom significantly slows the decay of the remaining photon which can either be lost irreversibly to the photon reservoir or be reversibly absorbed by the atom. Roughly speaking, the photon spends about half of its time inside the atom which shields it from the reservoir and so effectively doubles the photon lifetime in the cavity. This effect, of course, only occurs when the coupling g is large compared to the damping rate A. In Figs. 1 and 2 we discuss the decay of an initial number state of photons in more detail.

Note also that in the classical limit of a negligibly small coupling to the atom, $g\to 0$, all oscillating terms in (4.16) except for the one with the frequency $\Omega_0=(g^2-A^2/16)^{1/2}$ vanish, and since $\Omega_0\to iA/4$, this remaining term gives an additional damping factor $e^{-At/2}$ which combines with the factor already present to the correct classical exponential decay:

$$\left\{a_m + b_m\cos\left[2\left[g^2(n+1)-\frac{A^2}{16}\right]^{1/2} t\right] + c_m\sin\left[2\left[g^2(n+1)-\frac{A^2}{16}\right]^{1/2} t\right]\right\} \to \text{const}\times\delta_{m0}e^{-At/2} \tag{4.18}$$

so that $\langle a^\dagger a\rangle_t \to \langle a^\dagger a\rangle_0 e^{-At}$, as $g\to 0$.

So far we have been concerned with the eigenvalues of Eq. (4.1). To determine the corresponding eigenstates let us return to (4.5) and restrict the discussion, for notational transparence, to $k=0$ again. Let Γ be an eigenvalue of the master equation. Then the coordinate representation of the associated eigenstate P_Γ is given by simultaneous eigencolumns of the coupled system of matrix equations on the right-hand side of (4.5). These columns $\check{X}_n$ are iteratively computed from $\check{X}_N$, which is an eigencolumn of the matrix M_N. Thus

$$\begin{aligned} &M_N\check{X}_N = \Gamma\check{X}_N\,,\\ &\check{X}_{N-1} = (\Gamma - M_{N-1})^{-1}G\check{X}_N\,,\\ &\check{X}_n = (\Gamma-M_n)^{-1}G(\Gamma-M_{n+1})^{-1}G\cdots\\ &\qquad\times(\Gamma-M_{N-1})^{-1}G\check{X}_N\,. \end{aligned} \tag{4.19}$$

As the matrices M_n do have different eigenvalues for different n the inverse matrices appearing in (4.19) surely exist. The matrix G has only two nonzero entries, so that one can actually reduce (4.19) to the form

$$\check{X}_n = (\Gamma-M_n)^{-1}\check{G}_{n,N-1}\check{X}_N \quad \text{for } n\le N-2\,, \tag{4.20}$$

where $\check{G}_{n,N-1}$ is, apart from a multiplicative factor, identical with G:

$$\check{G}_{n,N-1} = g^{N-n-1}G\prod_{j=n+1}^{N-1}[(\Gamma-M_j)^{-1}_{23}+(\Gamma+M_j)^{-1}_{24}]\,. \tag{4.21}$$

The subdeterminants of $\Gamma-M_j$ can be used to evaluate the 23 and 24 matrix elements of the inverse matrices $(\Gamma-M_j)^{-1}$, and so one can give quite explicit expressions for the eigencolumns for any given eigenvalue [14].

After this discussion of right eigenstates of the Liouville operator $\mathcal{L}$ of (4.1) let us now turn to the left ones. For the left eigenstates the equations corresponding to (4.5) are deduced from (3.13) and read, for $k=0$,

$$\Gamma\check{P} = \check{\mathcal{L}}\check{P} = \check{P}\mathcal{L} \Longleftrightarrow \begin{cases} \Gamma X_n = X_n M_n + X_{n-1}G\,,\\ \qquad n = N+1, N+2, \ldots\\ \Gamma X_N = X_N M_N\,, \end{cases} \tag{4.22}$$

and similarly for $k\neq 0$. The rows X_n corresponding to the eigenstate $\check{P}_\Gamma$ for a given eigenvalue Γ are again iteratively calculated from X_N,

$$\begin{aligned} &X_N M_N = \Gamma X_N\,,\\ &X_{N+1} = X_N G(\Gamma-M_{N+1})^{-1}\\ &\quad\vdots\\ &X_n = X_N G(\Gamma-M_{N+1})^{-1}\cdots G(\Gamma-M_n)^{-1}\\ &\quad = X_N\tilde{G}_{N,n-1}(\Gamma-M_n)^{-1}\,, \end{aligned} \tag{4.23}$$

where the matrices $\tilde{G}_{N,n-1}$ are defined in (4.21). Note, however, that $n>N$ in (4.23) and the iteration does not terminate as in (4.20). Thus, in contrast to P_Γ, the coordinate representation of $\check{P}_\Gamma$ involves infinitely many columns.

We have thus given the analytical solution of cavity QED as described by the master equation (4.1). We now turn to the important and interesting application of this theory to the one-atom maser [15].

V. THE ONE-ATOM MASER ($\nu=0, s=0$)

In the situation of the one-atom maser the losses of the resonant photon field are balanced by excited atoms

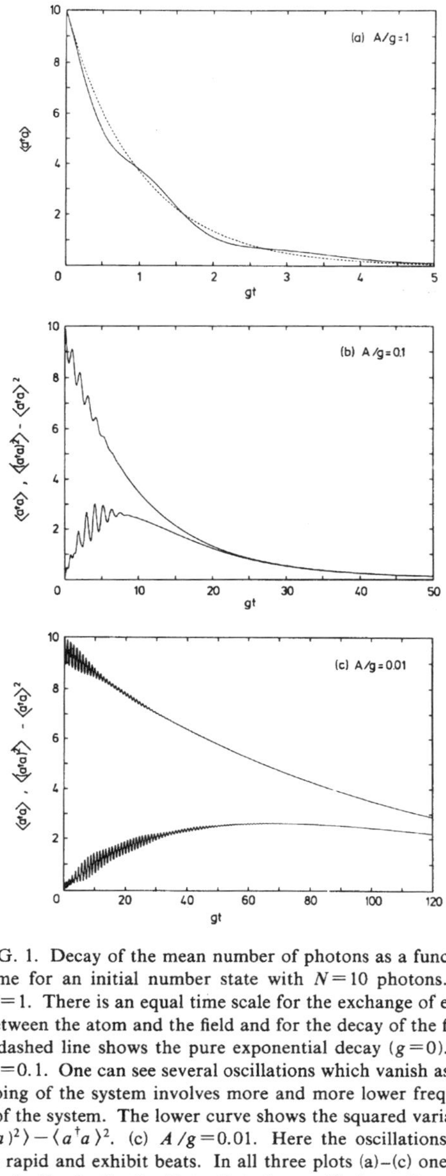

FIG. 1. Decay of the mean number of photons as a function of time for an initial number state with $N=10$ photons. (a) $A/g=1$. There is an equal time scale for the exchange of energy between the atom and the field and for the decay of the field. The dashed line shows the pure exponential decay ($g=0$). (b) $A/g=0.1$. One can see several oscillations which vanish as the damping of the system involves more and more lower frequencies of the system. The lower curve shows the squared variance $\langle (a^\dagger a)^2\rangle - \langle a^\dagger a\rangle^2$. (c) $A/g=0.01$. Here the oscillations are more rapid and exhibit beats. In all three plots (a)–(c) one has $\Delta/g=1$, $B=0$, $C=0$. Note that the oscillations die away once the damping has mixed states with photon numbers less than 10. For large t one can write $\langle a^\dagger a\rangle_t = \sum_{n=0}^{N-1}\sum_{\pm\pm} e^{\Gamma_n^{\pm\pm}t} c_n^{\pm\pm} \sim \sum_{\pm\pm} e^{\Gamma_0^{\pm\pm}t} c_0^{\pm\pm}$ where $\Gamma_0^{\pm\pm}$ are the smallest nonvanishing eigenvalues in (4.9) and $c_n^{\pm\pm}$ are some coefficients [cf. also (4.16) and (4.17) for $\Delta=0$]. Moreover, the coefficients $c_0^{\pm +}$ which multiply the oscillating terms are negligibly small compared to $c_0^{\pm -}$. Thus, an initial state with ten photons shows a very different behavior for large times than an initial state with one photon, although in both situations the relevant eigenvalues, $\Gamma_0^{\pm\pm}$, are the same (cf. Fig. 2). Similar curves are given in Refs. [12] and [18].

which pass through the cavity, depositing a photon in it with a certain probability that depends on the state of the field when the atom enters the cavity. Usually the atomic flux is so small that not more than a single atom is in the cavity at a time.

The dynamics of the photon field is described by cycles, see Fig. 3. A cycle begins, say, when an atom enters the cavity. The state of the composite system is then given by

$$P(0)=\rho_\sigma(0)\rho(0) , \tag{5.1}$$

where ρ_σ and ρ are the reduced states of the atom and the field, respectively. The atom will then interact for a certain time τ with the field. This is described by the master equation (4.1) or formally by

$$P(\tau)=e^{\mathcal{L}\tau}P(0) , \tag{5.2}$$

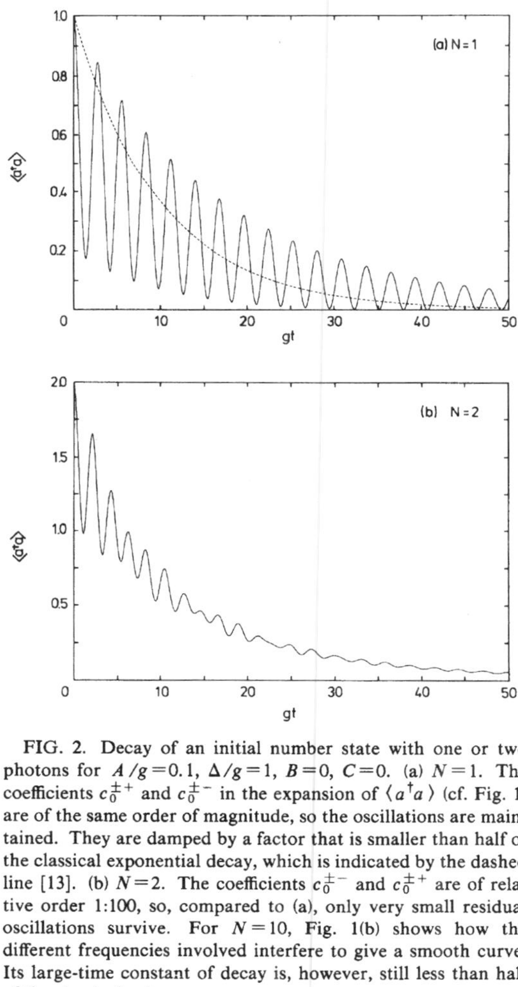

FIG. 2. Decay of an initial number state with one or two photons for $A/g=0.1$, $\Delta/g=1$, $B=0$, $C=0$. (a) $N=1$. The coefficients $c_0^{\pm +}$ and $c_0^{\pm -}$ in the expansion of $\langle a^\dagger a\rangle$ (cf. Fig. 1) are of the same order of magnitude, so the oscillations are maintained. They are damped by a factor that is smaller than half of the classical exponential decay, which is indicated by the dashed line [13]. (b) $N=2$. The coefficients $c_0^{\pm -}$ and $c_0^{\pm +}$ are of relative order 1:100, so, compared to (a), only very small residual oscillations survive. For $N=10$, Fig. 1(b) shows how the different frequencies involved interfere to give a smooth curve. Its large-time constant of decay is, however, still less than half of the classical value.

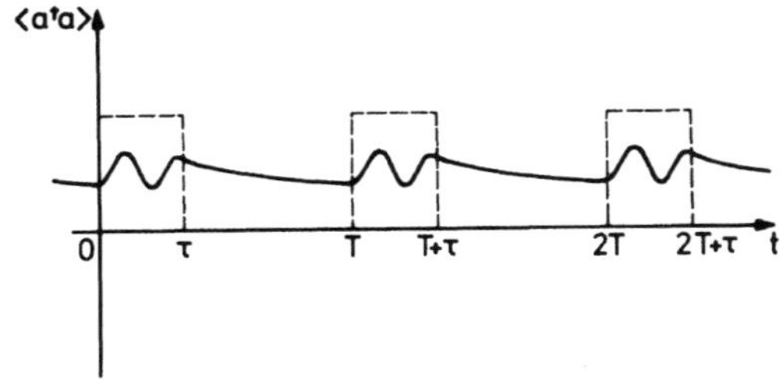

FIG. 3. Cyclically pumped one-atom maser. T is the pumping period, τ the time of interaction between the atom and the field, and $T-\tau$ is the duration of pure decay. To be more realistic we later include a finite probability of excitation for the incoming atoms and a fluctuating interaction time τ owing to a Fizeau-type velocity distribution for the atomic beam.

where $\mathcal{L}$ is the Liouville operator given by the right-hand side of Eq. (4.1). At the time $t=\tau$ the atom leaves the cavity and the reduced state of the photon field is then

$$\rho(\tau)=\mathrm{tr}_\sigma P(\tau)\ . \tag{5.3}$$

Subsequently, the field is subject to pure decay which is described by the master equation (4.1) for $g=0$ (and the trace taken over the atomic variables) or formally by

$$\rho(T)=e^{\mathcal{L}_0(T-\tau)}\rho(\tau)\ , \tag{5.4}$$

where

$$\mathcal{L}_0\rho=-i\omega[a^\dagger a,\rho]+L_a\rho \tag{5.5}$$

and $t=T\geq\tau$ is the time immediately before the next atom enters the cavity. For the following cycle $\rho(T)$ takes on the role of $\rho(0)$ in (5.1), and so on.

So the dynamics of a cycle is split into two parts, which are described by (5.2) and (5.4), respectively. The first part involves the time evolution operator $e^{\mathcal{L}\tau}$, whose precise meaning is given in (1.22) together with the coordinate representations (4.5) and (4.22) for the eigenstates of $\mathcal{L}$ and $\check{\mathcal{L}}$. The second part of the cycle follows (5.4). But this is, in our approach, a trivial special case of the dynamics described by the total $\mathcal{L}$, as the damping-basis states are eigenstates of $\mathcal{L}_0$, too.

Let us now apply the solutions of the eigenvalue problem (4.5) and (4.22) to the one-atom-maser dynamics. Equation (5.2) explicitly reads

$$P(\tau)=\sum_\Gamma e^{\Gamma\tau}P_\Gamma\mathrm{Tr}\{\check{P}_\Gamma P(0)\}\ , \tag{5.6}$$

which is Eq. (1.22). The coordinate representation of the eigenstates P_Γ is

$$\begin{aligned}P_\Gamma&=\rho^{(\Gamma)}\sigma_0+\rho_z^{(\Gamma)}\sigma_z+\rho_-^{(\Gamma)}\sigma_-+\rho_+^{(\Gamma)}\sigma_+\\&=\sum_n(\check{\alpha}_{nk}^{(\Gamma)}\sigma_0+\check{\beta}_{nk}^{(\Gamma)}\sigma_z+\check{\gamma}_{nk-1}^{(\Gamma)}\sigma_-+\check{\eta}_{nk+1}^{(\Gamma)}\sigma_+)\rho_n^{(k)}\ ,\end{aligned} \tag{5.7}$$

where the coefficients $\check{\alpha}_{nk}^{(\Gamma)},\check{\beta}_{nk}^{(\Gamma)},\check{\gamma}_{nk}^{(\Gamma)},\check{\eta}_{nk}^{(\Gamma)}$ make up the columns $\check{X}_{nk}^{(\Gamma)}$ associated with an eigencolumn $\check{X}_{Nk}^{(\Gamma)}$ of the matrix $M_N^{(k)}$, as in Eq. (4.19). Here we explicitly indicate the eigenvalue Γ which also serves as the index of summation in (5.6). Similarly the coordinates of the dual eigenstates are given by

$$\begin{aligned}\check{P}_\Gamma&=\check{\rho}^{(\Gamma)}\check{\sigma}_0+\check{\rho}_z^{(\Gamma)}\check{\sigma}_z+\check{\rho}_-^{(\Gamma)}\check{\sigma}_-+\check{\rho}_+^{(\Gamma)}\check{\sigma}_+\\&=\sum_n(\alpha_{nk}^{(\Gamma)}\check{\sigma}_0+\beta_{nk}^{(\Gamma)}\check{\sigma}_z+\gamma_{nk-1}^{(\Gamma)}\check{\sigma}_-+\eta_{nk+1}^{(\Gamma)}\check{\sigma}_+)\check{\rho}_n^{(k)}\ ,\end{aligned} \tag{5.8}$$

where the coefficients $\alpha_{nk}^{(\Gamma)}$, $\beta_{nk}^{(\Gamma)}$, $\gamma_{nk}^{(\Gamma)}$, $\eta_{nk}^{(\Gamma)}$ make up the rows $X_{nk}^{(\Gamma)}$ associated with an eigenrow $X_{Nk}^{(\Gamma)}$ of the matrix $M_N^{(k)}$. The values of N and k are fixed by the given value of Γ in (5.7) and (5.8).

If we finally write the coordinate representation of the reduced state of the photon field,

$$\rho(t)=\sum_{n,k}\check{\alpha}_{nk}(t)\rho_n^{(k)}\ , \tag{5.9}$$

the first part (5.2) and (5.3) of the one-atom-maser cycle reads

$$\begin{aligned}\rho(\tau)&=\sum_\Gamma e^{\Gamma\tau}\rho^{(\Gamma)}\mathrm{Tr}\{\check{P}_\Gamma P(0)\}\\&=\sum_n\sum_\Gamma e^{\Gamma\tau}\check{\alpha}_{nk}^{(\Gamma)}\mathrm{Tr}\{\check{P}_\Gamma\rho_\sigma(0)\rho(0)\}\rho_n^{(k)}\ .\end{aligned} \tag{5.10}$$

That is,

$$\check{\alpha}_{nk}(\tau)=\sum_\Gamma e^{\Gamma\tau}\check{\alpha}_{nk}^{(\Gamma)}\mathrm{Tr}\{\check{P}_\Gamma\rho_\sigma(0)\rho(0)\}\ , \tag{5.11}$$

which we proceed to express as a mapping of the coefficients $\check{\alpha}_{nk}(0)$ only. Now,

$$\mathrm{Tr}\{\check{P}_\Gamma\rho_\sigma(0)\rho(0)\}=\sum_{n',k'}\check{\alpha}_{n'k'}(0)\mathrm{Tr}\{\check{P}_\Gamma\rho_\sigma(0)\rho_{n'}^{(k')}\} \tag{5.12}$$

and, using (5.8) and the duality relations (2.27) and (2.7),

$$\begin{aligned}\mathrm{Tr}\{\check{P}_\Gamma\rho_\sigma(0)\rho_{n'}^{(k')}\}&=\alpha_{n'k'}^{(\Gamma)}\mathrm{tr}_\sigma\{\rho_\sigma(0)\check{\sigma}_0\}+\beta_{n'k'}^{(\Gamma)}\mathrm{tr}_\sigma\{\rho_\sigma(0)\check{\sigma}_z\}+\gamma_{n'k'-1}^{(\Gamma)}\mathrm{tr}_\sigma\{\rho_\sigma(0)\check{\sigma}_-\}+\eta_{n'k'+1}^{(\Gamma)}\mathrm{tr}_\sigma\{\rho_\sigma(0)\check{\sigma}_+\}\\&\equiv\check{c}_0\alpha_{n'k'}^{(\Gamma)}+\check{c}_z\beta_{n'k'}^{(\Gamma)}+\check{c}_-\gamma_{n'k'-1}^{(\Gamma)}+\check{c}_+\eta_{n'k'+1}^{(\Gamma)}\end{aligned} \tag{5.13}$$

if

$$\rho_\sigma(0)=\check{c}_0\sigma_0+\check{c}_z\sigma_z+\check{c}_-\sigma_-+\check{c}_+\sigma_+\ . \tag{5.14}$$

In this context, the coefficient $\check{c}_0$ always equals unity, of course. After inserting (5.12)–(5.14) into (5.11) we have

$$\check{\alpha}_{nk}(\tau)=\sum_{\Gamma,n',k'}e^{\Gamma\tau}\check{\alpha}_{nk}^{(\Gamma)}\{\check{c}_0\alpha_{n'k'}^{(\Gamma)}+\check{c}_z\beta_{n'k'}^{(\Gamma)}+\check{c}_-\gamma_{n'k'-1}^{(\Gamma)}+\check{c}_+\eta_{n'k'+1}^{(\Gamma)}\}\check{\alpha}_{n'k'}(0)\ . \tag{5.15}$$

The second part of the cycle, following (5.4), simply produces

$$\check{\alpha}_{nk}(T)=e^{-[i\omega k+A(n+|k|/2)](T-\tau)}\check{\alpha}_{nk}(\tau)\ , \tag{5.16}$$

where $-A(n+|k|/2)$ are the eigenvalues of the damping operator L_a.

So the state of the field at the end of a complete cycle,

$$\rho(T)=\sum_{n,k}\check{\alpha}_{nk}(T)\rho_n^{(k)}\ , \tag{5.17}$$

is given by the mapping [16]

$$\check{\alpha}_{nk}(T)=e^{-[i\omega k+A(n+|k|/2)]T}\sum_{\Gamma,n',k'}e^{[\Gamma+i\omega k+A(n+|k|/2)]\tau}\check{\alpha}_{nk}^{(\Gamma)}\{\check{c}_0\alpha_{n'k'}^{(\Gamma)}+\check{c}_z\beta_{n'k'}^{(\Gamma)}+\check{c}_-\gamma_{n'k'-1}^{(\Gamma)}+\check{c}_+\eta_{n'k'+1}^{(\Gamma)}\}\check{\alpha}_{n',k'}(0)\ . \tag{5.18}$$

All quantities in (5.18) are known, so it provides us with the analytical solution of the dynamics of the field in a one-atom maser including damping mechanisms for the atom and cavity leakage during the passage of the atom through the cavity.

Consider now the special case when all entering atoms are in the upper state,

$$\rho_\sigma(0)=\rho_\sigma(T)=\rho_\sigma(2T)=\cdots=\tfrac{1}{2}(1+\sigma_z)=\sigma_0+\sigma_z\ , \tag{5.19}$$

that is, $\check{c}_0=\check{c}_z=1$, $\check{c}_+=\check{c}_-=0$ in (5.14). In this situation a field that is successively loaded by passing atoms, starting from the vacuum state, remains diagonal. If we set, for the sake of additional notational simplicity, $C=B/2$, (5.18) takes on the form

$$\check{\alpha}_n(T)=e^{-AnT}\sum_{n'}\sum_{m}\sum_{\pm\pm}e^{\{-A(m-n+1/2)-B/2\pm 2i\Omega_m^{\pm}\}\tau}\check{\alpha}_n^{(m,\pm,\pm)}(\alpha_{n'}^{(m,\pm,\pm)}+\beta_{n'}^{(m,\pm,\pm)})\check{\alpha}_{n'}(0)\ , \tag{5.20}$$

with $\Omega_m^{\pm}=\Omega_{m0}^{\pm}$ from (4.9), and the coefficients $\check{\alpha}_n^{(m,\pm\pm)}$, $\alpha_{n'}^{(m,\pm\pm)}$, $\beta_{n'}^{(m,\pm\pm)}$ are the components of the eigencolumns and eigenrows of the matrix $M_m^{(0)}$, associated with the eigenvalues $\Gamma_{m0}^{\pm\pm}$ given in (4.9). Note that, according to (2.40) and (2.41), the mean number of photons and its variance are simply determined by the coefficients $\check{\alpha}_1(T)$ and $\check{\alpha}_2(T)$.

In the cyclically steady state one has $\check{\alpha}_n(T)=\check{\alpha}_n(0)=\check{\alpha}_n^{SS}$ for $n=0,1,2,\ldots$. So these coefficients $\check{\alpha}_n^{SS}$ are mapped onto zero by the matrix with the elements

$$M_{nn'}=e^{AnT}\sum_{m}\sum_{\pm\pm}e^{\{-A(m-n+1/2)-B/2\pm 2i\Omega_m^{\pm}\}\tau}\check{\alpha}_n^{(m,\pm\pm)}$$
$$\times(\alpha_{n'}^{(m,\pm\pm)}+\beta_{n'}^{(m,\pm\pm)})-\delta_{nn'}\ , \tag{5.21}$$

with $n,n'=0,1,2,\ldots$. Because $\check{\alpha}_{n+1}^{(m,\pm\pm)}=0$ if $n>m$ and similarly $\beta_{n'}^{(m,\pm\pm)}=0$ and $\alpha_{n'+1}^{(m,\pm\pm)}=0$ if $n'<m$, the m summation actually runs only from $n-1$ to n' in (5.21) or (5.20) and the coefficients $\check{\alpha}_n(T)$ depend only on initial coefficients $\check{\alpha}_{n'}(0)$ with $n'\geq n-1$.

So far we have assumed that all atoms are excited and have the same velocity when they enter the cavity. A more realistic description of the one-atom maser must take into account a spread in the velocity of the incoming atoms. This is done by integrating (5.18) over a (narrow) time interval $[\bar{\tau}-\delta\tau/2,\bar{\tau}+\delta\tau/2]$ as the Fizeau-type velocity selectors used in these experiments give approximately equal probabilities for atomic velocities v in a corresponding interval $[\bar{v}-\delta v/2,\bar{v}+\delta v/2]$. This results in replacing the factors

$$e^{[\Gamma+i\omega k+A(n+|k|/2)]\tau} \tag{5.22}$$

in (5.18) by

$$\frac{\sinh\{[\Gamma+i\omega k+A(n+|k|/2)](\delta\tau/2)\}}{[\Gamma+i\omega k+A(n+|k|/2)](\delta\tau/2)}$$
$$\times e^{[\Gamma+i\omega k+A(n+|k|/2)]\bar{\tau}}\ . \tag{5.23}$$

Further, one must take into account that the incoming atoms are only excited with a certain probability p. If they are excited, the dynamics follows (5.18). If they are *not* excited, the cavity field will be subject to a whole period of pure decay, which happens with the probability $1-p$. The inclusion of this pumping statistical parameter p into the above description modifies (5.18) to

$$\check{\alpha}_{nk}(T)=e^{-[i\omega k+A(n+|k|/2)]T}$$
$$\times\sum_{n',k'}\left\{p\sum_{\Gamma}e^{[\Gamma+i\omega k+A(n+|k|/2)]\tau}\check{\alpha}_{nk}^{(\Gamma)}(\check{c}_0\alpha_{n'k'}^{(\Gamma)}+\check{c}_z\beta_{n'k'}^{(\Gamma)}+\check{c}_-\gamma_{n'k'-1}^{(\Gamma)}+\check{c}_+\eta_{n'k'+1}^{(\Gamma)})+(1-p)\delta_{nn'}\delta_{kk'}\right\}\check{\alpha}_{n'k'}(0)\ . \tag{5.24}$$

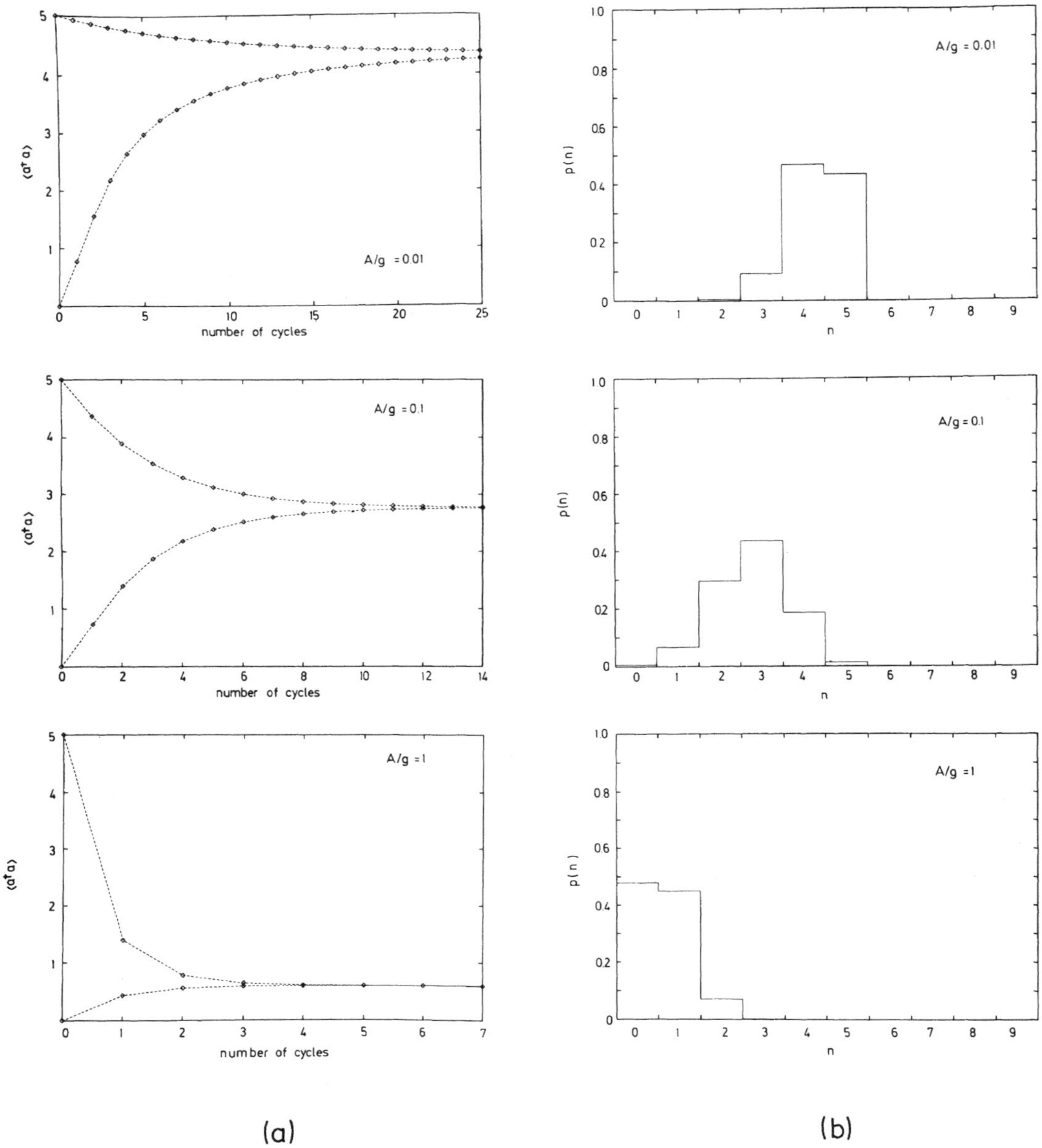

FIG. 4. (a) Stroboscopic evolution of the field in a one-atom maser towards the trapped state (5,1), given by $\Omega_5^+\tau=\pi$, for different values of A. The atoms enter the cavity at equidistant times $t=0,\tau,2\tau,3\tau,\ldots$, which means that there is always exactly one atom in the cavity so that there are no periods of pure relaxation. Both the vacuum and an initial state with five photons evolve towards a (cyclically) steady state with *not more* than five photons. The stroboscopic evolution is described by the quantum map (5.20) with $t=T=\pi/\Omega_5^+$. (b) In the (cyclically) steady state, (5.20) maps the coefficients $\check{\alpha}_n(0)$ onto themselves. Note that, for this trapping value for t the map (5.20) involves only the 6×6 matrix $M_{nn'}$ of (5.21) with n,n' running from 0 to 5. The probability distribution $p(n)$ for the final photon number is plotted for different values of A corresponding to (a). One can see that the final state depends very sensitively on the damping constant and only for very small damping can an approximate number state with $N=5$ photons be obtained.

Together with the replacement (5.22)→(5.23), the mapping (5.24) gives the analytical solution of the dynamics of the photon field in realistic one-atom-maser experiments. It includes detuning, cavity leakage effects, spontaneous decay mechanisms for the atoms, a Fizeau-type velocity distribution for the atomic beam, and a statistical parameter p for the excitation of the incoming atoms covering the limits $p \rightarrow 0$ for Poissonian pumping and $p \rightarrow 1$ for regular pumping. Incidentally, pumping with atomic coherence is also contained in (5.18) and (5.24) through the coefficients $\check{c}_+$ and $\check{c}_-$ of (5.14).

Numerical results of elementary one-atom-maser calculations are presented in Figs. 4 and 5. More details, including computations of the linewidth and the properties of the steady state with the aid of the quantum map (5.24), are planned to be reported shortly elsewhere.

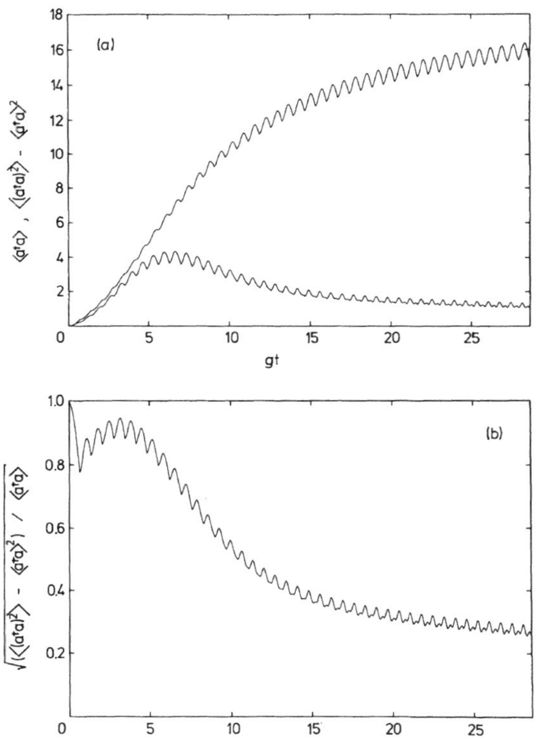

FIG. 5. Explicit evolution of the mean number of photons in a one-atom maser. The initial state is the vacuum and the system evolves towards the trapping state (19,1), which is given by $\Omega_{19}^{+}\tau = \pi$. Note that there are no periods of pure relaxation of the field since $\tau = T$, so an atom enters the cavity immediately after the preceding one has left it. The upper and the lower curves in (a) show the mean and the squared variance of the photon number, respectively. The curve in (b) gives the normalized variance. It shows that the field remains sub-Poissonian all the time while approaching the steady state. The parameters in (a) and (b) are $A/g = 0.001$, $D/g = 1$, $B = C = 0$.

VI. OUTLOOK: THE ONE-ATOM LASER

In the one-atom maser the cavity losses are balanced by excited atoms depositing photons while they pass through the cavity. If there is no atomic beam carrying energy into the resonator, but rather a single atom that stays in permanent interaction with the field, a steady state can only be achieved if the losses of the field are balanced by a permanent incoherent excitation of that atom. This is the standard pumping mechanism. In the master equation (1.1) it is taken into account by the term (1.6) with $s > 0$.

In this regime (1.1) describes a "one-atom laser," that is, a single permanently pumped atom in interaction with a field mode in a resonator with losses (mainly owing to the outgoing laser light). Its dynamics is given by Eqs. (3.6a)–(3.6d) or (3.7), respectively, where now $s > 0$, so that we really have to deal with the three-term recurrence relations (3.8) and (3.13) with matrices $F_n \neq 0$. The same situation arises at finite temperature ($\nu > 0$) of the photon reservoir.

These three-term recursions can be solved numerically by using the method of matrix continued fractions [17]. The results of this analysis are reported in [7].

ACKNOWLEDGMENTS

We would like to thank J. Bergou and Ch. Ginzel for valuable discussions as well as E. Figge, E. Krönauer, and U. Martini for technical assistance. H.J.B. thanks the Studienstiftung des deutschen Volkes for financial support.

APPENDIX

Inasmuch as the presence of the factor $\delta_{kk'}$ in the duality relation (2.27) for the $\rho_n^{(k)}$ of (2.23) and the $\check{\rho}_{n'}^{(k')}$ of (2.26) is immediately obvious, it suffices to derive this statement for $k = k'$. In addition, we can choose the common k value non-negative, because the sign of k is changed by taking the adjoint.

For $k \geq 0$, consider the generating functions

$$\rho^{(k)}(x) = \sum_{n=0}^{\infty} x^n \rho_n^{(k)} = \left[\frac{1}{1+\nu} \frac{1}{1+x} \right]^{k+1} a^{\dagger k} \!:\! \exp\left[-\frac{a^\dagger a}{(1+\nu)(1+x)} \right] \!: \tag{A1}$$

and

$$\check{\rho}^{(k)}(y) = \sum_{n=0}^{\infty} \frac{(n+k)!}{n!} y^n \check{\rho}_n^{(k)} = \left[\frac{1+\nu}{1+\nu(1+y)} \right]^{k+1} \!:\! \exp\left[\frac{y a^\dagger a}{1+\nu(1+y)} \right] \!:\! a^k , \tag{A2}$$

where the summations are performed with the aid of the identity

$$\left[\frac{1}{1+z}\right]^{k+1} \exp\left[\frac{xz}{1+z}\right] = \sum_{n=0}^{\infty} (-z)^n L_n^{(k)}(x) \; . \tag{A3}$$

Two other identities, viz.,

$$:\exp(za^\dagger a): = (1+z)^{a^\dagger a} \tag{A4}$$

and

$$a^{\dagger k} z^{a^\dagger a} a^k = \left[\frac{\partial}{\partial z}\right]^k z^{a^\dagger a} \; , \tag{A5}$$

are used in establishing

$$\rho^{(k)}(x)\check{\rho}^{(k)}(y) = \left[\frac{1}{1+x}\frac{1}{1+\nu(1+y)}\right]^{k+1} \times \left[\frac{\partial}{\partial\lambda}\right]^k \lambda^{a^\dagger a} \; , \tag{A6}$$

with

$$\lambda = \left[1 - \frac{1}{(1+\nu)(1+x)}\right]\left[1 + \frac{y}{1+\nu(1+y)}\right] . \tag{A7}$$

As a consequence we find

$$\begin{aligned} \mathrm{tr}\{\rho^{(k)}(x)\check{\rho}^{(k)}(y)\} &= \left[\frac{1}{1+x}\frac{1}{1+\nu(1+y)}\right] \times \left[\frac{\partial}{\partial\lambda}\right]^k \frac{1}{1-\lambda} \\ &= k!\left[\frac{1}{1+x}\frac{1}{1+\nu(1+y)}\frac{1}{1-\lambda}\right]^{k+1} \\ &= k!\left[\frac{1}{1-xy}\right]^{k+1} . \end{aligned} \tag{A8}$$

This implies

$$\sum_{n,n'=0}^{\infty} x^n \frac{(n'+k)!}{n'!} y^{n'} \mathrm{tr}\{\rho_n^{(k)}\check{\rho}_{n'}^{(k)}\} = \sum_{n=0}^{\infty} \frac{(n+k)!}{n!}(xy)^n \tag{A9}$$

or, equivalently,

$$\mathrm{tr}\{\rho_n^{(k)}\check{\rho}_{n'}^{(k)}\} = \delta_{nn'} \tag{A10}$$

and thus confirms the duality relation (2.27). In other words, the normalization factors in (2.23) and (2.26) are well chosen, indeed.

---

*Also at the Max-Planck-Institut für Quantenoptik, Ludwig-Prandtl-Straβe 10, D-8046 Garching, Germany.

[1] E. T. Jaynes and F. W. Cummings, Proc. IEEE **51**, 89 (1963).

[2] W. Weidlich, H. Risken, and H. Haken, Z. Phys. **201**, 396 (1967).

[3] H. Risken, in *Progress in Optics,* edited by E. Wolf (North-Holland, Amsterdam, London, 1970), Vol. 8.

[4] Ch. Ginzel, H.-J. Briegel, B.-G. Englert, and A. Schenzle (unpublished).

[5] Experiments with temperatures $T \lesssim 0.1$ K corresponding to a mean number $\nu \lesssim 10^{-5}$ of thermal photons in the cavity are in progress [H. Kosciessa, G. Raithel, *et al.* (private communication)].

[6] This dual conjugate $\check{\mathcal{L}}$ need not be identical with the Hermitian conjugate $\mathcal{L}^*$ whose defining property

$$(P_1, \mathcal{L}P_2) = (\mathcal{L}^* P_1, P_2)$$

(to hold for all states P_1 and P_2) is based upon the inner product

$$(P_1, P_2) = \mathrm{Tr}\{P_1^\dagger P_2\} \; .$$

The two conjugates are symmetrically related by

$$\mathcal{L}^* P = (\check{\mathcal{L}} P^\dagger)^\dagger \quad \text{and} \quad \check{\mathcal{L}} P = (\mathcal{L}^* P^\dagger)^\dagger \; .$$

For all semigroup generators

$$\mathcal{L}P = \frac{1}{i\hbar}[H,P] - \sum_j (V_j^\dagger V_j P + P V_j^\dagger V_j - 2V_j P V_j^\dagger) \; ,$$

however, for which (1.8) is an example, the dual conjugate $\check{\mathcal{L}}$ equals the Hermitian conjugate $\mathcal{L}^*$.

[7] Ch. Ginzel, H.-J. Briegel, U. Martini, B.-G. Englert, and A. Schenzle, Phys. Rev. A (to be published).

[8] In the notation of (3.7) it is understood that $\check{\alpha}_{-1,k}, \check{\beta}_{-1,k}, \check{\gamma}_{-1,k}, \check{\eta}_{-1,k} \equiv 0$ for any value of k. In addition, we apply the conventions

$$M_{-1}^{(0)} = 0, \quad G_{-1}^{(0)} = 0 \; ,$$

which incorporates (3.6d) into (3.6b), and

$$M_{-1}^{(k)} = \begin{bmatrix} -A\frac{k}{2} - i\omega k & 0 & 0 & -2g \\ 0 & 0 & 0 & 0 \\ 0 & 0 & 0 & 0 \\ -gk\left[s - \frac{1}{2}\right] & 0 & 0 & -A\left[\frac{k}{2} - \frac{1}{2}\right] - C - i\omega k + i\Delta \end{bmatrix} ,$$

$$G_{-1}^{(k)} = \begin{bmatrix} 0 & 0 & 0 & 0 \\ 0 & 0 & 0 & 0 \\ 0 & 0 & 0 & 0 \\ 0 & -gk & 0 & 0 \end{bmatrix}$$

for $k = 1,2,3,\ldots$, which incorporates (3.6c) into (3.6a). For negative k the complex conjugate equations have to be taken.

[9] In the one-atom maser these leakage effects are mainly owing to the holes in the cavity through which the atoms pass.

[10] According to (4.8) the decay rate of the electric field in a cavity is in particular reduced by the coupling of the field to an atom. This effect has also been discussed by G. S. Agarwal and R. R. Puri, Phys. Rev. A **33**, 1757 (1986) as well as by H. J. Carmichael, R. J. Brecha, M. G. Raizen, H. J. Kimble, and P. R. Rice, *ibid.* **40**, 5516 (1989).

[11] In the special case where $k = 0$, and without atomic damping, i.e., for $B = C = 0$, the eigenvalues (4.9) have independently been found, using "quasiprobability distributions," by Daeubler, Risken, and Schoendorf [12].

[12] B. Daeubler, H. Risken, and L. Schoendorf, Phys. Rev. A

46, 1654 (1992).

[13] See also the references in [10].

[14] The numerical computation is facilitated by the observation that, for $k=0$, the recurrence relation (4.5) can be turned into one involving the $\breve{\alpha}$ coefficients only:

$$\breve{\alpha}_n = -4g^2 n \frac{[\Gamma + A(n+1)][\Gamma + A(n-1)+B][\Gamma + A(n-\frac{1}{2})+C]}{\det(\Gamma - M_{n-1})(\Gamma + An + B)} \breve{\alpha}_{n+1},$$

with

$$\begin{aligned}\det(\Gamma - M_n) = {} & [\Gamma + A(n+1)](\Gamma + An + B) \\ & \times \{[\Gamma + A(n+\tfrac{1}{2})+C]^2 + \Delta^2\} \\ & + 4g^2(n+1)\left[\Gamma + A(n+\tfrac{1}{2}) + \frac{B}{2}\right] \\ & \times [\Gamma + A(n+\tfrac{1}{2})+C] .\end{aligned}$$

This is supplemented by the relations

$$\breve{\beta}_n = -\frac{\Gamma + A(n+1)}{\Gamma + An + B}\breve{\alpha}_{n+1},$$

$$\breve{\gamma}_n = \frac{\Gamma + A(n+1)}{4g(n+1)}\left[i + \frac{\Delta}{\Gamma + A(n+\frac{1}{2})+C}\right]\breve{\alpha}_{n+1},$$

$$\breve{\eta}_n = \frac{\Gamma + A(n+1)}{4g(n+1)}\left[-i + \frac{\Delta}{\Gamma + A(n+\frac{1}{2})+C}\right]\breve{\alpha}_{n+1}$$

for $\Delta \neq 0$ (and slightly modified in the degenerate case $\Delta = 0$) which express the other coefficients in terms of the $\breve{\alpha}_n$ thus found. So, for any given eigenvalue $\Gamma = \Gamma_{N,0}^{(\pm\pm)}$, starting from the maximal nonvanishing coefficient $\breve{\alpha}_{N+1}^{(\Gamma)}$, which is fixed by normalization, one iteratively obtains all coordinates of the eigenstate P_Γ through these relations. This method is also applicable for $\nu \neq 0$ and $s \neq 0$ as well as to the coordinates of the left eigenstates.

[15] D. Meschede, H. Walther, and N. Klein, Phys. Rev. Lett. **54**, 551 (1985).

[16] As every eigenvalue Γ fixes a corresponding diagonal with index k, the summation over the index k' in (4.41) is already settled by the identities

$$\alpha_{n'k'}^{(\Gamma)} = \delta_{k'k}\alpha_{n'k}^{(\Gamma)}, \quad \beta_{n'k'}^{(\Gamma)} = \delta_{k'k}\beta_{n'k}^{(\Gamma)},$$

$$\gamma_{n'k'-1}^{(\Gamma)} = \delta_{k'k+1}\gamma_{n'k}^{(\Gamma)}, \quad \eta_{n'k'+1}^{(\Gamma)} = \delta_{k'k-1}\eta_{n'k}^{(\Gamma)} .$$

[17] H. Risken, *The Fokker-Planck Equation* (Springer-Verlag, Berlin, 1984).

[18] J. Eiselt and H. Risken, Phys. Rev. A **43**, 346 (1991).

Errata:

1. The second ")" parenthesis in Eq. (2.3) should be omitted.
2. In Eq. (4.18) read $g^2(m+1)$ instead of $g^2(n+1)$, twice.
3. In Eq. (4.21) replace $\breve{G}_{n,N-1}$ by $\tilde{G}_{n,N-1}$.
4. The sum in Eq. (5.7) should read
   $\sum_n [(\breve{\alpha}_{nk}^{(\Gamma)}\sigma_0 + \breve{\beta}_{nk}^{(\Gamma)}\sigma_z)\rho_n^{(k)} + \breve{\gamma}_{nk}^{(\Gamma)}\sigma_-\rho_n^{(k+1)} + \breve{\eta}_{nk}^{(\Gamma)}\sigma_+\rho_n^{(k-1)}]$.
5. The sum in Eq. (5.8) should read
   $\sum_n [(\alpha_{nk}^{(\Gamma)}\breve{\sigma}_0 + \beta_{nk}^{(\Gamma)}\breve{\sigma}_z)\breve{\rho}_n^{(k)} + \gamma_{nk}^{(\Gamma)}\breve{\sigma}_-\breve{\rho}_n^{(k+1)} + \eta_{nk}^{(\Gamma)}\breve{\sigma}_+\breve{\rho}_n^{(k-1)}]$.
6. The first term on the right-hand side of Eq. (A8) should be taken to the $k+1$-th power.
7. The third author of Ref. [12] is L. Schoendorff.
8. The third author of Ref. [15] is G. Müller.
9. In Ref. [16] read (5.18) rather than (4.41).
10. In the caption to Fig. 4, the fifth line should begin with $\tau = T = \pi/\Omega_5^+$.
11. The abscissa of Fig. 5(b) should be labeled gt.

Michael Fleischhauer

Department of Physics and Research Center OPTIMAS,
University of Kaiserslautern, 67659 Kaiserslautern, Germany

Paper Title: **The Briegel-Englert equation for a maser with non-Poissonian injection statistics**

Reference: Original article (2014)

Personal message:

Berge Englert and the Micromaser

When learning Quantum Optics, the Micromaser was one of the things that fascinated me a lot—conceptually simple and clear, yet rich in fundamental physics. For me, one of the most virtuousic artists in this field is Berge Englert, whom I met when I joined the group in Munich. Although we never worked together on a specific project, I enjoyed many, many discussions with him. His lecture notes and the sometimes very special Englert homework problems are still used in my classes today. A great deal of my knowledge and most of my way of thinking about foundations of quantum mechanics, for which the micromaser is such a beautiful playground, are due to Berge. Thank you very much for this! One beautiful work from that time which he wrote together with Hans Briegel, was the theory of a micromaser with non-Poissonian injection statistics [Phys. Rev. A **52**, 2361 (1995)]. The following notes contain an attempt to derive the basic equation of that special micromaser as first suggested by Hans and Berge.

Happy Birthday, Berge!

The Briegel-Englert equation for a maser with non-Poissonian injection statistics

Michael Fleischhauer
Department of Physics and Research Center OPTIMAS
University of Kaiserslautern, 67659 Kaiserslautern, Germany

A derivation of the coarse-grained dynamical equation for a maser with periodic injection of atoms suggested by Briegel and Englert [Phys. Rev. A **52**, 2361 (1995)] from the microscopic master equation is presented.

The description of the dynamics of a maser with non-Poissonian injection statistics of atoms has been the subject of several discussions in the literature [1]. While a microscopic description of the field evolution by a sequence of kicks from the atoms and subsequent periods of free decay adequately models the maser dynamics [2], several approaches to derive a quasi-continuous master equation on a coarse-grained time-scale failed for non-Poissonian pumping [3]. Briegel and Englert suggested a macroscopic master equation which overcomes these problems [4]:

$$\frac{\partial}{\partial t}\bar{\rho}(t) = \mathcal{L}\bar{\rho}(t) + \mathcal{K}\frac{\mathcal{L}}{1-e^{-\mathcal{L}T}}\bar{\rho}(t). \tag{1}$$

The aim of this short note is to derive this equation from the microscopic master equation.

We here consider the following microscopic model of the maser: Some pumping mechanism periodically injects excited two-level atoms into a resonator. The number of atoms entering the cavity at the periodic times $t_j = jT$ $(j = 0, \pm 1, \ldots)$ can fluctuate. The probability for a k-atom event is denoted by p_k. If we assume that the transit time of the atom(s) is short compared to the time scale of interest, the effect of the atoms on the field can be described by a sequence of quasi instantaneous kicks,

$$\rho(jT+0) = (1+\mathcal{K})\rho(jT-0), \tag{2}$$

where

$$\mathcal{K} = \sum_{k=1}^{\infty} p_k \mathcal{M}_k \tag{3}$$

is the operator describing the average effect of the injected atoms on the field. $\mathcal{M}_k$ accounts for the change in the field resulting from a k-atom event and is not further specified here. It depends on the actual interaction process and the passage time. Denoting the Liouvillian that describes the coupling to the cavity reservoir by $\mathcal{L}$, the microscopic master equation for the field evolution in the interaction picture reads as

$$\frac{\partial}{\partial t}\rho(t) = \mathcal{L}\rho(t) + \lim_{\epsilon\to+0}\mathcal{K}\sum_j \delta(t-jT)\rho(t-\epsilon). \tag{4}$$

In order to derive an equation of motion on a coarse-grained time scale, we introduce a time-averaged density operator,

$$\bar{\rho}(t) = \int_{-T_0/2}^{T_0/2} d\tau\, \rho(t-\tau)f(\tau), \tag{5}$$

where $f(\tau)$ is a properly normalized, slowly varying function of time. The averaging interval T_0 is assumed to be larger, or of the order of, the injection period T. It is clear that we can not define a coarse-grained density operator in the immediate vicinity of the initial time $t = 0$. This is a generic feature of any coarse-graining approximation to an initial-value problem and the definition of $\bar{\rho}(t)$ makes only sense for times larger than the averaging interval $T_0/2$.

In this sense we may rewrite Eq. (5) in a form convenient for a Laplace transformation,

$$\bar{\rho}(t) = \int_0^t d\tau\, \rho(t-\tau)f(\tau). \tag{6}$$

We now use this equation as a definition of a coarse-grained density operator, noting that it has the correct properties of a density operator only for time $t \geq T_0$. The Laplace transform of $\bar{\rho}$ is then simply obtained from that of ρ via $\bar{\rho}(s) = \rho(s)f(s)$, where, for notational simplicity, we use the same symbols for the functions in Laplace space.

We proceed by transforming the microscopic master equation (4). The multiplication with the "filter function" $f(s)$ will allow some approximations which eventually yield the desired macroscopic master equation,

$$\begin{aligned}&(s-\mathcal{L})\rho(s) &(7)\\ &= \rho(t=0) + \lim_{\epsilon\to+0}\int_0^\infty dt \sum_j \delta(t-jT)e^{-st}\mathcal{K}\rho(t-\epsilon).\end{aligned}$$

The sum of delta-functions in Eq. (8) is equivalent to a sum of exponentials

$$\sum_{j=-\infty}^{\infty}\delta(t-jT) = \frac{1}{T}\sum_{\nu=-\infty}^{\infty} e^{2\pi i\nu t/T} \tag{8}$$

which yields

$$\begin{aligned}&(s-\mathcal{L})\rho(s) &(9)\\ &= \rho(t=0) + \lim_{\epsilon\to+0}\frac{\mathcal{K}}{T}\sum_\nu \int_0^\infty dt\; e^{-(s-2\pi i\nu/T)t}\rho(t-\epsilon).\end{aligned}$$

2

In Eq. (10) we can immediately identify the Laplace transform of ρ with a shifted argument:

$$(s-\mathcal{L})\rho(s) = \rho(t=0) + \lim_{\epsilon\to+0}\frac{\mathcal{K}}{T}\sum_{\nu} e^{-(s-2\pi i\nu/T)\epsilon}\rho(s-\frac{2\pi i\nu}{T}). \tag{10}$$

Since the r.h.s. of Eq. (11) is invariant under the transformation $s \to s+2\pi i n/T$ with $n = 0, \pm 1, \ldots$, the l.h.s is invariant as well. From this we infer

$$\rho(s+\frac{2\pi i\nu}{T}) = \left[s+\frac{2\pi i\nu}{T}-\mathcal{L}\right]^{-1}(s-\mathcal{L})\,\rho(s). \tag{11}$$

Inserting this result into Eq.(11) yields

$$\begin{aligned}(s-\mathcal{L})\rho(s)-\rho(t=0) &= \lim_{\epsilon\to+0}\frac{\mathcal{K}}{T}\sum_{\nu=-\infty}^{\infty}\frac{s-\mathcal{L}}{s+\frac{2\pi i\nu}{T}-\mathcal{L}}\,\rho(s)\,e^{-(s+2\pi i\nu/T)\epsilon}\\ &= \lim_{\epsilon\to+0}\frac{\mathcal{K}}{T}\sum_{\nu=-\infty}^{\infty}\left[1-\frac{i\frac{2\pi\nu}{T}}{s+\frac{2\pi i\nu}{T}-\mathcal{L}}\right]\rho(s)\,e^{-(s+2\pi i\nu/T)\epsilon}.\end{aligned} \tag{12}$$

So far, no approximations have been made. We now multiply Eq. (12) with the filter function $f(s)$, which rapidly decreases for $|s| > T_0^{-1}$. If we take an averaging interval long compared to the period of injection T, we may neglect s in the denominator of Eq. (12) as compared to $2\pi i\nu/T$. Note that the second term in the brackets vanishes for $\nu = 0$.

If we furthermore rewrite the sum over ν as a sum of integrals with the help of the Poisson summation formula [5], we obtain:

$$(s-\mathcal{L})\bar{\rho}(s) = \rho(t=0)f(s) + \lim_{\epsilon\to+0}\frac{\mathcal{K}}{T}\times \int_{-\infty}^{\infty}d\nu\sum_{l=-\infty}^{\infty}e^{-i2\pi\nu(\epsilon/T+l)}\left[1-\frac{\nu}{\nu+i\frac{\mathcal{L}T}{2\pi}}\right]\bar{\rho}(s). \tag{13}$$

The ν integration of the first term gives delta-functions $\delta(\epsilon/T+l)$ which vanish for $l = 0, \pm 1, \ldots$. The ν integration of the second term can be carried out by contour integration. Noting that the eigenvalues of $\mathcal{L}$ are zero or negative, we find that only terms with negative l values contribute. We thus have

$$\begin{aligned}(s-\mathcal{L})\bar{\rho}(s)-\rho(t=0)f(s) &= -\mathcal{K}\sum_{l=-\infty}^{-1}\mathcal{L}e^{-\mathcal{L}Tl}\bar{\rho}(s)\\ &= -\mathcal{K}\sum_{l=1}^{\infty}\mathcal{L}e^{\mathcal{L}Tl}\bar{\rho}(s)\\ &= \mathcal{K}\frac{\mathcal{L}}{1-e^{-\mathcal{L}T}}\,\bar{\rho}(s).\end{aligned} \tag{14}$$

A transformation into the time domain yields

$$\frac{\partial}{\partial t}\bar{\rho}(t) = \rho(t=0)f(t)-\bar{\rho}(t=0) + \mathcal{L}\bar{\rho}(t)+\mathcal{K}\frac{\mathcal{L}}{1-e^{-\mathcal{L}T}}\,\bar{\rho}(s). \tag{15}$$

Noting that according to Eq. (6), $\bar{\rho}(t=0) = 0$, and that for $t > T_0$, $f(t) \equiv 0$, we obtain the macroscopic master equation of Briegel and Englert [4]:

$$\frac{\partial}{\partial t}\bar{\rho}(t) = \mathcal{L}\bar{\rho}(t)+\mathcal{K}\frac{\mathcal{L}}{1-e^{-\mathcal{L}T}}\,\bar{\rho}(t). \tag{16}$$

Since Eq. (6) defines a correct coarse-grained density operator only for $t \geq T_0$, this equation is only true for these times. In order to find the correct initial value of $\bar{\rho}(t)$, one has to solve the microscopic equation for some small time interval and calculate $\bar{\rho}(T_0)$ according to Eq. (5). If the dynamics is, however, sufficiently slow, one may, to a good approximation, apply the macroscopic equation also in the initial time period and identify the macroscopic initial value with the microscopic one.

[1] Y.M. Golubev and I.V. Sokolov, Zh. Exp. Teor. Fiz. **87**, 408 (1984) [Sov. Phys. JETP **60**, 234 (1984)]; J. Bergou, L. Davidovich, M. Orszag, C. Benkert, M. Hillery, and M.O. Scully, Phys. Rev. A **40**, 5073 (1989).

[2] see for example: H.-J. Briegel, B.-G. Englert, C. Ginzel, and A. Schenzle, Phys. Rev. A **49**, 5019 (1994).

[3] L. Davidovich, S.-Y. Zhu, A.Z. Khoury, and C. Su, Phys. Rev. A **46**, 1630 (1992); C. Benkert and K. Rzążewski, Phys. Rev. A **40**, 5073 (1989).

[4] H.-J. Briegel and B.-G. Englert, Phys. Rev. A. **52**, 2361 (1995).

[5] R. Courant and D. Hilbert, *"Methods of Mathematical Physics"* (Interscience, New York, 1953).

Michael Thoss

Institut für Theoretische Physik, Friedrich-Alexander-Universität Erlangen-Nürnberg, Erlangen, Germany

Paper Title: **A Quantum Action Principle for Open Systems**

Reference: Letters in Mathematical Physics **37**, 293–308 (1996)

Personal message:

Dedicated to Professor Berthold-Georg Englert on the occasion of his 60th birthday.

With gratitude and very best wishes,
Michael Thoss

Letters in Mathematical Physics **37:** 293–308, 1996. 293

A Quantum Action Principle for Open Systems

MICHAEL THOSS[1] and BERTHOLD-GEORG ENGLERT[2]*
[1]*Institut für Physikalische und Theoretische Chemie, Technische Universität München, Lichtenbergstrasse 4, D-85748 Garching, Germany*
[2]*Max-Planck-Institut für Quantenoptik, Hans-Kopfermann-Strasse 1, D-85748 Garching, Germany. e-mail: bge@mpq.mpg.de*

(Received: 12 June 1995; revised version: 25 July 1995)

Abstract. Schwinger's quantum action principle is extended to the Liouville space and then to open systems with Markovian dynamics. This leads to a differential description of the Green function for Markovian master equations. For illustration, we apply the extended action principle to the driven, damped harmonic oscillator.

Mathematics Subject Classifications (1991): 49S05, 82C10, 81V80, 70H30.

Key words: Schwinger's quantum action principle, Liouville space, Markovian dynamics, quantum mechanics.

1. Introduction

Schwinger's quantum action principle (QAP) [1–3] allows for a very compact formulation of the laws of quantum mechanics [4]. The QAP is both of fundamental interest and a powerful tool for computations of transition amplitudes [5]. In its original form, the QAP is restricted to quantum systems with unitary time evolution, i.e. to closed systems.

Over the years, there has been a growing interest in open quantum systems. They are useful for the effective description of system-environment interactions or for modeling dissipative phenomena. As a rule, the dynamics in open systems is nonunitary and governed by master equations for the reduced density matrix, or by the Mori and Langevin equations for the system operators.

In this Letter, we investigate the question of how one can extend Schwinger's QAP to open systems [6]. We consider open systems whose time evolution is given by a Markovian master equation.

Here is a brief outline. In the next section, we review the essentials of Schwinger's QAP. Then, in Section 3, we formulate the QAP in Liouville space. One possibility to extend the action principle to open systems with Markovian dynamics is introduced in Section 4. As an illustration, we treat a driven, damped harmonic oscillator in Section 5. We close with a summary.

* Presently at Laboratoire de Physique des Lasers, Unité de recherche associée au CNRS (URA 282), Institut Galilée, Université Paris Nord, Av. J. B. Clément, F-93430 Villetaneuse, France.

This Letter is restricted to nonrelativistic quantum systems with continuous degrees of freedom. There exists also an action principle for discrete variables such as spin [7–9]. Owing to the discrete eigenvalues, one has to use other types of variations then. This would be beyond the scope of the present Letter.

2. Schwinger's Quantum Action Principle

To begin with, let us consider a quantum system with one continuous degree of freedom. Then the dynamical variables are the position and momentum operators $q(t)$ and $p(t)$ with improper eigenstates $|q',t\rangle$ and $|p',t\rangle$, respectively,

$$q(t)|q',t\rangle = |q',t\rangle q', \qquad p(t)|p',t\rangle = |p',t\rangle p', \tag{2.1}$$

and the equal-time commutator

$$[q(t),p(t)] = i\hbar. \tag{2.2}$$

The dynamics is determined by Heisenberg's equation of motion which involves the Hamilton operator $H\,(q(t),p(t),t)$.

Schwinger's QAP provides for a differential characterization of the transition amplitude, as exemplified by

$$\delta\langle q',t'|q'',t''\rangle = \frac{i}{\hbar}\langle q',t'|\delta W^{qq}_{t't''}|q'',t''\rangle. \tag{2.3}$$

It relates variations of the transition amplitude on the left-hand side to the corresponding matrix elements of the variation of an action operator on the right-hand side. The action operator that is appropriate for the transition (2.3) between position eigenstates reads

$$W^{qq}_{t't''} = \int_{t''}^{t'} \mathrm{d}t\,\left[p(t)\frac{\mathrm{d}}{\mathrm{d}t}q(t) - H\,(q(t),p(t),t)\right]. \tag{2.4}$$

Variations of two kinds are allowed in (2.3) with (2.4):

(1) Infinitesimal variations of the structure of the Hamilton operator, which we denote by $\overline{\delta}$,

$$\begin{aligned}\overline{\delta}\langle q',t'|q'',t''\rangle &= \frac{i}{\hbar}\langle q',t'|\overline{\delta}W^{qq}_{t't''}|q'',t''\rangle \\ &= -\frac{i}{\hbar}\langle q',t'|\int_{t''}^{t'}\mathrm{d}t\,\overline{\delta}H\,(q(t),p(t),t)\,|q'',t''\rangle.\end{aligned} \tag{2.5}$$

(2) Infinitesimal variations of the quantum numbers in the transition amplitude, denoted by $\overline{\overline{\delta}}$,

$$\overline{\overline{\delta}}\langle q',t'|q'',t''\rangle = \frac{i}{\hbar}\langle q',t'|\overline{\overline{\delta}}W^{qq}_{t',t''}|q'',t''\rangle \tag{2.6}$$

with

$$\bar{\bar{\delta}} W^{qq}_{t't''} = \int_{\tau''}^{\tau'} \mathrm{d}\tau \left(\frac{\mathrm{d}\tilde{t}}{\mathrm{d}\tau} \left[\tilde{p}(\tilde{t}) \frac{\mathrm{d}}{\mathrm{d}\tilde{t}} \tilde{q}(\tilde{t}) - H\left(\tilde{q}(\tilde{t}), \tilde{p}(\tilde{t}), \tilde{t}\right) \right] - \right.$$

$$\left. - \frac{\mathrm{d}t}{\mathrm{d}\tau} \left[p(t) \frac{\mathrm{d}}{\mathrm{d}t} q(t) - H\left(q(t), p(t), t\right) \right] \right) \tag{2.7}$$

and

$$\tilde{q}(\tilde{t}) = q(t) + \delta q(t), \qquad \tilde{p}(\tilde{t}) = p(t) + \delta p(t), \qquad \tilde{t}(t) = t + \delta t(t). \tag{2.8}$$

Here, $\delta q(t)$, $\delta p(t)$, and $\delta t(t)$ are real, infinitesimal test functions*, and t is a function of the parameter τ, where $t(\tau') = t'$ and $t(\tau'') = t''$.

For such variations $\bar{\bar{\delta}}$ that do not affect the structure of the Hamilton operator, one has

$$\bar{\bar{\delta}} \langle q', t' | q'', t'' \rangle = \frac{i}{\hbar} \langle q', t' | \left[G(t') - G(t'') \right] | q'', t'' \rangle \tag{2.9}$$

with the generator

$$G(t) = p(t)\delta q(t) - H(t)\delta t(t), \tag{2.10}$$

where $H(t)$ is a stand-in for $H(q(t), p(t), t)$. The QAP (2.6) then implies

$$\bar{\bar{\delta}} W^{qq}_{t't''} = G(t') - G(t''). \tag{2.11}$$

This is the principle of stationary action (PSA), which asserts that $\bar{\bar{\delta}} W^{qq}_{t't''}$ is independent of any variations at intermediate times. If we carry out the variations $\bar{\bar{\delta}} W^{qq}_{t't''}$ explicitly, the result is

$$\begin{aligned} &\bar{\bar{\delta}} W^{qq}_{t't''} \\ &= G(t') - G(t'') + \int_{t''}^{t'} \mathrm{d}t \left(\delta p(t) \left[\frac{\mathrm{d}}{\mathrm{d}t} q(t) - \frac{\partial}{\partial p(t)} H(t) \right] - \right. \\ &\left. - \delta q(t) \left[\frac{\mathrm{d}}{\mathrm{d}t} p(t) + \frac{\partial}{\partial q(t)} H(t) \right] + \delta t(t) \left[\frac{\mathrm{d}}{\mathrm{d}t} H(t) - \frac{\partial}{\partial t} H(t) \right] \right). \end{aligned} \tag{2.12}$$

Thus, if we postulate the QAP with the PSA, the integral must vanish and we obtain Heisenberg's equation of motion as an implication. Also, the commutator (2.2) and the Schrödinger equation can be derived from this principle [4, 10]. In other words, Schwinger's QAP offers a compact formulation of the basic laws of quantum mechanics.

* It is possible to consider more general variations $\delta q(t)$ and $\delta p(t)$ that are linear functions of $q(t)$ and $p(t)$, see [4, 5]. For the purpose of this Letter, these extensions are not needed, and we disregard this possibility.

Of course, the QAP is not restricted to one-dimensional systems. For N degrees of freedom, it reads

$$\begin{aligned}&\delta\langle q_1',\ldots,q_N',t'|q_1'',\ldots,q_N'',t''\rangle\\&\quad=\frac{i}{\hbar}\langle q_1',\ldots,q_N',t'|\delta W_{t't''}^{qq}|q_1'',\ldots,q_N'',t''\rangle,\end{aligned}\tag{2.13}$$

with the action

$$W_{t't''}^{qq}=\int_{t''}^{t'}\mathrm{d}t\left[\sum_{k=1}^{N}p_k(t)\frac{\mathrm{d}}{\mathrm{d}t}q_k(t)-H(t)\right].\tag{2.14}$$

Further, transition amplitudes between momentum eigenstates can be considered

$$\begin{aligned}&\delta\langle p_1',\ldots,p_N',t'|p_1'',\ldots,p_N'',t''\rangle\\&\quad=\frac{i}{\hbar}\langle p_1',\ldots,p_N',t'|\delta W_{t't''}^{pp}|p_1'',\ldots,p_N'',t''\rangle,\end{aligned}\tag{2.15}$$

or between position and momentum states,

$$\begin{aligned}&\delta\langle p_1',\ldots,p_N',t'|q_1'',\ldots,q_N'',t''\rangle\\&\quad=\frac{i}{\hbar}\langle p_1',\ldots,p_N',t'|\delta W_{t't''}^{pq}|q_1'',\ldots,q_N'',t''\rangle.\end{aligned}\tag{2.16}$$

The actions $W_{t't''}^{pp}$ and $W_{t't''}^{pq}$ differ from $W_{t't''}^{qq}$ only by terms at the initial and final stages

$$\begin{aligned}W_{t't''}^{pp}&=W_{t't''}^{qq}-\sum_{k=1}^{N}[p_k(t')q_k(t')-p(t'')q(t'')]\\&=\int_{t''}^{t'}\mathrm{d}t\left[-\sum_{k=1}^{N}q_k(t)\frac{\mathrm{d}}{\mathrm{d}t}p_k(t)-H(t)\right],\\W_{t't''}^{pq}&=W_{t't''}^{qq}-\sum_{k=1}^{N}p_k(t')q_k(t').\end{aligned}\tag{2.17}$$

We will also have a use for the Bargmann states $|a_k',t\rangle$ and $\langle a_k^{\dagger'},t|$ (see [11], for example). These are eigenstates of the ladder operators $a_k(t)$ and $a_k^\dagger(t)$, respectively:

$$a_k(t)|a_k',t\rangle=|a_k',t\rangle a_k',\qquad\langle a_k^{\dagger'},t|a_k^\dagger(t)=a_k^{\dagger'}\langle a_k^{\dagger'},t|,\tag{2.18}$$

$$a_k(t)=\frac{1}{\sqrt{2}}\left(\frac{q_k(t)}{\ell}+i\frac{p_k(t)}{\hbar/\ell}\right),\tag{2.19}$$

with complex eigenvalues a'_k and $a^{\dagger\prime}_k$. If one adopts, as we do, the usual convention, according to which $\langle a^{\dagger\prime}_k, t| = (|a'_k, t\rangle)^\dagger$ if $a^{\dagger\prime}_k = (a'_k)$, then the two eigenvalue equations (2.18) are adjoints of each other. The symbol ℓ in (2.19) denotes a typical length of the system under consideration; for instance, $\ell := \sqrt{\hbar/m\omega}$ is fitting for a harmonic oscillator of mass m and frequency ω. For transition amplitudes between Bargmann states, Schwinger's QAP reads

$$\begin{aligned}&\delta\langle a^{\dagger\prime}_1, \ldots, a^{\dagger\prime}_N, t'|a''_1, \ldots, a''_N, t''\rangle \\ &= \frac{i}{\hbar}\langle a^{\dagger\prime}_1, \ldots, a^{\dagger\prime}_N, t'|\delta W^{a^\dagger a}_{t't''}|a''_1, \ldots, a''_N, t''\rangle\end{aligned} \tag{2.20}$$

with the action

$$W^{a^\dagger a}_{t't''} = \int_{t''}^{t'} \mathrm{d}t \left[\sum_{k=1}^{N} i\hbar a^\dagger_k(t)\frac{\mathrm{d}}{\mathrm{d}t}a_k(t) - H(t)\right] - \sum_{k=1}^{N} i\hbar a^\dagger_k(t')a_k(t'). \tag{2.21}$$

Note the structural similarity with $W^{pq}_{t't''}$ of (2.17); indeed, $p_k \to i\hbar a^\dagger_k$, $q_k \to a_k$ is a similarity transformation that preserves commutation relations.

3. QAP in Liouville Space

Open systems must be described, in general, by statistical operators and their density matrices, not by state vectors and their wave functions. In a first step, we therefore extend the QAP to Liouville space. This space constitutes of all linear operators acting on the Hilbert space of state vectors under consideration. In this section, we still assume a unitary time evolution.

We use the following notational conventions (similiar to those of [12]). In analogy to the Dirac bras and kets, in Hilbert space we denote the elements of Liouville space by $|A\}$ and $\{A|$. Their relation to the linear operators acting on the state vectors is given by the scalar product, for which

$$\{A|B\} = \mathrm{tr}(A^\dagger B) \tag{3.1}$$

is a possible choice. In particular, we introduce the position element in Liouville space,

$$|q'_+, q'_-, t\} := |q'_+, t\rangle\langle q'_-, t|, \tag{3.2}$$

and its adjoint,

$$\{q'_+, q'_-, t| := \mathrm{tr}\left[|q'_-, t\rangle\langle q'_+, t|\,\cdot\right]. \tag{3.3}$$

The corresponding transition element,

$$\{q'_+, q'_-; t' \mid q''_+, q''_-, t''\} = \langle q'_+, t' \mid q''_+, t''\rangle\langle q''_-, t'' \mid q'_-, t'\rangle, \tag{3.4}$$

is the Green function of the density matrix in position representation,

$$\begin{aligned}\rho(q'_+, q'_-, t') &= \{q'_+, q'_-, t' \mid \rho(t'')\} \\ &= \int \mathrm{d}q''_+ \, \mathrm{d}q''_- \, \{q'_+, q'_-, t' \mid q''_+, q''_-, t''\}\rho(q''_+, q''_-, t'').\end{aligned} \tag{3.5}$$

The diagonal element equals the transition probability,

$$\{q'_+, q'_+, t' \mid q''_+, q''_+, t''\} = \langle q'_+, t' \mid q''_+, t''\rangle|^2. \tag{3.6}$$

Analogous expressions hold for momentum and Bargmann states.

Inasmuch as the operators that constitute the elements of the Liouville space do not commute, it is expedient to associate two operators on the Liouville space with each operator acting on the Hilbert space. Such linear operators that map elements of the Liouville space onto each other are frequently called superoperators. The pairs of superoperators associated with the operators $q(t), p(t), a(t)$, and $a^\dagger(t)$ are explicitly given by

$$\begin{aligned} q_+(t)A &:= q(t)A, & q_-(t)A &:= Aq(t), \\ p_+(t)A &:= p(t)A, & p_-(t)A &:= Ap(t), \\ a_+(t)A &:= a(t)A, & a_-(t)A &:= Aa(t), \\ a^\dagger_+(t)A &:= a^\dagger(t)A, & a^\dagger_-(t)A &:= Aa^\dagger(t), \end{aligned} \tag{3.7}$$

where A is any element in Liouville space. Thus, the plus sign denotes the multiplication from the left, and the minus sign that from the right. Obvious rules are exemplified by

$$(qp)_+ = q_+p_+, \qquad (qp)_- = p_-q_- = (q_-p_-)^\dagger; \tag{3.8}$$

generalizations to products with more than two factors are immediate. The position and momentum superoperators obey these commutation relations:

$$[q_+(t), p_+(t)] = i\hbar = [\, p_-(t), q_-(t)] \tag{3.9}$$

for equal times and

$$\begin{aligned} [q_+(t), q_-(\tau)] = 0 = [\, p_+(t), p_-(\tau)], \\ [q_+(t), p_-(\tau)] = 0 = [q_-(t), p_+(\tau)] \end{aligned} \tag{3.10}$$

for arbitrary times t and τ. The right and left eigenstates of $q_+(t)$ and $q_-(t)$ are

$$\begin{aligned} q_+(t)|q'_+, q'_-, t\} = q'_+|q'_+, q'_-, t\}, \quad q_-(t)|q'_+, q'_-, t\} = q'_-|q'_+, q'_-, t\}, \\ \{q'_+, q'_-, t|q_+(t) = q'_+\{q'_+, q'_-, t|, \quad \{q'_+, q'_-, t|q_-(t) = q'_-\{q'_+, q'_-, t| \end{aligned} \tag{3.11}$$

and analogous statements hold for the momentum superoperators. The temporal evolution in Liouville space is determined by the Liouville (super)operator

$$\mathcal{L}(t) = \mathcal{L}\left(q(t), p(t), t\right) = -\frac{i}{\hbar}[H\left(q(t), p(t), t\right)^{\cdot}, \cdot]. \tag{3.12}$$

In view of the commutator that enters the definition of $\mathcal{L}$, it is fitting to regard this Liouville operator as a function of $q_\pm$ and $p_\pm$. We emphasize this structure by writing

$$\begin{aligned}\mathcal{L}_{+-}(t) &:= \mathcal{L}_{+-}\left(q_+(t), p_+(t), q_-(t), p_-(t), t\right) \\ &:= -\frac{i}{\hbar} H\left(q_+(t), p_+(t), t\right) + \frac{i}{\hbar} H_-\left(q_-(t), p_-(t), t\right),\end{aligned} \tag{3.13}$$

where H_- is given by*

$$H_-(q_-, p_-, t) = \left[H(q_-, p_-, t)\right]^\dagger, \tag{3.14}$$

as implied by (3.8). The position Liouville elements obey the relations

$$\begin{aligned}\frac{\partial}{\partial t}|q'_+, q'_-, t\} &= -\mathcal{L}_{+-}(t)|q'_+, q'_-, t\}, \\ \frac{\partial}{\partial t}\{q'_+, q'_-, t| &= \{q'_+, q'_-, t|\mathcal{L}_{+-}(t),\end{aligned} \tag{3.15}$$

which essentially state Schrödinger's equation of motion. Heisenberg's equations of motion for q_+ and q_- take the form

$$\frac{\mathrm{d}}{\mathrm{d}t} q_\pm(t) = -[\mathcal{L}_{+-}(t), q_\pm(t)] \tag{3.16}$$

and analogously for the momentum operators p_+ and p_-. With the aid of these definitions, we can now formulate the QAP in Liouville space in a similiar form as in Hilbert space, viz.

$$\delta\{q'_+, q'_-, t' \mid q''_+, q''_-, t''\} = \frac{i}{\hbar}\{q'_+, q'_-, t'|\, \delta W^{q_+ q_-}_{t't''}\, |q''_+, q''_-, t''\} \tag{3.17}$$

with the action

$$\begin{aligned}W^{q_+ q_-}_{t't''} = \int_{t''}^{t'} \mathrm{d}t\, \Big[& p_+(t)\frac{\mathrm{d}}{\mathrm{d}t} q_+(t) - p_-(t)\frac{\mathrm{d}}{\mathrm{d}t} q_-(t) - \\ & - i\hbar\mathcal{L}_{+-}\left(q_+(t), p_+(t), q_-(t), p_-(t), t\right)\Big].\end{aligned} \tag{3.18}$$

This QAP can easily be derived from the Hilbert space form if one uses the product rule for the variations and the definitions of the plus/minus operators. Now, the four operators q_+, q_-, p_+, p_- and the time t are varied independently. As in Hilbert space, the PSA implies the Heisenberg equations of motion for q_+, q_-, p_+, p_-,

* This statement is correct only if H is expressed as a function of q and p, which are Hermitian. When H is given as a function of a and $a^\dagger$, which are Hermitian conjugates of each other, one should first re-express H in terms of q and p or combine (3.7) and (3.13) for a correct identification of H_-.

and $\mathcal{L}_{+-}$. Also, the QAP in Liouville space is, of course, not restricted to one-dimensional systems and can be formulated equally well for transition elements between momentum or Bargmann states.

The extension of the formalism from Hilbert to Liouville space seems to be artificial as long as the time evolution is unitary. But as we shall see in the next section, it is necessary for open systems with dissipation.

4. QAP for Open Systems with Markovian Time Evolution

In this section, we consider the situation of open quantum systems with Markovian time evolution. The dynamics of the statistical operator ρ is then determined by a Markovian master equation,

$$\frac{\partial}{\partial t}\rho = \mathcal{L}\rho = -\frac{i}{\hbar}[H(q,p,t),\rho] + L\rho. \tag{4.1}$$

We assume that the dissipative part L of the Liouville operator $\mathcal{L}$ is of the Lindblad form [13],

$$L\rho = \sum_j([V_j\rho, V_j^\dagger] + [V_j, \rho V_j^\dagger]). \tag{4.2}$$

For the sake of simplicity, we consider one degree of freedom described by the position and momentum operators q and p. Then the operators V_j and $V_j^\dagger$ are functions of q and p. The extension to N degrees of freedom is obvious.

The corresponding dynamics for the observables is generated by the dual operator of $\mathcal{L}$ [14]

$$\frac{\mathrm{d}}{\mathrm{d}t}A = \check{\mathcal{L}}A = \frac{i}{\hbar}[H(q,p,t),A] + \check{L}A, \tag{4.3}$$

where

$$\check{L}A = \sum_j([V_j^\dagger, A]V_j + V_j^\dagger[A, V_j]) \tag{4.4}$$

states the action of the dual partner $\check{L}$ of L.

In closed systems, one essential ingredient for the QAP was the formulation in the Heisenberg picture. But the Heisenberg picture generated by $\check{\mathcal{L}}$ cannot be used in a simple way for the derivation of an action principle because $\check{\mathcal{L}}$ does not fulfil the product rule. This is in conflict with the product rule of the numerical variations. Therefore, we have to introduce another Heisenberg picture in order to get a QAP similiar to (3.17).

The master equation (4.1) is formally solved by the time transformation operator $\Lambda_{t,0}$,

$$\rho(t) = \Lambda_{t,0}\rho(0), \qquad \Lambda_{t,0} = \left[\exp\Bigl(\int_0^t \mathrm{d}s\,\mathcal{L}(q,p,s)\Bigr)\right]^{(+)}, \tag{4.5}$$

where the plus sign denotes the time ordered exponential. A formal inverse of $\Lambda_{t,0}$ is

$$\Lambda_{t,0}^{-1} = \left[\exp\Big(-\int_0^t \mathrm{d}s\,\mathcal{L}(q,p,s)\Big)\right]^{(-)}, \tag{4.6}$$

an exponential with reversed time ordering. As a rule, the domain of $\Lambda_{t,0}^{-1}$ is not the whole Liouville space because $\mathcal{L}$ generates a semigroup only, not a group. For the applications that we have studied (harmonic and anharmonic oscillators) this limitation is not essential, however.

Using the notations introduced in Section 4, we cast the Liouville operator into the form

$$\begin{aligned}
&\mathcal{L}_{+-}(q_+,p_+,q_-,p_-,t)\\
&\quad= -\frac{i}{\hbar}H(q_+,p_+,t) + \frac{i}{\hbar}H_-(q_-,p_-,t) +\\
&\qquad+\sum_j \Big[2V_j(q_+,p_+)V_{j-}^\dagger(q_-,p_-) - V_j^\dagger(q_+,p_+)V_j(q_+,p_+) -\\
&\qquad\qquad - V_{j-}(q_-,p_-)V_{j-}^\dagger(q_-,p_-)\Big].
\end{aligned} \tag{4.7}$$

Here and in the following, the operators q_+,p_+,q_-,p_- refer to time $t=0$, for example, $q_+ = q_+(0)$. They are defined by (3.7) and obey the commutation relations (3.9), (3.10) with $t=\tau=0$. For later times $t\geqslant 0$, we define the Heisenberg operators in accordance with

$$q_\pm(t) := \Lambda_{t,0}^{-1}q_\pm(0)\Lambda_{t,0}, \qquad p_\pm(t) := \Lambda_{t,0}^{-1}p_\pm(0)\Lambda_{t,0}. \tag{4.8}$$

In the same way, one gets the Liouville operator

$$\begin{aligned}
\mathcal{L}_{+-}(t) &= \Lambda_{t,0}^{-1}\,\mathcal{L}_{+-}(q_+,p_+,q_-,p_-,t)\,\Lambda_{t,0}\\
&= \mathcal{L}_{+-}\left(q_+(t),p_+(t),q_-(t),p_-(t),t\right)\\
&= -\frac{i}{\hbar}H\left(q_+(t),p_+(t),t\right) + \frac{i}{\hbar}H_-\left(q_-(t),p_-(t),t\right) +\\
&\quad+\sum_j\Big[2V_j\left(q_+(t),p_+(t)\right)V_{j-}^\dagger\left(q_-(t),p_-(t)\right) -\\
&\qquad - V_j^\dagger\left(q_+(t),p_+(t)\right)V_j\left(q_+(t),p_+(t)\right) -\\
&\qquad - V_{j-}\left(q_-(t),p_-(t)\right)V_{j-}^\dagger\left(q_-(t),p_-(t)\right)\Big].
\end{aligned} \tag{4.9}$$

These operators obey the equation of motion (3.16), and the commutators (3.9), (3.10) hold at equal times $t = \tau$. For $t \neq \tau$, the commutation relations (3.10) may not be valid for the (super)operators (4.8). Because $\Lambda_{t,0}$ is typically nonunitary, the operators $q_\pm(t)$ and $p_\pm(t)$ are not Hermitian at times $t > 0$, but equations such as

$$\left(q_\pm(t)\right)^\dagger = \Lambda_{t,0}^\dagger\, q_\pm(0) \left(\Lambda_{t,0}^{-1}\right)^\dagger \tag{4.10}$$

hold. Therefore, these operators do not describe the physical observables at $t > 0$. They are just variables needed for the formulation of the action principle. If the generator $\mathcal{L}$ is not explicitly time-dependent, one can obtain the physical operators $q(t), p(t)$ from the variables $q_+(t), p_+(t), q_-(t), p_-(t)$. In this case,

$$\Lambda_{t,0} = \exp(\mathcal{L}t), \tag{4.11}$$

and for generators $\mathcal{L}$ of the Lindblad form (4.2) the dual $\check{\mathcal{L}}$ equals the adjoint $\mathcal{L}^\dagger$, so that

$$\left(\Lambda_{t,0}^{-1}\right)^\dagger = \exp(-\check{\mathcal{L}}t). \tag{4.12}$$

For the physical operators we then have

$$q(t) = \exp(\check{\mathcal{L}}t)q(0) = \left(q_\pm(t)\right)^\dagger \mathbf{1}, \tag{4.13}$$

where $\mathbf{1}$ denotes the identity in Liouville space.

The left and right eigenstates of $q_+(t)$ and $q_-(t)$ are

$$\{q'_+, q'_-, t| := \{q'_+, q'_-, 0|\Lambda_{t,0}, \qquad |q'_+, q'_-, t\} := \Lambda_{t,0}^{-1}|q'_+, q'_-, 0\}. \tag{4.14}$$

Equations (3.11) are then obeyed, but the expression (3.2) need not be correct at times $t > 0$ because a pure state may evolve into a mixed state.

With the aid of (4.14), we can represent the time transformation operator $\Lambda_{t,0}$ by position elements

$$\begin{aligned}&\Lambda_{t,0}(q'_+ q'_- | q''_+ q''_-)\\ &\quad = \{q'_+, q'_-, 0|\Lambda_{t,0}|q''_+, q''_-, 0\} = \{q'_+, q'_-, t|q''_+, q''_-, 0\}.\end{aligned} \tag{4.15}$$

The generalized QAP will state the response of such matrix elements to variations of the quantum numbers $q'_+, \ldots, q''_-$ and other parameters of the dynamics. For the derivation of the action principle, the generators for infinitesimal variations of the quantum numbers are particularly important. Upon applying the commutation relations (3.9), (3.10) at equal times $t = \tau$, we find

$$\begin{aligned}\delta_{q'_+}|q'_+, q'_-, t\} &= -\frac{i}{\hbar}\delta q'_+ p_+(t)|q'_+, q'_-, t\},\\ \delta_{q'_-}|q'_+, q'_-, t\} &= \frac{i}{\hbar}\delta q'_- p_-(t)|q'_+, q'_-, t\},\\ \delta_t|q'_+, q'_-, t\} &= -\delta t\mathcal{L}_{+-}(t)|q'_+, q'_-, t\},\end{aligned} \tag{4.16}$$

and

$$\begin{aligned}
\delta_{q'_+}\{q'_+, q'_-, t| &= \frac{i}{\hbar}\{q'_+, q'_-, t|p_+(t)\delta q'_+, \\
\delta_{q'_-}\{q'_+, q'_-, t| &= -\frac{i}{\hbar}\{q'_+, q'_-, t|p_-(t)\delta q'_-, \\
\delta_t\{q'_+, q'_-, t| = \{q'_+, q'_-, t|\mathcal{L}_{+-}(t)\delta t,
\end{aligned} \tag{4.17}$$

for the states defined in (4.14).

This enables us to formulate the QAP for propagators of Markovian master equations

$$\delta\{q'_+, q'_-, t'|q''_+, q''_-, t''\} = \frac{i}{\hbar}\{q'_+, q'_-, t'|\delta W^{q_+q_-}_{t't''}|q''_+, q''_-, t''\} \tag{4.18}$$

with the action

$$\begin{aligned}
W^{q_+q_-}_{t't''} = \int_{t''}^{t'} \mathrm{d}t \Big[& p_+(t)\frac{\mathrm{d}}{\mathrm{d}t}q_+(t) - p_-(t)\frac{\mathrm{d}}{\mathrm{d}t}q_-(t) - \\
& - i\hbar\mathcal{L}_{+-}\left(q_+(t), p_+(t), q_-(t), p_-(t), t\right)\Big].
\end{aligned} \tag{4.19}$$

A short derivation is given in the Appendix. As in Section 3, independent variations of the operators q_+, p_+, q_-, p_-, the time t and the structure of the Liouville operator are allowed. As usual, the PSA then implies the equations of motion for the operators,

$$\begin{aligned}
\delta p_+(t) &: \frac{\mathrm{d}}{\mathrm{d}t}q_+(t) = i\hbar\frac{\partial}{\partial p_+(t)}\mathcal{L}_{+-}(t) = -[\mathcal{L}_{+-}(t), q_+(t)], \\
\delta q_+(t) &: \frac{\mathrm{d}}{\mathrm{d}t}p_+(t) = -i\hbar\frac{\partial}{\partial q_+(t)}\mathcal{L}_{+-}(t) = -[\mathcal{L}_{+-}(t), p_+(t)], \\
\delta p_-(t) &: \frac{\mathrm{d}}{\mathrm{d}t}q_-(t) = -i\hbar\frac{\partial}{\partial p_-(t)}\mathcal{L}_{+-}(t) = -[\mathcal{L}_{+-}(t), q_-(t)], \\
\delta q_-(t) &: \frac{\mathrm{d}}{\mathrm{d}t}p_-(t) = i\hbar\frac{\partial}{\partial q_-(t)}\mathcal{L}_{+-}(t) = -[\mathcal{L}_{+-}(t), p_-(t)], \\
\delta t(t) &: \tfrac{\mathrm{d}}{\mathrm{d}t}\mathcal{L}_{+-}(t) = \tfrac{\partial}{\partial t}\mathcal{L}_{+-}(t).
\end{aligned} \tag{4.20}$$

The action (4.19) has the same appearance as (3.18). The difference lies in the particular form of the Liouville operator. This implies that the QAP (4.18) approaches the formulation for closed systems (3.17) in the limit of vanishing dissipation ($L \to 0$).

The action principle so derived represents a differential characterization of the Green function of Markovian master equations. Therefore, it should be an equally powerful computational tool, as is Schwinger's QAP for closed quantum systems. In particular, (4.18) may serve as the starting point of systematic approximations that generalize standard perturbation theory.

5. Driven Damped Harmonic Oscillator

An immediate application of the extended action principle concerns the calculation of transition probabilities. For illustration, we consider here the simple example of a driven damped harmonic oscillator at temperature $T = 0$. A technically more involved study of anharmonic oscillators will be reported elsewhere[15]. The dynamics of the oscillator is determined by the master equation

$$\frac{\partial}{\partial t}\rho = \mathcal{L}(t)\rho = -\frac{i}{\hbar}\left[H(a,a^\dagger,t),\rho\right] + \frac{\Gamma}{2}\left([a\rho,a^\dagger] + [a,\rho a^\dagger]\right) \tag{5.1}$$

with the Hamilton operator

$$H(t) = \hbar\omega\left(a^\dagger - \alpha^*(t)\right)(a - \alpha(t)) \,. \tag{5.2}$$

The complex function $\alpha(t)$ describes a classical driving force and Γ is the damping constant. The Liouville operator (5.1) is of the Lindblad form (4.2) with $V = \sqrt{\Gamma/2}\, a$. We are interested in the Green function $\{a', a^{\dagger\prime}, t|a'', a^{\dagger\prime\prime}, 0\}$ for the master equation (5.1). We proceed from the known answer for $\alpha \equiv 0 \equiv \alpha^*$, and obtain the result for $\alpha \neq 0$ by studying the response to variations of $\alpha(t), \alpha^*(t)$. The action principle states that this response is given by

$$\delta_{\alpha,\alpha^*}\{a', a^{\dagger\prime}, t|a'', a^{\dagger\prime\prime}, 0\} = \frac{i}{\hbar}\{a', a^{\dagger\prime}, t|\delta_{\alpha,\alpha^*}W_{t,0}^{a_+^\dagger a_-}|a'', a^{\dagger\prime\prime}, 0\}, \tag{5.3}$$

where

$$\begin{aligned} W_{t,0}^{a_+^\dagger a_-} &= \int_0^t \mathrm{d}s \left[i\hbar a_+^\dagger(s)\frac{\mathrm{d}}{\mathrm{d}s}a_+(s) - i\hbar a_-^\dagger(s)\frac{\mathrm{d}}{\mathrm{d}s}a_-(s) - i\hbar\mathcal{L}_{+-}(s)\right] - \\ &\quad - i\hbar a_+^\dagger(t)a_+(t) - i\hbar a_-^\dagger(0)a_-(0) \end{aligned} \tag{5.4}$$

is the action operator appropriate for Bargmann states [cf. (2.21)], and

$$\begin{aligned} \mathcal{L}_{+-}(s) &= -i\omega\left(a_+^\dagger(s) - \alpha^*(s)\right)(a_+(s) - \alpha(s)) + \\ &\quad + i\omega\,(a_-(s) - \alpha(s))\left(a_-^\dagger(s) - \alpha^*(s)\right) + \\ &\quad + \frac{\Gamma}{2}\left(2a_+(s)a_-^\dagger(s) - a_+^\dagger(s)a_+(s) - a_-(s)a_-^\dagger(s)\right) \end{aligned} \tag{5.5}$$

is the Liouville operator. Because only this Liouville operator depends explicitly on the functions α, α^*, their variation results in

$$\begin{aligned}&\delta_{\alpha,\alpha^*}\{a',a^{\dagger\prime},t|a'',a^{\dagger\prime\prime},0\}\\&\quad= i\omega\Big\{a',a^{\dagger\prime},t|\int_0^t \mathrm{d}s\,\Big[\big(a_+^\dagger(s)-a_-^\dagger(s)\big)\,\delta\alpha(s)+\\&\qquad\qquad+\big(a_+(s)-a_-(s)\big)\,\delta\alpha^*(s)\Big]\,|a'',a^{\dagger\prime\prime},0\Big\}. \qquad (5.6)\end{aligned}$$

Now we solve the equations of motion for the operators $a_+, a_-, a_+^\dagger, a_-^\dagger$ and insert these solutions into (5.6). After applying the operators to their eigenstates, we arrive at

$$\begin{aligned}&\frac{\delta_{\alpha,\alpha^*}\{a',a^{\dagger\prime},t|a'',a^{\dagger\prime\prime},0\}}{\{a',a^{\dagger\prime},t|a'',a^{\dagger\prime\prime},0\}}\\&\quad=\delta_{\alpha,\alpha^*}\Big[\beta(t)(a^{\dagger\prime}\mathrm{e}^{-(i\omega+\Gamma/2)\,t}-a^{\dagger\prime\prime}\mathrm{e}^{-\Gamma t})+\\&\qquad+\beta^*(t)(a'\mathrm{e}^{(i\omega-\Gamma/2)\,t}-a''\mathrm{e}^{-\Gamma t})-\mathrm{e}^{-\Gamma t}|\beta(t)|^2\Big], \qquad (5.7)\end{aligned}$$

where we have introduced the abbreviation

$$\beta(t) := i\omega\int_0^t \mathrm{d}s\,\mathrm{e}^{(i\omega+\Gamma/2)\,s}\alpha(s). \qquad (5.8)$$

With the result for $\alpha\equiv 0\equiv\alpha^*$, that is the damped harmonic oscillator without a driving force,

$$\begin{aligned}&\{a',a^{\dagger\prime},t|a'',a^{\dagger\prime\prime},0\}^{(\alpha=0)}\\&\quad=\exp\Big(-\Big[a^{\dagger\prime}-a^{\dagger\prime\prime}\mathrm{e}^{(i\omega-\Gamma/2)\,t}\Big]\times\\&\qquad\times\Big[a'-a''\mathrm{e}^{-(i\omega+\Gamma/2)\,t}\Big]+a^{\dagger\prime}a'+a^{\dagger\prime\prime}a''\Big), \qquad (5.9)\end{aligned}$$

we can integrate (5.7) and obtain the Green function

$$\begin{aligned}&\{a',a^{\dagger\prime},t|a'',a^{\dagger\prime\prime},0\}\\&\quad=\exp\Big(-\Big[a^{\dagger\prime}-\mathrm{e}^{(i\omega-\Gamma/2)t}\big(a^{\dagger\prime\prime}+\beta^*(t)\big)\Big]\times\\&\qquad\times\Big[a'-\mathrm{e}^{-(i\omega+\Gamma/2)t}(a''+\beta(t))\Big]+a^{\dagger\prime}a'+a^{\dagger\prime\prime}a''\Big) \qquad (5.10)\end{aligned}$$

In particular, (5.10) contains the transition probability between coherent states (these are normalized Bargmann states) with quantum numbers a' and a''

$$\begin{aligned}&\{a',(a')^*,t|a'',(a'')^*,0\}\,\mathrm{e}^{-|a'|^2-|a''|^2}\\&\quad=\exp\Big(-\Big|a'-\mathrm{e}^{-(i\omega+\Gamma/2)\,t}\,(a''+\beta(t))\,\Big|^2\Big). \qquad (5.11)\end{aligned}$$

This simple example shows how one can use the QAP for open systems as a computational tool.

6. Summary

We have extended Schwinger's quantum action principle from Hilbert space to Liouville space and then to open systems with Markovian dynamics. The new action principle provides for a differential characterization of the Green function for the master equation in a similar way as the original QAP does for the Green function of the Schrödinger equation. In contrast to the QAP for closed systems, the variables in the action are, in general, not the physical observables. One can get back the physical operators from the action variables if the Liouville operator is not explicitly time-dependent. The simple example of a driven, damped harmonic oscillator has been used for a demonstration of the extended action principle as a computational tool.

Appendix

Here we give a short heuristic derivation of the action principle (4.18). The method we use is similar to that of [7].

First we consider an infinitesimal time step $\mathrm{d}t$ and variations of all quantum numbers (type $\overline{\overline{\delta}}$). We assume $\delta q_+(t), \delta q_-(t), \delta p_+(t), \delta p_-(t)$ and $\delta t(t)$ to be real, infinitesimal test functions. Using the generators (4.16) for these variations and Heisenberg's equation of motion we obtain

$$\begin{aligned}
&\overline{\overline{\delta}}\{q'_+, q'_-, t'' + \mathrm{d}t | q''_+, q''_-, t''\} \\
&\quad = \{q'_+ + \delta q_+(t'' + \mathrm{d}t), q'_- + \delta q_-(t'' + \mathrm{d}t), t'' + \mathrm{d}t + \delta t(t'' + \mathrm{d}t)| \times \\
&\qquad \times q''_+ + \delta q_+(t''), q''_- + \delta q_-(t''), t''\} - \{q'_+, q'_-, t'' + \mathrm{d}t | q''_+, q''_-, t''\} \\
&\quad = \frac{i}{\hbar}\Big\{ q'_+, q'_-, t'' + \mathrm{d}t | \overline{\overline{\delta}} \times \\
&\qquad \times \Big(\mathrm{d}t \Big[p_+(t'')\frac{\mathrm{d}}{\mathrm{d}t''}q_+(t'') - p_-(t'')\frac{\mathrm{d}}{\mathrm{d}t''}q_-(t'') - \\
&\qquad\quad - i\hbar \mathcal{L}_{+-}(t'')\Big]\Big) | q''_+, q''_-, t'' \Big\}.
\end{aligned} \tag{A.1}$$

For variations of the structure of the Liouville operator (type $\overline{\delta}$), we have

$$\begin{aligned}
&\overline{\delta}\{q'_+, q'_-, t'' + \mathrm{d}t | q''_+, q''_-, t''\} \\
&\quad = \{q'_+, q'_-, t'' + \mathrm{d}t | \mathrm{d}t\, \overline{\delta}\mathcal{L}_{+-}(t'') | q''_+, q''_-, t''\},
\end{aligned} \tag{A.2}$$

which combines with $\overline{\overline{\delta}}\{\cdots\}$ to the general variations

$$\begin{aligned}
&\delta\{q'_+, q'_-, t'' + \mathrm{d}t | q''_+, q''_-, t''\} \\
&\quad = (\overline{\delta} + \overline{\overline{\delta}})\{q'_+, q'_-, t'' + \mathrm{d}t | q''_+, q''_-, t''\} \\
&\quad = \frac{i}{\hbar}\left\{q'_+, q'_-, t'' + \mathrm{d}t \middle| \delta\left(\mathrm{d}t\left[p_+(t'')\frac{\mathrm{d}}{\mathrm{d}t''}q_+(t'') - p_-(t'')\frac{\mathrm{d}}{\mathrm{d}t''}q_-(t'') - \right.\right.\right. \\
&\qquad\qquad \left.\left.\left. - i\hbar\mathcal{L}_{+-}(t'')\right]\right) \middle| q''_+, q''_-, t''\right\}. \qquad \text{(A.3)}
\end{aligned}$$

This result can easily be extended to n infinitesimal time steps

$$\begin{aligned}
&\delta\{q'_+, q'_-, t'' + n\mathrm{d}t | q''_+, q''_-, t''\} \\
&\quad = \frac{i}{\hbar}\left\{q'_+, q'_-, t'' + n\mathrm{d}t \middle| \delta \times \right. \\
&\qquad \times\left(\mathrm{d}t\sum_{k=0}^{n-1}\left[p_+(t'' + k\mathrm{d}t)\frac{\mathrm{d}}{\mathrm{d}t''}q_+(t'' + k\mathrm{d}t) - \right.\right. \\
&\qquad\qquad - p_-(t'' + k\mathrm{d}t)\frac{\mathrm{d}}{\mathrm{d}t''}q_-(t'' + k\mathrm{d}t) - \\
&\qquad\qquad \left.\left.\left. - i\hbar\mathcal{L}_{+-}(t'' + k\mathrm{d}t)\right]\right) \middle| q''_+, q''_-, t''\right\}. \qquad \text{(A.4)}
\end{aligned}$$

This produces the QAP of Equation (4.18) in the limit $n \to \infty$ with

$$\mathrm{d}t = (t' - t'')/n \to 0.$$

Acknowledgement

One of us (B.G.E.) is eternally grateful to the late Julian Schwinger for numerous unforgettable lectures on the quantum action principle.

Notes and References

1. Schwinger, J.: *Phys. Rev.* **82** (1951), 914.
2. Schwinger, J.: *Phys. Rev.* **91** (1953), 713.
3. Schwinger, J.: *Proc. Natl. Acad. Sci.* **46** (1960), 893.
4. Schwinger, J.: *Quantum Kinematics and Dynamics*, Benjamin, New York, 1970 (reprinted by Addison-Wesley, Redwood City, 1991).
5. Schwinger, J.: Quantum mechanics, lecture notes, University of California, Los Angeles 1984–85, unpublished.
6. Thoss, M.: Diplomarbeit, Universität München, 1994, unpublished.
7. Schwinger, J.: *Proc. Natl. Acad. Sci.* **47** (1961), 1075.

8. Schwinger, J.: *Proc. Natl. Acad. Sci.* **48** (1962), 603.
9. Volkov, D. V. and Peletminskii, S. V.: *Soviet Phys. JETP* **10** (1960), 121.
10. Finkelstein, R. J.: *Nonrelativistic Mechanics*, Benjamin, Reading, 1973.
11. Gardiner, C. W.: *Quantum Noise*, Springer, Berlin, 1991.
12. Fick, E. and Sauermann, G.: *Quantenstatistik Dynamischer Prozesse*, Bd. 2A, Harri Deutsch, Frankfurt am Main, 1986.
13. Lindblad, G.: *Comm. Math. Phys.* **48** (1976), 119.
14. Alicki, R.: *Quantum Dynamical Semigroups and Applications*, Lecture Notes in Phys. 286, Springer, Berlin, 1987.
15. Kleinle, J., Thoss, M., and Englert, B.-G.: in preparation.

Enrique Solano

Departamento de Química Física, Universidad del País Vasco UPV/EHU, Bilbao, Spain
IKERBASQUE, Basque Foundation for Science, Bilbao, Spain

Paper Title: **Measuring irreversible dynamics of a quantum harmonic oscillator**

Reference: Physical Review A **65**, 040102(R) (2002)

Personal message:

I met Englert, as people referred to him, when I started my postdoctoral research work back in January 2001 in the group of Herbert Walther in the Max Planck Institute for Quantum Optics (MPQ) in Garching. I arrived from my PhD studies in Rio de Janeiro, with a few-day stop in Berlin to see Anja, my future wife, and an urgent flight to Munich arranged by Mrs. Wild to enter directly in Schwabinger Krankenhaus due to a serious throat infection. As it happens to me even nowadays, I had left my coat inside the check-in bag and the t-shirt I was wearing was not enough to protect me during the long flight hours and a temperature gradient of almost 40 degrees. Everybody at MPQ knew my unfortunate story and I was greeted kindly at every corner. This was nice, but there are healthier techniques to becoming a celebrity in prestigious research institutions.

I was told rather early on that Englert was a solid scientist, a redoubtable debater, and almost inaccessible for superficial discussions. I was told much later that Englert's friends and colleagues called him Berge, that he was a strong chess player, that he worked with Julian Schwinger, and that he was even able to speak Polish. But what I realized on my own was that Berge was almost always outspoken and accessible for explaining, suggesting, and discussing, and always giving his generous and thorough opinion about almost everything with his constant assertive style, nodding with inquisitive open eyes and repeating the German "ja" at the end of each sentence. I used to go home in the U-Bahn repeating to myself several of his sentences that, in the beginning, did not make much sense to me, except for the "ja" at the end. After ruminating on them, I was able to grasp important theoretical concepts and understand how they were related to experimental methods. I just wanted to learn how Berge thought about physics, how he associated concepts, and the way in which he created the physical world in his mind. I arrived at MPQ from Rio de Janeiro, from the most important quantum optics theory group in Latin America. There, I had studied the minds of Nicim Zagury and Luiz Davidovich, two clever theorists working in cavity quantum electrodynamics (QED), which was not the cavity of Serge Haroche in Paris. With Berge, I learned how a lucid and abstract theorist understood the cavity of Herbert Walther in Garching, how different quantum technologies needed to be described with different theoretical and conceptual approaches. Furthermore, I realized that subtle technical differences could lead to rather different methods and concepts. At that time, all these were new to me and I had a wave-function approach in thinking about physics. Berge's algebraic methods to describe cavity QED with operators introduced a new dimension in my understanding of physics, on how the human mind can think differently about the quantum aspects of interacting light and matter. I was amazed that I could not think about physics like Berge, and I felt intellectually puzzled and more unprotected than walking in a t-shirt through the Garching winter. So I decided to study his abstract mind and I learnt somehow to describe cavity QED with operators instead of wave functions. How much I struggled with his lectures in Dresden! How much I was impressed by his capacity of giving lectures without notes! As time passed, I became more used to his language. Nevertheless, each discussion with him was still always a new lecture and intellectually challenging to me. One day I met him in the MPQ library, and he was reading an article directly from a heavy volume. I dared to ask him whether he was a regular visitor of the library. He told me he tried to read at least one different article each day—an answer that I have never forgotten and always tried to imitate, though frequently failed.

I know that anecdotes in our human and intellectual exchanges are mostly one-sided, and sometimes they are special for some of us while irrelevant for others. Fewer times we are conscious of how influential little details may be with students, with younger researchers, or with senior colleagues. The memories of my MPQ times are unavoidably tied to Berge's personal kindness and scientific support. I do not remember him saying a single impolite sentence to me or, probably, I have completely forgotten it. To know that he would move to Singapore was sad news for me, but he looked happy and that was enough. Later, I helped him with some book shipping issues between Germany and Singapore and, afterwards, we never saw each other again until a casual meeting in a Madrid congress. There, his attitude to me was even more friendly than in the past, introducing and recommending his students to me. I was positively surprised and feel honoured for being invited to this Berge Fest to celebrate his 60th Birthday in Singapore. I hope that when he reads this essay, he does not repeat one of his few disapproving sentences towards me. One day, during an MPQ lunch, I started to argue enthusiastically about testing some phenomena that are difficult to observe, such as the quantization of the atomic motion in cavities and relativistic effects in atomic entanglement. He disapproved of my insistence with a fast reaction: Kike, don't push it too much! I can confess now that, with respects to studying brilliant minds and developing ideas with passion, I have never stopped pushing it too much.

Sincerely,

Enrique Solano
Bilbao, 1 April, 2014

RAPID COMMUNICATIONS

PHYSICAL REVIEW A, VOLUME 65, 040102(R)

Measuring irreversible dynamics of a quantum harmonic oscillator

Giovanna Morigi,[1] Enrique Solano,[1,2] Berthold-Georg Englert,[1] and Herbert Walther[1]
[1]*Max-Planck-Institut für Quantenoptik, Hans-Kopfermann-Strasse 1, 85748 Garching, Germany*
[2]*Sección Física, Departamento de Ciencias, Pontificia Universidad Católica del Perú, Apartado 1761, Lima, Peru*
(Received 20 August 2001; published 10 April 2002)

We show that the unitary evolution of a harmonic oscillator coupled to a two-level system can be undone by a suitable manipulation of the two-level system—more specifically, by a quasi-instantaneous phase change. This enables us to isolate the dissipative evolution to which the oscillator may be exposed in addition. With this method we study the decoherence time of a photon mode in cavity QED, and that of the quantized harmonic motion of trapped ions. We comment on the relation to spin echoes and multipath interferometry.

DOI: 10.1103/PhysRevA.65.040102 PACS number(s): 03.65.Yz, 32.80.Qk, 42.50.Ct

Decoherence is currently under investigation theoretically and experimentally [1], thereby searching for the boundary between the quantum and classical realms [2,3]. From the point of view of technical applications, scalability of coherent control—like that needed in a quantum computer—is being pursued intensively and is limited mainly by the presence of decoherence [1,4]. With the tools of quantum optics, it has become possible to create quantum states of high purity [4] and to monitor their decay [5,6]. The measurement of a quantum state, however, can introduce additional coherence destruction, caused by the nonunitarity of the state reduction. Methods like quantum nondemolition measurements [7], partially solving this problem, have been implemented for studying nonclassical states of a harmonic oscillator and their decoherence [5]. Nevertheless, the decoherence of *arbitrary* initial states of the quantum oscillator deserves further studies.

In this contribution, we propose a method for measuring the decoherence time of a harmonic oscillator prepared in an arbitrary initial state by suitably coupling it to a stable two-level system (TLS), whose final level statistics carry the information of the oscillator decay. The composite system evolves following a Jaynes-Cummings (JC) interaction [8] for a total duration T. At an intermediate instant, after time τ has elapsed, the JC evolution is interrupted by a phase kick, that is, a fast unitary operation applied to the TLS that introduces a relative phase between the levels. We will show that, as a consequence of the phase kick, the coherent dynamics is effectively reversed, mimicking a time-reversal operation [9–11]. When $\tau = T/2$, the final state of the composite system matches the initial one, provided that the evolution is unitary. Any mismatch represents a measurement of changes in the oscillator state due to external actions and an estimation of the time in which decoherence occurs.

This method allows us to separate the irreversible decay of the oscillator state from the unitary JC dynamics. The scheme can be easily implemented in two representative scenarios in quantum optics, ion traps and cavity QED, for studying the irreversible decay of the quantized motion or the electromagnetic field, respectively. In cavity QED, the harmonic oscillator is a mode of the quantized radiation field, coupled to a resonant electronic transition of atoms sent through the resonator and undergoing JC dynamics [12]. In ion traps, the center-of-mass motion of the trapped ion is harmonic, and it couples to an internal atomic transition when the ion is irradiated by a laser [13]. We discuss below how the decay of these oscillators can be studied by means of the proposed scheme.

Let us consider the resonant interaction between a TLS, with lower state $|\mathrm{g}\rangle$ and upper state $|\mathrm{e}\rangle$, and a harmonic oscillator, with creation and annihilation operators $a^\dagger$ and a, respectively. The eigenstates of the number operator $a^\dagger a$ are denoted by $|n\rangle$ with $n=0,1,2,\ldots$. The transition energy of the TLS is $\hbar\omega$, equal to the energy of the oscillator quanta. The coherent dynamics is described by the Hamiltonian

$$H(t)=H_{\mathrm{JC}}+H_{\mathrm{kick}}=H_0+H_{\mathrm{int}}+H_{\mathrm{kick}}. \tag{1}$$

Here, H_{JC} describes the JC dynamics, composed of the free Hamiltonian

$$H_0=\hbar\omega(\sigma^\dagger\sigma+a^\dagger a), \tag{2}$$

with $\sigma=|\mathrm{g}\rangle\langle\mathrm{e}|$, $\sigma^\dagger=|\mathrm{e}\rangle\langle\mathrm{g}|$, and of the interaction term

$$H_{\mathrm{int}}=\hbar g(\sigma^\dagger a+\sigma a^\dagger), \tag{3}$$

where g is the coupling constant. H_{kick} generates the instantaneous phase kick at $t=\tau$ and is defined as

$$H_{\mathrm{kick}}=\hbar\pi\sigma\sigma^\dagger\delta(t-\tau). \tag{4}$$

The effect of the kick alone is thus given by

$$U_{\mathrm{kick}}=\exp\left(-\frac{i}{\hbar}\int_{\tau-0}^{\tau+0}dt\,H(t)=\sigma^\dagger\sigma-\sigma\sigma^\dagger=\sigma_z. \right. \tag{5}$$

After the duration T, the evolution operator for the whole process is then

$$\begin{aligned} U_\tau(T)&=e^{-(i/\hbar)H_{\mathrm{JC}}(T-\tau)}U_{\mathrm{kick}}e^{-(i/\hbar)H_{\mathrm{JC}}\tau}\\ &=\sigma_z\exp\left[\frac{i}{\hbar}\left(H_0+\frac{2\tau-T}{T}H_{\mathrm{int}}\right)T\right], \end{aligned} \tag{6}$$

where the second line exploits $[H_0,H_{\mathrm{int}}]=0$ and $\{H_{\mathrm{int}},\sigma_z\}=0$. So the net evolution amounts to propagation governed by a JC coupling of the effective strength

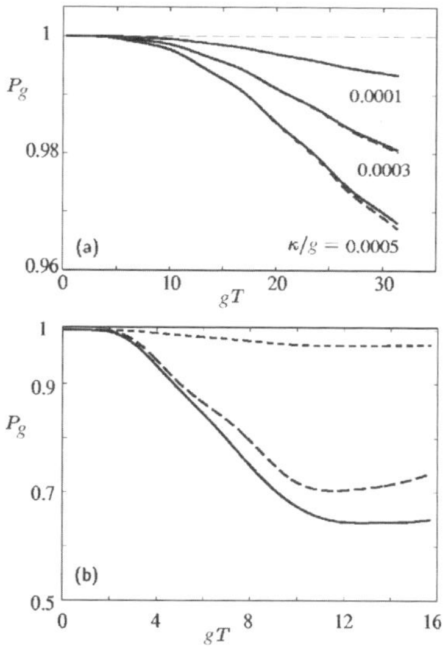

FIG. 1. (a) Numerical values [solid line, Eq. (8)] and approximate analytical values [dashed line, Eq. (10)] of P_g as a function of gT for different ratios κ/g and the field initially in the number state $|3\rangle$. (b) Numerical values of P_g for $\kappa=0.05g$ and initial cavity states $(|2\rangle+i|3\rangle)/\sqrt{2}$ (solid line), $|2\rangle$ (dashed line), coherent state $|\alpha\rangle$ with $\alpha=\exp(i\pi/4)/\sqrt{2}$ (dotted line).

$$g_{\text{eff}}=\frac{2\tau-T}{T}g \tag{7}$$

for time T, followed by the phase kick. Since the probability of finding the TLS in $|g\rangle$ or $|e\rangle$ is measured eventually, this final phase kick can in fact be ignored. For $\tau=T/2$, the effective interaction term in Eq. (6) vanishes and the final state is just the freely evolved initial state (except for the irrelevant phase kick). So, if the initial state were $|g\rangle\otimes|\psi_{\text{osc}}\rangle$, for example, the TLS would always be measured in the ground state after the interaction. The effect of the kick consists thus in symmetrizing the JC evolution [10], which is somewhat reminiscent of spin-echo techniques [14], where the evolution of spin-$\frac{1}{2}$ particles interacting with a *classical* field is undone. We remark that, with this scheme, the external action could come from any other coherent or even incoherent process, but not from the probing TLS. Nevertheless, it is its final statistics that carry the information about the effect of the environment on the oscillator.

We now turn to the application of this scheme to the particular scenario of cavity QED. The quantized electromagnetic field in a cavity is probed by atoms that fly through the resonator and whose dipole transition couples resonantly to a privileged field mode, undergoing the JC interaction described by H_{JC} in Eq. (1). The phase kick could be implemented by a laser that quasiresonantly couples the ground state of the atom to a third atomic level, thereby realizing a fast 2π pulse [11]. We consider the proposed scheme for measuring the decoherence time of the cavity field, coupled to a Markovian bath at zero temperature [12]. We assume that the atomic transition is not subject to dissipation by itself on the relevant time scale. The initial state $\rho(0)$ of the joint atom-field system is then evolved into the final state in accordance with

$$\rho(T)=e^{L(T-\tau)}Ke^{L\tau}\rho(0), \tag{8}$$

where $K\rho=U_{\text{kick}}\rho U_{\text{kick}}^{-1}=\sigma_z\rho\sigma_z$ is the effect of the kick at time τ, and the Liouville operator L is given by

$$L\rho=-\frac{i}{\hbar}[H_{\text{JC}},\rho]+\kappa\left(a\rho a^\dagger-\frac{1}{2}a^\dagger a\rho-\frac{1}{2}\rho a^\dagger a\right), \tag{9}$$

where κ is the field decay rate. This particular choice of the nonunitary part of L is taken for simplicity, but the method works just as well for other cases. An explicit form of P_g, the probability for measuring the atom at the exit in state $|g\rangle$, can be obtained from Eq. (8) in the limits $\kappa T\ll 1$ and $\kappa/g \ll 1$. At first order in these parameters, for $\tau=T/2$ and $\rho(0)=|g\rangle\langle g|\otimes\Sigma_{n,m}\rho_{nm}|n\rangle\langle m|$, it is

$$\begin{aligned}P_g=1-\sum_{n=2}^{\infty}\rho_{n,n}\Bigg\{&\frac{\kappa T}{4}(2n-1)\\&+\kappa\frac{\sin(gT\sqrt{n})}{4g\sqrt{n}}-\kappa\frac{\sin(gT\sqrt{n-1})}{4g\sqrt{n-1}}\\&-\frac{\kappa}{4g}\Big[\sqrt{n}(4n-3)\sin(gT\sqrt{n})\cos(gT\sqrt{n-1})\\&-\sqrt{n-1}(4n-1)\sin(gT\sqrt{n-1})\cos(gT\sqrt{n})\Big]\Bigg\}.\end{aligned} \tag{10}$$

The effect of the decay is evident, causing deviations of P_g from unity. These deviations are oscillatory functions of the duration T, resulting from the JC dynamics characterizing the system evolution. Note that in Eq. (10) the summation starts at $n=2$. In fact, for $\rho_{nn}=\delta_{n,1}$ one has $P_g=1$ in this perturbative limit, the deviations appearing at second order [15].

In Fig. 1, we plot P_g as a function of the duration T. Figure 1(a) compares the full numerical evaluation of Eq. (8) with the approximation (10) for different decay rates κ. In Fig. 1(b), the curves are plotted for different initial states. Note that for $\kappa T\gg 1$ the cavity state tends to the vacuum, and the probability of counting the atoms in the ground state approaches unity again. How big the deviation from $P_g=1$ should be for considering that the system has decohered is a question that will depend on each particular case or use.

The decay of the field phase can be studied by adding to the proposed scheme a Ramsey zone at the exit of the cavity, which performs the unitary transformation $\exp[-i\phi(\sigma e^{i\zeta}+\sigma^\dagger e^{-i\zeta})]$. In this way, the Ramsey zone and the detectors constitute a field-phase-sensitive device [16]. When only the coherent atom-field interaction takes place, we have $P_g=\cos^2\phi$, independent of the initial state of the field, and the decay of the field phase is measured through the deviations from this value. In Fig. 2, we plot the phase sensitive probability P_g, after the Ramsey zone with $\phi=\pi/4$, $\zeta=0$, as a

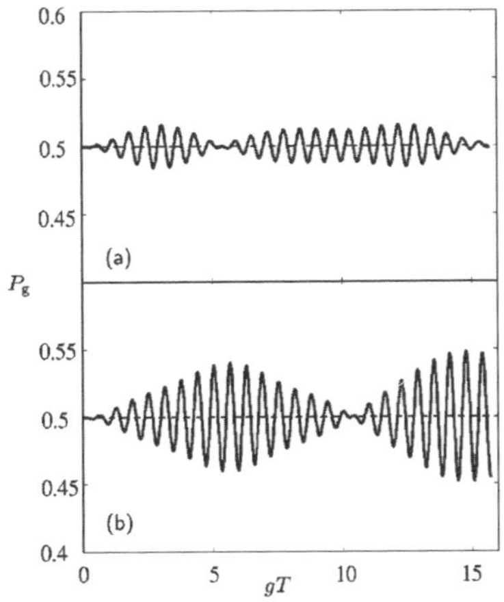

FIG. 2. Phase sensitive P_g as a function of gT for $\kappa=0.05g$ and $\omega=10g$. The initial field states are (a) $(|2\rangle+i|3\rangle)/\sqrt{2}$ and (b) coherent state $|\alpha\rangle$ with $\alpha=\exp(i\pi/4)/\sqrt{2}$. In both plots, the constant $P_g=0.5$ of an initial Fock state is plotted for reference.

function of the total interaction time for different initial field states. Both signals contain a high frequency modulation at the field frequency, chosen here for convenience as $\omega=10g$. For reference, we plot in both figures the curve corresponding to an initial state characterized only by a diagonal density matrix (dashed line)—a Fock state, say.

We now study the application of the phase-kick scheme in ion traps, where a JC dynamics between the internal transition of the ions and its center-of-mass motion is implemented by a laser. Now, the Hamilton operator (1) applies in the reference frame rotating at the laser frequency. The atomic phase kick can be realized here in a similar way as in the cavity QED case. Provided that the spontaneous emission from the atom can be neglected and that the motion is coupled to a Markovian bath at zero temperature, the evolution of the density matrix of the composite system is again given by Eq. (8), and the results reported above for the atom-cavity case are recovered. In the ion-trap situation, however, the phase kick is more easily implemented than in cavity QED. For example, we can switch off the laser system, after a fast initial coherent JC evolution and before the phase kick, and so interrupt the interaction during a convenient period of time T_{free}. By continuing with the phase kick and a fast final coherent JC evolution, the experiment is completed.

This variation of our scheme has features in common with Ramsey interferometry [17]. One could say that, during the initial short interaction time, interferometric paths are created in the ion-motion Hilbert space and that, after suffering "dephasing" during time T_{free}, these paths are recombined into the initial $g_{\mathrm{eff}}=0$ superposition. With the aid of an arbitrary phase kick, instead of the fixed 2π kick, we create a modulated output that is reminiscent of an interferometer pattern. This variation could also be implemented in the atom-cavity system, but the experimental setup would be much more demanding.

In general, the scheme can be applied to studying decoherence in the collective motion of a chain of N ions [18], i.e., a linear crystal of ions, achieved in linear ion traps at sufficiently low temperature. In this limit, the chain motion along its main axis is described by the collective excitations of its N normal modes with frequencies $\nu_1,\ldots,\nu_N$ [19]. A dynamics of the JC type can be designed between the jth ion internal state and the βth mode of the crystal, by laser irradiation analogous to the one-ion case. Then, the total Hamiltonian describing the interaction is given by $H_{\mathrm{JC}}^{(j,\beta)}=H_0^{(j,\beta)}+H_{\mathrm{int}}^{(j,\beta)}$ with

$$H_0^{(j,\beta)}=\hbar\nu_\beta(\sigma_j^\dagger\sigma_j+a_\beta^\dagger a_\beta) \tag{11}$$

and

$$H_{\mathrm{int}}^{(j,\beta)}=\hbar g_{j,\beta}(\sigma_j^\dagger a_\beta+\sigma_j a_\beta^\dagger), \tag{12}$$

where $g_{j,\beta}$ is the Rabi frequency for ion j and mode β, a_β, and $a_\beta^\dagger$ are the annihilation and creation operators of a vibrational quantum $\hbar\nu_\beta$, while ν_β is the detuning of the laser from the atomic transition and equal to the frequency of the addressed mode. The dynamics then follows the lines of that described for a single ion, except that here the σ_z kick is a laser pulse applied to ion j. In this way, we are able to study the irreversible decay of a chosen collective mode. Similarly, it is possible to measure the irreversible loss of coherence of an arbitrary motional state involving all modes. One would need to pair the ions with the modes, so that the corresponding Hamiltonian has the form $H=\Sigma_{(j,\beta)}H_{\mathrm{JC}}^{(j,\beta)}$, a sum over these pairs. One could then measure the decoherence time of each mode, and also of entangled states between different normal modes.

Here are some remarks on the experimental feasibility of the proposed scheme in relation to the systems discussed before. First, we note that the phase kick is crucial for the whole scheme, but it is not necessary to realize the ideal kick of Eq. (5). Indeed, the scheme is rather robust against experimental imperfections that would amount to multiplying H_{kick} of Eq. (4) by $1+\epsilon$. In particular, even with an error as large as $|\epsilon|\lesssim 7\%$, we would still cancel more than 99% of the JC dynamics.

In the case of cavity QED, our scheme could be implemented in the microwave and optical regime. For a test of the decoherence time associated with the population and the phase of the intracavity field, convenient ratios $\kappa/g\leqslant 1$ can be found in both regimes. In either case, atomic decay is irrelevant in a wide range of interaction times. Furthermore, the kick could be implemented by introducing a sheet of light—transversal to the mode of interest and crossed by the atomic trajectory at the center of the cavity—by shining a laser through a hole drilled at the side of the cavity (closed cavity), or by focusing laser beams (open cavity). Note that, for measuring decoherence in atom-cavity systems, a quality factor that is not too large may be desirable. Further, the use of an external kick for testing this scheme is not necessary, if the field mode in question changes sign at the center of the cavity. In this case, the odd spatial symmetry of the field naturally undoes the JC dynamics, producing the same effect as the phase kick for the fundamental mode.

Ion traps appear more fitting for testing and scaling the present scheme. For example, the atomic decay can be neglected by suitably coupling either a long-lived transition or two internal metastable states with a two-photon coherent process. Also, JC (or anti-JC) dynamics like that discussed here as well as pulses of well defined shape and duration are currently realized. Individual ions of a chain cannot be addressed by a laser in all experiments, but this is not necessary for the present application. It is always possible to illuminate the chain of ions homogeneously and, by giving an equally homogeneous phase kick, the JC dynamics could be "time reversed" and tested by measuring the global atomic ground state. In this case, we replace σ by $\Sigma_j \sigma_j$ in Eq. (12), and the phase-kick operator σ_z by $\Pi_j \sigma_z^j$.

In summary, we have discussed a method for measuring the decoherence time of a harmonic oscillator by coupling it to a two-level system and detecting its level statistics. The method is based on a phase kick applied to the two-level system during its interaction with the oscillator, thereby undoing the coherent interaction and isolating the effect of any external disturbance affecting the oscillator. We have discussed applications of the scheme for measuring decoherence times of the quantized field mode in cavity QED experiments and of the quantized motion of trapped atoms in harmonic traps. We think that the proposed scheme represents a useful tool for measuring decoherence times in different physical systems undergoing JC-like interactions. As a general feature, this method is sensitive to any deviation from the expected JC dynamics and could also be used, for instance, for measuring corrections to the rotating wave approximation, or for studying the anharmonic coupling among the normal modes of a chain of ions, or the influence of spectator atoms and spectator modes.

This work was partly supported by the European Commission (TMR networks ERB-FMRX-CT96-0077 and ERB-FMRX-CT96-0087).

[1] See for example, *Decoherence: Theoretical, Experimental and Conceptual Problems*, Vol. 538 of *Lecture Notes in Physics*, edited by P. Blanchard, D. Giulini, E. Joos, C. Kiefer, and I.-O. Stamatescu (Springer, Berlin, 2000).

[2] P. Bertet, S. Osnaghi, A. Rauschenbeutel, G. Nogues, A. Auffeves, M. Brune, J.M. Raimond, and S. Haroche, Nature (London) **411**, 166 (2001).

[3] M. Arndt, O. Nairz, J. Vos-Andreae, C. Keller, G. van der Zouw, and A. Zeilinger, Nature (London) **401**, 680 (1999); M. Arndt and A. Zeilinger, Phys. Bl. **56**, 69 (2000).

[4] See, for example, *The Physics of Quantum Information*, edited by D. Bouwmeester, A. Ekert, and A. Zeilinger (Springer, Berlin, 2000).

[5] M. Brune, E. Hagley, J. Dreyer, X. Maître, A. Maali, C. Wunderlich, J.M. Raimond, and S. Haroche, Phys. Rev. Lett. **77**, 4887 (1996).

[6] C.J. Myatt, B.E. King, Q.A. Turchette, C.A. Sackett, D. Kielpinski, W.M. Itano, C. Monroe, and D.J. Wineland, Nature (London) **403**, 269 (2000).

[7] V.B. Braginsky and F.Y. Khalili, in *Quantum Measurement*, edited by K.S. Thorne (Cambridge University Press, Cambridge, England, 1992).

[8] E.T. Jaynes and F.W. Cummings, Proc. IEEE **51**, 89 (1963).

[9] Phase kicks of this sort are at the basis of schemes for inhibiting decoherence of a small system coupled to a bath. See, for example, L. Viola and S. Lloyd, Phys. Rev. A **58**, 2733 (1998); D. Vitali and P. Tombesi, *ibid.* **59**, 4178 (1999); L. Viola, E. Knill, and S. Lloyd, Phys. Rev. Lett. **82**, 2417 (1999); L. Viola, S. Lloyd, and E. Knill, *ibid.* **83**, 4888 (1999).

[10] P. Zanardi, Phys. Lett. A **258**, 77 (1999).

[11] See, for example, G.S. Agarwal, M.O. Scully, and H. Walther, Phys. Rev. Lett. **86**, 4271 (2001).

[12] G. Raithel, C. Wagner, H. Walther, L.M. Narducci, and M.O. Scully, in *Advances in Atomic, Molecular and Optical Physics*, edited by P. Berman (Academic, New York, 1994), Suppl. 2.

[13] D.J. Wineland, C. Monroe, W.M. Itano, D. Leibfried, B.E. King, and D.M. Meekhof, J. Res. Natl. Inst. Stand. Technol. **103**, 259 (1998).

[14] See, for example, L. Allen, and J.H. Eberly, *Optical Resonance and Two-Level Atoms* (Wiley, London, 1975).

[15] For the initial state $\rho(0)=|g,1\rangle\langle g,1|$, one can evaluate P_g exactly:

$$P_g = 1 - 64\frac{g^2\kappa^2 e^{-\kappa T/2}}{(16g^2-\kappa^2)^2}\sin^4\left(\frac{1}{8}\sqrt{16g^2-\kappa^2}T\right),$$

which shows that the effect of the decay is $O(\kappa^2/g^2)$.

[16] B.-G. Englert, T. Gantsog, A. Schenzle, C. Wagner, and H. Walther, Phys. Rev. A **53**, 4386 (1996).

[17] N.F. Ramsey, *Molecular Beams* (Oxford University Press, New York, 1985).

[18] E. Solano, R.L. de Matos Filho, and N. Zagury, Phys. Rev. Lett. **87**, 060402 (2001).

[19] D.F.V. James, Appl. Phys. B: Lasers Opt. **66**, 181 (1998).

Ulrike Herzog

Nano-Optics, Institut für Physik, Humboldt-Universität zu Berlin, Berlin, Germany

Paper Title: **Statistics of photons and de-excited atoms in a micromaser with Poissonian pumping**

Reference: Physical Review A **50**, 783–786 (1994)

Personal message:

The paper selected for this Festschrift is related to one of Berge's early research fields, the quantum theory of the micromaser. The work reported in this paper had been done in parallel to work performed by Berge and his co-workers on a similar subject ["One-atom-maser: Statistics of detector clicks" by H.-J. Briegel, B.-G. Englert, and H. Walther, Phys. Rev. A **49**, 2962 (1994)]. I chose this paper because it marks the beginning of my acquaintance with him which soon turned into friendship, a friendship that would not have arisen without the German reunification.

In the first half of the nineties, Berge was working at the Max Planck Institute for Quantum Optics near Munich, where experiments with the one-atom maser were performed in Herbert Walther's group. At that time, I was a member of a temporary research group installed at the Humboldt University in Berlin by the Max Planck Society, thanks to the efforts of Herbert Walther to secure the survival of a small group of theoreticians from the former East German Academy of Sciences. Berge soon became a guest professor in Munich and invited me to give a talk there. During the visit, I met his nice family in the warm atmosphere of his home, his children being of about the same age as mine, and we became friends. After his position in Munich expired, Berge spent some time in Berlin in our then still existing Max-Planck group. During this period, I had the chance to enjoy his witty and funny remarks, and the straightforwardness and frankness of his comments on all kind of things, always based on profound knowledge. There was even the hope that Berge would become my future chief, since a new professorship had been advertised at the Humboldt University and Berge was elected as the number one in the list of three possible position holders. However, after we all had been waiting impatiently for his final appointment for more than two years, the prospective professorship was cancelled.

The German term Bergfest (mountain party) is used for a celebration after half of a certain period of time is over. Certainly it would be presumptuous to hope that this Berg(e)Fest celebrated in Singapore stands in the middle of Berge's active life as a scientist, but one can hope that the physics community will continue to profit for a long time from his ideas, and from his conscientious and dedicated work, always frankly telling the truth when it has to be told and writing papers in a style that is always a pleasure to read.

Thank you, Berge, for your long-time true friendship! On the occasion of your birthday, I wish you continuing health and creativeness. — Your friend Ulrike

PHYSICAL REVIEW A VOLUME 50, NUMBER 1 JULY 1994

Statistics of photons and de-excited atoms in a micromaser with Poissonian pumping

Ulrike Herzog

Arbeitsgruppe "Nichtklassische Strahlung" der Max-Planck-Gesellschaft an der Humboldt-Universität zu Berlin, Rudower Chaussee 5, 12484 Berlin, Germany

(Received 4 November 1993)

Explicit expressions are derived for calculating the two-time coincidence probability density, the Mandel Q parameter of the counting statistics, and the waiting-time distribution between detector clicks for the detection of de-excited atoms leaving a micromaser with Poissonian pumping. The procedure is illustrated by applying it to the simplest trapped micromaser state for which all results can be obtained analytically. Moreover, by making use of a symmetry relation which determines the properties of the time-dependent micromaser master equation solution, it is shown that in the stationary regime for negligible thermal photon number the normalized two-time intensity correlation function of the cavity field with respect to a given time difference is equal to the normalized two-time coincidence probability density of detecting two de-excited atoms with the same time difference at the exit of the micromaser.

PACS number(s): 42.52.+x, 42.50.Dv, 03.65.Bz

I. INTRODUCTION

In the micromaser a beam of excited Rydberg atoms interacts with the radiation field in a high-Q microwave cavity in such a way that at most one atom at a time is present in the cavity [1,2]. In order to get information about the radiation field, the energy state of the atoms leaving the cavity is measured in micromaser experiments [3]. For that reason the relation between atom and field statistics is a matter of particular interest [4,5,10].

In the present communication we consider the usual case that the injected pump atoms are statistically independent, therefore obeying Poissonian injection statistics. We show that the two-time coincidence probability density for the detection of two de-excited atoms at the exit of the micromaser at two given instants of time, notwithstanding the possible detection of other de-excited atoms at other times, can be calculated in a straightforward way from the well-known master equation for the cavity field. This quantity immediately yields the second factorial moment of the counting statistics of de-excited atoms. The knowledge of the whole set of all multitime coincidence probability densities would provide a complete description of the statistical properties from which all other quantities of interest, especially the elementary or exclusive two-time probability densities for two detection events at two given instants of time with no events in the interval in between, could be evaluated in principle [6]. On the other hand, we find that the latter quantity, which yields the waiting-time distribution between detection events, can be obtained in a much easier way from a linear differential equation for a non-normalized density matrix which is conditioned on the fact that no counts are observed in between. Our procedure is similar to the treatment of photodetection developed by Srinivas and Davies [7], which uses measurement theory as its basis.

II. BASIC EQUATIONS

To make our approach more transparent we first briefly sketch a derivation of the well-known micromaser master equation for the cavity field. Clearly, the change of the field density operator ρ during a time interval Δt arises both from the interaction with the atoms that traverse the cavity during this interval at instants which are randomly distributed according to the Poissonian injection statistics, and from cavity damping with loss constant γ between these instants. Mere cavity damping during an interval δt changes the density operator according to $\rho(t+\delta t)=\exp(L\delta t)\rho(t)$ where, in the photon-number representation with $\rho_{nn}=p_n$, the superoperator L acts as [8]

$$[L\rho]_{nn}=\gamma(n_b+1)[(n+1)p_{n+1}-np_n] -\gamma n_b[(n+1)p_n-np_{n-1}] . \tag{1}$$

Here n_b denotes the mean number of thermal photons. The transit of one atom through the cavity with a transit time $t_{\rm int}$, which is supposed to be small in comparison to all relevant time scales, causes a change of ρ which is formally described by a superoperator M as $\rho(t+t_{\rm int})=M(t_{\rm int})\rho(t)$ [8]. In the photon-number representation we have

$$[M\rho]_{nn}=(1-\beta_{n+1})p_n+\beta_n p_{n-1} , \tag{2}$$

where $\beta_n=\sin^2(\sqrt{n}\,gt_{\rm int})$ with g being the atom-field coupling constant. For a time interval Δt with $\Delta t \ll r^{-1}$, where r stands for the atomic injection rate, we may neglect the probability that two or more atoms arrive during Δt and approximate the probabilities that no atom or one atom arrives by $1-r\Delta t$ and $r\Delta t$, respectively. Moreover, if we require in addition that $n_{\max}\gamma\Delta T \ll 1$ with $n_{\max}$ denoting the maximum photon number n for which ρ_{nn} is noticeably different from zero, the evolution operator $\exp(L\delta t)$ describing the damping can be approximated by $(1+L\delta t)$ for all $\delta t \le \Delta t$. Therefore, for any interval of length $\Delta t \ll r^{-1}$, $(n_{\max}\gamma)^{-1}$, we obtain the lowest-order approximation

$$\rho(t+\Delta t)=(1-r\Delta t+L\Delta t)\rho(t)+r\Delta t M\rho(t) , \tag{3}$$

where the first and second terms on the right-hand side refer to the possibilities that zero or one atom could have traversed the cavity during Δt which have been taken into account with their corresponding probabilities. Replacing $[\rho(t+\Delta t)-\rho(t)]/\Delta t$ by $\dot{\rho}$ we arrive at the micromaser master equation [2]

$$\dot{\rho}=r(M-1)\rho+L\rho . \tag{4}$$

It should be kept in mind that this equation is time-coarse-grained in the sense that the condition $\Delta t \gg t_{\rm int}$ has to be fulfilled and, moreover, that it is based on the Markovian character, which is due to the fact that there are no correlations between the injected atoms. Introducing the evolution operator V we write the solution of Eq. (4) as

$$\rho(t_0+\tau)=V(\tau)\rho(t_0) . \tag{5}$$

If an atom leaving the cavity at time t is known to be in the de-excited state, the non-normalized field density matrix after interaction with this atom is given by the reduced version of Eq. (2) as $\beta_n p_{n-1}(t)\equiv[D\rho(t)]_{nn}$, where D may be interpreted as a deexcitation operator. The single-atom deexcitation probability follows to be [5]

$$W^{(1)}(t)=\sum_n \beta_n p_{n-1}(t)={\rm Tr}[D\rho(t)] . \tag{6}$$

Now we assume that the de-excited atoms are detected by an ionization detector with efficiency η, the single-atom detection probability therefore being given by $\eta W^{(1)}(t)$. Let us again refer to a time interval $\Delta t \ll r^{-1}$, $(n_{\max}\gamma)^{-1}$. The evolution of the field density matrix during this interval can be formally decomposed into the contributions which arise from the two possibilities that either zero or one de-excited atom is detected during Δt. Hence we introduce new evolution operators V_0 and V_1 which refer to these possibilities and write

$$\rho(t+\Delta t)=[V_0(\Delta t)+V_1(\Delta t)]\rho(t) . \tag{7}$$

The probability for the detection of zero or one atom during the interval Δt is then expressed by ${\rm Tr}[V_0(\Delta t)\rho(t)]$ and ${\rm Tr}[V_1(\Delta t)\rho(t)]$, respectively. On the other hand, with the help of Eq. (6) we find

$${\rm Tr}[V_1(\Delta t)\rho(t)]=\eta r\Delta t\,{\rm Tr}[D\rho(t)] , \tag{8}$$

since $r\Delta t$ is the probability that one atom arrives in the small interval Δt and since the change of ρ during Δt is negligible. Comparing Eqs. (3), (7), and (8) we may write

$$V_0(\Delta t)=1+\Delta t[r(M-1-\eta D)+L] . \tag{9}$$

Thus we arrive at the differential equation

$$\dot{V}_0=[r(M-1-\eta D)+L]V_0 \tag{10}$$

for the evolution operator V_0 that refers to the evolution of the non-normalized conditioned density operator $\sigma(t_0+\tau)=V_0(\tau)\rho(t_0)$, which describes the radiation field at time $t_0+\tau$ on condition that no de-excited atoms are detected during the time τ though the detector is active. With $\sigma_{nn}=q_n$ we get from Eq. (10) the modified master equation for the conditioned density matrix in the photon-number representation

$$\dot{q}_n=(1-\eta)\beta_n q_{n-1}-\beta_{n+1}q_n+[L\sigma]_{nn} , \tag{11}$$

where the last term is defined by replacing ρ and p by σ and q in Eq. (1).

In the following we refer to the stationary micromaser regime characterized by the steady-state density operator $\bar{\rho}$. First we consider the joint probability $W^{(2)}(\tau)$ that two atoms which have the time difference τ both leave the micromaser in the de-excited state, with an unknown number of excited or de-excited atoms in between them. In analogy to Eq. (6) this quantity can be written with the help of Eq. (5) in the symbolic form

$$\begin{aligned} W^{(2)}(\tau)&={\rm Tr}[DV(\tau)D\bar{\rho}]\\ &=\sum_n \beta_n \tilde{p}_{n-1}(\tau)|_{\tilde{p}_n(0)=\beta_n\bar{p}_{n-1}} , \end{aligned} \tag{12}$$

where the bar indicates that the (non-normalized) probabilities $\tilde{p}_n(\tau)$ $(n=0,1,\dots)$ have to be calculated by solving the master equation (4) in the photon-number representation on the given initial condition. For Poissonian injection statistics, the probability density for the arrival of two excited pump atoms with time difference τ does not depend on τ and is equal to r^2. Therefore the two-time coincidence probability density for detecting two de-excited atoms at the instants t_1 and t_2 using a detector of efficiency η follows from Eq. (12) to be

$$P^{(2)}(|t_2-t_1|)=\eta^2 r^2{\rm Tr}[DV(|t_2-t_1|)D\bar{\rho}] , \tag{13}$$

whereas the probability density for the detection of a single de-excited atom is given by $P^{(1)}=\eta r\,{\rm Tr}[D\bar{\rho}]$ because of Eq. (6). According to the theory of point processes the mean value and the second factorial moment of the number of de-excited atoms detected during a counting interval of length T can be obtained by using the equations [6] $\langle N\rangle_T=TP^{(1)}$ and

$$\langle N(N-1)\rangle_T=\int_0^T dt_1\int_0^T dt_2 P^{(2)}(|t_2-t_1|) , \tag{14}$$

which yield the Mandel Q parameter $Q_a(T)=\langle\Delta N^2\rangle_T/\langle N\rangle_T-1$. Another quantity characterizing the steady-state statistics of de-excited atoms is the waiting-time distribution $P_w(\tau)$ defined as the conditional probability density of detecting the next atom at time $t+\tau$, provided the first atom had been detected at time t. The two-time elementary or exclusive probability density $P_0^{(2)}(\tau)$ for detecting two de-excited atoms with time difference τ on condition that no de-excited atom is detected in between them can be written in analogy to Eqs. (13) and (12) as

$$\begin{aligned} P_0^{(2)}(\tau)&=\eta^2 r^2{\rm Tr}[DV_0(\tau)D\bar{\rho}]\\ &=\eta^2 r^2\sum_n \beta_n \tilde{q}_{n-1}(\tau)\|_{\tilde{q}_n(0)=\beta_n\bar{p}_{n-1}} , \end{aligned} \tag{15}$$

where the double bar now indicates that the non-normalized probabilities $\tilde{q}_n(\tau)$ $(n=0,1,\dots)$ have to be calculated by solving the master equation (11) for the conditioned density matrix on the given initial condition. From the above equation we get the waiting-time distribution

$$P_w(\tau)=\frac{P_0^{(2)}(\tau)}{P^{(1)}}=\eta r\frac{\mathrm{Tr}[DV_0(\tau)D\bar{\rho}]}{\mathrm{Tr}[D\bar{\rho}]} . \tag{16}$$

So far we considered the statistics of de-excited atoms. To characterize the statistical properties of the stationary radiation field inside the micromaser cavity its normalized two-time intensity correlation function

$$g^{(2)}(\tau)=\frac{\langle a^{\dagger}(0)a^{\dagger}(\tau)a(\tau)a(0)\rangle}{\langle n\rangle^2}$$

$$=\frac{1}{\langle n\rangle^2}\sum_{m=0}^{\infty} m\hat{p}_m(\tau)|_{\hat{p}_m(0)=(m+1)\bar{p}_{m+1}} \tag{17}$$

has been recently investigated numerically [9]. Here $a,a^{\dagger}$ and $\langle n\rangle$ denote the photon annihilation and creation operator and the mean photon number of the cavity field mode. As in Eq. (12), the bar indicates that the master equation (4) has to be used.

III. APPLICATIONS

In order to apply the above results one needs the time-dependent solutions of the master equations (4) and (11). In general, these solutions can be found only numerically. Nevertheless, by thoroughly inspecting the formal properties of the solution of Eq. (4) an interesting insight is gained. With the help of Eqs. (1) and (2) we write Eq. (4) in the photon-number representation as

$$\dot{\mathbf{p}}=\mathbf{A}\mathbf{p} , \tag{18}$$

with $\mathbf{p}=(p_0,p_1,\dots,p_n,\dots)^T$ being a column vector composed of the photon-number probabilities and $\mathbf{A}$ being a tridiagonal matrix, the nonzero elements of which are given by $A_{n+1,n}=b_{n+1}$, $A_{n,n}=-b_{n+1}-a_n$, and $A_{n,n+1}=a_{n+1}$, where $b_n=r\beta_n+\gamma n n_b$ and $a_n=\gamma n(1+n_b)$ $(n=0,1,\dots)$. The well-known steady-state solution for the intracavity photon-number distribution follows from the recursion relation [2]

$$\gamma n(1+n_b)\bar{p}_n=(r\beta_n+\gamma n n_b)\bar{p}_{n-1} . \tag{19}$$

Making use of the method of Laplace transformation we express the time-dependent solution of Eq. (18) with the help of the inverse matrix $\mathbf{B}(z)=(z\mathbf{1}-\mathbf{A})^{-1}$ as

$$p_n(t)=\sum_{\nu}\mathrm{Res}\left[e^{tz}\sum_{m=0}^{\infty}B_{nm}(z)p_m(0)\right]_{z=z_\nu} . \tag{20}$$

If we construct the inverse matrix $\mathbf{B}$ in the usual way from the subdeterminants of the matrix $z\mathbf{1}-\mathbf{A}$, thereby taking into account the special tridiagonal structure of $\mathbf{A}$, and if we utilize the properties of the stationary micromaser solution $\bar{p}_n$, we find that the elements of $\mathbf{B}$ obey the symmetry relation

$$\bar{p}_m B_{nm}(z)=\bar{p}_n B_{mn}(z) . \tag{21}$$

Now we assume that the temperature of the micromaser is low enough for the thermal photon number n_b to be neglected. From the recursion relation (19) with $n_b=0$ and from Eq. (6) we then obtain the well-known relation $rW^{(1)}=\gamma\langle n\rangle$ [5], which expresses the fact that in the stationary regime the rate of de-excited atoms at the exit of the cavity is equal to the rate of photons leaving the cavity due to the losses. Moreover, inserting the formal representation (20) of the time-dependent master-equation solution into Eqs. (17) and (12) and taking into account the symmetry relation (21) and the recursion relation (19) with $n_b=0$ as well as Eq. (13), we arrive at the result

$$g^{(2)}(\tau)=\frac{W^{(2)}(\tau)}{[W^{(1)}]^2}=\frac{P^{(2)}(\tau)}{[P^{(1)}]^2}\equiv w^{(2)}(\tau) , \tag{22}$$

which means that in the stationary regime of a micromaser with Poissonian pumping and negligible thermal photon number the normalized two-time intensity correlation function of the cavity field with respect to a time difference τ is equal to the normalized two-time coincidence probability density for detecting two de-excited atoms with the same time difference at the exit of the cavity. The above result is a generalization of the relation $g^{(2)}(0)=W^2(0)/[W^{(1)}]^2$ derived in [5]. Physically, it may be explained by the facts that, without thermal photons, every time a new photon emerges in the cavity an atom must be de-excited, and that the statistics of atomic injection and of cavity damping do not give rise to additional correlations.

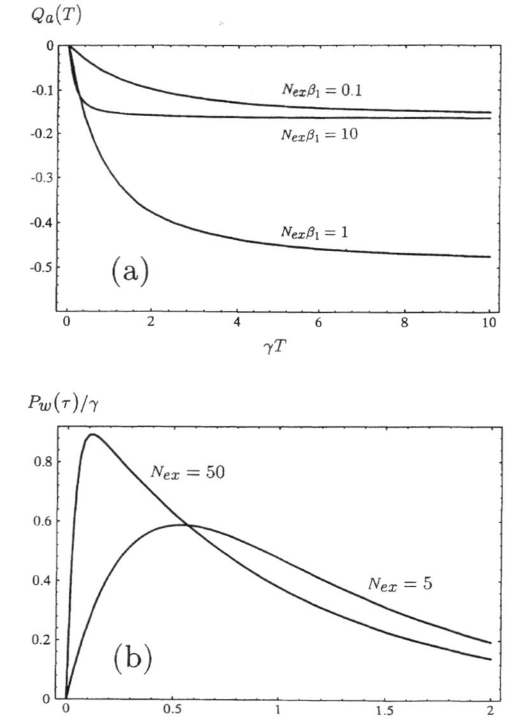

FIG. 1. Mandel Q parameter $Q_a(T)$ (a) and waiting-time distribution $P_w(\tau)$ (b) for the detection of de-excited atoms with efficiency $\eta=1$ at the exit of a micromaser in the trapped state at $gt_{\mathrm{int}}=\pi/\sqrt{2}$ for different values of $N_{\mathrm{ex}}=r/\gamma$ (b) and $N_{\mathrm{ex}}\beta_1=(r/\gamma)\sin^2(\pi/\sqrt{2})$ (a).

Finally we consider the stationary regime of a micromaser at zero temperature in the simplest trapped states arising from the trapping condition $gt_{\text{int}}=q\pi/\sqrt{2}$ ($q=1,2,3,\ldots$). In this case the photon-number distribution is truncated at $n=1$ and all results can be easily obtained analytically. By solving the master equation (18) we get the normalized two-time coincidence probability density $w^{(2)}(\tau)$ for the detection of the de-excited atoms [10] or the normalized two-time intensity correlation function of the cavity field

$$w^{(2)}(\tau)=g^{(2)}(\tau)=1-\exp\{-\gamma\tau(1+N_{\text{ex}}\beta_1)\}\ , \tag{23}$$

where $N_{\text{ex}}=r/\gamma$ and $\beta_1=\sin^2(q\pi/\sqrt{2})$. Obviously with growing N_{ex} the nonclassical character of the state manifesting itself in a normalized correlation function smaller than unity becomes more and more fragile. Furthermore, from the steady-state solution we find the Mandel Q parameter of the intracavity field $Q_f=\langle\Delta n^2\rangle/\langle n\rangle-1$ to be given by $Q_f=-N_{\text{ex}}\beta_1/(1+N_{\text{ex}}\beta_1)$. By using Eqs. (14), (22), and (23) we arrive at the expression

$$Q_a(T)=\frac{2\eta Q_f}{1+N_{\text{ex}}\beta_1}\left\{1+\frac{\exp[-\gamma T(1+N_{\text{ex}}\beta_1)]-1}{\gamma T(1+N_{\text{ex}}\beta_1)}\right\} \tag{24}$$

for the Mandel Q parameter characterizing the statistics of de-excited atoms with respect to a counting interval of length T in agreement with the result obtained independently in [10]. If $\gamma T \gg (1+N_{\text{ex}}\beta_1)^{-1}$ the parameter $Q_a(T)$ becomes independent of T; cf. Fig. 1(a). The waiting-time distribution is obtained with the help of the time-dependent solution of the master equation (11) as

$$P_w(\tau)=2\eta\beta_1 r\exp[-\tfrac{1}{2}\gamma\tau(1+N_{\text{ex}}\beta_1)] \times\frac{\sinh[\tfrac{1}{2}\gamma\tau\sqrt{N_{\text{ex}}^2\beta_1^2+2N_{\text{ex}}\beta_1(1-2\eta)+1}\,]}{\sqrt{N_{\text{ex}}^2\beta_1^2+2N_{\text{ex}}\beta_1(1-2\eta)+1}}\ ; \tag{25}$$

cf. Fig. 1(b). As it must be, the perfect antibunching expressed by $w^{(2)}(0)=0$ yields $P_w(0)=0$.

In conclusion, we remark that the method outlined in this paper can be easily extended to the treatment of the statistics of those atoms leaving the micromaser in the excited state or to the investigation of higher-order intensity correlation functions, multitime coincidence probabilities, higher-order moments of the number of counted atoms, or other statistical quantities. In particular, for $n_b=0$ in analogy to Eq. (22) also the normalized multitime intensity correlation function is found to be equal to the normalized multitime coincidence probability density of de-excited atoms.

ACKNOWLEDGMENTS

The author is indebted to Professor H. Paul, Dr. Th. Richter, and G. Kastelewicz for stimulating discussions.

[1] D. Meschede, H. Walther, and G. Müller, Phys. Rev. Lett. **54**, 551 (1985); G. Rempe, H. Walther, and N. Klein, *ibid* **58**, 353 (1987); J. Krause, M. O. Scully, and H. Walther, Phys. Rev. A **36**, 4547 (1987).

[2] P. Filipowic, J. Javanainen, and P. Meystre, Phys. Rev. A **34**, 3077 (1986); L. A. Lugiato, M. O. Scully, and H. Walther, *ibid.* **36**, 740 (1987).

[3] G. Rempe, F. Schmidt-Kaler, and H. Walther, Phys. Rev. Lett. **64**, 2783 (1990).

[4] G. Rempe and H. Walther, Phys. Rev. A **42**, 1650 (1990).

[5] H. Paul and Th. Richter, Opt. Commun. **85**, 508 (1991).

[6] See, e.g., B. Saleh, *Photoelectron Statistics* (Springer-Verlag, Berlin, 1978).

[7] M. D. Srinivas and E. B. Davies, Opt. Acta **28**, 981 (1981); see also C. W. Gardiner, *Quantum Noise* (Springer-Verlag, Berlin, 1991) and H. Carmichael, *An Open Systems Approach to Quantum Optics* (Springer-Verlag, Berlin, 1993).

[8] See, e.g., P. Meystre and M. Sargent III, *Elements of Quantum Optics* (Springer-Verlag, Berlin, 1990).

[9] T. Quang, Phys. Rev. A **46**, 682 (1992).

[10] H.-J. Briegel, B.-G. Englert, N. Sterpi, and H. Walther, Phys. Rev. A **49**, 2962 (1994).

Rui Han

Centre for Quantum Technologies, National University of Singapore, Singapore

Paper Title: **Raman transition without adiabatic elimination: a simple and accurate treatment**

Reference: Journal of Modern Optics **60**, 255–265 (2013)

Personal message:

I first met Berge in January 2005, at the beginning the second semester of my freshman year at NUS as an undergraduate student. By then, I have heard a lot from my seniors in the physics apartment about Professor Englert — the formal way I used to address him in those days — that he was one of the most outstanding physicists and lecturers in NUS. Thus, as a freshman, I decided to enroll in his module "Mathematical Methods in Physics I" designed for year-two physics students. That was a very tough module — partly also because I was too enthusiastic about and heavily involved in many school committees and activities, and didn't study hard enough. My memory of the "Mathematical Methods in Physics I" lectures has mostly faded away over time, and the reminder mainly associates with me sitting in a dark corner of the lecture room while Berge (in my mind there was even a spotlight on him, but that part of the memory might be twisted. . .) was doing magic tricks with vector calculations on the whiteboard. Yes, many of the tricks that he did on the whiteboard indeed appeared as magic to most of the students. I guess Berge does not remember me from those classes, as I was not an active student at all. Nevertheless, that was my first real encounter with him. To me, he was a brilliant physicist with an unimaginable amount of knowledge, and he also carried a serious look during lectures, which made me afraid of him.

I was upset about myself for not doing well for that module and decided to devote more time to study from my second year onwards. Subconsciously, we all wanted to show our true abilities to the people whom we respect the most, and, I guess, I was afraid of him because I had not managed to do so. Berge liked to frequently change the modules he taught, despite the extra work required to prepare new lectures. I was extremely lucky that his teaching plan had a large overlap with my module selections (or because I tried to adjust my schedule to take as many of his lectures as possible). During the four years of my undergraduate study, I had taken Mathematical Methods in Physics I and II, Quantum Mechanics II and III, Electrodynamics, Independent study, UROPS (NUS undergraduate research) and a Honours year project with Berge. Although his lectures were well-known among the students for being extremely challenging and I also struggled very often, probably, I gained much more knowledge from his classes than all my other modules combined.

I believe Berge started getting to know me from his Quantum Mechanics II module, as I concentrated as much as I could during those classes and could very occasionally point out a typo of his writing on the whiteboard. After gaining some confidence through this module, at the beginning of my first semester in the third university year, I introduced myself to him and asked him to be my supervisor for an independent-study module. Thus, in August 2006, Berge became my official supervisor. Honestly, I continued to be afraid of him for one more year. Every week when we met at Spinelli's in the Science canteen for discussion, I would show him my progress and where I had gotten stuck. And, very often, after a few minutes, he would say something like, "Oh, there is a much easier way of calculating this," or "Listen, we no longer use that old-fashioned method." Then he would show me, on a piece of A4 paper (sometimes a piece of napkin), an elegant way of going about the problem. I always marvelled.

In the summer of 2007, thanks to Berge, I had the opportunity of doing a two-month research internship in INLN, Nice, in Christian Miniatura's group. That was the start of the joint project between Singapore and France on cold fermions in the honeycomb lattice. Christian was there all the time, Berge stayed in Nice for a couple of weeks, and Benoît Grémaud also came to Nice for a few days. The discussions with all three of them were intensive and fruitful. During that summer, I also got to know what Berge was like outside physics, met his "headquarters" Ola (later known as Olanéni), started getting his humor, and enjoyed doing physics even more. That was the time I stopped being afraid of him.

At the end of my undergraduate study, Berge was trying to convince me that, for my own benefit and experience, I should consider pursuing a PhD overseas. The problem was that the undergraduate scholarship that I took did not allow me to leave Singapore within a few years after obtaining the Bachelor's degree. However, at the same time, I was very glad that I could stay in Singapore and continue learning from Berge. And, he continued supervising me tirelessly through my PhD. Occasionally, when I got stuck and felt frustrated, a chat with Berge always helped. He knew well how to reward students when they achieve something, to calm them down by repeatedly explaining the physical problems when they are confused, as well as to scold them when things are not done in the right manner (to me, there were many times that he had to point his index finger to the ceiling and say, "Scold! Scold!").

During the past few years, we went to a few conferences in Europe together. Berge often, very nicely, tried to "sell" me to his colleagues and friends during the conferences. Although we both joked that this was because I had been bothering him for too many years and he was eager to get rid of me, I fully understood his kindness. In October 2012, Berge and Ola, who treated me like their own daughter, finally met my parents in Xi'an, China. Although it was tiring to translate their conversations to each other all the time, I felt deeply how lucky I was.

One of the many things I learnt from Berge is to never take others' claims for granted before you manage to reproduce them, and to try as much as possible to reconstruct everything from scratch for a deeper understanding. The work in the paper I contribute is largely in this spirit. We were trying to use Rydberg transition, which is a two-photon Raman transition, to encode quantum information in the reference-frame-free qubit made of three atoms, and were not satisfied with the common approach of applying the adiabatic elimination process to obtain the effective Rabi frequency. Thus, we started from scratch and developed new techniques towards the problem. In this paper, we introduce a completely different method to solve the Raman transition and obtain the effective Rabi frequency. This new method is simple and gives much better results than the standard adiabatic elimination method.

To Berge: Wish you a very happy 60th birthday, and many more happy years to come! Continue to rock Physics in your own way!

Han Rui

Journal of Modern Optics, 2013
Vol. 60, No. 4, 255–265, http://dx.doi.org/10.1080/09500340.2013.770573

Taylor & Francis
Taylor & Francis Group

Raman transitions without adiabatic elimination: a simple and accurate treatment

Rui Han[a*], Hui Khoon Ng[a,b] and Berthold-Georg Englert[a,c]

[a] *Centre for Quantum Technologies, National University of Singapore, Singapore 117543, Singapore;* [b] *DSO National Laboratories, Singapore 118230, Singapore;* [c] *Department of Physics, National University of Singapore, Singapore 117542, Singapore*

(Received 17 November 2012; final version received 21 January 2013)

Driven Raman processes – nearly resonant two-photon transitions through an intermediate state that is non-resonantly coupled and does not acquire a sizeable population – are commonly treated with a simplified description in which the intermediate state is removed by adiabatic elimination. While the adiabatic-elimination approximation is reliable when the detuning of the intermediate state is quite large, it cannot be trusted in other situations, and it does not allow one to estimate the population in the eliminated state. We introduce an alternative method that keeps all states in the description, without increasing the complexity by much. An integro-differential equation of Lippmann–Schwinger type generates a hierarchy of approximations, but very accurate results are already obtained in the lowest order.

Keywords: Raman transition; adiabatic elimination; Rydberg excitation

1. Introduction

Atomic and molecular systems can be coupled in various ways, so that the atomic states evolve and the populations of states change. Of all the electromagnetic multipole couplings, the electric dipole is the strongest. Thus optically allowed atomic transitions are often stimulated by an electric-dipole coupling between the atom and a well-controlled laser.

Dipole-allowed transitions can be driven with optical lasers directly. It is, however, quite common that the desired transition is dipole forbidden or the transition frequency is outside of the popular optical range. Then some intermediate state can assist in an indirect transition. Examples for such processes include three-level Raman transitions [1], multi-level Raman transitions [2,3], and stimulated Raman adiabatic passage (STIRAP) [4,5]. They all have their useful applications, with three-level Raman transitions being perhaps the most widely used in a large variety of experiments.

In order to reduce the computational complexity of dealing with large Hilbert spaces, one can decrease the dimensionality of the system by eliminating states that are not populated much or not coupled strongly. For a typical three-level Raman transition, the intermediate state is far off-resonantly coupled to the relevant initial and target states. This enables one to perform the so-called *adiabatic elimination* that gets rid of the less relevant state and yields a two-level effective Hamiltonian. Although the procedure of adiabatic elimination is well understood – see [6] and[7], for instance – it gives a reliable approximation only when the detuning of the intermediate state is much larger than the Rabi frequencies for the coupling to the other two states.

In this paper, we introduce an alternative approach to the quantitative description of the three-level Raman transition. This new method does not rely on adiabatic elimination and gives a much more accurate solution while the computational complexity remains low. We set the stage in Section 2, where we briefly review driven three-level systems and state the notational conventions used throughout. Then, Section 3 deals with the usual adiabatic-elimination approach and comments on its limitations and problems. Our new approach is explained in Section 4: first we present the general methodology, then we show how it gives the exact solution to the resonant two-photon transition problem, and finally we generate reliable approximations for the situation of a non-resonant two-photon transition. We close with a summary and outlook.

2. The three-level system

A Raman transition is a two-photon process that gives an effective coupling between two states $|0\rangle$ and $|1\rangle$ via a far-detuned auxiliary state $|e\rangle$; see Figure 1. As mentioned above, Raman transitions are often used when the transition between levels $|0\rangle$ and $|1\rangle$ is dipole forbidden or has an inconvenient frequency. The Λ-type configuration of Figure 1(a) applies to transitions between different ground states via an excited state; the cascade-type configuration of

*Corresponding author. Email: han.rui@quantumlah.org

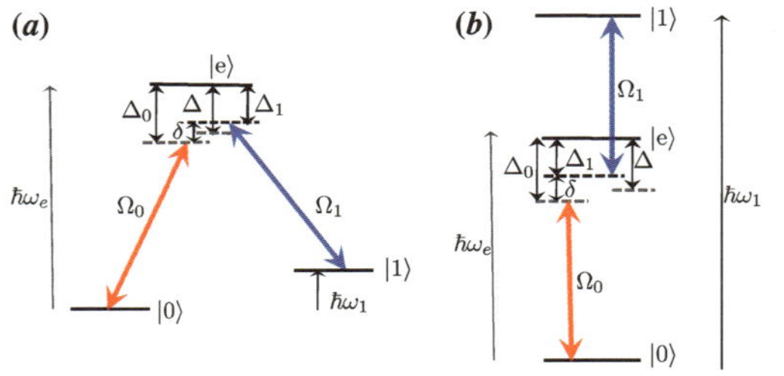

Figure 1. Level scheme of a typical Raman transition. Part (a) shows the level structure of a Λ-type Raman transition and (b) shows the level structure of a cascade-type Raman transition. Ω_0 and Ω_1 denote the Rabi frequencies of the individual two-level transitions, Δ denotes the detuning of the laser from the transition frequency of the excited state and δ is the detuning of the two-photon transition. The requirement is that the detuning Δ is much larger than the Rabi frequencies so that the excited state $|\mathrm{e}\rangle$ is not significantly populated. (The color version of this figure is included in the online version of the journal.)

Figure $1(b)$ can be used to achieve the transition between a ground state and a highly excited state, such as a Rydberg state. There is also a V-type transition with the level configuration similar to that of an inverted Λ-type transition. Although the level structures are different for these configurations, the underlying physics is essentially the same.

We will, therefore, restrict ourselves to treating the Λ-type Raman transition in detail. With reference to $|0\rangle$, $|1\rangle$, and $|\mathrm{e}\rangle$ in this order, the 3×3 matrix for the Hamiltonian of the system is

$$H = H_{\mathrm{Atom}} + H_{\mathrm{AL}}\,, \tag{1}$$

where

$$H_{\mathrm{Atom}} = \hbar\begin{pmatrix} 0 & 0 & 0 \\ 0 & \omega_1 & 0 \\ 0 & 0 & \omega_{\mathrm{e}} \end{pmatrix} \tag{2}$$

is the part for the atom by itself with the convention that the state $|0\rangle$ has energy zero, and

$$H_{\mathrm{AL}} = \frac{\hbar}{2}\begin{pmatrix} 0 & 0 & \Omega_0 \mathrm{e}^{\mathrm{i}\omega_{\mathrm{L}0}t} \\ 0 & 0 & \Omega_1 \mathrm{e}^{\mathrm{i}\omega_{\mathrm{L}1}t} \\ \Omega_0^* \mathrm{e}^{-\mathrm{i}\omega_{\mathrm{L}0}t} & \Omega_1^* \mathrm{e}^{-\mathrm{i}\omega_{\mathrm{L}1}t} & 0 \end{pmatrix} \tag{3}$$

accounts for the atom–laser interaction. The laser frequencies are denoted by $\omega_{\mathrm{L}0}$ and $\omega_{\mathrm{L}1}$, respectively. The Rabi frequencies for the electric-dipole transitions are

$$\Omega_0 = \frac{q_{\mathrm{el}}}{\hbar}\langle 0|\boldsymbol{r}\cdot\boldsymbol{E}_{\mathrm{L}0}|\mathrm{e}\rangle \quad \text{and}$$
$$\Omega_1 = \frac{q_{\mathrm{el}}}{\hbar}\langle 1|\boldsymbol{r}\cdot\boldsymbol{E}_{\mathrm{L}1}|\mathrm{e}\rangle\,, \tag{4}$$

where q_{el} is the electron charge, $\boldsymbol{E}_{\mathrm{L}0}$ and $\boldsymbol{E}_{\mathrm{L}1}$ are the amplitudes of the electric fields of the laser beams and $\boldsymbol{r}$ is the position vector of the atom. Each Rabi frequency depends on the intensity and polarization of the corresponding laser as well as the dipole matrix element between the two coupled states, and can be complex. Typically, the detuning Δ is designed to be large, i.e. $|\Delta| \gg |\Omega_0|, |\Omega_1|$, so that the auxiliary state $|\mathrm{e}\rangle$ does not get populated to avoid problems arising from uncontrolled spontaneous emission from $|\mathrm{e}\rangle$ to other states. Since the two-photon transition from $|0\rangle$ to $|1\rangle$ is nearly resonant, the overall detuning δ of the two-photon transition is small compared with the average detuning Δ, $|\delta/\Delta| \ll 1$. If the detunings between the lasers and the atomic frequencies are denoted by $\Delta_0 = \omega_{\mathrm{e}} - \omega_{\mathrm{L}0}$ and $\Delta_1 = \omega_{\mathrm{e}} - \omega_1 - \omega_{\mathrm{L}1}$, we have $\Delta = (\Delta_0 + \Delta_1)/2$ and $\delta = \Delta_0 - \Delta_1$; see Figure 1.

In view of the time-dependent phase factors in H_{AL}, it is expedient to switch to an interaction picture in which the Hamiltonian does not depend on time. This is achieved by identifying the 'free' Hamiltonian H_0 by a suitable splitting of the atomic Hamiltonian,

$$\begin{aligned} H_0 &= H_{\mathrm{Atom}} + \hbar\begin{pmatrix} \frac{1}{2}\delta & 0 & 0 \\ 0 & -\frac{1}{2}\delta & 0 \\ 0 & 0 & -\Delta \end{pmatrix} \\ &= \hbar\begin{pmatrix} \frac{1}{2}\delta & 0 & 0 \\ 0 & \omega_1 - \frac{1}{2}\delta & 0 \\ 0 & 0 & \omega_{\mathrm{e}} - \Delta \end{pmatrix}, \end{aligned} \tag{5}$$

for which we obtain the interaction-picture Hamiltonian

$$\begin{aligned} H_{\mathrm{I}} &= \exp(\mathrm{i}H_0 t/\hbar)(H - H_0)\exp(-\mathrm{i}H_0 t/\hbar) \\ &= \frac{\hbar}{2}\begin{pmatrix} -\delta & 0 & \Omega_0 \\ 0 & \delta & \Omega_1 \\ \Omega_0^* & \Omega_1^* & 2\Delta \end{pmatrix}. \end{aligned} \tag{6}$$

The evolution of the three-level system is then studied with the aid of the Schrödinger equation

$$\begin{aligned} \mathrm{i}\hbar\frac{\partial}{\partial t}\Psi_{\mathrm{I}}(t) &= H_{\mathrm{I}}\Psi_{\mathrm{I}}(t) \\ \text{with}\quad \Psi_{\mathrm{I}}(t) &= \begin{pmatrix} c_0(t) \\ c_1(t) \\ c_{\mathrm{e}}(t) \end{pmatrix} = \exp(\mathrm{i}H_0 t/\hbar)\Psi(t), \end{aligned} \tag{7}$$

where the components of $\Psi_{\mathrm{I}}(t)$ are the interaction-picture probability amplitudes for $|0\rangle$, $|1\rangle$, and $|\mathrm{e}\rangle$, related to the respective components of $\Psi(t)$, the amplitudes in the Schrödinger picture, by simple time-dependent phase factors. As a consequence of this simple relation between the components of $\Psi(t)$ and $\Psi_{\mathrm{I}}(t)$, we can simply square $c_0(t)$, $c_1(t)$, or $c_{\mathrm{e}}(t)$ to obtain the probability amplitudes for the respective atomic levels, as exemplified by the probability for $|0\rangle$,

$$\left|\begin{pmatrix}1\,0\,0\end{pmatrix}\Psi(t)\right|^2 = \left|\begin{pmatrix}1\,0\,0\end{pmatrix}\Psi_{\mathrm{I}}(t)\right|^2 = |c_0(t)|^2, \tag{8}$$

where, of course, $\begin{pmatrix}1\,0\,0\end{pmatrix}$ is the three-component row for $\langle 0|$.

To understand the system analytically, we can solve for the eigensystem of this time-independent Hamiltonian directly. However, very often, this is neither the most efficient way of getting the solution, nor the best method for obtaining a good physical insight into the problem. Various approaches have been developed, of which the adiabatic elimination and, for $\delta = 0$, the dark-state method [8] are particularly useful and popular.

3. Adiabatic elimination

3.1. The methodology

The standard textbook approach to the Raman-transition problem makes use of adiabatic elimination in accordance with the following line of reasoning. Since the excited state $|e\rangle$ is far-detuned by Δ, it will remain barely populated if it has no initial population. Thus, the change of the population in this state can be taken as approximately zero, $(\partial/\partial t)c_e(t) = 0$, so that

$$2i\frac{\partial}{\partial t}c_e(t) = \Omega_0^* c_0(t) + \Omega_1^* c_1(t) + 2\Delta c_e(t) = 0, \quad (9)$$

in view of the Schrödinger equation (7). We can now express $c_e(t)$ as a linear combination of $c_0(t)$ and $c_1(t)$, and so eliminate $c_e(t)$ from the equations of motion for $c_0(t)$ and $c_1(t)$. This gives us an effective 2×2 Hamiltonian for the evolution of the two relevant states,

$$\begin{aligned} H_{\text{eff}} &= -\frac{\hbar}{2}\begin{pmatrix} \delta + \frac{|\Omega_0|^2}{2\Delta} & \frac{\Omega_0\Omega_1^*}{2\Delta} \\ \frac{\Omega_1\Omega_0^*}{2\Delta} & -\delta + \frac{|\Omega_1|^2}{2\Delta} \end{pmatrix} \\ \text{for} \quad i\hbar\frac{\partial}{\partial t}\psi(t) &= H_{\text{eff}}\psi(t) \\ \text{with} \quad \psi(t) &= \begin{pmatrix} c_0(t) \\ c_1(t) \end{pmatrix}. \end{aligned} \quad (10)$$

As an immediate benefit of applying adiabatic elimination on the intermediate auxiliary state, the effective Hamiltonian is a simple 2×2 matrix, for which the eigenvalues and the projectors to the eigenspaces are readily available.

The eigenvalues of H_{eff} are

$$E_\pm = -\frac{\hbar}{8\Delta}\left(|\Omega_0|^2 + |\Omega_1|^2\right) \pm \frac{\hbar}{2}\Omega_R \quad (11)$$

with the positive frequency Ω_R given by

$$\Omega_R^2 = \frac{1}{(4\Delta)^2}\left(|\Omega_0|^2 + |\Omega_1|^2\right)^2 + \frac{\delta}{2\Delta}\left(|\Omega_0|^2 - |\Omega_1|^2\right) + \delta^2. \quad (12)$$

The projectors to the corresponding eigenspaces are

$$\frac{1}{2}(1 \pm \sigma_o),$$

where

$$\sigma_o = \frac{2H_{\text{eff}} - E_+ - E_-}{E_+ - E_-} \quad (13)$$

is a Pauli-type matrix. If the evolution starts with all the population in the ground state $|0\rangle$, i.e. $c_0(t = 0) = 1$, the population in state $|1\rangle$ at a later time t is

$$|c_1(t)|^2 = \frac{|\Omega_0|^2|\Omega_1|^2}{8\Delta^2\Omega_R^2}\left[1 - \cos(\Omega_R t)\right]. \quad (14)$$

This tells us the physical significance of

$$\Omega_R = (E_+ - E_-)/\hbar$$

: it is the effective Rabi frequency of the transition between states $|0\rangle$ and $|1\rangle$ via this Raman process.

In the situation of vanishing overall detuning, $\delta = 0$, we have

$$|c_1(t)|^2 = \frac{2|\Omega_0|^2|\Omega_1|^2}{\left(|\Omega_0|^2 + |\Omega_1|^2\right)^2} \times \left[1 - \cos\left(\frac{|\Omega_0|^2 + |\Omega_1|^2}{4\Delta}t\right)\right], \quad (15)$$

where the effective Rabi frequency is

$$\Omega_R = \left(|\Omega_0|^2 + |\Omega_1|^2\right)/(4|\Delta|)$$

and the amplitude of the Rabi oscillation is less than unity,

$$\frac{4|\Omega_0|^2|\Omega_1|^2}{\left(|\Omega_0|^2 + |\Omega_1|^2\right)^2} < 1, \quad (16)$$

unless $|\Omega_0|^2 = |\Omega_1|^2$. In other words, when $\delta = 0$, we get complete population transfer between state $|0\rangle$ and state $|1\rangle$ only if the two lasers drive the respective transitions equally strongly.

More generally, in this effective description, perfect population transfer from $|0\rangle$ to $|1\rangle$ is only possible if the two diagonal matrix elements of the effective Hamiltonian are identical, i.e. if there is no effective detuning after the adiabatic elimination. Then $|0\rangle$ and $|1\rangle$ are equal-weight superpositions of the eigenstates of H_{eff} and temporal evolution turns one into the other. For given laser intensities, and thus given Rabi frequencies Ω_0 and Ω_1, the experimenter can exploit the Zeeman or the Stark effect to adjust the overall detuning such that

$$\delta = \frac{|\Omega_1|^2 - |\Omega_0|^2}{4\Delta}. \quad (17)$$

This makes the effective detuning vanish and ensures perfect population transfer. Indeed, for this value of δ, the right-hand side of (14) simplifies,

$$|c_1(t)|^2 = \frac{1}{2}[1 - \cos(\Omega_R t)] \quad \text{with} \quad \Omega_R = \frac{|\Omega_0||\Omega_1|}{2|\Delta|}. \quad (18)$$

In experiments, specifically for two-photon population transfer from the ground state to a Rydberg state where the Rabi frequencies $|\Omega_0|$ and $|\Omega_1|$ can be an order of magnitude different in strength, adjusting the detuning in accordance with (17) is important [9].

3.2. Light shift

When an atomic transition is driven by an electromagnetic radiation field with detuning Δ, the dressed atomic levels are shifted. This is the so-called *light shift* of the atomic levels, which is a second-order correction to the eigenenergies of the Hamiltonian [10]. If an atomic transition is driven by Rabi frequency Ω with detuning Δ, the light shift of the ground state of the transition is $-|\Omega|^2/(4\Delta)$ and the light shift of the dressed excited state is of the same amount but opposite in sign. The overall light shift is a direct summation

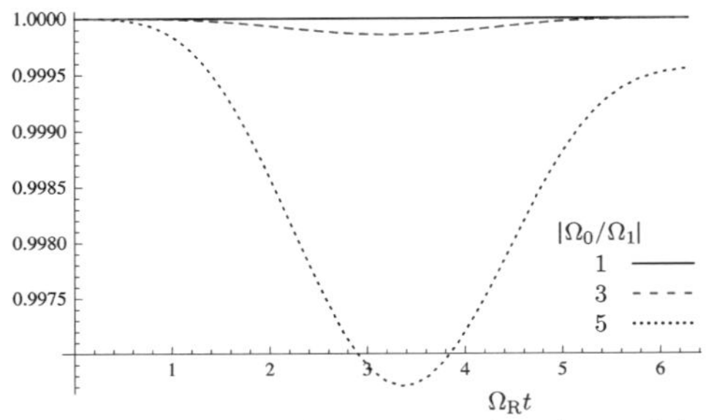

Figure 2. Fidelity between the two states at later time evolving with the exact Hamiltonian but for the two different δ values of (17) and (20). $\Delta = 400$ MHz and $|\Omega_1| = 40$ MHz are fixed and the different curves are for three different values of the ratio $|\Omega_0/\Omega_1|$. (The color version of this figure is included in the online version of the journal.)

of the light shifts arising from individual electromagnetic radiation fields, when the atomic level is addressed by multiple fields.

In the present context, the light shift of state $|0\rangle$ is $\hbar\delta_0 = -\hbar|\Omega_0|^2/(4\Delta_0)$ and the light shift of state $|1\rangle$ is $\hbar\delta_1 = -\hbar|\Omega_1|^2/(4\Delta_1)$. Thus, the value of δ that brings the two-photon transitions into resonance is determined by

$$\delta = \frac{|\Omega_1|^2}{4\Delta + 2\delta} - \frac{|\Omega_0|^2}{4\Delta - 2\delta}, \tag{19}$$

which is solved by

$$\delta = \frac{2\Delta\big(|\Omega_1|^2 - |\Omega_0|^2\big)}{8\Delta^2 + |\Omega_0|^2 + |\Omega_1|^2} + \cdots, \tag{20}$$

where the ellipsis stands for terms of relative size $|\Omega/\Delta|^2$ or smaller. The difference between the δ values obtained from (17) and (20) can be of the order of a percent ($\sim (|\Omega_0|^2 + |\Omega_1|^2)/(64\Delta^2)$) for typical experimental parameters. But the approximate solutions provided by the adiabatic-elimination method do not depend much on this small fractional difference in δ. Figure 2 shows the fidelity between the states at later times using the two different values of δ when the initial state is $|0\rangle$.

The effect of using these two different δ values is small when the difference between $|\Omega_0|$ and $|\Omega_1|$ is not too large. For the cases shown in Figure 2, the error from using (17) is a small fraction of a percent. This explains why (17) can serve as a good guidance for experiments, although (20) is more accurate.

3.3. *Problems with adiabatic elimination*

In physics, one speaks of an 'adiabatic process' if a relevant property evolves quite slowly, whereas other processes are fast – a clear separation of time scales is a defining element. In the context of adiabatic elimination, one invokes such a separation in the evolution of the components of $\Psi_{\rm I}(t)$, for which we have the Schrödinger equation in (7). One standard argument observes that, as a consequence of $|\Delta| \gg |\Omega_0|, |\Omega_1|$, the amplitude $c_{\rm e}(t)$ will undergo many oscillations during a period in which $c_0(t)$ and $c_1(t)$ do not change substantially. Then, if our interest is not in the short-time changes that would reveal the rapid oscillations of $c_{\rm e}(t)$ but predominantly in the longer-time changes of $c_0(t)$ and $c_1(t)$, the average change of $c_{\rm e}(t)$ over these longer periods is expected to be quite small. In the spirit of this reasoning, we should then read (9) as a statement about coarse-grained values of the probability amplitudes.

This is hardly a rigorous argument, and whether one regards it as convincing or not is largely a matter of taste. Clearly, though, a more solid argument would be welcome, and one has been provided in [7]. Indeed, the reasoning in [7] uses coarse graining in conjunction with a Markov approximation.

There are other problems that one needs to keep in mind when employing adiabatic elimination. We present them as four questions.

(i) Which is the correct interaction picture to use? The theory argues that the change of the (coarse-grained) population in the excited state is approximately zero and we arrived at (9) by using Schrödinger's equation of motion with the time-independent Hamiltonian $H_{\rm I}$ in (6). However, the choice of interaction picture is not unique and why do we apply the adiabatic elimination in this particular interaction picture instead of another? For example, by adding a constant term, the operator $H_{\rm I} + E$ describes the system equally well. But if we use an interaction picture with $E \neq 0$, the resultant two-level effective Hamilton operator also changes as Δ gets replaced by $\Delta + E/\hbar$ in (10) with consequential changes in the evolution of $\psi(t)$.

   This ambiguity in the choice of interaction picture was studied by Brion et al. with a Green's function formalism [6]. It would be premature, however, to claim that the case is closed. Further studies of the choice of interaction picture and a systematic way of performing the adiabatic elimination are presented in a companion paper [7].

(ii) Is it possible to estimate the population in the eliminated state? This does not have a simple answer because there are problems with the normalization of the wave function. Originally, we have $\Psi_{\rm I}(t)^\dagger\Psi_{\rm I}(t) = 1$. Then, since $H_{\rm eff}$ is hermitian, we also have $\psi(t)^\dagger\psi(t) = 1$. Combined with the initial condition $(c_0, c_1, c_{\rm e}) = (1, 0, 0)$ this implies $c_{\rm e}(t) = 0$ for all t, which contradicts the basic approximation in (9).

(iii) Is it possible to regard the adiabatic-elimination approximation as the first in a hierarchy of approximations? This is possible, indeed, as discussed in [7]. It turns out that the next approximation in the hierarchy gives a substantial improvement and better quantitative estimates. In addition, it answers the previous question inasmuch as the next approximation provides an estimate for the population in the auxiliary state.

(iv) Is it possible to avoid the adiabatic elimination without increasing the complexity much beyond the convenient two-level description of (10)? Yes, this is possible, as we demonstrate in Section 4.

4. Without adiabatic elimination

Alternative methods other than adiabatic elimination are also used to solve Raman-transition problems. The most direct way is, of course, to diagonalize the interaction-picture Hamiltonian of (6), and this can be done by hand since the dimensionality of the system is small. But the expressions for the eigenvalues and eigencolumns of H_I are quite involved and not transparent. An exception is the two-photon transition with vanishing overall detuning, $\delta = 0$, when one can identify a dark state and use it to reduce the three-level system to an effective two-level system, which can then be solved exactly rather simply. The dark states are particularly useful in the context of adiabatic population transfer and electromagnetically induced transparency. When the overall detuning is nonzero, however, there is no dark state. Methods of perturbation theory can then be used to find corrections for a small detuning δ, but the complexity grows quickly when high accuracy is required.

In this section, we provide a new way of solving the Raman-transition problem. Just like the dark-state method, it gives a compact exact solution for $\delta = 0$, and it can solve the $\delta \neq 0$ case with very high precision and rather little extra effort.

4.1. General methodology

It will be expedient to use a different interaction picture as the one of (5) and (6) that we used for the adiabatic elimination in Section 3.1. Instead, we choose

$$H_0 = \frac{\hbar}{2}\begin{pmatrix} \Delta+\delta & 0 & 0 \\ 0 & 2\omega_1+\Delta-\delta & 0 \\ 0 & 0 & 2\omega_\mathrm{e}-\Delta \end{pmatrix}, \tag{21}$$

and the Hamiltonian in this interaction picture is

$$H_\mathrm{I} = \frac{\hbar}{2}\begin{pmatrix} -\Delta-\delta & 0 & \Omega_0 \\ 0 & -\Delta+\delta & \Omega_1 \\ \Omega_0^* & \Omega_1^* & \Delta \end{pmatrix} = \frac{\hbar}{2}\begin{pmatrix} -(\Delta+\delta\sigma_3) & \Omega \\ \Omega^\dagger & \Delta \end{pmatrix}, \tag{22}$$

where the latter way of writing emphasizes the $3 = 2 + 1$ split into two relevant states and one auxiliary state; σ_3 is the standard third Pauli 2×2 matrix, and Ω and $\Omega^\dagger$ are the two-component column of Rabi frequencies,

$$\Omega = \begin{pmatrix} \Omega_0 \\ \Omega_1 \end{pmatrix}, \tag{23}$$

and its adjoint row. This change of interaction picture is equivalent to shifting the energy levels of the interaction-picture Hamiltonian in (6) by $-\hbar\Delta/2$, an example of the freedom of choice discussed in question (i) in Section 3.3.

We note that the square of H_I,

$$H_\mathrm{I}^2 = (\hbar M)^2 = \hbar^2(M_0^2 + \epsilon), \tag{24}$$

is the sum of a 'big' block-diagonal part,

$$M_0{}^2 = \frac{1}{4}\begin{pmatrix} (\Delta+\delta\sigma_3)^2 + \Omega\Omega^\dagger & 0 \\ 0 & \Delta^2 + \Omega^\dagger\Omega \end{pmatrix}, \tag{25}$$

and a 'small' off-diagonal part

$$\epsilon = -\frac{\delta}{2}\begin{pmatrix} 0 & \sigma_3\Omega \\ \Omega^\dagger\sigma_3 & 0 \end{pmatrix}. \tag{26}$$

Matrix M denotes a square root of the matrix $(H_\mathrm{I}/\hbar)^2$ and, since for each eigenvalue we have a choice of sign, there are many Ms that are equally good. Any one can be used as a replacement of $H_\mathrm{I}/\hbar$ in even functions of H_I; for example, we could choose $M > 0$ by convention. In particular, then, the unitary evolution matrix in this interaction picture can be written as

$$U(t) = \exp(-\mathrm{i}H_\mathrm{I}t/\hbar) = \cos(Mt) - \frac{\mathrm{i}}{\hbar}\frac{\sin(Mt)}{M}H_\mathrm{I}\,. \tag{27}$$

Except for the common physical approximations that enter the modeling of the atom–laser system by a driven three-level system described by the Hamiltonian of (1)–(3), this is an exact 3×3 matrix representing the evolution operator.

Instead of diagonalizing the interaction-picture Hamiltonian (22), we can determine the eigenvalues and eigencolumns of M^2, whose 'big plus small' structure, together with the block-diagonal form of M_0^2, facilitates approximations. This splitting of H_I^2 is possible because the two-photon transition is nearly resonant; we will see the advantage thereof shortly. Let us note that we can position the factor H_I in the second term of the right-hand side of (27) equally well to the left of $\sin(Mt)/M$, or break up $\sin(Mt)/M$ and sandwich H_I between even powers of M. Since $[H_\mathrm{I}, M^2] = 0$, such a change in the order of the matrices makes no difference in (27), but slightly different expressions are obtained when approximations are introduced for the trigonometric functions of M.

4.2. Vanishing overall detuning ($\delta = 0$) – exact solution

When $\delta = 0$, we have $\epsilon = 0$ and

$$M = M_0 = \frac{1}{2}\begin{pmatrix} -(\Delta^2 + \Omega\Omega^\dagger)^{1/2} & 0 \\ 0 & (\Delta^2 + \Omega^\dagger\Omega)^{1/2} \end{pmatrix}, \tag{28}$$

where the signs are chosen such that $M \to H_I/\hbar$ in the $\Omega \to 0$ limit. The resulting evolution operator (27) reads

$$U(t) = \cos(M_0 t) - \frac{\mathrm{i}}{\hbar}\frac{\sin(M_0 t)}{M_0}H_I\,. \tag{29}$$

Owing to the block-diagonal structure of M_0, the original 3×3 problem has been converted into an equivalent 2×2 problem without introducing any approximation. Clearly, the technical difficulty has been significantly reduced!

The column Ω is an eigencolumn of $\Delta^2 + \Omega\Omega^\dagger$ with eigenvalue $\Delta^2 + \Omega^\dagger\Omega$, so that

$$M_0^2 = \frac{1}{4}(\Delta^2 + \Omega^\dagger\Omega)\begin{pmatrix} \frac{\Omega\Omega^\dagger}{\Omega^\dagger\Omega} & 0 \\ 0 & 1 \end{pmatrix} + \frac{1}{4}\Delta^2\begin{pmatrix} 1-\frac{\Omega\Omega^\dagger}{\Omega^\dagger\Omega} & 0 \\ 0 & 0 \end{pmatrix} \tag{30}$$

is the spectral decomposition of M_0^2. This gives

$$\cos(M_0 t) = \cos\Big((\Delta^2 + \Omega^\dagger\Omega)^{1/2}\,t/2\Big)\begin{pmatrix} \frac{\Omega\Omega^\dagger}{\Omega^\dagger\Omega} & 0 \\ 0 & 1 \end{pmatrix} + \cos(\Delta t/2)\begin{pmatrix} 1-\frac{\Omega\Omega^\dagger}{\Omega^\dagger\Omega} & 0 \\ 0 & 0 \end{pmatrix} \tag{31}$$

and likewise for $\sin(M_0 t)$ in (29). Hence, the exact evolution of the system can be written out analytically. In particular, the population in the excited state $|\mathrm{e}\rangle$ is non-zero. With the atom initially in the ground state $|0\rangle$, it is

$$\begin{aligned}|c_\mathrm{e}(t)|^2 &= \left|(0\;0\;1)\,U(t)\begin{pmatrix}1\\0\\0\end{pmatrix}\right|^2 \\ &= \frac{|\Omega_0|^2}{\Delta^2+\Omega^\dagger\Omega}\sin\Big((\Delta^2+\Omega^\dagger\Omega)^{1/2}\,t/2\Big)^2.\end{aligned} \tag{32}$$

This exact expression shows that the population in the excited state oscillates with (angular) frequency

$$\frac{1}{2}(\Delta^2 + \Omega^\dagger\Omega)^{1/2}$$

and the oscillation amplitude can be non-negligible if $|\Omega_0|^2$ is a sizeable fraction of $\Delta^2 + \Omega^\dagger\Omega = \Delta^2 + |\Omega_0|^2 + |\Omega_1|^2$.

We note that the second projection matrix in (30) and (31) projects on the dark state, whose bra has the row $(-\Omega_1\;\Omega_0\;0)$ [8]. An atom prepared in this dark state stays in it, and there is no probability of finding the atom in the excited state at any time. The atom is essentially decoupled from the driving lasers under these circumstances. In this sense, one could regard (29) with (30) as the evolution matrix in the dark-state formalism but this is, in fact, not the case. In the dark-state approach, one diagonalizes H_I, reduced to a two-dimensional problem after putting the dark state aside, which amounts to choosing one particular square root of M_0^2 from the continuous family of square roots that the degenerate eigenvalue makes available, namely the square root whose eigenvalues and eigencolumns are those of $H_I/\hbar$. No such unique M_0 is needed in (29), nor is there any benefit from enforcing a unique square root of M_0^2 by imposing additional criteria. Although in the $\delta = 0$ case, this equivalence can be established between the new approach and the dark-state approach, the new approach offers more flexibility and it also has a clear advantage when dealing with the $\delta \neq 0$ case, as we shall see in Section 4.3.

For three sets of parameter values, Figure 3 shows the population $|c_\mathrm{e}(t)|^2$ of (32) as well as the populations $|c_0(t)|^2$ and $|c_1(t)|^2$ of the relevant states $|0\rangle$ and $|1\rangle$. The populations of $|0\rangle$ and $|1\rangle$ are oscillating with a slow frequency that goes with a large amplitude and a fast frequency that goes with a small amplitude. The population transfer is predominantly controlled by the slow frequency which, therefore, is the effective Rabi frequency of the system. Since M_0^2 has only two distinct eigenvalues, half the sum of their square roots gives the fast frequency and the difference gives the slow frequency. Thus, we have

$$\Omega_\mathrm{R} = \frac{1}{2}\Big((\Delta^2+\Omega^\dagger\Omega)^{1/2} - |\Delta|\Big) = \frac{\Omega^\dagger\Omega}{4|\Delta|} - \frac{(\Omega^\dagger\Omega)^2}{16|\Delta|^3} + \cdots \tag{33}$$

for the effective Rabi frequency. The expansion in powers of $|\Omega/\Delta| \ll 1$ permits a comparison of this exact expression with the approximation obtained by adiabatic elimination, the $\delta = 0$ version of (12): the adiabatic-elimination approximation gives the leading term, but does not reproduce any of the higher-order terms.

In summary, we find that the solution from the adiabatic elimination is indeed the zeroth-order approximation of the exact result in the expansion of $\Omega^\dagger\Omega/\Delta^2$. For example, in the zeroth order of $\Omega^\dagger\Omega/\Delta^2$, the excited state population vanishes as the oscillation amplitude is proportional to $\Omega^\dagger\Omega/\Delta^2 \sim 0$. We conclude that adiabatic elimination yields a reliable approximation only when $|\Omega_0|, |\Omega_1| \ll |\Delta|$. The difference between the solutions obtained by our method and the adiabatic-elimination approximation is demonstrated clearly in Figure 3.

The top left plot in Figure 3 shows the populations when $|\Omega_0| = |\Omega_1| = |\Delta|/10$ is small; in this parameter regime, we can already see the deviation of the adiabatic elimination from the exact result but the deviation is not significant. In the top-right plot, the population of state $|1\rangle$ only reaches a maximum of about 80% as $|\Omega_0| \neq |\Omega_1| \sim |\Delta|/10$ introduces an effective detuning for the two-photon transition; the adiabatic elimination gives a good approximation since we are still in the regime of $|\Omega_0|, |\Omega_1| \ll |\Delta|$. When the magnitudes of $|\Omega_0|$ and $|\Omega_1|$ are not much smaller than $|\Delta|$, the population of the excited state is no longer negligible, but the complete population transfer between states $|0\rangle$ and $|1\rangle$ is still achievable; see the bottom-left plot in Figure

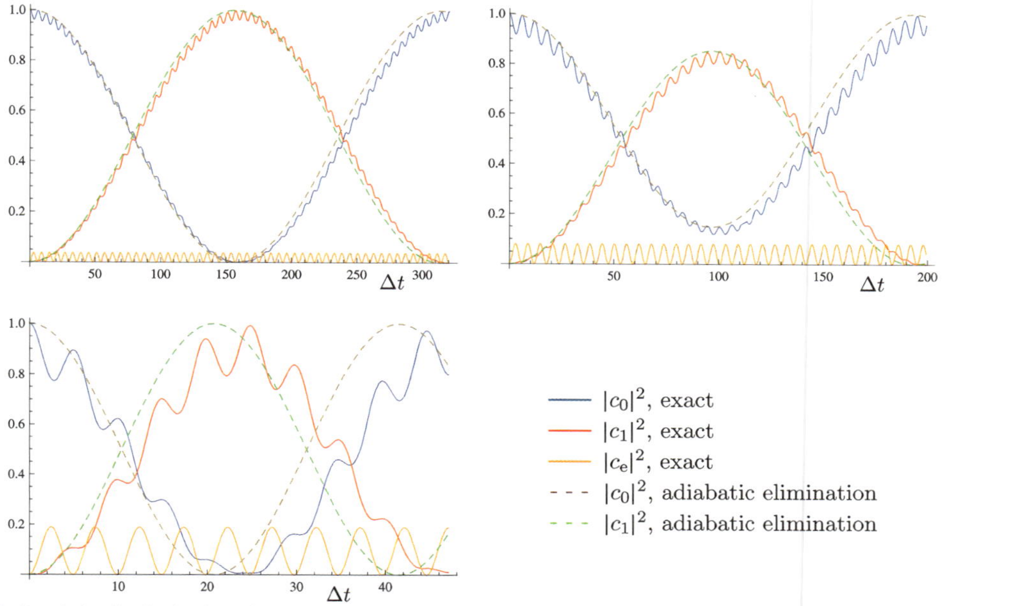

Figure 3. Population distribution for a single-atom Raman transition in time when $\delta = 0$. The solid curves show the exact solution, and the dashed curves the adiabatic-elimination approximation. The initial state is $|0\rangle$. The red curves are for the ground-state population $|c_0(t)|^2$, the blue curves for $|c_1(t)|^2$, and the orange curves report $|c_e(t)|^2$, the population in the excited state. The detuning is $\Delta = 400$ MHz for all plots; the top left plot is for $|\Omega_0| = |\Omega_1| = \Delta/10 = 40$ MHz; the top right plot is for $|\Omega_0| = \Delta/10 = 40$ MHz and $|\Omega_1| = \Delta/16 = 25$ MHz; the bottom plot is for $|\Omega_0| = |\Omega_1| = \Delta/4 = 100$ MHz. (The color version of this figure is included in the online version of the journal.)

3. The adiabatic elimination method fails in this case of stronger coupling between the relevant states $|0\rangle$, $|1\rangle$ and the excited state $|e\rangle$.

4.3. Non-zero overall detuning ($\delta \neq 0$)

As discussed in Section 3.3, one needs to adjust the detuning δ to compensate for the light shifts and achieve complete population transfer from $|0\rangle$ to $|1\rangle$. Besides this, there can also be other experimental reasons for choosing a particular δ value. Thus, the situation of $\delta \neq 0$ is of practical interest, and so we will regard the overall detuning δ as a free parameter that is small compared with the average detuning Δ. Typically, the ratios $|\delta/\Delta|$ and $\Omega^\dagger\Omega/\Delta^2$ are of the same small order, a few percent perhaps.

The splitting of H_I^2 in (24)–(26) has the diagonal blocks in M_0^2, including the δ-dependent contributions, whereas ϵ contains the off-diagonal parts linear in δ. We shall treat ϵ as a small quantity, without, however, regarding δ as an expansion parameter. Rather, the full δ-dependence of M_0^2 is taken into account. Then, the eigenvalues of M^2 agree with the eigenvalues of M_0^2 to first order in the perturbation of ϵ, and the leading correction will be of order ϵ^2.

The eigenvalues of M_0^2 in the subspace spanned by $|0\rangle$ and $|1\rangle$ are

$$\mu_\pm^2 = \frac{1}{4}(\Delta^2 + \delta^2) + \frac{1}{8}\Omega^\dagger\Omega \pm \frac{1}{8}\left((\Omega^\dagger\Omega)^2 + 8\delta\Delta\Omega^\dagger\sigma_3\Omega + (4\delta\Delta)^2\right)^{1/2}, \tag{34}$$

which we get from diagonalizing the upper 2×2 diagonal block $(\Delta + \delta\sigma_3)^2 + \Omega\Omega^\dagger$. The unnormalized eigencolumns are

$$\begin{pmatrix} [4\mu^2 - (\Delta + \delta\sigma_3)^2]\Omega \\ 0 \end{pmatrix} \quad \text{for } \mu^2 = \mu_\pm^2, \tag{35}$$

unless Ω is an eigencolumn of σ_3, which is a case of no interest. The eigenvalue $\frac{1}{4}(\Delta^2 + \Omega^\dagger\Omega)$ in the 1×1 block for $|e\rangle$ does not depend on δ. With its eigenvalues and eigencolumns at hand, all functions of M_0^2 are readily evaluated.

In passing, we observe that the difference between μ_+^2 and μ_-^2 is smallest, as a function of δ, when (17) holds, i.e. $4\delta\Delta = -\Omega^\dagger\sigma_3\Omega$. Then

$$\mu_+^2 - \mu_-^2 = \frac{1}{4}\left((\Omega^\dagger\Omega)^2 - (\Omega^\dagger\sigma_3\Omega)^2\right)^{1/2} = \frac{|\Omega_0|\,|\Omega_1|}{2}, \tag{36}$$

which is nonzero in all situations of interest.

For a systematic inclusion of correction of orders ϵ, ϵ^2, ϵ^3, ..., we do not use the perturbation theory for an approximation of the eigenvalues and eigencolumns of M^2 for use in (27). Rather, we generate approximations for the evolution matrix $U(t) = \exp(-\mathrm{i}H_\mathrm{I}t/\hbar)$ itself with the aid of an equation of Lippmann–Schwinger type,

$$U(t) = U_0^{(\mathrm{R})}(t) - \int_0^t \mathrm{d}t' \, \frac{\sin\bigl(M_0(t-t')\bigr)}{M_0}\epsilon U(t') \,, \quad (37)$$

where

$$U_0^{(\mathrm{R})}(t) = \cos(M_0 t) - \frac{\mathrm{i}}{\hbar}\frac{\sin(M_0 t)}{M_0}H_\mathrm{I} \quad (38)$$

differs from (29) by the inclusion of the δ dependent terms in M_0 and H_I.

As mentioned at the end of Section 4.1, the multiplication order of H_I and even powers of M^2 is irrelevant in (27) as they commute. In (38), however, the order does matter as $[H_\mathrm{I}, M_0^2] \neq 0$ when $\delta \neq 0$. In addition to (37) with (38) where H_I is on the right, we have, therefore, also an 'on the left' version,

$$U(t) = U_0^{(\mathrm{L})}(t) - \int_0^t \mathrm{d}t' \, U(t-t')\epsilon \frac{\sin(M_0 t')}{M_0} \quad (39)$$

with

$$U_0^{(\mathrm{L})}(t) = \cos(M_0 t) - \frac{\mathrm{i}}{\hbar}H_\mathrm{I}\frac{\sin(M_0 t)}{M_0} \,. \quad (40)$$

Half their sum gives a symmetrized version,

$$\begin{aligned} U(t) \; = \; & U_0^{(\mathrm{S})}(t) - \frac{1}{2}\int_0^t \mathrm{d}t' \, \frac{\sin\bigl(M_0(t-t')\bigr)}{M_0}\epsilon U(t') \\ & -\frac{1}{2}\int_0^t \mathrm{d}t' \, U(t-t')\epsilon \frac{\sin(M_0 t')}{M_0} \end{aligned} \quad (41)$$

with

$$U_0^{(\mathrm{S})}(t) = \cos(M_0 t) - \frac{\mathrm{i}}{2\hbar}\frac{\sin(M_0 t)}{M_0}H_\mathrm{I} - \frac{\mathrm{i}}{2\hbar}H_\mathrm{I}\frac{\sin(M_0 t)}{M_0} \,, \quad (42)$$

and there are many more variants that one could explore.

Each of the integral equations, (37), (39), and (41), provides a hierarchy of approximations by an iteration that commences with the respective zeroth-order approximation. This is, of course, the procedure by which one generates the Born series from the Lippmann–Schwinger equation. For the 'on the right' Equation (37), the kth-order approximation is

$$U_k^{(\mathrm{R})}(t) = U_{k-1}^{(\mathrm{R})}(t) - \int_0^t \mathrm{d}t' \, \frac{\sin\bigl(M_0(t-t')\bigr)}{M_0}\epsilon U_{k-1}^{(\mathrm{R})}(t') \,, \quad (43)$$

and analogous expressions apply to the 'on the left' version and the symmetrized variant. Note that $U_k^{(\mathrm{S})}(t)$ is not half the sum of $U_k^{(\mathrm{R})}(t)$ and $U_k^{(\mathrm{L})}(t)$ for $k \neq 0$; that arithmetic mean could also be taken as a valid kth-order approximation. For such a scheme to be useful in practice, the zeroth-order approximation should be quite good to begin with, the first-order approximation should be sufficient for many purposes, and it should not be necessary to go beyond the second order.

The various kth-order approximations differ from each other, but they are all accurate up to kth-order in $|\epsilon/M_0^2| \sim |\delta\Omega|/\Delta^2$. In the common case where $|\Omega/\Delta| \sim 10^{-1}$ and $|\delta/\Delta| \sim 10^{-2}$, we have $|\epsilon/M_0^2| \sim 10^{-3}$. Remember that the δ dependence in H_I and the ϵ dependence (which also depends on δ) are treated separately. Although ϵ goes to zero when δ vanishes, $|\epsilon/{M_0}^2|$ is one order of magnitude smaller than $|\delta/\Delta|$.

A technical point is the following. The approximate evolution matrices $U_k(t)$ are not unitary; rather, $U_k(t)^\dagger U_k(t)$ deviates from the unit matrix by an amount of order ϵ^{k+1}. One can cope with this in various ways. Perhaps the simplest is to ensure proper normalization by including a time-dependent factor that depends on the initial set of probability amplitudes, thereby arriving at an effectively unitary matrix $\widetilde{U}_k(t)$ which is suitable for the given initial column $\Psi_\mathrm{I}(0)$,

$$\widetilde{U}_k(t)\Psi_\mathrm{I}(0) = \frac{U_k(t)\Psi_\mathrm{I}(0)}{\bigl(\Psi_\mathrm{I}(0)^\dagger U_k(t)^\dagger U_k(t)\Psi_\mathrm{I}(0)\bigr)^{1/2}} \,. \quad (44)$$

In other words, we apply $U_k(t)$ to $\Psi_\mathrm{I}(0)$ and normalize the resulting column to unit length. This procedure worked fine for all examples that we studied, and more sophisticated alternative normalization methods can be found in [11].

As remarked above, the kth-order approximations

$$U_k^{(\mathrm{R})}(t),$$

$U_k^{(\mathrm{L})}(t)$, and $U_k^{(\mathrm{S})}(t)$ differ slightly and might not describe the system equally well. We discuss a few examples of the state populations as functions of time under the different approximations, and analyze their performance. We compare the results to the exact numerical answers. The value $\Delta = 400\,\mathrm{MHz}$ is taken for the average detuning, as it is of typical order for real experiments, and the values of the Rabi coupling strengths Ω_0 and Ω_1 can vary in a range of fractions of Δ. The overall detuning δ of the two-photon transition can be controlled to within 1 MHz accuracy in laboratory experiments. For the purpose of this analysis, then, we take the liberty of setting δ to any value we like.

Figure 4 gives an example that demonstrates the quality of the zeroth-order approximations given by $\widetilde{U}_0^{(\mathrm{R})}(t)$, $\widetilde{U}_0^{(\mathrm{L})}(t)$, and $\widetilde{U}_0^{(\mathrm{S})}(t)$. Since $U_0^{(\mathrm{S})}(t)$ is an average of $U_0^{(\mathrm{R})}(t)$ and $U_0^{(\mathrm{L})}(t)$, we expect the population curves for $\widetilde{U}_0^{(\mathrm{S})}(t)$ to lie between the other two curves and this can be seen quite clearly in Figure 4. Moreover, the solutions of the population in the initial state $|0\rangle$ do not depend much on which of the three zeroth-order approximations is used, and all of them are very close to the exact numerical solution. In this example, we use $\delta = (|\Omega_1|^2 - |\Omega_0|^2)/(4\Delta)$, the value of (17), which gives an effective resonant two-photon transition, and we can see that full population transfer from $|0\rangle$ to $|1\rangle$ can be achieved. In comparison, full population transfer cannot be achieved for other values of δ.

In Figure 4, the value of $|\delta|$ is about a few percent of $|\Delta|$. The plot shows that, in this parameter regime, $\widetilde{U}_0^{(\mathrm{S})}(t)$

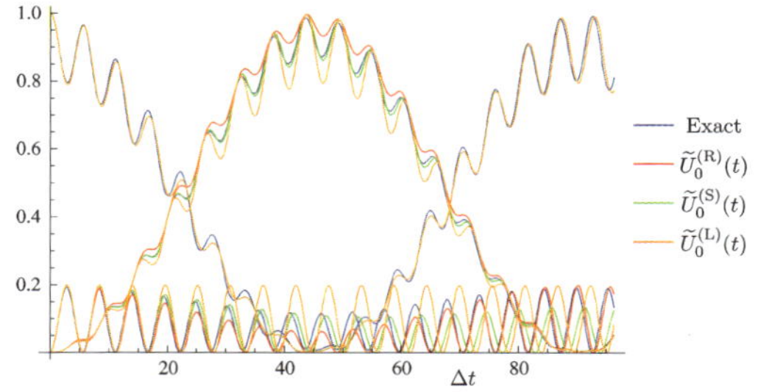

Figure 4. Plots of populations obtained from different zeroth-order solutions of the Lippmann–Schwinger equations. Blue curves give the exact results from numerical simulation using the Hamiltonian H_I; green curves show solutions from the symmetric approximation $\widetilde{U}_0^{(S)}(t)$; red curves show solutions from $\widetilde{U}_0^{(R)}(t)$; and orange curves show solutions from $\widetilde{U}_0^{(L)}(t)$. The parameters are $\Delta = 400\,\text{MHz}$, $|\Omega_0| = \Delta/2$, $|\Omega_1| = 3\Delta/10$ and $\delta = -\Omega^\dagger\sigma_3\Omega/(4\Delta) = (|\Omega_1|^2 - |\Omega_0|^2)/(4\Delta) = -16\,\text{MHz}$. The effective Rabi frequency is $\Omega_R = 27.8\,\text{MHz}$, about 7% of Δ. Initially, we have $c_0(0) = 1$ and $c_1(0) = c_e(0) = 0$. The curves starting at 1 show the approximations for $|c_0(t)|^2$; the curves that start at 0 and rise to 1 are for $|c_1(t)|^2$; and the curves that start at 0 and never exceed small values are for $|c_e(t)|^2$. (The color version of this figure is included in the online version of the journal.)

approximates the evolution of the probabilities for finding $|0\rangle$ and $|1\rangle$ quite well, and it certainly works best among the three different zeroth-order approximations shown here. The approximation for the excited state population also works well when t is short, i.e. during the first few fast oscillation periods, but the deviation grows quickly with time.

The accuracy is better for higher-order approximations. We compare the approximations of zeroth, first, and second order for the three different Lippmann–Schwinger equations in Figures 5(*a*), 5(*b*), and 6. Figure 5(*a*) shows that the deviation of the zeroth-order approximation $\widetilde{U}_0^{(R)}(t)$ from the exact numerical solution is large when about half of the effective Rabi cycle is completed, i.e. around $t = \pi/\Omega_R$ or $\Delta t \simeq 45$, and the deviation is smaller around a full Rabi cycle. The main deviation is in the size of the small-amplitude oscillations with short period, whereas the Rabi oscillation with longer period is reproduced equally well by $\widetilde{U}_k^{(R)}(t)$ with $k = 0, 1, 2$. The first-order approximation corrects part of the error in the zeroth-order approximation, and the second-order approximation improves matters further and gets the probabilities very close to their exact values. The same observations can be made about the corresponding 'on the left' approximations in Figure 5(*b*). How about the symmetric version $\widetilde{U}_k^{(S)}(t)$ whose zeroth-order approximation already works quite well?

Figure 6 shows that when the two-photon transition is resonant, the difference between $\widetilde{U}_0^{(S)}(t)$, $\widetilde{U}_1^{(S)}(t)$, and $\widetilde{U}_2^{(S)}(t)$ is difficult to detect. All three lowest-order approximations of (41) describe the evolution of the system well during the first Rabi cycle. The zeroth-order approximation works surprisingly well, at times it gives a better result than the higher-order approximations (see the bottom plot of the two blow-ups). The improvement offered by the higher-order approximations can be observed near the middle of the Rabi cycle (see the top plot of the blow-ups), and this improvement is more substantial when ϵ gets larger. For a practical application to experiments that aim at complete population transfer from $|0\rangle$ to $|1\rangle$, the approximation provided by $\widetilde{U}_0^{(S)}(t)$ is accurate enough to determine the parameter values reliably.

4.4. Discussion

To summarize our approach, we split $(H_I/\hbar)^2$, the square of the interaction Hamiltonian, into two parts: the dominant part M_0^2 and a small correction ϵ. With this splitting and any one of the Lippmann–Schwinger equations (37), (39), or (41), successive approximations to the unitary evolution matrix $U(t) = \exp(-\mathrm{i}H_I t/\hbar)$ can be obtained iteratively. If we use the approximations given by $\widetilde{U}_k^{(R)}(t)$ and $\widetilde{U}_k^{(L)}(t)$, we only need to do one iteration and stop at the first-order solution ($k = 1$) for a very good approximation. If we use the symmetric version $\widetilde{U}_k^{(S)}(t)$, we do not even need to go beyond the zeroth-order approximation since $\widetilde{U}_0^{(S)}(t)$ is already very close to the exact evolution for typical experimental parameters. Thus, we have

$$U(t) \simeq U_0^{(S)}(t) = \cos(M_0 t) - \frac{\mathrm{i}}{2\hbar}\frac{\sin(M_0 t)}{M_0}H_I - \frac{\mathrm{i}}{2\hbar}H_I\frac{\sin(M_0 t)}{M_0} \rightarrow \widetilde{U}_0^{(S)}(t), \quad (45)$$

where $\widetilde{U}_0^{(S)}(t)$ differs from $U_0^{(S)}(t)$ by the time-dependent factor of (44) that ensures $\Psi_I(t) = \widetilde{U}_0^{(S)}(t)\Psi_I(0)$ is properly normalized for the given initial column of probability amplitudes.

According to (45), the oscillation of the state populations of $|0\rangle$ and $|1\rangle$ are governed by the operator M_0^2. The effective Rabi oscillation frequency, to a very good approximation, only depends on the eigenvalues $\mu_\pm^2$ of the first diagonal block of M_0^2. Applying the same argument as in Section 4.2 for the case of $\delta = -\Omega^\dagger\sigma_3\Omega/4\Delta$ in (17), the effective Rabi frequency is (we take $\mu_\pm > 0$)

$$\Omega_R = \mu_+ - \mu_- = \frac{1}{2}\left[(\Delta^2+\delta^2) + \frac{1}{2}(|\Omega_0| + |\Omega_1|)^2\right]^{1/2} - \frac{1}{2}\left[(\Delta^2+\delta^2) + \frac{1}{2}(|\Omega_0| - |\Omega_1|)^2\right]^{1/2}. \quad (46)$$

With this particular choice of δ, the Rabi oscillation amplitude could reach unity; however, this might not be the real maximum that the population in state $|1\rangle$ can reach, because on top of this slow effective Rabi oscillation, the population also oscillates with a fast frequency. This fast

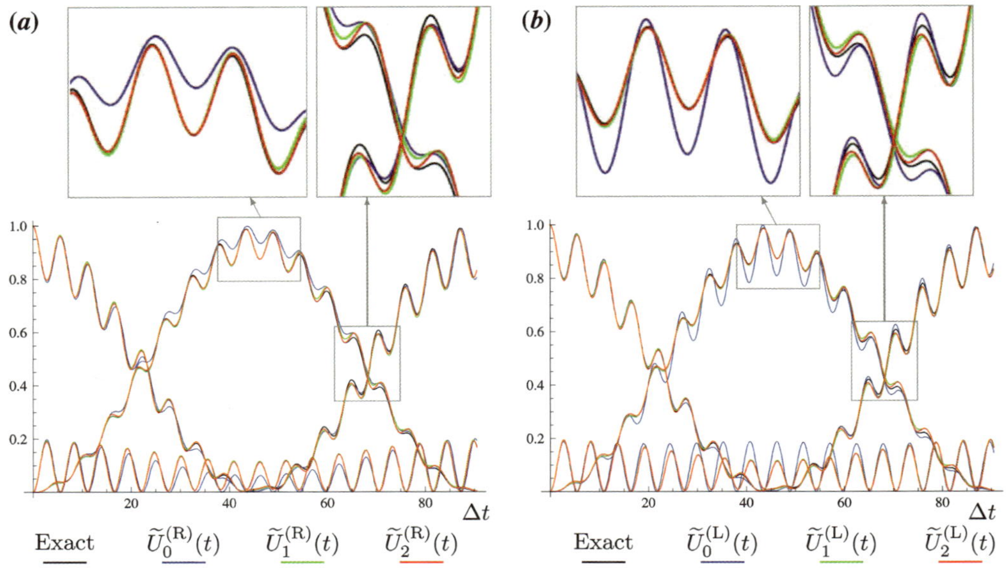

Figure 5. Comparison of the probabilities obtained from the zeroth-, first-, and second-order approximations of the Lippmann–Schwinger equations (37) and (39). In plot (a), the blue, green, and red curves show solutions from $\widetilde{U}_0^{(R)}(t)$, $\widetilde{U}_1^{(R)}(t)$, and $\widetilde{U}_2^{(R)}(t)$, respectively; and in plot (b), the blue, green, and red curves show solutions from from $\widetilde{U}_0^{(L)}(t)$, $\widetilde{U}_1^{(L)}(t)$, and $\widetilde{U}_2^{(L)}(t)$, respectively. The parameters and the initial state are the same as in Figure 4. (The color version of this figure is included in the online version of the journal.)

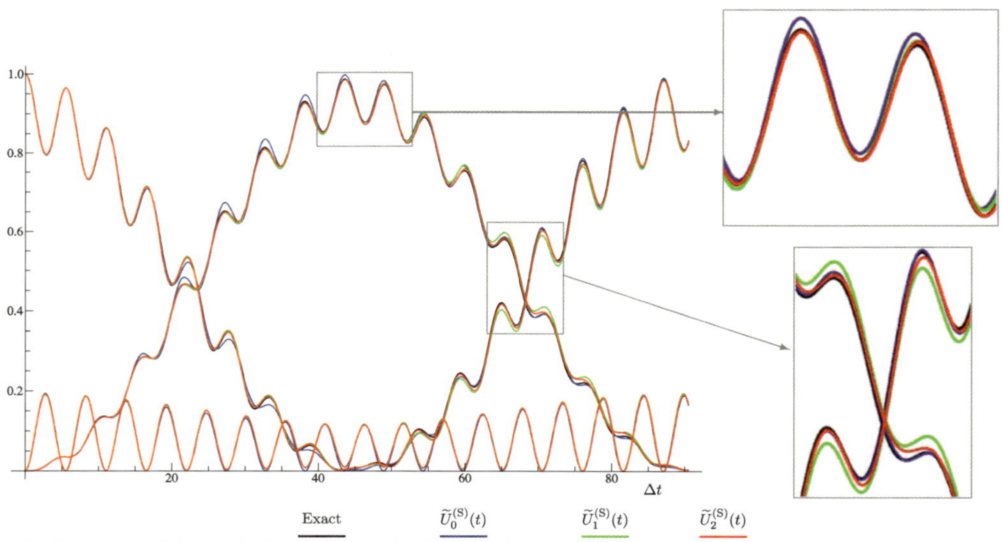

Figure 6. Comparison of the zeroth-, first-, and second-order approximations of the symmetric Lippmann–Schwinger equation (41). The parameter values and the initial state are the same as in Figure 4. (The color version of this figure is included in the online version of the journal.)

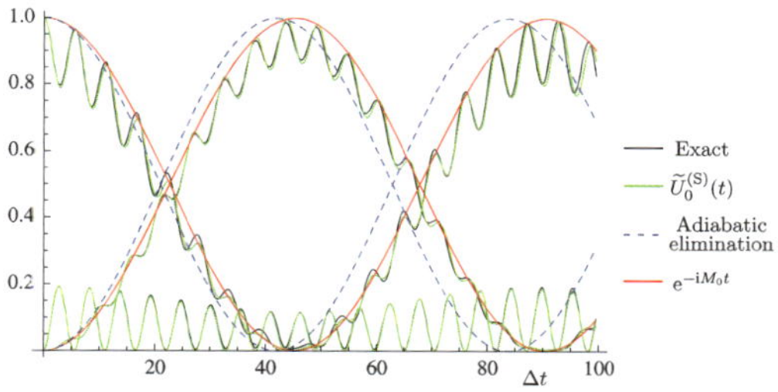

Figure 7. Improvement on the effective two-level Hamiltonian compared with adiabatic elimination. The black curves give the exact numerical solution; the green curves are for the symmetric zeroth-order approximation of $U(t)$; the blue dashed curves are for the adiabatic-elimination approximation; and the red curves result from taking M_0 as the effective Hamiltonian. The parameters are the same as in Figure 4. (The color version of this figure is included in the online version of the journal.)

oscillation goes roughly with the frequency $\mu_+ + \mu_- \simeq |\Delta|$. Nevertheless, up to linear order in δ/Δ, we find that the oscillation amplitude goes to unity, regardless of whether we choose δ in accordance with (17) or with (20). This shows why the evolution of the system is essentially the same for both δ values.

Moreover, since the effective Rabi oscillation of the two relevant states depends on the eigensystem of M_0^2 only, the effective Hamiltonian between states $|0\rangle$ and $|1\rangle$ is approximately given by the 2×2 upper diagonal block of $\hbar M_0$, if we exclude the excited state $|e\rangle$ from the evolution directly. That is

$$H_{\text{eff}} = -\frac{\hbar}{2}\left[(\Delta + \delta\sigma_3)^2 + \Omega\Omega^\dagger\right]^{1/2}. \tag{47}$$

The minus sign is chosen with the same reasoning as in (28). From this effective Hamiltonian, we can find the Rabi frequency directly and one of the special cases was already given in (46); the oscillation amplitude for an arbitrary δ is

$$P = 1 - \frac{(\Omega^\dagger\sigma_3\Omega + 4\delta\Delta)^2}{(\Omega^\dagger\Omega)^2 + 8\delta\Delta\Omega^\dagger\sigma_3\Omega + (4\delta\Delta)^2}, \tag{48}$$

and again we have $P = 1$ when $\delta = -\Omega^\dagger\sigma_3\Omega/4\Delta$. This approximate two-level effective Hamiltonian exhibits the same two-photon resonance condition in Equation (17) as given by the adiabatic elimination. Figure 7 shows that this effective Hamiltonian is much more accurate than that of the adiabatic-elimination approximation, inasmuch as the evolution by $\exp(-iM_0t)$ gives a very close envelope of the population oscillation.

5. Summary and outlook

After reviewing the standard adiabatic-elimination approximation, which reduces the theoretical description of a three-level Raman transition to an effective two-level problem, and identifying some of the shortcomings of this approach, we introduced an alternative approximation method. Similarly to adiabatic elimination, there is an essential two-level component in the new method without, however, eliminating the third auxiliary level. This makes the new method easy to use, inasmuch as one only needs to diagonalizea 2×2 matrix. Integro-differential equations of Lippmann–Schwinger type are the powerful tools that enable us to generate successive approximations. A particular one with high symmetry performs so well that the lowest-order approximation is all one needs for a highly reliable determination of experimental parameters.

For the sake of simplicity in presentation, the discussion was here limited to two-photon transitions in a single atom. Some generalization to multi-atom transitions can be found in [11]. The inclusion of spontaneous emission into the formalism is desirable; possibly, the methods developed in [12] are useful for this purpose.

Acknowledgements

This work is supported by the National Research Foundation and the Ministry of Education, Singapore. We are grateful for insightful discussions with Vanessa Paulisch, Antoine Browaeys, János Bergou, Gerd Leuchs and Björn Hessmo.

References

[1] Scully, M.O.; Zubairy, M.S. *Quantum Optics*; Cambridge University Press, Cambridge, UK, 1997.
[2] Kyrölä, E.; Lindberg, M. *Phys. Rev. A* **1987**, *35*, 4207–4225.
[3] Kyrölä, E.; Lindberg, M. *Opt. Commun.* **1983**, *48*, 284–286.
[4] Oreg, J.; Hioe, F.T.; Eberly, J.H. *Phys. Rev. A* **1984**, *29*, 690–697.
[5] Kuklinski, J.R.; Gaubatz, U.; Hioe, F.T.; Bergmann, K. *Phys. Rev. A* **1989**, *40*, 6741–6744.
[6] Brion, E.; Pedersen, L.H.; Mølmer, K.J. *Phys. A: Math. Theor.* **2007**, *40*, 1033–1043.
[7] Paulisch, V.; Han, R.; Ng, H.K.; Englert, B.-G. Raman Transitions: Adiabatic Elimination Revisited. 2012, arXiv:1209.6568 [quant-ph] e-Print archive. http://arxiv.org/abs/1209.6568 (accessed Nov 1, 2012)
[8] Fleischhauer, M.; Manka, A.S. Phys. Rev. A 1996, 54, 794–803.
[9] Miroshnychenko, Y.; Gaëtan, A.; Evellin, C.; Grangier, P.; Comparat, D.; Pillet, P.; Wilk, T.; Browaeys, A. *Phys. Rev. A* **2010**, *82*, 013405.
[10] Barratt, J.P.; Cohen-Tannoudji, C. J. Phys. Radium 1961, 22, 329, 443.
[11] Han, R. Robust Quantum Storage With Three Atoms. Ph.D. Thesis, National University of Singapore, 2012.
[12] Reiter, F.; Sørensen, A.S. *Phys. Rev. A* **2012**, *85*, 032111.

Miguel Orszag

Facultad de Física, Pontificia Universidad Católica de Chile, Santiago, Chile

Paper Title: **Thermal effects on sudden changes and freezing of correlations between remote atoms in a cavity quantum electrodynamics network**

Reference: Optics Letters **39**, 2668–2671 (2014)

Personal message:

All your buddies cheer
Berge, have a good year
We first met long time ago
in a small town in New Mexico
When the Humpty Dumpty egg broke
Marlan, Julian and you enjoyed the joke
You did a lot of thinking
without even blinking
You had fun with the quantum debate
Now your friends here celebrate
So, Berge, everybody cheers
have a happy birthday, my dear

2668 OPTICS LETTERS / Vol. 39, No. 9 / May 1, 2014

Thermal effects on sudden changes and freezing of correlations between remote atoms in a cavity quantum electrodynamics network

Vitalie Eremeev,[1,*] Nellu Ciobanu,[2] and Miguel Orszag[2]

[1]*Facultad de Ingeniería, Universidad Diego Portales, Santiago, Chile*

[2]*Facultad de Física, Pontificia Universidad Católica de Chile, Casilla 306, Santiago, Chile*

**Corresponding author: vitalie.eremeev@udp.cl*

Received February 17, 2014; revised March 31, 2014; accepted April 1, 2014; posted April 1, 2014 (Doc. ID 206115); published April 23, 2014

We investigate thermal effects on sudden changes and freezing of the quantum and classical correlations of remote qubits in a cavity quantum electrodynamics (CQED) network with losses. We find that the detrimental effect of thermal reservoirs on the freezing of correlations can be compensated via an efficient coupling of the fiber connecting the two cavities of the system. Furthermore, for certain initial conditions, we find a double sudden transition in the dynamics of Bures geometrical quantum discord. The second transition tends to disappear at a critical temperature, hence freezing the discord. Finally, we discuss the feasibility of the experimental realization of the present proposal.

OCIS codes: (270.5585) Quantum information and processing; (270.5580) Quantum electrodynamics.
http://dx.doi.org/10.1364/OL.39.002668

Quantum correlations play a fundamental role in quantum computation and quantum information processing [1], where entanglement is usually considered a popular measure of such correlations. However, more general measures such as quantum discord (QD) [2,3] and geometric quantum discord (GQD) with Bures distance [4–6] have been shown to capture nonclassical correlations, including completely separable systems, e.g., the deterministic quantum computation with one quantum bit (DQC1) model.

The unusual dynamics of classical and quantum decoherence originally reported in [7,8] and confirmed experimentally in [9–11] has stimulated high interest in the investigation of the phenomena of sudden changes in the correlations for different physical systems. Hence, during the past years, intensive effort has been focused on explaining the nature of sudden transitions and freezing effects of quantum correlations and the conditions under which such transitions occur. Also, from the perspective of applications, how efficiently one could engineer these phenomena in quantum technologies is also a matter of interest.

As has been shown in previous studies [6–16], the puzzling peculiarities of the sudden transitions and freezing phenomena are hidden in the structure of the density operator during the entire evolution of a bipartite quantum system for particular decoherence processes. Nevertheless, important questions remain open—how these fascinating effects are affected by the presence of noisy environments and if there are efficient mechanisms to control them in both nondissipative or dissipative decoherence models. The state-of-the-art research of cavity quantum electrodynamics (CQED) networks [17–20] has shown so far modest progress concerning the influence of thermal environments on the correlations, and the sudden transitions and freezing phenomena in particular [13,14,20–27]. This Letter presents interesting novel results in this line of research.

In Fig. 1(b) we show the time evolution of the classical and quantum correlations for a CQED network as discussed in [27] and applied to the case of two excitations, with the qubits initially prepared in a class of states as in Eq. (3) for all the reservoirs at zero temperature. We observe the quantum-classical sudden transition in the case of our CQED model similar to other studied systems [6–16]. Besides the classical correlations (CCs), entropic QD, and relative entropy of entanglement (REE), we have also studied two geometrical measures, geometric entanglement (GE) and GQD defined with Bures distance [4,5]. We notice that the Bures GQD and QD show similar behavior, having flat regions not affected by the dissipation processes during a particular time period, an effect known as freezing of QD. At the same time, CCs decay and meet QD at a point where a sudden change occurs. After this point, the CCs remain constant during another time period until another sudden change follows and so we observe periodic revival of the correlations [9,10,16]. On the other hand, the entanglements show different dynamics, evidencing effects of sudden death and birth not appearing in the QD and GQD for the given system.

In the following we present briefly the model proposed in [26,27], schematically shown in Fig. 1(a), and recall the analytical equations for a generalized model presented in this Letter. Basically, the model considers a quantum open system of two remote qubits (two-level atoms) where each qubit interacts independently with one of

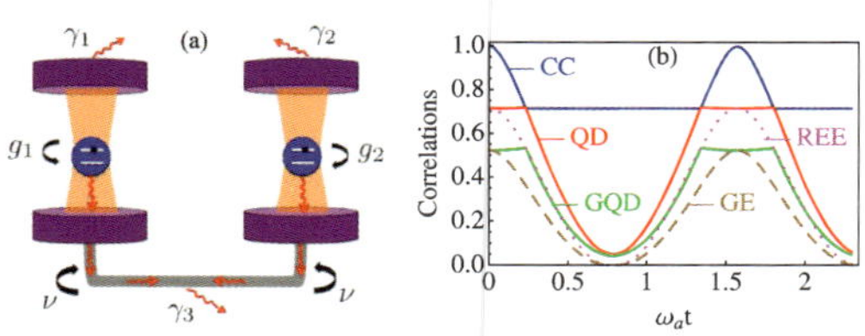

Fig. 1. (a) Remote atoms in the CQED network. (b) Time evolution of the correlations: CC (blue solid), QD (red solid), GQD (green solid), REE (magenta-dotted), and GE (brown-dashed) for the reservoirs at zero temperatures. The parameters are the same as in Fig. 2(a).

0146-9592/14/092668-04$15.00/0

the distant cavities coupled by a transmission line (e.g., fiber). For the sake of simplicity, one considers the approach of a short fiber limit: only one active mode of the fiber interacts with the cavity modes [20]. The whole system is open because the cavities and fiber exchange energy with their individual thermal baths; hence, we have a very general case of dissipative decoherence of the quantum system. The Hamiltonian of the system under the rotating-wave approximation in units of $\hbar$ reads

$$H_s = \omega_f a_3^\dagger a_3 + \sum_{j=1}^{2}(\omega_a S_{j,z} + \omega_0 a_j^\dagger a_j) + \sum_{j=1}^{2}(g_j S_j^+ a_j + \nu a_3 a_j^\dagger + \text{H.c.}), \quad (1)$$

where $a_1(a_2)$ and a_3 are the boson operator for cavity 1(2) and the fiber mode, respectively; ω_0, ω_f, and ω_a are the cavity, fiber, and atomic frequencies, respectively; g_j (ν) are the atom (fiber)-cavity coupling constants; and S_z and $S^\pm$ are the atomic inversion and ladder operators, respectively.

One of the important advances and the novelty of the present model compared to previous ones [26,27] is based on the generalization to large number of excitations in the system. To the best of our knowledge, this approach of many excitations in similar systems [17–20] is not common, and may be one of few existing studies.

To describe the evolution of an open quantum-optical system usually the approach of the Kossakowski–Lindblad phenomenological master equation is considered with the system Hamiltonian decomposed on the eigenstates of the field-free subsystems. However, sometimes a CQED system is much more realistically modeled based on the microscopic master equation (MME), developed in [28,29] where the system–reservoir interactions are described by a master equation with the system Hamiltonian mapped on the atom-field eigenstates, known as dressed states. The present system consists of two atoms within their own cavities connected by a fiber. We represent the leakage of the two cavities and the fiber via coupling to individual external environments, thus identifying three independent dissipation channels. Commonly, in CQED the main sources of dissipation originate from the leakage of the cavity photons due to the imperfect reflectivity of the cavity mirrors. Another mechanism of dissipation corresponds to the spontaneous emission of photons by the atom; however, this kind of loss is negligibly small in the CQED regime considered in our model and, consequently, is neglected. Hence, it is straightforward to bring the Hamiltonian H_s in Eq. (1) to a matrix representation in the atom-field eigenstates basis. To define a general state of the whole system we use the notation $|i\rangle = |A_1\rangle \otimes |A_2\rangle \otimes |C_1\rangle \otimes |C_2\rangle \otimes |F\rangle \equiv |A_1A_2C_1C_2F\rangle$, where $A_{1,2}$ correspond to the atomic states, which can be $e(g)$ the for excited (ground) state, while $C_{1,2}$ and F define the cavities and fiber states, respectively, which may correspond to 0, 1, … n photon states. Because the quantum system is dissipative, the excitations may leak to the reservoir degrees of freedom, hence the ground state of the system, $|0\rangle = |gg000\rangle$, should be also considered in the basis of the states. Therefore, in the case of N excitations in our system, the number of dressed states, $|i\rangle$, having a minimum of one excitation, i.e., excluding the ground state $|0\rangle$, is computed by a simple relation: $d_N = N + 2\sum_{k=1}^{N} k(k+1)$. For example, in the case of $N = 2$ excitations the Hamiltonian H_s in Eq. (1) is decomposed in a state-basis of dimension $1 + d_2$, i.e., it is a 19×19 matrix; for six excitations, H_s is represented by a 231×231 matrix, and so on. Hence it is evident that for large N the general problem becomes hard to solve even numerically. In the present work, we develop our calculations up to six excitations, which is an improvement as compared to some previous works with, e.g., two excitations [20].

Considering the above assumptions and following the approach of [28,29], the MME for the reduced density operator $\rho(t)$ of the system is defined in the form given by Eq. (2) in [27]. In the following we develop the equation for the density operator $\rho(t)$ mapped on the eigenstates basis, $\langle\phi_m|\rho(t)|\phi_n\rangle = \rho_{mn}$ for the case of N excitations in the system:

$$\dot{\rho}_{mn} = -i\bar{\omega}_{n,m}\rho_{mn} + \sum_{k=1}^{d_N}\left[\frac{\gamma_{k\to 0}}{2}(2\delta_{m0}\delta_{0n}\rho_{kk} - \delta_{mk}\rho_{kn} - \delta_{kn}\rho_{mk}) + \frac{\gamma_{0\to k}}{2}(2\delta_{mk}\delta_{kn}\rho_{00} - \delta_{m0}\rho_{0n} - \delta_{0n}\rho_{m0})\right], \quad (2)$$

where δ_{mn} is the Kronecker delta; the physical meaning of the damping coefficients $\gamma_{k\to 0}$ and $\gamma_{0\to k}$ refers to the rates of the transitions between the eigenfrequencies Ω_k and Ω_0 downward and upward, respectively, defined as follows: $\gamma_{k\to 0} = \sum_{j=1}^{3} c_i^2 \gamma_j(\bar{\omega}_{0,k})[\langle n(\bar{\omega}_{0,k})\rangle_{T_j} + 1]$, and by the Kubo–Martin–Schwinger condition we have $\gamma_j(-\bar{\omega}) = \exp(-\bar{\omega}/T_j)\gamma_j(\bar{\omega})$, where c_i are the elements of the transformation matrix from the states $\{|0\rangle, |1\rangle, \ldots, |d_N\rangle\}$ to the states $\{|\phi_0\rangle, |\phi_1\rangle, \ldots, |\phi_{d_N}\rangle\}$ {similar to Eq. (14) and Appendix A in [26]}. Here $\langle n(\bar{\omega}_{\alpha,\beta})\rangle_{T_j} = (e^{(\Omega_\beta - \Omega_\alpha)/T_j} - 1)^{-1}$ corresponds to the average number of thermal photons (with $k_B = 1$). The damping coefficients play a very important role in our model because their dependence on the reservoir temperatures imply a complex exchange mechanism between the elements of the system and the baths. Further, one solves numerically the coupled system of the first-order differential equations [Eq. (2)] and compute the evolution of different kinds of correlations between the two distant atoms, given some finite temperature of the reservoirs. In order to get the reduced density matrix for the atoms, one performs a measurement on the cavities and the fiber vacuum states, $|000\rangle = |0\rangle_{C1} \otimes |0\rangle_{C2} \otimes |0\rangle_F$; later we will explain how this task can be realized experimentally. We find that, after the projection, the reduced atomic density matrix has a X form and the correlations can be computed easily as developed in [4,30–32].

The system under consideration refers to atoms with long radiative lifetimes trapped in their own cavities and connected by a fiber. The cavities and fiber exchange

2670 OPTICS LETTERS / Vol. 39, No. 9 / May 1, 2014

their energy with individual reservoirs [Fig. 1(a)] which, for the sake of simplicity, are taken to have the same damping rate $\gamma_1 = \gamma_2 = \gamma_3 \equiv \gamma$. The transition frequency of the atom is considered a free parameter used to scale the rest of the dimensionless parameters, which we take as being similar to some experimental data [33], e.g., $\omega_a/2\pi = 10$ GHz. The atom–cavity couplings satisfy the constraint of the MME in a Markovian environment, i.e., $2g \gg \gamma$ [29] and we set the values $g_1 = g_2 = g = 10\gamma$ in Figs. 1 and 2. The values of γ and ν will be tuned to evidence the effects of the thermal baths. We find that the detunings do not have an important impact on the effects we discuss here, so we set the values $\omega_f = \omega_a$ and $\omega_a - \omega_0 = 0.1\omega_a$. In the following, we compute the time evolution of the atomic correlations, keeping in mind the main objective of this Letter, which is to find the influence of thermal baths on these correlations. To compute the general correlations—classical and quantum for the given system—we consider the concepts of mutual information, classical correlations, and entropic QD as defined and calculated in [2,3,30–32], as well the GQD with Bures distance, recently developed by one of us [4,5] and independently in [6]. Let us consider that the two atoms initially are prepared in a particular state as Bell-diagonal (BD), described by an X-type density matrix in Bloch form as

$$\rho(0) = [I \otimes I + \vec{c} \cdot (\vec{\sigma} \otimes \vec{\sigma})]/4, \tag{3}$$

where $\vec{\sigma} = (\sigma_1, \sigma_2, \sigma_3)$ is the vector given by Pauli matrices, I is the identity matrix, and the vector $\vec{c} = (c_1, c_2, c_3)$ defines completely the state with $-1 \leq c_i \leq 1$. It is very important to point out here that the majority of the works studying the sudden transition and freezing effects of the correlations in the bipartite quantum system consider the special decoherence mechanisms (noise channels known as bit flip, phase flip, and bit-phase flip [6,8,34]) so that during the time evolution the density matrix preserves the property of its initial state, i.e., $\rho_{11}(t) = \rho_{44}(t) = [1 + c_3(t)]/4$ and $\rho_{22}(t) = \rho_{33}(t) = [1 - c_3(t)]/4$. For this scenario the classical correlations and QD are easily computed with the help of the work of Luo in [30] or given explicitly in [8]. However, for quantum systems embedded in natural environments, like heat baths, the above-mentioned equality of the density matrix elements is no longer satisfied, as the system evolves in time, even if initially we have a BD state. Hence, it is of great interest to study the phenomena of sudden transition and freezing of the correlations for more general (realistic) dissipation models, such as our system under the MME approach.

Recently, Pinto *et al.* in [12] discussed the sensitivity of the sudden change of the QD to different initial conditions. In this context, the present work shed more light on this important subject. In particular, for the system under study, we find that the atomic density matrix preserves the initial BD form only for a short time under the action of the heat reservoirs. It very quickly evolves into a more general X-shaped non-BD form. Under these circumstances, we compute the QD as shown in Figs. 2 and 3 by using a more general algorithm developed in [31,32]. Furthermore, we also make use of an alternative, nonentropic measure of the quantum correlations, such as the GQD with Bures distance, proposed and calculated for a bipartite system in [4,5]. Hence, observing in Fig. 1(b) the effect of the sudden change and freezing of the correlations for the reservoirs at zero temperature, it is natural to inquire about the thermal effects on the classical and quantum correlations when the cavities and fiber are connected to reservoirs at finite (nonzero) temperatures. In our numerical analysis, we find that the freezing effects of the QD and GQD decay by increasing individually or collectively the temperatures of the cavities or the fiber. In Fig. 2(a) we show the effect of heating the fiber to four thermal photons and observe that the thermal effects act destructively on the freezing of both the entropic and geometric discords. However, the sudden transitions persist. The next question is: Could one recover from the damaging effects of the system being coupled to the thermal reservoirs? We find that we could, in principle, engineer such a recovery by a suitable increase in the fiber–cavity coupling. As a matter of fact, we show in Fig. 2(b) that, when we set the fiber's bath temperature to three thermal excitations, such recovery of the correlations, via the increase of the fiber–cavity coupling, is feasible, hence by this effect we understand the important role of the photon as the carrier of the quantum correlations between the remote qubits in such a network.

Recently in [34] the authors theoretically described another interesting class of sudden transitions and freezing of the quantum correlations, which later was observed experimentally in NMR setups [35]. They found the formation of an environment induced double transition of Schatten one-norm geometric quantum correlations

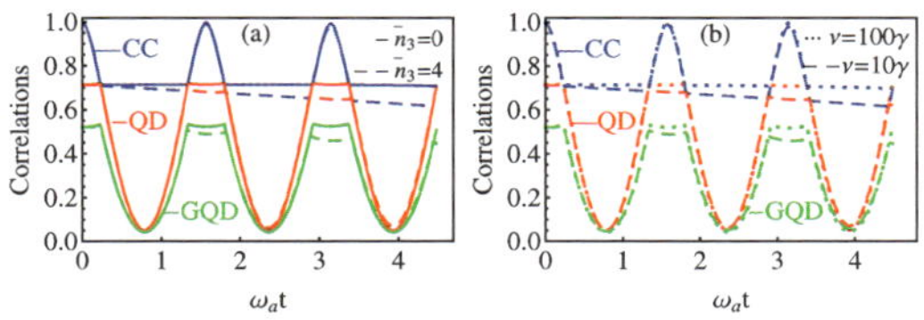

Fig. 2. Dynamics of the correlations: CC (blue), QD (red), and Bures GQD (green) for $\gamma = 0.008\omega_a$. (a) Varying the temperature of the fiber's reservoir given by the average number of the thermal photons, i.e., $\bar{n}_3 = 0$ (solid line) and $\bar{n}_3 = 4$ (dashed line) for constant cavity–fiber coupling $\nu = 10\gamma$. (b) Varying the cavity-fiber coupling, $\nu = 10\gamma$ (dashed line) and $\nu = 100\gamma$ (dotted line) for constant $\bar{n}_3 = 4$. The initial state is defined by $\vec{c} = (1, -0.9, 0.9)$ in Eq. (3).

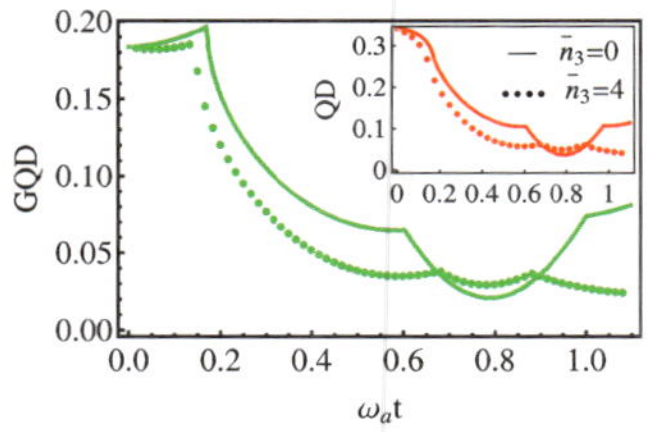

Fig. 3. Dynamics of GQD evidencing double sudden transitions (QD in inset). The parameters considered here are $\gamma = 0.1\omega_a$, $\nu = g = 5\gamma$ for $\bar{n}_3 = 0$ (solid line), and $\bar{n}_3 = 4$ (dotted line). The initial state is given by $\vec{c} = (0.85, -0.6, 0.36)$.

(GQD-1), which is not observed in the classical correlations, and is thus a truly quantum effect.

Motivated by this very recent finding, we simulate the dynamics of the Bures GQD for our model and find a type of double sudden transition somewhat different from the ones observed in [34,35]. To the best of our knowledge, this kind of double sudden transition and freezing effect for Bures GQD has not been reported in literature and by this result we come to an important conclusion that both the GQD-1 and Bures GQD show similar quantum effects. In Fig. 3 we see the double sudden changes in the dynamics of Bures GQD for the reservoirs at zero temperatures, while the QD suffers one sudden change. We observe the following interesting result: as we increase the temperature of the fiber's bath, there is a peculiar tendency to freeze the GQD and the second transition tends to disappear at a critical temperature.

The experimental realization of the present proposal hinges on the possibility of realizing quantum nondemolition (QND) measurements of the photon states in the fiber–cavities system. There is extensive literature on QND measurements in CQED; for a review see [33]. In our scheme we propose to measure the two-qubit density matrix in the condition that all the fields are in vacuum state, so it is feasible to monitor the probability of this state during the temporal evolution of the system [27].

In summary, we analyzed here the phenomena of sudden changes and freezing of correlations in a CQED network with thermal dissipation channels (Fig. 1). Double sudden transitions for Bures GQD are observed for the first time, to the best of our knowledge (Fig. 3). We conclude that by controlling the dissipation mechanisms one may engineer the quantum correlations with multiple sudden changes and freezing periods in the temporal evolution, effects which may have practical applications. A kind of thermal critical effect in this model is expected, as it is in other systems [36].

We acknowledge fruitful discussions with V. Montenegro, D. Spehner, and L. Morales-Molina. This work is supported by Chilean Fondecyt project no. 100039, and Programa de Investigacion Asociativa anillo ACT-1112.

References

1. M. A. Nielsen and I. L. Chuang, *Quantum Computation and Quantum Information* (Cambridge University, 2000).
2. L. Henderson and V. Vedral, J. Phys. A **34**, 6899 (2001).
3. H. Ollivier and W. H. Zurek, Phys. Rev. Lett. **88**, 017901 (2002).
4. D. Spehner and M. Orszag, New J. Phys. **15**, 103001 (2013).
5. D. Spehner and M. Orszag, J. Phys. A **47**, 035302 (2014).
6. B. Aaronson, R. L. Franco, and G. Adesso, Phys. Rev. A **88**, 012120 (2013).
7. J. Maziero, L. C. Céleri, R. M. Serra, and V. Vedral, Phys. Rev. A **80**, 044102 (2009).
8. L. Mazzola, J. Piilo, and S. Maniscalco, Phys. Rev. Lett. **104**, 200401 (2010).
9. J.-S. Xu, X.-Y. Xu, C.-F. Li, C.-J. Zhang, X.-B. Zou, and G.-C. Guo, Nat. Commun. **1**, 7 (2010).
10. J.-S. Xu, K. Sun, C.-F. Li, X.-Y. Xu, G.-C. Guo, E. Andersson, R. Lo Franco, and G. Compagno, Nat. Commun. **4**, 2851 (2013).
11. R. Auccaise, L. C. Céleri, D. O. Soares-Pinto, E. R. deAzevedo, J. Maziero, A. M. Souza, T. J. Bonagamba, R. S. Sarthour, I. S. Oliveira, and R. M. Serra, Phys. Rev. Lett. **107**, 140403 (2011).
12. J. P. G. Pinto, G. Karpat, and F. F. Fanchini, Phys. Rev. A **88**, 034304 (2013).
13. Q.-L. He, J.-B. Xu, D. X. Yao, and Y. Q. Zhang, Phys. Rev. A **84**, 022312 (2011).
14. Q. L. He and J.-B. Xu, J. Opt. Soc. Am. B **30**, 251 (2013).
15. B. You and L.-X. Cen, Phys. Rev. A **86**, 012102 (2012).
16. R. Lo Franco, B. Bellomo, E. Andersson, and G. Compagno, Phys. Rev. A **85**, 032318 (2012).
17. J. I. Cirac, P. Zoller, H. J. Kimble, and H. Mabuchi, Phys. Rev. Lett. **78**, 3221 (1997).
18. T. Pellizzari, Phys. Rev. Lett. **79**, 5242 (1997).
19. S. Ritter, C. Nölleke, C. Hahn, A. Reiserer, A. Neuzner, M. Uphoff, M. Mücke, E. Figueroa, J. Bochmann, and G. Rempe, Nature **484**, 195 (2012).
20. A. Serafini, S. Mancini, and S. Bose, Phys. Rev. Lett. **96**, 010503 (2006).
21. L.-T. Shen, Z.-B. Yang, H.-Z. Wu, X.-Y. Chen, and S.-B. Zheng, J. Opt. Soc. Am. B **29**, 2379 (2012).
22. S. Liu, J. Li, R. Yu, and Y. Wu, Phys. Rev. A **87**, 062316 (2013).
23. Z.-D. Hu and J.-B. Xu, Opt. Lett. **38**, 3639 (2013).
24. R. Coto and M. Orszag, J. Phys. B **46**, 175503 (2013).
25. R. Coto and M. Orszag, "Determination of the maximum global quantum discord via measurements of excitations in a cavity QED network," J. Phys. B, arXiv:1401.2934 (to be published).
26. V. Montenegro and M. Orszag, J. Phys. B **44**, 154019 (2011).
27. V. Eremeev, V. Montenegro, and M. Orszag, Phys. Rev. A **85**, 032315 (2012).
28. H. P. Breuer and F. Petruccione, *The Theory of Open Quantum Systems* (Oxford University, 2002).
29. M. Scala, B. Militello, A. Messina, J. Piilo, and S. Maniscalco, Phys. Rev. A **75**, 013811 (2007).
30. S. Luo, Phys. Rev. A **77**, 042303 (2008).
31. F. F. Fanchini, T. Werlang, C. A. Brasil, L. G. E. Arruda, and A. O. Caldeira, Phys. Rev. A **81**, 052107 (2010).
32. Y. Huang, Phys. Rev. A **88**, 014302 (2013).
33. S. Haroche and J. M. Raimond, *Exploring the Quantum: Atoms, Cavities and Photons* (Oxford University, 2006).
34. J. D. Montealegre, F. M. Paula, A. Saguia, and M. S. Sarandy, Phys. Rev. A **87**, 042115 (2013).
35. F. M. Paula, I. A. Silva, J. D. Montealegre, A. M. Souza, E. R. deAzevedo, R. S. Sarthour, A. Saguia, I. S. Oliveira, D. O. Soares-Pinto, G. Adesso, and M. S. Sarandy, Phys. Rev. Lett. **111**, 250401 (2013).
36. T. Werlang, C. Trippe, G. A. P. Ribeiro, and G. Rigolin, Phys. Rev. Lett. **105**, 095702 (2010).

Chapter 2

Wave-Particle Duality

Daniel Greenberger

City College of the City of New York, New York, USA

Paper Title: **Simultaneous wave and particle knowledge in a neutron interferometer**

Reference: Physics Letters A **128**, 391–394 (1988)

Personal message:

This is an article I wrote in 1988 with Allaine YaSin, a student of mine, who died of lupus shortly after she got her degree. The history is that Wootters and Zurek showed that you could get partial knowledge of momentum P and position X in a single experiment, and we produced a simple measure of how you could classify partial knowledge in a neutron interferometer. This was then taken up by Berge (together with Scully and Walther) sometime later and generalized. So this article is the inspiration for their much better known one.

Volume 128, number 8 PHYSICS LETTERS A 18 April 1988

SIMULTANEOUS WAVE AND PARTICLE KNOWLEDGE IN A NEUTRON INTERFEROMETER

Daniel M. GREENBERGER and Allaine YASIN
City College of the City of New York, New York, NY 10031, USA

Received 15 February 1988; accepted for publication 18 February 1988
Communicated by J.P. Vigier

We give a measure of particle knowledge in a neutron interferometer that reflects one's ability to predict in which beam a neutron is located. We can measure wave knowledge by contrast of the interference pattern. Then one's simultaneous knowledge of both is determined by a single parameter (not an uncertainty relation), running from full particle to full wave knowledge. We extend the discussion to partially coherent beams. Our measure of information is much simpler than the conventional one.

1. Fully coherent beams

Wootters and Zurek [1] caused some surprise when they analyzed the familiar two-slit diffraction experiment and showed that the results contained a significant amount of both particle and wave information, and also that one could simultaneously measure position and momentum to an unexpectedly high degree of precision, without violating the uncertainty relation. Specifically, if one performs an experiment to determine which beam the electron is in, it does not destroy all wave knowledge, i.e., there is still a discernable interference pattern.

We would like to point out here that a similar situation arises constantly in neutron interferometry, where it is obvious that a significant amount of both particle and wave knowledge is obtained, even if one does not make a direct measurement of which beam a particular particle is in. We shall introduce for this a measure of information that is qualitatively similar to the usual information measure, but much easier to handle analytically. (Mathematically, our situation is similar to that of the case of two unequal slits in the two-slit diffraction experiment).

As an example, consider the case of a three-eared interferometer (see fig. 1), where the beams ABD (beam I), and ACD (beam II), of equal intensity, interfere upon recombining and are counted by a detector located at E.

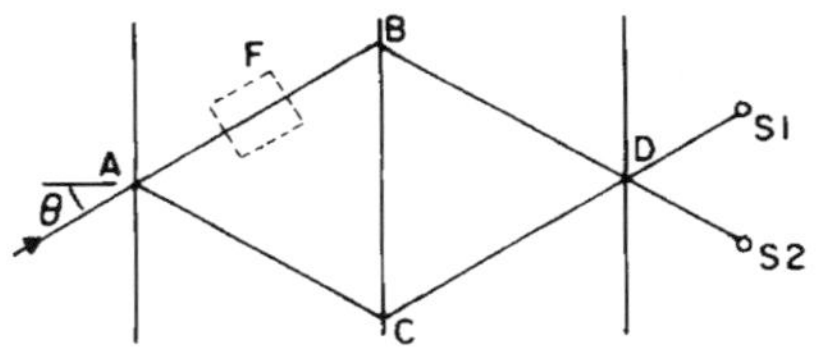

Fig. 1. Top view of a neutron interferometer. The incident beam enters at A and is split into beam I (ABD) and beam II (ACD), which are then recombined at D and detected at E. For our purposes, one can imagine a screen at B which records the interference pattern between the two beams. One can also introduce an absorber at F to partially absorb one of the beams.

If now an absorber is placed at F in beam I, such that it absorbs 99% of the intensity of beam I, there will still be a significant interference pattern at point E. In fact if the intensity of beam I is reduced from unity to ϵ^2, the contrast of the interference pattern will still be 2ϵ, or about 20% in our case, even though the intensity in beam I is only about 1% of that in beam II. This amazing result [2] testifies to the power of the superposition principle, and it has been clearly demonstrated experimentally.

We can use the contrast of the pattern,

$$W=\frac{I_{\max}-I_{\min}}{I_{\max}+I_{\min}} \tag{1}$$

as a measure of its wave properties (since it measures the amount of interference), so in this case

0375-9601/88/$ 03.50 © Elsevier Science Publishers B.V.
(North-Holland Physics Publishing Division)

Volume 128, number 8 PHYSICS LETTERS A 18 April 1988

about 20% of the wave knowledge is preserved in the beam.

On the other hand we know that 99% of the particles went via path II. Although we do not know which individual particle takes which path, we can predict that they will all take path II, and a measurement will prove us correct 99% of the time. So from a probability point of view we have almost complete knowledge of which beam the particle is in. (Of course if we actually measure which path a particular neutron is in, we will have 100% path knowledge for that particle, and it will not contribute at all to the interference pattern).

We can quantify both these features. We assume the beams are fully coherent, and that beam I hits the third ear with amplitude a, while beam II has amplitude b. Then if we imagine that there is a screen at D, we can write for the wave function there,

$$\psi=(a e^{+ik_x x}+b e^{i\varphi} e^{-ik_x x}) e^{ik_z z}, \tag{2}$$

where k_x, k_z are determined by the Bragg scattering condition with a and b real, and φ represents the phase difference between the beams, at some point. Then

$$|\psi|^2=a^2+b^2+2ab\cos(2k_x x+\varphi), \tag{3}$$

and the contrast will be

$$W=\frac{2ab}{a^2+b^2}. \tag{4}$$

If we write

$$a=R\cos\beta, \quad b=R\sin\beta, \tag{5}$$

then the contrast becomes

$$W=\sin 2\beta \tag{6}$$

and has its maximum at $\beta=45°$, when the beams are equal.

On the other hand, if we want a measure of the particle nature of the beam, we can take the following. We could take a measurement to see which beam each particle is actually in, by detecting it with a counter in each beam. We could also predict in advance the results of this measurement. The smartest strategy would be to predict that all the particles will be in the same beam, namely the more intense one. We then compare the success of this prediction to that obtained if we had no knowledge of which beam the particle is in (50–50 chance for each). This gives us a measure of our knowledge of which beam it is in, compared to the prediction of blind chance alone. If $|a|^2>|b|^2$, we can predict that all the particles will be in beam I, and we will be right a fraction $|a|^2/(|a|^2+|b|^2)$ of the time, which we compare to 1/2, and this gives us our particle measure (which we normalize),

$$P=\frac{|a|^2/(|a|^2+|b|^2)-1/2}{(1/2)} =\frac{|a|^2-|b|^2}{|a|^2+|b|^2}=\cos 2\beta. \tag{7}$$

(Of course, when we do not measure which beam a particle is in, quantum theory tells us that we do not have the right to attribute it to either beam.)

When the two beams are equal, we have no idea which beam a particular particle is in, since the two are equally likely, and $P=0$, while we have a maximum contrast, $W=1$. When all the particles are in one beam, we have full particle knowledge, and $P=1$, but then there will be no contrast, and $W=0$. In general, the more certain we are of which beam a particle will be found in, the less contrast the interference pattern will have, and this dual knowledge is controlled by the relationship

$$P^2+W^2=1. \tag{8}$$

So wave versus particle knowledge is not governed by an uncertainty principle, but rather by a single parameter (β), which varies from full particle to full wave knowledge, but in general one can have considerable knowledge of both!

Surprisingly, our measure of the contrast is not far from what a strict information-theoretical measure would give [3]. If p is the probability that the particle is in beam I, then we would have

$$p=\frac{|a|^2}{|a|^2+|b|^2}, \quad 1-p=\frac{|b|^2}{|a|^2+|b|^2},$$

$$I'(p)=-[p\ln p+(1-p)\ln(1-p)],$$

$$I'_0=I'(0.5)=0.693, \tag{9}$$

and a normalized measure of information would be

$$I(p)=I'(p)/I'_0. \tag{10}$$

This is to be contrasted to our W,

Volume 128, number 8 PHYSICS LETTERS A 18 April 1988

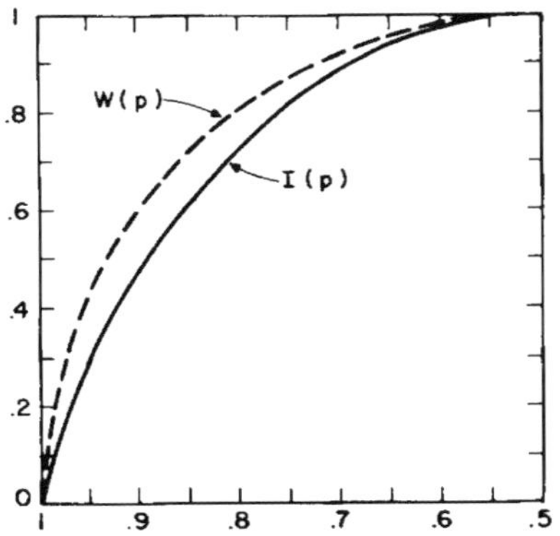

Fig. 2. Wave-like property of the neutron. $W(p)$, our measure of the wave-like property of the beam, determined by the contrast of the interference pattern, is the dotted curve. The solid curve represents an information theory measure of this information, eqs. (9) and (10). The two curves are quite similar. A plot of $W(p)^{1.5}$ would fit almost exactly on top of $I(p)$. Here p is the probability that the particle is in beam I.

$$W=2\sqrt{p(1-p)}\,, \tag{11}$$

and they are in fact quite close to each other (see fig. 2). In fact an even closer correspondence is given by $I(p)\approx W(p)^{1.5}$, which is accurate to within 0.02, and in fig. 2, one would hardly be able to distinguish them.

2. Partially coherent beams

If the beams I and II are not fully coherent, one can separate a and b into a coherent part a_1, b_1, and an incoherent part a_2, b_2. Then

$$|\psi|^2=a_1^2+b_1^2+a_2^2+b_2^2+2a_1b_1\cos(2k_xx+\varphi)\,,$$

$$W=\frac{2a_1b_1}{a_1^2+b_1^2+a_2^2+b_2^2}\,. \tag{12}$$

(Note: a and b refer to beams I and II, and the subscripts 1 and 2 refer to the coherent and incoherent parts.) If one introduces a coherent intensity, R,

$$a_1=R\cos\beta\,,\quad b_1=R\sin\beta\,, \tag{13}$$

and a degree of total coherence,

$$\frac{R}{A}=\cos\gamma\,,\quad A^2=R^2+a_2^2+b_2^2\,, \tag{14}$$

then from eq. (12) we have

$$W=\sin 2\beta\cos^2\gamma\,. \tag{15}$$

We have assumed here that both beams I and II have the same degree of coherence, so that all the incoherence arises from the initial input beam. In this case, which is not the most general one, eq. (13) gives

$$a_1=A\cos\gamma\cos\beta\,,\quad b_1=A\cos\gamma\sin\beta\,, \tag{16}$$

and we also have

$$a_2=A\sin\gamma\cos\beta\,,\quad b_2=A\sin\gamma\sin\beta\,, \tag{17}$$

so that the three parameters A, β, γ, completely characterize the beam. The parameter P is given by

$$P=\frac{a_1^2+a_2^2-(b_1^2+b_2^2)}{A^2}=\cos 2\beta\,, \tag{18}$$

while W is determined by eq. (15). So incoherence in the beam decreases the wavelike information, while leaving unchanged the particle information, which one would expect from the fact that coherence and wave-like behavior are non-classical concepts for a particle.

However if the experiments that are performed are capable of producing extra incoherence independently into each of the two beams, then one will need an extra parameter, δ, to describe the situation. Thus, in this case,

$$a_1=A\cos\beta\cos\gamma\,,\quad \alpha_2=A\cos\beta\sin\gamma\,,$$
$$b_1=A\sin\beta\cos\delta\,,\quad b_2=A\sin\beta\sin\delta\,, \tag{19}$$

To determine the degree of coherence of each beam separately, one would have to perform an interference experiment on each beam, by splitting and recombining it, and using the above results. Then the extra loss of contrast in each beam, W_I and W_II would determine γ and δ. One could introduce a degree of coherence for each beam as before, by introducing R_I and R_II,

$$a_1=R_\mathrm{I}\cos\beta\,,\quad R_\mathrm{I}=A\cos\gamma\,,$$
$$b_1=R_\mathrm{II}\sin\beta\,,\quad R_\mathrm{II}=A\cos\delta\,. \tag{20}$$

Then eq. (14) would generalize to

Volume 128, number 8 PHYSICS LETTERS A 18 April 1988

$$\frac{R_I R_{II}}{A^2} = \cos\gamma\cos\delta\,. \tag{21}$$

The four amplitudes a_1, a_2, b_1, b_2, form a four-dimensional euclidean space. If one identifies the vector (a_1, a_2, b_1, b_2) with position in that space, then for a given total intensity, A, the angles β, γ, δ sweep out the surface of a unit sphere. One can characterize any beam by a rotation from the polar axis $(1, 0, 0, 0) \equiv \hat{e}$, as follows. First, rotate (R_1) in the a_1, b_1 plane by the angle β,

$$R_1\hat{e} = \begin{pmatrix} \cos\beta & 0 & -\sin\beta & 0 \\ 0 & 1 & 0 & 0 \\ \sin\beta & 0 & \cos\beta & 0 \\ 0 & 0 & 0 & 1 \end{pmatrix}\begin{pmatrix} 1 \\ 0 \\ 0 \\ 0 \end{pmatrix} = \begin{pmatrix} \cos\beta \\ 0 \\ \sin\beta \\ 0 \end{pmatrix}, \tag{22}$$

then rotate (R_2) in the a_1, a_2 plane by γ, and in the b_1, b_2 plane by δ, (R_3) to get a general beam,

$$R_3R_2R_1\hat{e} = \begin{pmatrix} \cos\gamma & -\sin\gamma & 0 & 0 \\ \sin\gamma & \cos\gamma & 0 & 0 \\ 0 & 0 & \cos\delta & \sin\delta \\ 0 & 0 & \sin\delta & \cos\delta \end{pmatrix}\begin{pmatrix} \cos\beta \\ 0 \\ \sin\beta \\ 0 \end{pmatrix} = \begin{pmatrix} a_1 \\ a_2 \\ b_1 \\ b_2 \end{pmatrix}. \tag{23}$$

One can then manipulate between beam configurations by arbitrary rotations in this space.

One can also set up a density matrix formulation for a partially coherent beam. If one defines

$$Z = \begin{pmatrix} a_2+b_2 & a_1-\mathrm{i}b_1 \\ a_1+\mathrm{i}b_1 & a_2-b_2 \end{pmatrix},$$

$$Z' = \begin{pmatrix} a_2-b_2 & a_1+\mathrm{i}b_1 \\ a_1-\mathrm{i}b_1 & a_2+b_2 \end{pmatrix}, \tag{24}$$

then

$$Z' = \sigma_x Z \sigma_x\,,$$

$$\tfrac{1}{2}\,\mathrm{Tr}\,Z^2 = \tfrac{1}{2}\,\mathrm{Tr}(Z')^2 = A^2\,,$$

$$P = \mathrm{Tr}(ZZ' + Z'Z)/\mathrm{Tr}\,Z^2\,,$$

$$\mathrm{i}W = \mathrm{Tr}[(Z'Z - ZZ')\sigma_z]/\mathrm{Tr}\,Z^2\,, \tag{25}$$

where W is still determined by eq. (12), and P by the first part of eq. (18), and the σ are the Pauli matrices.

The complementarity equation (8) generalizes in the partially coherent case to the inequality

$$P^2 + W^2 \leqslant 1\,, \tag{26}$$

or more explicitly, to the equality

$$P^2 + \frac{W^2}{\cos^2\gamma\cos^2\delta} = 1\,. \tag{27}$$

(The intermediate case described is covered by the above equation with $\delta = \gamma$.)

Acknowledgement

We would like to thank Professor Cliff Shull and the members of the Neutron Diffraction Laboratory at MIT for their hospitality and interesting discussions about this work [4]. This research was supported in part by grants from the National Science Foundation, the Department of Energy, and the CUNY-FRAP program of the City University of New York.

References

[1] W.K. Wootters and W.H. Zurek, Phys. Rev. D 19 (1979) 473;
L.S. Bartell, Phys. Rev. D 21 (1980) 1698 [these papers are also reprinted in: J.A. Wheeler and W.H. Zurek, eds, Quantum theory and measurement (Princeton, 1983)].

[2] H. Rauch and K. Summhammer, Phys. Lett. A 104 (1984) 44.

[3] A. Zeilinger, Physica B 137 (1986) 235.

[4] A preliminary version of this work was presented at the conference on New techniques and ideas in quantum measurement theory, ed. D. Greenberger, Ann. New York Acad. Sci. No. 480 (1987).

Helmut Rauch

Vienna University of Technology, Atominstitut, Vienna, Austria
Institut Laue-Langevin, Grenoble, France

Paper Title: **Wave-particle duality and quantum erasure in polarized-neutron interferometry**

Reference: Nuclear Instruments and Methods in Physics Research A **440**, 562–567 (2000)

Personal message:

Berge Englert's guest professorship at our Institute (1998 and 2000/01); guest professor at the University of Technology in Vienna) was very effective for our experimental group working in neutron quantum optics because he has a theoretician's special talent to listen to experimentalists. We learned about Wigner representation of quantum states and about particle-wave duality. His contact with several students stimulated dedicated work on particle-wave phenomena mainly in cooperation with Michael Zawisky. Berge also motivated many students to attend his lectures on experimental quantum physics. He was very helpful in bridging between theory and experiment.

ELSEVIER Nuclear Instruments and Methods in Physics Research A 440 (2000) 562–567

NUCLEAR INSTRUMENTS & METHODS IN PHYSICS RESEARCH Section A

www.elsevier.nl/locate/nima

Wave–particle duality and quantum erasure in polarized–neutron interferometry

G. Badurek[a,*], R.J. Buchelt[a], B.-G. Englert[b,1], H. Rauch[c]

[a]*Institut für Kernphysik, Technische Universität Wien, Stadionallee 2, A-1020 Wien, Austria*
[b]*Max-Planck-Institut für Quantenoptik, D-85748 Garching, Germany*
[c]*Atominstitut der Österreichischen Universitäten, A-1020 Wien, Austria*

Abstract

An interference experiment with polarized neutrons is proposed to check the duality relation between the fringe visibility and the which-way information for massive fermions with macroscopic spatial separation of the interfering sub-beams. The connection between polarization-sensitive post-selection of sub-ensembles and the concept of quantum erasure is discussed. This neutron experiment will supplement those recently performed with photons and atoms.

Keywords: Wave–particle duality; Quantum erasure; Neutron interferometry

1. Introduction

Wave–particle duality (WPD) dates back to the very early days of quantum mechanics, to Einstein's seminal paper on the photoelectric effect [1], and is a striking manifestation of Bohr's principle of complementarity [2,3]. The familiar phrase "each experiment must be described either in terms of particles or in terms of waves" emphasizes the extreme cases and disregards the intermediate situations, in which particle aspects and wave aspects are present simultaneously. Theoretical investigations [4,11], supplemented by a few experimental studies [12–14], have led to a quantitative formulation of WPD, namely the duality relation of Eq. (3) below. Very recently, the validity of the duality relation was tested with a Mach–Zehnder-type *photon interferometer* [15] and an *atom interferometer* [16,17] that mimics some of the essential features of a Young double-slit setup. The experiment we are proposing here will test the duality relation in the realm of *neutron interferometry*.

Since its invention in 1974 [18] perfect crystal neutron interferometry has become a powerful and in many respects unique tool to test and to demonstrate fundamental principles of quantum mechanics with massive particles on a macroscopic space–time scale. This is particularly true if the spin-$\frac{1}{2}$ property of the neutron is explicitly taken into account as it will be the case in our proposal. The list of successfully performed experiments belonging to the spinor character of the neutron

* Corresponding author. Tel.: + 43-1-72701-229; fax: + 43-1-72-89-220.

E-mail address: badurek@ati.ac.at (G. Badurek)

[1] Presently at: Department of Physics, Texas A&M University, College Station, TX 77843-4242, USA.

0168-9002/00/$ - see front matter
PII: S0168-9002(99)01037-2

comprises topics like the first explicit verification of the 4π-periodicity of spinor wave functions [19–21], demonstration of the quantum mechanical principle of linear superposition of states [22,23], macroscopic quantum beating of the neutron wave function [24], multiple photon exchange between the neutron and an applied electromagnetic field [25], neutron interferometric separation of geometric and dynamical phases [26], as well as interferometric [27] and polarimetric [28] verification of the topological character of the scalar Aharonov–Bohm effect. A general state-of-the-art survey about neutron interferometry is found in Refs. [29,30]. Unlike any other matter wave interference technique developed thus far with either electrons, atoms or ions, perfect-crystal neutron interferometry allows a coherent separation of quantum states of massive particles over macroscopic distances of several centimeters followed by subsequent coherent recombination. It is therefore ideally suited to test the validity of the duality relation.

For a quantification of WPD we need quantitative, measurable characteristics for the wave aspects and the particle aspects. The former are naturally quantified by the *visibility* $\mathscr{V}$ of the observed interference fringes. How about the latter? Another familiar phrase, "a wave is in both arms of an interferometer simultaneously, a particle is in one arm or the other", gives a hint. The better we can tell which way has been taken by a particular neutron, the more pronounced are its particle aspects. Their quantification is, accordingly, based on the *likelihood* $\mathscr{L}$ of guessing the way right. Since a random guess gives a likelihood of $\mathscr{L} = \frac{1}{2}$, whereas $\mathscr{L} = 1$ indicates that we know the way with certainty, the actual which-way (WW) *knowledge* $\mathscr{K}$ is given by

$$\mathscr{L} = \tfrac{1}{2} + \tfrac{1}{2}\mathscr{K} \quad \text{with } 0 \leq \mathscr{K} \leq 1. \tag{1}$$

The value of $\mathscr{K}$ depends on the "betting strategy" we are employing; the optimal strategy maximizes $\mathscr{K}$ and this identifies the *distinguishability* $\mathscr{D}$ of the ways,

$$\mathscr{D} = \max\{\mathscr{K}\}. \tag{2}$$

In an asymmetric interferometer, one way is more likely than the other to begin with, so that we would know something about the way beforehand; WW information of this kind is fittingly called *predictability* (denoted by $\mathscr{P}$), and the inequalities $\mathscr{P} \leq \mathscr{K} \leq \mathscr{D}$ state an obvious hierarchy because WW detection cannot decrease the betting odds.

The central quantitative statement about WWD is the *duality relation*

$$\mathscr{D}^2 + \mathscr{V}^2 \leq 1 \tag{3}$$

that has been found recently [10,11]; there are examples, in which the equality sign holds. The implied statement $\mathscr{P}^2 + \mathscr{V}^2 \leq 1$ has been known for some time, implicitly or explicitly, in a variety of physical contexts [4–9,12–14]. All experimental tests refer necessarily to the version

$$\mathscr{K}^2 + \mathscr{V}^2 \leq 1 \tag{4}$$

with the knowledge $\mathscr{K}$ optimized under the constraints of the setup, hoping for $\mathscr{K}_{\max} \approx \mathscr{D}$. In the experiment we are proposing, we will usually have $\mathscr{P} = 0$ and it should be possible to extract (almost) all the WW information that becomes potentially available, so that the acquired WW knowledge $\mathscr{K}$ is expected to come very close to its theoretical upper bound $\mathscr{D}$.

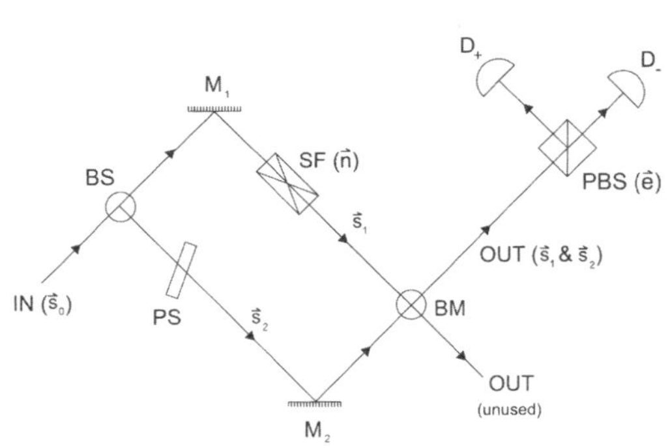

Fig. 1. Schematic sketch of the proposed neutron interferometer with which-way marking (see text).

2. Scheme of the experiment

Fig. 1 shows a schematic sketch of the proposed experiment. It is the neutron analog of the photon experiment of Ref. [15]. A standard Mach–Zehnder-type neutron interferometer is supplemented by devices for (i) encoding WW information in the neutron's polarization and (ii) extracting WW knowledge later on. For the sake of brevity here we are concentrating on the general concepts and not on the actual technical realization.

In the setup of Fig. 1, polarized neutrons enter from the left at the beam splitter BS. After being reflected at one of the mirrors (M_1 or M_2) the amplitudes are recombined at the beam merger BM. The neutrons exiting at the symmetric output port are eventually counted by the detectors D_+ and D_-. Those emerging at the asymmetric output are not used for the measurement itself, only for collecting control data.

2.1. Determining the fringe visibility $\mathscr{V}$

The phase shifter PS in the lower way ($BS\text{–}M_2\text{–}BM$) enables us to change the relative phase ϕ of the two partial amplitudes, and so we can scan through the interference pattern. The total count rate of D_+ and D_- will be proportional to $1 + \mathscr{V}\cos\phi$, where $\mathscr{V}$ is the fringe visibility. Since we are observing in the symmetric output, the maximal value $\mathscr{V} = 1$ is achievable if the partially transparent mirrors that are used for the BS and the BM have identical properties.

2.2. Labeling the neutrons and acquiring which-way knowledge $\mathscr{K}$

In the upper way ($BS\text{–}M_1\text{–}BM$) a spin flipper SF can change the initial polarization to another one, so that the neutrons acquire a polarization label that marks their way through the interferometer. This information is read out with the aid of the polarizing beam splitter PBS that sorts the neutrons, so that D_+ counts the neutrons polarized in the $+\boldsymbol{e}$ direction and D_- those polarized in the (orthogonal) $-\boldsymbol{e}$ direction. Here, the unit vector $\boldsymbol{e}$ characterizes the PBS.

Table 1

	D_+	D_-
M_1-way	$1 + \boldsymbol{e}\cdot\boldsymbol{s}_1$	$1 - \boldsymbol{e}\cdot\boldsymbol{s}_1$
M_2-way	$1 + \boldsymbol{e}\cdot\boldsymbol{s}_2$	$1 - \boldsymbol{e}\cdot\boldsymbol{s}_2$

The Pauli vector $\boldsymbol{s}_0 = \langle\boldsymbol{\sigma}\rangle_0$, which is the expectation value of the spin vector operator $\boldsymbol{\sigma}$ in the initial state, specifies the spin state of the neutrons entering the interferometer at the BS. Neutrons coming along the M_1-way pass through the (unitary) SF which rotates their spin vector around the axis $\boldsymbol{n}$ and so turns $\boldsymbol{s}_0$ into $\boldsymbol{s}_1$; the neutrons of the M_2-way have $\boldsymbol{s}_2 = \boldsymbol{s}_0$. If no depolarization occurs, the lengths of $\boldsymbol{s}_1$ and $\boldsymbol{s}_2$ remain the same (identical to the length of $\boldsymbol{s}_0$) and they differ only in their direction.

In general, the count rates of D_+ and D_- will be different for the two ways. Their relative sizes are summarized in Table 1.

The experimental values corresponding to Table 1 entries are obtained by measuring the count rates in each row with the other way blocked.

The betting odds for guessing the way right are most favorable when we always bet on the way that contributes most to the probability of triggering the detector that actually fired. This results in a measured WW knowledge given by [10,11]

$$\mathscr{K} = \frac{|r_{1+} - r_{2+}| + |r_{1-} - r_{2-}|}{r_{1+} + r_{2+} + r_{1-} + r_{2-}} = \frac{1}{2}|\boldsymbol{e}\cdot(\boldsymbol{s}_1 - \boldsymbol{s}_2)| \tag{5}$$

where, for example, r_{1+} is the count rate of D_+ for M_1-way neutrons. Clearly, theory predicts that $\boldsymbol{e}$ should be chosen parallel to $\boldsymbol{s}_1 - \boldsymbol{s}_2$ for maximal WW knowledge, and that $\mathscr{K} = 1$ is only possible if (i) $\boldsymbol{s}_1 = -\boldsymbol{s}_2$ and (ii) both are unit vectors. In operational term: The neutrons must be fully polarized in both ways and these polarization states must be orthogonal.

For the equal sign to hold in Eq. (3), condition (ii) suffices. It is then possible to study the trade-off between fringe visibility and WW knowledge under ideal circumstances.

2.3. Post-selection, quantum erasure

The polarization-sensitive detection of the neutrons sorts them into two sub-ensembles. Depending on the choice for the characterizing unit vector $\boldsymbol{e}$, different sortings are possible. For example, the WW sorting just described is given by $\boldsymbol{e}\|(\boldsymbol{s}_1 - \boldsymbol{s}_2)$. Inasmuch as the choice of sorting is undecided until the neutrons reach the PBS in the setup of Fig. 1, that is, until they have left the interferometer proper, the sub-ensembles are *post-selected.*

A sorting of a particularly interesting kind has features in common with *quantum erasure* (QE) [31,32]. It identifies sub-ensembles that exhibit perfectly visible fringes, even when the fringe visibility $\mathscr{V}$ of the unsorted totality of counted neutrons is reduced, possibly to $\mathscr{V} = 0$. This is achieved by setting $\boldsymbol{e}$ equal to the unit vector $\boldsymbol{n}$ that specifies the direction around which the SF rotates the neutron's spin vector.

It is clear that intermediate situations are realized by choosing $\boldsymbol{e}$ different from both the WW choice and the QE choice. This amounts to an experimental realization of *partial QE.*

It is remarkable that the initial polarization $\boldsymbol{s}_0$ is irrelevant for QE. Irrespective of the length and the direction of $\boldsymbol{s}_0$ the sorting characterized by $\boldsymbol{e} = \boldsymbol{n}$ always identifies two sub-ensembles that have fringes with unit visibility. In the extreme case of totally unpolarized neutrons (that is $\boldsymbol{s}_0 = 0$) the SF does not produce a WW label, but will reduce the fringe visibility to zero if the rotation angle is 180°. The fringes are then lost, but no WW information is available [$\mathscr{D} = \mathscr{V} = 0$ in Eq. (3)]. Nevertheless, one can do the QE sorting and recover the fringes. An experimental demonstration of the latter property of QE would confirm that the availability of WW information is not a precondition for successful QE.

3. Experimental essentials

There is no room to present the details of the experimental setup that will be required to realize our proposal. Hence, we completely omit a description of the interferometer itself and focus instead on the methods to prepare the input and to analyze the final neutron state, since they are of particular importance for such kind of experiment.

3.1. Input state preparation

The extremely narrow angular width of perfect-crystal Bragg or Laue reflections enables us to exploit the spin-dependent birefringence of neutrons upon passing through prismatically shaped magnetic fields. The angular separation of the two spin states of an initially unpolarized beam of neutrons with energy E which passes at an asymmetry angle ε (see Fig. 2) through a magnetic field prism of apex angle Φ is given by [33]

$$\delta = \frac{2\mu B}{E} \frac{\sin(\Phi)}{\cos(\Phi) + \cos(2\varepsilon)}. \tag{6}$$

Recently, we could develop for the first time a polarizer of this kind consisting of a permanent magnetic yoke with prismatically shaped air gap of 1 cm height and apex angle $\Phi = 116°$. Without producing any heat a field of about 0.9 T is achieved. Fig. 2 shows the angular splitting measured with a high-resolution double-crystal small-angle camera for symmetric passage through one and two of these field prisms. Owing to the influence of stray fields the experimentally observed

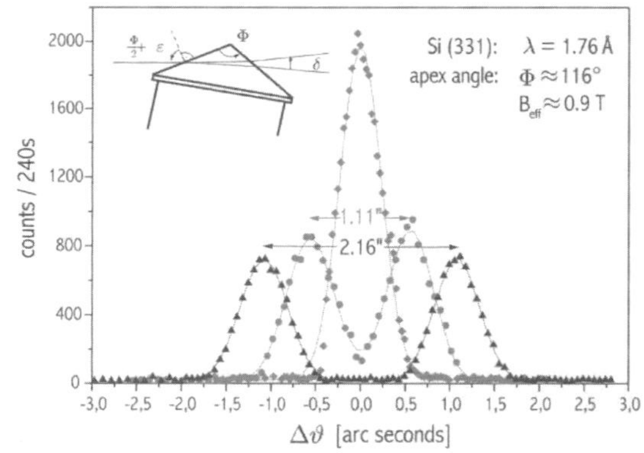

Fig. 2. Angular splitting of the two spin states upon transmission through *one* (●) and *two* (▲) magnetic field prisms. The central curve (◆) displays the resolution curve of the double-crystal small-angle instrument.

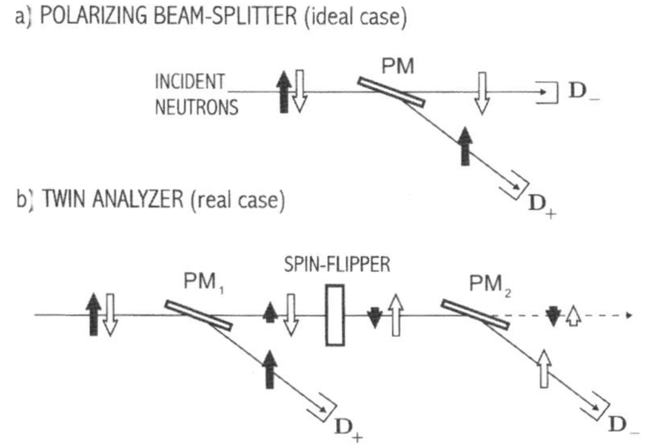

Fig. 3. (a) Ideal polarizing beam splitter (PBS) with 100% reflectivity of the spin-*up* state. (b) Approximate realization of a PBS by placing a high performance spin flipper between two polarizing mirrors with reduced reflectivity

splitting is about 20% smaller than theoretically expected. Nevertheless, it is sufficiently large compared to the reflection width of perfect Si crystals to guarantee fully polarized beams inside the interferometer.

3.2. *Spin state analysis*

WW detection crucially depends on the quality of the polarization sensitive detection. In the ideal case a polarizing beam splitter sorts each neutron according to its spin state either in detector D_+ or D_- (Fig. 3). Evidently, the reflectivity of such an ideal beam splitter must reach a value very close to 100% in order to avoid a contamination of the transmitted beam with the "wrong" spin state. At least for cold neutrons the performance of supermirror polarizers is quite close to this ideal situation. But even in the case of reduced reflectivity a combination of a pair of such imperfect polarizers should allow for an almost perfect spin-dependent neutron sorting, provided a spin-flip device with an efficiency close to 100% is placed between them, as indicated in Fig. 3. There is no particular difficulty in fabricating neutron spin flippers of the required quality, but it is essential that the two polarizers neither attenuate nor depolarize the transmitted beam. Then a low reflectivity only causes a poor neutron economy, but it has no influence on the significance of the spin state sorting.

References

[1] A. Einstein, Ann. Phys. (Leipzig) 17 (1905) 132.
[2] N. Bohr, Naturwissenschaften 16 (1928) 245 [English version: Nature (London) 121 (1928) 580].
[3] M.O. Scully, B.-G. Englert, H. Walther, Nature (London) 351 (1991) 111.
[4] W.K. Wootters, W.H. Zurek, Phys. Rev. D 19 (1979) 473.
[5] R. Glauber, Ann. New York Acad. Sci. 480 (1986) 336.
[6] Ph. Grangier, Doctoral Thesis, Université de Paris-Sud, Orsay, 1986.
[7] D.M. Greenberger, A. Yasin, Phys. Lett. A 128 (1988) 391.
[8] L. Mandel, Opt. Lett. 16 (1991) 1882.
[9] G. Jaeger, A. Shimony, L. Vaidman, Phys. Rev. A 51 (1995) 54.
[10] B.-G. Englert, Phys. Rev. Lett. 77 (1996) 2154.
[11] B.-G. Englert, Acta Phys. Slov. 46 (1996) 249; in: D. Han, J. Janszky, Y.S. Kim, V.I. Man'ko (Eds.), Proceedings of the Fifth International Conference on Squeezed States and Uncertainty Relations, Balatonfüred 1997, NASA/CP-1998-206855, Greenbelt, 1998, pp. 603–608.
[12] H. Rauch, J. Summhammer, Phys. Lett. A 104 (1984) 44.
[13] J. Summhammer, H. Rauch, D. Tuppinger, Phys. Rev. A 36 (1987) 4447.
[14] P. Mittelstaedt, A. Prieur, R. Schieder, Found. Phys. 17 (1987) 891.
[15] P.D.D. Schwindt, P.G. Kwiat, B.-G. Englert, Nature (London) submitted for publication.
[16] S. Dürr, T. Nonn, G. Rempe, Nature (London) 395 (1998) 33.
[17] S. Dürr, T. Nonn, G. Rempe, Phys. Rev. Lett., submitted for publication.
[18] H. Rauch, W. Treimer, U. Bonse, Phys. Lett. 47 A (1974) 369.
[19] H. Rauch, A. Zeilinger, G. Badurek, A. Wilfing, W. Bauspiess, U. Bonse, Phys. Lett. 54 A (1975) 425.
[20] S.A. Werner, R. Colella, A.W. Overhauser, C.F. Eagen, Phys. Rev. Lett. 35 (1975) 1053.
[21] P. Fischer, F. Mezei, A. Ioffe, Physica B 241–243 (1998) 117.
[22] J. Summhammer, G. Badurek, H. Rauch, U. Kischko, A. Zeilinger, Phys. Rev. A 27 (1983) 2523.
[23] G. Badurek, H. Rauch, J. Summhammer, Phys. Rev. Lett. 51 (1983) 1015.
[24] G. Badurek, H. Rauch, D. Tuppinger, Phys. Rev. A 34 (1986) 2600.
[25] J. Summhammer, K.A. Hamacher, H. Kaiser, H. Weinfurter, D.L. Jacobson, S.A. Werner, Phys. Rev. Lett. 75 (1995) 3206.
[26] A.G. Wagh, V.C. Rakhecha, J. Summhammer, G. Badurek, H. Weinfurter, B.E. Allman, H. Kaiser, K. Hamacher, D.L. Jacobson, S.A. Werner, Phys. Rev. Lett. 78 (1997) 755.
[27] B.E. Allman, A. Cimmino, A.G. Klein, G.I. Opat, H. Kaiser, S.A. Werner, Phys. Rev. Lett. 68 (1992) 2409.
[28] G. Badurek, H. Weinfurter, R. Gähler, A. Kollmar, S. Wehinger, A. Zeilinger, Phys. Rev. Lett. 71 (1993) 307.

[29] G. Badurek, H. Rauch, A. Zeilinger (Eds.), Proceeding of the International Workshop Matter Wave Interferometry, Wien, 1987; Physica B 151.
[30] S. Kawano, T. Kawai, A. Kawaguchi (Eds.), Proceedings of the International Symposium on Neutron Optics and Related Research Facilities, Kumatori 1996; J. Phys. Soc. Japan 65 Suppl. A (1996).
[31] M.O. Scully, K. Drühl, Phys. Rev. A 25 (1982) 2208.
[32] B.-G. Englert, M.O. Scully, H. Walther, Sci. Am. 271 (6) (1994) 56.
[33] G. Badurek, H. Rauch, A. Wilfing, U. Bonse, W. Graeff, J. Appl. Crystallogr. 12 (1979) 186.

Leong Chuan Kwek

Centre for Quantum Technologies, National University of Singapore, Singapore
National Institute of Education, Nanyang Technological University, Singapore

Paper Title: **Information-Theoretic Approach to Single-Particle and Two-Particle Interference in Multipath Interferometers**

Reference: Physical Review Letters **91**, 037901 (2003)

Personal message:

This was my first collaboration with Berge. Marek Żukowski visited us sometime in 2003 and I think it was Dagomir Kaszlikowski's idea to generalize the complementarity principle to qutrit and higher-dimensional systems. We had a lot of laughter and fun with the project. During a casual conversation, we revived one of the popular puzzle questions that Artur Ekert raised. This was about putting red dots on the forehead of dwarves and how these dwarves were called each morning to an assembly and to raise their hand if they know that they have a red dot on their forehead. Marek was not convinced that the problem was tractable, so Berge suggested that we did an experiment! The experiment was conducted between the four of us, with Marek as the Giant (who gave the red dots), and perhaps a few others who were in the room at that time. Each time, we succeeded in getting the correct "dwarves" with red dots on their forehead, much to Marek's anguish or "amazement."

Paper Title: **Quantum cryptography: Security criteria reexamined**

Reference: Physical Review A **70**, 032306 (2004)

Personal message:

We were fascinated with quantum cryptography at that time. Ajay joined the "quantum lah" group as the laboratory/administration assistant and also as my Masters student. Following Peter Shor's paper on the security of quantum cryptography with square-root measurements, we attempted to re-examine the issue for generalized measurements. We found that, indeed, there was a large class of positive operator-valued measurements (POVM) that provide better security bounds. Ajay was good at programming and so he did some of the numerics. Berge, I think, did his own version. Berge had much fun with the project as he dabbled with the dimensions and the number of particles. We were rather optimistic about the results, and we decided at some stage to go for a *Nature* publication. It was unfortunately rejected despite the hyped-up title.

VOLUME 91, NUMBER 3 PHYSICAL REVIEW LETTERS week ending 18 JULY 2003

Information-Theoretic Approach to Single-Particle and Two-Particle Interference in Multipath Interferometers

Dagomir Kaszlikowski,[1] Leong Chuan Kwek,[2,1] Marek Żukowski,[3] and Berthold-Georg Englert[1]

[1]*Department of Physics, National University of Singapore, Singapore 117542, Singapore*

[2]*National Institute of Education, Nanyang Technological University, Singapore 637616, Singapore*

[3]*Instytut Fizyki Teoretycznej i Astrofizyki, Uniwersytet Gdański, 80-952 Gdańsk, Poland*

(Received 18 February 2003; published 17 July 2003)

We propose entropic measures for the strength of single-particle and two-particle interference in interferometric experiments where each particle of a pair traverses a multipath interferometer. Optimal single-particle interference excludes any two-particle interference, and vice versa. We report an inequality that states the compromises allowed by quantum mechanics in intermediate situations, and identify a class of two-particle states for which the upper bound is reached. Our approach is applicable to symmetric two-partite systems of any finite dimension.

DOI: 10.1103/PhysRevLett.91.037901 PACS numbers: 03.67.Mn, 07.60.Ly

Interference effects of two kinds can be observed in quantum processes that involve pairs of particles. There are the usual single-particle interference fringes and there are phase-dependent coincidence probabilities that constitute two-particle interferences (see, e.g., [1,2]). As Horne and Zeilinger noted already in 1985 [1], there is a certain complementarity between the two types of interference: Optimal interference of one kind excludes any interference of the other.

The various quantitative studies of the possible trade-off between single-particle and two-particle interference that are on record (see [3,4] in particular) are dealing with two-path interferometers for the single particles. It is then possible to measure the strength of the single-particle interference by the familiar Michelson fringe visibility—the difference of maximal and minimal probabilities in one output divided by their sum—and introduce a matching measure for the two-particle interference.

There is, however, no unique analog for multipath interferometers, even for single-particle interference [5,6]. Dürr has recently compiled a list of desirable properties of any multipath generalization of Michelson's two-path visibility [7], but some proposals by others do not have all of these properties (see, e.g., [8]). Yet, quite a few of equally plausible definitions do meet Dürr's criteria [9], and the extension to two-particle interference is largely unexplored territory.

In an attempt to close this gap, at least partly, we introduce here an information-theoretic definition of the measure of the strength of two-particle interference. It exploits the mutual information contained in the coincidence probabilities. The corresponding single-particle measure is, in some sense, the fringe-contrast analog of the entropic measure for path knowledge that was first used by Wootters and Żurek [10] in 1979 in the context of a two-path interferometer (Young's double slit), and later also by others (see, e.g., [11,12]). But entropic measures of this kind have never been popular and have been criticized occasionally [4,7].

Yet, we think that they possess certain advantages. In particular, there is the immediate benefit that the dimensionality of the systems is irrelevant (although we assume here that the two systems are of the same dimension). Further, two-particle interferometers are building blocks of certain schemes for quantum cryptography [13]. The security criterion for some of them can be recast as a statement about entropic measures for single-particle and two-particle visibilities. These matters are, however, beyond the scope of the present Letter.

Let us close these introductory remarks by mentioning that the problem studied here is very closely linked with the relations between single-particle fringe visibility and which-path knowledge in quantitative studies of wave-particle duality. Going back to Einstein's 1905 paper on the photoelectric effect, this intriguing aspect of quantum mechanics is as old as quantum theory itself. Recent studies include [3,4,7,14–16], where also the earlier literature can be found.

Consider an interferometric experiment of the kind that is sketched in Fig. 1. A source emits pairs of entangled particles, one of each pair propagating to observer Alice (A) on the left, the other to Bob (B) on the right, N paths being available on either side. Before its registration by the respective detectors, each particle undergoes a unitary transformation consisting of the action of phase shifters (phases $\phi_k^{(A)}$ and $\phi_l^{(B)}$, respectively) and an unbiased symmetric multiport beam splitter. The latter effects the unitary discrete Fourier transformation, F, that is specified by the matrix elements $F_{mn} = \gamma_N^{mn}/\sqrt{N}$, where $\gamma_N = \exp(2\pi i/N)$ is the basic Nth root of unity. Depending on the two-particle state emitted by the source, the resulting paired detector clicks may exhibit correlations that originate in the path interference that is tested by the Fourier transformation in conjunction with the phase shifters. It is our objective here to quantify the

 0031-9007/03/91(3)/037901(4)$20.00

VOLUME 91, NUMBER 3 PHYSICAL REVIEW LETTERS week ending 18 JULY 2003

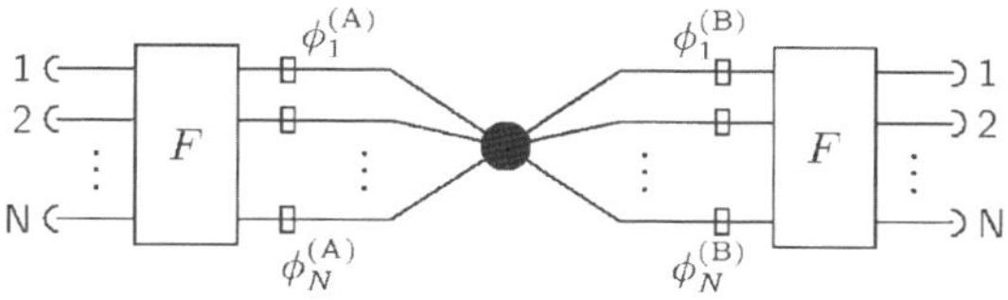

FIG. 1. Sketch of the interferometric setup. The source at the center emits paired particles. One of them propagates to the left, the other to the right, whereby both have a choice between N paths. After traversing the respective arrays of phase shifters (phases $\phi_k^{(A)}$ on the left, $\phi_l^{(B)}$ on the right), the particles pass through N-port beam splitters that effect a discrete Fourier transformation (F) of the probability amplitudes for the paths, and are then detected.

strength of the interference observed in experiments of this kind.

The experimental data are the N^2 coincidence probabilities $p_{mn}^{(AB)} = p_{mn}^{(AB)}(\phi_1^{(A)}, \ldots, \phi_N^{(A)}; \phi_1^{(B)}, \ldots, \phi_N^{(B)})$, of unit sum, between Alice's mth detector and Bob's nth detector ($m, n = 1, 2, \ldots, N$). The relative frequencies of Alice's and Bob's individual detection events are given by the marginal sums of $p_{mn}^{(AB)}$:

$$\begin{aligned} p_m^{(A)}(\phi_1^{(A)}, \ldots, \phi_N^{(A)}) &= \sum_{n=1}^{N} p_{mn}^{(AB)}, \\ p_n^{(B)}(\phi_1^{(B)}, \ldots, \phi_N^{(B)}) &= \sum_{m=1}^{N} p_{mn}^{(AB)}. \end{aligned} \tag{1}$$

In their dependence on the local interferometric phases, $p_m^{(A)}$ and $p_n^{(B)}$ contain the pertinent information about the phase relations between the N paths on Alice's and Bob's side, respectively, and information about the N^2 path pairs can be extracted from the coincidence probabilities $p_{mn}^{(AB)}$.

We have mentioned above that Michelson's standard definition of the fringe visibility for two-path interferometers has no unique generalization for multipath interferometers. This is so for a simple reason: A single number cannot do full justice to the complex pattern displayed by $p_m^{(A)}(\phi_1^{(A)}, \ldots, \phi_N^{(A)})$ or $p_n^{(B)}(\phi_1^{(B)}, \ldots, \phi_N^{(B)})$. Thus, one must make a judicious choice among the various possible definitions that meet the reasonably obvious criteria on Dürr's list [7].

Here we opt for measuring the various interference strengths in terms of the respective Shannon entropies. We shall continue to speak of "visibility" although the analogy with the usual two-path visibility is a bit remote. Alice's single-particle visibility $V^{(A)}$ is then defined in accordance with

$$V^{(A)} = \max_{\phi_1^{(A)}, \ldots, \phi_N^{(A)}} [1 - H_N(p^{(A)})], \tag{2}$$

where

$$H_N(p^{(A)}) = -\sum_{m=1}^{N} p_m^{(A)} \log_N p_m^{(A)} \tag{3}$$

is the normalized Shannon entropy of her marginal probability distribution, and Bob's single-particle visibility $V^{(B)}$ is defined analogously. For convenient normalization, we take the logarithm to base N so that, by construction, both $H_N(p^{(A)})$ and $V^{(A)}$ are nonnegative numbers that do not exceed unity.

The definition of $V^{(A)}$ in (2) exploits the fact that path knowledge after a Fourier transformation corresponds to interference strength before the transformation [9]. The largest amount of terminal path knowledge that is potentially available for optimally set phase shifters is thus a self-suggesting measure for the interference strength at the input, and the Shannon entropy is one natural quantitative measure of knowledge about a probability distribution — the one we find particularly useful here. This *a priori* justification of our definition of $V^{(A)}$ is, of course, no more than an appeal at plausibility; the definition is ultimately justified by the consequences that we report below.

We note that $V^{(A)}$ of (2) reaches its upper bound of unity only for certain pure states of Alice's particle, namely, those for which it is possible to adjust the phases $\phi_k^{(A)}$ such that the particle is surely directed to one particular detector, which is to say that the arriving particle must follow each path with the same probability of $1/N$ and that there must be definite relative phases between the paths. It is equally important that $V^{(A)}$ vanishes if Alice's particle arrives along one particular path, in which situation there is no room for relative phases between the paths. Accordingly, $V^{(A)}$ vanishes also if the statistical operator $\rho^{(A)}$ of Alice's particle is a convex sum of pure states that refer to definite paths. In the case of the totally mixed state $\rho^{(A)} = 1/N$, we face the extreme situation of utter ignorance in which nothing is predictable about the particle's path before or after the Fourier transformation.

The single-particle visibility of (2) also meets the indispensable criterion of convexity: The visibility of a convex sum of states cannot exceed the convex sum of the individual visibilities. This is an immediate consequence of the concavity of the Shannon entropy.

As a preparation for the definition of the two-particle visibility $V^{(AB)}$ in (5) below, we now consider the interferometric experiment from the perspective of information theory. Imagine that the source and beam splitters are contained in a black box and only the detectors are accessible. We observe paired clicks of the detectors with the coincidence probabilities $p_{mn}^{(AB)}$ and, on either side, individual clicks with the marginal probabilities $p_m^{(A)}$ and $p_n^{(B)}$, respectively. We then wonder about the strength of correlations between clicks of pairs of detectors at opposite sides of the experiment. Obviously, the quantitative measure of two-particle interference should vanish if there are no correlations between the clicks for

VOLUME 91, NUMBER 3 PHYSICAL REVIEW LETTERS week ending 18 JULY 2003

any setting of the phase shifters, and it should be maximal if perfect correlations can be achieved for some setting.

A good candidate for a measure having these properties, and the natural one to accompany the single-particle visibility of (2), is the maximal value of the mutual information $I(p^{(AB)})$ contained in the $p_{mn}^{(AB)}$, which is the relative Shannon entropy between the probability distributions $p_{mn}^{(AB)}$ and $p_n^{(A)}p_m^{(B)}$; that is, between the actual coincidence probabilities and the corresponding products of the marginal probabilities.

For a particular setting of the phase shifters, we have

$$\begin{aligned} I(p^{(AB)}) &= H_N(p^{(A)}) + H_N(p^{(B)}) - H_N(p^{(AB)}) \\ &= \sum_{m,n=1}^{N} p_{mn}^{(AB)} \log_N \frac{p_{mn}^{(AB)}}{p_n^{(A)} p_m^{(B)}}. \end{aligned} \tag{4}$$

If the two distributions happen to be equal, $p_{mn}^{(AB)} = p_n^{(A)}p_m^{(B)}$, then the stochastic variables n and m are statistically independent, the detector clicks are not correlated, and the mutual information vanishes accordingly. Since it is also bounded, $0 \le I(p^{(AB)}) \le 1$, we are invited to define the two-particle visibility as

$$V^{(AB)} = \max_{\phi_k^{(A)},\phi_l^{(B)}} I(p^{(AB)}). \tag{5}$$

Before proceeding we should not fail to mention the alternative proposal by Jaeger, Horne, and Shimony [3]. In the context of two-path interferometers, they define a two-particle visibility by a Michelson-type formula for the extremal values of the *difference* $p_{mn}^{(AB)} - p_n^{(A)}p_m^{(B)}$, and it appears that a systematic study of the possible generalization to multipath interferometers is lacking. By contrast, our definition involves (the logarithm of) the *ratio* $p_{mn}^{(AB)}/(p_n^{(A)}p_m^{(B)})$ and is immediately applicable to multipath interferometers.

We note that the two-particle visibility (5) is maximal, $V^{(AB)} = 1$, if the source emits a maximally entangled state such as

$$\rho_{\text{max ent}}^{(AB)} = \frac{1}{N} \sum_{j,k=1}^{N} |jj\rangle\langle kk|, \tag{6}$$

where $|jk\rangle$ is the ket vector for one particle in Alice's jth path and the other in Bob's kth path. If the source emits a product state,

$$\begin{aligned} \rho^{(AB)} &= \rho^{(A)}\rho^{(B)}, \qquad \rho^{(A)} = \mathrm{tr}_B \rho^{(AB)}, \\ \rho^{(B)} &= \mathrm{tr}_A \rho^{(AB)}, \end{aligned} \tag{7}$$

then $p_{mn}^{(AB)} = p_n^{(A)}p_m^{(B)}$ and $V^{(AB)} = 0$. This limit is also reached in the case of the maximally mixed, or chaotic, state

$$\rho_{\text{chaos}}^{(AB)} = \frac{1}{N^2} \sum_{j,k=1}^{N} |jk\rangle\langle jk| = \frac{1}{N^2}. \tag{8}$$

All these features are in complete agreement with what one would expect from a reasonably defined two-particle visibility.

The stage is now set for presenting the central result: For correlations generated by quantum mechanics (arbitrary joint probabilities $p_{mn}^{(AB)}$ do not have this property), the inequalities

$$V^{(A)} + V^{(AB)} \le 1, \qquad V^{(B)} + V^{(AB)} \le 1 \tag{9}$$

are obeyed by the single-particle visibilities $V^{(A)}$, $V^{(B)}$ and the two-particle visibility $V^{(AB)}$, and there are families of pure two-particle states, $\rho_\lambda^{(AB)} = |\lambda\rangle\langle\lambda|$ with $0 \le \lambda \le 1$, for which the equal signs hold with each visibility covering the whole range from 0 to 1. In other words, the inequalities can be saturated.

In view of the complete symmetry between Alice and Bob, it suffices to prove one of the inequalities, say, the first one. Consider the experiment of Fig. 2, where we have two observers who can apply arbitrary $U(N)$ transformations, U_A and U_B, to their respective incoming particles, which arrive in some state $\rho^{(AB)}$ that could be pure or mixed. Since there is no restriction here to Fourier transformations, the more general single-particle visibility $\tilde{V}^{(A)}$ bounds $V^{(A)}$ from above,

$$V^{(A)} \le \tilde{V}^{(A)} = \max_{U_A}[1 - H_N(p^{(A)})] = 1 - S_N(\rho^{(A)}), \tag{10}$$

where $S_N(\rho) = -\mathrm{tr}\{\rho \log_N \rho\}$ is the scaled von Neumann entropy of the reduced state $\rho^{(A)}$. The maximum in (10) is reached for those U_A that diagonalize $\rho^{(A)}$.

To find a corresponding upper bound for the two-particle visibility, we exploit an information-theoretic argument. When Bob performs his unitary transformation U_B and then registers a click of his nth detector, it is *as if* he prepared Alice's particle in the state described by

$$\rho_n^{(A)} = \frac{1}{p_n^{(B)}} \mathrm{tr}_B\{P_n^{(B)} U_B^\dagger \rho^{(AB)} U_B\}, \tag{11}$$

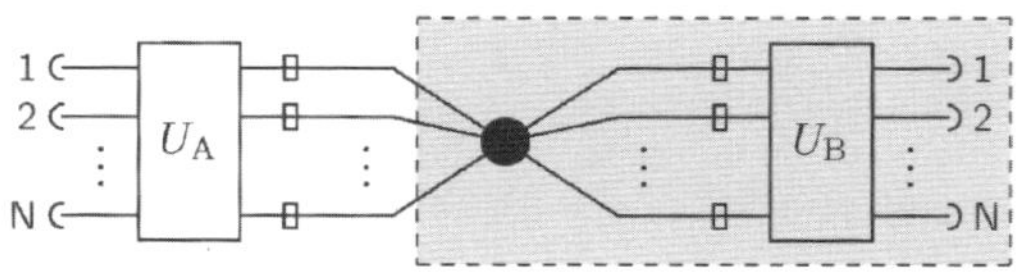

FIG. 2. Another view of the experiment. It is *as if* Bob prepared Alice's particle on the left in a particular state by detecting his particle on the right. Upon regarding the components inside the shaded area as Bob's effective source, the setup can be viewed as realizing a certain communication protocol, and can thus be interpreted and assessed from the point of view of information theory.

VOLUME 91, NUMBER 3 PHYSICAL REVIEW LETTERS week ending 18 JULY 2003

where $P_n^{(B)}$ is the projector corresponding to a click of his nth detector. In effect, Alice applies her U_A to $\rho_n^{(A)}$ and then detects her particle.

The process of finding the two-particle visibility in the experiment can now be understood as a communication protocol where Alice and Bob collaborate to obtain the maximal mutual information in the B-to-A quantum channel with the restriction that only projective measurements are allowed. The well-known Holevo bound [17] limits the available mutual information in situations of this kind. In the particular case under consideration, the Holevo bound reads

$$I(p^{(AB)}) \le S_N(\rho^{(A)}) - \min_{U_B}\left\{\sum_{n=1}^{N} p_n^{(B)} S_N(\rho_n^{(A)})\right\}. \quad (12)$$

Since the subtracted term is nonnegative, it follows that $V^{(AB)} \le S_N(\rho^{(A)})$, and the argument concludes as above.

To prove the second assertion at (9), namely that the upper bound for the sums of single-particle and two-particle visibilities in (9) is tight, we consider the following one-parameter family of pure states:

$$\rho_\lambda^{(AB)} = |\lambda\rangle\langle\lambda|, \qquad |\lambda\rangle = \frac{|0\rangle(1-\lambda) + |1\rangle\lambda\sqrt{N}}{\sqrt{1+(N-1)\lambda^2}}, \quad (13)$$

where we have, for $\lambda = 0$, the ket vector,

$$|0\rangle = \frac{1}{\sqrt{N}}\sum_{k=1}^{N} |kk\rangle, \quad (14)$$

of the maximally entangled state (6) and, for $\lambda = 1$, the ket vector,

$$|1\rangle = \frac{1}{N}\sum_{j,k=1}^{N} |jk\rangle = (F_A F_B)^{-1}|NN\rangle, \quad (15)$$

of the product state that is obtained from $|NN\rangle$ by twofold inverse Fourier transformation. The normalizing denominator in (13) takes $\langle 0|1\rangle = 1/\sqrt{N}$ into account.

If the source emits the two-particle state $\rho_\lambda^{(AB)}$ and all phases $\phi_k^{(A)}$, $\phi_l^{(B)}$ are set to zero, then the coincidence probabilities are $p_{mn}^{(AB)} = |\langle mn|\lambda\rangle|^2$. The ingredients of the probability amplitude are

$$\begin{aligned}\langle mn|F_A F_B|0\rangle &= \frac{\delta_{m+n,N}^{(N)}}{\sqrt{N}}, \\ \langle mn|F_A F_B|1\rangle &= \langle mn|NN\rangle = \delta_{m,N}^{(N)}\delta_{n,N}^{(N)},\end{aligned} \quad (16)$$

where the periodic Kronecker delta $\delta_{j,k}^{(N)}$ is 1 if $j = k$ modulo N, and 0 otherwise. The resulting probabilities,

$$p_{mn}^{(AB)} = \frac{[(1-\lambda)\delta_{m+n,N}^{(N)} + N\lambda\delta_{m,N}^{(N)}\delta_{n,N}^{(N)}]^2}{N[1+(N-1)\lambda^2]}, \quad (17)$$

exhibit perfect correlations inasmuch as for each m value there is only one n value for which $p_{mn}^{(AB)} \neq 0$. As a consequence, all summands vanish in the double sum of

$$\begin{aligned}I(p^{(AB)}) - H_N(p^{(A)}) &= H_N(p^{(B)}) - H_N(p^{(AB)}) \\ &= \sum_{m,n=1}^{N} p_{mn}^{(AB)} \log_N \frac{p_{mn}^{(AB)}}{\sum_{m'} p_{m'n}^{(AB)}} = 0. \end{aligned} \quad (18)$$

This in turn implies

$$1 = [1 - H_N(p^{(A)})] + I(p^{(AB)}) \le V^{(A)} + V^{(AB)} \le 1, \quad (19)$$

so that equal signs must hold throughout. Indeed, the upper bound in (9) is reached for all λ, with $V^{(AB)} = 1$ for $\lambda = 0$ and $V^{(AB)} = 0$ for $\lambda = 1$, and all intermediate values in between.

We thank Thomas Durt for very valuable discussions. M. Ż. and B. G. E. gratefully acknowledge the splendid hospitality extended to them at the National University of Singapore. This work was supported by A *Star Grant No. 012-104-0040, by KBN Project No. 5 PO3B 088 20.

[1] M. A. Horne and A. Zeilinger, *Proceedings of the Symposium on the Foundations of Modern Physics*, edited by P. Lahti and P. Mittelstaedt (World Scientific, Singapore, 1985).
[2] D. M. Greenberger, M. A. Horne, and A. Zeilinger, Phys. Today **46**, No. 8, 22 (1993).
[3] G. Jaeger, M. A. Horne, and A. Shimony, Phys. Rev. A **48**, 1023 (1993).
[4] G. Jaeger, A. Shimony, and L. Vaidman, Phys. Rev. A **51**, 54 (1995).
[5] G. Weihs, M. Reck, H. Weinfurter, and A. Zeilinger, Phys. Rev. A **54**, 893 (1996).
[6] M. Żukowski, A. Zeilinger, and M. Horne, Phys. Rev. A **55**, 2564 (1997).
[7] S. Dürr, Phys. Rev. A **64**, 042113 (2001).
[8] A. Luis, J. Phys. A **34**, 8597 (2001).
[9] B.-G. Englert, D. Kaszlikowski, and L. C. Kwek (unpublished).
[10] W. K. Wootters and W. Żurek, Phys. Rev. D **19**, 473 (1979).
[11] P. Mittelstaedt, A. Prieur, and R. Schieder, Found. Phys. **17**, 891 (1987).
[12] P. J. Lahti, P. Busch, and P. Mittelstaedt, J. Math. Phys. (N.Y.) **32**, 2770 (1991).
[13] A. Ekert, Phys. Rev. Lett. **67**, 661 (1991).
[14] B.-G. Englert, Phys. Rev. Lett. **77**, 2154 (1996).
[15] B.-G. Englert and J. Bergou, Opt. Commun. **179**, 337 (2000).
[16] M. Jakob and J. A. Bergou, arXiv/quant-ph/0302075.
[17] A. S. Holevo, in *Proceedings of the Second Japan–USSR Symposium on Probability Theory*, edited by G. Maruyama and J. V. Prokhorov, Lecture Notes in Mathematics Vol. 330 (Springer-Verlag, Berlin, 1973), pp. 104–119.

PHYSICAL REVIEW A **70**, 032306 (2004)

Quantum cryptography: Security criteria reexamined

Dagomir Kaszlikowski,[1] Ajay Gopinathan,[1,2] Yeong Cherng Liang,[1] L. C. Kwek,[1,2] and Berthold-Georg Englert[1]
[1]*Department of Physics, National University of Singapore, Singapore 117 542, Singapore*
[2]*National Institute of Education, Nanyang Technological University, Singapore 637616, Singapore*
(Received 23 October 2003; revised manuscript received 26 May 2004; published 8 September 2004)

We find that the generally accepted security criteria are flawed for a whole class of protocols for quantum cryptography. This is so because a standard assumption of the security analysis, namely that the so-called square-root measurement is optimal for eavesdropping purposes, is not true in general. There are rather large parameter regimes in which the optimal measurement extracts substantially more information than the square-root measurement.

DOI: 10.1103/PhysRevA.70.032306 PACS number(s): 03.67.Dd, 03.67.Hk

I. INTRODUCTION

All practical implementations of protocols for quantum cryptography have to deal with the unavoidable noise in the transmission lines, and possibly the intervention of an eavesdropper, that degrade the correlations in the raw-key data of the communicating parties, namely Alice and Bob. They then face a double task: First, they must establish how much Eve, the evildoing eavesdropper, can possibly know about their data; and second, they must extract a secure noise-free key sequence from the insecure noisy raw data.

The second task of key generation is solved by exploiting the findings and methods of classical information theory, in particular the lesson of the seminal work by Csiszár and Körner [1]. They demonstrated that Alice and Bob can always generate a secure key, provided that the mutual information between them exceeds the mutual information between either one of them and Eve.

The first task of determining how much Eve knows thus amounts to figuring out the maximally attainable mutual information between her and either Alice or Bob. There are two different, but equivalent, lines of reasoning that one can choose to follow, depending on how one pictures the communication between Alice and Bob, and Eve's tampering with it.

One scenario is that of the 1984 protocol by Bennett and Brassard (BB84 [2]), in which Alice sends quantum-information carriers to Bob through an appropriate, authenticated quantum channel. Eve intercepts each carrier in transmission and keeps an imperfect copy, obtained by operating a quantum-cloning machine, before forwarding the carrier to Bob. The quest is then for the best cloning machine—best for *this* purpose—in conjunction with the best way of extracting information from the clones.

The other scenario is that of the 1991 protocol by Ekert (E91 [3]), in which a source distributes entangled pairs of carriers to Alice and Bob, who make statistically independent measurements on them, thereby effectively establishing a quantum channel between themselves. Eve is given full control of the source. She keeps a quantum record of what is sent in the form of auxiliary quantum systems, usually termed *ancillas*, that she entangles with the paired carriers. Here the quest is for the best ancilla states in conjunction with the best way of extracting information from the ancillas.

Because of the lack of superior alternatives, the standard analysis of protocols of BB84 type invokes unproven assumptions about optimal cloning machines; see, for example, Refs. [4,5] and the recent paper by Acín *et al.* [6]. Likewise, there is a common assumption in the analysis of E91-type protocols, namely that the so-called square-root measurement (SRM [7]) is optimal for Eve's processing of the ancillas; see the recent paper by Liang *et al.* [8], for example. The established equivalence of the BB84 and E91 scenarios [9], and the fully equivalent security criteria thus found, is strong circumstantial evidence that these assumptions—about Eve's best intercept strategy and her best way of processing the ancillas, respectively—are equivalent as well.

It is the objective of this article to demonstrate that the SRM is *not* optimal for a whole class of quantum cryptography protocols, the tomographic protocols of Refs. [8,10]; it may very well not be optimal for other protocols, too. The equivalence stated above then implies the well-founded conjecture that there are also better intercept strategies than those usually regarded as best. We offer some remarks about the connection of this work with intercept strategies in the Appendix.

II. THE PYRAMID OF ANCILLA STATES

We build on the work of Ref. [8], where the protocols are phrased as generalizations of the E91 scenario to N letter alphabets ($N=2,3,\ldots$). The source controlled by Eve would emit pairs of qubits for $N=2$, pairs of qutrits for $N=3,\ldots$, and pairs of *qunits* in the general case. After everything is said and done, Eve knows that her ancilla is in the state described by ket $|E_k\rangle$ if Alice obtains value k for her qunit of the respective pair (with $k=0,1,\ldots,N-1$). Since there is a common (real) angle between every pair of ancilla states,

$$\langle E_k|E_l\rangle = \lambda + (1-\lambda)\delta_{kl} = \left\{\begin{array}{ll} 1 & \text{if } k=l \\ \lambda & \text{if } k\neq l \end{array}\right\} = r_0 - r_1 + N r_1 \delta_{kl}, \tag{1}$$

the N ancilla kets can be regarded as the edges of an N-dimensional pyramid [11]; see Fig. 1 for an illustration of the case of $N=3$. The average ancilla ket

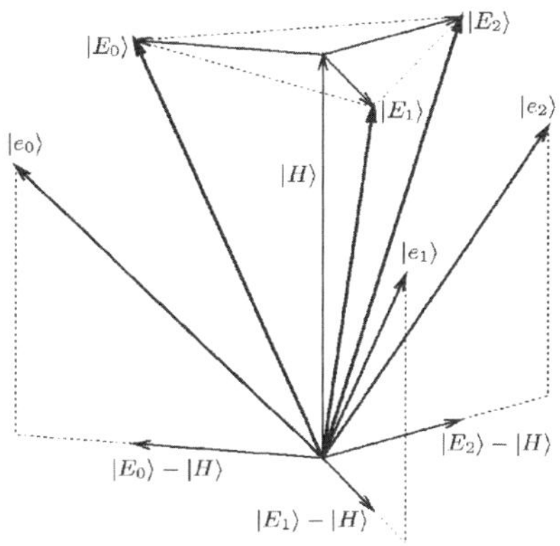

FIG. 1. Pyramid geometry for $N=3$. The ancilla kets $|E_k\rangle$, of unit length, are the edges of the ancilla pyramid. Its shape is determined by the parameter λ of Eq. (1), the cosine of the acute angle between any pair of edges. The height ket $|H\rangle$ of Eq. (2) points from the tip of the pyramid to the center of its base; its length is $\sqrt{r_0}$. The kets $|E_k\rangle-|H\rangle$, of length $\sqrt{1-r_0}$, point from the center of the pyramid base to its corners. The SRM kets $|e_k\rangle$ of Eq. (9), of unit length, define the SRM pyramid, which has right angles between its edge kets. The SRM pyramid is wider than, but not as high as, the ancilla pyramid.

$$|H\rangle = \frac{1}{N}\sum_{k=0}^{N-1} |E_k\rangle \tag{2}$$

points from the tip of the pyramid to the center of its $(N-1)$-dimensional base [12], so that the length of $|H\rangle$, $\sqrt{\langle H|H\rangle}=\sqrt{r_0}$, is the height of the pyramid. The pyramid volume is given by $(1/N!)(Nr_0)^{1/2}(Nr_1)^{(N-1)/2}$; it is largest for $\lambda=0$, $r_0=r_1=1/N$ when the pyramid is a corner of an N-dimensional cube.

Geometry restricts λ to the range $-1/(N-1)\leqslant\lambda\leqslant 1$, where both limits correspond to degenerate pyramids that have no N-dimensional volume. For $\lambda=1$, we have a single ancilla state and the pyramid is just a line, a pyramid of unit height and no base; and for $\lambda=-1/(N-1)$ we have linearly dependent ancilla kets that span an $(N-1)$-dimensional subspace, so that the pyramid has no height. In the context of quantum cryptography, however, only non-negative λ values are relevant, for which $r_0\geqslant r_1$. In other words, the pyramids of interest are acute, in the sense that the common angle between each pair of their edges is acute.

Alice gets each k value with probability $1/N$, so that

$$\rho = \frac{1}{N}\sum_{k=0}^{N-1} |E_k\rangle\langle E_k| \tag{3}$$

is the statistical operator for Eve's ancillas. The height ket $|H\rangle$ of Eq. (2) is the eigenket of ρ to the eigenvalue r_0 and all kets orthogonal to $|H\rangle$ are eigenkets to the $(N-1)$-fold degenerate eigenvalue $r_1=r_0-\lambda=(1-\lambda)/N$.

The N kets $|E_k\rangle-|H\rangle$, each of length $\sqrt{1-r_0}=\sqrt{(N-1)r_1}$, point from the center of the ancilla-pyramid base to its corners. They span the $(N-1)$-dimensional subspace to eigenvalue r_1.

III. WHICH EDGE OF THE PYRAMID?

A. The pretty good square-root measurement

Eve extracts information out of ρ with the aid of a generalized measurement, a positive-operator-valued measure (POVM), specified by a decomposition of the identity in the N-dimensional ancilla space into M non-negative operators,

$$1 = \sum_{m=0}^{M-1} P_m, \quad P_m \geqslant 0. \tag{4}$$

The mutual information between Alice and Eve,

$$I = \sum_{n=0}^{N-1}\sum_{m=0}^{M-1} p_{nm}\log_N\frac{p_{nm}}{p_{n\cdot}p_{\cdot m}}, \tag{5}$$

is then computable from the joint probabilities

$$p_{nm} = \frac{1}{N}\langle E_n|P_m|E_n\rangle \tag{6}$$

and their marginals

$$p_{n\cdot} = \sum_{m=0}^{M-1} p_{nm} = \frac{1}{N}, \quad p_{\cdot m} = \sum_{n=0}^{N-1} p_{nm}. \tag{7}$$

For convenient normalization, the logarithm in Eq. (5) is taken to base N, so that $I\leqslant 1$ with the maximum achieved for uniform perfect correlations, that is, for $M=N$ and $p_{nm}=\delta_{nm}/N$.

The POVM for the SRM is specified by setting $M=N$ and

$$P_m = (N\rho)^{-1/2}|E_m\rangle\langle E_m|(N\rho)^{-1/2} \equiv |e_m\rangle\langle e_m| \tag{8}$$

with

$$|e_m\rangle = (|E_m\rangle - |H\rangle)\frac{1}{\sqrt{Nr_1}} + |H\rangle\frac{1}{\sqrt{Nr_0}}. \tag{9}$$

The resulting joint probabilities are

$$p_{nm} = \frac{1}{N}|\langle E_n|e_m\rangle|^2 = \frac{1}{N}[\eta_1 + (\eta_0 - \eta_1)\delta_{nm}], \tag{10}$$

where

$$\sqrt{\eta_0} - \sqrt{\eta_1} = \sqrt{Nr_1} \quad \text{and} \quad \eta_0 + (N-1)\eta_1 = 1. \tag{11}$$

We note that the SRM thus associated with the ancilla pyramid happens to be a standard von Neumann measurement, not a POVM proper, because the projectors in Eq. (8) are pairwise orthogonal, $\mathrm{tr}\{P_mP_{m'}\}=\delta_{mm'}$. The mutual information acquired by performing the SRM,

$$I^{(\mathrm{SRM})} = \eta_0\log_N(N\eta_0) + (N-1)\eta_1\log_N(N\eta_1), \tag{12}$$

is shown in Fig. 2 for $N=2,3,5,10,20,100$.

B. Better than pretty good

Whereas the SRM is known to be "pretty good" as a rule [13], it is also known that it does not always optimize the

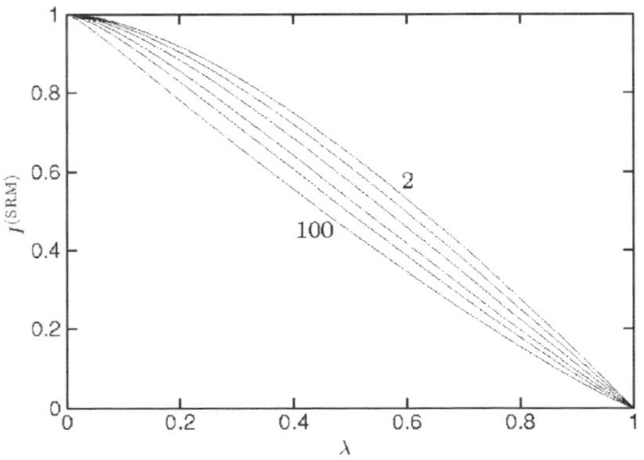

FIG. 2. Mutual information between Alice and Eve if Eve performs the square-root measurement. The curves refer to $N=2,3,5,10,20$, and 100, and the plot covers the range $0\leq\lambda\leq 1$ that is relevant for quantum cryptography.

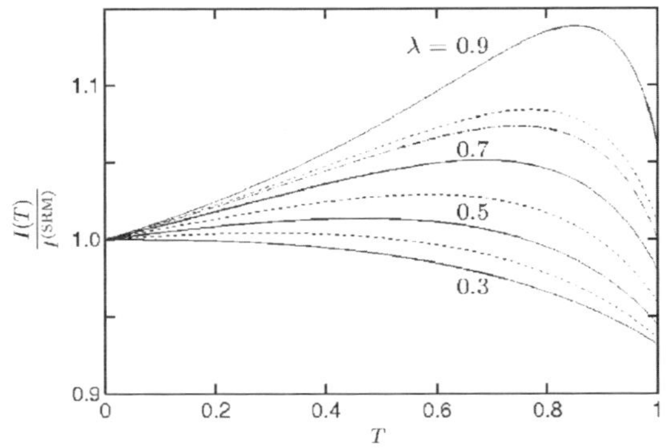

FIG. 3. Mutual information for the POVM of Eq. (13) relative to that of the SRM. For $N=10$, the plot shows the ratio of $I(T)/I^{(\mathrm{SRM})}$ as a function of T for $\lambda=0.9,0.7,0.5,0.3$ (solid lines) and for $\lambda=0.8,0.6,0.4$ (dashed lines). The left end ($T=0$) refers to the SRM, the right end ($T=1$) to the MUD. For $\lambda=0.772\,76$ (dash-dotted line), both give the same mutual information.

mutual information. In particular, Shor has pointed out that there are superior POVMs for $N=3$ and some $\lambda<0$, and has conjectured that there is also a $\lambda>0$ range in which other POVMs could be better [14]. Shor's explicit example for $\lambda<0$ is interesting in its own right but does not seem to have any bearing on the security analysis of quantum-cryptography protocols. By contrast, the $\lambda>0$ examples reported below are of immediate relevance, as they invalidate, at least partly, established security criteria.

Consider the one-parametric family of POVMs defined by $M=N+1$ and $P_m=|\overline{e_m}\rangle\langle\overline{e_m}|$ with

$$|\overline{e_m}\rangle = (|E_m\rangle - |H\rangle)\frac{1}{\sqrt{Nr_1}} + |H\rangle\frac{t}{\sqrt{Nr_0}}, \quad \text{for } m<N,$$

$$|\overline{e_N}\rangle = |H\rangle\sqrt{\frac{1-t^2}{r_0}}, \tag{13}$$

where $0\leq t\leq 1$. The SRM kets of Eq. (9) obtain for $t=1$.

For $t<1$, the measurement pyramid, which has the kets $|\overline{e_0}\rangle,\ldots,|\overline{e_{N-1}}\rangle$ for its edges, has the same base area as the SRM pyramid, but is of smaller height and therefore obtuse. Since the angle between any such given $|\overline{e_m}\rangle$ and the ancilla kets $|E_n\rangle$ with $n\neq m$ increases as t decreases from $t=1$, the sector of $m<N$ will have increased mutual information. But this comes at a price: When Eve finds $|\overline{e_N}\rangle\propto|H\rangle$, she has no clue about Alice's value; the sector $m=N$ is inconclusive and provides no contribution at all to the mutual information. Accordingly, the optimal choice of t is such that the increase of mutual information in the $m<N$ sector is balanced against the increase in the probability of the inconclusive result; this probability equals $(1-t^2)r_0$.

For $t=\sqrt{r_1/r_0}$, the POVM specified by Eq. (13) is the "measurement for unambiguous discrimination" (MUD [7]), for which $\langle E_n|\overline{e_m}\rangle=0$ if $n\neq m<N$, so that there are perfect correlations, and thus full mutual information, in the $m<N$ sector. The cost for this perfection is, however, so high that the MUD never maximizes the mutual information, although it can outperform the SRM. The optimal choice for t is always in the range $\sqrt{r_1/r_0}<t\leq 1$. This observation is illustrated in Fig. 3 for $N=10$ and various values of λ, including $\lambda=0.772\,76$, for which the MUD and the SRM give the same mutual information. The plot shows only the t range of interest, conveniently reparametrized in terms of T, a scaled version of t, introduced in accordance with

$$t = 1 - T + T\sqrt{r_1/r_0}. \tag{14}$$

Thus, $T=0$ refers to the SRM, and $T=1$ to the MUD.

The mutual information for the POVMs specified by Eqs. (13) is given by

$$I(T) = \overline{\eta_0}\log_N\frac{N\overline{\eta_0}}{\overline{\eta_0}+(N-1)\overline{\eta_1}} + (N-1)\overline{\eta_1}\log_N\frac{N\overline{\eta_1}}{\overline{\eta_0}+(N-1)\overline{\eta_1}}, \tag{15}$$

where

$$\overline{\eta_0} = (\sqrt{\eta_0} - T\sqrt{\eta_1})^2, \quad \overline{\eta_1} = (1-T)^2\eta_1 \tag{16}$$

are the T-dependent versions of η_0, η_1. For ancilla pyramids with a large volume, $0<\lambda<(3-4/N)/(N-1)\equiv\Lambda$, the maximum of $I(T)$ is obtained for $T=0$, which is to say that the SRM is optimal in this range of small λ values. By contrast, for ancilla pyramids with a rather small volume, $\Lambda<\lambda<1$, the maximum of $I(T)$ is reached for $T=1-(\sqrt{\eta_0/\eta_1}-1)/(N-2)$, that is, when the arguments of the two logarithms in Eq. (15) equal $N-1$ and $1/(N-1)$, respectively. Then, the measurement pyramid is obtuse.

In summary, we have

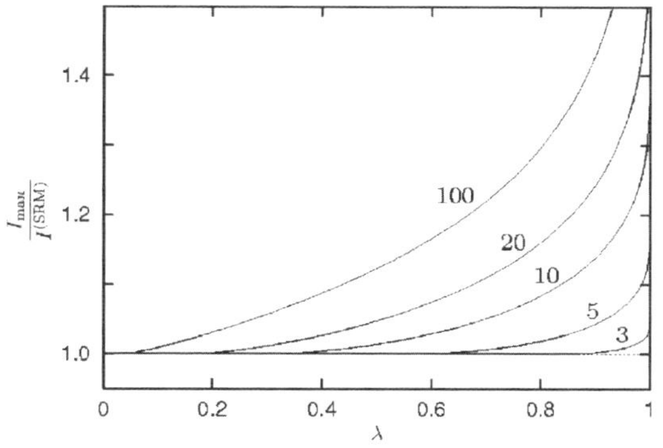

FIG. 4. Ratio of the maximal mutual information I_{max} and the SRM value $I^{(SRM)}$, for $N=3,5,10,20,100$, as a function of λ.

$$I_{max} \equiv \max_T I(T) = \begin{cases} I^{(SRM)} \text{ of Eq. (12)} & \text{if } 0 \leq \lambda \leq \Lambda = \dfrac{3N-4}{N(N-1)}, \\ (1-\lambda)\dfrac{N-1}{N-2}\log_N(N-1) & \text{if } \Lambda \leq \lambda \leq 1. \end{cases} \quad (17)$$

This is our central result.

For λ values that exceed the threshold value of Λ substantially, the optimal POVM from the family (13) gives significantly more mutual information than the SRM. This can be seen by plotting the ratio $I_{max}/I^{(SRM)}$ as a function of λ; see Fig. 4. The $\lambda \to 1$ limit,

$$\frac{I_{max}}{I^{(SRM)}} \to \frac{N/2}{N-2}\ln(N-1) \quad \text{as } \lambda \to 1, \quad (18)$$

shows that the optimal POVM provides much more information than the SRM if N is large, and then the range $0 \leq \lambda < \Lambda \simeq 3/N$ is small in addition.

IV. SUMMARY AND DISCUSSION

In summary, there are POVMs that outperform the SRM for $\lambda > \Lambda$, and we know the optimal POVM of the sort defined by Eq. (13) quite explicitly. We are, in fact, quite sure that it is the global optimum because an extensive numerical search failed to find any better POVM.

A first search covered a large class of POVMs that respect the geometry of the ancilla pyramid: We took parameter t to be complex; we rotated around the symmetry axis specified by ket $|H\rangle$; and we considered weighted sums of several such POVMs, with different t parameters and different rotations. For all of the many N and λ values, for which the numerical investigation was performed, the optimal POVM was always of the kind described above.

A second search, not restricted by geometrical or other constraints, confirmed these findings. It used the numerical method of Ref. [15], which is a fix-point iteration that converges monotonically toward the optimal POVM.

TABLE I. Threshold values for the disturbance below which the Csiszár-Körner theorem ensures that a secure key can be extracted from the noisy raw data. The second column gives the critical disturbance, that is, $(N-2)^2/[(N-2)^2+N]$, above which the SRM is optimal, as implied by Eq. (17). The third column repeats the values of Refs. [5,8], where Eve extracts information with the aid of the SRM. The true threshold values of the fourth column are obtained for the optimal POVM.

	Critical	Csiszár-Körner thresholds	
N	value	SRM	true
2	0.0%	15.6373%	15.6373%
3	25.0%	22.6714%	22.6707%
4	50.0%	26.6561%	26.5989%
5	64.3%	29.2303%	29.1038%
10	86.5%	34.9713%	34.7051%
30	96.3%	39.8403%	39.6259%
50	97.9%	41.1886%	41.0284%
100	99.0%	42.5282%	42.4295%
∞	100.0%	50.0000%	50.0000%

We note further that the large relative difference shown in Fig. 4 occurs where both I_{max} and $I^{(SRM)}$ are small, and so the absolute difference is rather small (see the figure in Ref. [16]). Therefore, the SRM threshold values given in Table I of Ref. [8] are quite good approximations for the true threshold values, as shown by the numerical values in Table I.

The "disturbance" values listed in this table are the quantities denoted by D_{d+1}^{ind} in Ref. [5] and by $1-\beta_0$ in Ref. [8], respectively. There is no difference for $N=2$, of course, but for all $N>2$ the true threshold is noticeably lower than the SRM threshold. In addition to this shift of the threshold, there is also a reduced efficiency inside the Csiszár-Körner regime (below the threshold) and this must be taken into account when extracting the secure key sequence from the noisy raw data. Fortunately, however, almost all of the practical quantum cryptography scheme presently implemented use qubits ($N=2$), and then the SRM *is* optimal. Also, the optimal POVMs have no bearing on the threshold for classical advantage distillation [6,10], because the SRM remains optimal in the relevant limit, even for coherent eavesdropping attacks [17].

In the spirit of Shor's investigation of obtuse pyramids, the eavesdropping procedure presented here can be viewed as a quantum communication channel, in which Alice transmits nonorthogonal and equally distributed signal states to Eve. The amount of information about the sequence of states sent by Alice, maximized over all possible POVMs, is then the *accessible information* of this quantum channel. Therefore, the maximal mutual information (17) between Alice and Eve gives us also this accessible information for $0 \leq \lambda$, which supplements, for $N=3$, Shor's $\lambda < 0$ result.

ACKNOWLEDGMENTS

We wish to thank Antonio Acín, Thomas Durt, and Jaroslav Řeháček for valuable discussions. We gratefully acknowledge the financial support from A*Star Grant No. 012-104-0040 and from NUS Grant No. WBS: R-144-000-089-112.

APPENDIX: INTERCEPT ATTACKS

Here are a few remarks about the connection with intercept attacks on qunits sent through an authenticated quantum channel. We make use of the notational conventions of Ref. [8] without explaining them anew, and refer to Eq. (12), say, of Ref. [8] by ([8]-12).

The geometry of the unnormalized ancilla states $|\widetilde{E}_{kl}^{(m)}\rangle$ is completely determined, for a given m value, by the inner products of Eq. ([8]-6), and Eq. ([8]-7) states the transformation law between ancilla states to different m values. It follows from this equation that the k index of $|\widetilde{E}_{kl}^{(m)}\rangle$ is analogous to that in $|\overline{m_k}\rangle$, and the l index to that in $|m_l\rangle$. Therefore, it is expedient to regard the $|\widetilde{E}_{kl}^{(m)}\rangle$'s as the kets of two-qunit states that are superpositions of basis kets of the $|\overline{m_k}m_l\rangle$ kind. They then acquire the strikingly simple explicit form

$$|\widetilde{E}_{kl}^{(m)}\rangle = |\overline{\psi}\rangle\delta_{kl}\frac{a}{\sqrt{N}} + |\overline{m_k}m_l\rangle\frac{b}{N}, \tag{A1}$$

where

$$\overline{\psi} = \frac{1}{\sqrt{N}}\sum_k |\overline{m_k}m_k\rangle \quad (\text{any} \quad m \quad \text{value}) \tag{A2}$$

is the maximally entangled state that is conjugate to $|\psi\rangle$ of Eq. ([8]-2). This ansatz for $|\widetilde{E}_{kl}^{(m)}\rangle$ is consistent with Eq. ([8]-6) if the complex amplitudes a,b obey

$$\left|a+\frac{1}{N}b\right|^2 = \beta_0 - \frac{N-1}{N}\beta_1, \quad |b|^2 = N\beta_1, \tag{A3}$$

but no other restrictions apply, so that $a=\sqrt{\beta_0-\beta_1}$, $b=i\sqrt{N\beta_1}$ is a permissible choice.

The entangled pure state $|\Psi\rangle$ of Eq. ([8]-5) that is prepared by Eve is then of the compact form

$$|\Psi\rangle = |\psi_{12}\overline{\psi_{34}}\rangle a + |\psi_{13}\overline{\psi_{24}}\rangle b, \tag{A4}$$

where qunit 1 is sent to Alice, qunit 2 is sent to Bob, and qunits 3 and 4 make up Eve's ancilla. We note that this is the generic form of $|\Psi\rangle$ because *all* alternatives are obtained from this $|\Psi\rangle$ by unitary transformations on the ancilla.

Now, the "asymmetric universal quantum cloning machines" [18], generalizations of the symmetric ones introduced by Bužek and Hillery [19], that are employed in Refs. [4,5] for the analysis of intercept attacks on the qunit in transmission from Alice to Bob, are characterized by a four-qunit state of the form (A4). The resulting states of the clone-anticlone pair are thus fully analogous to the ancilla states $|\widetilde{E}_{kl}^{(m)}\rangle$ in Eq. (A1). Of those, the ones with $k\neq l$ are orthogonal among themselves and orthogonal to those with $k=l$, and the latter form the pyramid of ancilla states described in Sec. II. Accordingly, Eve can extract more information if she applies the optimal POVM of Sec. III B to the clone-anticlone pair, rather than submitting them to the usual SRM.

---

[1] I. Csiszár and J. Körner, IEEE Trans. Inf. Theory **24**, 339 (1978).

[2] C. H. Bennett and G. Brassard, in *IEEE Conference on Computers, Systems, and Signal Processing, Bangalore, India* (IEEE, New York, 1984), p. 175.

[3] A. K. Ekert, Phys. Rev. Lett. **67**, 661 (1991).

[4] M. Bourennane, A. Karlsson, G. Björk, N. Gisin, and N. J. Cerf, J. Phys. A **35**, 10 065 (2002).

[5] N. J. Cerf, M. Bourennane, A. Karlsson, and N. Gisin, Phys. Rev. Lett. **88**, 127902 (2002).

[6] A. Acín, N. Gisin, and V. Scarani, Quantum Inf. Comput. **3**, 563 (2003).

[7] See, e.g., the review article by A. Chefles, Contemp. Phys. **41**, 401 (2000), and the pertinent references therein.

[8] Y. C. Liang, D. Kaszlikowski, B.-G. Englert, L. C. Kwek, and C. H. Oh, Phys. Rev. A **68**, 022324 (2003).

[9] C. H. Bennett, G. Brassard, and N. D. Mermin, Phys. Rev. Lett. **68**, 557 (1992).

[10] D. Bruß, M. Christandl, A. Ekert, B.-G. Englert, D. Kaszlikowski, and C. Macchiavello, Phys. Rev. Lett. **91**, 097901 (2003).

[11] Put $N\rightarrow n$, $|E_k\rangle\rightarrow|E_{kk}\rangle$, and $\lambda\rightarrow 1-\beta_1/\beta_0$ to convert to the notational conventions of Ref. [8] . Other symbols, in particular r_0, r_1 and η_0, η_1, have the same significance.

[12] Geometrically speaking, the base of the N-dimensional pyramid is itself an $(N-1)$-dimensional pyramid with edges of length $\sqrt{2Nr_1}$ and an angle of 60° between them.

[13] P. Hausladen and W. K. Wootters, J. Mod. Opt. **41**, 2385 (1994).

[14] P. W. Shor, IBM J. Res. Dev. **48**, 115 (2004).

[15] J. Řeháček, B.-G. Englert, and D. Kaszlikowski, e-print quant-ph/0408134.

[16] D. Kaszlikowski, A. Gopinathan, Y. C. Liang, L. C. Kwek, and B.-G. Englert, e-print quant-ph/0307086.

[17] D. Kaszlikowski, J. Y. Lim, L. C. Kwek, and B.-G. Englert, e-print quant-ph/0312172.

[18] N. J. Cerf, Acta Phys. Slov. **48**, 115 (1998); J. Mod. Opt. **47**, 187 (2000); Phys. Rev. Lett. **84**, 4497 (2000).

[19] V. Bužek and M. Hillery, Phys. Rev. Lett. **81**, 5003 (1998).

Paul Kwiat

Department of Physics, University of Illinois at Urbana-Champaign, Urbana, Illinois, USA

Paper Title: **Quantum erasing the nature of reality: or, perhaps, the reality of nature?**

Reference: Chapter 15, pp. 306–328 of the book *"Science and ultimate reality: quantum theory, cosmology, and complexity"*, edited by John D. Barrow, Paul C. W. Davies, and Charles L. Harper, Jr., Cambridge University Press (2004).

Personal message:

According to my email records, I've known Berge Englert for nearly two decades, but it hardly seemed so few. I recall many enlightening, and some 'endarkening' discussions with him, some of which led to the manuscript presented here, a summary of our understanding of the relationship between two not-so-distant cousins – quantum erasure and quantum entanglement. The paper explains in a very clear way (I can say this, since Berge wrote most of it!) how these are simply two different descriptions of the same phenomenon. Once this correspondence is elucidated, the paper goes on to discuss various methods to prepare mixed states, despite the fact that there are those who deny their very existence! Berge's superb scientific skills are evident throughout the article, as is his fabulous sense of humor, which produced the title, as well as numerous footnotes including references to cats, dogs, Humpy Dumpty, and the Ghost of Bohr. Of the articles I've been lucky to write in my life, this is simultaneously one of my favorite (and not just because Berge wrote most of it), and one of the least read (ditto).

15

Quantum erasing the nature of reality: or, perhaps, the reality of nature?

Paul G. Kwiat
University of Illinois–Urbana

Berthold-Georg Englert
National University of Singapore

Introduction

WHY must I treat the measuring device classically? What will happen to me if I don't?!
Eugene Wigner

The quantum measurement problem – how does an apparently "classical" definite world arise out of the random world of quantum superpositions – was and continues to be one of the fundamental philosophical issues in quantum mechanics. What we mean by a classical world, for example, is one in which macroscopic objects are not in superposition states of being simultaneously in several locations at once, and cats are never in coherent superpositions of being alive and dead. This lack of coherence is actually a *loss* of the coherence that exists at the level of the isolated quanta, but somehow does not survive the transition to the classical level of measuring apparatus. Such incoherent states are known as mixed states. Therefore, to study the quantum–classical interface, or even to investigate whether such an interface exists at all other than in the minds of "classical sympathizers,"[1] one should look carefully at mixed states, how they arise and how they behave. Here we describe a set of experiments, both real and *gedanken*, investigating the subtleties of quantum interference when mixed states are involved. We start by describing the well-known double-slit experiment, and the loss of interference when which-path information can be had. Next we discuss the principle of the "quantum eraser," by which one can make measurements on the which-path detector in such a way as to remove the path-information; by selecting only a particular set of results one can recover the interference fringes. We will see that this is true even if there was no which-path

[1] (At least) one of us secretly belongs to that camp!

Science and Ultimate Reality, eds. J. D. Barrow, P. C. W. Davies and C. L. Harper, Jr. Published by Cambridge University Press.

information available at the outset because the which-path quantum system was in a mixed state. Because the notion of mixture is strongly correlated to the idea of entanglement with some unobserved system, we describe (pp. 315–20) an analogous experiment performed using entangled photon pairs. We can then show clearly the relationship between the two central quantum mechanical "mysteries" (to quote Feynman): interference and entanglement. We then describe a variety of methods to produced mixed states. Although formally these have the same outcome, physically the underlying interpretations are quite different, and may stimulate some insight into the nature of reality, and the reality of nature. Final thoughts and remarks are gathered in the summary section.

Quantum eraser for pure and mixed states

Interference is at the heart of quantum mechanics. In reality it contains the *only* mystery. We cannot make the mystery go away by 'explaining' how it works.

(*Richard Feynman*)

The complementary nature of wavelike and particlelike behavior is frequently interpreted as follows: as a consequence of the uncertainty principle, any attempt to measure the position (particle aspect) of a quantum leads to an uncontrollable, irreversible disturbance in its momentum, thereby washing out any interference pattern (wave aspect) (Bohr 1983; Feynman *et al.* 1965). This picture is incomplete though; no "state reduction," or "collapse," is necessary to destroy interference, and measurements which do not involve reduction can be reversible. One must view the loss of coherence as arising from an *entanglement* of the system with the which-way marker (WWM), which may be another degree of freedom or another quantum system entirely. Previously interfering alternatives can thereby become distinguishable, such that no interference is observed. Interference may be regained, however, if one manages to "erase" the distinguishing information. This is the physical content of *quantum erasure* (QE) (Scully and Drühl 1982; Hillery and Scully 1983; Scully *et al.* 1991) (though as we shall see below, this simple physical picture cannot explain the results when nonpure WWM states are considered).

The double slit

As the simplest possible example, consider the archetypal double-slit experiment as indicated in Fig. 15.1. A photon is directed from a (relatively) far away source to a wall with two slits, behind which sits (again, at a relatively large spacing) an observation screen, consisting of some sort of classical mechanism for indicating the arrival point of the single photon. As is very well known, the likelihood for the

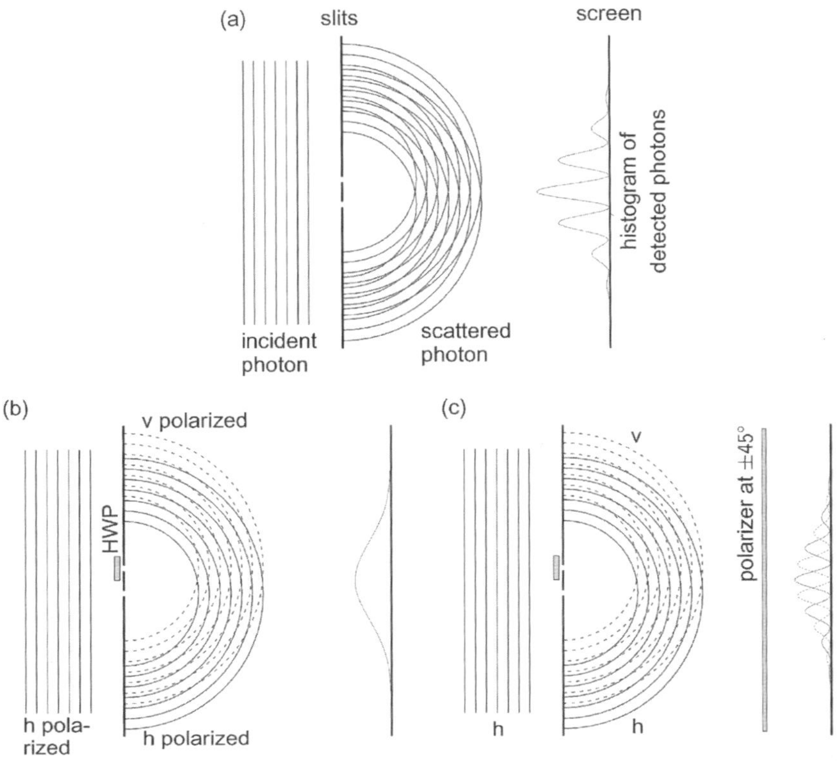

Figure 15.1. The "mysterious" double-slit experiment. (a) Photons are directed at a pair of slits. They arrive with well-defined wavelength and are scattered at the two slits. At the instant under consideration, one photon is approaching the slits, and its immediate predecessor has just been scattered but has not reached the screen as yet. The histogram built up by photons that have been detected on that far screen already shows that they are observed to be more likely to arrive in some places, and completely unlikely to arrive at others. This interference results from the indistinguishability – the in-principle unknowability – of which slit a photon traversed. (b) A half-wave plate (HWP) in front of the top slit rotates the initial horizontal polarization (h) to vertical (v). Accordingly, the scattered spherical wave centered at the top slit is v polarized (dashed half-circles), whereas the one from the bottom slit carries the original h polarization (solid half-circles). In this situation, no fringes are observable on the screen – the which-path information stored in the photon polarization distinguishes the top-slit/bottom-slit processes. (c) The distinguishing which-path information may be erased by analyzing the polarization in a particular way. For example, if we place a polarizer at 45° in front of the screen, we will recover the original fringe pattern of (a), only reduced in intensity by half (solid curve). If instead we use a −45° polarizer, we will again see fringes, but now shifted in phase by 180° (dashed curve). The sum of the two histograms in (c) equals the fringeless pattern of (b). Likewise, we could set the polarizer such that only v (or only h) polarized photons reach the screen, and they would all have passed through the top slit (or all through the bottom slit, respectively).

photon to arrive at any given point on the screen is determined by the coherent sum of the probability amplitudes for the photon to have traveled through the top slit or the bottom slit. Although it is not necessary for the experiment just described, we will now assume that the incident photon is horizontally polarized, i.e., in the polarization state $|\mathbf{h}\rangle$. Next we use a (very small) half-wave plate (HWP) to rotate the polarization of the photon passing through the upper slit by 90° to $|\mathbf{v}\rangle$, the state of vertical polarization. Again, it is well established that no interference will be observed at the screen (Fig. 15.1b). Of course, according to classical physics, the loss of interference occurs because the electric fields $\vec{E}_{\text{top}} = \vec{E}_{\mathsf{v}}$ and $\vec{E}_{\text{bottom}} = \vec{E}_{\mathsf{h}}$ corresponding to the electromagnetic wave passing through the top and bottom slits, respectively, are orthogonal; therefore, the interference cross term, which is proportional to $\vec{E}_{\text{top}} \cdot \vec{E}_{\text{bottom}}$, is zero.

A more modern and more useful perspective is that the process of passing through the double slits with the wave plate covering one of them results in an entanglement of the path degree of freedom with the polarization degree of freedom of the photon. The polarization acts as the WWM here. The interference is determined by the reduced density matrix ρ_{red} that one obtains by tracing over the quantum state of the WWM, since it is not being measured. The off-diagonal elements of ρ_{red} are proportional to the overlap of the two WWM quantum states associated with the top and bottom slits. If these WWM states are orthogonal, as in our example above, then the off-diagonal elements in ρ_{red} will vanish, and no interference will be observed. Note that it is absolutely not necessary to actually measure the polarization; the mere fact that such which-way information in principle *could* be obtained is enough to destroy the interference.

The interference may be recovered, however, if we can find some way to "erase" the distinguishing information being carried in the WWM quantum state. In this example, we can perform this erasing operation by simply passing the photons through a final analyzer at ±45° before the screen (Fig. 15.1c). This effectively removes the distinguishability of the two paths – any photon which is transmitted through an analyzer at 45° is equally likely to have originally been v-polarized (upper slit) or h-polarized (lower slit). This is an example of QE. Note that the presence of the polarizer at 45° merely serves to partition the final data into two subsets: those photons which were transmitted through the polarizer, and those which were absorbed by it. If, instead, we have the polarizer at −45°, 100%-visibility fringes will again be observed, but now they will be shifted 180° with respect to the fringes obtained with the 45° polarizer. Note that we could have instead decided to partition the ensemble of data in such a way that our polarization measurements yielded directly the information of which path the photon chose.

The primary lesson is that one must consider the total physical state, including any WWM with which the interfering quantum has become entangled. If the coherence of the WWM is maintained, then interference may be recovered by effectively post-selecting particular subensembles. In general there are two distinct possibilities – either one can measure the WWM in such a way as to recover the which-way (WW) information, a particlelike characteristic; or one can measure the WWM in such a way as to recover interference, a wavelike characteristic. The familiar phrase "each experiment must be described either in terms of particles or in terms of waves" emphasizes the extreme cases and disregards the intermediate situations, in which particle aspects and wave aspects are present simultaneously.[2] Remarkably, if the WWM is a separate system from the interfering system, one can even make the choice of how to measure the WWM *after* the other quantum has been detected, as we discuss below. This is an extension of the original delayed-choice experiment popularized by Wheeler (1979).[3] In all cases though, it is necessary to correlate the results of the measurements. Only after this postselection do the fringes reappear.[4]

Set-up and procedure

In contrast to many interference situations where the WW information may be inaccessible, for our measurements the WW labels are easily manipulated: we used a photon in a simple Mach–Zehnder interferometer as the interfering system, and its polarization as the WWM (see Fig. 15.2) (Kwiat *et al.* 1999a; Schwindt *et al.* 1999).[5] By changing the polarization in one path of the interferometer, we entangle the spatial mode with the polarization and so partially or completely "label" the path followed by the photon. This is enough to make the paths partially or completely distinguishable, resulting in a reduced visibility *even if the polarization is not measured.*

Photons at 670 nm from the output of a single-mode fiber were directed into our Mach–Zehnder interferometer. The entire interferometer was compressed from the usual rectangular configuration (i.e., the angle of incidence on the beam-splitter was set to 10°) in order to minimize polarization variations in the reflection and

[2] Nature allows compromises between the particle aspects and the wave aspects of a quantum object. In the context of two-path interferometers, there are various inequalities that quantify these compromises. These matters are reviewed in Englert and Bergou (2000) where the interested reader will also find extended remarks on the history of the subject, in which the theoretical developments of Wootters and Zurek (1979), Glauber (1986), Greenberger and Yasin (1988), Mandel (1991), Jaeger *et al.* (1995), and Englert (1996) and the experiments reported in Rauch and Summhammer (1984), Summhammer *et al.* (1987), Mittelstaedt *et al.* (1987), Dürr *et al.* (1998a, b, c), and Schwindt *et al.* (1999) play a central role.

[3] The first discussion of delayed choice was given in von Weizsäcker (1941). The first experiments on Wheeler's conception appeared in Hellmuth *et al.* (1987), Alley *et al.* (1987), and Baldzuhn *et al.* (1989).

[4] It is precisely the requirement of correlating the results – necessarily requiring the transmission of classical information at or below the speed of light – which prevents this procedure from enabling superluminal communication.

[5] A similar experiment was carried out much earlier by Baldzuhn and Martienssen (1991) – they used a fast electrooptic switch to decide between measuring fringes or which-path information. However, they focused exclusively on pure input states.

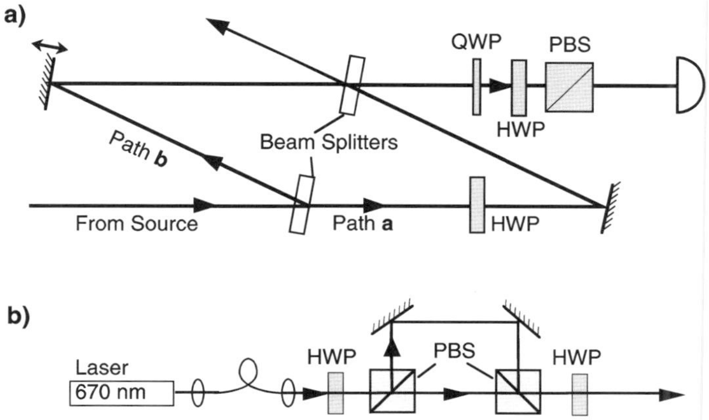

Figure 15.2. Experimental setup to investigate the phenomenon of quantum erasure (QE). (a) The half-wave plate (HWP) in path **a** of the interferometer generates varying amounts of which-way information, depending on the input polarization state. The quarter-wave plate (QWP), HWP, and Rochon prism (polarizing beamsplitter PBS) in the output port allow analysis in an arbitrary polarization basis, for determining which-way information or to enable QE of it. (b) One method to prepare mixed and partially mixed quantum states. By varying the input polarization to the polarizing Mach–Zehnder interferometer (which is unbalanced by more than the coherence length of the diode laser), one can make an arbitrarily mixed polarization state.

transmission amplitudes. The interferometer arms were adjusted to have equal lengths after the desired polarization-transforming elements were inserted into one or both arms. For the results presented below, an adjustable HWP in path a of the interferometer was used to vary the amount of polarization "labeling" of the photons.

Our analysis system consisted of a Rochon calcite prism preceded by adjustable half- and quarter-wave plates, allowing the polarization of the photons to be measured in any arbitrary polarization basis. The photons were detected using a Geiger-mode avalanche photodiode – a Single Photon Counting Module (EG&G #SPCM-AQ), with detection efficiency ~40%. The input source, described below, was greatly attenuated so that the maximum detection rates were always less than 50 000 s^{-1}; for the interferometer passage time of 1 ns, this means that on average there were fewer than 10^{-4} photons in the interferometer at any time. This one-photon-at-a-time operation is essential to allow sensible discussion of the likely path taken by an individual light quantum.[6] For visibility measurements the

6 It is, therefore, not true that all features of an experiment of this kind can be understood in terms of classical physics, which is a misplaced objection that has been raised occasionally, most recently by Trifonov *et al.* (2002). They, and others, perhaps have in mind a "semi-classical" description of the phenomenon, consisting of a classical treatment of the electromagnetic field and a quantum treatment of matter (to account for the "clicking" of detectors that indicates the arrival of a single light quantum). It is indeed well known that single-photon interference patterns are identical with the classical intensity patterns (Fermi 1929). But it is equally well known

maximum and minimum count rates at the detector were measured as the length of path b was adjusted using a piezoelectric transducer on the mirror. After subtracting out the separately measured detector background (i.e., the count rate when the input to the interferometer was blocked, typically 100–400 s^{-1}), the fringe visibility is calculated in the standard manner:

$$V = (\text{Max} - \text{Min})/(\text{Max} + \text{Min}).$$

In the limiting case that the paths are completely distinguishable, no interference is observed and the visibility vanishes, $V = 0$. This occurs, for example, when the HWP in path a of the interferometer is used to rotate the polarization in that path by 90°. In fact, with the HWP in place the visibility will be zero *even when the input is in a mixed state of polarization*. This is less intuitive, since there is no WW information to distinguish the paths, but can be understood by examining the behavior of orthogonal *pure* states, with no definite phase relationship between them. In the basis where the HWP rotates the states by 90°, the orthogonal polarizations from paths a and b cannot interfere; in the basis aligned with the HWP's axes, each polarization individually interferes, but the interference patterns are shifted relatively by 180° (due to the birefringence of the HWP), so the sum is a fringeless constant.

Now that we have made interference disappear, we may consider methods to recover it. If we have initially a pure state of the polarization (or, more generally, of the quantum system comprising the WWM), the lack of interference may be directly attributed to the WW information stored in the polarization. By making a suitable analysis of the WWM we can read out this information, e.g., by detecting photons in a state that is orthogonal to the polarization associated with path a, we can know with certainty that they must have traveled via path b, and vice versa. Or we can analyze the polarization so that a detected photon is equally likely to have come from either path in the interferometer. There is actually a whole class of measurements that achieve this: if the polarizations from two paths are θ_{a} and θ_{b}, then analyzing in any basis of the form $(|\theta_{\mathsf{a}}\rangle + e^{i\phi}|\theta_{\mathsf{b}}\rangle)/\sqrt{2}$ will suffice to recover complete interference, i.e., $V = 1$. For example, if the polarizations from the paths a and b are horizontal (h) and vertical (v), respectively, analysis at +45° will recover complete fringes, while analysis at −45° will recover complete *anti*-fringes (i.e., $V = 1$, but shifted by 180°).

that such a semi-classical description has its limitations and is not generally valid, as has been demonstrated, e.g., in Clauser (1974) and Grangier *et al.* (1986). Light really consists of photons, whether or not a given experiment directly proves that fact. Put most simply, the notion of taking this path or that path through an interferometer is utterly meaningless in Maxwellian electromagnetism where there is always intensity in both paths.

In this context it should be mentioned that this erroneous reasoning would also imply that quantum cryptography with faint pulses is somehow completely classical. There are, however, cryptography protocols with attenuated coherent states, rather than true single-photon states, that are *provably* secure (Gisin *et al.* 2002). Somewhat paradoxically, it is the classical "click" of a photon counter that collapses the quantum state of a weak pulse (described by a low-amplitude coherent state) into a very close approximation of a true single-photon state, albeit "posthumously."

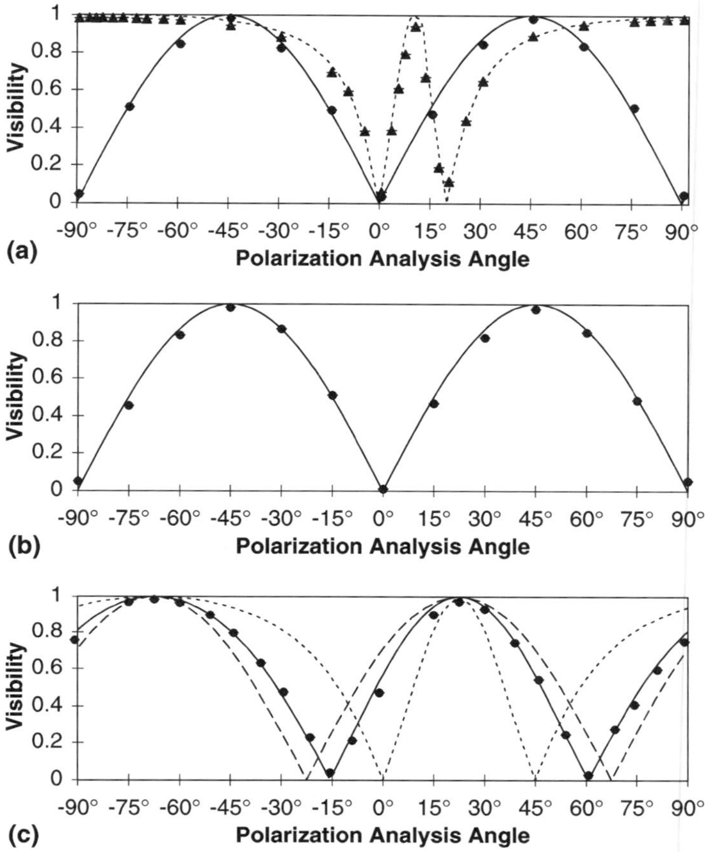

Figure 15.3. QE visibility data and theory curves for various input states, with a HWP in path a of the interferometer: (a) a purely vertically polarized input ($\equiv 90°$), with the polarization rotated by the HWP in path a by 90° (circles, solid line; $\theta_{\mathrm{HWP}} = 45°$) or 20° (triangles, dashed line; $\theta_{\mathrm{HWP}} = 10°$); (b) a completely-mixed state, with $\theta_{\mathrm{HWP}} = 45°$; and (c) a partially mixed state (1:2 pure to mixed; circles, solid line), with the HWP at $\theta_{\mathrm{HWP}} = 22.5°$. For comparison, the dotted and dashed curves show the corresponding theoretical predictions for pure and completely mixed states, respectively.

Pure and mixed states of the which-way marker

Figure 15.3a shows the results when a pure vertical-polarization state was input to the interferometer, and rotated by the HWP in path a by either 90° or 20°, so that the intrinsic visibilities without QE were 0 and 0.94, respectively. We see that the visibility *after* the analyzer can assume any value between 0 and 1. In the latter case, we have a complete QE of WW information. The minima on the curves correspond to analysis that transmits light from only one or the other path; the maxima fall midway between these minima. The maximum measured values of V are slightly lower than 100% because the intrinsic visibility of the interferometer (even without the HWP in path a) is only ~98%, due to imperfect optics.

Next we investigated QE for non-pure WWM states. If the WWM is initially in a mixed or partially mixed state, formally the key to recovering full fringe visibility is to measure in the basis for which the Hamiltonian coupling the interferometer path to the WWM is diagonal. With polarization as the WWM, this means analyzing along one of the two eigenmodes of the optical element responsible for rotating the polarization in one of the paths.[7] The effect is to select out one of the two perfect-visibility subensembles.

To generate photons in arbitrary mixed and partially mixed states of polarization, we used a "tunable source" (see Fig. 15.2b). It consisted of a 670 nm diode laser – which was spectrally filtered with a narrow-band interference filter (1.5 nm FWHM) and spatially filtered via a single-mode optical fiber – whose horizontal and vertical components were separated by much more than the laser's ~1-cm coherence length, using an asymmetric Mach–Zehnder interferometer with polarizing beam-splitters (we will discuss below (pp. 321–25) other possible preparation methods). By rotating the (pure linear) polarization input to the first polarizing beam-splitter, one can control the relative contribution of h and v components. For example, for incident photons at 45°, one has equal amplitudes of h and v which are then added together with a random and rapidly varying phase to produce an effectively completely mixed state of polarization. If one has two times more vertical than horizontal, the state is then 1/3 pure to 2/3 completely mixed.[8]

In Fig. 15.3b we show that even for a completely mixed state, it is still possible to recover interference. Since there is no WW information to erase, this *nonerasing QE* may seem quite remarkable at first sight. However, as discussed above, the essential feature of QE is not that it destroys the possibly available WW information,[9] but that it enables us to sort the photons into subensembles, each exhibiting high-visibility fringes. Complete interference is recoverable by analyzing along the eigenmodes of the internal HWP – along one axis we see fringes, and along the other we see anti-fringes, shifted by 180°. The amount of this shift is precisely that imparted to light polarized along the fast axis of the HWP relative to light polarized along the slow axis. Finally, Fig. 15.3c shows the results for a partially mixed state. Note

[7] Throughout we are implicitly assuming a *unitary* transformation of the WWM; if this were not the case, e.g., if a polarizer were placed in one of the interferometer arms, the behavior would be much more complex; see Englert and Bergou (2000) for such complications.

[8] In terms of density matrices (referring to the h/v basis):

$$\frac{1}{3}\begin{pmatrix}1 & 0\\ 0 & 0\end{pmatrix}+\frac{2}{3}\begin{pmatrix}0 & 0\\ 0 & 1\end{pmatrix}=\frac{1}{3}\begin{pmatrix}1 & 0\\ 0 & 2\end{pmatrix}=\frac{1}{3}\begin{pmatrix}0 & 0\\ 0 & 1\end{pmatrix}+\frac{2}{3}\begin{pmatrix}\frac{1}{2} & 0\\ 0 & \frac{1}{2}\end{pmatrix}.$$

[9] In hindsight, that is merely a side effect, but historically it used to be emphasized much; see Englert and Bergou (2000) and pp. 315–20 below.

that the analysis angles yielding zero visibility for the partially mixed state fall between those for pure and mixed states. Specifically, if the fractional purity is s, the angles are at $\theta_{\mathrm{HWP}} \pm 1/2 \arccos[s \cos(2\theta_{\mathrm{HWP}})]$, where θ_{HWP} is the HWP angle. For example, in Fig. 15.3c we have $s = 1/3$ and $\theta_{\mathrm{HWP}} = 22.5°$, so that $V = 0$ is observed at $60.7°$ and $-15.7°$.

The WW labeling in the experiment of Fig. 15.2 arose from an *entanglement* between the photon's spatial mode and polarization state (a similar result was obtained in Baldzuhn and Martienssen (1991)). As discussed at length below, it could just as well have been with another photon altogether (as was the case in the experiments reported in Zajonc *et al.* (1991), Kwiat *et al.* (1992), Herzog *et al.* (1995), Monken *et al.* (1995), and Kim *et al.* (2000)), or even with a totally different kind of quantum system (Eichmann *et al.* 1993). Analogous outcomes are predicted – our results are relevant as long as the WWM can be mapped onto a two-state system. For example, instead of the photon's polarization we could use a two-level atom as the WWM. The analogy to unpolarized light would then be an atom in a quantum mechanical mixture of ground and excited state, perhaps because it is entangled to yet another system. More generally, our findings are extendable to analogous experiments with different kinds of interfering quanta, such as interferometers with electrons (Buks *et al.* 1998), neutrons (Badurek *et al.* 1988, 2000; Rauch 1995), or atoms (Dürr *et al.* 1998a, b, c).

Sorting at a distance: what does a quantum eraser erase?

[Entanglement is] the characteristic trait of quantum mechanics, the one that enforces its entire departure from classical lines of thought.

(Erwin Schrödinger)

Consider the experimental set-up sketched in Fig. 15.4.[10] Paired photons are emitted by the source, one of each pair to Alice and the other to Bob. For the moment, assume that the source emits the paired photons in the polarization-entangled state

$$|\Psi_1\rangle \equiv \frac{1}{\sqrt{2}}(|\mathrm{vh}\rangle - |\mathrm{hv}\rangle) = \frac{1}{\sqrt{2}}(|-+\rangle - |+-\rangle), \tag{15.1}$$

where $|\mathrm{vh}\rangle$, for instance, symbolizes the state in which Alice's photon is vertically polarized and Bob's horizontally. The states labeled by $\pm$,

$$|+\rangle \equiv \frac{1}{\sqrt{2}}(|\mathrm{v}\rangle + |\mathrm{h}\rangle) \quad \text{and} \quad |-\rangle \equiv \frac{1}{\sqrt{2}}(|\mathrm{v}\rangle - |\mathrm{h}\rangle), \tag{15.2}$$

are linearly polarized at $\pm 45°$.

[10] A simpler version was discussed in problem 9-6 in Ballentine (1998); this extended variant was introduced in Englert (1999).

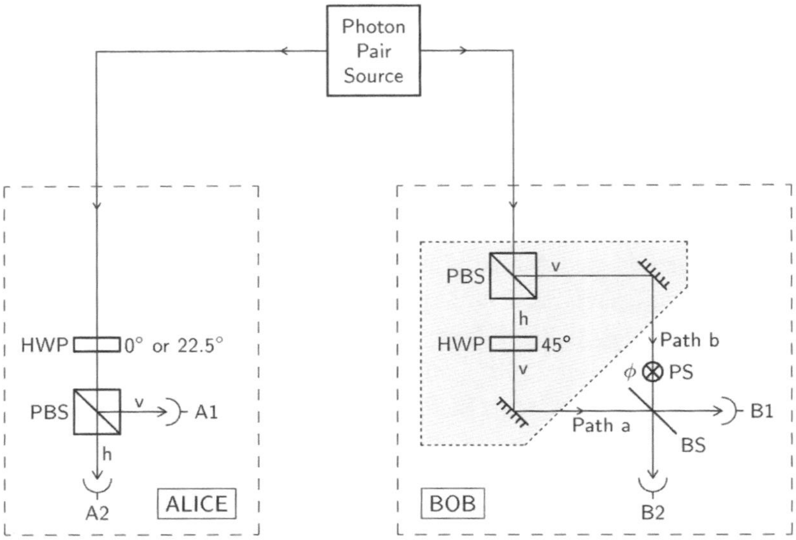

Figure 15.4. Scheme of a photon-pair experiment illustrating various aspects of wave–particle duality. A source emits paired photons whose polarization qubits are entangled. Alice sends her photon through a HWP, set either at 0° or at 22.5°, and then through a PBS to detectors A1 and A2. With the HWP set at 0°, a click of A1 or A2 indicates vertical polarization (v) or horizontal polarization (h), respectively. When the HWP is set at 22.5°, however, a click of A1 means linear polarization at +45° of the incoming photon, and at −45° for a click of A2. Bob uses a PBS in conjunction with a HWP set at 45° to convert the polarization qubit of his photon into a spatial-mode qubit. This conversion happens in the shaded area of Bob's apparatus: photons that arrive h polarized emerge v polarized in path a, and those arriving with v polarization are deflected into path b. An adjustable phase shifter (PS) and a 50–50 beam-splitter (BS) enable Bob to measure the fringe visibility of the Mach–Zehnder interference pattern. Alternatively, Bob can determine the path of the photon by removing BS altogether, so that path a only goes to B1 and path b only goes to B2.

Alice performs one of two polarization measurements on her photons, either distinguishing vertical from horizontal polarization – detecting the polarization states $|\mathsf{v}\rangle$ or $|\mathsf{h}\rangle$ – or distinguishing between $|+\rangle$ and $|-\rangle$. She observes that the outcome is utterly unpredictable: both outcomes are obtained equally frequently in the v/h mode of operation, and also in the $+/-$ mode.

Bob does not perform a polarization measurement (at least, not a direct one). Rather, he converts the polarization qubit of his photon into the spatial-mode qubit of a Mach–Zehnder interferometer.[11] Then he measures the visibility of the

[11] Within the context of the new field of quantum information, Bob's transformations also have another interpretation. Namely, the HWP in path a of the interferometer is actually implementing a controlled-not (CNOT) gate, with the spatial mode as the control qubit and the polarization as the target qubit: only if the control is in state

interferometer fringes, thereby probing for definite phase relations between the amplitudes referring to the two paths. He observes no fringes at all – each of his detectors, B1 and B2, registers approximately 50% of the photons.

The conversion in the shaded area of Bob's apparatus in Fig. 15.4 is summarized by

$$|\mathbf{h}\rangle \to |\mathbf{a}\rangle, \qquad |\mathbf{v}\rangle \to |\mathbf{b}\rangle, \tag{15.3}$$

and its effect on $|\Psi_1\rangle$ is given by

$$|\Psi_1\rangle \to \frac{1}{\sqrt{2}}(|\mathbf{va}\rangle - |\mathbf{hb}\rangle). \tag{15.4}$$

The statistical operator of the reduced state that applies to Alice's polarization qubit is then

$$\rho_1^{(\mathrm{A})} = \frac{1}{2}(|\mathbf{v}\rangle\langle\mathbf{v}| + |\mathbf{h}\rangle\langle\mathbf{h}|) = \frac{1}{2}(|+\rangle\langle+| + |-\rangle\langle-|). \tag{15.5}$$

Bob in turn accounts for the spatial properties of his photon by the statistical operator

$$\rho_1^{(\mathrm{B})} = \frac{1}{2}(|\mathbf{a}\rangle\langle\mathbf{a}| + |\mathbf{b}\rangle\langle\mathbf{b}|). \tag{15.6}$$

Expressions (15.5) and (15.6) are obtained by tracing $|\Psi_1\rangle\langle\Psi_1|$ over Bob's spatial-mode qubit or Alice's polarization qubit, respectively.

In more sophisticated correlation experiments, Alice and Bob can exploit the entanglement between the two qubits. For example, by performing the v/h distinction, Alice can easily determine the trajectory followed by each of Bob's photons: as indicated in (15.4), path a is strictly correlated with v polarization, and path b with h polarization. This is summarized by the conditional density matrices

$$\rho_{1\mathsf{v}}^{(\mathrm{B})} = |\mathsf{a}\rangle\langle\mathsf{a}| \quad \text{and} \quad \rho_{1\mathsf{h}}^{(\mathrm{B})} = |\mathsf{b}\rangle\langle\mathsf{b}| \tag{15.7}$$

which apply to Bob's subensembles that are identified, and labeled, by the polarization detected by Alice.[12] This sorting of Bob's photons is fittingly called *WW sorting*.

a is the target qubit flipped. A polarizing beam-splitter realizes the other CNOT gate, in which the polarization qubit controls the spatial mode qubit. It is now well known that in fact all quantum algorithms can be synthesized using only linear optics, e.g., waveplates, beam-splitters, etc. (Cerf *et al.* 1998); however, there is a price to pay – for more than ~5 qubits, the number of optical elements needed to implement a particular algorithm grows exponentially with the number of qubits required (Kwiat *et al.* 2000). For instance, a 10-qubit algorithm would require interferometers of order 2^{10} spatial paths, or cavities with 2^{10} resolvable modes (Bhattacharya *et al.* 2002). Remarkably, it has recently been shown theoretically how to achieve *scalable* quantum computation using the same linear optics (and true single-photon sources), combined with the implicit effective nonlinearity of the photon detection process itself (Knill *et al.* 2001).

12 The superscript identifies this as the state of Bob's photon; the first subscript refers to the state $|\Psi_1\rangle$ produced by the source; and the second subscript identifies the results Alice obtained from her measurement.

If Alice instead performs the $+/-$ distinction, the density matrices for the corresponding subensembles of Bob's photons are then given by

$$\rho_{1+}^{(\mathrm{B})} = \frac{1}{2}(|\mathsf{a}\rangle\langle\mathsf{a}| + |\mathsf{b}\rangle\langle\mathsf{b}| - |\mathsf{a}\rangle\langle\mathsf{b}| - |\mathsf{b}\rangle\langle\mathsf{a}|) = \frac{|\mathsf{a}\rangle - |\mathsf{b}\rangle}{\sqrt{2}}\frac{\langle\mathsf{a}| - \langle\mathsf{b}|}{\sqrt{2}} \tag{15.8}$$

and

$$\rho_{1-}^{(\mathrm{B})} = \frac{1}{2}(|\mathsf{a}\rangle\langle\mathsf{a}| + |\mathsf{b}\rangle\langle\mathsf{b}| + |\mathsf{a}\rangle\langle\mathsf{b}| + |\mathsf{b}\rangle\langle\mathsf{a}|) = \frac{|\mathsf{a}\rangle + |\mathsf{b}\rangle}{\sqrt{2}}\frac{\langle\mathsf{a}| + \langle\mathsf{b}|}{\sqrt{2}}. \tag{15.9}$$

The probability that the photon is detected by B1, say, is then

$$\left|\frac{(\langle\mathsf{a}| + \mathrm{e}^{\mathrm{i}\phi}\langle\mathsf{b}|)}{\sqrt{2}}\frac{(|\mathsf{a}\rangle - |\mathsf{b}\rangle)}{\sqrt{2}}\right|^2 = \frac{1}{2}(1 - \cos\phi) \tag{15.10}$$

if it is in the 1+ subensemble (i.e., Alice measured the entangled partner to be polarized at $+45°$), and

$$\left|\frac{(\langle\mathsf{a}| + \mathrm{e}^{\mathrm{i}\phi}\langle\mathsf{b}|)}{\sqrt{2}}\frac{(|\mathsf{a}\rangle + |\mathsf{b}\rangle)}{\sqrt{2}}\right|^2 = \frac{1}{2}(1 + \cos\phi) \tag{15.11}$$

if it is in the 1− subensemble (Alice's photon polarized at $-45°$). Accordingly, Bob finds unit visibility for the fringes of these individual subensembles, and we note that this is the "fringes and anti-fringes situation" that we encountered already (p. 312), and earlier in Fig. 15.1c. For historical reasons, one speaks of the *QE sorting* here. This terminology derives from the fact that, as soon as she has performed the $+/-$ polarization distinction, Alice can no longer determine the path of Bob's photon by the v/h distinguishing measurement. She has "erased" the WW information that was latently available before.

But, one could equivalently argue that the WW sorting erases the latently available phase information that could be revealed by QE sorting (whether the relative phase between $|\mathsf{a}\rangle$ and $|\mathsf{b}\rangle$ is 0 or π as in (9) and (8), respectively). The bias in favor of WW information, i.e., in favor of the photon's particle aspect, is very natural – given the local nature of all physical interactions, the state reduction associated with generic measurements tends to leave the observed system in a well-localized state and thus with attributes that are analogous to those of a classical particle.[13] On a more abstract level, however, one cannot fail to note that in experiments of the kind we are discussing here it is hardly justifiable to regard the photon's particle nature as more fundamental than its wave nature, or vice versa.

We now address the question: what does a quantum eraser erase? Or, in more technical terms, is the availability of WW information a precondition for QE? The

[13] This natural bias in favor of particles rather than waves is ubiquitous. For instance, it makes us prefer the term "elementary particle physics" over "elementary wave physics."

answer is no, which can be seen as follows. First, consider that the source instead of $|\Psi_1\rangle$ emits photon pairs in the entangled state

$$|\Psi_2\rangle \equiv \frac{1}{\sqrt{2}}(|\mathbf{vv}\rangle - |\mathbf{hh}\rangle) = \frac{1}{\sqrt{2}}(|-+\rangle + |+-\rangle). \tag{15.12}$$

After Bob's conversion,

$$|\Psi_2\rangle \rightarrow \frac{1}{\sqrt{2}}(|\mathbf{vb}\rangle - |\mathbf{ha}\rangle), \tag{15.13}$$

so that the which-path sorting results in the subensembles specified by

$$\rho_{2\mathrm{v}}^{(\mathrm{B})} = |\mathrm{b}\rangle\langle\mathrm{b}| \quad \text{and} \quad \rho_{2\mathrm{h}}^{(\mathrm{B})} = |\mathrm{a}\rangle\langle\mathrm{a}| \tag{15.14}$$

and QE produces

$$\rho_{2+}^{(\mathrm{B})} = \frac{1}{2}(|\mathrm{a}\rangle\langle\mathrm{a}| + |\mathrm{b}\rangle\langle\mathrm{b}| - |\mathrm{a}\rangle\langle\mathrm{b}| - |\mathrm{b}\rangle\langle\mathrm{a}|) = \frac{|\mathrm{a}\rangle - |\mathrm{b}\rangle}{\sqrt{2}}\frac{\langle\mathrm{a}| - \langle\mathrm{b}|}{\sqrt{2}} \tag{15.15}$$

and

$$\rho_{2-}^{(\mathrm{B})} = \frac{1}{2}(|\mathrm{a}\rangle\langle\mathrm{a}| + |\mathrm{b}\rangle\langle\mathrm{b}| + |\mathrm{a}\rangle\langle\mathrm{b}| + |\mathrm{b}\rangle\langle\mathrm{a}|) = \frac{|\mathrm{a}\rangle + |\mathrm{b}\rangle}{\sqrt{2}}\frac{\langle\mathrm{a}| + \langle\mathrm{b}|}{\sqrt{2}}. \tag{15.16}$$

In short, we have $\rho_{2\mathrm{v}}^{(\mathrm{B})} = \rho_{1\mathrm{h}}^{(\mathrm{B})}$, $\rho_{2\mathrm{h}}^{(\mathrm{B})} = \rho_{1\mathrm{v}}^{(\mathrm{B})}$ and $\rho_{2+}^{(\mathrm{B})} = \rho_{1+}^{(\mathrm{B})}$, $\rho_{2-}^{(\mathrm{B})} = \rho_{1-}^{(\mathrm{B})}$. Just as before, the WW sorting yields full path knowledge for each photon in question, and the QE sorting gives subensembles with unit fringe visibility.

But now consider the situation in which the source emits with equal probability photon pairs either in state $|\Psi_1\rangle$ or in state $|\Psi_2\rangle$, randomly alternating between the two possibilities. Bob's subensembles are then described by

$$\rho_{3\mathrm{v}}^{(\mathrm{B})} = \rho_{3\mathrm{h}}^{(\mathrm{B})} = \frac{1}{2}(|\mathrm{a}\rangle\langle\mathrm{a}| + |\mathrm{b}\rangle\langle\mathrm{b}|) \tag{15.17}$$

and

$$\begin{aligned} \rho_{3+}^{(\mathrm{B})} &= \frac{1}{2}(|\mathrm{a}\rangle\langle\mathrm{a}| + |\mathrm{b}\rangle\langle\mathrm{b}| - |\mathrm{a}\rangle\langle\mathrm{b}| - |\mathrm{b}\rangle\langle\mathrm{a}|) = \frac{|\mathrm{a}\rangle - |\mathrm{b}\rangle}{\sqrt{2}}\frac{\langle\mathrm{a}| - \langle\mathrm{b}|}{\sqrt{2}}, \\ \rho_{3-}^{(\mathrm{B})} &= \frac{1}{2}(|\mathrm{a}\rangle\langle\mathrm{a}| + |\mathrm{b}\rangle\langle\mathrm{b}| + |\mathrm{a}\rangle\langle\mathrm{b}| + |\mathrm{b}\rangle\langle\mathrm{a}|) = \frac{|\mathrm{a}\rangle + |\mathrm{b}\rangle}{\sqrt{2}}\frac{\langle\mathrm{a}| + \langle\mathrm{b}|}{\sqrt{2}}. \end{aligned} \tag{15.18}$$

No WW information is available at all – nothing that Alice can do to her photon would enable her to guess the path of Bob's photon – and yet the QE sorting still identifies subensembles with unit fringe visibility. Clearly, no WW information is erased by Alice's $+/-$ polarization measurement here, because there is simply no such information present to begin with. This is, of course, exactly the situation of Fig. 15.3b, and the oxymoron *nonerasing QE* summarizes the matter quite appropriately.

Similarly, one can easily construct situations in which Alice cannot identify *any* subensembles that display interference fringes. For example, if the source instead emits with equal probability photons in states $|\Psi_1\rangle$ and

$$\frac{1}{\sqrt{2}}(|\mathrm{vh}\rangle + |\mathrm{hv}\rangle) = \frac{1}{\sqrt{2}}(|{+}{+}\rangle - |{-}{-}\rangle) \rightarrow \frac{1}{\sqrt{2}}(|\mathbf{va}\rangle + |\mathbf{hb}\rangle), \quad (15.19)$$

then we have, in this fourth case, $\rho_{4\mathsf{v}}^{(\mathrm{B})} = |\mathsf{a}\rangle\langle\mathsf{a}|$, $\rho_{4\mathsf{h}}^{(\mathrm{B})} = |\mathsf{b}\rangle\langle\mathsf{b}|$ so that full WW information can be acquired, and

$$\rho_{4+}^{(\mathrm{B})} = \rho_{4-}^{(\mathrm{B})} = \frac{1}{2}(|\mathsf{a}\rangle\langle\mathsf{a}| + |\mathsf{b}\rangle\langle\mathsf{b}|), \quad (15.20)$$

so that no fringes can be recovered.

As a final remark concerning the scheme of Fig. 15.4, we note that the temporal order in which Alice and Bob perform their measurements is of no relevance. For instance, one could incorporate delay lines (e.g., many loops of optical fiber guiding the photons) such that Bob completes his measurements before Alice does hers. Thus, in the best Wheelerian tradition (Wheeler 1979), Alice could delay her choice of measuring v vs. h or performing the $+/-$ distinction.[14] Likewise, Bob could delay his choice between the wave experiment (BS in place) and the particle experiment (BS removed) until *after* Alice has finished her polarization measurements. No delay of either kind has any bearing on what is said above; in particular, none of the various conditional density matrices depends on whether Alice measures first, or Bob, or both measurements are made in spacelike separated regions. No superluminal signaling is possible, despite that the correlations concommitant with the state $|\Psi_1\rangle$ (and also with $|\Psi_2\rangle$) can be shown to be nonlocal.

Although the experiment of Fig. 15.4 has not yet been performed, the various scenarios discussed have been realized in spirit in the experiment by Schwindt *et al.* (1999) (see Fig. 15.2), which differs from the set-up sketched in Fig. 15.4 mainly by the circumstance that Alice's polarization qubit and Bob's spatial-mode qubit are carried by one and the same photon, rather than by the two photons of a pair. By contrast, photon pairs are actually used in the experiments of Trifonov *et al.* (2002) or Walborn *et al.* (2002). The latter experiment combines elements described above, inasmuch as the setup is similar to Fig. 15.4 but with the double slit of Fig. 15.1b replacing Bob's Mach–Zehnder interferometer.

[14] There is no paradox here that Bob's photons seem to display interference "after the fact" – no matter how Alice measures (or even whether she measures at all), his results as a whole do *not* display fringes. It is only *after* he receives the classical information from Alice on how to partition his data into subensembles that interference may be recovered.

The four (or five!) types of mixed states

Contraria non contradictoria sed complementa sunt.

(*Niels Bohr*)

As alluded to above, it is certainly well known that *if* system A is entangled with system B, then tracing over system B leaves system A in a mixed or partially mixed state. In fact, this is precisely one method of quantifying the amount of entanglement for bipartite pure states, namely, calculating the entropy of the reduced density matrix after tracing over one of the subsystems: $E = S(\rho^{(A)})$, where $S(\rho) = -\mathrm{Tr}(\rho \ln \rho)$, and $\rho^{(A)} = \mathrm{Tr}_B(\rho^{(A\&B)})$. The reverse question is perhaps more profound, and potentially has very important physical and philosophical implications – if the system A is in a mixed state, is it necessarily the case that it is entangled with another system? Hiding behind this seemingly innocuous question is the much deeper issue of whether or not a classical world actually exists. For, if indeed quantum mechanics does describe every possible interaction[15] between two systems – if quantum mechanics is "the whole story" – then indeed one might believe that the universe is described by one humongous entangled state of all the constituent parts.[16]

To highlight this question, we can identify four or five arguably distinct methods for producing a completely mixed (i.e., unpolarized) photon (see Fig. 15.5).

(a) In Fig. 15.5a we depict a situation where the mixture comes from the entanglement of the polarization to another degree of freedom (e.g., frequency) of the same photon, as was used in the experiment described above (pp. 310–15) (see Fig. 15.2b). The PBS transmits the horizontal component of the light directly, while deflecting the vertical component into a long delay line. The phase accumulated by light in that delay line depends on the frequency ω: $\Delta\phi = \omega L/c$, where L is the optical path length of the delay line, and c is the speed of light. If L is much greater than the coherence length of the light, in turn determined by the spread in frequencies, the resultant reduced density matrix of the polarization will be in a completely mixed state (assuming the incident light had equal horizontal and vertical polarization components).[17]

[15] We are actually restricting ourselves to a fairly limited scope of the word "interaction." For example, we do not necessarily mean to include gravity, as there does not yet exist a quantum theory of gravity. However, it should not be dismissed that such a theory may have tremendous implications for the questions we are addressing, e.g., such a theory could in principle completely resolve the semi-paradox of the emergence of a seemingly classical world from solely quantum mechanical foundations. Interesting discussions on the connection between quantum mechanics and gravity can be read elsewhere in this book.

[16] We can almost hear the Ghost of Bohr, acting as the custodian of the Copenhagen spirit, speak up: "Some of us (including one of you!) do not take kindly to your invoking the 'wave function of the universe': a quantum state is merely a mathematical entity that summarizes the conditions under which we make statistical predictions about the quantum system in question." Of course, there are those who might wish to give more meaning to Schrödinger's wave function, but if they do, then they are responsible for the consequences, as van Kampen (1988) has rightly emphasized.

[17] *Individual* photons thus manipulated are unpolarized; for a discussion of unpolarized multi-photon states see (Lehner *et al.* 1996).

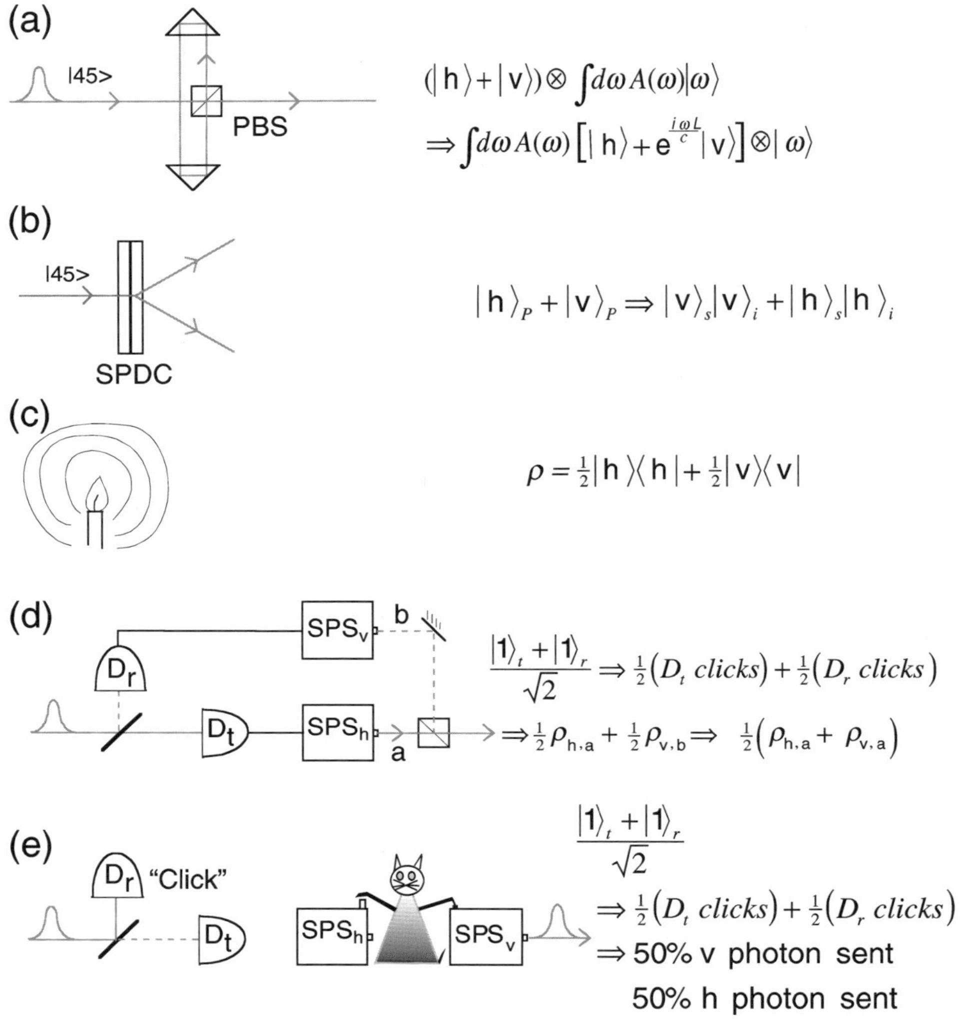

Figure 15.5. Several ways of producing a mixed polarization quantum state of a single photon. The transformations of the incident state are indicated on the right.

(b) Figure 15.5b illustrates one method that has been used to generate polarization entangled pairs of photons (Kwiat *et al.* 1999b): an ultraviolet pump photon is directed into two adjacent nonlinear crystals. Crystal 1 is oriented such that a horizontally polarized pump photon can down-convert into two vertically polarized photons (each with approximately half the energy of the pump), while the second crystal allows for the possibility of a vertically polarized pump photon to down-convert into two horizontally polarized photons. By coherently driving the crystals with the pump polarization of 45°

we arrive at the entangled state of $|vv\rangle + |hh\rangle$. As discussed above (p. 317), tracing over one of the photons leaves the other in a completely mixed state.[18]

(c) Figure 15.5c is meant to illustrate some natural random process that emits photons, such as a candle, a star, or simply some fluorescing atoms in a trap. In this case the polarization is seen to be completely random on all timescales long compared to the coherence time of the light. If we look at a classical light source of this type, at any instance in time there is a definite polarization, but it wanders rapidly over all possible states. At the single photon level, the polarization of any given photon is completely undetermined and, perhaps more importantly, *indeterminable*.

(d) Figure 15.5d shows a slightly more sophisticated method of making an unpolarized single photon state. A photon is directed at a 50–50 beam-splitter, with detectors in each output. If the photon is transmitted (and detected), then a SPS (gedanken single-photon source) is automatically fired, producing an h-polarized photon; if the original photon is reflected (and detected), then a different SPS is automatically fired, producing a v-polarized photon. Because it is intrinsically random which way the initial photon travels at the beam-splitter, it is unpredictable which of the two detectors will fire.[19] Finally, the two modes are combined using a polarizing beam-splitter.

(e) Figure 15.5e shows an extension of the previous scheme. The difference is that now a sentient observer is required to actually activate one of the single photon sources, depending on which detector registers the initial photon.[20] Although interpretations of quantum mechanics that rely on consciousness to collapse the wave function are no longer in vogue,[21] by this example we mean to highlight explicitly the role that complexity of the measuring system may play in reducing the state from a pure superposition to a mixed state.

Let us now briefly discuss the five scenarios from the point of view of entanglement and reversibility. The systems are presented in order of increasing size and complexity. In each case, at the very least we start with some entanglement. This is most obvious for the case (b), where we have explicitly written down the entangled state of one photon with the other. In system (a), the entanglement is with another degree of freedom carried by the same quantum particle. Although it is true that such a system could never be used in tests of nonlocality, at least formally there

[18] Of course, one could also use the technique of Fig. 15.5a to prepare the pump itself in a mixed state, which would then prepare the partially mixed, but correlated state $\rho \propto |vv\rangle\langle vv| + |hh\rangle\langle hh|$. No quantum erasure is possible in this case.

[19] In fact, this very system has been used to generate random numbers for quantum cryptography (Rarity *et al.* 1994; Gisin *et al.* 2002), and for a rapid-switching test of Bell's inequalities (Weihs *et al.* 1998).

[20] We hope the reader will accept this at least as a thought-experiment. We fully admit the unlikelihood that cats, even if sentient, could ever be trained to reliably perform such a task (a dog might be better suited (Bergou and Englert 1998)). On the other hand, a cat's own fickleness is likely to lead to outcomes just as random as the photon at the beam-splitter!

[21] The Ghost of Bohr interjects: "And for good reasons!"

seems to be no difference between entanglement of two degrees of freedom on separate particles and on the same particle.[22]

For scenario (c) we envision an atom in a gas, or part of a much larger ensemble of similar atoms. If there are enough of them – especially if they are more closely spaced than the wavelength of the emitted photon – it becomes impossible ever to determine which emitted the photon. One can further identify two subcategories: (1) it is impossible *in principle* to make the determination; (2) it is only impossible *in practice*. Note that if we reduce the number of emitters in the system to only a few, then it becomes possible in principle and perhaps even in practice to distinguish which emitter emitted the photon, i.e., by observing the motions of the emitters and seeing which one recoiled. In this case we have returned to a situation completely analogous to that in scenario (b).

In variation (d) the superposition of the initial photon is rapidly converted into an entangled state. Assuming, as in Fig. 15.5d, that the path for the photon to reach one detector is longer than the path to reach the other, and that the initial photon arrives at a well-specified time, the state of the whole system when the photon has had time to reach detector D_r but not yet D_t is the entangled state whose two terms are: (photon-in-the-longer-path and all-detector-atoms-quiescent) and (no-photon-in-the-long-path and first-few-detector-electrons-in-an-excited-state), which state will eventually, through various amplification processes, lead to a "click." Note that by this stage it is pretty clear that we have made the transition to the classical realm, apparently somewhere in the multi-step amplification process from the single-photon input to the electrical pulse containing perhaps hundreds or thousands of electrons, each of them interacting with its own local environment. Of course, if one accepts that explanation, then there is hardly a need to list scenario (e). However, if one instead takes the viewpoint that our inability to disentangle a complicated entangled state of the photon and detector is merely a consequence of our *technical* inability to sufficiently isolate the detector from the rest of the world, then it *does* make sense to ask whether or not there might be another natural dividing line at the level of some form of consciousness or life.[23]

We can also consider our various mixed-state creation schemes in terms of reversibility. When does it become impossible to "unscramble" the mixture? For example, the entangling process in system (a) may be very easily reversed simply by sending the photon through another delay line which now delays the horizontal

[22] Actually, the entangled state in system (a) tends to be more robust to decoherence, since the two quantum labels (polarization and frequency), are carried by the same entity, and therefore interact in most scenarios with the same local environment (Remarkably, recent experimental results seem to provide evidence for the spatial separation of the charge and spin wave functions of an electron in a one-dimensional conductor (Lorenz *et al.* 2002).) In contrast, for system (b) the two photons are able to interact with other quantum systems which are not located at the same spacetime coordinates, in general leading to faster decoherence.

[23] Clearly, again, there are numerous levels that one could consider, ranging from a single cell bacterium (which might indeed be physically smaller than the detector element in Fig. 15.5d) all the way up to human consciousness.

component with respect to the vertical, thereby "stitching" the wave function back together in its original productlike form. Scenario (b) is somewhat more difficult to reverse, in that it requires strong nonlinear process to up-convert the two photons.[24] As mentioned above, in the third scenario, as long as we were dealing with only a few atoms, it might be possible to engineer the reversibility of the system. However, it is clear that at some point it will become effectively impossible to achieve this (and maybe even impossible in principle). Apparently, by the time we are at the "click" of the detectors in scenarios (d) and (e), there is no possibility of reversing the procedure and returning to the pure initial state.[25]

Summary and discussion

This isn't right. This isn't even wrong.

(*Wolfgang Pauli*)

It was absolutely marvelous working for Pauli. You could ask him anything. There was no worry that he would think a particular question was stupid, since he thought all questions were stupid.

(*Victor Weisskopf*)

Now let us make a connection to the procedure of QE and harken back to the opening theme of the Introduction. In order for QE to recover interference fringes, the entangled quantum system plus WWM must remain in an entangled state, not irreversibly decohere into a mixed state. More specifically, the quantum phase coherence between the different terms of the entangled state must be maintained if that phase is to be mapped onto the subensemble corresponding to a particular measurement outcome of the WWM. For example, the quantum phase coherence implied by (15.8) depends entirely on the fact that (15.4) also possesses phase coherence between the two terms. If the joint state of Alice's and Bob's photons was instead a mixed state (i.e., $\rho^{(A\&B)} \propto |\mathsf{va}\rangle\langle\mathsf{va}| + |\mathsf{hb}\rangle\langle\mathsf{hb}|$), then no recovery of fringes would be possible, even in principle. We see, then, that QE is in fact a very useful tool for establishing whether or not the system has evolved out of the reversible quantum realm into the classical world of definite reality. As discussed in sections "Quantum eraser for pure and mixed states" and "Sorting at a distance," the sources described by Fig. 15.5a and 15.5b clearly possess this erasable feature.

24 Although the presence of a bright phase-coherent pump can dramatically improve the efficiency of the up-conversion process (Herzog *et al.* 1994; Resch *et al.* 2001), in this case there is no way to pick out the particular photon that resulted from the pair.

25 On the macroscopic scale, we are used to the fact that changes are not fully reversible, as witnessed by the folk wisdom of the Humpty Dumpty nursery rhyme: all the king's men cannot put a broken egg together again. This has its quantum analog – a silver atom split in two by a Stern–Gerlach magnet cannot be brought back to its initial spin state (Schwinger *et al.* 1988), and this observation contains a lesson about the reversibility of a quantum evolution (Englert 1997).

In this sense we would not say that the state of the photon prior to measurement was classical.

On the other extreme, to the extent that we believe that cats (Fig. 15.5e), or even photon counters (Fig. 15.5d), do not exist in quantum mechanical superposition states,[26] then there is, even in principle, no possibility of ever undoing the mixture of the polarization state of the finally emitted photon. There could be a record kept at the source and, therefore, each photon is h or v polarized in an objective sense.[27] No such objective reality can be ascribed to a photon emerging from the sources in Figs. 15.5a and 15.5b.

Figure 15.5c in some sense could be thought to cover the transition between the two extremes. If, for example, there are only a few atoms that might be emitting the photon, then – at least in principle, if not in practice – one could envision carefully measuring the atoms before and after emission to see which of them produced the photon. If we do this before the atom has had a chance to interact with other parts of the environment, then we might be able to measure the atom in such a way as to erase the which-polarization information. On the other hand, when sufficient numbers of atoms are involved, with sufficient numbers of degrees of freedom to degrade the quantum coherence, it rapidly becomes impossible – at least in practice, if not in principle – to perform QE.

As we continue to improve our experimental technology, we will undoubtedly be able to recover fringes in still larger systems; but since the difficulties of such an experiment seem to scale exponentially with the number of degrees of freedom we have to keep track of and control, the most likely scenario is that the in-practice constraints will greatly dominate over any possible in-principle constraint like those we alluded to here. In other words, we may never be able to tell whether the unrecoverable nature of the fringes is due to an actual irreversible transition to a classical mixture, or whether no such transition ever occurs but our measurement apparatus nevertheless are not able to reverse the complicated string of interactions. Quantum mechanics with its fundamental indeterminism sets the ultimate limits to how well we can make measurements. But these measurements, curiously, are our window back into the quantum realm: somewhat paradoxically, the only way to experimentally probe the underlying quantum mechanical reality is by making the transition – through amplification-enabled measurement – to the classical world:

No elementary quantum phenomenon is a phenomenon until it is a registered phenomenon.
(*John A. Wheeler*)

[26] There is certainly an *effective* nonexistence of such macroscopic superpositions because there are no known phenomena that are sensitive to the relative phases between the superposed states: how would one distinguish $|\text{dead cat}\rangle + |\text{live cat}\rangle$ from $|\text{dead cat}\rangle - |\text{live cat}\rangle$? And perhaps, as the Ghost of Bohr endorses, this effective nonexistence is all that is needed for the apparent emergence of the classical world.

[27] To the experimenter who has no knowledge of the preparation method, it is, for all purposes, as if the photon were h or v polarized without him knowing which one is the case or, equivalently, + or − polarized. In fact, many such "as if" realities are consistent with the experimenter's data.

References

Alley, C O, Jakubowicz, O G, and Wickes, W C (1987) In *Proc. 2nd Int. Symposium on Foundations of Quantum Mechanics*, Tokyo 1986, ed. M. Namiki, M. *et al.*, p. 36. Tokyo: Physical Society of Japan.
Badurek, G, Rauch, H, and Summhammer, J (1988) *Physica B&C* **151**, 82.
Badurek, G, Buchelt, R J, Englert, B-G, *et al.* (2000) *Nucl. Instr. Meth. Nucl. Res.* **A440**, 562.
Baldzuhn, J, and Martienssen, W (1991) *Zeits. Phys.* **B82**, 309.
Baldzuhn, J, Mohler, E, and Martienssen, W (1989) *Zeits. Phys.* **B77**, 347.
Ballentine, L E (1998) *Quantum Mechanics*, 2nd edn. Singapore: World Scientific Press.
Bergou, J A and Englert, B-G (1998) *J. Mod. Opt.* **45**, 701.
Bhattacharya, N, van Linden van den Heuvell, H B, and Spreeuw, R J C (2002) *Phys. Rev. Lett.* **88**, 137901.
Bohr, N (1983) In *Quantum Theory and Measurement*, ed. J. A. Wheeler and W. H. Zurek, p. 9. Princeton NJ: Princeton University Press.
Buks, E, Schuster, R, Heiblum, M, *et al.* (1998) *Nature* **391**, 871.
Cerf, N J, Adami, C, and Kwiat, P G (1998) *Phys. Rev.* **A57**, R1477.
Clauser, J F (1974) *Phys. Rev.* **D9**, 853.
Dürr, S, Nonn, T, and Rempe, G (1998a) *Nature* **395**, 33.
(1998b) *Phys. Rev. Lett.* **81**, 5705.
(1998c) *Phys. Rev.* **A57**, R1477.
Eichmann, U, Bergquist, J C, Bollinger, J J, *et al.* (1993) *Phys. Rev. Lett.* **70**, 2359.
Englert, B-G (1996) *Phys. Rev. Lett.* **77**, 2154.
(1997) *Zeits. Naturforsch.* **52a**, 13.
(1999) *Zeits. Naturforsch.* **54a**, 11.
Englert, B-G, and Bergou, J A (2000) *Opt. Commun.* **179**, 337.
Fermi, E (1929) *Rend. Lincei* **10**, 72.
Feynman, R P, Leighton, R B, and Sands, M (1965) *The Feynman Lectures on Physics*, Reading, MA: Addison-Wesley.
Gisin, N, Ribordy, G, Tittel, W, *et al.* (2002) *Rev. Mod. Phys.* **74**, 145.
Glauber, R (1986) *Ann. N. Y. Acad. Sci.* **480**, 336.
Grangier, P, Roger, G, and Aspect, A (1986) *Europhys. Lett.* **1**, 173.
Greenberger, D M, and Yasin, A (1988) *Phys. Lett.* **A128**, 391.
Hellmuth, T, Walther, H, Zajonc, A, *et al.* (1987) *Phys. Rev.* **A35**, 2532.
Herzog, T J, Rarity, J G, Weinfurter, H, *et al.* (1994) *Phys. Rev. Lett.* **72**, 629.
Herzog, T J, Kwiat, P G, Weinfurter, H, *et al.* (1995) *Phys. Rev. Lett.* **75**, 3034.
Hillery, M and Scully, M O (1983) In *Quantum Optics, Experimental Gravitation, and Measurement Theory*, ed. P. Meystre and M. O. Scully, p. 65. New York: Plenum Press.
Jaeger, G, Shimony, A, and Vaidman, L (1995) *Phys. Rev.* **A51**, 54.
Kim, Y-H, Yu, R, Kulik, S, *et al.* (2000) *Phys. Rev. Lett.* **84**, 1.
Knill, E, Laflamme, R, and Milburn, G (2001) *Nature* **409**, 46.
Kwiat, P G, Steinberg, A M, and Chiao, R Y (1992) *Phys. Rev.* **A45**, 7729.
Kwiat, P G, Schwindt, P D D, and Englert, B-G (1999a) In *Mysteries, Puzzles, and Paradoxes in Quantum Mechanics*, ed. R. Bonifacio, p. 69. New York: The American Institute of Physics.
Kwiat, P G, Waks, E, White, A G, *et al.* (1999b) *Phys. Rev.* **A60**, R773.
Kwiat, P G, Mitchell, J R, Schwindt, P D D, *et al.* (2000) *J. Mod. Opt.* **47**, 257.
Lehner, J, Leonhardt, U, and Paul, H (1996) *Phys. Rev.* **A53**, 2727.
Lorenz, T, Hofmann, M, Grüninger, M, *et al.* (2002) *Nature* **418**, 614.
Mandel, L (1991) *Opt. Lett.* **16**, 1882.

Mittelstaedt, P, Prieur, A, and Schieder, R (1987) *Found. Phys.* **17**, 891.

Monken, C H, Branning, D, and Mandel, L (1995) In *Coherence and Quantum Optics VII*, ed. J. H. Eberly, L. Mandel, and E. Wolf, p. 701. New York: Plenum Press.

Rarity, J G, Owens, P C M, and Tapster, P R (1994) *J. Mod. Opt.* **41**, 2435.

Rauch, H (1995) *Ann. N. Y. Acad. Sci.* **755**, 263.

Rauch, H and Summhammer, J (1984) *Phys. Lett.* **A104**, 44.

Resch, K J, Lundeen, J S, and Steinberg, A M (2001) *Phys. Rev. Lett.* **87**, 123603.

Schwindt, P D D, Kwiat, P G, and Englert, B-G (1999), *Phys. Rev.* **A60**, 4285.

Schwinger, J, Scully, M O, and Englert, B-G (1988) *Zeits Phys.* **D10**, 135.

Scully, M O, and Drühl, K (1982) *Phys. Rev.* **A25**, 2208.

Scully, M O, Englert, B-G, and Walther, H (1991) *Nature* **351**, 111.

Summhammer, J, Rauch, H, and Tuppinger, D (1987) *Phys. Rev.* **A36**, 4447.

Trifonov, A, Björk, G, Söderholm, J, *et al.* (2002) *Eur. Phys. J.* **D18**, 251.

van Kampen, N G (1988) *Physica* **A153**, 97.

von Weizsäcker, C F (1941) *Zeits. Phys.* **118**, 489.

Walborn, S P, Terra Cunha, M O, Pádua, S, *et al.* (2002) *Phys. Rev.* **A65**, 033818.

Weihs, G, Jennewein, T, Simon, C, *et al.* (1998) *Phys. Rev. Lett.* **81**, 5039.

Wheeler, J A (1979), In *Problems in the Formulation of Physics*, ed. G. T. diFrancia, p. 395. Amsterdam: North-Holland.

Wootters, W K and Zurek, W H (1979) *Phys. Rev.* **D19**, 473.

Zajonc, A G, Wang, L J, Zou, X Y, *et al.* (1991) *Nature* **353**, 507.

Björn Hessmo

Centre for Quantum Technologies, National University of Singapore, Singapore

Paper Title: **Quantum Entanglement and Partial Path Information for a Double Slit**

Reference: Unpublished (1998)

Personal message:

To Berge:

First of all, I wish you a very happy birthday and a joyous celebration! I would also like to thank you for starting my interests in quantum optics. When I was a graduate student, I did my first experiment because I was fascinated by a paper of yours on path information and interference visibility. For some reason, this experiment was never submitted for publication until now. We didn't meet until our paths crossed here in Singapore, and it is equally nice to be inspired by you these days as it was during my student years reading your papers.

Quantum Entanglement and Partial Path Information for a Double Slit

Björn Hessmo[1,2,*] and Alois Mair[2,†]

[1]Centre for Quantum Technologies, National University of Singapore, Singapore
[2]Institut für Experimentalphysik, Universität Innsbruck, Technikerstraße 25, A-6020 Innsbruck, Austria

This paper investigates relationships between non-perfect entanglement and path information, and how they influence interference patterns for a Young's double slit experiment. We illuminate a double slit with photons produced in a parametric down-conversion source. Diffraction in the source is used to modify the degree of entanglement between the photons, making it possible to adjust the amount of path information available.

PACS numbers: PACS number(s): 03.75.Be, 03.65.Nk

Quantum mechanics explains the double-slit interference as a consequence of the superposition principle. Quantum complementarity says, if you have path information you cannot have interference, and vice versa [1]. In this paper, we experimentally examine how interference patterns change with access to and erasure of path information. A quantitative theoretical description of this is given by Englert [2] and Jaeger, *et al.* [3, 4]. The probability to find a particle at position x on a screen far behind two narrow slits separated by a distance b is given by the equation

$$P \approx |a_1 e^{ikL_1} + a_2 e^{ikL_2}|^2 \approx 1 + 2|a_1||a_2| \cos(kb\theta)\,, \quad (1)$$

where a_1 and a_2 are the probability amplitudes for passing through slit 1 and 2 respectively. L_1 and L_2 are the lengths of the two paths from the slits to the detection plane, and k is the wave number of the particle. The angle θ refers to the direction in which the particle travels after the slits. To classify the interference patterns, it is convenient to use the visibilities [5] of fringe patterns. The visibility is defined as

$$V = \frac{P_{max} - P_{min}}{P_{max} + P_{min}} = 2|a_1||a_2|\,, \quad (2)$$

where the last equality is for the special case when P is given by Eq. (1). A convenient measure of path information (or path distinguishability) is given by

$$D \;=\; \left|(|a_1|^2 - |a_2|^2)\right|, \quad (3)$$

where $|a_1|^2$ ($|a_2|^2$) is the probability of passing through slit 1 (slit 2). It is easy to see that visibility and path information are related by the equation,

$$D^2 + V^2 = 1\,. \quad (4)$$

In the case of the complete lack of path information, that is when $D = 0$ corresponding to $|a_1| = |a_2|$, the visibility of the interference fringes is one. In all other situations, the visibility is less than unity.

Consider a double slit which is illuminated by light generated in a parametric down-conversion process (see Fig. 1). In the case of perfect entanglement, the two generated photons form the Bell state

$$|\Psi\rangle \;=\; \frac{1}{\sqrt{2}}\left(|U\rangle_i|L\rangle_s + |L\rangle_i|U\rangle_s\right) \quad (5)$$

where U (L) refers to the upper (lower) path in Fig. 1. If the idler photon (subscript i) is found in the upper arm, then the corresponding signal photon (subscript s) will, with certainty, be found in the opposite lower arm, and vice versa. In this situation, no interference should be observed. This is due to the fact that it is possible to make a measurement on the idler photon to determine which path the signal photon took through the slits.

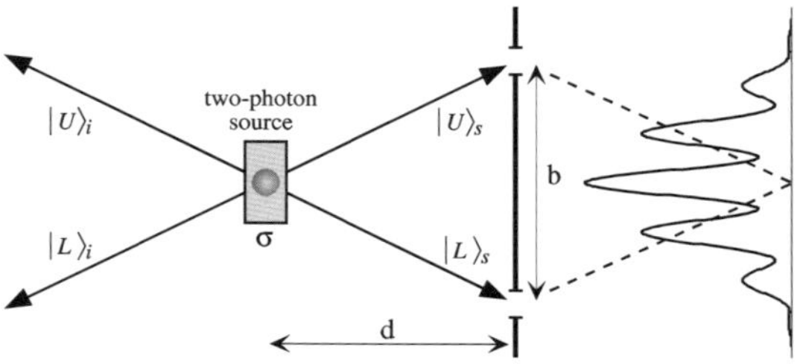

Figure 1: This is a schematic of the experimental setup. Two photons, the idler $|\cdot\rangle_i$ and the signal $|\cdot\rangle_s$, leave the two-photon source in the center of the figure. The signal photon travels a (variable) distance d through two slits separated a distance b from each other. Behind the double sit, an interference pattern may be formed.

If, on the other hand, the two-photon state is in a non-maximally entangled state,

$$\begin{aligned}|\Psi\rangle \;=\; &\frac{1}{\sqrt{2}}\big[\,|U\rangle_i\,(\cos\theta|L\rangle_s + \sin\theta|U\rangle_s)\\ &+ |L\rangle_i\,(\sin\theta|L\rangle_s + \cos\theta|U\rangle_s)\,\big],\end{aligned} \quad (6)$$

there is no longer a perfect correlation between the idler and signal paths. Detection of the idler photon in the upper arm will not allow us to infer with certainty that the signal photon is in the opposite lower arm. Instead, one finds that it has amplitude $\cos\theta$ ($\sin\theta$) for the lower (upper) arm. This implies that the fringe pattern behind the double slit has visibility $V = \sin 2\theta$ if position is measured on the idler photon. The path distinguishability is $D = \cos 2\theta$.

By rewriting the state in Eq. (6) in the basis

$$|U\rangle_k = \frac{1}{\sqrt{2}}\left(|A\rangle_k + |A^{\perp}\rangle_k\right), \quad (7)$$

$$|L\rangle_k = \frac{1}{\sqrt{2}}\left(|A\rangle_k - |A^{\perp}\rangle_k\right), \quad (8)$$

which is rotated by 45 degrees in Hilbert space, $|\Psi\rangle$ transforms into its Schmidt decomposition,

$$\begin{aligned} |\Psi\rangle &= \frac{\sin\theta + \cos\theta}{\sqrt{2}}|A\rangle_i|A\rangle_s \\ &+ \frac{\sin\theta - \cos\theta}{\sqrt{2}}|A^{\perp}\rangle_i|A^{\perp}\rangle_s. \end{aligned} \quad (9)$$

This is a transformation into the eigenstates of an observable complementary to the observable which has the position states $|U\rangle$ and $|L\rangle$ as its eigenstates. If $|A\rangle$ or $|A^{\perp}\rangle$ is measured on the idler, it will be impossible to get path information (which concerns U and L) for the idler. The measurement will also project the signal photon into the same state as the idler photon due to the nature of the state described by Eq. (9). Since the amplitudes for U and L are equal in the $|A\rangle$- or $|A^{\perp}\rangle$-states, interference pattern recorded in coincidence should be fully recovered, with $V = 1$.

Experimentally, the state given by Eq. (6) is obtained by changing the source width, or the distance d between source and double slit in the setup illustrated by Fig. 1. The source has a Gaussian intensity distribution with beam waist w_0, determined by the pump beam. For a finite source, the two photons (signal and idler) diffract from the source where they are produced. When the idler photon is detected at position U (L), the signal photon is *not* found at position L (U) with certainty. Instead, one should consider the diffraction pattern to obtain the correct probability distribution for the signal photon. The probability amplitude for finding the signal photon at position x (with the center of path L at $x = 0$) on the screen containing the double slit is given by

$$A(x) \propto e^{-\left(\frac{2\pi w_0 x}{\lambda d}\right)^2}, \quad (10)$$

where λ is the optical wavelength, w_0 is the pump beam width, and d is the distance between the source and the plane containing the double slit. This formula is valid only in the Fraunhofer limit. Let us consider two narrow slits separated by a distance b from each other. If the idler photon is detected at U, the (normalized) probability amplitude for the signal photon to be found at L is

$$a_{\mathrm{UL}} = \frac{1}{\sqrt{1+\exp[-2(2\pi w_0 b/\lambda d)^2]}} = \cos\theta, \quad (11)$$

and the amplitude for the other slit is

$$a_{\mathrm{UU}} = \frac{\exp[-(2\pi w_0 b/\lambda d)^2]}{\sqrt{1+\exp[-2(2\pi w_0 b/\lambda d)^2]}} = \sin\theta. \quad (12)$$

Also, if the idler is detected at L, the amplitudes are

$$a_{\mathrm{LU}} = \frac{1}{\sqrt{1+\exp[-2(2\pi w_0 b/\lambda d)^2]}} = \cos\theta, \quad (13)$$

$$\text{and } a_{\mathrm{LL}} = \frac{\exp[-(2\pi w_0 b/\lambda d)^2]}{\sqrt{1+\exp[-2(2\pi w_0 b/\lambda d)^2]}} = \sin\theta. \quad (14)$$

Equations (11) to (14) give

$$\begin{aligned} |\Psi\rangle &= a_{\mathrm{UL}}|U\rangle_i|L\rangle_s + a_{\mathrm{UU}}|U\rangle_i|U\rangle_s \\ &+ a_{\mathrm{LL}}|L\rangle_i|L\rangle_s + a_{\mathrm{LU}}|L\rangle_i|U\rangle_s, \end{aligned} \quad (15)$$

which is identical to Eq. (6). For small w_0/d, the photons are diffracted strongly, corresponding to $\theta = \pi/4$. This situation gives a non-entangled product state. To obtain a state with full entanglement, w_0/d must be large in order to minimize diffraction. This corresponds to $\theta = 0$. Practically, in a down-conversion experiment, the effective size can be changed either by changing the pump beam diameter at the nonlinear crystal or by translating the crystal with respect to the double slit. In our experimental setup, we use the second strategy, keeping $w_0 \sim 0.25$ mm and varying the distance d.

The next step is to project the idler photon onto the state $|A\rangle$ or $|A^{\perp}\rangle$, as described by Eq. (9). One way to do this is by introducing a second double slit in the idler path. When the idler is detected behind this second double slit, it is not possible to tell which slit it went through (assuming that the signal photon also passes through its double slit and is not useful for extracting path information for the idler). Because of the entanglement, the signal photon is projected onto a state with equal amplitudes on its two paths. This setup has several advantages. For more details, we refer to [6] and [7]. In our experiment, we however lack the two-photon intensity necessary for using two double slits due to poor mode-matching.

An alternative method to the above procedure is to measure the momentum of the idler photon. A momentum measurement is realized by introducing a lens in the idler path. Different momenta are mapped to different points in the focal plane of the lens.

The experiment employs a pump beam coming from an Argon-ion laser with wavelength $\lambda = 351$nm and an output power of 300 mW. This beam pumps a BBO (β-Barium borate) crystal. One of the down-converted photons (the signal photon) goes to a double slit. Behind the double slit, a monomode fiber is mounted in a fiber coupler pointing at the center of the double slit. The fiber coupler consists of a lens with the optical fiber located in the focal plane of the lens. The optical fiber has a core diameter of $\approx 5\mu$m. This device can be rotated around the double slit by a motor which defines the x-axis in Fig. 2. The other photon (the idler photon) goes to an identical fiber coupler pointing at the BBO crystal, realising the momentum measurement. The fibers guide the light through an interference filter selecting light with wavelength 702 nm to an avalanche photodiode. The signals registered by the photodiodes are sent to a computer which stores the number of photons detected per point (defined by the motor behind the double slit). In addition, the coincidence between the

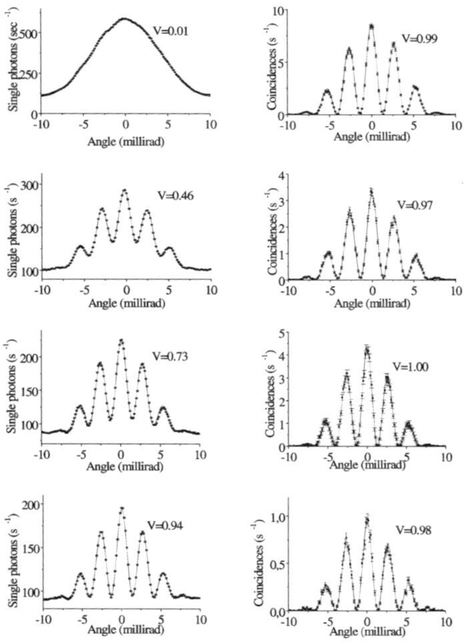

Figure 2: The first column shows patterns collected behind the double slit if the idler photon is ignored. The second column shows patterns recorded in coincidence with a momentum measurement on the idler photon. For the first row of measurements, the source is close to the double slit, and in the last row, the distance is large. The solid curves are from theory assuming the visibilities indicated on each graph.

two detectors is also registered by the computer. The distance between the BBO-crystal and the double slit can be changed from 0.23 m to 1 m. This changes the apparent source size. The results for four different distances are shown in Fig. 2. The first column shows the signal photon distribution behind the double slit when the idler photon is ignored. The second column shows the signal photon distribution collected in coincidence with the momentum measurement on the idler photon. In the first row of Fig. 2, the source is also large and interference can be seen only after erasing the path information. The following two rows show how the single-particle interference increases as the distance between source and slits increases. The last row shows a nearly perfect interference pattern (visibility 0.94), and the idler photon is of little use for extracting path information for the photon passing through the double slit.

The conclusion of this paper is that it is possible to change the amount of entanglement in a two-photon state generated in a down-conversion process by using diffraction effects in the source. For a very large source, the state has nearly perfect entanglement (a Bell state) and for a very small source, the entanglement is lost, yielding a product state. One of the photons in a non-maximally entangled photon pair is sent through a double slit, and the other one can be used to extract path information. This makes the visibility of the interference pattern dependent on the degree of entanglement. Performing a momentum measurement on the idler photon renders it useless for extracting path information. This recovers full interference visibility. It is seen that the fringe visibility of the one-photon pattern increases when the source is placed further from the double slit, whereas the pattern recorded in coincidence with a momentum measurement yields a constant and high visibility. This raises the question of whether it is possible to make a long distance Bell experiment relying on position/momentum entanglement, since the entanglement degrades with propagation.

We want to thank A. Zeilinger for his support during this experiment. We also wish to thank M. Zukowski, M. Horne, and C. Simon for many valuable discussions during this experiment.

Note: This experiment was carried out in 1998 and intentionally left unpublished until Berge's 60th birthday. Since then, related experiments have been carried out along the same direction, most notably the one by M. d'Angelo *et al.* [8]. Also, the experiment of B. Dofler appeared in the literature, which was the starting point of our experiment [9]. This was also reproduced in a later experiment [10]. Most of these experiments relate to the work of B.-G. Englert.

---

* Electronic address: bjorn.hessmo@quantumlah.org
† Electronic address: amair@gmx.at

[1] N. Bohr, in Albert Einstein: Philosopher-Scientist, edited by P.A. Schlipp (The Library of Living Philosophers, Evanston, IL, 1949), p. 216.
[2] B. G. Englert, Phys. Rev. Lett. **77**, 2154 (1996).
[3] G. Jaeger, M. A. Horne, and A. Shimony, Phys. Rev. A **48**, 1023 (1993).
[4] G. Jaeger, A. Shimony, and L. Vaidman, Phys. Rev. A **51**, 54 (1995).
[5] A. A. Michelson, Phil. Mag. **30**, 1 (1890).
[6] D. M. Greenberger, M. A. Horne, and A. Zeilinger, Physics Today **46**, 22 (1993).
[7] M. Horne, in *Experimental Metaphysics* (ed. P.A. Schlipp, Kluwer Academic Publisher, 1997), pp. 109–119.
[8] M. d'Angelo, M. Chekhova, and Y. Shih, Phys. Rev. Lett. **87**, 013602 (2001).
[9] A. Zeilinger, Rev. Mod. Phys. **71**, S288 (1999).
[10] M. d'Angelo, Y.-H. Kim, S. P. Kulik, and Y. Shih, Phys. Rev. Lett. **92**, 233601 (2004).

Gerhard Rempe

Max-Planck-Institut für Quantenoptik, Garching, Germany

Paper Title: **Origin of quantum-mechanical complementarity probed by a 'which-way' experiment in an atom interferometer**

Reference: Nature **395**, 33–37 (1998)

Paper Title: **Fringe visibility and which-way information in an atom interferometer**

Reference: Physical Review Letters **81**, 5705–5709 (1998)

Further reference: *Can wave-particle duality be based on the uncertainty relation?*, S. Dürr and G. Rempe, American Journal of Physics **68**, 1021 (2000).

Personal message:

Dear Berge,

Please accept the best birthday wishes from a long-standing admirer of your work. As unfortunately I am not able to congratulate you in person, I take the opportunity and the liberty to write to you in order to praise your timeless work that you did together with Marlan Scully and Herbert Walther. That work concerned an old physics problem, namely the connection between Bohr's complementarity principle and Heisenberg's uncertainty relation. When I first heard about your theoretical proposal for a micromaser test experiment I was surprised and sceptical, largely because of the technological challenges of the proposed experiment. Others were sceptical, too, but for other reasons, and formulated extensive comments, thus starting a lively scientific debate in one of the most prestigious and most-read journals. Our experiments from a few years later, performed with atomic spin states instead of micromaser fields, were able to clarify several issues of the so-far purely academic debate. I vividly remember several inspiring discussions with you during that time. These discussions helped me a lot to better understand the subtle issues concerning one of the most fundamental questions of quantum physics. Looking back, I find those years very memorable, exciting and stimulating! Thank you so much for sharing your insights with me and, again, happy birthday and good health for the years to come!

Best wishes to you and, of course, your family!

G. Rempe

Origin of quantum-mechanical complementarity probed by a 'which-way' experiment in an atom interferometer

S. Dürr, T. Nonn & G. Rempe

Fakultät für Physik, Universität Konstanz, 78457 Konstanz, Germany

The principle of complementarity refers to the ability of quantum-mechanical entities to behave as particles or waves under different experimental conditions. For example, in the famous double-slit experiment, a single electron can apparently pass through both apertures simultaneously, forming an interference pattern. But if a 'which-way' detector is employed to determine the particle's path, the interference pattern is destroyed. This is usually explained in terms of Heisenberg's uncertainty principle, in which the acquisition of spatial information increases the uncertainty in the particle's momentum, thus destroying the interference. Here we report a which-way experiment in an atom interferometer in which the 'back action' of path detection on the atom's momentum is too small to explain the disappearance of the interference pattern. We attribute it instead to correlations between the which-way detector and the atomic motion, rather than to the uncertainty principle.

In classical physics, a particle moves along a well-defined trajectory. A quantum object, however, reveals its wave character in interference experiments in which the object seems to move from one place to another along several different paths simultaneously. It is essential that these ways are indistinguishable, because any attempt to observe which way the object actually took unavoidably destroys the interference pattern.

The usual explanation for the loss of interference in a which-way experiment is based on Heisenberg's position–momentum uncertainty relation. This has been illustrated in famous *gedanken* experiments like Einstein's recoiling slit[1] or Feynman's light microscope[2]. In the light microscope, electrons are illuminated with light immediately after they have passed through a double slit with slit separation d. A scattered photon localizes the electron with a position uncertainty of the order of the light wavelength, $\Delta z \approx \lambda_{light}$. Owing to Heisenberg's position–momentum uncertainty relation, this localization must produce a momentum uncertainty of the order of $\Delta p_z \approx h/\lambda_{light}$. This momentum uncertainty arises from the momentum kick transferred by the scattered photon. For $\lambda_{light} < d$, which-way information is obtained, but the momentum kick is so large that it completely washes out the spatial interference pattern.

However, Scully *et al.*[3] have recently proposed a new *gedanken* experiment, where the loss of the interference pattern in an atomic beam is not related to Heisenberg's position–momentum uncertainty relation. Instead, the correlations between the which-way detector and the atomic beams are responsible for the loss of interference fringes.

Such correlations had already been studied experimentally. They are, for example, responsible for the lack of ground-state quantum beats in time-resolved fluorescence spectroscopy[4]. Other examples are neutron interferometers, where which-way information can be stored by selectively flipping the neutron spin in one arm of the interferometer[5,6].

Nevertheless, the *gedanken* experiment of Scully *et al.* was criticized by Storey *et al.*[7], who argued that the uncertainty relation always enforces recoil kicks sufficient to wash out the fringes. This started a controversial discussion[8–11] about the following question: "Is complementarity more fundamental than the uncertainty principle?"[10]. This motivated Wiseman *et al.*[10,11] to investigate what constitutes a momentum transfer in a double-slit experiment. They call the usual momentum transfer, like that in Feynman's light microscope, a "classical" kick; in addition, they define the concept of "quantum" momentum transfer. They find that the loss of interference need not be due to "classical" kicks. In this case the "quantum" momentum transfer cannot be less than that required by the uncertainty principle, so that these "quantum" kicks wash out the fringes.

In this context, Eichmann *et al.*[12] performed an experiment with a light interferometer, where the double slit is replaced by two trapped ions, which can store which-way information in internal states. This scheme was criticized[9], because the ions play a double role: they act as sources of elementary waves (just like a double slit) and simultaneously as a which-way detector. Hence the momentum transfer from the double slit and the which-way detector cannot be separated.

Here we report on a which-way experiment with an atom interferometer. A microwave field is used to store the which-way information in internal atomic states. We study the mechanical effect of the which-way detection on the atomic centre-of-mass motion separately, and find that the "classical" momentum kicks are much too small to wash out the interference pattern. Instead, correlations between the which-way detector and the atomic motion destroy the interference fringes. We show that the back action onto the atomic momentum implied by Heisenberg's position–momentum uncertainty relation cannot explain the loss of interference.

The atom interferometer

Figure 1 shows a scheme of our atom interferometer. An incoming beam of atoms passes through two separated standing wave light beams. The detuning of the light frequency from the atomic resonance, $\Delta = \omega_{light} - \omega_{atom}$, is large so that spontaneous emission can be neglected. The light fields each create a conservative potential U for the atoms, the so-called light shift, with $U \propto I/\Delta$, where I is the light intensity (see, for example, ref. 13). In a standing wave the light intensity is a function of position, $I(z) = I_0\cos^2(k_{light}z)$, where k_{light} is the wavevector of the light. Hence the light shift potential takes the form $U(z) = U_0\cos^2(k_{light}z)$, with $U_0 \propto I_0/\Delta$.

The atoms are Bragg-reflected from this periodic potential, if they enter the standing light wave at a Bragg angle (see, for example, ref. 14). This process is similar to Bragg reflection of X-rays from the periodic structure of a solid-state crystal, but with the role of matter and light exchanged. In our experiment, the light creates the periodic structure, from which the matter wave is reflected.

The standing light wave splits the incoming atomic beam A (see Fig. 1) into two beams, a transmitted beam C and a first-order Bragg-reflected beam B. The angle between the beams B and C corresponds to a momentum transfer of exactly $2\hbar k_{light}$, as determined by the spatial period of $U(z)$. By varying the light intensity, the fraction of reflected atoms can be adjusted to any arbitrary value. In our experiment, the reflectivity of the beam splitter is tuned to ~50%.

After switching off the first standing light wave, the two beams are allowed to propagate freely for a time interval t_{sep}. During this time, beam B moves a horizontal distance $d/2$ to the left and beam C moves $d/2$ to the right. The longitudinal velocities (vertical in Fig. 1) of the two beams are not affected by the light field. Then a second standing light wave is switched on, which also serves as a 50% beam splitter. Now two atomic beams D and E are travelling to the left, while beams F and G are travelling to the right. In the far field, each pair of overlapping beams produces a spatial interference pattern. The fringe period is the same as in a double-slit experiment with slit separation d. The relevant wavelength is the de Broglie wavelength associated with the momentum of the atoms. The envelope of the fringe pattern is given by the collimation properties of the initial atomic beam A.

The experiment is performed with the apparatus described in ref. 15: ^{85}Rb atoms are loaded into a magneto-optical trap (MOT). After trapping and cooling, the cloud of atoms is released and falls freely through the apparatus. The resulting pulsed atomic beam is collimated with a mechanical slit 20 cm below the MOT. The atoms then pass the interaction region with the standing light wave inside a microwave resonator. In the far field of the interaction region, 45 cm below the MOT, the atomic position distribution is observed by exciting the atoms with a resonant laser beam and detecting the fluorescence photons. The interaction time t_{Bragg} of the atoms with the standing light wave is controlled by switching the light on and off. The small atomic velocity of $2\,\mathrm{m\,s^{-1}}$ in the interaction region allows us to perform the whole interferometer experiment with only one standing light wave, which is switched on and off twice. As compared to ref. 15, only a few changes have been made. The width of the collimation slit was enlarged to 450 μm. In order to improve the position resolution, the horizontal waist of the detection laser beam was reduced to $w = 50\,\mu m$, and a second collimation slit with a width of 100 μm was added 1 cm below the MOT. Finally, the preparation of the internal atomic state was improved by removing atoms in wrong Zeeman sublevels.

Figure 2 shows the spatial fringe pattern in the far field for two different values of t_{sep}. We note that the observed far-field position distribution is a picture of the atomic transverse momentum distribution after the interaction.

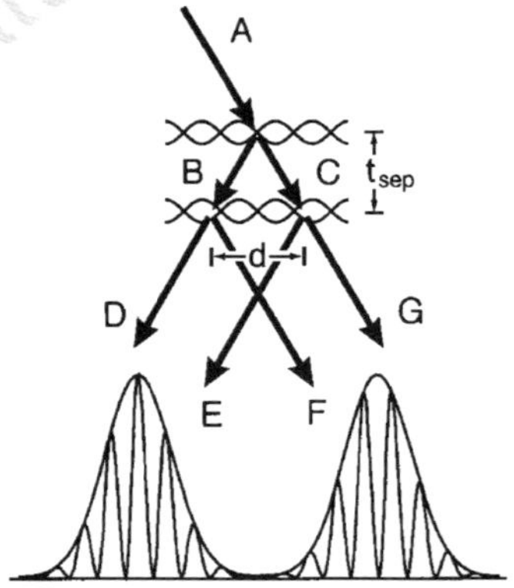

Figure 1 Scheme of the atom interferometer. The incoming atomic beam A is split into two beams: beam C is transmitted and beam B is Bragg-reflected from a standing light wave. The beams are not exactly vertical, because a Bragg condition must be fulfilled. After free propagation for a time t_{sep} the beams are displaced by a distance d. Then the beams are split again with a second standing light wave. In the far field, a spatial interference pattern is observed.

Storing which-way information

A second quantum system is now added to the interferometer in order to store the information whether the atom moved along way B

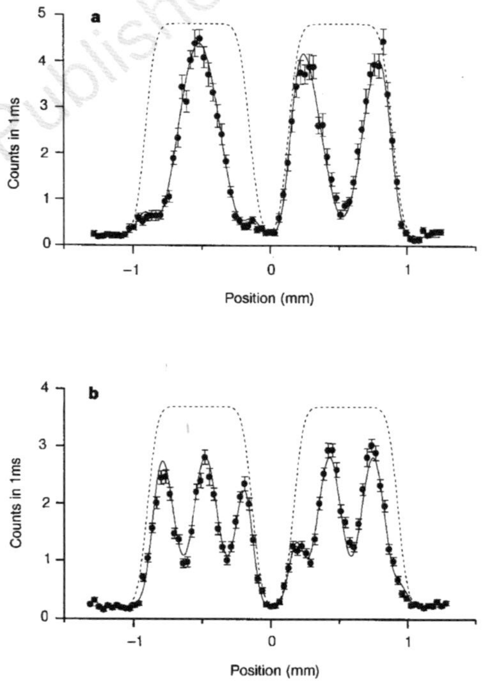

Figure 2 Spatial fringe pattern in the far field of the interferometer. The data were obtained with $t_{Bragg} = 45\,\mu s$, and the 50% beam splitter was realized with $|U_0| = \hbar\pi/t_{Bragg}$. We chose $t_{sep} = 105\,\mu s$ with $d = 1.3\,\mu m$ (**a**), and $t_{sep} = 255\,\mu s$ with $d = 3.1\,\mu m$ (**b**). In both cases, the fringe period is in good agreement with the theoretical expectation. Each solid line represents a fit to the experimental data. The best-fit values for the visibilities are (75 ± 1)% and (44 ± 1)%, respectively. The reduced visibility for the case of the narrow fringes is due to the finite position resolution of our apparatus. The dashed lines represent the independently measured beam envelope, which consists of two broad peaks. The right peak is due to beams F and G (see Fig. 1), with a shape determined by the momentum distribution of the initial beam A. The left peak is a combination of beams D and E. It is a Bragg-reflected picture of the right peak. The fringe patterns under these two broad peaks are complementary, that is, the interference maxima in the left peak correspond to interference minima in the right peak, and vice versa.

or C. Two internal electronic states of the atom are used as a which-way detector system. A simplified level scheme of ^{85}Rb is shown in Fig. 3a. Rabi oscillations between states $|2\rangle$ and $|3\rangle$ can be induced by applying a microwave field at ~3 GHz. To describe the information storing process, we first investigate the properties of one single Bragg beam splitter, using a simple model, whose validity will be discussed later.

The frequency of the standing light wave, ω_{light}, is tuned halfway between the $|2\rangle \rightarrow |e\rangle$ and $|3\rangle \rightarrow |e\rangle$ transitions. Hence the detunings from these transitions, Δ_{2e} and Δ_{3e}, have the same absolute value but opposite sign. The reflectivity of the beam splitter, that is, the probability of reflecting an atom, depends on $t_{Bragg} \times |U_0|$, and it is independent of the internal state.

However, the amplitude of the wavefunction experiences a phase shift which depends on the internal atomic state. A simple analogy for this phase shift can be found in light optics: a light wave reflected from an optically thicker medium experiences a phase shift of π, while reflection from an optically thinner medium or transmission into an arbitrary medium does not cause any phase shift. This argument also applies in atom optics: in our experiment, an atom in $|2\rangle$ sees a negative light shift potential (because $\Delta_{2e} < 0$), corresponding to an optically thicker medium, while an atom in $|3\rangle$ sees a positive potential (because $\Delta_{3e} < 0$), corresponding to an optically thinner medium. Hence an atom will experience a π phase shift only if it is reflected in $|2\rangle$.

This phase shift can be converted into a population difference between the hyperfine levels. For that purpose two microwave $\pi/2$ pulses resonant with the hyperfine transition are applied. They form a Ramsey scheme as shown in Fig. 3b. The atom is initially prepared in state $|2\rangle$. Then a microwave $\pi/2$ pulse is applied, converting the internal state to the superposition state $(|3\rangle + |2\rangle)/\sqrt{2}$. After this, the atom interacts with the standing light wave. As explained above, the atom will experience a π phase shift only if it is reflected and in state $|2\rangle$. Thus the internal state of the reflected beam is changed to $(|3\rangle + |2\rangle)/\sqrt{2}$, while the internal state of the transmitted beam is not affected. As a result, there is an entanglement created between the internal and the external degree of freedom of the atom. The state vector of the system becomes:

$$|\psi\rangle \propto |\psi_B\rangle \otimes (|3\rangle - |2\rangle) + |\psi_C\rangle \otimes (|3\rangle + |2\rangle) \quad (1)$$

where $|\psi_B\rangle$ and $|\psi_C\rangle$ describe the centre-of-mass motion for the reflected and transmitted beams (see Fig. 1), respectively. This entanglement is the crucial point for the storage of information. The second microwave pulse acting on both beams (the transmitted and the reflected), converts the internal state of the transmitted beam to state $|3\rangle$, while the reflected beam is converted to state $-|2\rangle$. Thus, the state vector after the pulse sequence shown in Fig. 3b becomes:

$$|\psi\rangle \propto -|\psi_B\rangle \otimes |2\rangle + |\psi_C\rangle \otimes |3\rangle \quad (2)$$

We note that momentum transfer from the microwave slightly changes $|\psi_B\rangle$ and $|\psi_C\rangle$, but this has negligible effects, as will be discussed below.

Equation (2) shows that the internal state is correlated with the way taken by the atom. The which-way information can be read out later by performing a measurement of the internal atomic state. The result of this measurement reveals which way the atom took: if the internal state is found to be $|2\rangle$, the atom moved along beam B, otherwise it moved along beam C.

A detailed calculation of the beam splitter reveals an additional phase shift, not discussed so far. It arises because the atoms travel in the light shift potential during t_{Bragg}. This creates a phase shift for the atoms proportional to $U_0 \times t_{Bragg}$, resulting in a relative phase shift between atoms in states $|2\rangle$ and $|3\rangle$. Fortunately, this phase shift is identical for the transmitted and reflected beams and therefore does not affect the storing process in an essential manner. Moreover, a small detuning of the microwave frequency from the atomic resonance allows us to compensate for this effect, so that the simple model discussed above is valid.

Interferometer with which-way information

After considering a single beam slitter, we now return to the complete interferometer. Sandwiching the first Bragg beam splitter between two microwave $\pi/2$ pulses stores the which-way information in the internal atomic state, as described above. We note that the second Bragg beam splitter does not change the internal state.

Will there still be interference fringes in the far field, when the which-way information is stored? The experimental result is shown in Fig. 4: there are no fringes. The data were recorded with the same parameters as in Fig. 2a. The only difference is that two microwave pulses were added to store the which-way information. Atoms in both hyperfine states were detected, so that the which-way detector was not read out. The mere fact that which-way information is stored in the detector and *could* be read out already destroys the interference pattern. We have verified experimentally that the fringes also disappear when the which-way detector is read out, that is, when only atoms in state $|2\rangle$ or only atoms in state $|3\rangle$ are detected. Of course, the absolute size of the signal is reduced by a factor of two in these cases.

Mechanical effects

We now discuss whether the loss of interference can be explained by mechanical effects of the which-way detector on the atomic

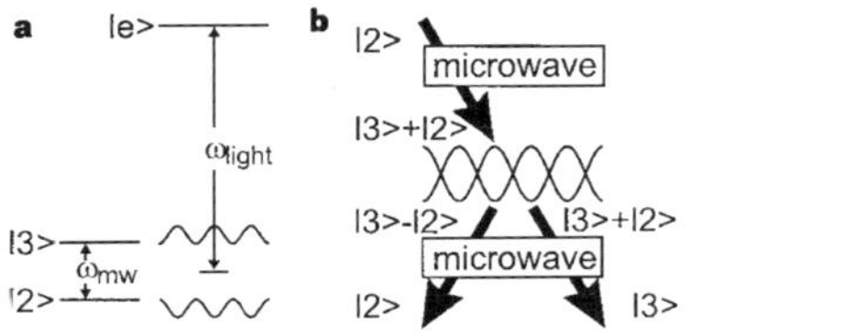

Figure 3 Storage of which-way information. **a**, Left, simplified level scheme of ^{85}Rb. The excited state ($5^2P_{3/2}$) is labelled $|e\rangle$. The ground state ($5^2S_{1/2}$) is split into two hyperfine states with total angular momentum $F = 2$ and $F = 3$, which are labelled $|2\rangle$ and $|3\rangle$, respectively. Right, the standing light wave with angular frequency ω_{light} induces a light shift for both ground states which is drawn as a function of position. **b**, The beam splitter produces a phase shift that depends on the internal and external degree of freedom. A Ramsey scheme, consisting of two microwave $\pi/2$ pulses, converts this phase shift into a population difference (see text).

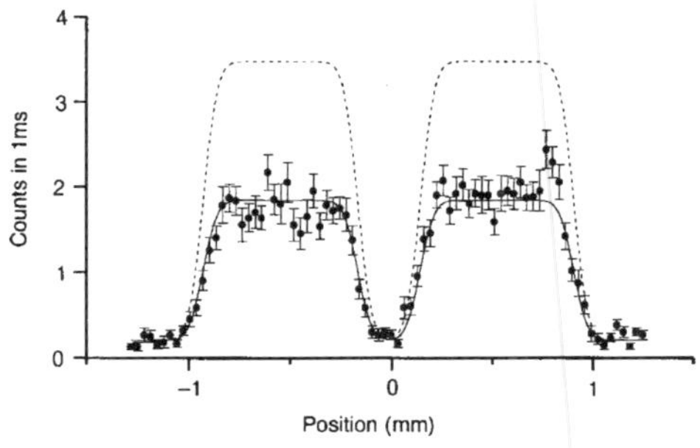

Figure 4 Same as Fig. 2a, but with which-way information stored in the internal atomic state. The interference fringes are lost due to the storage of which-way information.

centre-of-mass motion. Therefore the transverse momentum transfer in the microwave field and the longitudinal displacement of the atomic wavefunction must be investigated.

To wash out the interference fringes, the transverse momentum transfer must have a distribution whose spread must correspond to at least half a fringe period. Such a "classical" momentum transfer distribution would also broaden the envelope of the fringe pattern by the same amount[11]. Comparing the experimental data in Fig. 2a and Fig. 4, it is obvious that the width of the envelope is not changed. This experimental result clearly shows that there is no significant transfer of transverse momentum in the microwave field.

The momentum transfer during the interaction with the microwave field can also be estimated theoretically. The microwave field is a standing wave diffracting the atomic beam. Because the atoms are travelling much less than a microwave wavelength during the interaction, the Raman–Nath approximation is valid. For a plane atomic wave, the probability to pick up n photon momenta during a $\pi/2$ pulse is $J_n^2(\pi/2)$, where J_n is the nth order Bessel function[16]. Hence the probability of transferring more than two microwave photon momenta is less than 1%. The absorption of a single microwave photon shifts the position of the atom in the detector plane by 5 nm. It follows that the mechanical recoils during the interaction with the microwave field can shift the pattern by at most ± 10 nm—too small to be observed. Of course, the transverse width of the atomic beam is much less than a microwave wavelength, so that the atomic beam is not a plane wave. However, the atomic beam is a superposition of plane waves. The above Raman–Nath calculation applies for each plane wave component, so that the spread of the atomic beam cannot be larger than for a plane wave.

The second mechanical effect that could explain the loss of interference is a longitudinal displacement of the atomic wavefunction, as discussed in ref. 7. In our experiment, this argument does not apply, because the whole interaction sequence is pulsed. The atom's potential energy varies as a function of time, not as a function of longitudinal position, so that the interaction does not create any longitudinal forces and displacements.

We conclude that the "classical" mechanical effects of the which-way detector on the atomic centre-of-mass motion are negligible, so that some other mechanism must enforce the loss of interference fringes.

Correlations destroy interference

In order to investigate why the interference is lost, we consider the state vector for the interaction sequence used in Fig. 4. The state vector after the interaction with the first beam splitter sandwiched between the two microwave pulses is given in equation (2). The second beam splitter transforms this state vector into:

$$|\psi\rangle \propto -|\psi_D\rangle \otimes |2\rangle + |\psi_E\rangle \otimes |3\rangle + |\psi_F\rangle \otimes |2\rangle + |\psi_G\rangle \otimes |3\rangle \quad (3)$$

The sign of $|\psi_F\rangle$ is positive due to the π phase shift during the reflection from the second beam splitter.

In the far field, the atomic position distribution under the left peak of the envelope is given by:

$$P(z) \propto |\psi_D(z)|^2 + |\psi_E(z)|^2 - \psi_D^*(z)\psi_E(z)\langle 2|3\rangle - \psi_E^*(z)\psi_D(z)\langle 3|2\rangle \quad (4)$$

because here the spatial wavefunctions $\psi_F(z)$ and $\psi_G(z)$ vanish. The first two terms describe the mean intensity under the envelope. Interference could only be created by the last two terms, but they vanish because $\langle 2|3\rangle = 0$. Precisely the same entanglement that was required to store the which-way information is now responsible for the loss of interference. In other words: the correlations between the which-way detector and the atomic motion destroy the interference, as discussed in ref. 3.

This loss of interference manifests itself as a dramatic change in the momentum distribution when adding the microwave fields to the interferometer, even though the microwave itself does not transfer enough momentum to the atom to wash out the fringes. However, the addition of the microwave fields modifies the probability for momentum transfer by the light fields. This modification of the momentum transfer probability is due to the correlations between the which-way detector and the atomic motion.

Correlations between the interfering particle and the detector system are produced in any which-way scheme, for example, in the previously mentioned *gedanken* experiments of Einstein's recoiling slit and Feynman's light microscope. But in these experiments "classical" mechanical effects of the detector on the particle's motion can explain the loss of interference as well, so that the effect of the correlations is hidden.

So far, the microwave field has been treated as a classical field, that is, not as a quantized field. At first glance this might seem to be unjustified, because the population difference between the two hyperfine states corresponds to the absorption of one microwave photon, so that the system also becomes entangled with the microwave field. Hence the interference terms in equation (4) must include an additional factor $\langle\alpha|\beta\rangle$, where $|\alpha\rangle$ denotes the initial state of the microwave field, which changes to $|\beta\rangle$ due to the absorption of one photon. In our experiment, the initial state $|\alpha\rangle$ is a coherent state with a large mean photon number, and therefore the spread of the photon number is also large. It follows that $\langle\alpha|\beta\rangle \approx 1$, so that the entanglement with the microwave field has negligible effects, as has already been pointed out in ref. 3.

Uncertainty relation

We now discuss the role of Heisenberg's position–momentum uncertainty relation in our experiment. For this discussion it is essential that the two ways through the interferometer (beams B and C in Fig. 1) are never separated in transverse position space. This is because the beams have a transverse width of 450 μm (as determined by the width of the lower collimation slit), but are shifted transversely only by a few micrometres (given by d) with respect to each other. This has an important consequence: the storage of which-way information does not imply any storage of transverse position information.

Moreover, the atom is not localized with a precision of the order of d at any stage of the experiment. In particular, the atom stays delocalized during its whole passage through the interaction region, regardless of whether the microwave is on or off. Heisenberg's position–momentum uncertainty relation could only be invoked if the atom were localized. Hence the uncertainty relation does not imply any back action onto the transverse momentum, either "classical" or "quantum".

Of course, in every real experiment there are localization effects due to the finite size of the apparatus. This localization leads to a back action onto the momentum. But this back action is not at all related to the fringe separation. For example, in our set-up the atoms are localized within one wavelength of the microwave. The corresponding back action onto the transverse momentum implied by the uncertainty relation is of the order of one microwave photon recoil and has been analysed above. This back action is four orders of magnitude smaller than the fringe separation, so that it cannot explain the loss of interference. Hence correlations can explain the loss of interference, while the uncertainty relation cannot. This answers the controversial question cited in the introduction: complementarity is not enforced by the uncertainty relation.

This result is not in conflict with the results of Wiseman *et al.*[11], who considered only "experiments in which double-slit interference patterns are destroyed by making a position measurement"[11]. In our experiment, no double slit is used and no position measurement is performed, so that the results of ref. 11 do not apply. It is an open question whether the concept of "quantum" momentum transfer can be generalized to schemes without a mechanical double slit. Such a generalization would have to take into account the fact that in our experiment, the amount of momentum transferred by the light fields is always either zero or exactly $2\hbar k_{light}$.

Although beams B and C are never separated in transverse position space, they are separated in transverse momentum space. The separation is $2\hbar k_{\mathrm{light}}$, as is required for first-order Bragg reflection. Storing which-way information therefore corresponds to storing transverse momentum information with an accuracy of the order of $\Delta p_z \approx \hbar k_{\mathrm{light}}$. So the uncertainty relation implies that the storing process must include a back action onto the transverse position of the order of $\Delta z \approx \lambda_{\mathrm{light}}$. This back action is due to the following effect. In the Bragg regime, the interaction time with the standing light wave, t_{Bragg}, is so long that the atoms move at least a transverse distance of the order of $\lambda_{\mathrm{light}}/2$ within t_{Bragg}. In a naive picture, the atoms can be Bragg-reflected at the beginning or at the end of this interaction, which implies a transverse position uncertainty of the order of $\Delta z \approx \lambda_{\mathrm{light}}$. But this back action onto the near-field position cannot destroy the far-field fringe pattern. The interference pattern created in the interferometer is a pattern in momentum space, not in position space. The far-field position distribution is simply a picture of the final momentum distribution. □

Received 3 February; accepted 15 June 1998.

1. Bohr, N. in *Albert Einstein: Philosopher-Scientist* (ed. Schilpp, P. A.) 200–241 (Library of Living Philosophers, Evanston, 1949); reprinted in *Quantum Theory and Measurement* (eds Wheeler, J. A. & Zurek, W. H.) 9–49 (Princeton Univ. Press, 1983).
2. Feynman, R., Leighton, R. & Sands, M. in *The Feynman Lectures on Physics* Vol. III, Ch. I (Addison Wesley, Reading, 1965).
3. Scully, M. O., Englert, B. G. & Walther, H. Quantum optical tests of complementarity. *Nature* **351**, 111–116 (1991).
4. Haroche, S. in *High-resolution Laser Spectroscopy* (ed. Shimoda, K.) 253–313 (Topics in Applied Physics, Vol. 13, Springer, 1976).
5. Rauch, H. *et al.* Verification of coherent spinor rotation of fermions. *Phys. Lett. A* **54**, 425–427 (1975).
6. Badurek, G., Rauch, H. & Tuppinger, D. Neutron interferometric double-resonance experiment. *Phys. Rev. A* **34**, 2600–2608 (1986).
7. Storey, P., Tan, S., Collett, M. & Walls, D. Path detection and the uncertainty principle. *Nature* **367**, 626–628 (1994).
8. Englert, B. G., Scully, M. O. & Walther, H. Complementarity and uncertainty. *Nature* **375**, 367–368 (1995).
9. Storey, E. P., Tan, S. M., Collett, M. J. & Walls, D. F. Complementarity and uncertainty. *Nature* **375**, 368 (1995).
10. Wiseman, H. & Harrison, F. Uncertainty over complementarity? *Nature* **377**, 584 (1995).
11. Wiseman, H. M. *et al.* Nonlocal momentum transfer in welcher weg measurements. *Phys. Rev. A* **56**, 55–75 (1997).
12. Eichmann, U. *et al.* Young's interference experiment with light scattered from two atoms. *Phys. Rev. Lett.* **70**, 2359–2362 (1993).
13. Cohen-Tannoudji, C. Effect of non-resonant irradiation on atomic energy levels. *Metrologia* **13**, 161–166 (1977); reprinted in Cohen-Tannoudji, C. *Atoms in Electromagnetic Fields* 343–348 (World Scientific, London, 1994).
14. Kunze, S., Dürr, S. & Rempe, G. Bragg scattering of slow atoms from a standing light wave. *Europhys. Lett.* **34**, 343–348 (1996).
15. Kunze, S. *et al.* Standing wave diffraction with a beam of slow atoms. *J. Mod. Opt.* **44**, 1863–1881 (1997).
16. Bernhardt, A. F. & Shore, B. W. Coherent atomic deflection by resonant standing waves. *Phys. Rev. A* **23**, 1290–1301 (1981).

Acknowledgements. We thank S. Kunze for discussions. This work was supported by the Deutsche Forschungsgemeinschaft.

Correspondence and requests for materials should be addressed to G.R. (e-mail: gerhard.rempe@uni-konstanz.de).

PHYSICAL REVIEW LETTERS

VOLUME 81 28 DECEMBER 1998 NUMBER 26

Fringe Visibility and Which-Way Information in an Atom Interferometer

S. Dürr, T. Nonn, and G. Rempe

Fakultät für Physik, Universität Konstanz, 78457 Konstanz, Germany

(Received 19 June 1998)

We experimentally investigate the reduction of the fringe visibility in an atom interferometer due to the storage of which-way information. We focus on the case of incomplete which-way information and use the distinguishability D to quantify how much information is stored. For a given value of D, the fringe visibility V is limited by the duality relation $D^2 + V^2 \leq 1$. We have measured D and V independently. Combining the results, we find good agreement with the duality relation. [S0031-9007(98)08074-0]

PACS numbers: 03.65.Bz, 03.75.Dg, 32.80.−t, 42.50.Vk

Complementarity expresses the fact that every quantum system has at least two properties, which cannot be observed simultaneously. This is often illustrated by means of interferometers, where "the observation of an interference pattern and the acquisition of which-way information are mutually exclusive" [1]. These two properties are associated with a wave and a particle picture, respectively. Two extreme cases of this wave-particle duality are well known from textbook examples: In the absence of which-way (WW) information, a fringe pattern with perfect visibility can be observed, while in the presence of full WW information, there are no fringes. In this paper we consider intermediate situations in which one obtains incomplete WW information and retains interference fringes with a reduced visibility. One of the most interesting questions in this intermediate regime is the following: How much WW information can be obtained for a given value of the visibility? To answer this question, a quantitative measure for WW information is required. We follow the definitions of Englert [1], who distinguishes between two different methods to obtain WW information in two-way interferometers.

The first method is to set up the interferometer such that the particle flux along the two ways differs. This leads to an *a priori* WW knowledge due to the difference between the probabilities, w_+ and w_-, that the particle takes one way or the other. The magnitude of this difference, $P = |w_+ - w_-|$, is the predictability. For a given P, the visibility V of the interference fringes is limited by $P^2 + V^2 \leq 1$ (see Ref. [1]). Here $V = (I_{\max} - I_{\min})/(I_{\max} + I_{\min})$ is determined from the maximum and minimum intensities of the interference fringes, $I_{\max}$ and $I_{\min}$, respectively. Equivalent expressions for this limit were derived in Refs. [2–4]. Experimental results obtained with neutron interferometers [5] are in good agreement with this limit. We therefore restrict the following considerations to symmetric interferometers, where $w_+ = w_- = 1/2$.

The second method to obtain WW information is to add a second quantum system, the WW detector, to the setup. When the particle passes through the interferometer, an appropriate interaction changes the state of the WW detector depending on the particle's way. Hence WW information is stored and can be read out by measuring a suitable observable W of the WW detector. Let $p(W_i, +)$ and $p(W_i, -)$ denote the joint probabilities that the eigenvalue W_i of W is found and that the interfering object took way "+" or way "−." If W_i is found, the best guess about the way one can make is to opt for way + if $p(W_i, +) > p(W_i, -)$, and for way − otherwise. This yields the "likelihood for guessing the right way" [1]

$$L_W = \sum_i \max\{p(W_i, +), p(W_i, -)\}. \tag{1}$$

L_W depends on the choice of the observable W, as this determines the fraction of the stored WW information which is read out. Consider, e.g., the case where full WW information is stored: choosing W carefully, one can

0031-9007/98/81(26)/5705(5)$15.00

VOLUME 81, NUMBER 26 PHYSICAL REVIEW LETTERS 28 DECEMBER 1998

reach $L_W = 1$, while an unfortunate choice of W could result in $L_W = 1/2$, so that one could just as well toss a coin. To measure quantitatively how much WW information is *stored,* the arbitrariness of the *readout* process must be eliminated. This motivates the definition of the "distinguishability of the ways" [1,6]

$$D = -1 + 2\max_{W}\{L_W\}, \tag{2}$$

which uses the maximum of L_W that is reached for the best choice of W. The ways cannot be distinguished at all if $D = 0$, and they can be held apart completely if $D = 1$. For a given value of D the visibility V is limited by the duality relation

$$D^2 + V^2 \leq 1. \tag{3}$$

The equal sign in Eq. (3) holds if the WW detector is initially prepared in a pure state. The duality relation makes a quantitative test of wave-particle duality possible. It has been derived only recently by Jaeger *et al.* [6] and independently by Englert [1], but it has not been tested experimentally. So far, only the reduction of the visibility due to storage of incomplete WW information has been observed in experiments using an optical interferometer [7] or a Ramsey interferometer [8]. However, the distinguishability has not been measured.

In this Letter, we report on measurements of V and D in an atom interferometer, where WW information is stored in internal atomic states. By varying the parameters of the experiment, D can be adjusted continuously. We determined D by an independent measurement of the atom's way and observed the reduction of the visibility. Combining the results for D and V, we find good agreement with the duality relation.

We start with a brief description of the atom interferometer, which has been presented in more detail in Refs. [9,10]. Figure 1 shows a scheme of the setup, which uses Bragg reflection of atoms from standing light waves. A first standing light wave splits the incoming atomic beam, A, into two beams, B and C, of equal atomic flux. After free propagation, a second standing light wave splits each atomic beam into two components. Now two beams, D and E, are traveling to the left, while beams F and G are traveling to the right. In the far field, each pair of overlapping beams produces a spatial interference pattern. The envelope of the interference pattern is given by the collimation properties of the incoming atomic beam. For detection, the atoms are illuminated with laser light and the fluorescence photons are observed.

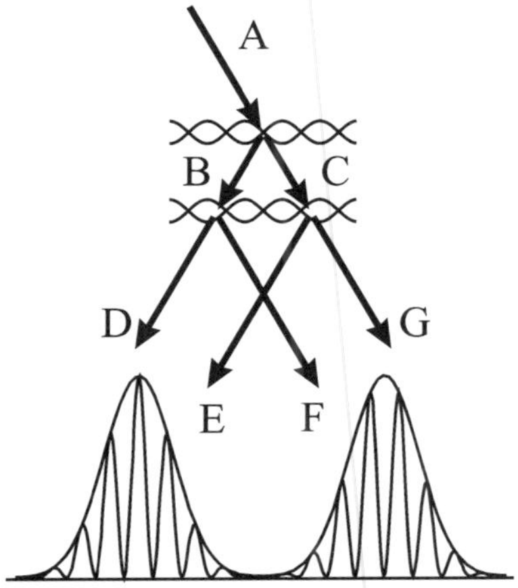

FIG. 1. Scheme of the atom interferometer. Bragg reflection from a standing light wave splits the incoming atomic beam, A, into two beams, B and C. A second standing light wave splits the beams again. In the far field, a spatial interference pattern is observed.

The experiment employs a beam of ^{85}Rb atoms, whose ground state is split into two hyperfine components with total angular momentum $F = 2$ and $F = 3$, labeled $|2\rangle$ and $|3\rangle$, respectively. In these long-lived internal states, WW information can be stored. For that purpose, the frequency of the standing light wave, ω_{light}, is tuned halfway between the $|2\rangle \leftrightarrow |e\rangle$ and $|3\rangle \leftrightarrow |e\rangle$ transition (see Fig. 2). Hence atoms in state $|2\rangle$ ($|3\rangle$) see a red (blue) detuned light field creating a negative (positive) ac-Stark shift potential, corresponding to an optically thicker (thinner) medium. In analogy to light optics one therefore expects [10] that the wave experiences a π phase shift if reflected from an optically thicker medium, i.e., if an atom is Bragg reflected in state $|2\rangle$. However, a detailed calculation [11] shows that here this π phase shift occurs if an atom is transmitted in state $|2\rangle$.

This phase shift is converted into a population difference between states $|2\rangle$ and $|3\rangle$ by applying a microwave field, at frequency ω_{mw}, resonant with the $|2\rangle \leftrightarrow |3\rangle$ transition. Two $\pi/2$ pulses of the microwave are required. They form a Ramsey scheme: one is applied before and one after one of the two standing light waves. We first consider the case, where the first standing light wave is sandwiched between the microwave pulses. The internal atomic state is initially prepared in state $|2\rangle$ and then converted to the superposition state $(|2\rangle + |3\rangle)/\sqrt{2}$ by the first microwave pulse. Next, the standing light wave splits the beam, so

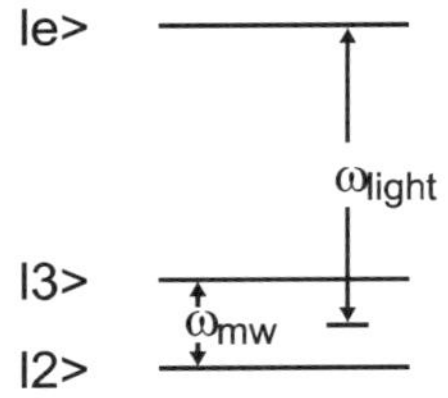

FIG. 2. Level scheme. The ground state is split into two components $|2\rangle$ and $|3\rangle$. The light frequency is chosen midway between the corresponding resonances to the excited state $|e\rangle$.

VOLUME 81, NUMBER 26 PHYSICAL REVIEW LETTERS 28 DECEMBER 1998

that the state vector is changed to

$$|\psi\rangle \propto |\psi_B\rangle \otimes (|3\rangle + |2\rangle) + |\psi_C\rangle \otimes (|3\rangle - |2\rangle), \quad (4)$$

where the minus sign is due to the π phase shift, and $|\psi_B\rangle$ and $|\psi_C\rangle$ denote the state vectors of the center-of-mass motion for the reflected and transmitted beams (B and C in Fig. 1), respectively. The second microwave pulse, acting on both beams (B and C), converts the state vector to

$$|\psi\rangle \propto |\psi_B\rangle \otimes |3\rangle - |\psi_C\rangle \otimes |2\rangle. \quad (5)$$

Obviously, the atom's internal and external degrees of freedom are entangled. This entanglement is the key point for the storage of WW information. If later a measurement of the internal state is performed, the result of this measurement reveals in which of the beams the atom is: if the internal state is found to be $|3\rangle$, the atom was Bragg reflected, otherwise transmitted. Equation (5) indicates that full WW information is stored.

Generalizing this scheme to arbitrary microwave pulse areas allows us to store incomplete WW information. It is sufficient to consider the case where the areas φ of both microwave pulses are identical. In this case, the state vector after the second microwave pulse is

$$|\psi\rangle \propto |\psi_B\rangle \otimes (\cos\varphi|2\rangle + \sin\varphi|3\rangle) - |\psi_C\rangle \otimes |2\rangle. \quad (6)$$

For $\varphi = 0$ no WW information is stored, while for $\varphi = \pi/2$ full WW information is obtained. For intermediate values of φ, incomplete WW information is stored.

Alternatively, the *second* (instead of the *first*) standing light wave can be sandwiched between the two microwave pulses. Then the internal state contains the information whether the atom was reflected or transmitted in the *second* standing light wave. There is no essential difference between these two schemes: both produce two pairs of beams which overlap in the far field thereby creating a spatial interference pattern with visibility

$$V(\varphi) = |\cos\varphi|. \quad (7)$$

In order to measure the distinguishability D, a suitable observable W_{opt} has to be found which maximizes L_{W}. It can be shown that this observable must have the eigenvectors

$$\cos\left(\frac{\pi}{4} - \frac{\varphi}{2}\right)|2\rangle - \sin\left(\frac{\pi}{4} - \frac{\varphi}{2}\right)|3\rangle, \quad (8)$$

$$\sin\left(\frac{\pi}{4} - \frac{\varphi}{2}\right)|2\rangle + \cos\left(\frac{\pi}{4} - \frac{\varphi}{2}\right)|3\rangle, \quad (9)$$

and that a measurement of this observable yields $L_{\mathrm{W}} = (1 + |\sin\varphi|)/2$, so that

$$D(\varphi) = |\sin\varphi|. \quad (10)$$

To measure the likelihood L_{W}, we perform the experiment with only one standing light wave sandwiched between the two microwave pulses, while the other standing light wave is removed. This allows us to measure the atom's external and internal states simultaneously in the following way: first, the far-field position of an atom now reveals whether the atom is in beam B or C. Second, it is possible to detect atoms in either state $|2\rangle$ or state $|3\rangle$ by an appropriate choice of the frequency of the detection laser. This frequency is tuned into resonance with either a transition $|2\rangle \leftrightarrow |e\rangle$ or a transition $|3\rangle \leftrightarrow |e\rangle$, where $|e\rangle$ denotes an excited level. This measurement yields the joint probabilities $p(2,B)$, $p(3,B)$, $p(2,C)$, and $p(3,C)$ which determine L_{W}. Note that a position measurement without internal state detection is also possible. For that purpose the atoms are illuminated with light at both frequencies simultaneously, so that all atoms are detected.

The internal state measurement discussed so far is a measurement of an observable W_0 with eigenvectors $|2\rangle$ and $|3\rangle$. In order to measure an arbitrary observable W of the internal state, a third microwave pulse with suitable parameters is applied before measuring W_0 [12]. For example, to measure W_{opt} according to Eqs. (8) and (9), the area of the third microwave pulse must be $\pi/2 - \varphi$ (or $3\pi/2 - \varphi$, etc.). This allows us to measure the probabilities required to determine L_{W}, from which D can be inferred.

In Fig. 3, the measured values of D (dots) are displayed as a function of the area φ of the first two microwave pulses. Simultaneously, the area of the third microwave pulse was varied, in order to keep it equal to the optimum value $\pi/2 - \varphi$ (or $3\pi/2 - \varphi$, etc.). Hence W_{opt} is measured and the measured data maximize L_{W}. We checked experimentally that no larger value of L_{W} could be obtained by varying the parameters of the third microwave pulse.

According to Eq. (10), D should be unity at $\varphi = \pi/2$. In the experiment, we find $D_{\max} = 0.81 \pm 0.02$. This reduction is mainly due to two effects: first, the presence of background counts in the fluorescence detection reduces D by a factor of 0.90. Second, the standing light wave is not a perfect plane wave but has a transverse profile

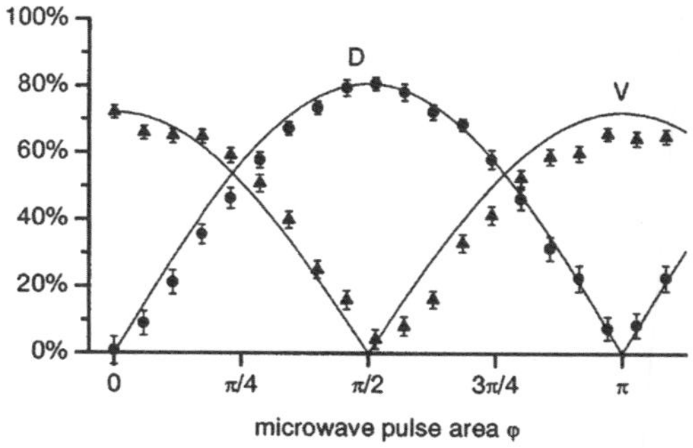

FIG. 3. Visibility V (triangles) and distinguishability D (dots) as a function of the microwave pulse area φ. The solid lines are the theoretical expectations Eqs. (11) and (12).

which exhibits intensity wiggles caused by imperfections in the optical elements. We measured the relative intensity variations and obtained $\sigma_I \approx 12\%$ (rms). Hence the light intensity I seen by an individual atom depends on its transverse position. A detailed calculation shows that the amount of WW information stored is independent of I, but that the parameters of the third microwave pulse must depend on I in order to maximize L_W. As I is not known for each atom, the best one can do is to choose the parameters of the third microwave pulse such that the ensemble average of L_W is maximized. This reduces D by a constant factor $1 - (\pi\sigma_I)^2/2 = 0.93$. Multiplication of both factors gives $D = 0.84$ at $\varphi = \pi/2$, which is in reasonable agreement with the experimental result $D_{max} = 0.81 \pm 0.02$.

Because the background as well as the intensity variations reduce the measured value of D by a constant factor, independent of φ, the measured quantity is

$$D(\varphi) = D_{max}|\sin\varphi| \tag{11}$$

instead of Eq. (10). Equation (11) with $D_{max} = 0.81$ is plotted as a solid line in Fig. 3.

In order to observe an interference pattern, both standing light waves are turned on. The second standing light wave is sandwiched between the two microwave pulses. The triangles in Fig. 3 show the measured values of the fringe visibility V, when all atoms are detected. V was determined by measuring the atomic flux at a maximum and a minimum of the interference pattern. We observed a maximum visibility of $V_{max} = 0.72 \pm 0.02$ at $\varphi = 0$. The reduction from unity can be explained by two effects: first, background counts reduce the visibility by a constant factor 0.92. Second, the detection laser beam and the atom source have a finite size, so that the fringes wash out. This reduces the visibility by a constant factor 0.82. Both factors together give $V = 0.75$ at $\varphi = 0$, which is in reasonable agreement with the experimental result $V_{max} = 0.72 \pm 0.02$. Again, the two factors are independent of φ. Hence Eq. (7) should be modified to

$$V(\varphi) = V_{max}|\cos\varphi|\,. \tag{12}$$

The result is shown as a solid line in Fig. 3 with $V_{max} = 0.72$.

We now discuss whether the variations of the light intensity influence the visibility. As was mentioned above, there is no influence onto the amount of WW information stored. Hence one might assume that the intensity variations do not influence the visibility either. However, it is easy to see that the intensity variations actually reduce the visibility in the right part of the interference pattern which is formed by beams F and G (see Fig. 1). This reduction is due to the fact that the probability for Bragg reflection of an atom from the standing light wave depends on the light intensity. Beam F is reflected twice, while beam G is transmitted twice. This leads to different atomic fluxes in beams F and G and reduces the visibility in the right part of the interference pattern.

This argument does not apply in the left part of the interference pattern, because beams D and E are transmitted and reflected once, so that any deviations from the exact 50:50 beam-splitter ratio compensate each other. Hence the atomic fluxes in beams D and E are equal and the visibility is unaffected. For this reason, the results for V displayed in Fig. 3 were determined in the left part of the interference pattern, namely, from the atomic flux at the central maximum and at one of the neighboring minima.

We conclude that the reduction of the measured distinguishability D and visibility V is well understood. In order to test the duality relation, it is therefore justified to divide the measured data from Fig. 3 by D_{max} and V_{max}, respectively. The result is shown in Fig. 4, where

$$\left(\frac{D(\varphi)}{D_{max}}\right)^2 + \left(\frac{V(\varphi)}{V_{max}}\right)^2 \tag{13}$$

is plotted as a function of φ. The data at $\varphi = 0$ and $\varphi = \pi/2$ (open circles) are close to unity by definition. All other data (full circles) are below unity which means that we find no violation of the duality relation.

The initial state of the WW detector is a pure state, so that all data in Fig. 4 should reach unity. The data for $\varphi < \pi/2$ agree well with this prediction. The deviations for $\varphi > \pi/2$ are due to the fact that the measured visibility is smaller here than for $\varphi < \pi/2$ (see Fig. 3). This reduction of V can be explained by noting that the contrast of the interference pattern is reversed for $\varphi > \pi/2$; i.e., the positions of interference maxima and minima are exchanged. As a consequence, the height of the interference maximum for $\varphi > \pi/2$ is smaller than for $\varphi < \pi/2$ because of the finite width of the beam envelope. Hence background counts and the finite position resolution have a slightly larger effect on V for $\varphi > \pi/2$.

To summarize, we performed the first quantitative test of the recently published duality relation with an atom interferometer. The duality relation sustained the test.

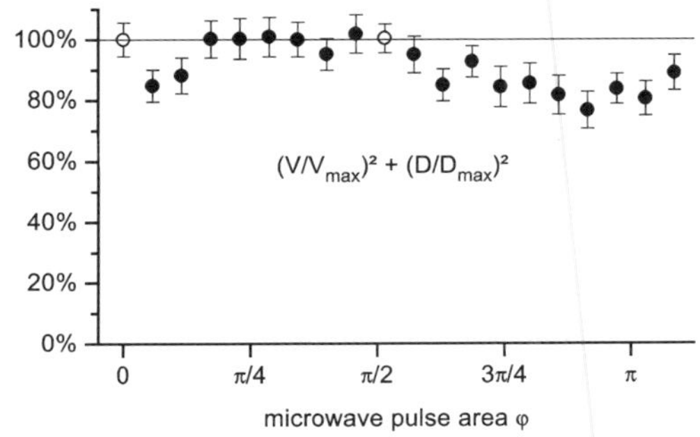

FIG. 4. Experimental test of the duality relation based on the data from Fig. 3. $[D(\varphi)/D_{max}]^2 + [V(\varphi)/V_{max}]^2$ is plotted as a function of φ. According to the duality relation the data points may not exceed unity.

We finally mention that a similar experiment could be performed with a light interferometer, where a half-wave plate is used to rotate the light polarization in one interferometer arm. However, such an experiment can be described in terms of classical electrodynamics, without any quantum theory. Hence, it could not be considered as a test of quantum-mechanical complementarity. In contrast to this, our experiment employs *massive* particles, whose interference phenomena can be explained only in quantum theory.

This work was supported by the Deutsche Forschungsgemeinschaft and the European Union TMR-Network "Coherent Matter Wave Interactions."

[1] B.-G. Englert, Phys. Rev. Lett. **77**, 2154 (1996).
[2] W.K. Wootters and W.H. Zurek, Phys. Rev. D **19**, 473 (1979).
[3] D.M. Greenberger and A. Yasin, Phys. Lett. A **128**, 391 (1988).
[4] L. Mandel, Opt. Lett. **16**, 1882 (1991).
[5] J. Summhammer, H. Rauch, and D. Tuppinger, Phys. Rev. A **36**, 4447 (1987).
[6] G. Jaeger, A. Shimony, and L. Vaidman, Phys. Rev. A **51**, 54 (1995).
[7] X.Y. Zou, L.J. Wang, and L. Mandel, Phys. Rev. Lett. **67**, 318 (1991).
[8] M. Brune *et al.,* Phys. Rev. Lett. **77**, 4887 (1996).
[9] S. Kunze *et al.,* J. Mod. Opt. **44**, 1863 (1997).
[10] S. Dürr, T. Nonn, and G. Rempe, Nature (London) **395**, 33 (1998).
[11] S. Dürr and G. Rempe, in "Advances in Atomic, Molecular, and Optical Physics," edited by B. Bederson and H. Walther (Academic Press, New York, to be published).
[12] For technical reasons, it is difficult to detect only atoms in state $|2\rangle$. The use of the third microwave pulse solves this problem, because for a pulse area of π, the subsequent detection of atoms in state $|3\rangle$ corresponds to the direct detection of atoms in state $|2\rangle$.

Chapter 3

Mutually Unbiased Bases and the Mean King's Problem

Karol Życzkowski

Instytut Fizyki Uniwersytetu Jagiellońskiego, Kraków, Poland
Centrum Fizyki Teoretycznej PAN, Warszawa, Poland

Thomas Durt

École Centrale de Marseille, Institut Fresnel, Domaine Universitaire de Saint-Jérôme, Marseille, France

Ingemar Bengtsson

Stockholms Universitet, Fysikum, Stockholm, Sweden

Article Title: **On mutually unbiased bases**

Reference: International Journal of Quantum Information **8**, 535–640 (2010)

Editors' note: Due to the length of the article, only the abstract and the contents pages are reprinted here.

Personal messages:

Berge, unbiased bases and positive biased encounters

Karol Życzkowski Kraków, March 7, 2014

During the conference *Quantum Optics V* organized in Zakopane (the Polish Tatra Mountains) in June 2001, I met Berge Englert for the very first time. He delivered there a nice talk on communication with single-photon two-qubit states, but at that time this subject was not so interesting for me. However, as I was then asked by the editors of a journal of the Polish Physical Society to translate from German an article on teaching quantum mechanics, I approached Berge asking him about some sophisticated, long expressions in the German text. Not only did he offer immediate help and provided several good translation hints, he also amazed me with his truly exceptional command of my native language. Not so often can one meet a German Physikprofessor with such a deep knowledge of Polish!

A few years later Berge visited Cracow and I had a chance to know him better. Only then I learned that Berge stands for Berthold-Georg, and that the owner of this name plays chess and bridge well! As he visited us at the Institute of Physics of the Jagiellonian University, we started to discuss various methods to find mutually unbiased bases and complex Hadamard matrices — unitary matrices with equal moduli for all their entries.

As we were aware of some statements concerning the existence of such matrices for orders 2–5, and showed Berge some nice examples of such matrices of size $N = 6$, he encouraged us to collect a-ll cases known and put it on the web. It took us several years to complete this task and to update this unusual zoo by some new species, but eventually an online catalog of complex Hadamard matrices was constructed, and is there, available to all people interested, `http://chaos.if.uj.edu.pl/~karol/hadamard`, due to a direct inspiration and encouragement of Berge.

$$\begin{bmatrix} 1 & d & d & 1 & d & 1 & 1 \\ 1 & 1 & d & d & 1 & d & 1 \\ 1 & 1 & 1 & d & d & 1 & d \\ d & 1 & 1 & 1 & d & d & 1 \\ 1 & d & 1 & 1 & 1 & d & d \\ d & 1 & d & 1 & 1 & 1 & d \\ d & d & 1 & d & 1 & 1 & 1 \end{bmatrix}$$

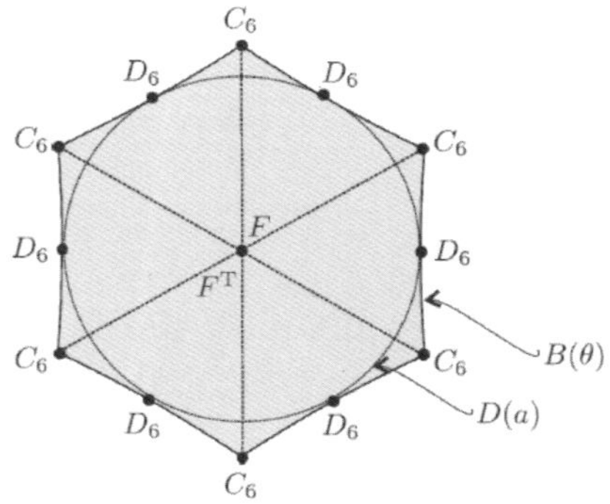

Figure. On the left, a circular complex Hadamard matrix C_7 of order 7, with $d = \frac{1}{4}(i\sqrt{7} - 3)$; on the right, a family of $N = 6$ complex Hadamard matrices — the final plot by B.-G. E.

Early 2005, Berge invited me for a conference to Singapore. I was working then on the structure of sets of quantum states and quantum maps, but I met there Thomas Durt, who convinced me to continue to work on mutually unbiased bases. Furthermore, Thomas and Berge invited me then to write a joint review on MUBs and complex Hadamard matrices. As we already knew a lot about these issues, the project looked rather innocent, and originally we could hardly imagine that it would take us more than five years to complete it. As Ingemar Bengtsson from Stockholm decided to join the team, the four of us tarted to work together, and kept exchanging by e-mail the chapters, tex-files and figures between four different places. The number of pages, sections and references of our opus grew fast, and Berge was the only person who was in control of the entire complex process. During all these years, I was really amazed by the unbelievable patience and understanding with which Berge was treating our neverending list of wishes, suggestions and corrections concerning the review. Moreover, Berge cared a lot, to improve and expand all the results and subsections he got from us. For instance, as I suggested to add to the paper an appendix with a prime distinguishing function (equal to zero for prime arguments, and different from zero elsewhere), related to the fact that the standard MUB's construction works for prime dimensions only, Berge started to work on the subject himself and found quickly some amazing properties of this function.

It was a great experience for me to participate in the project and to work together with Berge. I am always pleased to meet him from time to time here or there, as it is always a pleasure and fun to talk to a person with such a broad knowledge of different issues as Berthold-Georg.

Hence Berge: Happy Birthday, Sto lat and ***Alles Gute zum Geburtstag.*** **KŻ**

Thomas Durt Marseille, March 5, 2014

I met Berge some years ago during a visit to Singapore. At that time, the Q-Lah was in its pioneering phase, a seed of what would become the Centre for Quantum Technologies some years later. Berge was in a sense cruel with me: He explained to me the so-called King's problem (see below), and let me gently marinate in my own juices for at least one day.

The King's Problem:
A mean king challenges a physicist, who got stranded on the remote island ruled by the king, to prepare a spin-1/2 atom in any state of her choosing and to perform a control measurement of her liking. Between her preparation and her measurement, the king's men determine the value of either the spin along X, Y or Z. Only after she completed her control measurement, the physicist is told which spin component has been measured, and she must then state the result of that intermediate measurement correctly. How does she do it?

Needless to say, I did not find the answer, in which, by the way, entanglement plays a prior role. Later I investigated more in depth the role played by Galois fields in this problem (and also in the so-called MUB problem), and Berge proposed to me to write together a kind of review paper about these questions. This was the beginning of a very long process, which resulted in a very long paper too, during which two coauthors Karol Życzkowski and Ingemar Bengtsson also played an important role (I will come back to that). Our work around this paper was punctuated by yearly visits that I paid to NUS in that period, and also by the exchange of many emails.

First of all, I am thankful to Berge and our coauthors who proposed me to be the first author. I still believe that, to some extent, if I enjoy today a full professor position in France, this paper contributed to it. Besides, this long collaboration let me discover many aspects of Berge's personality as I will explain now. The first one is rigor and sense of detail (which are not my first qualities I am afraid); among others I discovered by then, despite the fact that I was not so young when we worked together, there exist some widespread conventions which had previously escaped my attention. For instance, internal references to equations are always under parentheses (:-)), the odd pages of a book always lie on the pages on the right of the open book, and so on. Berge also completed and generalized certain proofs, and he systematically dived deeper than me in the underlying maths, to the extent that I was quickly convinced that when he had corrected and edited a section of our paper, there was nothing to add. To the great despair of his wife Ola, Berge devoted a lot of his weekends and holiday periods to work on the paper which became a kind of humorous subject between us. John Bell pretended to be a quantum engineer during his weekly work at CERN "but to have principles on Sundays"; Berge was in my eyes a teacher and group leader during the week and an accomplished mathematical physicist on Sundays.

Our patience was also challenged by our two coauthors Karol and Ingemar who grandly contributed to the part of the paper devoted to the problem of MUB in a six-dimensional Hilbert space. The "problem" was that there exists an active community of mathematicians and physicists who permanently work on this topic and that every six months this chapter had to be rewritten due to new developments in the field. This also became a humorous subject between Berge and me.

Actually, I nearly never saw Berge impatient or angry; I saw him nevertheless ironical in several opportunities, mostly in relation to the foundational and interpretational problems. Roughly, my feeling is that Berge does not like "charlatanism", I mean hereby empty discussions and vague speculation. His approach to the Quantum Theory is in my eyes, a pragmatic approach, not very distant from Bohr's view: Even if it is not clear where the border line of the theory would be exactly located, nothing prevents us from making use of it for solving problems, realizing experiments, developing new, elegant and clever tools. This attitude also characterizes in my eyes the generation of quantum physicists (quantum opticians, atomic physicists) to which he belongs, a generation which constitutes a bridge between Feynman and Schwinger's QED and the more recent developments of Quantum Information. This generation was very creative and productive if we think about major breakthroughs such as the laser, laser cooling, BECs, QED cavities, ion traps, entanglement related technologies and so on. To conclude, concerning the foundations of the Quantum Theory, I think that Berge constitutes a consistent "incarnation" of Bohrian complementarity: He respects conventions without being a formalist, he continues tradition without being a traditionalist, and he is actively involved in fundamental physics without being a fundamentalist.

Last but not least, I salute the work of Berge as a teacher, having for instance in mind the third volume of his lectures on quantum mechanics, that was to some of my master and Ph.D students what mother milk is to newborn babies. Enfin, it is my pleasure to contribute to this Festschrift, and to say...

Bon anniversaire mister Englert! TD

Ingemar Bengtsson Stockholm, March 6, 2014

I first met Berge in a very small meeting that Karol arranged in Kraków. There were only five participants and two talks, in one of which I expressed my surprise that the MUB existence problem in six dimensions had not been solved. I understood only later that "Berge" is short for "B.-G. Englert", so I was puzzled by the fact that one participant knew such a lot. The collaboration we had was just what Tom says it was. Occasionally I would disagree with the others, but then there would come long, carefully argued emails from Berge setting out in detail why I was wrong, so no harm was done. Afterwards

I came to know also some of Berge's students, and they have clearly inherited deep understanding and broad knowledge from their supervisor. I think this has something to do with the pleasant atmosphere he has created in Singapore. And incidentally, although papers like that of Raynal, Lü, and Englert have made it clear what to expect, **the MUB existence problem in dimension six is still, strictly speaking, unsolved. IB**

International Journal of Quantum Information
Vol. 8, No. 4 (2010) 535–640

DOI: 10.1142/S0219749910006502

ON MUTUALLY UNBIASED BASES

THOMAS DURT

TONA Vrije Universiteit Brussel,
Pleinlaan 2, B-1050 Brussels, Belgium
thomdurt@vub.ac.be

BERTHOLD-GEORG ENGLERT

Centre for Quantum Technologies,
National University of Singapore,
3 Science Drive 2, Singapore 117543, Singapore

Department of Physics, National University of Singapore,
2 Science Drive 3, Singapore 117542, Singapore
cqtebg@nus.edu.sg

INGEMAR BENGTSSON

Stockholms Universitet, Fysikum, Alba Nova,
106 91 Stockholm, Sweden
ingemar@physto.se

KAROL ŻYCZKOWSKI

Instytut Fizyki Uniwersytetu Jagiellońskiego,
ul. Reymonta 4, 30-059 Kraków, Poland

Centrum Fizyki Teoretycznej PAN, Al. Lotników 32/44,
02-668 Warszawa, Poland
karol@tatry.if.uj.edu.pl

Received 20 April 2010

Mutually unbiased bases for quantum degrees of freedom are central to all theoretical investigations and practical exploitations of complementary properties. Much is known about mutually unbiased bases, but there are also a fair number of important questions that have not been answered in full as yet. In particular, one can find maximal sets of $N+1$ mutually unbiased bases in Hilbert spaces of prime-power dimension $N = p^{\mathrm{M}}$, with p prime and M a positive integer, and there is a continuum of mutually unbiased bases for a continuous degree of freedom, such as motion along a line. But not a single example of a maximal set is known if the dimension is another composite number ($N = 6, 10, 12, \ldots$).

In this review, we present a unified approach in which the basis states are labeled by numbers $0, 1, 2, \ldots, N-1$ that are both elements of a Galois field and ordinary integers. This dual nature permits a compact systematic construction of maximal sets of mutually unbiased bases when they are known to exist but throws no light on the open existence problem in other cases. We show how to use the thus constructed mutually unbiased bases in quantum-informatics applications, including dense coding, teleportation, entanglement swapping, covariant cloning, and state tomography, all of which rely on an explicit set of maximally entangled states (generalizations of the familiar two−q-bit Bell states) that are related to the mutually unbiased bases.

There is a link to the mathematics of finite affine planes. We also exploit the one-to-one correspondence between unbiased bases and the complex Hadamard matrices that turn the bases into each other. The ultimate hope, not yet fulfilled, is that open questions about mutually unbiased bases can be related to open questions about Hadamard matrices or affine planes, in particular the notorious existence problem for dimensions that are not a power of a prime.

The Hadamard-matrix approach is instrumental in the very recent advance, surveyed here, of our understanding of the $N = 6$ situation. All evidence indicates that a maximal set of seven mutually unbiased bases does not exist — one can find no more than three pairwise unbiased bases — although there is currently no clear-cut demonstration of the case.

Keywords: Mutually unbiased bases; complex Hadamard matrices; generalized Bell states.

Contents

Yakir Aharonov

School of Physics and Astronomy, Tel-Aviv University, Tel-Aviv, Israel
Institute for Quantum Studies, Chapman University, Orange, California, USA

Paper Title: **The mean king's problem: Spin 1**

Reference: Zeitschrift für Naturforschung **56a**, 16–19 (2001)

Paper Title: **The mean king's problem: prime degrees of freedom**

Reference: Physics Letters A **284**, 1–5 (2001)

Personal message:

The first time I met Berge, he was still at the Max Planck Institute in Munich. Already at this meeting, I was very impressed with his deep insights regarding quantum mechanics, as well as his outstanding mathematical abilities. Since then, we met many times and collaborated in a number of works. In those collaborations, he played a major role in developing the ideas and solving the difficult problems we encountered. I find particularly important Berge's generalization of the Mean King's Problem (see the two paper contributions). I wish him a very happy birthday and many more productive years.

The Mean King's Problem: Spin 1

Yakir Aharonov[a,b] and Berthold-Georg Englert[c,d]

[a] School of Physics and Astronomy, Tel-Aviv University, Tel-Aviv 69978, Israel
[b] Physics Department, University of South Carolina, Columbia SC 29208, USA
[c] Atominstitut der Österreichischen Universitäten, Stadionallee 2, 1020 Wien, Austria
[d] Max-Planck-Institut für Quantenoptik, Hans-Kopfermann-Straße 1, 85748 Garching, Germany

Reprint requests to Dr. B.-G. E.; E-mail: bge@mpq.mpg.de

Z. Naturforsch. **56 a**, 16–19 (2001); received January 12, 2001

Presented at the 3rd Workshop on Mysteries, Puzzles and Paradoxes in Quantum Mechanics, Gargnano, Italy, September 17 - 23, 2000.

We show how one can ascertain the values of four mutually complementary observables of a spin-1 degree of freedom. – Pacs: 03.65.Bz

Key words: Complementary Observables; Complete Sets of Observables; Entanglement.

1. Introduction

About a dozen years ago, one of us (YA) co-authored a paper [1] with the somewhat provocative title "How to ascertain the values of σ_x, σ_y, and σ_z of a spin-$\frac{1}{2}$ particle". It reports the solution of what has later become known as *The King's Problem*:

> A ship-wrecked physicist gets stranded on a far-away island that is ruled by a mean king who loves cats and hates physicists since the day when he first heard what happened to Schrödinger's cat. A similar fate is awaiting the stranded physicist. Yet, mean as he is, the king enjoys defeating physicists on their own turf, and therefore he maliciously offers an apparently virtual chance of rescue.
>
> He takes the physicist to the royal laboratory, a splendid place where experiments of any kind can be performed perfectly. There the king invites the physicist to prepare a certain silver atom in any state she likes. The king's men will then measure one of the three cartesian spin components of this atom – they'll either measure σ_x, σ_y, or σ_z without, however, telling the physicist which one of these measurements is actually done. Then it is again the physicist's turn, and she can perform any experiment of her choosing. Only after she's finished with it, the king will tell her which spin component had been measured by his men. To save her neck, the physicist must then state correctly the measurement result that the king's men had obtained.
>
> Much to the king's frustration, the physicist rises to the challenge – and not just by sheer luck: She gets the right answer any time the whole procedure is repeated. How does she do it?

Readers who don't know the answer should try to figure it out themselves rather than consult the said reference. There is a lesson here about the wonderful things entanglement can do for you.

It is worth mentioning that this thought experiment of 1987 has not been realized as yet. Very recently, however, a quantum-optical analog has been formulated [2], and it is hoped that experimental data will be at hand shortly.

The present paper deals with a generalization of the king's problem. Instead of the traditional spin-$\frac{1}{2}$ atom, we consider the situation of a spin-1 atom. The two main questions are then: What are the appropriate spin-1 analogs of the spin-$\frac{1}{2}$ observables σ_x, σ_y, σ_z? And, how does the physicist save her neck now?

The first question is answered in Sect. 2 in terms of a complete set of mutually complementary observables. The answer to the second question is given in Sect. 3; it employs essentially the same strategy that works in the spin-$\frac{1}{2}$ case, so that we have a genuine generalization indeed. Further generalizations to even higher spins will be discussed in [3].

2. Mutually Complementary Observables

The three spin-$\frac{1}{2}$ observables σ_x, σ_y, σ_z are *complete* in the sense that the probabilities for finding their eigenvalues as the results of measurements specify uniquely the statistical operator that characterizes the spin-$\frac{1}{2}$ degree of freedom of the ensemble under consideration. They are not overcomplete because this unique specification is not ensured if one of the spin components is left out.

In addition to being complete, the observables σ_x, σ_y, σ_z are also pairwise *complementary*, which is to say that in a state where one of them has a definite value, all measurement results for the other ones are equally probable. For example, if $\sigma_x = 1$ specifies the ensemble, say, then the results of σ_y measurements are utterly unpredictable: +1 and -1 are found with equal frequency; and the same applies to σ_z measurements.

What is essential here are not the eigenvalues of σ_x, σ_y, σ_z, but their sets of eigenstates. It is familiar that they are related to each other by

$$\begin{aligned} |\sigma_x = \pm 1\rangle &= 2^{-\frac{1}{2}}\left(|\sigma_z = +1\rangle \pm |\sigma_z = -1\rangle\right), \\ |\sigma_y = \pm 1\rangle &= 2^{-\frac{1}{2}}\left(|\sigma_z = +1\rangle \pm i|\sigma_z = -1\rangle\right), \end{aligned} \tag{1}$$

if the usual phase conventions are adopted. The fact that the transition probabilities

$$\begin{aligned} \left|\langle\sigma_x = \pm 1|\sigma_y = \pm 1\rangle\right|^2 &= \frac{1}{2}, \\ \left|\langle\sigma_y = \pm 1|\sigma_z = \pm 1\rangle\right|^2 &= \frac{1}{2}, \\ \left|\langle\sigma_z = \pm 1|\sigma_x = \pm 1\rangle\right|^2 &= \frac{1}{2}, \end{aligned} \tag{2}$$

do not depend on the quantum numbers ± 1, is the statement of the pairwise complementary nature of σ_x, σ_y, and σ_z. Their algebraic completeness is then an immediate consequence of the insight that a spin-$\frac{1}{2}$ degree of freedom can have at most three mutually complementary observables [4].

Analogously, there can be no more than four such observables for a spin-1 degree of freedom. Let's call them A_0, A_1, A_2, and A_3, and to be specific, we take their eigenvalues to be 0, 1, and 2. We denote by $|m_k\rangle$ the eigenstate of A_m to eigenvalue k, and we express the eigenstates of A_1, A_2, A_3 in terms of those of A_0. With

$$x = e^{i2\pi/3}, \tag{3}$$

the basic cubic root of unity, it is a matter of inspection to verify that the choice

$$\begin{aligned} \left(|1_0\rangle, |1_1\rangle, |1_2\rangle\right) &= \left(|0_0\rangle, |0_1\rangle, |0_2\rangle\right)\frac{1}{\sqrt{3}}\begin{pmatrix} x & 1 & 1 \\ 1 & x & 1 \\ 1 & 1 & x \end{pmatrix}, \\ \left(|2_0\rangle, |2_1\rangle, |2_2\rangle\right) &= \left(|0_0\rangle, |0_1\rangle, |0_2\rangle\right)\frac{1}{\sqrt{3}}\begin{pmatrix} x^2 & 1 & 1 \\ 1 & x^2 & 1 \\ 1 & 1 & x^2 \end{pmatrix}, \\ \left(|3_0\rangle, |3_1\rangle, |3_2\rangle\right) &= \left(|0_0\rangle, |0_1\rangle, |0_2\rangle\right)\frac{1}{\sqrt{3}}\begin{pmatrix} 1 & 1 & 1 \\ 1 & x & x^2 \\ 1 & x^2 & x \end{pmatrix}, \end{aligned} \tag{4}$$

is indeed such that

$$\left|\langle m_k|m'_{k'}\rangle\right|^2 = \begin{cases} \delta_{kk'} & \text{if } m = m', \\ \frac{1}{3} & \text{if } m \neq m', \end{cases} \tag{5}$$

so that each set consists of 3 orthonormal states, as it should, and any two different sets are complementary.

Repeated measurements of the observables A_m (on identically prepared spin-1 systems) eventually determine the probabilities $p_k^{(m)}$ for finding their eigenstates $|m_k\rangle$. As a consequence of their mutual complementarity, knowledge of the probabilities for one A_m contains no information whatsoever about the probabilities for any other one. These 12 probabilities represent 8 parameters in total, since $p_0^{(m)} + p_1^{(m)} + p_2^{(m)} = 1$ for each of the four A_ms. The statistical operator that characterizes the ensemble of identically prepared spin-1 systems,

$$\rho = \sum_{m=0}^{3}\sum_{k=0}^{2}|m_k\rangle\left(p_k^{(m)} - \frac{1}{4}\right)\langle m_k|, \tag{6}$$

is therefore uniquely determined by the probabilities $p_k^{(m)} = \langle m_k|\rho|m_k\rangle$. Indeed, the A_ms constitute a complete set of mutually complementary observables for the spin-1 degree of freedom.

3. Spin-1 Version of the Mean King's Problem

Accordingly, in the spin-1 version of *The King's Problem* either one of A_0, A_1, A_2, or A_3 is measured by the mean king's men, on a spin-1 atom suitably prepared by the physicist. Without knowing which measurement was done actually, the physicist performs a subsequent measurement of her own, and – after then being told which A_m was measured by the king's men – she has to state correctly what they found: 0 or 1 or 2.

The physicist solves the problem by first preparing a state $|\Psi_0\rangle$ in which the given spin-1 atom is entangled with another, auxiliary, spin-1 atom. Two-atom states in which the given atom is in state $|m_k\rangle$ and the auxiliary atom in $|m'_{k'}\rangle$ are denoted by $|m_k m'_{k'}\rangle$. Then

$$\begin{aligned} |\Psi_0\rangle &= 3^{-\frac{1}{2}}\bigl(|0_00_0\rangle + |0_10_1\rangle + |0_20_2\rangle\bigr) \\ &= 3^{-\frac{1}{2}}\bigl(|1_02_0\rangle + |1_12_1\rangle + |1_22_2\rangle\bigr) \\ &= 3^{-\frac{1}{2}}\bigl(|2_01_0\rangle + |2_11_1\rangle + |2_21_2\rangle\bigr) \\ &= 3^{-\frac{1}{2}}\bigl(|3_03_0\rangle + |3_13_2\rangle + |3_23_1\rangle\bigr) \qquad (7) \end{aligned}$$

are alternative ways of writing the state she prepares. Their equivalence is easily verified with the aid of the transformation laws (4).

If the king's men then measure A_m on the given atom and find the value k, the resulting two-atom state is the respective $|m_k m'_{k'}\rangle$ component of $|\Psi_0\rangle$. After their measurement, there are thus all together 4 trios of possible two-atom states. We write them compactly as

$$\begin{aligned} \bigl(|0_00_0\rangle, |0_10_1\rangle, |0_20_2\rangle\bigr) &= \bigl(|\Psi_0\rangle, |\Psi_1\rangle, |\Psi_2\rangle\bigr)\mathcal{U}, \\ \bigl(|1_02_0\rangle, |1_12_1\rangle, |1_22_2\rangle\bigr) &= \bigl(|\Psi_0\rangle, |\Psi_3\rangle, |\Psi_4\rangle\bigr)\mathcal{U}, \\ \bigl(|2_01_0\rangle, |2_11_1\rangle, |2_21_2\rangle\bigr) &= \bigl(|\Psi_0\rangle, |\Psi_5\rangle, |\Psi_6\rangle\bigr)\mathcal{U}, \\ \bigl(|3_03_0\rangle, |3_13_2\rangle, |3_23_1\rangle\bigr) &= \bigl(|\Psi_0\rangle, |\Psi_7\rangle, |\Psi_8\rangle\bigr)\mathcal{U}, \quad (8) \end{aligned}$$

where the 3-rows on the right are multiplied by the unitary 3×3 matrix

$$\mathcal{U} = \frac{1}{\sqrt{3}}\begin{pmatrix} 1 & 1 & 1 \\ 1 & x & x^2 \\ 1 & x^2 & x \end{pmatrix}, \qquad (9)$$

which we met in (4) as well. Since the members of each trio are orthogonal to each other, the 8 two-atom states $|\Psi_1\rangle, \ldots, |\Psi_8\rangle$ introduced here are orthogonal to $|\Psi_0\rangle$ by construction. It is equally immediate that the paired states $|\Psi_{2m+1}\rangle, |\Psi_{2m+2}\rangle$ are orthogonal to each other for $m = 0, 1, 2, 3$. That, more generally, the orthonormality relation

$$\langle\Psi_j|\Psi_k\rangle = \delta_{jk} \quad \text{for } j, k = 0, \ldots, 8 \qquad (10)$$

holds also for states from different trios can be checked explicitly (or one recognizes a special case of a more general statement [3]).

The physicist will be able to state correctly the measurement result found by the king's men if she can find a two-atom observable P with a set of eigenstates $|P_0\rangle, \ldots, |P_8\rangle$ such that each $|P_k\rangle$ is orthogonal to two members each of the four trios on the left of (8). It is convenient to specify such states by indicating which members they are *not* orthogonal to, so that

$$|[k_0k_1k_2k_3]\rangle \qquad (11)$$

has the defining property of being orthogonal to the two-atom states that result when measurements of A_m do *not* give the eigenvalue k_m.

In order to see how this enables her to infer the measured value, suppose the physicist finds the two-atom system in state $|[1012]\rangle$. She then knows that if the king's men had measured A_0, A_1, A_2, or A_3, the respective results must have been 1, 0, 1, and 2, because she would never find $|[1012]\rangle$ for other measurement results.

Accordingly, all that is needed to complete the solution of the spin-1 version of the mean king's problem is the demonstration that we can have a complete orthonormal set of two-atom states of the kind (11). First note that the expansion of $|[k_0k_1k_2k_3]\rangle$ in the $|\Psi_j\rangle$ basis is given by

$$\begin{aligned} |[k_0k_1k_2k_3]\rangle &= \frac{1}{3}|\Psi_0\rangle \qquad (12) \\ &\quad + \frac{1}{3}\sum_{m=0}^{3}\bigl(|\Psi_{2m+1}\rangle x^{k_m} + |\Psi_{2m+2}\rangle x^{-k_m}\bigr). \end{aligned}$$

Then observe that

$$\langle[k_0k_1k_2k_3]|[k'_0k'_1k'_2k'_3]\rangle = \frac{1}{3}\sum_{m=0}^{3}\delta_{k_m,k'_m} - \frac{1}{3}, \quad (13)$$

so that two such states are orthogonal if $k_m = k'_m$ for one and only one m value. Therefore, a possible choice of basis states for the physicist's final measurement is

$$|P_0\rangle = |[0000]\rangle,\ |P_1\rangle = |[0111]\rangle,\ |P_2\rangle = |[0222]\rangle,$$
$$|P_3\rangle = |[1012]\rangle,\ |P_4\rangle = |[1120]\rangle,\ |P_5\rangle = |[1201]\rangle,$$
$$|P_6\rangle = |[2021]\rangle,\ |P_7\rangle = |[2102]\rangle,\ |P_8\rangle = |[2210]\rangle. \tag{14}$$

After being told which measurement the king's men performed on the given atom, she can then infer their measurement result correctly, and with certainty, in the manner described above for $|P_3\rangle = |[1012]\rangle$.

Acknowledgements

YA gratefully acknowledges the kind hospitality extended to him by Herbert Walther during a visit at the MPI für Quantenoptik in Garching. BGE would like to thank Wolfgang Schleich and the University of Ulm for financial support while part of this work was done. YA's research is supported in part by Grant No. 471/98 of the Basic Research Foundation (administered by the Israel Academy of Sciences and Humanities) and NSF Grant No. PHY-9971005.

[1] L. Vaidman, Y. Aharonov, and D. Z. Albert, Phys. Rev. Lett. **58**, 1385 (1987).

[2] B.-G. Englert, Ch. Kurtsiefer, and H. Weinfurter, Universal unitary gate for single-photon 2-qubit states, Phys. Rev. A (in print).

[3] Y. Aharonov and B.-G. Englert, The mean king's problem: Prime degrees of freedom, in preparation.

[4] W. K. Wootters and B. D. Fields, Ann. Phys. (NY) **191**, 363 (1989).

ELSEVIER

28 May 2001

Physics Letters A 284 (2001) 1–5

PHYSICS LETTERS A

www.elsevier.nl/locate/pla

The mean king's problem: prime degrees of freedom

Berthold-Georg Englert [a,b,*], Yakir Aharonov [c,d]

[a] *Atominstitut, Technische Universität Wien, Stadionallee 2, 1020 Wien, Austria*
[b] *Max-Planck-Institut für Quantenoptik, Hans-Kopfermann-Strasse 1, 85748 Garching, Germany*
[c] *School of Physics and Astronomy, Tel-Aviv University, Tel-Aviv 69978, Israel*
[d] *Physics Department, University of South Carolina, Columbia, SC 29208, USA*

Received 27 March 2001; accepted 17 April 2001
Communicated by P.R. Holland

Abstract

We show how one can ascertain the values of a complete set of mutually complementary observables of a prime degree of freedom.

Keywords: Quantum kinematics; Complementary observables; Complete sets of observables; Entanglement

1. Introduction

In 1987, one of us (Y.A.) co-authored a paper [1] with the somewhat provocative title "How to ascertain the values of σ_x, σ_y, and σ_z of a spin-1/2 particle". It reports the solution of what later became known as *The King's Problem*: A mean king challenges a physicist, who got stranded on the remote island ruled by the king, to prepare a spin-1/2 atom in any state of her choosing and to perform a control measurement of her liking. Between her preparation and her measurement, the king's men determine the value of either σ_x, or σ_y, or σ_z. Only after she completed her control measurement, the physicist is told which spin component has been measured, and she must then state the result of that intermediate measurement correctly. How does she do it?

This thought experiment has not been realized as yet. But recently an optical analog has been formulated [2], and experimental data should be at hand shortly. Somewhat unexpectedly, and rather rewardingly, the photon version of the king's problem suggested a new scheme for quantum cryptography [3].

Also very recently, we reported a generalization of the king's problem [4] where, instead of the traditional spin-1/2 atom, a spin-1 atom is used. This generalization required answers to two questions: What are the appropriate spin-1 analogs of the spin-1/2 observables $\sigma_x, \sigma_y, \sigma_z$? And, how does the physicist rise to the challenge now?

In the present Letter we deal with the further generalization to arbitrary prime degrees of freedom, where measurements can have at most p different outcomes, p being any prime number. Of course, the situations of Refs. [1] and [4], spin-1/2 ($p = 2$) and spin-1 ($p = 3$), respectively, are particular realizations of the prime case. We believe that this extension of the idea of Ref. [1] teaches us a potentially important lesson about the mathematical structure of quantum kinematics.

* Corresponding author.
E-mail address: bge@mpq.mpg.de (B.-G. Englert).

0375-9601/01/$ – see front matter
PII: S0375-9601(01)00271-7

In Section 2 we answer the general-prime version of the first question asked above. The analogs of the three spin-1/2 observables are identified as complete sets of mutually complementary observables. Then the answer to the second question is given in Section 3; it employs essentially the same strategy that works in the cases of spin-1/2 and spin-1, so that we have a genuine generalization indeed. We leave it as a moot point whether generalizations to non-prime degrees of freedom are possible, or if there are analogs of the variants of the spin-1/2 problem that were found by Ben-Menahem [5] and Mermin [6]. Also, we do not address the intriguing question of whether the geometrical reasoning that works so well in the spin-1/2 case [7] lends itself to generalizations for spin-1 or richer degrees of freedom.

2. Pairwise complementary observables

The three spin-1/2 observables $\sigma_x, \sigma_y, \sigma_z$ are *complete* in the sense that the probabilities for finding their eigenvalues as the results of measurements specify uniquely the statistical operator that characterizes the spin-1/2 degree of freedom of the ensemble under consideration. They are not overcomplete because this unique specification is not ensured if one of the spin components is left out.

In addition to being complete, the observables $\sigma_x, \sigma_y, \sigma_z$ are also pairwise *complementary*, which is to say that in a state where one of them has a definite value, all measurement results for the other ones are equally probable. For example, if $\sigma_x = 1$ specifies the ensemble, say, then the results of σ_y measurements are utterly unpredictable: $+1$ and -1 are found with equal frequency; and the same is true for σ_z measurements.

What is essential here are not the eigenvalues of $\sigma_x, \sigma_y, \sigma_z$, but their sets of eigenstates. In technical terms, the fact that the transition probabilities

$$\begin{aligned} &\left|\langle\sigma_x=\pm1|\sigma_y=\pm1\rangle\right|^2=\frac{1}{2},\\ &\left|\langle\sigma_y=\pm1|\sigma_z=\pm1\rangle\right|^2=\frac{1}{2},\\ &\left|\langle\sigma_z=\pm1|\sigma_x=\pm1\rangle\right|^2=\frac{1}{2} \end{aligned} \tag{1}$$

do not depend on the quantum numbers ±1, is the statement of the pairwise complementary nature of σ_x, σ_y, and σ_z. Their algebraic completeness is then an immediate consequence of the insight that a spin-1/2 degree of freedom can have at most three mutually complementary observables.

More generally, there can be no more than $p+1$ such observables for a degree of freedom with a p-dimensional space of state vectors [8]. Following Weyl [9,10] and Schwinger [11–13], we shall find it convenient to deal with unitary operators, rather than the Hermitian operators to which they would be closely related. Thus the $p+1$ observables $U_0, U_1, \ldots, U_p$ are unitary and of period p,

$$\begin{aligned} &U_m^p=1,\\ &U_m^r\neq 1 \quad \text{if } r=1,2,\ldots,p-1, \end{aligned} \tag{2}$$

for $m=0,1,\ldots,p$. The eigenvalues of each U_m are powers of

$$q\equiv e^{2\pi i/p}, \tag{3}$$

the basic pth root of unity, and we denote by $|m_k\rangle$ the kth eigenstate of U_m, so that

$$U_m|m_k\rangle=|m_k\rangle q^k \tag{4}$$

for $m=0,1,\ldots,p$ and $k=1,2,\ldots,p$.

Both the orthonormality of the $|m_k\rangle$'s for each m and the mutual complementarity for different m's are summarized in

$$\begin{aligned} \left|\langle m_k|m'_{k'}\rangle\right|^2&=\delta_{mm'}\delta_{kk'}+\frac{1}{p}(1-\delta_{mm'})\\ &=\begin{cases}\delta_{kk'} & \text{if } m=m',\\ p^{-1} & \text{if } m\neq m',\end{cases} \end{aligned} \tag{5}$$

for $m,m'=0,1,\ldots,p$ and $k,k'=1,2,\ldots,p$. With

$$U_m=\sum_{k=1}^{p}|m_k\rangle q^k\langle m_k| \tag{6}$$

this implies

$$p^{-1}\,\mathrm{tr}\{U_m^r U_{m'}^s\}=\delta_{mm'}\delta_{r,-s}^{(p)}+(1-\delta_{mm'})\delta_{r,0}^{(p)}\delta_{s,0}^{(p)}, \tag{7}$$

where $m,m'=0,1,\ldots,p$ and $r,s=0,\pm1,\pm2,\ldots$, and

$$\delta_{rs}^{(p)}\equiv\left\{\begin{matrix}1 & \text{if } q^r=q^s\\ 0 & \text{otherwise}\end{matrix}\right\}=\frac{1}{p}\sum_{k=1}^{p}q^{(r-s)k} \tag{8}$$

is the appropriate p-periodic version of Kronecker's delta symbol. The reverse is also true: (7) implies (5), as can be shown with the aid of

$$|m_k\rangle\langle m_k| = \frac{1}{p}\sum_{r=1}^{p}\left(q^{-k}U_m\right)^r. \tag{9}$$

Thus, given a set of $p+1$ unitary operators of period p, we can verify the defining property (5) of their pairwise complementarity by demonstrating that (7) holds.

Repeated measurements of the observables U_m (on identically prepared systems) eventually determine the probabilities $w_k^{(m)}$ for finding their eigenstates $|m_k\rangle$. As a consequence of their mutual complementarity, knowledge of the probabilities for one U_m contains no information whatsoever about the probabilities for any other one. These $(p+1)\times p$ probabilities represent p^2-1 parameters in total, since

$$\sum_{k=1}^{p} w_k^{(m)} = 1 \tag{10}$$

for each of the $p+1$ measurements. The statistical operator that characterizes the ensemble of identically prepared systems,

$$\rho = \sum_{m=0}^{p}\sum_{k=1}^{p}|m_k\rangle\left(w_k^{(m)} - \frac{1}{p+1}\right)\langle m_k|, \tag{11}$$

is therefore uniquely determined by the probabilities $w_k^{(m)} = \langle m_k|\rho|m_k\rangle$. Indeed, the U_m's constitute a complete set of pairwise complementary observables for the prime degree of freedom under consideration.

Actually, the prime nature of p has not been significant so far, but it is for the explicit construction of the set $U_0, U_1, \ldots, U_p$ that we turn to now. We pick an arbitrary period-p unitary operator for U_0. The unitary operator that permutes the eigenvectors of U_0 cyclically is used for U_p. Its eigenvectors in turn are cyclically permuted by U_0, so that U_0 and U_p are jointly characterized by

$$\langle 0_k|U_p = \langle 0_{k+1}|, \qquad U_0|p_k\rangle = |p_{k+1}\rangle \tag{12}$$

for $k = 1, 2, \ldots, p-1$ and, to complete the cycle,

$$\langle 0_p|U_p = \langle 0_1|, \qquad U_0|p_p\rangle = |p_1\rangle. \tag{13}$$

The fundamental Weyl commutation relation

$$U_0U_p = q^{-1}U_pU_0 \tag{14}$$

is an immediate consequence of this reciprocal definition of U_0 and U_p. The other U_m's are chosen as

$$U_m = U_0^m U_p. \tag{15}$$

Since p is a prime — what follows is not true for composite numbers; try $p=6$, for instance, to see what goes wrong — the powers of the U_m's that appear in (9) comprise all products of powers of U_0 and U_p, and since the unitary operators

$$U_0^r U_p^s \quad \text{with } r, s = 1, 2, \ldots, p, \tag{16}$$

which are p^2 in number, are a basis in the p^2-dimensional operator algebra [9–13], the p^2-1 unitary operators

$$U_m^r \quad \text{with } r = 1, 2, \ldots, p-1, \tag{17}$$

supplemented by $1 = U_0^p = U_1^p = \cdots = U_p^p$ are also such an operator basis. As it should be, these bases are complete, but not overcomplete; none of the basis operators is superfluous.

As a consequence of (12) and (13), all operators in (16) are traceless with the sole exception of the identity operator that obtains for $r = s = p$. It is then a matter of inspection to verify that the U_m's thus constructed obey (7) and are, therefore, a set of pairwise complementary observables, indeed. From the point of view of the information-theoretical approach to quantum mechanics that is being developed by Brukner and Zeilinger [14], the U_m's form a complete set of mutually complementary propositions.

3. The mean king's problem generalized

In the generalized version of the king's problem then, either one of the observables $U_0, U_1, \ldots, U_p$ is measured by the mean king's men, on a p-system suitably prepared by the physicist. Without knowing which measurement was done actually, the physicist performs a subsequent measurement of her own, and — after then being told which U_m was measured by the king's men — she has to state correctly what they found: $|m_1\rangle$, or $|m_2\rangle$, ..., or $|m_p\rangle$.

The physicist solves the problem by first preparing a state $|\Psi_0\rangle$ in which the given p-system, the *object*, is entangled with an auxiliary p-system, the *ancilla*, whose operators and states are barred for distinction.

For the ancilla, there are analogs $\overline{U}_0$ and $\overline{U}_p$ of the fundamental Weyl operators U_0 and U_p that we have for the object. It is advantageous, however, to interchange the roles of $\overline{U}_0$ and $\overline{U}_p$ in their reciprocal definition. So, rather than just copying the object relations (12) and (13), we write for the ancilla

$$\overline{U}_p|\bar{0}_k\rangle = |\bar{0}_{k+1}\rangle, \qquad \langle\bar{p}_k|\overline{U}_0 = \langle\bar{p}_{k+1}| \tag{18}$$

for $k = 1, 2, \ldots, p-1$ and

$$\overline{U}_p|\bar{0}_p\rangle = |\bar{0}_1\rangle, \qquad \langle\bar{p}_p|\overline{U}_0 = \langle\bar{p}_1|, \tag{19}$$

and the corresponding analog of (15) is

$$\overline{U}_m = \overline{U}_p\overline{U}_0^m \tag{20}$$

for $m = 1, \ldots, p-1$. Then the transition amplitudes $\langle 0_j|m_k\rangle$ and $\langle\bar{0}_j|\bar{m}_k\rangle$ between the eigenstates of U_0 and U_m and between those of $\overline{U}_0$ and $\overline{U}_m$, respectively, obey recurrence relations

$$\frac{\langle 0_{j+1}|m_k\rangle}{\langle 0_j|m_k\rangle} = q^{-jm+k}, \qquad \frac{\langle\bar{0}_{j+1}|\bar{m}_k\rangle}{\langle\bar{0}_j|\bar{m}_k\rangle} = q^{jm-k} \tag{21}$$

(for $m \neq 0$, of course), which allow and invite to choose the phase conventions such that

$$\langle 0_j|m_k\rangle = \langle\bar{m}_k|\bar{0}_j\rangle. \tag{22}$$

We note in passing that the $|m_k\rangle$'s, or the $|\bar{m}_k\rangle$'s, are essentially identical with the states found by Wootters and Fields [8] if one opts for the solutions

$$\langle 0_j|m_k\rangle = p^{-1/2}q^{jk-j(j-1)m/2} = \langle\bar{0}_j|\bar{m}_k\rangle^* \tag{23}$$

of recursions (21).

Joint states in which the object is in $|m'_{k'}\rangle$ and the ancilla in $|\bar{m}_k\rangle$ are denoted by $|m'_{k'}\bar{m}_k\rangle$. Then

$$|\Psi_0\rangle = p^{-1/2}\sum_{k=1}^{p}|m_k\bar{m}_k\rangle \tag{24}$$

is the entangled object-ancilla state that the physicist prepares. Thanks to the phase conventions (22), the m dependence is only apparent. For either value of $m = 0, 1, \ldots, p$ we get the same $|\Psi_0\rangle$.

If the king's men then measure the object observable U_m and find the eigenvalue q^k, the resulting object-ancilla state is $|m_k\bar{m}_k\rangle$. After their measurement, there are thus all together $p+1$ sets (labeled by m) of p possible object-ancilla states each. These $(p+1)\times p$ states cannot be linearly independent because the state space is only p^2-dimensional. Indeed, each of the p-dimensional subspaces spanned by the $p+1$ sets contains $|\Psi_0\rangle$ by construction. In addition, there are $(p+1)\times(p-1) = p^2-1$ other states, and we now proceed to show that they *are* linearly independent.

Consider $m \neq m'$ and any pair of values for k and k'. Then

$$\langle m_k\bar{m}_k|m'_{k'}\bar{m}'_{k'}\rangle = p^{-1} \tag{25}$$

as a consequence of (5) and (22), and definition (24) of $|\Psi_0\rangle$ implies

$$\langle\Psi_0|m_k\bar{m}_k\rangle = p^{-1/2} = \langle\Psi_0|m'_{k'}\bar{m}'_{k'}\rangle. \tag{26}$$

The two vectors

$$|m_k\bar{m}_k\rangle - p^{-1/2}|\Psi_0\rangle,$$
$$|m'_{k'}\bar{m}'_{k'}\rangle - p^{-1/2}|\Psi_0\rangle \tag{27}$$

are therefore orthogonal to $|\Psi_0\rangle$ and orthogonal to each other. Accordingly, $|\Psi_0\rangle$ together with the p^2-1 vectors $|\Psi_1\rangle, |\Psi_2\rangle, \ldots, |\Psi_{p^2-1}\rangle$ that are defined by

$$|\Psi_{(p-1)m+j}\rangle = p^{-1/2}\sum_{k=1}^{p}|m_k\bar{m}_k\rangle q^{-jk} \tag{28}$$

for $m = 0, 1, 2, \ldots, p$ and $j = 1, 2, \ldots, p-1$ constitute an orthonormal basis,

$$\langle\Psi_n|\Psi_{n'}\rangle = \delta_{nn'} \quad \text{for } n, n' = 0, 1, \ldots, p^2-1, \tag{29}$$

in the p^2-dimensional object-ancilla state space.

Two $|\Psi_n\rangle$'s that have the same m value in (28) are orthogonal by construction. And if $|\Psi_n\rangle$ and $|\Psi_{n'}\rangle$ belong to different m values, their orthogonality follows immediately as soon as one replaces $|m_k\bar{m}_k\rangle$ in (28) by the difference of (27), which does not alter the value of the sum.

Let us now see how all of this helps the physicist to meet the mean king's challenge. She will be able to state correctly the measurement result found by the king's men if she can find an object-ancilla observable P with eigenstates $|P_1\rangle, \ldots, |P_{p^2}\rangle$ such that each $|P_n\rangle$ is orthogonal to $p-1$ members each of the $p+1$ sets of states that are potentially the case after the measurement by the king's men. We characterize the looked-for eigenstates of P by an ordered set of numbers $k_0, k_1, \ldots, k_p$ that indicate which members they are *not* orthogonal to, so that

$$\big|[k_0k_1\dots k_p]\big\rangle \tag{30}$$

has the defining property of being orthogonal to the object-ancilla states that result when measurements of U_m do *not* give the eigenvalue q^{k_m}.

Suppose the physicist finds the state $|[325\dots7]\rangle$. She then knows that if the king's men had measured U_0, U_1, U_2, or U_p, the respective results must have been q^3, q^2, q^5, and q^7, because she would never find $|[325\dots7]\rangle$ for other measurement results.

Accordingly, all that is needed to complete the solution of the generalized mean king's problem is the demonstration that we can have a complete orthonormal set of object-ancilla states of the kind (30). First note that the expansion of $|[k_0k_1\dots k_p]\rangle$ in the $|\Psi_n\rangle$ basis is given by

$$\big|[k_0k_1\dots k_p]\big\rangle = \frac{1}{p}\left(|\Psi_0\rangle + \sum_{m=0}^{p}\sum_{j=1}^{p-1} q^{jk_m}|\Psi_{(p-1)m+j}\rangle\right). \tag{31}$$

Then observe that

$$\big\langle[k_0k_1\dots k_p]\big|\big[k'_0k'_1\dots k'_p\big]\big\rangle = \frac{1}{p}\sum_{m=0}^{p}\delta_{k_m,k'_m} - \frac{1}{p}, \tag{32}$$

so that two such states are orthogonal if $k_m = k'_m$ for one and only one m value. Therefore, a possible choice of basis states for the physicist's final measurement is given by those p^2 states for which $k_0, k_1 = 1, 2, \dots, p$ and

$$k_m = (m-1)k_0 + k_1 \pmod{p} \tag{33}$$

for $m = 2, 3, \dots, p$. The prime nature of p is crucial for the otherwise straightforward demonstration of the orthogonality of two such states that differ in their values of k_0, or k_1, or both.

So, the physicist just has to choose her object-ancilla observable P such that it distinguishes the states specified in (33). After being told which measurement the king's men performed on the object, she can then infer their measurement result correctly, and with certainty, in the manner described above for $|[325\dots7]\rangle$.

Acknowledgements

Y.A. gratefully acknowledges the kind hospitality extended to him by Herbert Walther during a visit at the MPI für Quantenoptik in Garching. B.G.E. would like to thank Wolfgang Schleich and the University of Ulm for financial support while part of this work was done, and he expresses his sincere gratitude for the hospitable environment provided by Gerald Badurek and Helmut Rauch at the Atominstitut and for the financial support by the Technical University of Vienna. Y.A.'s research is supported in part by Grant No. 471/98 of the Basic Research Foundation (administered by the Israel Academy of Sciences and Humanities), by NSF Grant No. PHY-9971005, and by ONR Grant No. N00014-00-1-0383.

References

[1] L. Vaidman, Y. Aharonov, D.Z. Albert, Phys. Rev. Lett. 58 (1987) 1385.
[2] B.-G. Englert, Ch. Kurtsiefer, H. Weinfurter, Phys. Rev. A 63 (2001) 032303.
[3] A. Beige, B.-G. Englert, Ch. Kurtsiefer, H. Weinfurter, quant-ph/0101066, Phys. Rev. Lett., submitted.
[4] Y. Aharonov, B.-G. Englert, Z. Naturforsch. 56a (2001) 16.
[5] S. Ben-Menahem, Phys. Rev. A 39 (1989) 1621.
[6] N.D. Mermin, Phys. Rev. Lett. 74 (1995) 831.
[7] Y. Aharonov, L. Vaidman, J. Phys. A: Math. Gen. 24 (1991) 2315.
[8] W.K. Wootters, B.D. Fields, Ann. Phys. (NY) 191 (1989) 363.
[9] H. Weyl, Z. Phys. 46 (1927) 1.
[10] H. Weyl, Gruppentheorie und Quantenmechanik, Hirzel, Leipzig, 1928;
English translation by H.P. Robertson, Theory of Groups and Quantum Mechanics, E.P. Dutton, New York, 1932.
[11] J. Schwinger, Proc. Natl. Acad. Sci. USA 46 (1960) 570.
[12] J. Schwinger, in: W. Deppert, K. Hübner, A. Oberschelp, V. Weidemann (Eds.), Exact Sciences and Their Philosophical Foundations, Verlag Peter Lang, Frankfurt am Main, 1985.
[13] J. Schwinger, Quantum Mechanics. Symbolism of Atomic Measurements, Springer, Berlin, 2001.
[14] Č. Brukner, A. Zeilinger, Phys. Rev. Lett. 83 (1999) 3354.

Ady Mann

Department of Physics, Technion–Israel Institute of Technology, Haifa, Israel

Amir Kalev

Centre for Quantum Technologies, National University of Singapore, Singapore
Center for Quantum Information and Control, University of New Mexico, Albuquerque, New Mexico, USA

Michael Revzen

Department of Physics, Technion–Israel Institute of Technology, Haifa, Israel

Paper Title: **Choice of Measurement as the Signal**

Reference: Physical Review Letters **110**, 260502 (2013)

Paper Title: **Quantum-mechanical retrodiction through an extended mean king problem**

Reference: EPL (Europhysics Letters) **104**, 50008 (2013)

Personal message:

These two papers were inspired by and built upon the insightful work of a mentor and colleague, Berge Englert, and dedicated to him for his 60th birthday.

PRL **110,** 260502 (2013) PHYSICAL REVIEW LETTERS week ending 28 JUNE 2013

Choice of Measurement as the Signal

Amir Kalev

Centre for Quantum Technologies, National University of Singapore, 3 Science Drive 2, Singapore 117543, Singapore
Center for Quantum Information and Control, University of New Mexico, Albuquerque, New Mexico 87131-0001, USA

Ady Mann and Michael Revzen

Department of Physics, Technion—Israel Institute of Technology, Haifa 32000, Israel
(Received 5 April 2013; published 25 June 2013)

In classical mechanics, performing a measurement without reading the measurement outcome is equivalent to not exploiting the measurement at all. A nonselective measurement in the classical realm carries no information. Here we show that the situation is remarkably different when quantum mechanical systems are concerned. A nonselective measurement on one part of a maximally entangled pair can allow communication between two parties. In the proposed protocol, the signal is encoded in the *choice* of the measurement basis of one of the communicating parties, while the outcomes of the measurement are irrelevant for the communication and therefore may be discarded. Different choices for the (nonselective) measurement correspond to different signals. The implication of the study of measurements in quantum mechanics is considered. The scheme is studied in a Hilbert space of prime dimension.

DOI: 10.1103/PhysRevLett.110.260502 PACS numbers: 03.67.Hk, 03.65.Ta

An important distinction between classical and quantum measurement is that the latter implies an inevitable disturbance to the measured system. In the present work, we show that this disturbance is trackable to the extent that it may be used for communication. Thus we study nonselective measurements where the outcomes are not recorded. Such measurements within the classical theory do not carry information and hence cannot be used for communication [1,2]. Here we show how nonselective measurements on one part of a quantum system of a maximally entangled pair can be used to encode and eventually communicate information. In the proposed protocol the basis, i.e., the choice, of the (nonselective) measurement is the signal. The outcomes of the measurement are totally irrelevant. The trackable choice of measurement (rather than its outcome) analysis allows a novel interpretation of quantum measurements. Thus a quantum state does, in general [3–5], imply contextual values for measurable dynamical variables. Hence it is attractive to interpret our result as suggesting that quantum measurement is built up of two stages. The first stage, to be associated with the nonselective measurement, elevates a particular set of dynamical variables (those labeling the basis in our case) to reality (i.e., having a prescribed value). The state after this stage is, in general, mixed. The second stage involves the determination of the value of the dynamical variable. This stage has clear classical attributes.

Confining our study at the moment to a Hilbert space of odd prime dimension d, we consider as alternative choices for measurements the alternative mutual unbiased bases (MUB). For prime dimension there are $d + 1$ MUB [6–12]. A possible set of $d + 1$ MUB can be defined as follows. The first basis is the computational basis $\{|n\rangle\}_{n=0}^{d-1}$, composed of the d orthonormal eigenstates of the generalized Pauli operator $\hat{Z}$, $\hat{Z}|n\rangle = \omega^n|n\rangle$, $|n+d\rangle = |n\rangle$, $\omega = e^{i(2\pi/d)}$. The other d orthonormal bases are parametrized by $b = 0, 1, \ldots, d-1$. The kets that compose the d remaining bases are given in terms of the computational basis by [9]

$$|m; b\rangle = \frac{1}{\sqrt{d}} \sum_{n=0}^{d-1} |n\rangle \omega^{bn^2 - 2nm}; \quad b, m = 0, 1, \ldots, d-1. \tag{1}$$

We shall designate the computational basis by $b = \ddot{0}$, and depending on the context we may also denote the kets of the computational basis $|m\rangle$ by $|m; \ddot{0}\rangle$. Thus, the $d + 1$ bases are labeled by $b = \ddot{0}, 0, 1, \ldots, d-1$.

The proposed communication protocol is described in what follows. We assume that the two communicating parties, Alice and Bob, agree beforehand upon a code, associating messages with the parameters, b, specifying the MUB. There is no classical communication between Alice and Bob beyond this point. The protocol involves a two d-level system (qudit) entangled state prepared by Alice with one qudit available to Bob, who wishes to communicate a message, $b = \ddot{0}, 0, 1, \ldots, d-1$, to Alice. To this end, Bob measures the part of the system that is available to him in the basis parametrized with the b of his message. He must complete the measurement yet may ignore its outcome and then return the qudit to Alice. This step renders values to a class of dynamical variables: the complete set of commuting operators $(XZ^b)^n$; $n = 0, 1, \ldots, d-1$. Now Alice measures the two-qudit resultant state, deduces, almost always, the basis b used by Bob, and, hence, decodes the message. The procedure is quantal in that the signal corresponds to the basis of Bob's

PRL **110,** 260502 (2013) PHYSICAL REVIEW LETTERS week ending 28 JUNE 2013

measurement, that is, to the "alignment" of his instrument, and in that the measurement outcomes are irrelevant and may be unrecorded.

Choice of measurement basis as signal.—To establish a communication channel, let Alice prepare one of the following d^3 two-qudit maximally entangled states [13,14]:

$$|c, r; s\rangle_{1,2} = \frac{1}{\sqrt{d}} \sum_{n=0}^{d-1} |n\rangle_1 |c - n\rangle_2 \omega^{sn^2-2rn}, \qquad (2)$$

with $c, r, s = 0, 1, \ldots, d-1$, and send one of the qudits, say, the one labeled by 1, to Bob. We note that, for a given s value, these states form an orthonormal, maximally entangled, basis for the Hilbert space of the two qudits. Thus, s labels the basis, and c and r label the d^2 orthonormal states within a basis. The reduced state for Bob's qudit is the completely mixed state.

To communicate a message to Alice, Bob measures his qudit in one of the MUB labeled by $b = \ddot{0}, 0, 1, \ldots, d-1$. The message is his choice of the basis used for the measurement. Bob may or may not record the measurement outcome. This is of no relevance to the protocol. After completing his nonselective measurement, Bob sends the qudit back to Alice. The two-qudit state is described now by

$$\rho_{1,2} = \sum_{m=0}^{d-1} |m; b\rangle_1 \langle m; b|c, r; s\rangle_{1,2} \langle c, r; s|m; b\rangle_1 \langle m; b|. \qquad (3)$$

We note in passing that making a nonselective measurement in basis b is equivalent to performing a random unitary transformation which is diagonal in the b basis [1]. To retrieve the message, Alice now measures the two qudits in the basis of preparation, $\{|c', r'; s\rangle_{1,2}\}_{c',r'=0}^{d-1}$ of Eq. (2). The probability to obtain an outcome which corresponds to the basis state $|c', r'; s\rangle_{1,2}$ is

$$\langle c', r'; s|\rho_{1,2}|c', r'; s\rangle_{1,2} = \frac{1}{d} \begin{cases} \delta_{c,c'} & \text{for } b = \ddot{0}, \\ \delta_{(b-s)c+r,(b-s)c'+r'} & \text{for } b = 0, 1, 2, \ldots, d-1. \end{cases} \qquad (4)$$

The arithmetics is modulo d. According to the above equation, based on the outcome of her measurement, Alice can decode the message sent from Bob, that is, the basis of his measurement. If the outcome corresponds to a state $|c', r'; s\rangle_{1,2}$ with $c \neq c'$, Alice infers that $b = s + (r' - r)/(c - c')$. Since she knows the values of c, r, c', r', and s, she can calculate the message b. If, on the other hand, $c = c'$ and $r \neq r'$, Alice infers that $b = \ddot{0}$. The case of $c = c'$ and $r = r'$ is inconclusive for Alice. The inconclusive outcome occurs with probability $1/d$. In that case, the preparation state and the detection state of the two qudits is the same, and she does not gain any information about Bob's message. Hence the decoding table is

$$\begin{aligned} c \neq c' \rightarrow b &= s + \frac{r - r'}{c' - c}, \\ r \neq r',\ c = c' \rightarrow b &= \ddot{0}, \\ r = r',\ c = c' \rightarrow &\ \text{inconclusive}. \end{aligned} \qquad (5)$$

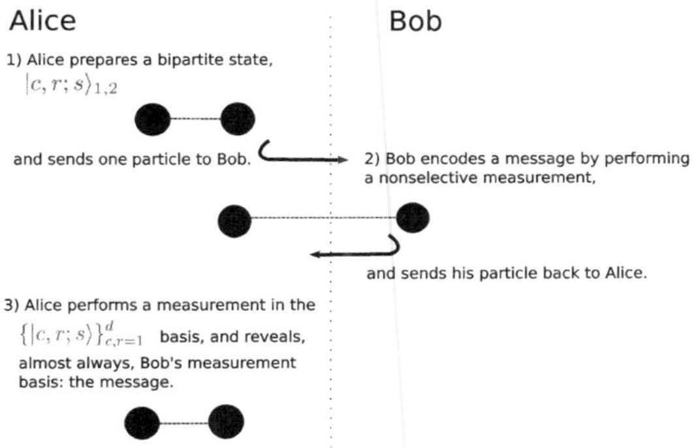

FIG. 1. The scheme of the communication protocol.

For the even prime dimension $d = 2$, by plugging the imaginary unit i instead of ω in all of the above equations, one retrieves the same decoding table (5). The protocol is schematically drawn in Fig. 1.

Conclusions and remarks.—In conclusion, we showed how nonselective measurements in MUB on one part of an entangled pair could be used to encode information. The scheme uniquely utilizes quantum features of the system, since performing nonselective measurements on classical systems (no matter how correlated they are) cannot carry or manipulate information [1,2]. Alternatively, the trackability of the nonselective measurement allows a novel view of quantum measurement. Thus we may view a quantum measurement as a two-stage process. The first, to be associated with the nonselective part, involves the promotion of a set of dynamical variables (labeled, in our case, by the basis b) to reality (i.e., [5] having a definite value). After this stage, in general, the state is a mixed state. The second stage involves the determination of the outcome among these values. This stage allows a classical interpretation: The experiment determines a possible preassigned value. Dealing, as we do in this work, with an entangled state renders the stages separable: Bob's measurement (unknown to Alice) consummates a first stage *for the two-particle system*. The sequential measurement, by Alice, selects possible values of the two-particle system allowed by the (mixed) state that resulted from Bob's measurement.

PRL **110,** 260502 (2013) PHYSICAL REVIEW LETTERS week ending 28 JUNE 2013

In the considered protocol, by sending a qudit, Bob is able to transfer, on average, more than $\log_2 d$ bits of information to Alice. This is, in some respect, a form of dense coding. We note for comparison that superdense coding [15] achieves $2\log_2 d$ bits per qudit sent from Bob to Alice. However, in the superdense coding scheme specific unitary transformations are used for the encoding, while here non-selective measurements, or, equivalently, random unitary transformations, are utilized. Though, in its present form, the protocol cannot be used for secure communication, we leave it as an open question whether one could consider variations of the present protocol that would render it suitable for cryptography tasks. Preliminary study indicates that the proposed scheme can be generalized to encompass prime-powers dimensions. Finally, this protocol exemplifies how tasks which seem impossible by classical reasoning are realized in quantum systems.

A. K. thanks Professor B.-G. Englert for fruitful discussions and for his insightful comments. The Centre for Quantum Technologies is a Research Centre of Excellence funded by the Ministry of Education and by the National Research Foundation of Singapore. This research was supported in part by NSF Grant No. PHY-1212445.

[1] J. Schwinger, *Quantum Mechanics: Symbolism of Atomic Measurements*, edited by B.-G. Englert (Springer, Berlin, 2001).

[2] L. Diósi, *A Short Course in Quantum Information Theory*, Lecture Notes in Physics Vol. 827 (Springer, Berlin, 2011), 2nd ed.

[3] S. Kochen and E. P. Specker, J. Math. Mech. **17**, 59 (1967).

[4] J. S. Bell, Rev. Mod. Phys. **38**, 447 (1966).

[5] N. D. Mermin, Phys. Rev. Lett. **65**, 3373 (1990); Rev. Mod. Phys. **65**, 803 (1993).

[6] I. D. Ivanovic, J. Phys. A: Math. Gen. **14**, 3241 (1981).

[7] A. Vourdas, Rep. Math. Phys. **40**, 367 (1997).

[8] W. K. Wootters and B. D. Fields, Ann. Phys. (N.Y.) **191**, 363 (1989).

[9] S. Bandyopadhyay, P. O. Boykin, V. Roychowdhury, and F. Vatan, Algorithmica **34**, 512 (2002).

[10] K. S. Gibbons, M. J. Hoffman, and W. K. Wootters, Phys. Rev. A **70**, 062101 (2004).

[11] A. Vourdas, Rep. Prog. Phys. **67**, 267 (2004).

[12] T. Durt, B.-G. Englert, I. Bengtsson, and K. Życzkowski, Int. J. Quantum. Inform. **08**, 535 (2010).

[13] M. Revzen, Phys. Rev. A **81**, 012113 (2010).

[14] M. Revzen, J. Phys. A: Math. Theor. **46**, 075303 (2013).

[15] C. H. Bennett and S. J. Wiesner, Phys. Rev. Lett. **69**, 2881 (1992).

A LETTERS JOURNAL EXPLORING THE FRONTIERS OF PHYSICS

December 2013

EPL, **104** (2013) 50008
doi: 10.1209/0295-5075/104/50008

www.epljournal.org

Quantum-mechanical retrodiction through an extended mean king problem

A. KALEV[1], A. MANN[2] and M. REVZEN[2]

[1] *Center for Quantum Information and Control, MSC07-4220, University of New Mexico Albuquerque, NM 87131-0001, USA*
[2] *Department of Physics, Technion, Israel Institute of Technology - Haifa 32000, Israel*

received 14 October 2013; accepted 5 December 2013
published online 2 January 2014

PACS 03.65.Ta – Foundations of quantum mechanics; measurement theory
PACS 03.67.Hk – Quantum communication

Abstract – The mean king problem is a conditional retrodiction problem. In this problem Alice prepares a two prime-dimensional particles state and avails one of the particles to the king who measures its state in one of mutually unbiased bases of his choice. The king *tells* Alice his choice of basis *after* she completes a control measurement on his particle. Conditioned on this knowledge, she now infers the state observed by the king by utilizing the outcome of her control measurement. In the *extended* mean king problem, studied in this paper, the king does not tell Alice his measurement basis, but instead both the king and Alice repeat their measurements. Proper ordering of these allows Alice to deduce *both* the basis used by the king and the outcome of his *first* measurement, with the king reticent throughout, *i.e.*, this protocol effects a (almost) complete retrodiction of the king's first measurement.

Introduction. – Retrodiction, viz. assigning the state of a system in the past, is direct in deterministic classical physics: Knowledge of the present state, *e.g.* via measurements, allows both prediction and retrodiction. Within quantum theory, the status of this issue is not as clear as the theory is fundamentally a probabilistic theory. Thus *e.g.* does the theory retrodict the observations or their probabilistic attributes? The fundamental difficulty involved was noted almost at the inception of quantum mechanics [1]: "The principles of quantum mechanics actually involve an uncertainty in the description of past events which is analogous to the uncertainty in the prediction of future events". More recent studies pin the difficulty in quantum retrodiction on the contextuality of quantum observables, suggested by Bell-Kochen-Specker theorem [2].

The mean king problem (MKP), for which the present work is an extension, is a quantum-mechanical retrodiction problem [2,3]. It originated with [4] for spin-half particles and generalized to prime dimensionality in [5] and power of prime in [6–8]. Further generalizations were given in [3]. The MKP involves a two-particle state prepared by Alice, availing one of the constituent particles to the king who measures the particle's state, m, in one of mutually unbiased bases of his choice, b (unknown to Alice). Subsequent control measurement by Alice of the two particles allows her to determine —*upon being informed* by the king of the basis, b, he used— the outcome, m, of his measurement, *i.e.*, the state retrodiction is "conditional" on further information: the king's measurement basis, b. Thus Alice is *seemingly* [2] violating a fundamental quantum tenant by being able to assign (to one system) outcome values for non-commuting (distinct bases) observables.

In [9,10] we considered a deterioration in the king's mood: Alice is *not* being told the basis, b (nor the outcome, m) of the king's measurement; we showed that by an appropriate state preparation and measurements, she nonetheless is able to unravel the basis, b. In this paper, we consider an extension of the MKP which, essentially, completes the retrodiction. In the extended problem the king does not reveal his measurement basis to Alice, but repeats his measurement (*i.e.* uses the same basis of the first measurement). Alice, likewise, repeats her (control) measurement. Based on her measurement results alone, Alice now retrodicts the king's *first* measurement: both outcome and basis. The protocol may be used as a key distribution with an extra authentication scheme.

Mutually unbiased bases (MUB) play a central role in the analysis [5,9,10]. In what follows we list some MUB characteristics that are utilized in the present work. Two sets of orthonormal maximally entangled states studied in [9] provide the means for solving the retrodiction problem. The full protocol involved in retrodicting the king's measurement is then spelled out. Furthermore we outline a possible variant of the retrodiction protocol. The last part contains our conclusions inclusive of possible implication of the results in the quantum measurement theory.

Brief review on mutually unbiased bases. – In a d-dimensional Hilbert space, two complete, orthonormal bases, $\mathcal{B}_1, \mathcal{B}_2$, are said to be MUB if and only if ($\mathcal{B}_1 \neq \mathcal{B}_2$)

$$\forall |u\rangle \in \mathcal{B}_1 \text{ and } \forall |v\rangle \in \mathcal{B}_2, \quad |\langle u|v\rangle| = 1/\sqrt{d}. \tag{1}$$

The physical meaning of this is that knowledge that a system is in a particular state in one basis implies complete ignorance of its state in the other basis.

Ivanovic [11] proved that there are at most $d+1$ MUB in a d-dimensional Hilbert space and gave an explicit formula for the $d+1$ bases in the case of $d =$ prime number. Wootters and Fields [12] constructed such $d+1$ bases for power of prime dimensions. Variety of methods for construction of the $d+1$ bases for power of prime dimensions are now known [13–17].

Our present study is confined to $d =$ prime number. We now give explicitly the MUB states in conjunction with the Weyl-Schwinger pair Z, X [18–20]. Thus we label the d orthonormal states spanning the Hilbert space, termed the computational basis, by $|n\rangle$, $n = 0, \ldots, d-1$, and $|n+d\rangle = |n\rangle$,

$$\begin{aligned} Z|n\rangle &= \omega^n|n\rangle, \ \omega = e^{i2\pi/d}, \\ X|n\rangle &= |n+1\rangle. \end{aligned} \tag{2}$$

The above relation, together with the commutation relation

$$ZX = \omega XZ, \tag{3}$$

imply that the pair Z, X completely specify the degree of freedom. The d states in each of the $d+1$ MUB bases [13,19] are the computational basis and the d bases:

$$|m;b\rangle = \frac{1}{\sqrt{d}}\sum_{0}^{d-1}\omega^{\frac{b}{2}n(n-1)-nm}|n\rangle, \quad b, m = 0, \ldots, d-1, \tag{4}$$

where b labels the bases, and m labels the states within a basis. Each basis corresponds to Weyl-Schwinger pair Z, X by [13]

$$XZ^b|m;b\rangle = \omega^m|m;b\rangle. \tag{5}$$

For later reference we shall refer to the computational basis by $b = \ddot{0}$. We denote $|m;\ddot{0}\rangle$ by $|m\rangle$ when no confusion should arise. Thus the $d+1$ bases are labeled as $b = \ddot{0}, 0, 1, \ldots, d-1$. We have of course,

$$|\langle m;b|m';b'\rangle| = \delta_{m,m'}\delta_{b,b'} + (1-\delta_{b,b'})\frac{1}{\sqrt{d}}. \tag{6}$$

This completes our discussion of single particle MUB.

Several studies [8,21–24] considered the entanglement of two d-dimensional particles via MUB state labeling. Our presentation is based on [9,10]. In [9,10] we considered maximally entangled states that are *product* states in the particles' collective (*i.e.*, "center of mass" and "relative") coordinates. Based on these latter representative states one obtains [9] orthonormal bases that span the d^2-dimensional Hilbert space with maximally entangled states. In the next section we use two such bases sets.

Retrodiction of the king's measurement. – Retrodiction within quantum mechanics involves the specification of past measurement outcome. To effect such retrodiction the extended MKP includes the following bases:

$$\begin{aligned} &\{|m;b\rangle|m = 0,\ldots,d-1\} \text{ single particle MUB}, \\ &\{|u,v;-\rangle|u,v = 0,\ldots,d-1\} \text{ two-particle basis}, \\ &\{|u,v;+\rangle|u,v = 0,\ldots,d-1\} \text{ two-particle basis}. \end{aligned} \tag{7}$$

The last two bases (the "minus" and "plus" bases) are manifestly orthonormal sets of maximally entangled states for the two d-dimensional particles [9,10]:

$$\begin{aligned} |u,v;-\rangle &= \frac{1}{\sqrt{d}}\sum_{n=0}^{d-1}|n\rangle_1 X_2^{2u}Z_2^{v}|-n\rangle_2, \\ |u,v;+\rangle &= \frac{1}{\sqrt{d}}\sum_{n=0}^{d-1}|n\rangle_1 X_2^{-2u}Z_2^{-v}|n\rangle_2. \end{aligned} \tag{8}$$

A possible protocol is as follows. Alice prepares a state from the "plus" basis, say $|u,v;+\rangle$, cf. eq. (8), and sends particle 1 to the king. As promised, the king measures the particle in one of the $d+1$ MUB of his choice. His outcome is, say, m, and he returns the particle to Alice. Now Alice measures the two particles in the "minus" basis, $\{|c,r;-\rangle\ |c,r = 0,\ldots,d-1\}$ and observes, say, an outcome labelled by u' and v'. Thus, by direct evaluation of

$${}_{1,2}\langle u',v';-|m;b\rangle_1\langle m;b|u,v;+\rangle_{1,2} \neq 0, \tag{9}$$

she obtains

$$m = \begin{cases} u'+u, & \text{for } b = \ddot{0}, \\ \dfrac{v'+v}{2} + b(u'+u) - \dfrac{b}{2}, & \text{for } b = 0,\ldots,d-1. \end{cases} \tag{10}$$

Had Alice been informed of the basis used by the king, b, she could have deduced his outcome, m. However in this protocol the king does not disclose his measurement basis. Instead Alice now sends particle 1 again to the king who repeats his measurement (*i.e.*, uses the same measurement basis b), obtains some result, say m' and returns the particle to Alice. Actually, the outcome of his second measurement is immaterial for the protocol [10]. The king might as well use a non-selective measurement in the b basis [10,25]. At this point, Alice repeats her measurement in the "minus" basis, and obtains some result, u'', v''. This allows her to deduce the basis used (twice) by the king. Thus, the condition

$$_{1,2}\langle u'', v''; -|m'; b\rangle_1 \langle m'; b|u', v'; -\rangle_{1,2} \neq 0, \tag{11}$$

implies that

$$b = \begin{cases} \ddot{0}, & \text{for } u'' = u', \\ \dfrac{v' - v''}{2(u' - u'')}, & \text{for } u'' \neq u'. \end{cases} \tag{12}$$

We note that the above procedure fails when *both* $u'' = u'$ *and* $v'' = v'$, in which case the basis b is undetermined and Alice is unable to retrodict the king's observation. This case happens with probability $1/d$. Upon learning the measurement basis, Alice uses eq. (10) to find the outcome of the king's first measurement as well. This completes the retrodiction.

An obvious alternative protocol would be a retrodiction of the king's *second* measurement (assumed *selective* in this case). This requires a simpler procedure: Alice prepares her state, *e.g.* $|u, v; +\rangle$, the king measures (non-selectively) in one of the MUB. Now Alice measures in the "plus" basis. This allows her to deduce [10] the basis, b, used by the king. The king's repeated measurement leaves particle 1 in the state $|m, b\rangle$, thus allowing Alice to determine m simply by measuring this particle in the same basis used by the king.

The above protocol was formulated for an odd prime dimension. The mathematical reason for that may be traced back to eqs. (6) and (8). We can, however, include the dimension two as well. In this case, the first part of the protocol follows the original protocol of [4], by which Alice obtains the information $m(b)$, that is the outcome of the king's measurement conditioned on his measurement basis. Here, as before, the king does not reveal his measurement basis, but instead Alice prepares another state, say $|u', v'; +\rangle$ in the "plus" basis, formed by the states

$$|u, v; +\rangle = \frac{1}{\sqrt{2}} \sum_{n=0,1} |n\rangle_1 X_2^u Z_2^v |n\rangle_2, \ u, v = 0, 1, \tag{13}$$

and sends particle 1 to the king. He, in turn, measures the particle in the same basis he used in his first measurement and returns the particle to Alice, who measures the two particles in the "plus" basis of eq. (13). From her measurement results u'', v'' she can deduce the measurement basis used by the king according to

$$b = \begin{cases} \ddot{0}, & \text{for } u'' = u', \\ \dfrac{v' - v''}{u' - u''}, & \text{for } u'' \neq u'. \end{cases} \tag{14}$$

The procedure fails when *both* $u' = u''$ *and* $v' = v''$, in which case the basis b is undermined and Alice is unable to retrodict the king's observation. This happens with probability $1/2$. (Hence, considering n two-level systems, the probability for the undetermined case would be $1/2^n$.)

Discussion and conclusions. – Classical physics being deterministic and time symmetric allows direct retrodiction: measuring an observable implies its value at an earlier time. Quantum mechanics, though time symmetric, having a distinct approach to measurements and being a probabilistic theory cannot directly be classified as allowing retrodiction. Thus, *e.g.*, Bell-Kochen-Specker theorem seems to require contextuality among the observables, straining retrodiction. We demonstrated above that, using entanglement and extension of a protocol used for tracking measurement bases, it is possible to retrodict, *i.e.*, to give the basis and outcome of past measurement given present experimental values under certain specified conditions.

The study may be summarized as follows: The king measured some observables associated with a measurement setting, b (in our case $b = \ddot{0}, 0, 1, \ldots, d-1$) and obtained an outcome m (in our case $m = 0, \ldots, d-1$); Alice via her control measurement can tabulate the correspondence $m(b)$ for all b. Thus, within classical physics, being informed of b she can stipulate (for both past and future) the king's outcome, m. We may refer to this as "conditional" retrodiction, it being conditioned on knowing the "alignment", b. On the other hand, quantum mechanics precludes such definite outcome values in general and particularly when distinct alignments correspond to non-commuting observables (as is in our case the MUB b correspond to XZ^b, $[XZ^{b'}, XZ^b] \neq 0$, $b \neq b'$). Nonetheless, the corresponding tabulation is provided within the solution to the MKP. This is possible only as "conditional" retrodiction with the *understanding* that the tabulated value, $m(b)$, has a meaning only for the actually chosen alignment b. The other are meaningless (this point was stressed in [2]).

Now in the extended MKP studied in this work the king repeats his measurement after Alice undertook her (first) control measurement. Within classical physics this does *not* provide any extra information to Alice, in particular as the king is allowed to ignore the outcome of his second measurement (*i.e.*, he can perform a non-selective measurement). However, within quantum mechanics this may, as is shown in this work, provide enough extra information to allow (almost) complete retrodiction of his (first) measurement, that is, it allows Alice to deduce *both* the "alignment", b, used by the king and the outcome, m, of his measurement.

The study may be viewed as providing a novel means for a key distribution within a cryptographic protocol. Thus the parties may agree on having both/either the outcome of the king (first) measurement and/or the basis he used as forming their key and use the other observable for checking on the security of their communication.

* * *

AK would like to acknowledge the support of NSF grant No. PHY-1212445.

REFERENCES

[1] Einstein A., Tolman R. C. and Podolsky B., *Phys. Rev.*, **37** (1931) 780.
[2] Mermin N. D., *Phys. Rev. Lett.*, **65** (1990) 3373.
[3] Reimpell M. and Werner R., *Phys. Rev. A*, **75** (2007) 062334.
[4] Vaidman L., Aharonov Y. and Albert D., *Phys. Rev. Lett.*, **58** (1987) 1385.
[5] Englert B.-G. and Aharonov Y., *Phys. Lett. A*, **284** (2001) 1.
[6] Aravind P. K., *Z. Naturforsch.*, **58a** (2003) 85.
[7] Durt T., *Int. J. Mod. Phys. B*, **20** (2006) 1742.
[8] Durt T., Englert B.-G., Bengtsson I. and Życzkowski K., *Int. J. Quantum Inf.*, **8** (2010) 535.
[9] Revzen M., *J. Phys. A: Math. Theor.*, **46** (2013) 075303.
[10] Kalev A., Mann A. and Revzen M., *Phys. Rev. Lett.*, **110** (2013) 260502.
[11] Ivanovic I. D., *J. Phys. A: Math. Gen.*, **14** (1981) 3241.
[12] Wootters W. K. and Fields B. D, *Ann. Phys. (N.Y.)*, **191** (1989) 363.
[13] Bandyopadhyay S., Boykin P. O., Roychowdhury V. and Vatan F., *Algorithmica*, **34** (2002) 512.
[14] Gibbons K. S., Hoffman M. J. and Wootters W. K., *Phys. Rev. A*, **70** (2004) 062101.
[15] Vourdas A., *Rep. Math. Phys.*, **40** (1997) 367.
[16] Vourdas A., *Rep. Prog. Phys.*, **467** (2004) 267.
[17] Klimov A. B., Romero J. L., Björk G. and Sànchez-Soto L. L., *J. Phys. A: Math. Theor.*, **40** (2007) 3987.
[18] Schwinger J., *Proc. Natl. Acad. Sci. U.S.A.*, **46** (1960) 560.
[19] Kalev A., Revzen M. and Khanna F. C., *Phys. Rev. A*, **80** (2009) 022112.
[20] Revzen M., *EPL*, **98** (2012) 10001.
[21] Fivel D. I., *Phys. Rev. Lett.*, **74** (1995) 835.
[22] Nagler B. and Durt T., *Phys. Rev. A*, **66** (2003) 042323.
[23] Klimov A. B., Sych D., Sanches-Soto L. L. and Leuchs G., *Phys. Rev. A*, **79** (2009) 052101.
[24] Revzen M., *Phys. Rev. A*, **81** (2010) 012113.
[25] Diósi L., *A Short Course in Quantum Information Theory, Lect. Notes Phys.*, Vol. **827** (Springer, Berlin, Heidelberg) 2011, p. 11.

Chapter 4

Foundations of Quantum Mechanics

János Bergou

Department of Physics and Astronomy, Hunter College of the City University of New York, New York, USA

Paper Title: **Heisenberg's dog and quantum computing**

Reference: Journal of Modern Optics **45**, 701–711 (1998)

Personal message:

I first met Berge many years ago in Albuquerque, New Mexico, of all places. It was at the time a rare occurrence that a German and a Hungarian scientist first met in the middle of American nowhere, not the least because the two of us had very different backgrounds: We came from different sides of the then-still-existing iron curtain. Berge came to Albuquerque to give a series of extremely polished lectures on Schwinger's take on high-energy physics. It was the first of many occasions that I could enjoy his impeccable lecturing skills. But it had to wait for a second encounter before we became friends. That time he returned to Albuquerque together with his family for an extended stay. Very quickly, we discovered that his two children were born in the same years and same months as our two older children, and his wife Ola also came from the wrong side of the curtain. So, we had a lot in common besides physics, and I got to know Berge's non-physics side very well. Not just the two of us but also the two families became very close friends. We went on field trips, the kids played together, we played chess and just had a great time all along. I think we also gave the Californian and German wine industries a boost in the process. Since then they visited us in our two homes, in East Brunswick and Budapest, and we returned the favor by visiting them in their two homes, in Ismaning and Singapore. It is truly a global friendship.

Of course, Berge being Berge, one cannot escape writing some papers together, so we, too, produced some. We had some very light-hearted papers on the notorious but nevertheless non-existent phase operator, and on duality. But a round anniversary is serious business, so it deserves that our most serious work be included here. It offers a vast generalization of the Schrödinger's cat, elevating it to a whole new level and firmly establishing a new concept that we appropriately termed as the Heisenberg's dog. Alas, the physics community is biased. We, that is, both authors, are convinced that this is our most read paper; unfortunately, it is certainly not one of our most cited ones. To remedy the situation, I am re-issuing the paper for the betterment of our community at large. The paper is truly paradigm changing. For example, we submitted to the journal three reports by the anonymous referees alongside the manuscript. Imagine, in our busy times of overburdened editors, how much this would ease their lives. The editor, of course, was very appreciative of the idea: Not only was the paper accepted immediately but excerpts from the reports were published adjacent to the paper. Since both Berge and I are editors of some journals, we speak from experience. We believe the entire community should follow suit.

To Berge:

Since you speak every conceivable language and all of them well, I restrict it to three: I wish you Happy Anniversary, Herzlichen Glückwunsch zum Geburtstag and Boldog Születésnapot! Und auch viel Spass in der Zukunft und danke für die Freundschaft!

JOURNAL OF MODERN OPTICS, 1 APRIL 1998, VOL. 45, NO. 4, 701–711

Un-topical review
Heisenberg's dog and quantum computing

JÁNOS A. BERGOU
Department of Physics and Astronomy, Hunter College of the City University of New York, 695 Park Avenue, New York, NY 10021, USA

and BERTHOLD-GEORG ENGLERT
Sektion Physik, Universität München, Theresienstrasse 37, 80333 München, Germany

(*Received 13 August 1997*)

Abstract. A Heisenberg's dog is the central unit in a quantum computer that factors prime numbers. We report the outcome of a proof-of-principle *Gedankenexperiment* in which the number 3 was factored and good agreement was found both with an analytical theoretical prediction and with the results of a quantum Las Vegas calculation. Other future applications of Heisenberg's dog are hinted at. The anonymous reports of the referees and our reply to them are reproduced in appendix A and appendix B respectively.

1. Introduction

Ever since its birth in 1935, Schrödinger's [1] cat has received much and well deserved attention.[1] The original, very complicated set-up that consisted of a radioactive sample, a Geiger counter, a hammer, a cell of poisonous gas, etc.—in addition to the cat itself—gradually evolved into arrangements that are considerably simpler and experimentally more accessible. At the present state of cat evolution it has become reduced to the superposition of two slightly excited coherent states of opposite phase. The amount of theoretical work behind this reduction cannot be overestimated!

Let us break a lance for historical justice and recall that this extraordinary theoretical achievement was triggered by two developments.[2] First, it has become increasingly difficult to obtain permission for experiments on live objects [5].

[1] For recent developments see [2, 3].

[2] We suspect, but cannot prove the case, that Lamb's [4] verdict about Schrödinger's cat, namely

> It is a messy, rather pointless problem, and in bad taste for several different reasons. (1) The contemplated treatment of the cat is inhumane. (2) The possible death of the cat serves no useful scientific purpose. (3) Cats often have nine lives. (4) A different kind of bad taste involves proposing a complicated problem when a simple one will suffice to illustrate the physical principles involved.

provided an additional impetus.

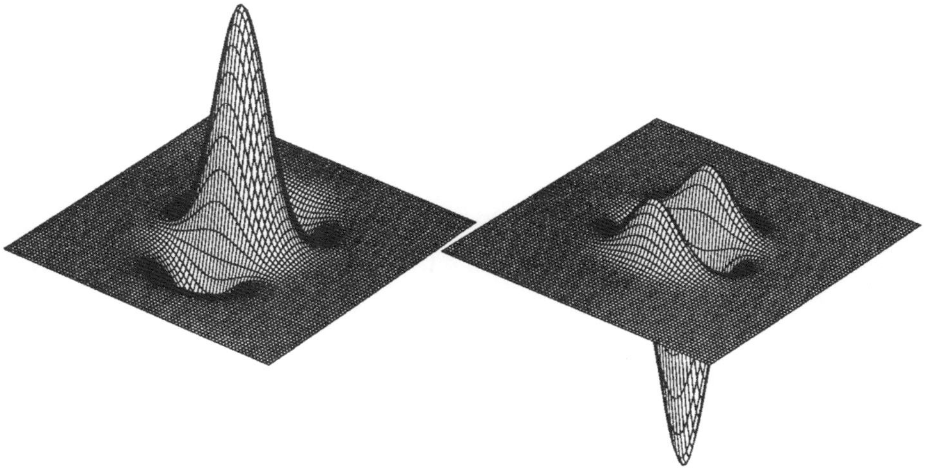

Figure 1. A male Schrödinger's cat (on the left) eyeing a female Schrödinger's cat (on the right). Only the essential parts are shown. (Plots supplied by J. Janszky.)

Second, at the current level of funding for fundamental physical phenomena, it is impossible to cover the fee requested by the veterinarian, who is needed to carry out the measurement at the final stage of any Schrödinger's cat experiment.

It is remarkable, however, that, together with the said simplification, the cat, originally a rather neutral object, has developed two sexes (figure 1). This discovery, generally credited to Janszky *et al.* [6], brought about the possibility of mating these objects, and the offsprings at their initial state are termed Schrödinger's kittens. It is also very interesting that, when mature Schrödinger's cats are superimposed, there is a tendency towards constructive interference in same-sex superpositions and toward largely unpredictable, but often destructive interference in opposite-sex superpositions. According to Janszky, the leading expert [6] on cat sexology, the implications of this disturbing observation are far from being fully understood. Domokos and Janszky [7] have observed that the state resulting from an opposite-sex superposition strongly depends on whether the Schrödinger's cats are parallel or perpendicular. Further studies are called for, but they are beyond the scope of the present contribution.

The ageing of Schrödinger's cats has also been at the focus of theoretical and experimental studies [8], so that all aspects of an individual cat's life have been covered. It has been noticed that the sexual differences disappear as Schrödinger's cats grow old, and senior cats are indistinguishable from the point of view of maleness and femaleness. This phenomenon, the existence of which in other mammals is vehemently denied by those involved in practical studies, therefore is surely a quantum phenomenon. The experts speak of decoherence [9, 10]. For more vulgar minds it is the loss of ability to form a superposition state.

In marked contrast with the scientific scrutiny enjoyed by Schrödinger's cat, Heisenberg's dog[3] (HD) attracted considerably less attention. Its discovery is just

[3] Gajzenberga Gund in the Russian literature; see the textbook series by Landau and Lifshitz. Admittedly, we do not know which volume should be consulted, but our Russian colleagues are positive that everything can be found in Landau and Lifshitz.

as well documented[4], and so it will suffice to recall briefly the physical properties of HD. It is so fuzzy that, when Professor Heisenberg walked his dog at the natural speed of the Planck stroll, that is 2 ft s^{-1} [11] or, for senior citizens, 9.48 in s^{-1}[12], with an uncertainty of $10^{-?}$ ft s^{-1} (the exponent is fuzzy), the dog's position became so uncertain that it was impossible to tell the dog from its fuzz.[5] The velocity uncertainty originates, of course, in the random up-and-down and side-to-side motion of the dog's tail (incidentally, an early manifestation of quantum chaos [14]), for which the quantum optics community has adopted the expression 'waving its tail'. The momentum uncertainty is thus the measure of the dog's waviness whereas the position uncertainty is clearly related to fuzziness.

The recent advance in the art of quantum computing resulted from the observation that Schrödinger's cats are good at doing arithmetic. Indeed, all quantum computing schemes employ Schrödinger's cats of various races, sizes and shapes. They promise the realization of one of the oldest dreams of mankind: the fast factoring of large numbers into primes.

The technical problems faced by those building quantum computers are formidable [15] and the field is highly competitive. There is not only strong competition among various groups sharing the same goal, but also the little-spoken-about race against research on other topics. The least-spoken-about issue concerns the next industrial revolution: what will first be found in every household, a quantum computer or a quantum teleporter?

Inasmuch as HD is complementary to Schrödinger's cat, it is hardly surprising that HDs will eventually solve another problem of great cultural value and utmost practical importance, namely the factorization of small prime numbers into nonintegers. A first step towards this goal is reported in the present contribution.

Although it is beyond the scope of the current review, we should mention, for completeness, that research is in progress on a complete history of HD with special attention to its cultural conditioning.[6]

Here is an outline of the paper. In section 2 we report the set-up of the *Gedankenexperiment* used for our studies of HD. For reasons analogous to those that led to the great simplification of the concept of Schrödinger's cat mentioned above, we went through much trouble to reduce the HD concept to something more mundane. The utter simplicity that characterizes present-day Schrödinger's cats has not been achieved as yet, but we are confident that the readers will appreciate the progress that has been made. This is followed, in section 3, by a comparison of the experimental results with theoretical predictions. Good agreement was reached both with an analytical formula and with a quantum Las Vegas calculation. We hint at possible future applications (section 4) and close with a

[4] It can be traced to some serious after-dinner conversations between perfectly sober physicists, among them the current authors and some others who, owing to an overdose of modesty, prefer to remain anonymous.

[5] To the best of our knowledge there is no photographic document showing the dog and its fuzz. We proudly report, however, that Heisenberg's dog was heard barking in Hungary recently [13].

[6] It has recently been pointed out by Sokal [16] that the findings of scientific studies are much dependent on the cultural context in which the investigations are carried out. A striking example is the value of Planck's constant; physicists of one cultural background insist that $\hbar = 10^{-27}$ erg s, and others are convinced that $\hbar = 1$.

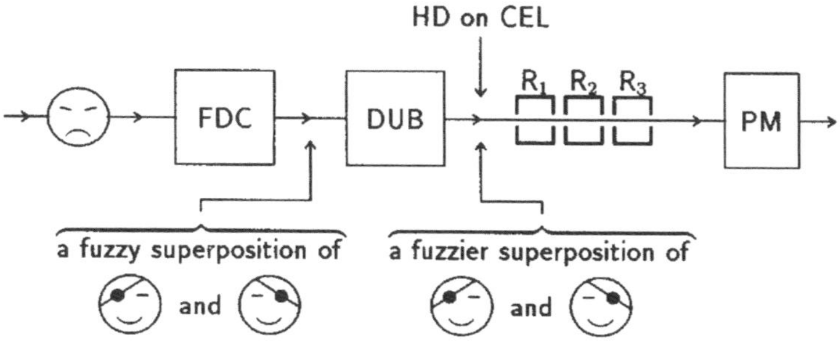

Figure 2. Experimental set-up for factoring the number 3 into two factors.

summary. Appendix A and appendix B contain the referees' reports and our reply to them respectively.

2. Experimental set-up

We shall now describe our experimental set-up (figure 2). It is well known that one-level atoms and two-level atoms are not sufficiently fuzzy to allow for the construction of superposition states of the HD kind. Fortunately, three-level atoms of the V type are perfectly suited but, since we need essentially a two-level system at one stage of the experiment, we must dress the atom so that one of the excited states is turned into a dark state when forming the HD. Thus we need a device that effects the transition

 $\longrightarrow$ fuzzy superposition of  and . (1)

Note the appearance of the two orthogonal pirate states.[7] Ödmann [17] has shown recently that the desired transition is obtained when the three-level atom is passed through a Fermi–Dirac condenser (FDC), operated in reverse mode.[8] The fuzzy superposition resulting from the action of the FDC does not, however, possess all the essential properties of a HD. Therefore, we pass it through a down-under booster (DUB), which transfers non-local momentum without applying a force [18]. As a consequence of the force-free momentum transfer, the three-level atom has now really become a HD or, more precisely, a HD on a conveniently extended leash (HD on CEL).[9]

The stage is now set for the factorization experiment. More specifically, in this

[7] We have no idea why dressed three-level atoms with one dark excited state are said to be in a 'pirate state'; it is one of the many mysteries of the jargon of quantum optics.

[8] The original FDC would put two spin-$\frac{1}{2}$ particles into the same orbital state. It effects, for instance, the formation of a helium atom in its ground state if two electrons and an α particle happen to be at hand. Ödmann discovered that reversed-mode operation enables the experimenter to put one particle into a fuzzy superposition of two states. In the case of a three-level atom these are obviously the pirate states.

[9] Any similarity to other pre-existing acronyms, such as HD for Humpty Dumpty or CEL for correlated emission laser, are unavoidable and not unintended. The possibility of CEL was recognized for the first time by Scully [19]. A medieval poem on a HD is on record (see [20]).

proof-of-principle *Gedankenexperiment* we attempt to factor the number 3 into two factors. The two-factor situation requires a two-state HD which is prepared at the initial stage of the experiment, as described in the preceding paragraph. Now, since the number 3 is to be factored, the HD must play the quantum *Hütchenspiel* with three *Hütchens*.[10] Extensions to factoring larger numbers into more factors, such as 137 into 44 factors, are straightforward. Here, the three *Hütchens* are realized by three consecutive microwave resonators R_1, R_2 and R_3 with a photon stored in one of them, which one being unknown to the experimenter. The HD runs through the resonators and finds the photon with the aid of a quite normally destructive (QND) measurement. QND measurements have been reviewed recently by Braginsky and Khalili [23].

At the final stage of the experiment, the state of the HD is determined by a protective measurement (PM) [24].[11] The result of this measurement is one of the factors,[12] the other factor is given by the amount of non-local momentum that the DUB has transferred.

Finally, we recycle the three-level atom for another run of the experiment. For the sake of reproducible initial conditions, the HD must not remember the previous run. We therefore apply a quantum eraser (QE) (introduced in [29]; not shown in figure 2) to the three-level atom. Control *Gedankenexperiments* have shown that complete amnesia can be achieved.

3. Results

Our results are summarized in figure 3. We should perhaps remind the reader that one outcome of the *Gedankenexperiment* (displayed as a cross (+)) corresponds to the generation of the x and the y coordinate. We note that the agreement with theory (solid line) is marvellous. The solid line represents our analytical result, namely[13]

$$yx = 3; \tag{2}$$

according to a back-of-the-envelope calculation it should be well approximated by a straight line in a log–log plot.

Somewhat less satisfactory is the comparison with the x, y pairs ($\otimes$) produced by a quantum Las Vegas calculation.[14] We employed two personal computers, each armed with a 586 processor. For the purpose of parallel computing, their keyboards were meticulously aligned on an optical bench. It appears that the

[10] We recall that in the classical *Hütchenspiel* (English translation, shell game) a little ball is hidden under one of three thimbles. Whereas the classical variant is characterized by the property that the ball is never under the thimble lifted by the player (this statement is supported by experimental data in abundance [21]), the quantum *Hütchenspiel* exhibits the opposite behaviour, that is the photon is always in the resonator in which you look for it [22].

[11] The PM is known as a 'quantum condom' (QC) in the quantum optics jargon [25]; we disapprove of such frivolous abuse of the language. Some people claim that the QCs have big holes [26–28], but we prefer to ignore these findings.

[12] It is a familiar fact that a PM recognizes the dog by its paw (*ex ungue canem*).

[13] This is a quantum-mechanical result. Remember to read it from right to left: 3 equals x times y.

[14] According to Mebody *et al.* [30], a quantum Las Vegas calculation is similar, in spirit, to a quantum Monte Carlo calculation [31, 32]. The latter was not applicable here because we could not incorporate the FDC and the DUB into the existing schemes.

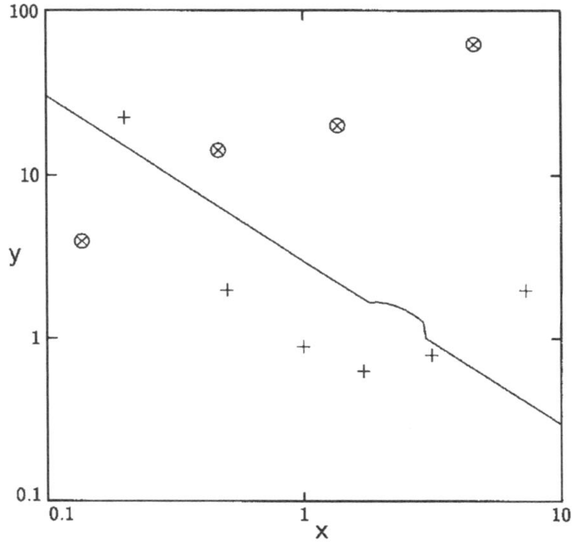

Figure 3. Comparison of the outcomes of our *Gedankenexperiment* (+) with an analytical approximation (——) and the results of a quantum Las Vegas calculation (⊗). The analytical approximation is actually a rule of thumb. (The solid line was drawn by running a pen along a ruler; the boss shows where the pen hit the thumb.)

quantum Las Vegas data are suffering from a small systematic error of unknown origin (grossly exaggerated in this log–log plot). We have excluded all known sources of error (tidal forces, anisotropy in the 3 K background radiation, fluctuations in the solar-neutrino flux, misalignment of the keyboards, etc.) and are at present considering the unlikely possibility of a malfunction of the 586 processors.

We have not indicated any statistical error bars in figure 3, simply because they are not available. After the action of the FDC the system is subject to the *Pauliverbot* so that each outcome can occur once at most, and we refuse to do a statistical analysis under these circumstances.

4. Future applications

In addition to the factorization of small prime numbers, HDs are also very good at detecting loopholes hiding vainly (LHVs) in theories aimed at replacing standard quantum theory. A crucial test is whether the *Bellungleichung* (German for 'bark inequality')[15]

$$\mathcal{D}^2 + |\mathcal{C}|^2 \leqslant 1, \tag{3}$$

is obeyed. This inequality [33] sets bounds on the dogness $\mathcal{D}$ that measures the amount of the Heisenberg picture present in a given quantum-theoretical

[15] Traditionally, many nouns referring to fundamental issues are taken over from the German language in which they are defined most easily. Translations into other languages are unavoidably imprecise. Perhaps unnecessarily we remind the reader of the best-known example; a *Welcherweggedankenexperiment* is an obvious notion, but who could make sense of 'which way thought experiment' without professional training?

treatment, and the (complex) catness $\mathcal{C}$ that quantifies the relative presence of the Schrödinger picture. An illustrating example is Dirac's interaction picture, where one would ideally have $\mathcal{D} = |\mathcal{C}| = 1/2^{1/2}$. Needless to say, HDs can bark and are therefore predestined for testing the validity of the *Bellungleichung*. We are at present involved in experimental studies of these matters and shall report results in due course.

5. Summary

We have shown that a three-level atom in a HD state can be used for a reliable factorization of small prime numbers. Extensions to large prime numbers are straightforward. As soon as the Schrödinger's-cat-based quantum computers are equally successful in factoring large numbers into primes, the complete factorization problem will be solved. Further, we have pointed out that HDs—a race traditionally bred in the mist of Copenhagen—are particularly well suited to testing alternatives to standard quantum theory.

Acknowledgments

We are grateful for the many insights gained in numerous enjoyable discussions with H.-J. Briegel, M. Fleischhauer, R. Glauber, U. Herzog, J. Janszky and A. Schenzle. Without their encouragement this unf(o)unded research project would never have been completed. We thank B. L. Ödmann for sharing his discovery with us before [17] was published, and we thank J. Janszky for supplying the plots of figure 1.

Appendix A: Anonymous referee reports by Axel Schenzle and Michael Fleischhauer

A.1. (*Excerpts from*) *the first referee's report*

Since the end of the last century the evolution of natural sciences has marched on a street of success without precedent. The reason is that science has devoted its undivided attention to every problem (small or large, important or seemingly dull) that has attracted its attention. Erwin Schrödinger with his unappeasable interest for pussy-cats opened a new line of reseach to the so far undeservedly neglected animal kingdom.

In view of this success story of the twentieth century, it is entirely inexcusable that important other creatures have been so pitilessly neglected. Where is the quantum theory of non-localizable pigs or quantum-jumping sheep? Where is the theoretical progress on man's favourite companion, the dog? The existence of HD has remained entirely in the dark for half a century. It has always been argued that it is the enormous fuzziness of this peculiar dog that has made its investigation so extremely difficult, but these are nothing but evasions which should not divert from the incompetence and unwillingness of scientists to deal with this problem so far.

The ultimate reduction of Schrödinger's cat to two particles by Bohm is an epoch-making triumph of science. It cannot be held against the present paper that HD has not yet been simplified to its very essence. Unfortunately Bohm's holistic disciples, who meanwhile worship the 'hiddenness' of physical concepts, are of no help in simplifying the entangled-dog concept either. The HD is a miraculously aesthetic concept by itself that satisfies all theoretical desires but, in addition, it has the enormous potential for revolutionizing mathematics. Fellow mathematicians, *cave canem*! While Schrödinger's cat can be trained to do various useless things such as factoring large numbers into primes, the HD will be able to do quite the opposite; it will be able to factor small primes into large numbers.

The authors demonstrate unexpected talents and a long-standing experience in interpreting experimental and numerical results. Their convincing and ingenious interpretation of figure 3 opens our eyes and makes it immediately obvious that the agreement is an overwhelming proof of the presented theory and not only a rule(r) of thumb.

At this point it cannot be swept under the rug that the paper is also plagued by a number of serious shortcomings.

(i) While the manuscript is written mostly in English, it uses an abundance of technical terminology and quantum-optical jargon mostly borrowed from German. This is entirely inexcusable and makes the paper very hard to read, especially to those who understand neither of the two languages. The authors should eventually stick to a single language, most preferably to Finno-Ugric since it cannot be comprehended by a sizeable number of people and therefore gives nobody an unfair advantage.

(ii) The authors have trundled into the mud by advocating the use of contraceptive means—not to use the vulgar terminology of the paper. We thoroughly refuse the use of such means in complete posterior agreement with the *Encyclica Humanae Vitae*.

(iii) The lack of moral standards of the authors also becomes obvious in the vague and suggestive discussion of superposition states. Parental guidance must be ensured before general distribution of the paper.

(iv) The ridiculous discussion on the value of fundamental constants such as $\hbar$ and c does not fit the high scientific quality of the paper. Everybody has known since childhood that $\hbar = c = 1$. However, attempts to prove that also $2\pi/6 = 1$ have failed badly, in spite of the fact that it had been defined this way by legislative action in the State of Indiana in 1897 [34].

(v) By mentioning the name of the present referee in the acknowledgments of the paper, the authors have attempted to influence the referee and his independent opinion in an undue way. Here they meet with a man of immutable and uncorruptible standards who is not to be affected by such a blunt attempt. In any case, however, I include (on a separate sheet[16]) my bank account number and suggestions for a generous contribution to the funding of my otherwise badly supported research. Depending on the generosity of the transactions, I will suggest immediate publication or rejection of the paper.

I hope that I have been of help to the Editor in deciding this matter. If there are any further questions, call on somebody else. In case the Editor decides not to follow my well substantiated advice, he is entirely responsible for the consequences.

[16] Not reproduced here.

A.2. (*Excerpts from*) *the second referee's report*

I very much enjoyed reading this manuscript. It is sound and I believe that it is a very important contribution to theoretical quantum optics and will open up new areas of research. I strongly recommend the publication of this interesting and very original work.

The authors present for the first time a proof-of-principle *Gedankenexperiment* to factor the prime number 3 and in this way boldly go where no one has gone before. To this end they propose and study a novel quantum computer based on HD states. The rule-over-the-thumb result together with the analytical calculatons showing the different factorizations of 3 in figure 3 are convincing. Only the numerical Montevideo simulations are somewhat dissatisfying. Nevertheless, I believe that also the Monteverdi simulations are in support of the authors' results. I am convinced that the HD states will soon gain the same growing attention which is now paid to the famous cat states of Schrödinger, which are complementary to the first.

Finally I would like to remark on a seemingly side aspect of the paper, the pirate states. Astonishingly these important non-classical states of the radiation field have not been discussed in the quantum optics literature so far. Not only is it because according to Ödmann [17] they are produced in the reverse-mode operation of a FDC and because they are fundamental blocks for the HD states that makes them appealing, but also they are interesting in their own right. I am convinced that the discussion of the pirate states in the present manuscript will trigger a large number of further publications studying the properties of these states. For example, I believe that an analysis of the Q-, the Wigner- and other s-parametrized phase-space distributions will yield interesting and potentially useful insights into these fascinating states and will help us to understand their subtle nature. It may lead to a whole new field, to 'quantum piracy'. But of course, these are all things which go much beyond the scope of the present paper and should be understood as suggestions for future work.

All in all, despite the small weaknesses in the Montessori simulations, this is a very nice paper, which should definitely be published.

Appendix B: Authors' reply to the referees

We are grateful for the referees' efforts and are pleased by their positive recommendations. To their minor objections we respond as follows.

To the first referee,

(i) we have naturally adopted the lingua franca of modern science, namely broken English as canonized by Casimir [35].

(ii) and (iv) We are bewildered. How could a person of such high moral standards, as witnessed by his support of the *Encyclica Humanae Vitae*, argue against the usage of the biblical value of π?[17]

(iii) *Honi soit qui mal y pense.*

[17] 'And he made a molten sea, ten qubits from one brim to the other; it was round all about, and his height was five qubits; and a line of thirty qubits did compass it round about.' (I Kings, 7, 23 [36]) implies that $\pi = 3$. We have replaced the archaic spelling 'cubit' by the more appropriate modern spelling.

(v) We value the anonymity of the referees very highly. In other words, we cannot afford to support everybody mentioned in the acknowledgments.

To the second referee, Montevideo? Monteverdi? Montessori?—is the referee implying something?

We think that these brief remarks are properly responsive to the referees' mild objections and see no need for modifications of or additions to the text as submitted originally. We would like, however, to express our disappointment about the absence of any comment by the referees on our novel method of performing parallel computing. Therefore, we shall, for the time being, not reveal how we can do orthogonal computing.

References

[1] Schrödinger, E., 1935, *Naturwissenschaften*, **23**, 807, 823, 844; English transl., Trimmer, J. D., 1980, *Proc. Am. Phil. Soc.*, **124**, 323; Reprinted in J. A. Wheeler and W. H. Zurek (editors), 1983, Quantum Theory and Measurement (Princeton University Press), 152–167.

[2] Kleppner, D., November 1996, *Phys. Today*, **49**, 11.

[3] *Phys. Today*, April 1997, letters to the editor by J. H. Hetherington, by R. J. von Gutfeld, by W. R. Hitchens, and by Th. W. Hänsch.

[4] Lamb, Jr, W. E., 1987, in *Reminiscences about a Great Physicist: Paul Adrien Maurice Dirac*, edited by B. N. Kursunoglu and E. P. Wigner (Cambridge University Press), pp. 249–261.

[5] *National Science Foundation Grant Proposal Manual*, 1995, chapter 7, section 710: Protection of Living Organisms, subsection 713: Animal Welfare Requirements. For further information consult http://www.nsf.gov/nsf/nsfpubs/gpm95/ch.7ht#ch7-23.

[6] Janszky, J., Földesi, I., Szabo, S., Adam, P., and Kim, M. S., 1996, *Acta Phys. Slov.*, **46**, 269, and references therein.

[7] Domokos, P., and Janszky, J., 1994, *Phys. Lett.* A, **186**, 289.

[8] Brune, M., Hagley, E., Dreyer, J., Maali, A., Wunderlich, C., Raimond, J. M., and Haroche, S., 1996, *Phys. Rev. Lett.*, **77**, 4887.

[9] Zurek, W. H., October 1991, *Phys. Today*, **44**, 36; ***Phys. Today***, April 1993, letters to the Editor.

[10] Omnès, R., 1994, *The Interpretation of Quantum Mechanics* (Princeton University Press).

[11] Zeilinger, A., 1990, *Am. J. Phys.*, **58**, 103.

[12] Wadlinger, R. L. P., and Hunter, G., 1990, *Am. J. Phys.*, **58**, 903.

[13] Scully, M. O., 1997, private communication.

[14] Gutzwiller, M. C., 1990, *Chaos in Classical and Quantum Mechanics* (Berlin: Springer).

[15] Haroche, S., and Raimond, J.-M., August 1996, *Phys. Today*, **49**, 51; ***Phys. Today***, November 1996, letters to the Editor.

[16] Sokal, A., 1996, *Transgressing the Boundaries: Towards a Transformative Hermeneutics of Quantum Gravity*, Social Text Nos. 46/47, pp. 217–252. For further information consult http://www.physics.nyu.edu/faculty/sokal/.

[17] Ödmann, B. L., 1997, ***Sitzungsber. Akad. Wiss. Schilda***, **1**, 1.

[18] Wiseman, H., and Harrison, F., 1995, *Nature*, **377**, 584; Wiseman, H. M., Harrison, F. E., Collett, M. J., Tan, S. M., Walls, D. F., and Killip, R. B., 1997, *Phys. Rev.* A, **56**, 55.

[19] Scully, M. O., 1985, *Phys. Rev. Lett.*, **55**, 2802.

[20] Schwinger, J., Scully, M. O., and Englert, B.-G., 1988, *Z. Phys.* D, **10**, 135.

[21] Götz, G., 1996, private communication.

[22] Bergou, J. A., 1997, *Acta Phys. Slov.*, **47**, 185.

[23] Braginsky, V. B., and Khalili, F. Ya., 1996, *Rev. mod. Phys.*, **68**, 1.

[24] Aharonov, Y., Anandan, J., and Vaidman, L., 1993, *Phys. Rev.* A, **47,** 4616.
[25] Robert, J., 1995, private communication.
[26] Schwinger, J., 1993, *Science,* **262,** 826.
[27] Rovelli, C., 1994, *Phys. Rev.* A, **50,** 2788.
[28] Alter, O., and Yamamoto, Y., 1995, *Phys. Rev. Lett.*, **74,** 4106; 1996, *Phys. Rev.* A, **53,** R2911.
[29] Scully, M. O., and Drühl, K., 1986, *Phys. Rev.* A, **25,** 2208.
[30] Mebody, S. O., Other, A. N., and Hird, A. T., 1997, *Prog. Par. Quant. Comp.* (to be published).
[31] Dalibard, J., Castin, Y., and Mølmer, K., 1992, *Phys. Rev. Lett.*, **68,** 580.
[32] Dum, R., Zoller, P., and Ritsch, H., 1992, *Phys. Rev.* A, **45,** 4879.
[33] Englert, B.-G., 1996, *Phys. Rev. Lett.*, **77,** 2154.
[34] Beckmann, P., 1971, *A History of* π (New York: Golem Press).
[35] Casimir, H. B. G., 1983, *Haphazard Reality* (New York: Harper & Row).
[36] *The Bible.*

Lev Vaidman

Raymond and Beverly Sackler School of Physics and Astronomy, Tel-Aviv University, Tel-Aviv, Israel

Paper Title: **Surrealistic trajectories**

Reference: Original article (2014)

Paper Title: **Surrealistic Bohm trajectories**

Reference: Zeitschrift für Naturforschung **47a**, 1175–1186 (1992)

Surrealistic trajectories

March 3, 2014

Berge Englert wrote many interesting and important papers. The one which left the strongest impact on me is his work on Surrealistic Bohm trajectories [1]. I am apparently the first reader of this paper and I continue to work on this topic until today. After reading it, I could not believe that the result is correct. I do not remember any other case in which I was absolutely sure that it cannot be true, but now I understand that it is a correct result. The paper created a controversy, but it still awaits a proper acknowledgement of its importance.

I know that Berge's and my views on foundations of quantum mechanics are different. My understanding is that Berge belongs to the orthodox camp. He is ready to accept the position of the founders of quantum theory according to which we just have to learn how to calculate the results of experiments and that there is no reality beyond this. Berge wrote [2] that quantum mechanics is a theory about events, "but an answer to the question *Why are there events?* cannot be given by quantum theory." I am not ready to give up trying to know what happens in between the "events". Although my preferred picture is a "pure" quantum mechanics in which everything is a wave (which forces me to accept numerous parallel worlds), I am looking for other options, and for me Bohmian mechanics is the best candidate for a complete final theory of Nature within a single world. It is deterministic and elegant (in the pilot wave form). I did study it carefully, discussed it with Bohm himself (who considered it only as an intermediate step toward a deeper theory) and was trying to persuade Yakir Aharonov that it is interesting. I thought I understood it well.

I was not particularly successful in making Yakir interested in the Bohmian mechanics. So, when he received Berge's paper on this topic to referee, he asked me to look at it. I felt that I understood well the measurement process in Bohmian theory. I was confident that Bohmian mechanics treats well all cases of quantum measurements; yet Berge's paper claimed the opposite. I re-read it several times, but saw no error. I was still not convinced by the "no crossing argument". I calculated the result myself without relying on the no-crossing theorem and, indeed, Bohmian mechanics failed to behave in an expected way. With apparently good position detectors, the particle left a trace where the Bohmian particle was not present!

Some Bohmians received Berge's result with disbelief [3]. But it seems that today there is a consensus that it is true [4]. I think that surrealistic trajectories represent a weakness of the theory. We should admit that perfectly legitimate experiments in the framework of standard quantum mechanics "fool" the observer that accepts the Bohmian mechanics. I also find it a fascinating phenomenon and I enjoyed analyzing what exactly are the conditions when surrealistic trajectories appear [5,6]. I asked Berge to help with this analysis, but he did not share with me the excitement about this project. Berge apparently does not appreciate the importance of his paper. He did not mention it in his "Scientific Curriculum". He did not care to make it available online. I understand that this follows from his orthodox approach to quantum mechanics and his belief that

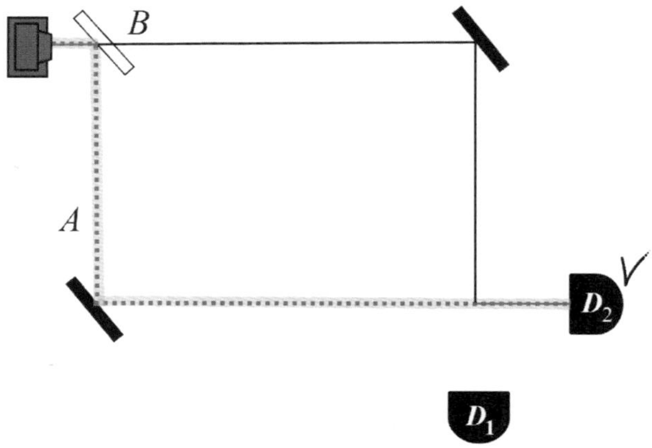

Figure 1: The particle emitted by the source is detected by detector D_2. The dotted line shows its trace as shown by recordings of Berge's slow nondemolition particle detectors. The continuous line shows Bohmian trajectory. The grey wide line shows the weak trace left by the particle. The weak trace remains the same when the slow detectors are not present.

"there is no measurement problem" [2]. He did, however, join in the analysis of Bohmian trajectories in the setup of protective measurements [7] where surrealistic trajectories appear as well.

I feel that this work led me to analyze possible trajectories of a quantum particle outside the frame of Bohmian mechanics. I found another surprising example in which the particle leaves a trace in an unexpected place [8]. In Berge's work on surrealistic Bohmian trajectories, a particle passes through a Mach-Zehnder interferometer (MZI) in which the second beamsplitter was removed, transforming it into a "which-path experiment"; see Fig. 1. Berge considered special nondemolition detectors in which the position of their pointers separated according to the results of the measurement after the particle passed the location of the missing beam splitter. Berge and his collaborators showed that a trace of these slowly operating detectors is left on arm A while the Bohmian particle goes through arm B.

Bohmian mechanics predictions for the results of experiments are identical to those of standard quantum mechanics, so an experiment cannot distinguish between them. In order to find some way to decide which interpretation has an advantage, I suggested to define the past of a quantum particle (a concept which has no clear definition in the standard interpretation) as the place where it leaves a weak trace. In Berge's setup, it happens in arm A even if his slow operating detectors are not present. So, I have another reason to name Bohmian trajectories surrealistic.

Slow detectors and the weak trace show the same trajectory. This is also a common-sense trajectory adopted by Wheeler: It had to be this trajectory since the particle could not have come any other way (according to the standard quantum evolution equations).

Instead of removing the second beam splitter in the MZI, I transformed the experiment into a which-path detector by deflecting the particle in path B using an additional properly tuned nested MZI; see Fig. 2. Now again, slowly operating detectors

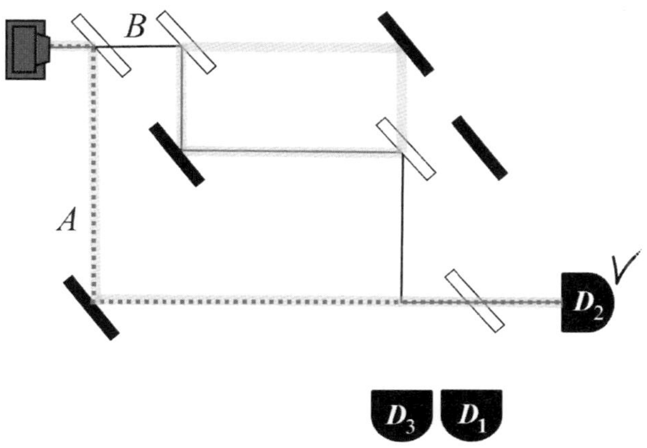

Figure 2: The nested MZI is tuned for destructive interference for continuation to the beam splitter of the external MZI. The particle emitted by the source is detected by detector D_2. The dotted line shows its trace as shown by recordings of Berge's slow nondemolition particle detectors. The continuous line shows Bohmian trajectory. The grey wide line shows the weak trace left by the particle when the slow detectors are not present.

and the weak trace show the path in arm A agreeing with the common-sense argument, while the Bohmian trajectory takes the arm B passing through one of the arms of the nested interferometer.

An additional surprise happens when we consider the weak trace in this setup without slowly operating detectors. As the common-sense argument predicts, we do see a weak trace in arm A, but we also observe a weak trace of a comparable strength in the part of arm B inside the nested interferometer. The photons reaching detector D_2 leave a trace in the interferometer through which they could not pass. And they do not leave the weak trace on the way in and on the way out of this interferometer. The weak trace cannot be explained by any set of continuous trajectories.

In spite of our different approaches to the interpretation of quantum mechanics, my and Berge's research paths kept intersecting. As I followed Berge on the issue of surrealistic trajectories, Berge developed the Aharonov's Mean King Problem [9–11] which does appear in his Scientific Curriculum.

I learned a lot from Berge and enjoyed very much our intertwined research. I hope we will continue to stimulate each other's research for many years to come.

Bibliography

[1] B.-G. Englert, M. O. Scully, G. Süssmann, and H. Walther, "Surrealistic Bohm trajectories," Z. Naturforsch. A **47**, 1175 (1992).

[2] B.-G. Englert, "On Quantum Theory", Eur. Phys. Jour. D **67**, 238 (2013).

[3] B. J. Hiley and R. E. Callaghan, "Delayed-choice experiments and the Bohm approach," Phys. Scripta **74**, 336 (2006).

[4] `http://youtu.be/CCW93koLNYY`, "What are the surrealistic trajectories in Bohmian Mechanics?"

[5] L. Vaidman, "The reality in Bohmian quantum mechanics or can you kill with an empty wave bullet?" Found. Phys. **35**, 299 (2005).

[6] G. Naaman-Marom, N. Erez and L. Vaidman, "Position Measurements in the de Broglie-Bohm Interpretation of Quantum Mechanics," Ann. Phys. **327**, 2522 (2012).

[7] Y. Aharonov, B.-G. Englert, and M. Scully, "Protective measurements and Bohm trajectories," Phys. Lett. A **263**, 137 (1999).

[8] L. Vaidman, "Past of a quantum particle," Phys. Rev. A **87**, 052104 (2013).

[9] L. Vaidman, Y. Aharonov, and D. Z. Albert, "How to ascertain the values of σ_x, σ_y, and σ_z of a spin-1/2 particle?" Phys. Rev. Lett. **58**, 1385 (1987).

[10] Y. Aharonov and B.-G. Englert, "The mean king's problem: Spin 1," Z. Naturforsch. **56a**, 16 (2001).

[11] B.-G. Englert and Y. Aharonov, "The mean king's problem: Prime degrees of freedom," Phys. Lett. A **284**, 1 (2001).

Surrealistic Bohm Trajectories

Berthold-Georg Englert[1,2], Marlan O. Scully[3], Georg Süssmann, and Herbert Walther[1,2]

[1] Sektion Physik, Universität München, Am Coulombwall 1, D-8046 Garching, Germany
[2] Max-Planck-Institut für Quantenoptik, Ludwig-Prandtl-Straße 10, W-8046 Garching.
[3] Department of Physics, Texas A & M University, College Station, TX 77843-4242.

Z. Naturforsch. **47a,** 1175–1186 (1992); received September 22, 1992

A study of interferometers with one-bit which-way detectors demonstrates that the trajectories, which David Bohm invented in his attempt at a *realistic* interpretation of quantum mechanics, are in fact *surrealistic*, because they may be macroscopically at variance with the observed track of the particle. We consider a two-slit interferometer and an incomplete Stern-Gerlach interferometer, and propose an experimentum crucis based on the latter.

Introduction

In Bohmian mechanics [1–3], ordinary quantum theory [4] is supplemented by classical particle trajectories. These trajectories are usually unobserved – indeed, they are unobservable in the face of Heisenberg's uncertainty relation for position and momentum – but nevertheless they are regarded as realistic, similar to the tracks seen in a bubble chamber. Yes, such tracks supposedly exhibit Bohm trajectories within the limitations set by uncertainty relations. More common, however, is the situation in which a particle is detected without being observed earlier, and then the particle's Bohm trajectory can be *retro*dicted. This retrodiction assigns reality to the Bohm trajectories in an operational sense, the more so because the actual trajectory cannot be *pre*dicted owing to the fundamental ignorance of the actual initial position. Predictions in Bohmian mechanics are limited by the probabilistic knowledge about initial conditions, which knowledge comes from quantal probability distributions. Therefore, the predictive power of Bohmian mechanics does not exceed that of ordinary quantum theory, and so the alleged superiority of Bohmian mechanics over ordinary quantum theory is of a purely philosophical nature.

A particle traverses a bubble chamber and is then detected. Does the retrodicted Bohm trajectory always agree with the observed track? Our answer is: No. For, our considerations show that the Bohm trajectory may be macroscopically at variance with the recorded track. For the sake of simplicity, we do, however, not study the complicated motion of a charged particle through a bubble chamber, where many degrees of freedom are involved. Instead, we look at atom interferometers, and a few which-way detectors replacing the plethora of bubbles. These one-bit which-way detectors are placed such that they enable us to distinguish one class of tracks from another, macroscopically different one. It turns out that the atom's Bohm trajectory may not belong to the observed class of tracks. We are thus led to the conclusion that the Bohm trajectories, originally introduced with the aim of arriving at a "realistic interpretation" of quantum theory, are in fact surrealistic.

The next section contains a concise review of Bohmian mechanics. Then we turn to gedanken experiments in which atoms pass through double slits, eventually supplemented by which-way detectors capable of recording through which slit the atom went without thereby disturbing the atom's center-of-mass motion. We find that there are events when the Bohm trajectory goes through one slit, but the atom through the other. This is followed by a detailed treatment of magnetic atoms traversing an incomplete Stern-Gerlach interferometer. Here the track recorded by the which-way detectors is *always* macroscopically different from the corresponding Bohm trajectories. We are thus proposing an experimentum crucis which, according to our quantum theoretical prediction, will clearly demonstrate that the reality attributed to Bohm trajectories is rather metaphysical than physical.

In Appendix A we supply a brief description of quantum-optical which-way detectors, which have been discussed in detail elsewhere [5].

Reprint requests to Prof. Dr. G. Süssmann, Sektion Physik, Universität München, Am Coulombwall 1, W-8046 Garching bei München.

0932-0784 / 92 / 1200-1175 $ 01.30/0. – Please order a reprint rather than making your own copy.

The one-bit which-way detectors may evolve dynamically in Bohmian mechanics, too. The description is then changed quantitatively but not qualitatively. A large fraction of atomic Bohm trajectories remains macroscopically at variance with the recorded track. This situation is discussed in Appendix B.

Bohmian Mechanics in a Nutshell

In Bohmian mechanics [1–3] the continuity equation

$$\frac{\partial}{\partial t}\varrho(t, \boldsymbol{r}) + \boldsymbol{\nabla}\cdot\boldsymbol{j}(t, \boldsymbol{r}) = 0\,, \tag{1}$$

obeyed by the probability density ϱ for finding the particle in the vicinity of the point $\boldsymbol{r}$ and the corresponding probability current density $\boldsymbol{j}$, is taken as the starting point for defining, and eventually computing, trajectories $\boldsymbol{R}(t)$. These Bohm trajectories are interpreted as the possible paths along which the particle could have propagated. The actual trajectory is then selected by the (initial or rather) final position that is somehow observed. The trajectories are determined by Bohm's velocity field $\boldsymbol{v}(t, \boldsymbol{r})$ that is given by

$$\boldsymbol{v}(t, \boldsymbol{r}) = \boldsymbol{j}(t, \boldsymbol{r})/\varrho(t, \boldsymbol{r})\,. \tag{2}$$

Fortunately, knowledge of $\boldsymbol{v}$ where $\varrho = 0$ is never needed. The differential equation

$$\frac{\mathrm{d}}{\mathrm{d}t}\boldsymbol{R}(t) = \boldsymbol{v}(t, \boldsymbol{R}(t)) \tag{3}$$

in conjunction with a known position at some instant – usually when the particle is finally detected – then supplies the actual Bohm trajectory. The probabilistic predictions of ordinary quantum theory remain unaltered because an ensemble of trajectories that mimicks ϱ at one time will equally well mimick ϱ at any later time. So, with the understanding that in repeated experiments initial positions are realized according to the probabilities implied by ϱ, an ensemble average over the corresponding trajectories will agree with the quantum theoretical predictions. In addition, before the velocity field (2) is available, one must first solve the Schrödinger equation for the underlying wave function from which ϱ and $\boldsymbol{j}$ are then calculated. Thus, the Bohm trajectories cannot provide us with more information than what is carried by the wave function.

In the sequel we shall confine the discussion to nonrelativistic situations involving one particle of mass m. The dynamics will by governed by a Hamilton operator of the form

$$\mathbf{H} = \frac{1}{2m}\boldsymbol{p}^2 + V(t, \boldsymbol{r}; \mathbf{A})\,, \tag{4}$$

where $\boldsymbol{r}$ and $\boldsymbol{p}$ are the position and momentum operators and $\mathbf{A}$ symbolizes additional degrees of freedom like spin. The wave function may then posses a number of components $\psi_\alpha(t, \boldsymbol{r})$ labelled by a subscript α. From these, ϱ and $\boldsymbol{j}$ are constructed in the familiar way:

$$\varrho(t, \boldsymbol{r}) = \sum_\alpha |\psi_\alpha(t, \boldsymbol{r})|^2\,, \tag{5}$$

$$\boldsymbol{j}(t, \boldsymbol{r}) = \frac{1}{m}\,\mathrm{Re}\sum_\alpha \psi_\alpha^*(t, \boldsymbol{r})\,\frac{\hbar}{i}\boldsymbol{\nabla}\psi_\alpha(t, \boldsymbol{r})\,,$$

of which the latter applies only if the $\boldsymbol{p}$-dependence of $\mathbf{H}$ does not differ from that in (4).

In Bohmian mechanics, a particle has a position and nothing else, as is particularly emphasized in Bell's [6] formulation (see also [7]). The properties associated with $\mathbf{A}$ in (4) (or the label α in (5)), such like spin, are possessed and carried by the wave function only. Nevertheless, after the trajectory $\boldsymbol{R}(t)$ has been found, the values of the $\psi_\alpha(t, \boldsymbol{R}(t))$ can be employed to specify what – in a conscious departure from standard Bohmian mechanics – could be regarded as the actual values of the quantities $\mathbf{A}$. For example, if $\psi_\pm$ are the two spin-$\frac{1}{2}$ components referring to the z direction, then the unit spin vector at time t is

$$\boldsymbol{S}(t) = \frac{1}{\varrho}\begin{pmatrix} 2\,\mathrm{Re}\,\psi_+^*\,\psi_- \\ 2\,\mathrm{Im}\,\psi_+^*\,\psi_- \\ |\psi_+|^2 - |\psi_-|^2 \end{pmatrix} \tag{6}$$

with $\varrho = |\psi_+|^2 + |\psi_-|^2$ (see (5)), where $\psi_\pm$ are evaluated at $\boldsymbol{r} = \boldsymbol{R}(t)$.

Double-Silt Interferometer

Consider the symmetric double-slit set-up in Figure 1. The particles coming in from the left all have the same velocity perpendicular to the slit plate. Thus the incident wave function is a plane wave with the planes parallel to the slit plate. In the interference region between the slit plate and the screen, the wave function is the coherent superposition of the two contributions from the slits,

$$\psi(t, \boldsymbol{r}) = \psi_>(t, \boldsymbol{r}) + \psi_<(t, \boldsymbol{r})\,. \tag{7}$$

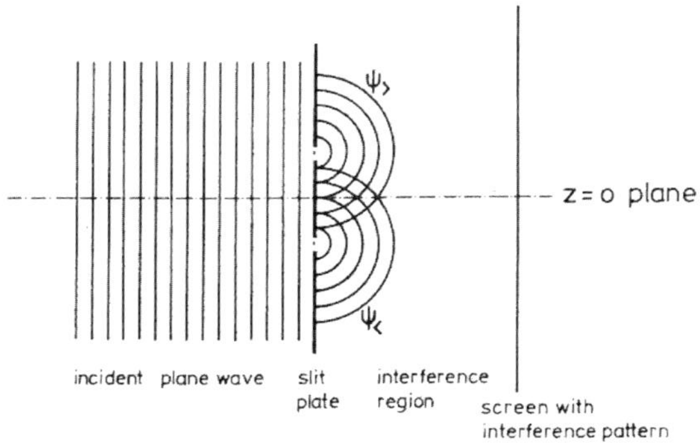

Fig. 1. Double-slit interferometer with indication of phase fronts for incident plane wave and scattered partial waves $\psi_>$ and $\psi_<$ which interfere.

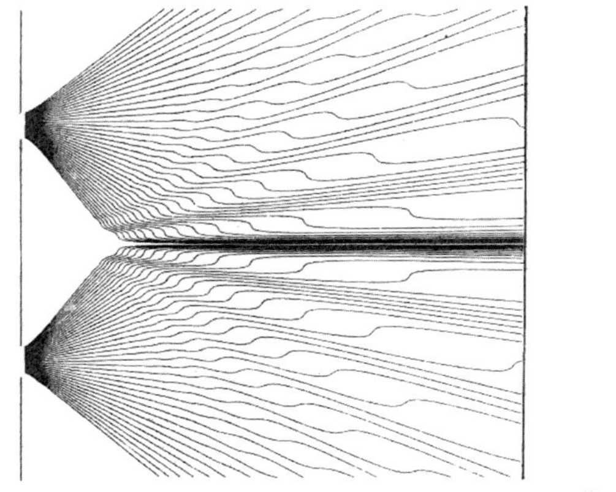

Fig. 2. Bohm trajectories in the interference region of a double-slit interferometer; adapted from [8].

In view of the geometrical symmetry of the arrangement, $\psi_<$ is obtained from $\psi_>$ by reflection at the $z=0$ plane,

$$\psi_<(t, x, y, z) = \psi_>(t, x, y, -z) . \tag{8}$$

Consequently, the probability density $\varrho = |\psi|^2$ as well as the x- and y-components of the current vector $j = \frac{1}{m} \operatorname{Re} \psi^* \frac{\hbar}{i} \nabla \psi$ are even functions of z, in contrast to the z-component of $\boldsymbol{j}$, which is odd. Therefore, the z-component of the velocity field (2) is also an odd

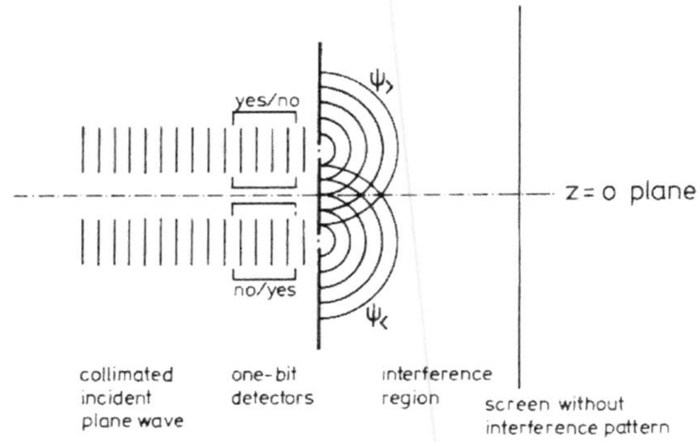

Fig. 3. Double-slit interferometer with one-bit detectors through which the collimated incident plane wave reaches the slit plate. With which-way information available, the screen no longer displays the double-slit interference pattern.

function of z, so that this component vanishes on the $z=0$ plane of symmetry. This has the immediate implication that the Bohm trajectories do not cross the $z=0$ plane [6]. In other words, when a particle has hit the upper half of the screen (where $z>0$), then its retrodicted Bohm trajectory goes through the upper slit, and those particles arriving on the lower half of the screen (where $z<0$) have, according to Bohmian mechanics, passed through the lower slit. These symmetry considerations are confirmed by the Bohm trajectories computed by Philippidis, Dewdney, and Hiley [8], see Figure 2.

In ordinary quantum mechanics, the statement that the particle went through one slit and not the other is, of course, utterly meaningless, as long as no corresponding observation is performed (as exemplified by the string of bubbles in a bubble chamber). Quite consciously, no definite classical history is assigned to a single event. An adherent of Bohmian mechanics, however, would proudly announce that he does not have to give up the concept of a classical particle trajectory when studying quantal phenomena. And really, what is the argument about? The particle is not *observed* while it traverses the apparatus in Figure 1. So why can't we all accept the Bohm trajectories as real, their empirical reality originating in retrodiction rather than direct observation?

The advanced techniques of modern quantum optics enable the experimenter to build one-bit detectors sensitive to the passage of a single atom. For a description consult Appendix A. The details of the detection mechanism are not relevant for the present discussions, so we shall symbolize the one-bit recording as a transition from "no" to "yes". It is important to realize that this transition happens with virtual certainty and that the atom's center-of-mass wave function is not altered noticeably in the process.

The set-up of Fig. 1 is now supplemented by two such one-bit detectors, one for each slit as in Figure 3. The detectors will supply us with which-way information, not quite in the detailed sense of registering a whole track (like that seen in a bubble chamber), but by clearly distinguishing the class of tracks through one slit from the class through the other. The two-slit interference pattern on the screen is lost, of course, as soon as we have this which-way information available, but that is not the issue here.

With the one-bit which-way detectors in place, the wave function in the interference region is now

$$\Psi = \psi_> \left|{}^{\rm yes}_{\rm no}\right\rangle + \psi_< \left|{}^{\rm no}_{\rm yes}\right\rangle , \tag{9}$$

in which the contribution of the upper slit is correlated to the which-way information documented in the upper detector, and likewise for the lower slit *and the lower detector*. In contrast to (7), the functions $\psi_>$ and $\psi_<$ in (9) are components of the two-component wave function Ψ. The density ϱ and current $\boldsymbol{j}$ associated with Ψ are, therefore, different from what is obtained from (8). In particular, the terms responsible for the two-slit interference pattern are absent now, as it must be. However, most important for the matter of interest, the symmetry properties of ϱ and $\boldsymbol{j}$ under reflections at the $z=0$ plane are the same as before: ϱ as well as the x- and y-component of $\boldsymbol{j}$ and $\boldsymbol{v}$ are unaffected by the transformation $z \to -z$, whereas the z-components of $\boldsymbol{j}$ and $\boldsymbol{v}$ change sign and accordingly vanish on the symmetry plane $z=0$. Thus it is still true that when the atom is found on the upper half of the screen its retrodicted Bohm trajectory goes through the upper slit, and the trajectory of an atom found on the lower half passes through the lower slit.

But through which slit did the atom come? Suppose that the upper detector says "yes", the lower "no". Then the probability for finding the atom somewhere on the screen is proportional to $|\psi_>|^2$, which does *not* vanish on the lower half of the screen. Consequently, there will be events when the *atom goes through* the upper detector and therefore through *the upper slit* and then hits the lower half of the screen, so that the corresponding *Bohm trajectory goes through the lower slit*. In other words: the Bohm trajectory is here macroscopically at variance with the actual, that is: observed, track. Tersely: Bohm trajectories are not realistic, they are surrealistic.

As a result of these considerations we disagree with Bell [6] who says (words adapted to the present discussion): "The naive classical picture [of ordinary quantum mechanics] has the particle, arriving on a given half of the screen, going through the wrong slit." Per definition, the "right slit" is, for Bell, the one traversed by the Bohm trajectory. To state it once more clearly: for us, the right slit is the one through which the atom is actually observed going, never mind the naive unobservable Bohm trajectory. And if one does not observe through which slit the atom goes, then the notion of the right or the wrong slit is meaningless.

Bell [6] has more to say about double-slit interferometers with which-way detectors. His detectors, however, are not of the one-bit type, but consist of very many particles, of which a good fraction gets macroscopically displaced. Bell considers treating these particles also according to the rules of Bohmian mechanics and arrives at the conclusion that, in effect, either $\psi_>$ or $\psi_<$ becomes irrelevant. Thereafter the symmetry argument no longer applies and the Bohm trajectory passes always through the same slit as the atom's track. Bell's reasoning does not apply to the present scheme, in which no macroscopic displacements occur until the which-way information stored in the one-bit detectors is finally read off. This reading-off is done (long) after the atom has hit the screen, so the relevant Bohm velocity field is, indeed, determined by the two-component wave function (9). There the yes/no degree of freedom of the one-bit detectors appears just as a label, like α in (5). One cannot associate continuous position variables with such discrete degrees of freedom and, therefore, they are not dynamical in Bohmian mechanics, in contrast to the motional degrees of freedom of the many particles that constitute Bell's macroscopic detector.

One could object that the yes/no label of the one-bit detector need not really refer to two orthogonal states of a spin-$\frac{1}{2}$-type degree of freedom. It could just as well indicate that some physical system – a harmonic oscillator, say – is either in its ground state ("no") or in its first excited state ("yes"). Then one does have a corresponding continuous position variable with its own Bohmian dynamics. And the Bohm velocity field for the atom no longer has a definite symmetry, it changes according to the evolution of the detector variable. Consequently, now the atom trajectory may cross the $z=0$ plane. But, as discussed in

Appendix B, there remains a considerable number of trajectories that do not pass through the same slit as the atom's track.

The one-bit detectors possess yet another essential property that is lacking in macroscopic detectors. Before they are actually read off, nothing irreversible happens. One can even erase the which-way information carefully and recover the two-slit interference pattern in the process. For details consult [9].

There is one more lesson in Bell's article, a psychological one. Namely that this tentative supporter of Bohmian mechanics very much *wants* the Bohm trajectory to pass through the same slit as the observed track of the atom. Nature, however, does not grant this favor.

Incomplete Stern-Gerlach Interferometer

The double-slit interferometer for atoms with which-way detectors is very difficult, if not impossible, to realize experimentally. We have discussed its features above mainly for pedagogical reasons which include the contact made with [6] and [8]. We now turn to the much more realistic experiments in which atoms traverse three quarters of a Stern-Gerlach interferometer, see Figure 4. The missing fourth quarter, necessary for reuniting the partial beams, is absent. Therefore, entering spin-up atoms are first deflected up, then down, and finally hit the screen well below the $z=0$ plane; conversely, spin-down atoms are first deflected down, then up, and hit the screen well above the $z=0$ plane. We emphasize that both the intermediate and final separation between the beams are macroscopic and, in particular, large compared with the individual beam widths.

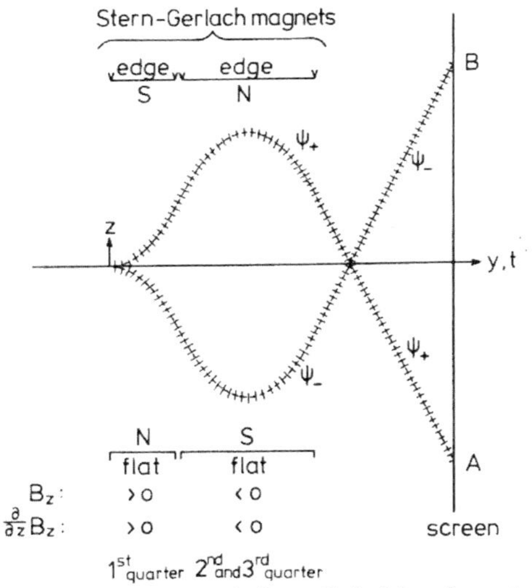

Fig. 4. Three quarters of a Stern-Gerlach interferometer. The dashed curves represent $\pm\Delta z$ of (15) along which the centers of the wave function components $\psi_\pm$ propagate. After crossing the $z=0$ plane, atoms arrive on the screen either in region A ("spin up") or in region B ("spin down").

The subtleties studied in [10], which are essential for the coherent reunion of the two partial beams in a complete (four quarters) Stern-Gerlach interferometer, are irrelevant here. We can, therefore, adopt the strategy of [11] and disregard the x-motion totally while regarding the y-motion as classical. Then y is replaced by vt and the spatial y-dependence of the magnetic field is thereby effectively turned into a time-dependence. For the z-motion and the spin evolution we then have, as in [1], the Hamilton operator

$$\mathbf{H} = \frac{1}{2m}\mathbf{p}^2 + \mathcal{E}(t)\,\sigma_z - F(t)\,z\,\sigma_z\,, \tag{10}$$

which has the structure (4) with $\mathbf{A}$ standing for the spin vector operator. The term $\mathcal{E}(t)\,\sigma_z$ is the magnetic energy at the center of the magnets (where $z=0$) and $F(t)\,\sigma_z$ is the force produced by the z-inhomogeneity of the magnetic field.

The net momentum transferred to a spin-up atom is

$$\Delta p(t) = \int_0^t \mathrm{d}t'\,F(t')\,, \tag{11}$$

and its accumulated z-displacement becomes

$$\Delta z(t) = \int_0^t \mathrm{d}t'\,\Delta p(t')/m = \int_0^t \mathrm{d}t'\,\frac{t-t'}{m}\,F(t')\,, \tag{12}$$

where $t=0$ is the instant when the atom enters the magnetic field. For example, the idealized force

$$F(t) = \begin{cases} F_0 & \text{for} \quad 0<t<T_0\,, \\ -F_0 & \text{for} \quad T_0<t<3\,T_0\,, \\ 0 & \text{for} \quad 3\,T_0<t\,, \end{cases} \tag{13}$$

where T_0 is the time to traverse one of the three quarters, yields

$$\Delta p(t) = \begin{cases} F_0\,t & \text{for} \quad 0<t<T_0\,, \\ F_0(2\,T_0-t) & \text{for} \quad T_0<t<3\,T_0\,, \\ -F_0\,T_0 & \text{for} \quad 3\,T_0<t \end{cases} \tag{14}$$

and

$$\Delta z(t) = \frac{F_0}{2m}\begin{cases} t^2 & \text{for} \quad 0 < t < T_0\,, \\ 2T_0^2 - (t-2T_0)^2 & \text{for} \quad T_0 < t < 3T_0\,, \\ (7T_0 - 2t)\,T_0 & \text{for} \quad 3T_0 < t\,. \end{cases} \tag{15}$$

Hence the up and down beams intersect around time $t = \frac{7}{2}T_0$. At time $t = 2T_0$ the (macroscopic) distance between the two beams is $2|\Delta z(2T_0)| = 2F_0T_0^2/m$, and if the position of the screen corresponds to $t = 5T_0$, the spots produced by the spin-up and spin-down atoms are $2|\Delta z(5T_0)| = 3F_0T_0^2/m$ apart.

At the initial time $t=0$ the magnetic atom is supposed to be in a minimum uncertainty state. (This is an innocuous simplifying assumption; in an atomic beam minimum uncertainty states are hard to come by.) To take advantage of the symmetry of the set-up, we take the atom initially polarized in the x-direction and choose the spatial wave function symmetric to the $z=0$ plane. Thus the two components referring to the eigenvalues $\sigma_z = \pm 1$ of the spin operator σ_z are initially

$$\begin{aligned} \psi_+(0,z) &= \psi_-(0,z) \\ &= (2\pi)^{-1/4}(2\,\delta z_0)^{-1/2}\exp\left[-\left(\frac{z}{2\,\delta z_0}\right)^2\right], \end{aligned} \tag{16}$$

where δz_0 is the initial spread in z, which uncertainty is small compared to the maximal separation of the beams. The corresponding spread in momentum is $\delta p = \frac{1}{2}\hbar/\delta z_0$. In a reasonable experiment one must avoid that the natural spreading happens too fast. In the present context this requires that the characteristic time $m\,\delta z_0/\delta p$ is at most of the order of T_0.

At later times we get, by solving the Schrödinger equation,

$$\begin{aligned} \psi_\pm(t,z) &= (2\pi)^{-1/4}\left[2\left(\delta z_0 + i\frac{t}{m}\,\delta p\right)\right]^{-1/2} \\ &\cdot \exp\left[-\frac{1}{4}\,\frac{(z \mp \Delta z(t))^2}{\delta z_0(\delta z_0 + i\,t\,\delta p/m)} \pm \frac{i}{\hbar}\,z\,\Delta p(t) \mp \frac{i}{2}\,\Phi(t)\right] \end{aligned} \tag{17}$$

up to an irrelevant common z-independent phase factor. The quantity

$$\Phi(t) = \frac{2}{\hbar}\int_0^t dt'\,\mathscr{E}(t') \tag{18}$$

is the accumulated Larmor precession angle. It equals

$$\Phi(t) = \begin{cases} \Omega t & \text{for} \quad 0 < t < T_0\,, \\ \Omega(2T_0 - t) & \text{for} \quad T_0 < t < 3T_0\,, \\ -\Omega T_0 & \text{for} \quad 3T_0 < t \end{cases} \tag{19}$$

in the case of

$$\mathscr{E}(t) = \begin{cases} \frac{\hbar}{2}\Omega & \text{for} \quad 0 < t < T_0\,, \\ -\frac{\hbar}{2}\Omega & \text{for} \quad T_0 < t < 3T_0\,, \\ 0 & \text{for} \quad 3T_0 < t\,. \end{cases} \tag{20}$$

The two components (17) interchange under the spatial reflection according to

$$\psi_+(t,z) = \psi_-(t,-z)\,, \tag{21}$$

which is the essential symmetry also possessed by the components $\psi_<$ and $\psi_>$ in (9). Consequently, the probability density

$$\varrho(t,z) = |\psi_+(t,z)|^2 + |\psi_-(t,z)|^2 \tag{22}$$

and the z-component

$$j(t,z) = \frac{1}{m}\,\mathrm{Re}\left[\psi_+^*\,\frac{\hbar}{i}\frac{\partial}{\partial z}\psi_+ + \psi_-^*\,\frac{\hbar}{i}\frac{\partial}{\partial z}\psi_-\right](t,z) \tag{23}$$

of the current are symmetric and antisymmetric, respectively:

$$\varrho(t,-z) = \varrho(t,z)\,, \qquad j(t,-z) = -j(t,z)\,. \tag{24}$$

The resulting Bohm velocity field (z-component again) is then antisymmetric, too,

$$\begin{aligned} v(t,z) &= t\,z\left(\frac{\delta p}{m\,\delta z(t)}\right)^2 \\ &+ \left[\frac{\Delta p(t)}{m} - t\,\Delta z(t)\left(\frac{\delta p}{m\,\delta z(t)}\right)^2\right]\tanh\left(\frac{z\,\Delta z(t)}{(\delta z(t))^2}\right) \\ &= -v(t,-z)\,. \end{aligned} \tag{25}$$

Thus it vanishes on the $z=0$ plane with the now familiar consequence that the Bohm trajectories do not cross this plane. In (25), $\delta z(t) = \sqrt{(\delta z_0)^2 + (t\,\delta p/m)^2}$ is the spread in z at time t.

The Bohm-mechanical equation of motion for the trajectory $Z(t)$, *viz.*

$$\frac{d}{dt}Z(t) = v(t, Z(t))\,, \tag{26}$$

is implicitly solved by

$$\int_{-\infty}^{Z(t)} dz\,\varrho(t,z) = \int_{-\infty}^{Z_0} dz\,\varrho(0,z)\,, \tag{27}$$

where $Z_0 = Z(0)$ is the initial condition. A final condition could be employed analogously and would, indeed, emphasize the retrodictive nature. In view of

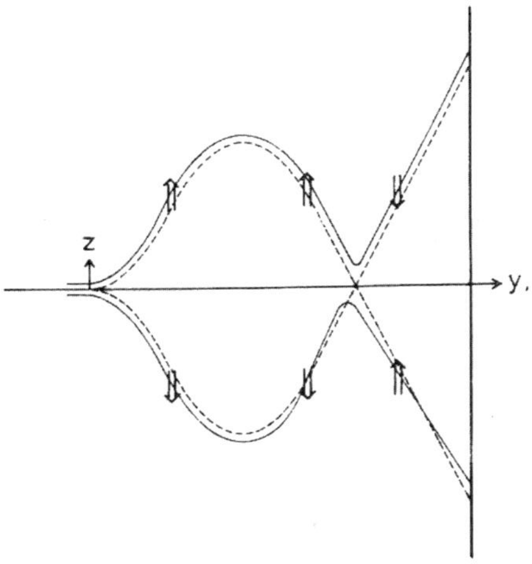

Fig. 5. Bohm trajectories through the incomplete Stern-Gerlach interferometer (magnets not drawn). The dashed curves represent $\pm\Delta z$ of (15); the solid curves are typical Bohm trajectories, which do not cross the $z=0$ plane. The arrows indicate the spin vectors of (29).

(22) with (17), (27) is equivalent to

$$\int\limits_{\Delta z(t)-Z(t)}^{\Delta z(t)+Z(t)} \frac{dz}{\delta z(t)} \exp\left[-\frac{1}{2}\left(\frac{z}{\delta z(t)}\right)^2\right]$$

$$= \int\limits_{-Z_0}^{Z_0} \frac{dz}{\delta z_0} \exp\left[-\frac{1}{2}\left(\frac{z}{\delta z_0}\right)^2\right]. \quad (28)$$

These integrals can, of course, be evaluated in terms of the error function, but that hardly adds transparence. Note that both integrands are even functions of z, which has the immediate consequence that the sign of $Z(t)$ is that of Z_0, quite independent of the sign of $\Delta z(t)$. So the explicit expression (28) confirms the result of the symmetry consideration: no Bohm trajectory crosses the $z=0$ plane. Further, if Z_0 does not exceed δz_0 by much, as is the most probable situation, and if, in addition, $\Delta z(t)$ is much larger than $\delta z(t)$, signifying a macroscopic separation between ψ_+ and ψ_-, then $|Z(t)|$ does not differ markedly from $|\Delta z(t)|$. Thus, whereas the spatial wave function components $\psi_\pm$ intersect where $\Delta z(t)$ changes sign, each Bohm trajectories first (roughly) follows the center of *one* of the components up to this cross-over region, beyond which they follow the *other*. These statements are illustrated in Figure 5.

The spin vectors indicated in this figure represent what is produced by (6) and (17),

$$S(t) = \left[\cosh\left(\frac{Z(t)\,\Delta z(t)}{(\delta z(t))^2}\right)\right]^{-1} \begin{pmatrix} \cos\varphi(t) \\ \sin\varphi(t) \\ \sinh\left(\frac{Z(t)\,\Delta z(t)}{(\delta z(t))^2}\right) \end{pmatrix} \quad (29)$$

with the effective Larmor angle

$$\varphi(t) = \Phi(t) - \frac{2}{\hbar} Z(t)\left[\Delta p(t) - \frac{t}{m}\left(\frac{\delta p}{\delta z(t)}\right)^2 \Delta z(t)\right]. \quad (30)$$

Outside the cross-over region, the argument of the hyperbolic functions is large, so that, practically speaking,

$$S = \begin{pmatrix} 0 \\ 0 \\ (\operatorname{sgn} Z_0)(\operatorname{sgn}\Delta z) \end{pmatrix} \quad (31)$$

there. In the cross-over region, the direction of S is reversed, rapidly though continuously.

We now supplement the incomplete Stern-Gerlach interferometer of Figure 4 with two one-bit which-way detectors located where $|\Delta z|$ is largest, that is just after two of the three quarters, see Figure 6. When the detectors are included into the formulation, we have to replace the two-component wave function

$$\Psi = \begin{pmatrix} \psi_+ \\ \psi_- \end{pmatrix} \quad (32)$$

by a four-component one, i.e.

$$\Psi_{\text{before}} = \begin{pmatrix} \psi_+\left|{}^{\text{no}}_{\text{no}}\right\rangle \\ \psi_-\left|{}^{\text{no}}_{\text{no}}\right\rangle \end{pmatrix} \quad (33)$$

at the times before the atom traversed (one of) the detectors, and by

$$\Psi_{\text{after}} = \begin{pmatrix} \psi_+\left|{}^{\text{yes}}_{\text{no}}\right\rangle \\ \psi_-\left|{}^{\text{no}}_{\text{yes}}\right\rangle \end{pmatrix} \quad (34)$$

after the transition. Since the which-way detectors have no noticeable influence on the center-of-mass wave functions $\psi_\pm$, both the density ϱ and the current j as well as the velocity field v of (25) remain unaltered. Consequently, the (retrodicted) Bohm trajectories are identical with the ones found without the which-way detectors installed. There is only one difference. The

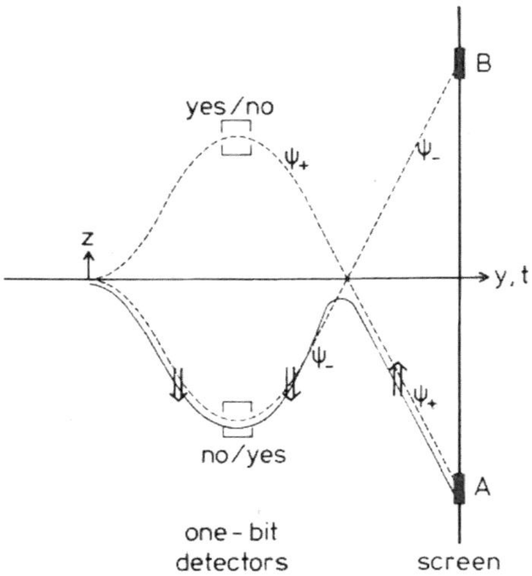

Fig. 6. Incomplete Stern-Gerlach interferometer with one-bit which-way detectors (magnets not drawn). The dashed curves represent $\pm\Delta z(t)$ of (15). The solid curve shows a Bohm trajectory through the *lower* detector, corresponding to an atom that reached the lower region A through the *upper* detector. The arrows indicate the spin vectors of (29) and (35), respectively.

spin vector "after" is not given by (6) and (29) but by

$$\boldsymbol{S}_{\text{after}} = \begin{pmatrix} 0 \\ 0 \\ \operatorname{sgn}(|\psi_+|^2 - |\psi_-|^2) \end{pmatrix} = \begin{pmatrix} 0 \\ 0 \\ (\operatorname{sgn} Z_0)(\operatorname{sgn} \Delta z) \end{pmatrix}, \tag{35}$$

which now applies in the cross-over region, too. Thus, spin-flip now appears to happen instantaneously at the moment when Δz changes sign. For an adherent of Bohmian mechanics this is nothing to worry about, though, because for him spin is a property of the wave function and not carried by the atom itself.

With the which-way detectors in place we can again compare the Bohm trajectory with what is known about the track of the atom. In the lower region A of Figure 6. ψ_- vanishes effectively and only ψ_+ contributes to the density ϱ. Therefore, as implied by (35), if an atom hits the lower screen in region A, it has left a trace in the upper detector. Its Bohm trajectory, in contrast, goes through the lower detector. Likewise, an atom arriving in the upper region B went through the lower detector, but its Bohm trajectory through the upper one. The actual tracks and the Bohm trajectories could not be more at variance than that.

A supporter of Bohmian mechanics woulds insist that the atom went along its Bohm trajectory through one of the detectors, but left its mark in the other one. Thus, he concludes, the which-way detectors do not deserve their name. The usual which-way detection, he declares, is an illusion. But, in arguing this way, is he not putting away with the basics of Bohmian mechanics? For, on the one hand, the retrodicted Bohm trajectories are supposed to be "truly realistic", on the other hand, checking them out is forbidden or, at least, impossible. We conclude: The reality attributed to Bohm trajectories is not physical, it is metaphysical.

A real experiment along these lines appears feasible, as all components of the apparatus have been realized separately; for details consult Appendix A. The purpose of the proposed experiment is to verify the predicted quantum theoretical correlations: Whenever an atom hits the screen in the lower region A, the upper detector says "yes" – and likewise for the upper region B and the lower detector. Since we know that (some of) the retrodicted Bohm trajectories pass through the other detector, this verification suffices to demonstrate the asserted metaphysical character of the Bohm trajectories.

Acknowledgements

BGE and GS are grateful for the hospitality experienced at the Center for Advanced Studies at UNM. This work is partially supported by the Office of Naval Research.

Appendix A

The central instrument needed for the construction of the quantum-optical which-way detector is the micromaser. This [12] consists of a resonator cavity with superconducting walls that enclose a volume of a few cubic centimeters. It is traversed by atoms that are suitably excited prior to entering the cavity. The resonator can be tuned very precisely to transitions between rather neighboring Rydberg states. In the ongoing micromaser experiments, one exemplary transition is between the $63\,\mathrm{p}_{3/2}$ and the $61\,\mathrm{d}_{5/2}$ states of Rubidium at a frequency of about 21.5 GHz, corresponding to the size of the resonator, which has a cylindrical shape and is about 24 mm both in length and in diameter.

By cooling the cavity to very low temperatures one reduces the number of thermal photons practically to

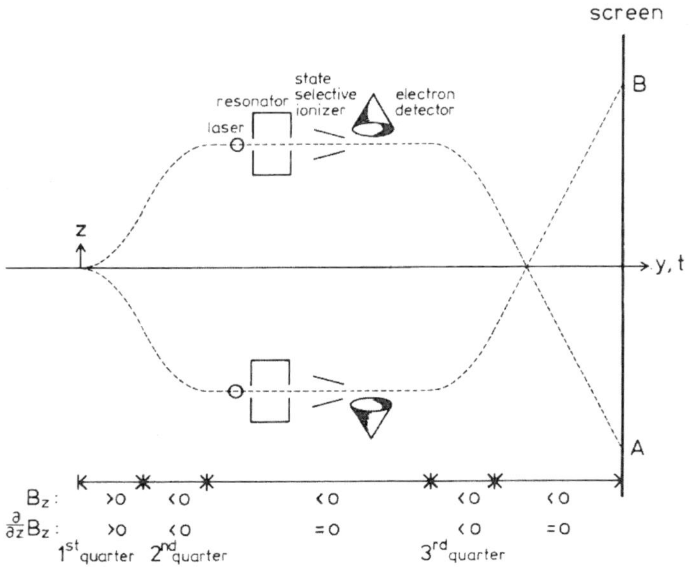

Fig. 7. Schematic set-up of the proposed experimentum crucis. An incomplete Stern-Gerlach interferometer is supplemented – between the second and third quarter – by quantum-optical one-bit which-way detectors consisting of laser beams for the excitation to the Rydberg states, micromaser resonators for storing the tell-tale photons, electrostatic condensors for the state-selective ionization of the Rydberg atoms, and electron detectors. A measurement cycle begins with a laser pulse that excites an atom into the upper Rydberg state. After this atom has been detected on the screen in regions A or B, a second laser pulse excites a reader atom into the lower Rydberg state. This reader atom picks up the tell-tale photon and is detected by the state-selective field ionization.

zero. When an atom, excited to the upper states of the Rydberg transition, enters this empty cavity, it will emit a single microwave photon with virtual certainty, provided that the interaction time is chosen appropriately by controlling the atom's velocity (~ 300 m/s). The emerging atom is then in the lower one or the two relevant Rydberg states. Thus the presence or absence of a photon in the resonator indicates whether an atom went through it or not. The micromaser can, therefore, be used as a one-bit detector sensitive to the passage of a single atom.

The lifetime of the tell-tale photon in the ultracold cavity is as long as a few hundred milliseconds. This is more then enough time for reading off the stored information. For this purpose, a "reader atom" in the lower state of the Rydberg transition is sent through the cavity with the right velocity, so that it absorbs the photon with certainty, if it is there. By state-selective field ionization one can then find out if the emerging reader atom is in the upper or the lower Rydberg state. Thereafter one knows for sure whether a tell-tale photon was stored in the cavity or not.

The interaction of the atom with the microwave cavity field does not appreciably disturb the atom's center-of-mass wave function [13], although the exchange of energy may change the internal atomic state. The center-of-mass motion is most naturally discussed in terms of kinetic and potential energy. In this language, the coupling between the atom and the quantized photon field appears as a very small potential energy whose sign and magnitude depend on the internal atomic and the photonic quantum numbers. The wave function then consists of two components, one exposed to a weak attractive potential and the other to a repulsive one; the dynamical difference between attraction and repulsion effects the internal atomic transition accompanied by the emission or absorption of a photon. After the atom has left the cavity, these potential energies are again zero and the kinetic energy has its initial value. Thus no net momentum is transferred to the atom during the interaction with the cavity field and, accordingly, its center-of-mass wave function is not affected markedly. Perhaps we should also mention that the localized cavity photon does not possess a definite momentum, and therefore the atom-photon interaction calculations cannot be simply carried out on the basis of momentum transfer and the recoil associated with it.

In the incomplete Stern-Gerlach interferometer of Figs. 4–6, the separation of the two partial beams after two of the three quarters can be made so large that enough room is available for the two micromaser resonators, see Figure 7. A sufficiently long stretch with a homogeneous magnetic field between the second and third provides the required longitudinal space. The atoms move parallel to the y-axis through this region. The Rb atoms are in their $5s_{1/2}$ ground state when leaving the second quarter with magnetic quantum numbers $m_j = +1/2$ or $m_j = -1/2$, respectively.

Prior to entering either one of the micromaser cavities, a laser beam exites the atom to the $63\,p_{3/2}$ Rydberg state. It is expedient to use a laser beam propagating along the x-direction and linearly polarized along the z-direction, that is: parallel to the magnetic field. In the π-transitions thus achieved the magnetic quantum number is not changed, so that only the Zeeman levels with $m_j = +1/2$ in the upper partial beam and $m_j = -1/2$ in the lower one get populated. For the same reason, the resonators are oriented such that the polarization of the photon mode is also along the z-direction. Thus, when entering the third quarter of the incomplete Stern-Gerlach interferometer, the Rb atoms are in the $61\,d_{5/2}$ Rydberg states with $m_j = +1/2$ or $m_j = -1/2$, respectively. Ideally, these atoms are deflected as indicated in Figure 6. In reality, however, the inhomogeneous magnetic field induces transitions to neighboring Rydberg levels with different magnetic properties. As a consequence, the two partial beams acquire a substructure, and a more complicated pattern emerges on the screen. This is not damaging for our purpose because all we need are identifiable regions on the lower/upper half of the screen such that, whenever an atom arrives there, the tell-tale photon is found in the upper/lower resonator. These regions are surely there.

To maintain the essential symmetry of the set-up the two laser beams in Fig. 7 have to be *equivalent* partial beams produced by sending the output of a single laser through a beam splitter. The laser frequency can be so adjusted that either the upper or the lower one of the two relevant Rydberg states is excited. In each measurement cycle, the laser is first tuned to the upper Rydberg state and switched on for a period so short that one atom at most is excited. After this atom has reached the screen, the laser is again activated for a short while, this time tuned to the lower Rydberg state. The second Rydberg atom possibly thus produced will absorb the tell-tale photon left behind by the first atom, thereby reading-off the which-way information stored in the respective resonator. The "reader atom" is then in the upper Rydberg state. It is ionized in an inhomogeneous static electric field whose strength is such that it cannot ionize atoms in the lower Rydberg state. The electron stripped off the reader atom is then detected. After waiting long enough to ensure that an unnoticed tell-tale photon has faded away, the next cycle is started. Each cycle in which one of the electron detectors responds supplies us with one atom whose track through the incomplete Stern-Gerlach interferometer is known.

Appendix B

The yes/no label attached to the quantum-optical which-way detectors of Appendix A refers to the presence or absence of one photon only. The photon modes in question can be regarded as harmonic oscillators which are either in their respective ground states ("no") or first excited states ("yes"). Rather than using discrete photon numbers, one could thus employ (dimensionless) position variables, q_1 and q_2, and the corresponding harmonic oscillator wave functions to represent the state of the system. In Bohmian mechanics the two harmonic oscillators are also subject to a dynamical evolution, although their waves remain in their ground or first excited state, respectively.

The four-component wave function $\Psi(t, z)$ of (34) is then replaced by a two-component one, *viz.*

$$\Psi(t, z; q_1, q_2) = \begin{pmatrix} \Psi_+ \\ \Psi_- \end{pmatrix} = \sqrt{2/\pi}\,\exp(-\tfrac{1}{2}q_1^2 - \tfrac{1}{2}q_2^2)\begin{pmatrix} q_1\,\psi_+(t, z) \\ q_2\,\psi_-(t, z) \end{pmatrix} \tag{36}$$

with $\psi_\pm(t, z)$ from (17). For notational simplicity we drop the subscript "after". The probability density is now

$$\varrho(t, z; q_1, q_2) = |\Psi_+|^2 + |\Psi_-|^2 = \frac{2}{\pi}\exp(-q_1^2 - q_2^2)\,[q_1^2|\psi_+(z)|^2 + q_2^2|\psi_-(t, z)|^2] \tag{37}$$

and the current components are given by

$$j_z(t, z; q_1, q_2) = \frac{\hbar}{m}\,\mathrm{Im}\left\{\Psi_+^*\frac{\partial}{\partial z}\Psi_+ + \Psi_-^*\frac{\partial}{\partial z}\Psi_-\right\} = \frac{2}{\pi}\frac{\hbar}{m}\exp(-q_1^2 - q_2^2)\left[q_1^2\,\mathrm{Im}\left\{\psi_+^*\frac{\partial}{\partial z}\psi_+\right\} + q_2^2\,\mathrm{Im}\left\{\psi_-^*\frac{\partial}{\partial z}\psi_-\right\}\right] \tag{38}$$

as well as

$$j_1(t,z;q_1,q_2)=\omega\,\mathrm{Im}\left\{\Psi_+^*\frac{\partial}{\partial q_1}\Psi_+ +\Psi_-^*\frac{\partial}{\partial q_1}\Psi_-\right\}=0,$$
$$j_2(t,z;q_1,q_2)=\omega\,\mathrm{Im}\left\{\Psi_+^*\frac{\partial}{\partial q_2}\Psi_+ +\Psi_-^*\frac{\partial}{\partial q_2}\Psi_-\right\}=0, \tag{39}$$

where ω denotes the common natural frequency of the oscillators. The velocity fields

$$v_1=j_1/\varrho,\qquad v_2=j_2/\varrho, \tag{40}$$

for the $q_{1,2}$ motions, vanish so that the oscillators do not actually move. The velocity field for the atom's z-motion,

$$\begin{aligned} v_z(t,z;q_1,q_2)&=j_z/\varrho \\ &=t\,z\left(\frac{\delta p}{m\,\delta z(t)}\right)^2+\left[\frac{\Delta p(t)}{m}-t\,\Delta z(t)\left(\frac{\delta p}{m\,\delta z(t)}\right)^2\right] \\ &\quad\cdot\frac{q_1^2\exp\left(\frac{z\,\Delta z(t)}{(\delta z(t))^2}\right)-q_2^2\exp\left(-\frac{z\,\Delta z(t)}{(\delta z(t))^2}\right)}{q_1^2\exp\left(\frac{z\,\Delta z(t)}{(\delta z(t))^2}\right)+q_2^2\exp\left(-\frac{z\,\Delta z(t)}{(\delta z(t))^2}\right)} \end{aligned} \tag{41}$$

is different from (25) unless $q_1^2=q_2^2$. Consequently, the implied trajectories $Z(t)$ are not the ones generated by (25). In particular, the new trajectories *may* cross the $z=0$ plane, and some of them do so.

The Bohm trajectories stay now in the three-dimensional z, q_1, q_2-space: $Z(t)$, $Q_1(t)$, $Q_2(t)$. They are determined by

$$\begin{aligned} \frac{\mathrm{d}}{\mathrm{d}t}Z&=v_z(t,Z;Q_1,Q_2), \\ \frac{\mathrm{d}}{\mathrm{d}t}Q_1&=v_1(t,Z;Q_1,Q_2), \\ \frac{\mathrm{d}}{\mathrm{d}t}Q_2&=v_2(t,Z;Q_1,Q_2). \end{aligned} \tag{42}$$

In view of $v_1=v_2=0$ there is a set of $Z(t)$ trajectories to each pair of initial Q_1, Q_2 values, and two trajectories from the same set do not intersect in the t,z plane. Therefore, to calculate the probability that the atomic $Z(t)$ trajectory crosses the $z=0$ plane we first compare, for given Q_1 and Q_2, the probabilities for finding the atom below the $z=0$ plane at the initial and final times t_i and t_f, and then integrate over all Q_1 and Q_2 values. The probability for crossing is thus

$$P_{\text{crossing}}=\int_{-\infty}^{\infty}\mathrm{d}Q_1\int_{-\infty}^{\infty}\mathrm{d}Q_2\left|\int_{-\infty}^{0}\mathrm{d}z\,[\varrho(t_i,z;Q_1,Q_2)-\varrho(t_f,z;Q_1,Q_2)]\right|=\sqrt{2/\pi^3}\left|\int_{(\Delta z/\delta z)_f}^{(\Delta z/\delta z)_i}\mathrm{d}s\,e^{-\frac{1}{2}s^2}\right|. \tag{43}$$

This is to be applied to the situation in which $\Delta z/\delta z\gg 1$ initially and $-\Delta z/\delta z\gg 1$ finally. Then the s integral equals $\sqrt{2\pi}$ and the crossing probability is $2/\pi=64\%$. The remaining 36% of the atomic trajectories do not cross the $z=0$ plane, so they do not pass through the same detector as the atom's track.

The spin vector associated with the Z, Q_1, Q_2 trajectory is

$$\boldsymbol{S}=\left[Q_1^2\exp\left(\frac{Z\,\Delta z}{(\delta z)^2}\right)+Q_2^2\exp\left(-\frac{Z\,\Delta z}{(\delta z)^2}\right)\right]^{-1}\cdot\begin{pmatrix}2Q_1Q_2\cos\varphi\\ 2Q_1Q_2\sin\varphi\\ Q_1^2\exp\left(\frac{Z\,\Delta z}{(\delta z)^2}\right)-Q_2^2\exp\left(-\frac{Z\,\Delta z}{(\delta z)^2}\right)\end{pmatrix}, \tag{44}$$

which is different from both (35) and (29). It has more in common with the latter, which is contained as the symmetric $Q_1=Q_2$ case. Along those $Z(t)$ trajectories that do not cross the $z=0$ plane, the direction of $\boldsymbol{S}(t)$ is reversed in the cross-over region, rapidly though continuously. The instantaneous spin flip observed for (35) would require $Q_1=0$ or $Q_2=0$, in which situation practically all $Z(t)$ trajectories cross the $z=0$ plane and $\boldsymbol{S}(t)$ is time independent. Accordingly, that instantaneous spin flip does not occur.

For the double-slit interferometer with which-way detectors considered by Bell [6], the Z, Q_1, Q_2 trajectories are more complicated because Q_1 and Q_2 possess a genuine time dependence. Analogously to (36), we replace the two-component wave function (9) by the single function

$$\begin{aligned} \psi(t,\boldsymbol{r};q_1,q_2)&\\ =\sqrt{2/\pi}\exp(-\tfrac{1}{2}q_1^2-\tfrac{1}{2}q_2^2)\,[q_1\psi_>(t,\boldsymbol{r})+q_2\psi_<(t,\boldsymbol{r})]. \end{aligned} \tag{45}$$

There are now interference terms in the resulting probability density

$$\begin{aligned} \varrho(t,\boldsymbol{r};q_1,q_2)=|\psi|^2=\frac{2}{\pi}\exp(-q_1^2-q_2^2)& \\ \cdot\,[q_1^2|\psi_>|^2+q_2^2|\psi_<|^2+2q_1q_2\,\mathrm{Re}\{\psi_>^*\psi_<\}]& \end{aligned} \tag{46}$$

as well as in the currents and Bohm velocity fields

$$j(t,r;q_1,q_2)=\varrho\, v(t,r;q_1,q_2)=\frac{\hbar}{m}\,\mathrm{Im}\{\psi^*\nabla\psi\}$$
$$=\frac{2}{\pi}\frac{\hbar}{m}\exp(-q_1^2-q_2^2)$$
$$\cdot\,\mathrm{Im}\{q_1^2\psi_>^*\nabla\psi_>+q_2^2\psi_<^*\nabla\psi_<+q_1q_2(\psi_>^*\nabla\psi_<+\psi_<^*\nabla\psi_>)\},$$

$$j_1(t,r;q_1,q_2)=\varrho\, v_1(t,r;q_1,q_2)=\omega\,\mathrm{Im}\left\{\psi^*\frac{\partial}{\partial q_1}\psi\right\}$$
$$=\omega\frac{2}{\pi}\exp(-q_1^2-q_2^2)\,q_2\,\mathrm{Im}\{\psi_<^*\psi_>\},\tag{47}$$
$$j_2(t,r;q_1,q_2)=\varrho\, v_2(t,r;q_1,q_2)=\omega\,\mathrm{Im}\left\{\psi^*\frac{\partial}{\partial q_2}\psi\right\}$$
$$=\omega\frac{2}{\pi}\exp(-q_1^2-q_2^2)\,q_1\,\mathrm{Im}\{\psi_>^*\psi_<\}.$$

In contrast to (39), the velocities for the $q_{1,2}$ motions do not vanish here. But there is still a constant of motion, namely $Q_1^2+Q_2^2$, as implied by

$$q_1j_1+q_2j_2=0\,.\tag{48}$$

This invites the use of polar coordinates in the q_1, q_2-plane for the parameterization of $Q_1(t)$ and $Q_2(t)$,

$$Q_1(t)=Q\cos\vartheta(t),\quad Q_2(t)=Q\sin\vartheta(t),\tag{49}$$

where $Q>0$ is time independent.

The Bohm mechanical time dependence of $\vartheta(t)$ and of the atom trajectory $R(t)$ are determined by

$$\frac{d}{dt}R=v(t,R;Q\cos\vartheta,Q\sin\vartheta)\,,\tag{50}$$

where the right-hand side is actually independent of Q, and

$$\frac{d}{dt}\vartheta=\frac{\omega}{Q^2}\frac{\mathrm{Im}\{\psi_>^*(t,R)\,\psi_<(t,R)\}}{|\psi_>(t,R)\cos\vartheta+\psi_<(t,R)\sin\vartheta|^2}\,.\tag{51}$$

After restricting the atomic motion to the z-direction for simplicity and modeling the double-slit wave functions $\psi_>$ and $\psi_<$ by ψ_+ and ψ_- from (17) with $\phi\equiv0$, we have studied these equations numerically for the relevant period between $t=3T_0$ and $t=4T_0$. When the same order of magnitude was chosen for the time scales of the evolution of $Z(t)$ and $\vartheta(t)$, as expressed by $\omega T_0\cong1\cdots10$, we found that the Bohm trajectories depend very sensitively on the initial values of Z, ϑ, and Q. We note that this range of oscillator frequencies is, of course, much smaller than is physically reasonable. Since an analysis like the one that produced (43) above appears impossible here, we have resorted to computing a few hundred trajectories with their initial values selected at random. About 20% of them do not cross ther $z=0$ plane. So here, too, a substantial fraction of the atomic Bohm trajectories do not pass through the same detector as the atom's track.

When, in contrast, very different time scales were chosen, such as by the physically reasonable value of $\omega T_0=10^5$, we did not find a single trajectory that crosses the $z=0$ plane. Here, very few Bohm trajectories, at best, pass through the same detector as the atom's track. This behavior results from the different frequencies. For, the angle $\vartheta(t)$ grows so fast that $\sin\vartheta$ and $\cos\vartheta$ in (50) oscillate extremely rapidly, so that v of (50) can be effectively replaced by its ϑ-average,

$$\frac{d}{dt}R=v_{\mathrm{av}}(t,R)=\int_{(2\pi)}\frac{d\vartheta}{2\pi}\,v(t,R;Q\cos\vartheta,Q\sin\vartheta)\,.\tag{52}$$

The resulting averaged velocity field $v_{\mathrm{av}}(t,r)$ has the required symmetry property: its z-component changes sign under the reflection $r=(x,y,z)\to(x,y,-z)$. Consequently, the Bohm trajectories generated by v_{av} must stay on one side of the $z=0$ plane.

[1] D. Bohm, Phys. Rev. **85,** 166 (1952) and **85,** 180 (1952); both reprinted in: Quantum Theory and Measurement (J. A. Wheeler and W. H. Zurek, eds.), Princeton University Press 1983.
[2] D. Bohm and B. J. Hiley, Found. Phys. **5,** 93 (1975).
[3] D. Bohm, J. Hiley, and P. N. Kaloyerou, Phys. Rep. **144,** 321 (1987).
[4] D. Bohm, Quantum Theory, Prentice-Hall, Englewood Cliffs 1951.
[5] M. O. Scully, B.-G. Englert, and J. Schwinger, Phys. Rev. A **40,** 1775 (1989).
[6] J. S. Bell, Int. J. Quant. Chem. **14,** 155 (1980); reprinted in: Speakable and unspeakable in quantum mechanics, Cambridge University Press 1987.
[7] D. Dürr, S. Goldstein, and N. Zanghi, J. Stat. Phys. **67,** 843 (1992).
[8] C. Philippidies, C. Dewdney, and B. J. Hiley, Nuovo Cim. **52 B,** 15 (1979).
[9] M. O. Scully, B.-G. Englert, and H. Walther, Nature London **351,** 111 (1991).
[10] J. Schwinger, M. O. Scully, and B.-G. Englert, Z. Phys. D **10,** 135 (1988).
[11] B.-G. Englert, J. Schwinger, and M. O. Scully, Found Phys. **18,** 1045 (1988).
[12] For a review of the Garching micromaser work see: G. Rempe, M. O. Scully, and H. Walther, Physica Scripta **T34,** 5 (1991).
[13] B.-G. Englert, J. Schwinger, and M. O. Scully, in: New Frontiers in Quantum Electrodynamics (A. O. Barut, ed.) Plenum, New York 1990, pp. 513–519.

Artur Ekert

Centre for Quantum Technologies, National University of Singapore, Singapore
Mathematical Institute, University of Oxford, Oxford, United Kingdom

Paper Title: **Creative procrastination**

Reference: Original article (2014)

Personal message:

Dear Berge—Wishing you many creative procrastinations and an endless supply of great wines. It has always been a pleasure discussing science with you, even though after few drinks I have always seen few more universes than you. Artur.

CREATIVE PROCRASTINATION

ARTUR EKERT

On a drowsy Singapore afternoon, protected from the scorching sun by the shadow of a verandah, an armchair physicist was ready for a siesta. A glass of chilled wine, a remnant of a sumptuous lunch, added to his quiet satisfaction with life. He watched for a moment how the sun reflected off the glass in bright sparkles of light and smiled. Scattered light had no mystery to him. Admittedly, photons were somewhat weird, but so was quantum theory. Well, perhaps not as weird as many of his colleagues wanted to believe, but weird nonetheless. Probabilities became less fundamental than some complex numbers called amplitudes, for particles, such as photons, couldn't care less about the additivity axiom of probability theory. Indeed, there is no reason why they should care, he thought. The rules of quantum theory, this beautiful generalisation of probability theory, were deceptively simple. He believed that they could have been invented years ago, without any input from experiment. And now he was ready to give it a try. He poured himself more wine, took a sip, leaned back and closed his eyes.

Classical probability is a good starting point – he thought – let us see where it leads and how it fails. Assume that any physical system can be prepared in some finite number of distinguishable states. Introduce the state vector which tabulates probabilities of the system being in a particular state and make sure that admissible transformations map these probability vectors into probability vectors. What are these admissible transformations? Let us keep it simple, keep it linear and use transition matrices. The admissible transformations are then represented by stochastic matrices P – they have nonnegative elements such that $\sum_m P_{mn} = 1$, i.e., entries in each column add up to one. The matrix element P_{mn} is the probability that the system initially in state labeled by n evolves over a prescribed period of time into the state labeled by m. The probability vector with components p_n evolves as $p_n \mapsto \sum_n P_{mn}\, p_n$.

Let us take a sequence of two *independent* evolutions, P followed by Q. What is the probability that the system initially in state n evolves over a prescribed period of time into the state m via some intermediate state k? The evolutions are independent so for any particular k the probability is $Q_{mk}P_{kn}$. But there are several intermediate perfectly distinguishable states k, thus there are several mutually exclusive ways to get from n to m. Following the Kolmogorov additivity axiom, we add up the constituent probabilities, $\sum_k Q_{mk}P_{kn}$, and discover that the matrix multiplication QP in one swoop takes care of the multiplication and addition of probabilities. Products of stochastic matrices are stochastic matrices, so far so good, at least on the basis of logical consistency.

Now we add one more requirement – *continuity* of evolution. Any product of stochastic matrices gives a stochastic matrix but now we are asking for more. It should be possible to view any evolution as a sequence of *independent* evolutions over shorter periods of time. In particular, we should be able to take the square root, or the cube root, or any root of any transition matrix and obtain a valid transition matrix. Take, for example, a physical system with two states, say a physical bit, and consider a transformation which swaps the two states; a logical

`NOT` if you wish, represented by the stochastic matrix with zeros on the diagonals and ones on the off-diagonals. Let us take the square root. The two eigenvalues of this matrix are ± 1 so we have to end up with a matrix with complex entries; indeed,

$$\begin{pmatrix} 0 & 1 \\ 1 & 0 \end{pmatrix} = \frac{1}{2}\begin{pmatrix} 1+i & 1-i \\ 1-i & 1+i \end{pmatrix} \frac{1}{2}\begin{pmatrix} 1+i & 1-i \\ 1-i & 1+i \end{pmatrix}. \tag{1}$$

Square roots of stochastic matrices are, usually, not stochastic matrices. In other words – by adding the continuity requirement, we have gracefully thrashed the classical theory. Another sip of wine is in order and let us try again.

Let us keep state vectors and transition matrices T, but let them have complex entries, simply because they pop up as soon as you start taking roots. Hopefully, we will be able to relate complex numbers to probabilities later on. Let us take the continuity requirement seriously and parameterise transition matrices $T(t)$ with some real parameter t, that we may as well call time. Require that $T(t+s) = T(t)T(s)$, for any two time intervals t and s, and set $T(0) = \mathbb{I}$. Ha! This clearly points toward an exponential map $T(t) = \exp(tX)$, where X is any complex matrix. Now, taking the nth roots or inverses is a breeze: $T(t)^{1/n} = T(t/n)$ and $T(t)^{-1} = T(-t)$. We also recall that any matrix can be written in its polar form $T = RU$ where R is a positive matrix and U is unitary, it is analogous to writing a complex number in the polar form or viewing linear transformation T as the "stretching" R and the "rotation" U. But the exponential increase of stretching with time does not look good, we do not want to have exponential divergencies in our theory, so we had better drop R. Now, we are left with a unitary evolution of the form $U(t) = \exp(itX)$, where X is Hermitian. It looks good, decent periodic evolution, no exponential divergences. Unitary matrices act on complex vectors and rotate them continuously. But how do we get probabilities out of these vectors?

Now, we have to follow our hunch – probability should be preserved under the admissible evolution, so what is it that remains invariant under unitary operations... Eureka! The length of a vector! The Euclidian norm or the 2-norm, if you wish. Hence the squares of absolute values of complex components are probabilities. Indeed, suppose probabilities were given by the absolute values of some complex numbers raised to power p. Then we would be looking for operations that preserve the p-norms. Recall that the p-norm of vector v, with components $v_1, v_2, ..., v_n$, is defined as $\sqrt[p]{|v_1|^p + |v_2|^p + ...|v_n|^p}$. When $p = 2$, the norm-preserving operations, i.e., isometries, are unitary matrices. In all other cases, the only admissible operations are permutation of the basis vectors and multiplication by phase factors. He could almost see this result, as he imagined unit circles in different p norms, e.g. for $p = 1, 2, 42$, and ∞.

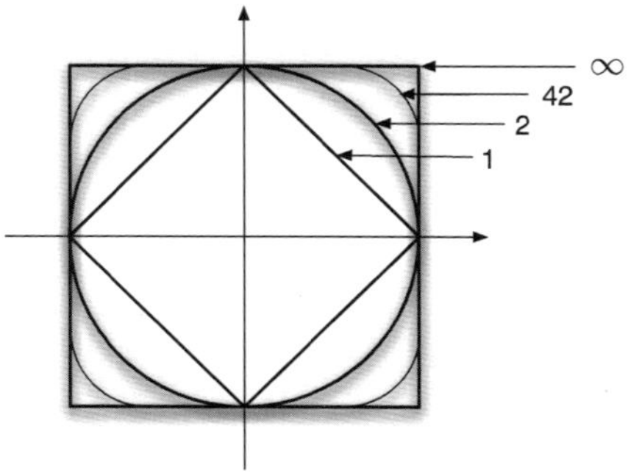

He could see the 2-norm being special because of its rotational invariance – the probability measure picks out no preferred basis in the space of state vectors.

And so he had it all – state vectors with complex components, unitary transition matrices, and a simple rule on how to get probabilities out of the complex numbers. He congratulated himself. He derived quantum theory without moving the butt from the armchair. Well, almost, there are a few holes in this plausibility argument, he thought, but they can be fixed, with some more wine, of course. He decided to skip this philosophical mumbo-jumbo and deal with the measurement problem properly the following day. It has to to be sorted out once and for all, he thought as he slumbered away.

Endnote

Needless to say, our armchair physicist is well read and knows that more refined arguments can be found in a number of papers, in particular, in a very readable exposition by Lucien Hardy (arXiv:quant-ph/0101012). Lucien argues, very convincingly, that if we try to construct a good statistical theory from a few (actually five) very reasonable axioms, then once we request *continuity* of admissible evolutions, we will end up with quantum theory, and if this requirement is dropped, we obtain classical probability theory.

Gerd Leuchs

Max-Planck Institute for the Science of Light, Erlangen, Germany
Department of Physics, University of Erlangen-Nuremberg, Erlangen, Germany
Department of Physics, University of Ottawa, Ontario, Canada

Paper Title: **The physical reality of the quantum wave function**

Reference: Original article (2014) (arXiv: 1501.07199)

Personal message:

Dear Berge,

I very much appreciate being part of your 60th anniversary party. I always enjoyed discussing with you, be it physics or any other topic, and am hoping for many more such occasions. Happy birthday and all the best for the years to come!

Gerd Leuchs

¿ The physical reality of the quantum wave function ?

Gerd Leuchs
Max Planck Institute for the Science of Light, Erlangen
Department of Physics, University of Erlangen-Nuremberg
Department of Physics, University of Ottawa

Abstract

Can it possibly be that the proper interpretation of quantum theory, which allows a quantum system to be in a superposition of mutually exclusive states, is itself a super–position of seemingly contradicting interpretations?

The quantum theory is amazingly successful and we all know how to use it. For a given experimental situation we can predict the outcome of the measurement. Often the prediction tells us the probability with which the detector will give a certain result. When dealing with a single quantum system allowing for several simultaneous measurements on sub-systems, then quantum physics can make precise predictions about correlations of the measured data if the system is in an entangled quantum state. David Mermin described the situation as follows: *"The only proper subjects of physics are correlations among different parts of the physical world. Correlations are fundamental, irreducible, and objective. They constitute the full content of physical reality"* [1]. The numerous discussions on the topic are well reflected in his American Journal of Physics article [2, see also 3]. One particular question is what happens in the process of measuring a quantum system, when quantum state reduction takes place. In his wonderful essay 'On quantum theory' Berge Englert stresses that there is no measurement problem. One by one he goes through all the different topics, some of us find cumbersome. With regard to the measurement process he writes: "*... the collapse of the state" or "wave function collapse" are popular synonyms for state reduction. The connotation that the transition* (associated with state reduction) *is a dramatic dynamical process, as if the physical system were evolving, is clearly misleading*" [4]. And indeed it would be quite troublesome if a single photon or even a single massive particle going through a beam splitter would simultaneously exit through both output ports, only to "collapse" upon a measurement at one or the other output port [5]. This would be

INTERPRETATION	remark	wave function	**you should not ask**
Copenhagen	knowledge collapses, not the wave function	not real	what happens inside the apparatus?
many worlds	different worlds evolve by measurement	real	where are all these worlds, exponentially growing in number?
Ghirardi-Rimini-Weber	measurement collapses the wave function	real	for the physical dynamics of wave function collapse (?)
QBism [6]	uses Bayesian probabilities	not real	how can nature be properly described by personalized probability amplitudes?

Table 1: ***There is no free lunch.***
This list of interpretations is not complete and just shows examples. The different interpretations listed here agree in the predictions of experimental results. But all interpretations have their cumbersome aspects, some find difficult to accept. These cumbersome aspects are merely moved around. This teaches us to accept a cumbersome aspect, one or the other way.

troublesome because energy or mass would have to move instantaneously over finite distances. But not being allowed to ask what happens inside the apparatus is also somewhat cumbersome, at least for experimenters. Therefore, attempts were made to avoid this cumbersome aspect by searching for alternative interpretations of quantum physics, but then a cumbersome aspect is bound to pop up somewhere else - there does not seem to be a free lunch.

Some scientists find the interpretation most appealing, which interprets the wave function and the quantum physical predictions as statements about the information we have about a system rather than as descriptions of the physical reality [4]. Thus there may be a deep underlying connection between quantum physics and information theory [see e.g. 5, 7, 8]. Maybe the long lasting discussion about interpretations tells us something. The different combinations of theory and interpretation seem to be equally valid although the different interpretations by scientists seem to be contradicting. So maybe the link between quantum physics and information theory is deeper than some of us think.

The experimenter will not be able to provide evidence for one interpretation being correct and the others having limited applicability as long as the different interpretations - or rather the theories behind them - do not lead to contradicting predictions for the outcome of an experiment. So it is no surprise that again and again experiments are performed to keep quantum theory on the test bench. In the following a selection of experiments providing such tests are reviewed and a new experiments are proposed.

Violation of Bell's inequalities

Regarding a possible reality associated with the quantum wave function there had been ample speculations about the existence of hidden variables. Early on Heisenberg argued qualitatively, that hidden variables couldn't exist, because the quantum classical boundary is not well defined but depends on our ability to control the system (see [9] p. 506, 507). Then John Bell proposed his celebrated inequality to quantitatively test for local hidden variables [10], requiring systems in entangled quantum states, leading to a series of subsequent measurements ever improving with regard to potential loop holes, see e.g. [11-16]. Today it is clear that there is less and less room for local hidden variables in quantum theory. We note in passing, that John Bell's inequality is currently studied also in the classical regime [17], i.e. in the context of classically excited non-separable mode functions, such as cylindrical vector beams [18-21]. It may be too early to make final remarks but the violation may be associated with classical correlations being arbitrarily strong though not non-local. In any case applications in metrology were already proposed [22].

Wheeler's delayed choice experiment

Coming back to the single photon impinging on a beam splitter, one has a choice to either measure at which output port the photon appears, or one can let the two output beams interfere at a second beam splitter closing the geometry to form an interferometer and then measure the interference of the photon. Clearly, in either case only one detector will fire. Without the recombining beam splitter both detectors will fire stochastically, the relative percentage reflecting the splitting ratio. However, with the beam recombiner and for the appropriate path difference the photon will always exit at one output port of the recombining beam-splitter, or for a properly modified phase change at the other output port and only one detector will fire the other one being silent

[23]. In this latter case the photon obviously[1] seems to exit via both outputs of the first beam splitter in order for the wave function to interfere via the two paths. Needless to say, both these different experimental scenarios and the corresponding measurement results are properly described by quantum theory. Yet it is meaningless to ask whether the photon is really split or not, unless one goes and looks, i.e. measures in the two interferometer arms - but then one is back to the first scenario. John Archibald Wheeler had the idea to refine the apparatus and decide whether or not to insert the beam recombiner only after the photon has passed the first beam splitter [24]. This proposal is feasible if the photon travel distances are so large that the associated travel times are longer than the time constant of the switch with which one alternates between the which-path (i.e. 'welcher-Weg') and the interference measurement. The result is that one sees no difference as compared to the static cases. Switching back and forth before the photon arrives does not have any influence on the measurement results [25-28]. Further studies of the topic use weak measurements in order to gain more information [7, 23, 29] and the discussion about it [30].

The "01 - 10" state

Little more than ten years ago there was a fierce debate about whether the quantum state of the single photon is entangled or not after passing through the beam splitter. There is no simpler wave function than that of a single photon underlining the relevance of this discussion. The opponents argued that by a mere transformation to other modes the state, which looked like displaying entanglement between a single photon state and the vacuum ground state, would become separable. Hence, this quantum state is not entangled. The proponents on the other hand said that what looks entangled is entangled. Stephen van Enk nicely summarized the discussion [31]. A major conceptual advance was made by Björk, Jonsson and Sanchez-Soto who proposed keeping the beams at the beam splitter output separate and interfering each of them with a local oscillator before detection [32, see also 33]. In the experiments, click detectors were used [34] but one could also use homodyne detection instead [35] or a combination of the two [36]. In view of the above, this measurement is exciting because it shows that the single photon splitting in the beam splitter leads to simultaneous signals in the two detectors - a correlation caused by the one single photon[2]. Furthermore, the transfer of single-photon entanglement in and out of an atomic system serving as a quantum memory has been demonstrated [37]. A Bell test was performed with this one particle state [38-39] and quantum teleportation was demonstrated [40-41]. It is interesting to note that the state $|\Psi\rangle = 2^{-1/2}(|01\rangle - |10\rangle)$ has a certain relation to the classical entanglement mentioned above, in that in both cases the relation between the mathematically non-separable mode function and the notion of quantum entanglement was discussed. It will be interesting to devise new experiments also in this context.

Looking into complementarity using entanglement

Quantum optics allows for designing tests of complementarity [42]. In 2012 Menzel et al. published inspiring novel results [43] on the double slit experiment illuminated with one of the photons of an entangled pair. They studied experimentally and theoretically in how far the visibility of the interference fringes in the signal beam is affected by a which-slit measurement on the idler beam [43-44]. They emphasize that duality is well

---

[1] To quote the experimenter

[2] Whatever this means in the quest for the physical reality of the quantum wave function. The least on can deduce is that the single photon is not local.

defined only for a single particle. Jakob and Bergou extended the concept of complementarity to bi-partite systems [45-46], which yet has to be applied to the two-photon double-slit experiment. Bolduc et al. pointed out that the auxiliary transverse variable provides an additional degree of freedom [47], which affects the interpretation of the experiment by Menzel et al. [43-44].

The new experiments

If new experiments can be done to further test a physical theory they should be done. The possibility to measure a single photon at several different locations opens a window for novel investigations The next steps exploring single photon entanglement seem obvious, with compelling goals:

(1) perform a delayed choice experiment with the $|\Psi\rangle = 2^{-1/2}(|01\rangle - |10\rangle)$ state using two simultaneous measurements.

(2) repeat the two-slit experiment with an entangled single photon.

(3) extend the studies to measurements at more than two points in space-time.

Instead of using the combination of a local oscillator and click detection, it may be advantageous to replace the click detector by a homodyne detector [35,36], or even better with an eight-port homodyne detector allowing for simultaneously measuring two conjugate field quadratures [48-50], keeping in mind that these different detectors correspond respectively to normal, symmetric and anti-normal ordering of the field operators and the associated quasi probability functions in phase space. For this task one can use integrated multiport splitters [51-53], which are interferometrically stable, making the proposed experiments viable. An interesting set-up uses a bi-prism [54], which may be combined with these different detector scenarios. Non-separable mode functions sometimes referred to as 'classical entanglement' [18-23] is closely related to single photon entanglement. The connection between these two areas and the potential synergy will be further explored.

References

[1] N.D. Mermin, *arXiv*:quant-ph/**9609013** (1996)

[2] N.D. Mermin, *Am. J. Phys.* **66**, 753 (1998)

[3] P.C.W. Davies and J.R. Brown, 'The Ghost in the Atom – a dicussion oft he mysteries of quantum physics', Cambridge University Press (1986)

[4] B.-G. Englert, *Eur. Phys. J. D* **67**, 238 (2013)

[5] M.F. Pusey, J. Barrett and T. Rudolph, *Nature Phys.* **8**, 475 (2012); see also [54] p. 19

[6] C.A. Fuchs, N.D. Mermin, R. Schack, *arXiv:***1311.5253** (2013); It is interesting to watch how a scientist having a strong opinion on which interpretation to prefer can convert to a different interpretation. The ambivalence in the scientific community may exist also in a single person. My experimentalist's soul is still waiting for things to check in the lab.

[7] K.W. Murch, S.J. Weber, C. Macklin and I. Siddiqi, *Nature* **502**, 211 (2013)

[8] J.A. Wheeler, *Annals New York Academy of Sciences* **655**, 349 (1992)

[9] K.F. von Weizsäcker, *Zeitschrift für Physik* **118**, 489 (1941)

[10] J.S. Bell, *Physics (N.Y.)* **1**, 195 (1964)

[11] J.F. Clauser, M.A. Horne, M.A. Shimony and R.A. Holt, "Proposed Experiment to Test Local Hidden-Variable Theories", *Phys. Rev. Lett.* **23**, 880 (1969)

[12] A. Aspect, P. Grangier and G. Roger, *Phys. Rev. Lett.* **49**, 91 (1982)

[13] M. A. Rowe, D. Kielpinski, V. Meyer, C. A. Sackett, W. M. Itano, C. Monroe and D. J. Wineland, *Nature* **409**, 791 (2000)

[14] D.N. Matsukevich, P. Maunz, D.L. Moehring, S. Olmschenk and C.Monroe, *Phys. Rev. Lett.* **100**, 150404 (2008)

[15] M. Giustina, A. Mech, S. Ramelow, B. Wittmann, J. Kofler, J. Beyer, A. Lita, B. Calkins, T. Gerrits, S.W. Nam, R. Ursin and A. Zeilinger, *Nature* **497**, 227 (2013)

[16] B.G. Christensen, K.T. McCusker, J.B. Altepeter, B. Calkins, T. Gerrits, A.E. Lita, A. Miller, L.K. Shalm, Y. Zhang, S.W. Nam, N. Brunner, C.C.W. Lim, N. Gisin, P.G. Kwiat, *Phys. Rev. Lett.* **111**, 130406 (2013)

[17] C. V. S. Borges, M. Hor-Meyll, J.A.O. Huguenin and A.Z. Khoury, *Phys. Rev. A* **82**, 033833 (2010)

[18] R.J.C. Spreeuw, *Foundations of Physics* **28**, 361 (1998)

[19] A. Holleczek, A. Aiello, Ch. Gabriel, Ch. Marquardt and G. Leuchs, *arXiv:***1007.2528** (2010)

[20] X.-F. Qian and J.H. Eberly, *Opt. Lett.* **36**, 4110 (2011)

[21] A. Holleczek, A. Aiello, Ch. Gabriel, Ch. Marquardt and G. Leuchs, *Opt. Expr.* **19**, 9714 (2011)

[22] F. Töppel, A. Aiello, Ch. Marquardt, E. Giacobino and G. Leuchs, *New J. Phys.* **16**, 073019 (2014)

[23] L. Vaidman, Phys. Rev. A **87**, 052104 (2013)

[24] J.A. Wheeler, in 'Quantum Theory and Measurement', edited by J.A. Wheeler and W.H. Zurek, pp.182-213, Princeton Univ. Press, Princeton, NJ (1984)

[25] T. Hellmuth, H. Walther, A.G. Zajonc and W.P. Schleich, *Phys. Rev. A* **72**, 2533 (1987)

[26] G. Weihs, Th. Jennewein, Ch. Simon, H. Weinfurter, and A. Zeilinger, *Phys. Rev. Lett.* **81**, 5039 (1998)

[27] V. Jacques, E. Wu, F. Grosshans, F. Treussart, P. Grangier, A. Aspect and J.-F. Roch, *Science* **315**, 966 (2007)

[28] A. Peruzzo, P. Shadbolt, N. Brunner, S. Popescu, and J.L. O'Brien, *Science* **338**, 634 (2012)

[29] A. Danan, D. Farfurnik, S. Bar-Ad and L. Vaidman, *Phys. Rev. Lett.* **111**, 240402 (2013)

[30] Z.-H. Li, M. Al-Amri and M.S. Zubairy, *Phys. Rev. A* **88**, 046102 (2013)

[31] S.J. van Enk, *Phys. Rev. A* **72**, 064306 (2005)

[32] G. Björk, P. Jonsson and L.L. Sanchez-Soto, *Phys. Rev. A* **64**, 042106 (2001)

[33] S.M. Tan, D.F. Walls and M.J. Collett, *Phys. Rev. Lett.* **66**, 252 (1991)

[34] B. Hessmo, P. Usachev, H. Heydari and G. Björk, *Phys. Rev. Lett.* **92**, 180401 (2004)

[35] M. Fuwa, S. Takeda, M. Zwierz, H.M. Wiseman and A. Furusawa, arXiv:1412.7790 (2014)

[36] A.I. Lvovsky, H. Hansen, T. Aichele, O. Benson, J. Mlynek and S. Schiller, *Phys. Rev. Lett.* **87**, 050402 (2001)

[37] K. S. Choi, H. Deng, J. Laurat and H. J. Kimble, *Nature* **452**, 67 (2008)

[38] S.A. Babichev, J. Appel and A.I. Lvovsky, *Phys. Rev. Lett.* **92**, 193601 (2004)

[39] M. D'Angelo, A. Zavatta, V. Parigi and M. Bellini, *Phys. Rev. A* **74**, 052114 (2006)

[40] S. Giacomini, F. Sciarrino, E. Lombardi and F. De Martini, *Phys. Rev. A* **66**, 030302 (2002)

[41] H. de Riedmatten, I. Marcikic, W. Tittel, H. Zbinden, D. Collins and N. Gisin, *Phys. Rev. Lett.* **92**, 047904 (2004)

[42] M.O. Scully, B.-G. Englert and H. Walther, *Nature* **351**, 111 (1991)

[43] R. Menzel, A. Heuer, D. Puhlmann and W.P. Schleich, *PNAS (U.S.A.)* **109**, 9314 (2012)

[44] R. Menzel, A. Heuer, D. Puhlmann, K. Dechoum, M. Hillery, M.J.A. Spähn and W.P. Schleich, *J. Mod. Opt.* **60**, 86 (2013)

[45] M. Jakob and J.A. Bergou, *Phys. Rev. A* **76**, 052107 (2007)

[46] M. Jakob and J.A. Bergou, *Opt. Commun.* **283**, 827 (2010)

[47] E. Bolduc, J. Leach, F.M. Miatto, G. Leuchs and R.W. Boyd, *PNAS (U.S.A.)* **111**, 12337 (2014)

[48] M.J. Collett, R. Loudon and C.W. Gardiner, *J. Mod. Opt.* **34**, 881, (1987)

[49] M. Freyberger, K. Vogel and W.P. Schleich, *Phys. Lett. A* **176**, 41 (1993); see also W.P. Schleich, "Quantum optics in phase space", Wiley-VCH (2001)

[50] W. Vogel and D.-G. Welsch, "Quantum optics", John Wiley & Sons (2006)

[51] R. Ulrich and G. Ankele, *Appl. Phys. Lett.* **27**, 337 (1975)

[52] P. Roth and G. Voirin, *Proc. SPIE* **1014**, *Micro-Optics* 35 (1989); doi:10.1117/12.949392

[53] A. Peruzzo, A. Laing, A. Politi, T. Rudolph and J.L. O'Brien, *arXiv*: **1205.4926** (2012)

[54] V. Jacques, E. Wu, T. Toury, F. Treussart, A. Aspect, P. Grangier, and J.-F. Roch, *Eur. Phys. J. D* **35** , 561 (2005)

[55] P. Wallden, V. Dunjko and E. Andersson, *J. Phys. A: Math. Theor.* **47**, 125303 (2014)

Hui Khoon Ng

Yale-NUS College, Singapore
Centre for Quantum Technologies, National University of Singapore, Singapore

Paper Title: **On quantum theory**

Reference: The European Physical Journal D **67**, 238 (2013)

Personal message:

Paths Intertwined: A Guiding Light

I first met Berge in early 2004—the precise date has by now escaped my memory, but that it was nearly exactly ten years ago is certain! At that time, I was back in Singapore for a year, in between completing my undergraduate study and commencing on a PhD programme in the US. That one year gap was required by the scholarship board (DSO/DSTA) that sponsored my undergrad degree: I was to return to Singapore to work for a year in the Applied Physics Lab in DSO National Laboratories, apparently to get a better feel of what area of research in my PhD would better fit in with the demands of DSO. In reality, I had by then already decided to pursue a PhD in quantum information/quantum computation. That was the same time that the Quantum Information Technology (QIT) Lab (predecessor of today's CQT) was being set up in NUS, and I had the good fortune of having a liberal boss in DSO who thought it a good idea for me to intern at the QIT Lab for six months. Thus, I was introduced to Berge and I soon started working in his group as a research assistant, then very much an ignorant newbie.

Those were the early days of quantum information in Singapore. The so-called QIT Lab really only occupied two adjoining rooms, one room being the experimental lab, the other being the theory space—together about the size of today's Quantum Cafe and the seminar room. Berge had just joined NUS the year before; Christian Kurtsiefer was just setting up his lab (I recall him unwrapping all the new lab equipment he'd purchased!); Antia had just arrived as a new assistant professor; Dagomir was there before all of us; Prof Oh was still the Physics Head then. Everything was new, lively, and bubbling with excitement. From those days, I also remember Alex Ling (a research assistant then in Christian's group), Darwin Gosal (who was at that time working on quantum information research), Tey Meng Khoon (a student in the group, whom I saw again briefly in Innsbruck years later), Tin Kah Min and Goh Choon Guan (two undergraduates working on ellipsometry under Berge and in Christian's new lab, with me trying to offer some — futile — help). And, of course, Janet Anders, who had just started her PhD in Berge's group, with whom I shared many long days poring over

QKD security calculations for what later became known as the Singapore protocol. We were a tiny bunch, in those days, compared to CQT today.

I stayed for six memorable months in the QIT Lab, and then left for the US to begin my PhD in Caltech. Even during those years away, I kept contact with Berge. I remember, in particular, occasions when Berge visited me in Los Angeles, when he was there for the Schwinger Foundation board meetings. The intention then, even from DSO's perspective (I had a 5-year scholarship bond to fulfill with DSO after my PhD), was for me to re-join Berge's group as a postdoc after my studies. I was always very grateful to know that I had something wonderful to come back to.

As it turned out, however, when I did return to Singapore in late 2009, things were rather more complicated with DSO, and the original plan of me re-joining Berge's group fell through. But I was determined (probably much to DSO's despair) to at least spend a part of my time in CQT. CQT, by that time, had already blossomed into pretty much what you see today, gathering together a prominent group of researchers and offering one of the best research environments in Singapore. DSO was, however, somewhat rigid (quite understandably so) in terms of how its employees spent their time. It took some six months of creative maneuvering and not a small amount of luck, helped along, most assuredly, by Berge's influence and charm, before I was finally allowed to spend a portion of my time in CQT, alongside my position in DSO.

Since then, I've spent an increasing amount of time within Berge's group in CQT. Berge gave me the chance of being a part of many different research projects, allowed me ample opportunities in learning about mentoring and guiding students, and just simply provided me with a warm home within his group. Today, having left DSO, I still maintain a joint position in CQT in his group, while also being a part of the new Yale-NUS College. I owe my current position as an assistant professor in Yale-NUS in no small part to Berge's constant guidance and support through the years.

So, what did I learn from Berge through these years? Many things. Who could avoid learning to do a few integrals and gaining a few tricks (or techniques, as Berge will correct me, I'm sure: Tricks only work once; techniques are general approaches that work in many situations!) while watching him dazzle everyone with his elegant calculations? My recent paper with him on a simple solution to the 1D localization problem [Phys. Rev. B **88**, 054201 (2013)] is an example of me benefitting from Berge's brilliant analytical skills. But, the most important lesson Berge has taught me has to be his no-nonsense, completely honest attitude towards science and research: One needs to do meaningful and sensible scientific work, to value honest and sincere research, and to not mystify or sensationalize good old genuine science—good science is itself intriguing enough to not need any added hype.

As a tribute to that last point, having noticed that no one else has suggested it yet, I would like to include in this collection Berge's most recent article on EPJ-D, *On Quantum Theory*. For those who don't know the history of this paper: This article started as a talk Berge gave in CQT about a year ago, with an amusing, and I'm sure, infuriating to many, title of *Quantum mechanics is local, and there is no measurement problem*. That was a memorable occasion. Announced as a regular seminar talk, the usual CQT seminar room was the standard venue. The overwhelming number of attendees, however, quickly made it clear that the (mobile) dividing walls between the seminar room and the Quantum Cafe had to be removed to accommodate more people. Thence followed an intense session where Berge reminded everyone of basic quantum mechanics that we all are, in some way or another, guilty of having forgotten or neglected. This neglect has culminated in perhaps many of the somewhat overly-hyped topics discussed today under the heading of foundations of QM. Following that talk was a stream of visitors to Berge's office: Some came to argue against his points; others came to thank him for finally standing up and saying (in a most satisfying manner) what many have long wanted to point out. I remember commenting to Benoît during lunch those few days that we really shouldn't get Berge started again on the topic, or he just doesn't stop! Since then, Berge had composed his thoughts carefully into an article, and was subsequently convinced by the editor of EPJ-D to submit it as a journal article. Despite Berge's emphasis that he is really telling us nothing new, the article has since garnered quite some attention in the community, including being mentioned in various science news articles. Some variant of that talk has also since been repeated in a few other places and occasions. Hyped up? At least it is good and true science.

Finally, Berge, here's Happy 60th Birthday and Good Health! May we continue to share many more days together, and many more drives home!

Eur. Phys. J. D (2013) 67: 238
DOI: 10.1140/epjd/e2013-40486-5

THE EUROPEAN PHYSICAL JOURNAL D

Colloquium

On quantum theory*

Berthold-Georg Englert[a]

Centre for Quantum Technologies and Department of Physics, National University of Singapore, Singapore

Received 9 August 2013 / Received in final form 24 August 2013
Published online 22 November 2013 – © EDP Sciences, Società Italiana di Fisica, Springer-Verlag 2013

Abstract. Quantum theory is a well-defined local theory with a clear interpretation. No "measurement problem" or any other foundational matters are waiting to be settled.

Quantum theory had essentially taken its final shape by the end of the 1920s and, in the more than eighty years since then, has been spectacularly successful and reliable – there is no experimental fact, not a single one, that contradicts a quantum-theoretical prediction. Yet, there is a steady stream of publications that are motivated by alleged fundamental problems: We are told that quantum theory is ill-defined, that its interpretation is unclear, that it is nonlocal, that there is an unresolved "measurement problem", and so forth[1].

It may, therefore, be worth reviewing what quantum theory is and what it is about. While it is neither possible nor desirable to cover the ground exhaustingly, we shall deal with the central issues and answer questions such as these:

- Is quantum theory well defined?
- Is the interpretation of quantum theory clear?
- Is quantum theory local?
- Is quantum evolution reversible?
- Do wave functions collapse?
- Is there instant action at a distance?
- Where is Heisenberg's cut?
- Is Schrödinger's cat half dead and half alive?
- Is there a "measurement problem"?

1 Physical theories: phenomena, formalism, interpretation; preexisting concepts

In physics, a *theory* has three defining constituents: the physical phenomena, the mathematical formalism, and the interpretation. The *phenomena* are the empirical evidence about physical objects gathered by passive observation, typical for astronomy and meteorology, or by active experimentation in the laboratory. The *formalism* provides the adequate mathematical description of the phenomena and enables the physicist to make precise quantitative predictions about the results of future experiments. The *interpretation* is the link between the formalism and the phenomena.

Each physical theory relies on preexisting concepts, which reflect elementary experiences and without which the theory could not be stated. In Newton's classical mechanics [2], these concepts include mass and force, and the theory deals with the motion of massive bodies under the influence of forces. Classical mechanics cannot provide any insight why there are masses and forces.

Likewise, electric charge is a preexisting concept in Maxwell's electromagnetism [3], which accounts for the forces exerted by the electromagnetic field on charged bodies and how, in turn, the charges and their currents give rise to the electromagnetic field. Maxwell's theory cannot tell us why there are electricity and magnetism or why there are charges and fields, nor is this the purpose of the theory.

One preexisting concept of quantum theory is the event, such as the emission of a photon by an atom, the radioactive decay of a nucleus, or the ionization of a molecule in a bubble chamber. The formalism of quantum theory has the power to predict the probabilities that the events occur, whereby Born's rule [4] is the link between formalism and phenomenon. But an answer to the question *Why are there events?* cannot be given by quantum theory.

In addition, there are preexisting concepts that are common to all physical theories: the three-dimensional space in which the natural phenomena happen, and the time whose flow distinguishes the past from the future.

* In sincere gratitude for many instructive discussions, I dedicate this essay to Professor Rudolf Haag on the occasion of his 90th birthday.

[a] e-mail: cqtebg@nus.edu.sg

[1] No references are given here or later; why point a finger at a particular representative of one or the other community? The pertinent essays in the 2009 *Compendium of Quantum Physics* [1] cite the relevant papers. This no-finger-pointing policy applies throughout the chapter. In particular, we note that the quotes in (53), (60), (62), and (63) are not made up. They are actual quotes from existing sources. The sources are irrelevant, however, the quotes are representatives of many similar statements.

We must also acknowledge a fundamental philosophical concept of science – the conviction that it is possible, with means accessible to the human mind, to give a systematic account of the natural phenomena.

2 Events

The formalism of quantum theory can be used to predict probabilities – probabilities for *events*. In the classic Stern-Gerlach experiment [5][2], the silver atom eventually hits the glass plate, and we can calculate the probability that the next atom will land in a certain region. The hitting of the glass plate is localized in space and time and has an element of irreversibility, *videlicet* the emission of long-wavelength photons during the violent deceleration of the atom, which photons are irretrievably lost. Another example of an event is the radioactive α-decay of a nucleus, which is invariably accompanied by the bremsstrahlung associated with the escaping α particle and the two now-superfluous electrons that are left behind. Yet another example is the absorption of a photon by a semiconductor detector. These examples illustrate the defining features of an event.

First, an event is well localized in space and time. Haag [8] emphasized, and rightly so, that short-range interactions are crucial for the localization of an event and that events are linked by particles, so that a causal history evolves – think, for instance, of the sequence of ionization events in a bubble chamber. By their spatial-temporal relations, the events give meaning to the space-time structure, the great stage for all natural phenomena.

Second, an event is irreversible, it leaves a document behind, a definite trace. It has often been emphasized that an amplification is necessary to bring the event to the attention of the experimenter and that the amplification is an irreversible process; this is true, but it downplays (or even ignores) the irreversibility of the event itself, which is independent of human observation. If you don't have an event in the first place, there is nothing you can amplify. In the bubble chamber situation above, the amplification turns the sequence of ionization events into a string of bubbles that reveal the track of the charged particle to the human eye.

Third, events are randomly realized. Before an event occurs, there are usually several, perhaps many, different events that could happen, each of them with a certain probability of occurrence. Eventually, a definite one of the possible events will be realized. We cannot predict which one will be the case – because this is unknowable, not because we are missing a crucial piece of information.

Quantum theory can be very helpful in understanding quite a lot of the details of, say, the ionization of a molecule by a charge that passes by. We can calculate, with reasonable accuracy, the probability of ionization, the probability that the molecule gets energetically excited without being ionized and subsequently emits one or more photons, and the probability that the molecule remains unaffected by the charged particle. Whether the ionization event happens or not, we cannot predict – the *principle of random realization* (Haag in [9]) ensures that the events do happen in accordance with their probabilities of occurrence.

Born's rule tells us these probabilities – the probability for the ionization event, the probability for the photon-emission event, and the no-event probability for the molecule remaining unaffected. For that to have any meaning, the existence of events must be accepted in the first place. In this sense, then, the event is a preexisting concept of quantum theory. We cannot formulate the theory without this concept.

3 Measurements, Born's rule

3.1 An example

Consider the set-up sketched in Figure 1 where photons are incident on a half-transparent mirror that serves as a symmetric beam splitter (BS). Whether transmitted or reflected, the photons then encounter a polarizing beam splitter (PBS) that reflects vertically (V) polarized photons and transmits horizontally (H) polarized photons. For simplicity, we assume that the BS is symmetric for all polarizations, so that the next photon to arrive has an equal chance of being reflected or transmitted at the BS, irrespective of its polarization. The photons that are reflected pass through a quarter-wave plate (QWP), which converts right-handed (R) circular polarization to vertical polarization and left-handed (L) polarization to horizontal polarization. The four detectors $D_1, \ldots, D_4$ in the output ports of the PBSs, therefore, distinguish between H and V polarization or between R and L polarization of the incoming photon.

The construction of the single-photon detector for a particular outcome is largely irrelevant, except that its efficiency depends on which physical process is used for absorbing the photon and how the absorption event is amplified to finally make the detector "click". With photons arriving one by one, separated in time by more than the dead-time of the detectors, the measurement result consists of a sequence of detector clicks. Quantum theory cannot predict which sequence will be recorded for the next, say, one hundred photons but, for the next photon to arrive, it tells us reliably the probability of clicking for each detector, provided we know the polarization of this incoming photon.

Which one of the four detectors will be triggered by the next photon, which one of the four possible events will be the actual one, this is not just unknown, it is unknowable. One of the events will be randomly realized.

That quantum theory cannot predict which detector will click for the next photon, does not imply that quantum theory is incomplete. There is nothing missing; it

[2] The accounts by Friedrich and Herschbach [6] and by Bernstein [7] of this experiment's history are recommended reading.

Eur. Phys. J. D (2013) 67: 238

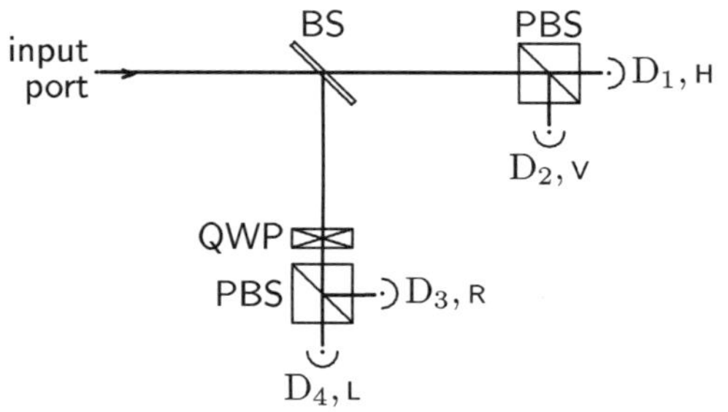

Fig. 1. A four-outcome measurement of photon polarization. An incident photon has equal chance of transmission or reflection at the beam splitter (BS). The polarizing beam splitters (PBS) reflect vertically polarized photons and transmit horizontally polarized photons. Photons transmitted at the BS are probed for horizontal (H) and vertical (V) polarization and detected by detectors D_1 and D_2, respectively. The combination of quarter-wave plate (QWP) and PBS probes the photons reflected at the BS for right-handed (R) or left-handed (L) circular polarization, with respective detectors D_3 and D_4.

is not possible to add further elements to the quantum-theoretical formalism or its interpretation and then have the power of making such predictions. There are no consistent modifications of quantum theory that will achieve this without getting wrong predictions in other situations. The question *Can Quantum-Mechanical Description of Physical Reality be Considered Complete?* that Einstein, Podolsky, and Rosen asked in 1935 [10][3] has a clear answer: Yes, quantum theory gives a complete description of the phenomena. Bohr's reply [12,13][4] hit the mark.

In the experiment of Figure 1, it is in fact impossible to prepare the photon such that a particular detector will certainly click. Yes, we can make sure that one of the detectors will not click – H-polarized photons will never give rise to a click of D_2, for instance, but then the photon has probabilities of $\frac{1}{2}$, $\frac{1}{4}$, $\frac{1}{4}$ of entering one of the other three detectors. Of course, what we have here is just an example of the fundamental probabilistic nature of quantum processes – an illustration of the principle of random realization at work.

It should be noted that the reflection or transmission of the photon at the BS does not constitute an event. The photon needs a short-range interaction with a partner to give rise to an event. If no such partner is available, we could complete a Mach-Zehnder interferometer by adding mirrors and another BS, after which it would be unknowable if the exiting photon was reflected or transmitted at the BSs.

3.2 Mathematical description

We represent the kets for the polarization state of a photon by two-component columns, for which we use these conventions:

$$|\mathsf{V}\rangle \mathrel{\widehat{=}} \begin{pmatrix}1\\0\end{pmatrix}, \quad |\mathsf{H}\rangle \mathrel{\widehat{=}} \begin{pmatrix}0\\1\end{pmatrix},$$
$$|\mathsf{R}\rangle \mathrel{\widehat{=}} \frac{1}{\sqrt{2}}\begin{pmatrix}1\\\mathrm{i}\end{pmatrix}, \quad |\mathsf{L}\rangle \mathrel{\widehat{=}} \frac{1}{\sqrt{2}}\begin{pmatrix}\mathrm{i}\\1\end{pmatrix}. \tag{1}$$

Accordingly, the statistical operator for the polarization degree of freedom for the incoming photon is represented by a 2×2 matrix,

$$\rho = \begin{pmatrix}|\mathsf{V}\rangle & |\mathsf{H}\rangle\end{pmatrix}\begin{pmatrix}\rho_{\mathsf{VV}} & \rho_{\mathsf{VH}}\\ \rho_{\mathsf{HV}} & \rho_{\mathsf{HH}}\end{pmatrix}\begin{pmatrix}\langle\mathsf{V}|\\ \langle\mathsf{H}|\end{pmatrix} \mathrel{\widehat{=}} \begin{pmatrix}\rho_{\mathsf{VV}} & \rho_{\mathsf{VH}}\\ \rho_{\mathsf{HV}} & \rho_{\mathsf{HH}}\end{pmatrix}, \tag{2}$$

and the probability that detector D_1 will click for the next photon is

$$p_1 = \frac{1}{2}\eta_1\rho_{\mathsf{HH}} = \frac{1}{2}\eta_1\langle\mathsf{H}|\rho|\mathsf{H}\rangle = \mathrm{tr}\{\rho\Pi_1\} \tag{3}$$

with the probability operator

$$\Pi_1 = |\mathsf{H}\rangle\frac{1}{2}\eta_1\langle\mathsf{H}| \mathrel{\widehat{=}} \frac{1}{2}\eta_1\begin{pmatrix}0 & 0\\ 0 & 1\end{pmatrix}. \tag{4}$$

Here, the factor of $\frac{1}{2}$ is the transmission probability of the BS and η_1 is the detection efficiency of D_1, the conditional probability that a click is triggered by an incident photon.

Similarly, the probability operators for the other three detectors are[5]

$$\Pi_2 = |\mathsf{V}\rangle\frac{1}{2}\eta_2\langle\mathsf{V}| \mathrel{\widehat{=}} \frac{1}{2}\eta_2\begin{pmatrix}1 & 0\\ 0 & 0\end{pmatrix},$$
$$\Pi_3 = |\mathsf{R}\rangle\frac{1}{2}\eta_3\langle\mathsf{R}| \mathrel{\widehat{=}} \frac{1}{4}\eta_3\begin{pmatrix}1 & -\mathrm{i}\\ \mathrm{i} & 1\end{pmatrix},$$
$$\Pi_4 = |\mathsf{L}\rangle\frac{1}{2}\eta_4\langle\mathsf{L}| \mathrel{\widehat{=}} \frac{1}{4}\eta_4\begin{pmatrix}1 & \mathrm{i}\\ -\mathrm{i} & 1\end{pmatrix}. \tag{5}$$

These need to be supplemented by the probability operator Π_0 for the possibility that the photon escapes detection (no click),

$$\Pi_0 = 1 - (\Pi_1 + \Pi_2 + \Pi_3 + \Pi_4), \tag{6}$$

to make up the probability-operator measurement (POM[6]) for the set-up of Figure 1.

With the probability operators Π_k at hand, the respective probabilities p_k are obtained by *Born's rule*,

$$p_k = \mathrm{tr}\{\rho\Pi_k\}. \tag{7}$$

[3] Reprinted in [11].

[4] Reference [13] reprinted, with pages in wrong order, in [11].

[5] For simplicity, we ignore here and in (3) the possibility of "dark counts" – spontaneous detector clicks not triggered by an incident photon.

[6] POM, with its emphasis on "probability" and "measurement", is preferable quantum physics jargon. The mathematical term POVM (positive-operator-valued measure) makes reference to measure theory. Note that "theory" here identifies a mathematical discipline and does not have the meaning of Section 1.

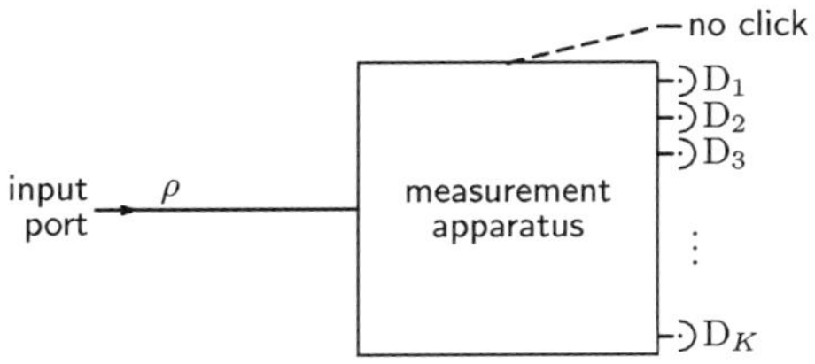

Fig. 2. A generic measurement. A physical object (photon, electron, atom, molecule, ...), whose relevant quantum degrees of freedom are described by the statistical operator ρ, enters the measurement apparatus and triggers a click of one of the K detectors D_1, D_2, ..., D_K, or it escapes detection (no click).

As already noted in Section 1 above, Born's rule is a defining element of the interpretation of quantum theory, it links symbols of the mathematical formalism with the phenomenology: The statistical operator, which summarizes what we know about the polarization state of the incoming photon, and the probability operator, which represents the apparatus, with the probability of a detector click, which we can determine in an experiment by observing sufficiently many repetitions of the situation "single incident photon triggers a detector click".

3.3 Generic measurements

The particular example of Figure 1 illustrates the general measurement scenario. Let us briefly recall the generic properties of a measurement in the quantum realm, as depicted in Figure 2. The relevant degrees of freedom of the quantum system are described by the statistical operator ρ, and the K outcomes of the POM have probability operators Π_1, Π_2, ..., Π_K.

The statistical operator is positive and normalized to unit trace,

$$\rho \geq 0, \quad \mathrm{tr}\{\rho\} = 1, \tag{8}$$

and the probability operators are positive and have unit sum,

$$\Pi_k \geq 0, \quad \sum_{k=0}^{K} \Pi_k = 1, \tag{9}$$

where we include Π_0, the probability operator for the "no click" case. The probability that the kth detector clicks is given by Born's rule (7), which also applies for the no-click probability p_0. As a basic check of consistency, one can verify without effort that the probabilities are positive and have unit sum,

$$p_k \geq 0, \quad \sum_{k=0}^{K} p_k = 1. \tag{10}$$

The probabilities p_k are conditional probabilities – they are conditioned on what we know about the physical system, which information is encoded in the statistical operator that we use for our probabilistic predictions. Someone else may have different knowledge about the system and, therefore, use a different statistical operator and arrive at different values of the probabilities p_k. Yet, her predictions and ours, though different, will both be correct and verifiable.

4 Statistical operators

4.1 Mixtures, blends, and as-if realities

The general form of a statistical operator for the polarization degree of freedom of a photon is given in (2), where the matrix elements are restricted by

$$\rho_{\mathsf{AB}} = \rho_{\mathsf{BA}}{}^* \quad \text{for} \quad \mathsf{A}, \mathsf{B} = \mathsf{H}, \mathsf{V},$$
$$\rho_{\mathsf{VV}} + \rho_{\mathsf{HH}} = 1, \quad \text{and} \quad \rho_{\mathsf{VV}}\rho_{\mathsf{HH}} \geq \rho_{\mathsf{VH}}\rho_{\mathsf{HV}}. \tag{11}$$

Let us consider a ρ with a positive nonmaximal determinant,

$$\frac{1}{4} > \det\{\rho\} = \rho_{\mathsf{VV}}\rho_{\mathsf{HH}} - \rho_{\mathsf{VH}}\rho_{\mathsf{HV}} > 0, \tag{12}$$

so that the polarization state is impure ($\rho^2 \neq \rho$) and ρ has the two positive eigenvalues

$$\left.\begin{matrix} r_1 \\ r_2 \end{matrix}\right\} = \frac{1}{2} \pm \frac{1}{2}\sqrt{1 - 4\det\{\rho\}}, \quad 1 > r_1 > r_2 > 0. \tag{13}$$

The rank-one projectors to the respective eigenspaces are

$$|r_1\rangle\langle r_1| = \frac{\rho - r_2}{r_1 - r_2} \quad \text{and} \quad |r_2\rangle\langle r_2| = \frac{r_1 - \rho}{r_1 - r_2}, \tag{14}$$

and

$$\rho = |r_1\rangle r_1 \langle r_1| + |r_2\rangle r_2 \langle r_2| \tag{15}$$

is the spectral decomposition of ρ.

If ρ refers to a large ensemble of photons, we could say that it is *as if* ρ came about by having fractions r_1 and r_2 of photons in the polarization states associated with kets $|r_1\rangle$ and $|r_2\rangle$, respectively. Speaking of the one photon of interest, regarded as the representative of a Gibbs ensemble of photons, it is *as if* that photon had respective chances r_1 and r_2 of being of the $|r_1\rangle$ or $|r_2\rangle$ kind. This is but one of very many *as-if realities* for the one statistical operator ρ. For example, we can also write

$$\rho = |r_+\rangle\frac{1}{2}\langle r_+| + |r_-\rangle\frac{1}{2}\langle r_-| \tag{16}$$

with

$$|r_\pm\rangle = |r_1\rangle\sqrt{r_1} \pm |r_2\rangle\sqrt{r_2}. \tag{17}$$

Now it is as if the photon had equal chances of being in the polarization states for kets $|r_+\rangle$ and $|r_-\rangle$. Hereby, each choice for the relative phase between $|r_1\rangle$ and $|r_2\rangle$, when factoring the projectors in (14) into ket-bra products, gives different $r_\pm$ kets in (17) and, therefore, different projectors on the right-hand side of (16) and a different as-if reality to go with them. Any convex sum of (15) and (16) identifies yet another as-if reality.

We have the same mixed state ρ, the same *mixture*, and different *blends* for it[7], each of them offering a different as-if reality. All that matters is the mixture, not how it is blended as a convex sum of pure states. Born's rule (7) makes no reference to the blend and its as-if reality, it only cares about the mixture ρ. There is no way of distinguishing between the different blends of (15) and (16) or of their many convex sums; no experiment can tell whether this as-if reality or that one is "what really is the case". Attempts at discriminating between blends founder on fundamental principles, such as Einsteinian causality [15,16]. All ways of blending a mixture are equally good, all as-if realities are equally virtual. In particular, the spectral decomposition in (15) and the as-if reality associated with it have no special significance.

Of course, these observations about mixtures, blends, and as-if realities – here illustrated in the simple context of photon polarization – carry over to more general situations. Whenever the statistical operator is not a rank-one projector, there will be a plethora of blends for that mixture and a corresponding abundance of as-if realities.

4.2 Purification

A mixture that is blended from J rank-one ingredients,

$$\rho = \sum_{j=1}^{J} |a_j\rangle w_j \langle a_j| = \sum_{j,j'=1}^{J} |a_j\rangle \sqrt{w_j}\, \delta_{j,j'}\, \sqrt{w_{j'}} \langle a_{j'}| \tag{18}$$

with positive weights of unit sum ($w_j > 0$, $\sum_j w_j = 1$), can be written as the partial trace of a rank-one projector of a bipartite system,

$$\rho = \mathrm{tr}_2\{|\,\rangle\langle\,|\} \tag{19}$$

with

$$|\,\rangle = \sum_{j=1}^{J} |a_j\, b_j\rangle \sqrt{w_j}, \tag{20}$$

where the b kets are orthonormal, $\langle b_j | b_{j'}\rangle = \delta_{j,j'}$. Since the blend that we start with in (18) is not unique and any orthonormal set of b kets of any auxiliary second system will serve the purpose, there are many different ket-bras $|\,\rangle\langle\,|$ that one can use in (19) equally well.

In particular situations, such as the one discussed in Section 7, we are really dealing with a bipartite system in a pure state (or nearly so), and then a physically meaningful ket $|\,\rangle$ may be available. In the generic case, however, we have our statistical operator ρ for the physical system of interest and the *purification* of (19) with (20) is just a mathematical procedure that might be helpful in performing a certain calculation but has no physical significance whatsoever.

Yet, the "church of the larger Hilbert space" offers membership to those who believe that there is always such a larger physical system in a pure state. Now, it is already quite challenging to acquire sufficient information about a few degrees of freedom to justify, as a good approximation, the use of a rank-one projector as the statistical operator. How, then, can we have a pure state for the unknown degrees of freedom of a fictitious larger system that exists only in our fantasy? Of course, we cannot.

It appears that sloppy language is one source of this misconception. One often speaks of "the wave function of the electron" or "the statistical operator of the photon" rather than "our wave function for the electron" or "our statistical operator for the photon". The precise "our ... for" phrases leave no doubt that we are describing the electron by this wave function or the photon by that statistical operator, whereas the sloppy "the ... of" wordings wrongly suggest an independent existence of the wave function or the statistical operator. When this suggestion is taken seriously, the church gates stand wide open.

Is there not the one largest system, the universe, with the physical system of interest as a small part of it? Yes, there is the universe, but no-one has a wave function *for* the universe, and there is no wave function *of* the universe.

5 Evolution

5.1 Equations of motion

In Born's rule (7), both the statistical operator ρ and the probability operator Π_k are functions of the dynamical variables – collectively symbolized by $Z(t)$ – with, possibly, a parametric time dependence as well. *All* such operator-valued functions evolve in accordance with *Heisenberg's equation of motion*,

$$\frac{\mathrm{d}}{\mathrm{d}t} f(Z(t),t) = \frac{\partial}{\partial t} f(Z(t),t) + \frac{1}{\mathrm{i}\hbar}\left[f(Z(t),t), H(Z(t),t)\right], \tag{21}$$

the quantum analog of Hamilton's equation of motion for phase-space functions. In (21), $H\big(Z(t),t\big)$ is the Hamilton operator, itself a function of this kind, and the commutator term accounts for the dynamical time dependence of $f\big(Z(t),t\big)$. Whereas Π_k and H may or may not have a parametric time dependence, the parametric time dependence of the statistical operator compensates for its dynamical time dependence because ρ is constant,

$$\begin{aligned} \frac{\mathrm{d}}{\mathrm{d}t}\rho(Z(t),t) &= 0, \\ \frac{\partial}{\partial t}\rho(Z(t),t) &= \frac{1}{\mathrm{i}\hbar}\left[H(Z(t),t), \rho(Z(t),t)\right]. \end{aligned} \tag{22}$$

The latter equation, really a special case of (21), is known as *von Neumann's equation of motion*, the quantum analog of Liouville's equation of motion for phase-space densities.

Since the von Neumann equation concerns solely the parametric time dependence of ρ, it does not matter to which common time t_0 the dynamical operators refer,

$$\frac{\partial}{\partial t}\rho(Z(t_0),t) = \frac{1}{\mathrm{i}\hbar}\left[H(Z(t_0),t), \rho(Z(t_0),t)\right], \tag{23}$$

[7] We are adopting Süßmann's fitting terminology who distinguishes between *Gemisch* (mixture) and *Gemenge* (blend) [14].

and since a unitary transformation relates the dynamical operators at time t to those at time t_0, we also have

$$\begin{aligned} p_k(t) &= \operatorname{tr}\{\Pi_k(Z(t),t)\rho(Z(t),t)\} \\ &= \operatorname{tr}\{\Pi_k(Z(t_0),t)\rho(Z(t_0),t)\}. \end{aligned} \tag{24}$$

According to standard terminology, these two expressions for $p_k(t)$ are "in the Heisenberg picture" and "in the Schrödinger picture", respectively. The central von Neumann equation in (22) or (23), however, is picture independent. The parametric time dependence of the statistical operator is what it is.

The product $\Pi_k\rho$ is also an operator that obeys Heisenberg's equation of motion (21), and the two versions of (24) give

$$\frac{\mathrm{d}}{\mathrm{d}t}p_k(t) = \operatorname{tr}\left\{\frac{\mathrm{d}}{\mathrm{d}t}(\Pi_k\rho)\right\} = \operatorname{tr}\left\{\frac{\partial}{\partial t}(\Pi_k\rho)\right\} \tag{25}$$

for the time derivative of the kth probability. The difference of the two traces is the vanishing trace of the commutator of $\Pi_k\rho$ with the Hamilton operator. In this sense, then, the Heisenberg picture pays attention to the full time dependence, whereas the Schrödinger picture cares only about the parametric time dependence. This is particularly well visible when the probability operator Π_k has no parametric time dependence (the same set-up is used at different times), so that (25) becomes

$$\frac{\mathrm{d}}{\mathrm{d}t}p_k(t) = \operatorname{tr}\left\{\frac{\mathrm{d}\Pi_k}{\mathrm{d}t}\rho\right\} = \operatorname{tr}\left\{\Pi_k\frac{\partial\rho}{\partial t}\right\} \tag{26}$$

after the product rule is applied,

$$\frac{\mathrm{d}}{\mathrm{d}t}(\Pi_k\rho) = \frac{\mathrm{d}\Pi_k}{\mathrm{d}t}\rho, \quad \frac{\partial}{\partial t}(\Pi_k\rho) = \Pi_k\frac{\partial\rho}{\partial t}. \tag{27}$$

The first of these statements is an identity that is generally valid, the second holds only when $\frac{\partial}{\partial t}\Pi_k = 0$.

If we denote by $\langle a,t|$ the common eigenbras of a maximal set $A(t)$ of commuting dynamical variables, a subset of the collection $Z(t)$, with a symbolizing the set of eigenvalues, $\langle a,t|A(t) = a\langle a,t|$, then we have the *Schrödinger equation*

$$\mathrm{i}\hbar\frac{\partial}{\partial t}\langle a,t| = \langle a,t|H(Z(t),t) \tag{28}$$

for the relation between $\langle a,t+\mathrm{d}t|$ and $\langle a,t|$. The matrix elements of a rank-one statistical operator, $\rho = |\,\rangle\langle\,|$, in this a basis,

$$\langle a,t|\rho|a',t\rangle = \psi(a,t)\psi(a',t)^*, \tag{29}$$

are products of the probability amplitudes

$$\psi(a,t) = \langle a,t|\,\rangle \tag{30}$$

and their complex conjugates. The von Neumann equation (22) is then equivalent to

$$\mathrm{i}\hbar\frac{\partial}{\partial t}\psi(a,t) = \sum_{a'}\langle a,t|H(Z(t),t)|a',t\rangle\psi(a',t) \tag{31}$$

or

$$\mathrm{i}\hbar\frac{\partial}{\partial t}\Psi(t) = \mathcal{H}\Psi(t), \tag{32}$$

if we collect the probability amplitudes $\psi(a,t)$ in the column $\Psi(t)$, the *wave function*, and the matrix elements of H in the matrix $\mathcal{H}$. The corresponding matrix version of the von Neumann equation itself is then

$$\frac{\partial}{\partial t}\varrho = \frac{1}{\mathrm{i}\hbar}[\mathcal{H},\varrho], \tag{33}$$

where $\varrho = \Psi\Psi^\dagger$ is the matrix that represents the statistical operator $\rho = |\,\rangle\langle\,|$. Of course, (32) is the most familiar version of the Schrödinger equation, as it applies to a wave function.

Expressions such as (29) may convey the false impression that the statistical operator depends on time. Yes, its matrix elements do depend on time, whereas we have $\frac{\mathrm{d}}{\mathrm{d}t}\rho = 0$ for the time dependence of ρ. It is the time-dependent basis of bras that gives time dependence to the wave function in (30), whereas the state ket $|\,\rangle$ is time independent. In

$$|\,\rangle = \sum_a |a,t\rangle\psi(a,t) = \sum_{a'}|a',t'\rangle\psi(a',t'), \tag{34}$$

which holds for any time t or t', the basis kets $|a,t\rangle$ and $|a',t'\rangle$ as well as the probability amplitudes $\psi(a,t)$ and $\psi(a',t')$ change in time, but $|\,\rangle$ does not. This is, of course, as it should be because $\rho = |\,\rangle\langle\,|$ represents our knowledge about the preparation of the system and that makes reference to some definite instant in the past when this knowledge was acquired.

In passing, we note that some textbook authors have failed to distinguish clearly between the state ket $|\,\rangle$ (an abstract object) and its wave function Ψ (a basis-dependent numerical representation of $|\,\rangle$), between the Hamilton operator H and its matrix $\mathcal{H}$, and between the statistical operator ρ and its matrix ϱ. In their books, then, (32) is written as $\mathrm{i}\hbar\frac{\partial}{\partial t}|\Psi(t)\rangle = H|\Psi(t)\rangle$, as if there were a time-dependent state ket $|\Psi(t)\rangle$ that obeys a Schrödinger equation. As a consequence, it then seems that we have a time-dependent statistical operator $\rho = |\Psi(t)\rangle\langle\Psi(t)|$ although, in fact, ρ is constant. The related failure of distinguishing between the total and the parametric time dependence then gives rise to the misleading statement that equations such as (26) demonstrate that "in the Schrödinger picture ρ depends on time and Π_k is constant, whereas ρ is constant and Π_k depends on time in the Heisenberg picture" and the like.

5.2 Time-reversal symmetry and irreversible evolution[8]

Let us consider the Schrödinger equation for the position wave function of a particle of mass M that moves along the x axis under the influence of the forces associated with the potential energy $V(x)$,

$$\left(\frac{\partial}{\partial t} + \mathrm{i}\Omega\right)\psi(x,t) = 0, \tag{35}$$

[8] This section follows [17] closely.

where the differential operator Ω is given by

$$\hbar\Omega = \frac{1}{2M}\left(\frac{\hbar}{\mathrm{i}}\frac{\partial}{\partial x}\right)^2 + V(x) > 0. \quad (36)$$

Here, for simplicity, the physical requirement that the energy is bounded from below is equivalently replaced by insisting on the positivity of Ω. The time-reversed wave function

$$(\Theta_T\psi)(x,t) = \psi(x, 2T-t)^* \quad (37)$$

is also a solution of the Schrödinger equation (35),

$$\left(\frac{\partial}{\partial t} + \mathrm{i}\Omega\right)\psi(x, 2T-t)^* = 0. \quad (38)$$

This *time-reversal symmetry* of the Schrödinger equation does not imply, however, that quantum evolution is reversible.

One can imagine the undoing of all changes in $\psi(x,t)$ that have occurred between, say, $t=0$ and $t=T$ by first applying Θ_T, followed by evolution under the same dynamics for $T<t<2T$, and final application of Θ_{2T}, so that an instant τ after $t=2T$ one has the same wave function as at time $t=\tau$, that is

$$\begin{aligned}\psi(x, 2T+\tau) &= \mathrm{e}^{-\mathrm{i}\Omega\tau}\Theta_{2T}\mathrm{e}^{-\mathrm{i}\Omega T}\Theta_T\mathrm{e}^{-\mathrm{i}\Omega T}\psi(x,0)\\ &= \psi(x,\tau) = \mathrm{e}^{-\mathrm{i}\Omega\tau}\psi(x,0)\end{aligned} \quad (39)$$

for all x. But this cannot be achieved in any real sense.

The changes symbolized by Θ_T and Θ_{2T} must be of dynamical origin themselves, which they cannot be because these operators are not unitary; they are anti-unitary,

$$\Theta_T(\lambda_1\psi_1 + \lambda_2\psi_2) = \lambda_1^*\,\Theta_T\psi_1 + \lambda_2^*\,\Theta_T\psi_2. \quad (40)$$

It is true that all that matters is the product

$$\Theta_{2T}\mathrm{e}^{-\mathrm{i}\Omega T}\Theta_T = \mathrm{e}^{\mathrm{i}\Omega T}, \quad (41)$$

and this is unitary. But for most Ω the right-hand side is *not* of the form $\mathrm{e}^{-\mathrm{i}\widetilde{\Omega}T}$ with a physical, positive $\widetilde{\Omega}$. It follows that the undoing of all evolutionary changes is impossible, at least if one just wants to exploit the time-reversal symmetry of the Schrödinger equation.

A more general scheme, not so tightly bound to elementary nonrelativistic quantum mechanics, is what one could call the *Humpty-Dumpty problem*, which name refers to the rhymed riddle (solution: egg) that succinctly expresses folk wisdom about irreversibility[9]. Here, then, is the

> **Humpty-Dumpty problem**
> Given a statistical operator ρ and a Hamilton operator $H_1(t)>0$, acting from $t=0$ to $t=T$, find $\widetilde{T}$ and $H_2(t)>0$, acting from $t=T$ to $t=\widetilde{T}$, such that the unitary evolution operators $U(T,0)$ and $U(\widetilde{T},T)$, constructed as time-ordered exponentials from $H_1(t)$ and $H_2(t)$, respectively, give an over-all evolution operator $U(\widetilde{T},0) = U(\widetilde{T},T)U(T,0)$ that has no effect on ρ: $U(\widetilde{T},0)\,\rho = \rho\,U(\widetilde{T},0)$.[10] (42)

[9] You can find the riddle in [18].

A solution is found when we have

$$\begin{aligned}U(\widetilde{T},T)^\dagger &= \exp\left(+\frac{\mathrm{i}}{\hbar}\int_T^{\widetilde{T}}\mathrm{d}t\,H_2(t)\right)_<\\ &\doteq \exp\left(-\frac{\mathrm{i}}{\hbar}\int_0^T\mathrm{d}t\,H_1(t)\right)_> = U(T,0)\end{aligned} \quad (43)$$

with standard time ordering in the second line and the reverse order in the first, and the dotted equal sign means "same effect on ρ", that is

$$U(\widetilde{T},T)^\dagger\,\rho\,U(\widetilde{T},T) = U(T,0)\,\rho\,U(T,0)^\dagger. \quad (44)$$

In view of the positivity of both $H_1(t)$ and $H_2(t)$ this can only be realized under exceptional circumstances. The question is then: For which ρ and $H_1(t)$ is it possible at all? There are, of course, two very important rules in this game, namely

(a) only real physical interactions are considered;
(b) over-idealizations are not allowed. (45)

It can safely be conjectured that for generic statistical operators ρ and Hamilton operators $H_1(t)$, the Humpty-Dumpty problem has no solution. The examples of Section 2 are to the point.

Maybe part of the evolutionary changes can be undone at least? Consistent with rule (45a) we ask more specifically: Can the two partial beams of a Stern-Gerlach apparatus (SGA) be reunited with such precision that the initial spin state is recovered? That is, we make no attempt at undoing in full the changes in the center-of-mass state of the atom. References [18,19] deal with the details of such a Stern-Gerlach interferometer; it is found that one needs three more SGAs for the beam reunion. Ideally, the four SGAs would be perfectly identical. With due respect to rule (45b), however, we have to ask how large a mismatch between the SGAs can be tolerated.

Two relevant quantities are the net transverse momentum transferred to either partial beam and their net transverse displacement, respectively measured by

$$\Delta p = \int\mathrm{d}t\,F(t) \quad\text{and}\quad \Delta z = -\int\mathrm{d}t\,F(t)\frac{t}{M}, \quad (46)$$

where $F(t)$ is the force on the up-component, for instance, produced by the inhomogeneity of the magnetic field. Thus, Δp and Δz are properties of the macroscopic SGAs; the deviations from the ideal values $\Delta p=0$, $\Delta z=0$ are

[10] More generally, one could also consider ancilla-assisted evolution between T and $\widetilde{T}$, so that the net effect on ρ would come from a completely positive map that may not be describable by a single unitary operator $U(\widetilde{T},T)$. The main conclusion that, as a rule, quantum evolution is irreversible is not affected, however.

resulting from the lack of control over the SGAs. It turns out that to maintain spin coherence one must, at least, ensure that

$$\left(\delta z\,\Delta p/\hbar\right)^2 + \left(\delta p\,\Delta z/\hbar\right)^2 \ll 1 \tag{47}$$

holds, where δz and δp are the natural spreads of the beam prior to entering the first SGA. Therefore, one can tolerate only such variations in the magnetic field that both $|\Delta z| \ll \hbar/\delta p$ and $|\Delta p| \ll \hbar/\delta z$, with the consequence

$$\frac{|\Delta z|\,|\Delta p|}{\hbar} \ll \frac{\hbar}{\delta z\,\delta p}. \tag{48}$$

This states that the *macro*scopic magnetic field must be controlled with *sub-micro*scopic precision – an impossible feat. The spin state is so severely affected by the inhomogeneous magnetic field of the first SGA that it shares Humpty Dumpty's fate: One cannot undo the damage.

6 State reduction

Our statistical operator ρ represents our knowledge about the preparation of the system, and this knowledge does not change until we acquire additional information. When that happens, we must update the statistical operator to account for the change in the conditions under which we predict the results of future experiments. *State reduction* is the technical term for this updating.

State reduction is not particular to quantum theory, it is part and parcel of all statistical formalisms. Bookmakers know about state reduction, even if they never heard the term and have not studied quantum theory.

Here is an example from the world of chess, showing two cases of state reduction[11]: After eleven rounds of the 2013 candidates tournament in London that would determine the challenger of world champion Anand, a prediction based on past results among the participants, their ratings, and perhaps other data gave a chance of 87% to Carlsen to win the competition, a chance of 7% to Kramnik, and a chance of 6% to Aronian. In the twelfth round, Carlsen lost against Ivanchuk and Kramnik won against Aronian, which resulted in updated chances of 65% for Kramnik and 35% for Carlsen, with Aronian out of the race. Then, in the thirteenth round, Carlsen won against Radjabov and Kramnik drew his game with Gelfand, and the probabilities for winning the tournament became 84% for Carlsen and 16% for Kramnik. The highly improbable happened in the final fourteenth round: Both Carlsen and Kramnik lost against Svidler and Ivanchuk, respectively, and Carlsen took first place.

After the kth detector of a POM clicked, we update the statistical operator in accordance with

$$\rho \to \frac{K_k \rho K_k^\dagger}{p_k}, \tag{49}$$

where K_k is the appropriate *Kraus operator* [20] for the probability operator Π_k,

$$\Pi_k = K_k^\dagger K_k. \tag{50}$$

Depending on how the POM is implemented, there can be quite different Kraus operators for the probability operator of interest. In the example of Section 3, we have

$$K_1 = |\textsc{vac}\rangle \sqrt{\eta_1/2}\,\langle \textsc{h}| \tag{51}$$

for detector D_1 and similar expressions for D_2, D_3, and D_4. Irrespective of which detector clicks, we get

$$\rho \to |\textsc{vac}\rangle\langle\textsc{vac}| \tag{52}$$

upon reducing the state, thereby correctly stating that the photon has been absorbed and we are left with the "no more photon" situation of the photon vacuum, symbolized by the state ket $|\textsc{vac}\rangle$.

More complicated cases are possible, such as a sum of several $K^\dagger K$ terms instead of the single term in (50). The folklore that "a measurement leaves the system in the relevant eigenstate of the observable" applies only to over-idealized projective measurements (meaning that the K_ks are pairwise orthogonal projectors). It is puzzling that some textbook authors consider it good pedagogy to elevate this folklore to an "axiom" of quantum theory.

"Collapse of the state" or "wave function collapse" are popular synonyms for state reduction. The connotation that the transition in (49) is a dramatic dynamical process, as if the physical system were evolving, is clearly misleading. The various mathematical models that, exactly or approximately, accomplish the state reduction (49) as a process in time result from a category mistake: The statistical operator is not a physical object or a property of a physical object, it *describes* the object by encoding what we know about it, and state reduction is the bookkeeping device for updating the description. An electron carries its mass and charge around, but not "its" statistical operator. It is worth recalling here that different physicists may very well use different, equally correct, statistical operators for predictions about that same electron.

7 No action at a distance

Here is a quote from a recent news item:

> "Quantum theory makes the distinctive prediction that non-local correlations are instant: for example, a measurement of the polarization of one of a pair of quantum-entangled photons should immediately set the polarization of the other, no matter how far the photons are apart, without either photon's polarization being in any way predetermined." (53)

In fact, quantum theory makes no such prediction. There is no instant action at a distance.

The situation alluded to is of the kind depicted in Figure 3. A source of entangled photon pairs (SEPP) emits

[11] This is a slightly simplified account; for more details, see the news items on www.chessbase.com, 29–31 March 2013.

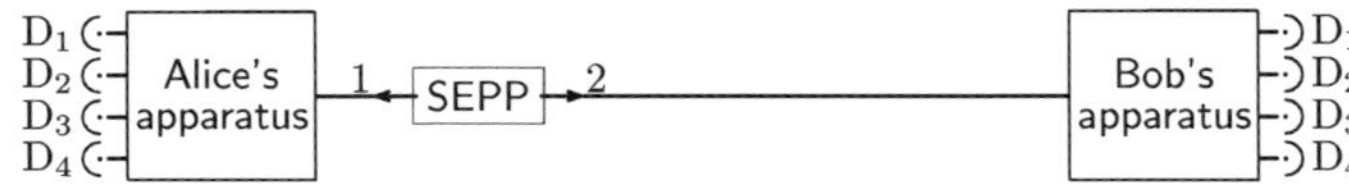

Fig. 3. A source of entangled photon pairs (SEPP) sends photon 1 of each pair to Alice's laboratory and photon 2 to Bob's laboratory. Alice's apparatus detects photon 1 before photon 2 reaches Bob's apparatus.

correlated photons in pairs, with photon 1 sent to Alice and photon 2 to Bob who, for the sake of the argument, is assumed to be much farther away from the SEPP than she is. For the purpose of this discussion, we allow ourselves the idealization that the photon pairs are described by a rank-one statistical operator, specifically

$$\rho_{1\&2} = |\mathrm{EPRB}\rangle\langle\mathrm{EPRB}| \tag{54}$$

with an Einstein-Podolsky-Rosen [10] state ket in Bohm's version [21],

$$|\mathrm{EPRB}\rangle = \frac{1}{\sqrt{2}}\big(|\mathsf{HV}\rangle - |\mathsf{VH}\rangle\big) = \frac{1}{\sqrt{2}}\big(|\mathsf{RL}\rangle - |\mathsf{LR}\rangle\big), \tag{55}$$

where, for instance, the ket $|\mathsf{HV}\rangle$ stands for "photon 1 is H-polarized, photon 2 is V-polarized". Alice and Bob measure their respective photons by four-outcome POMs of the construction shown in Figure 1, with her detector clicking much earlier than his.

Bob's statistical operator for photon 2 is that of a completely mixed polarization state,

$$\rho_2^{(\mathrm{B})} = \mathrm{tr}_1\{\rho_{1\&2}\} = \frac{1}{2}\big(|\mathsf{H}\rangle\langle\mathsf{H}| + |\mathsf{V}\rangle\langle\mathsf{V}|\big) = \frac{1}{2}\big(|\mathsf{R}\rangle\langle\mathsf{R}| + |\mathsf{L}\rangle\langle\mathsf{L}|\big). \tag{56}$$

This is also Alice's statistical operator for photon 2 prior to measuring photon 1.

Now suppose Alice's detector D_1 clicks for photon 1. Then, her reduced state for the photon pair is:

$$\rho^{(\mathrm{A})} = |\mathrm{VAC}\ \mathsf{V}\rangle\langle\mathrm{VAC}\ \mathsf{V}|, \tag{57}$$

meaning that really only photon 2 is left, and Alice's updated statistical operator for that photon is:

$$\rho_2^{(\mathrm{A})} = \mathrm{tr}_1\left\{\rho^{(\mathrm{A})}\right\} = |\mathsf{V}\rangle\langle\mathsf{V}|. \tag{58}$$

Has anything happened to photon 2 when Alice detected photon 1 with her apparatus? Has "the measurement... immediately set the polarization" of photon 2? Of course not. The transition

$$\frac{1}{2}\big(|\mathsf{H}\rangle\langle\mathsf{H}| + |\mathsf{V}\rangle\langle\mathsf{V}|\big) \to |\mathsf{V}\rangle\langle\mathsf{V}| \tag{59}$$

is Alice's bookkeeping, it is not a physical process that photon 2 is undergoing, and it happens when she updates her statistical operator.

Incidentally, we have here an example of the situation mentioned twice above, at the end of Section 3.3 and at the end of Section 6: Alice and Bob use different statistical operators for predictions about photon 2, reflecting the different knowledge they have about the situation. Alice knows the outcome of the measurement on photon 1, Bob does not. And we must not fall into the trap of concluding that Alice's statistical operator is better than Bob's; both her $\rho_2^{(\mathrm{A})}$ and his $\rho_2^{(\mathrm{B})}$ give correct statistical predictions, as can be verified by many repetitions of the experiment[12].

So, what about the quote (53)? It is another instance of the category mistake noted in Section 6. A change in Alice's description of photon 2 is misunderstood as a physical process affecting photon 2, and then *ex falso sequitur quodlibet*, namely the false assertion about instantaneous nonlocal correlations, exemplified by the immediate setting of the polarization of a photon that can be arbitrarily far away.

8 No nonlocality

Nonlocal correlations are a familiar consequence of local actions: A timing signal from a radio station synchronizes clocks at large distances from each other. These nonlocal correlations are clearly not instant; they have a common local cause. But then there are claims that quantum systems can exhibit correlations of nonlocal origin.

Recall, for instance, the abstract of a recent paper, which begins with these words:

> "Bell's 1964 theorem, which states that the predictions of quantum theory cannot be accounted for by any local theory, represents one of the most profound developments in the foundations of physics." (60)

We submit: Doesn't quantum theory itself, which is a local theory, account for its own predictions?

As the authors of this quote know very well, experimental data contradict Bell's theorem [23,24][13], which implies that – as a statement about physical systems – the theorem is wrong[14]. Since there is no error in the reasoning that establishes the theorem from its assumptions, the flaw must be in the assumptions. Specifically, it is the assumption that a mechanism exists that determines which detector will click for the next photon registered by an apparatus of the kind depicted in Figure 1. There is no such deterministic mechanism – quantum processes

[12] A more complicated example is discussed in Section 4.3 of [22].

[13] Reference [23] reprinted in [11].

[14] We leave it as a moot point whether the theorem is "one of the most profound developments in the foundations of physics".

are fundamentally probabilistic, events are randomly realized – and the violation of Bell's theorem by actual data confirms that.

Insights of this kind predate Bell's work by decades. Recall, for example, the matter-of-fact statement in Schrödinger's essay of 1935 [25–27], which reads

> "[If I wish to ascribe] to *all* determining parts definite (merely not exactly known to me) numerical values, then there is no supposition as to these numerical values *to be imagined* that would not conflict with some portion of quantum theoretical assertions." (61)

in Trimmer's translation [28][15].

In a derivation of Bell's theorem, or any of the many variants on record, the formalism of quantum theory is not used. Instead, one relies on common-sense arguments of a certain plausibility in an attempt at describing quantitatively the properties of the joint probabilities in experiments similar to that of Figure 3. Part of that common sense is a certain notion of locality: If the beam splitter in Figure 1 is replaced by a mirror that the experimenter may or may not put in place, the free-will decisions by Alice should have no influence on the click frequencies that Bob records in his experiment, and *vice versa*.

The actual joint probabilities – which are experimentally determined by measuring many pairs emitted by the SEPP and are correctly accounted for by quantum theory – do not obey the restrictions that follow from that common-sensible adhockery, and why should they? Findings of an inadequate nonquantum formalism are irrelevant for quantum physics. If the findings are at variance with the experimental data, as is the case here, we are reminded of the inappropriateness of the reasoning. It follows that common sense of that sort does not apply in the quantum realm.

Rather disturbingly, though, it has become acceptable to turn the argument into its opposite. It is taken for granted that quantum physics *should* obey such common sense, but then that inadequate nonquantum formalism needs nonlocal features – or so it seems[16]. The conclusion that

$$\text{“Quantum theory is nonlocal.”} \tag{62}$$

appears inevitable: So reads the opening line of a paper that reports a violation of Bell's theorem by experimental data.

In fact, however, there is no nonlocality in an experiment such as the one sketched in Figure 3 or any other quantum-physics experiment on record. The SEPP uses well-localized short-range interactions for the creation of the photon pairs and the same is true for the processes by which the photon absorption in one of the detectors gives rise to the observed click. Quantum theory is a *local theory* in exactly the same sense as all other physical theories: All interactions come about by local couplings; a continuity equation states that probability is locally conserved; energy, momentum, angular momentum, and other properties are transferred between particles by highly localized scattering events, and so forth. And just like the other local theories do, quantum theory predicts nonlocal correlations that originate in local processes.

It is true that joint probabilities that result from quantum processes can have stronger correlations than those available by nonquantum simulations; a violation of Bell's theorem or one of its variants tells us when such a situation is at hand. This can be exploited for various purposes, including quantum key distribution [29], and experimental realizations [30] have no nonlocal elements. Accordingly, there are real-life practical applications of Bell's theorem, and they give Bell's work lasting value.

We close this subject with noting that a very basic notion of *Local Quantum Physics* (the title of Haag's book [31]) is precisely the locality concept of Bell's theorem, namely that "the free-will decisions by Alice should have no influence on the click frequencies that Bob records in his experiment, and vice versa". In the technical terms of relativistic quantum field theory, this is the requirement that the local observables in Alice's space-time region commute with the local observables in Bob's region, which is at a space-like separation from Alice's. Since the notion of locality that enters Bell's theorem is exactly the notion of locality in quantum field theory, the observed violation of Bell's theorem by quantum systems cannot support the claim (62).

9 Heisenberg's cut

9.1 The cut is where you put it

Alice and Bob have conducted the experiment of Figure 3 on a certain number of entangled photon pairs, independently and identically prepared by the SEPP, and now they report their measurement results to Charlie. What they communicate to him is the sequence of detector clicks observed and the times at which the clicks occurred. These data enable Charlie to pair Alice's clicks with Bob's and so verify that the click statistics observed is consistent with the predictions obtained from applying Born's rule to this situation.

Alice and Bob will surely not trouble Charlie with accounts of their minds, their retinas (or eardrums), or other irrelevant information. Nor will Charlie bother with such matters when computing probabilities. Yet, exactly this is what certain dogmatists insist must be done *in principle*, although *in practice* they themselves never do it[17].

[15] Reprinted in [11].

[16] One could also conclude that there is no deterministic mechanism, were this conclusion not prevented by that common sense.

[17] Statements like "In principle, I could solve the Schrödinger equation to predict the next solar eclipse." are empty unless you can do it in practice. Something can be *impossible in principle*, such as drawing a triangular circle or building a so-called Popescu-Rohrlich box [32], or *possible in practice*. But to declare that something is *possible in principle* can at best mean that one is not aware of any reason that would make it impossible in principle.

The BS, the QWP, the PBSs of the apparatus in Figure 1 are made of atoms with which the incoming photon interacts. The detectors $D_1, \ldots, D_4$ are made of atoms, too, and so is the experimenter's retina and her brain. If we apply quantum theory to the photon, we should also apply it to the atoms of the optical elements, the detectors, ..., the experimenter's brain. And why stop here? There are also the atoms of your (the reader's of these lines) retina, optical nerve, and brain. A truly complete account of the experiment should include all that – as a matter of principle, that is, not of practice.

No, it shouldn't. The experiment of Figure 3 investigates the polarization properties of entangled photon pairs, and the phenomena of interest do not depend on which material is used to realize the BS and the other components of the four-outcome POM of Figure 1. Yes, if one wishes to do so, one can extend the quantum description beyond the crucial degrees of freedom of the photons, but nothing relevant will be added to Charlie's analysis of the data.

In practice, the formalism of quantum theory is always applied to the relevant quantum variables (almost always these are few) and the inclusion of irrelevant ones increases the effort only but not the benefit[18]. The identification of the relevant variables is usually unproblematic and, if in doubt, one can take more quantum degrees of freedom into account than might be necessary. There are always plenty of technical details that are important for the success of the experiment without entering the quantum-theoretical description of the physical system that is studied. For example, the experimenter has to calibrate the detectors and determine their efficiencies, perhaps exploiting a mathematical model of some crucial components, but in the end the only significant number is the efficiency η_k for the kth detector. When Charlie analyzes the data, he must be able to trust the efficiencies reported to him, but what kind of detector is actually used does not matter to him.

The imagined demarcation line between the degrees of freedom that are given a full quantum-theoretical description and the "rest of the universe" (to use a pompous expression) is known as *Heisenberg's cut*, and the inclusion of a few more variables into the quantum-theoretical description than really necessary is the *shift of Heisenberg's cut*.

In a manner of speaking, one can regard Heisenberg's cut as separating the microscopic realm of quantum physics from the macroscopic world of classical physics, but one should not be carried away by the connotations that the micro/macro distinction may bring about. We are always referring to a particular context, such as Alice and Bob taking data and Charlie analyzing them, not to a generally valid demarcation line. When applying the quantum-theoretical formalism to an experiment in Munich, the polarization degrees of freedom of photon pairs emitted by a SEPP in Singapore are on the macroscopic side of Heisenberg's cut, whereas they are on the microscopic side when the Singapore experiment is evaluated.

In his posthumous eloquent polemic *Against 'measurement'* [33][19] and also on earlier occasions, Bell[20] talks about Heisenberg's cut as if it were a universal border between quantum physics and classical physics. Having made this mistake[21], he then concludes that quantum theory is ill-defined because it does not specify the location of the cut with the mathematical precision that he demands it should have.

Bell is missing three essential points. First, there is a Heisenberg's cut whenever the quantum-theoretical formalism is applied in a particular situation: There is no such universal border. Second, the cut is imprecise by its nature, it can be shifted to include as many quantum degrees of freedom as one can handle in a calculation or control in an experiment: Heisenberg's cut is where you put it. Third, the distinction between microscopic phenomena and macroscopic phenomena is meaningful even if one cannot give rigorous criteria for the micro/macro distinction.

We illustrate this remark by the following analogy. The concepts of land and sea are meaningful although you cannot draw, in a unique and natural way, a mathematical line on the beach that has the sea on one side and the land on the other. You can draw lines such that all of the sea is assuredly on one side, or all of the land is on the other, but there is no line that accomplishes both.

A fourth point is this: Alice and Bob rely on the macroscopic world around them, with its well-defined properties, when they record their data and communicate them to Charlie. One cannot even speak about quantum phenomena without reference to the rather robust classical-physics environment, in which all human activity happens. Denying that there is this robust environment, is denying the obvious.

In summary, then, Heisenberg's cut is needed and, by its nature, it is imprecise, and this is just fine.

9.2 Decoherence

Whenever we apply the formalism of quantum theory to a particular physical situation, we identify the relevant degrees of freedom and ignore all the rest. In other words, we put Heisenberg's cut where we find it appropriate. This involves a physical approximation because we disregard all dynamical effects that arise from interactions between the quantum degrees of freedom that are treated in full and those on the other side of the cut, often referred to as the *environment*. As a consequence, the predictions we make about future measurements on the system are not

[18] This is also true for any application of Newton's classical mechanics, Maxwell's electromagnetism, or any other physical theory, without the dogmatists crying foul.

[19] The replies by van Kampen [34,35], Peierls [36], and Gottfried [37] must not be missed.

[20] This one exception from the no-finger-pointing policy of footnote 1 is unavoidable.

[21] Bell also makes that category mistake of regarding the wave function as a physical object, and he misunderstands state reduction as a physical process.

absolutely reliable and, since small errors accumulate, the predictions are less reliable for the far future than the near future.

In terms of the parametric time dependence of the statistical operator, this means that the von Neumann equation (23) applies only during the initial period when the residual interactions with the environment can be ignored. If we wish to extrapolate from the initial time t_0 to much later times, the uncertainty that originates in our lack of knowledge about the system-environment interactions across Heisenberg's cut has to be taken into account. This leads to a degradation of our statistical operator at later times t, usually noticeable as an increase of the entropy. There is a rich literature about this so-called *decoherence*, and many ways of modeling the environment and the system-environment interactions have been proposed and studied (a simple example of "phase decoherence" appears in Sect. 11.1). When a suitable model is adopted, the resulting modification of the von Neumann equation (23) can give quite an accurate account of the loss in precision of our predictions about the future properties of the quantum system.

10 No murky interpretation

The introduction of a recent paper contains the following sentences:

> "[...] a century after the discovery of quantum mechanics, it seems that we are no closer to a consensus about its interpretation than we were in the beginning. The collapse of the quantum state occurring during the process of measurement [...] does not have an unambiguous definition and a reasonable explanation. [...] some radical changes in our classical understanding of reality have to be made; e.g. constructing a physical process of collapse, accepting the existence of parallel worlds, or adding nonlocal hidden variables." (63)

Of course, the authors of these lines use Born's rule for the calculation of probabilities, associate the wavelengths of spectral lines with differences of energy eigenvalues, and use all other standard links between the mathematical formalism and the physical phenomena. In other words, these authors' interpretation of quantum theory is the usual one[22]. Why, then, is there this blunt assertion that "there is no consensus about the interpretation"?

Clearly, the authors of (63) have a nonstandard meaning of the word "interpretation" in mind, and so have others[23]. The actual interpretation, that is: The battle-tested link between the formalism and the phenomena, is not what they are questioning. They would refer to that as the "minimal interpretation" and that is not enough for them. They want to know "what is really going on", thereby requiring answers to questions that quantum theory cannot provide.

These are questions about the *mind-body problem*, the relation between our conscious human experiences and our free will on one side, and the reality of natural phenomena on the other – philosophical issues that are outside of physics. Yes, it is important to understand the philosophical implications of the lessons of quantum physics, but one must not confuse physics with philosophy.

The "lack of consensus" noted in (63) is not about the interpretation of quantum theory in the strict sense of Section 1, but about its bearing on philosophy. The use of the term "interpretation" for the more or less successful reconciliation of philosophical preconceptions with the quantum-physical phenomena is most unfortunate, and there is little hope that this practice will change. Fuchs's and Peres's verdict *Quantum Theory Needs No 'Interpretation'* (the title of their essay [42]) should be understood in this context: Quantum theory has its interpretation and does not need a philosophical *Überbau*.

Of course, philosophers should be encouraged to study the lessons of quantum theory – for example, the lesson that the future is not predetermined by the past[24] – but their philosophical debates are really irrelevant for quantum theory as a physical theory, although they are likely to be of great interest to physicists. We could leave it at that, were there not the widespread habit of the debaters to endow the mathematical symbols of the formalism with more meaning than they have. In particular, there is a shared desire to regard the Schrödinger wave function as a physical object itself after forgetting, or refusing to accept, that it is merely a mathematical tool that we use for a description of the physical object (electron, atom, photon, ...). The regrettable textbook practice of invoking analogies between Schrödinger's probability-amplitude waves and sound waves, water-surface waves, or electromagnetic waves provokes and feeds that desire. The fundamental difference is then ignored: Sound waves, water waves, electromagnetic waves are physical objects with mechanical properties, they carry energy, momentum, angular momentum from here to there, whereas the Schrödinger waves are mathematical symbols[25].

The reification of the wave function requires that the "collapse of the quantum state" is a physical process, but since it isn't, there cannot be "an unambiguous definition", and a "reasonable explanation" is not possible. We submit: With such incorrect meaning given to

[22] It is simply *the* interpretation of quantum theory. Calling it the standard interpretation or using any other identifier is unnecessary and, indeed, harmful because "this sounds as if there were several interpretations of quantum mechanics. There is only one". (Peierls as quoted by Davies and Brown [38].)

[23] The *Compendium of Quantum Physics* [1] has entries on more than ten "interpretations", and two recent polls [39,40] gave about that many "interpretations" to choose from, while Tegmark's poll of 1998 [41] had only half as many choices.

[24] However the next photon is prepared, there is no way of knowing which of the four detectors in the experiment of Figure 1 will click.

[25] Also, only very special cases of wave functions are probability-amplitude waves in the three-dimensional space of physical positions.

the wave function and to the process of state reduction, one no longer deals with quantum theory, but with something else. The conclusions that are then reached about philosophical implications are of no consequence.

Indeed, the alleged implications ("parallel worlds") are bewildering and prompt us to paraphrase van Kampen's Theorem IV [43]: If you endow the mathematical symbols with more meaning than they have, you yourself are responsible for the consequences, and you must not blame quantum theory when you get into dire straits.

The interpretation of quantum theory is not murky, it is absolutely clear; there is no lack of consensus about it. It is one of the three defining constituents of quantum theory (Sect. 1), and all practicing quantum physicists rely on it all the time.

11 You cannot see what you wouldn't recognize

11.1 No measurement problem

We can use the set-up of Figure 1, see Sections 3.1 and 3.2, for a discussion of the so-called "measurement problem". A typical argument would develop along the following lines. At the initial time t_{ini}, we have the photon approaching the BS with a polarization-state ket

$$|\wp\rangle = |\mathsf{V}\rangle\alpha + |\mathsf{H}\rangle\beta \quad \text{with} \quad |\alpha|^2 + |\beta|^2 = 1, \tag{64}$$

and the four detectors are in the "ready" state

$$|\mathrm{R}\rangle = |\mathrm{R}_1\mathrm{R}_2\mathrm{R}_3\mathrm{R}_4\rangle, \tag{65}$$

so that we write

$$|\,\rangle = |\wp\,\mathrm{R}, t_{\text{ini}}\rangle \tag{66}$$

for the state ket as characterized by properties referring to the initial time. Expressed with reference to the final time t_{fin}, this state ket is

$$|\,\rangle = |\mathrm{VAC\ NC}, t_{\text{fin}}\rangle\sqrt{p_0} + \sum_{k=1}^{4} |\mathrm{VAC\ C}_k, t_{\text{fin}}\rangle\sqrt{p_k}, \tag{67}$$

where C_k stands for "the kth detector has clicked and the other three have not" and NC denotes the no-click situation. At the final time, the photon has been absorbed and we have the photon vacuum state for all detector states, which allows us to focus on the final state of the detectors,

$$|\,\rangle = |\mathrm{VAC\ D}, t_{\text{fin}}\rangle \tag{68}$$

with

$$|\mathrm{D}\rangle = |\mathrm{NC}\rangle\sqrt{p_0} + \sum_{k=1}^{4} |\mathrm{C}_k\rangle\sqrt{p_k}. \tag{69}$$

This is the point where some start to worry: The detectors are in a superposition of this one having clicked or that one having clicked, then how does one account for a definite click of one of the detectors and why do we never see superpositions such as $\big(|\mathrm{C}_1\rangle + |\mathrm{C}_2\rangle\big)/\sqrt{2}$ with the two detectors having half-clicked?

Although these questions seem to be addressing profound issues, they are not formulating a serious "measurement problem". At best, they are rephrasing matters already discussed.

First, we note that the state ket $|\,\rangle$ in (68) is the same as the one in (66). The two versions of $|\,\rangle$ are particular examples of the general statement in (34). Both versions of $|\,\rangle$ encode what we know about the physical situation. Why, then, are all those questions asked about (68), but none about (66)?

Second, the superposition $\big(|\mathrm{C}_1\rangle + |\mathrm{C}_2\rangle\big)/\sqrt{2}$ does not describe a situation where detectors D_1 and D_2 each have "half-clicked" (whatever that would mean) but an entirely new situation that is outside our experience. Remember that $\big(|\mathsf{V}\rangle + |\mathsf{H}\rangle\mathrm{i}\big)/\sqrt{2}$ is not a polarization state that is "half-vertical" and "half-horizontal" but one that is entirely right-circular, phenomenologically very different from both $|\mathsf{V}\rangle$ and $|\mathsf{H}\rangle$. We know what $\big(|\mathsf{V}\rangle+|\mathsf{H}\rangle\mathrm{e}^{\mathrm{i}\varphi}\big)/\sqrt{2}$ means (as a statement about a photon's polarization) for any value of the phase φ, but we have no clue about $\big(|\mathrm{C}_1\rangle+|\mathrm{C}_2\rangle\mathrm{e}^{\mathrm{i}\varphi}\big)/\sqrt{2}$. So, we never see these superpositions because we cannot recognize them[26]. If we could, either version of $|\,\rangle$ would enable us to calculate the probability of seeing these superpositions.

Third, despite strong assertions that it does, decoherence does not come to the rescue. The asserters consider the statistical operator associated with ket $|\mathrm{D}\rangle$,

$$\begin{aligned}\rho_{\mathrm{D}} &= |\mathrm{D}\rangle\langle\mathrm{D}| \\ &= |\mathrm{NC}\rangle p_0\langle\mathrm{NC}| + \sum_{k=1}^{4}|\mathrm{C}_k\rangle p_k\langle\mathrm{C}_k| + \{\text{cross terms}\},\end{aligned} \tag{70}$$

where the cross terms collect all ket-bra products of two different outcomes,

$$\begin{aligned}\{\text{cross terms}\} &= \sum_{k=1}^{4}\big(|\mathrm{NC}\rangle\sqrt{p_0p_k}\langle\mathrm{C}_k| + |\mathrm{C}_k\rangle\sqrt{p_0p_k}\langle\mathrm{NC}|\big) \\ &\quad + \sum_{j\neq k}|\mathrm{C}_j\rangle\sqrt{p_jp_k}\langle\mathrm{C}_k|.\end{aligned} \tag{71}$$

At this point, one remembers that $|\mathrm{C}_k\rangle$ symbolizes a state of very many quantum degrees of freedom (all those nuclei and electrons that make up the optical elements and the detectors) and, therefore, one is suddenly unsure about details such as the over-all phase assigned to $|\mathrm{C}_k\rangle$, and so it appears natural to refine the quantum-theoretical description by the replacements $|\mathrm{C}_k\rangle \to |\mathrm{C}_k\rangle\mathrm{e}^{\mathrm{i}\phi_k}$ with random phases ϕ_k. Here, "random" simply means that the

[26] He who interjects "But, in principle, I could certainly measure the observables described by the hermitian operators $|\mathrm{C}_1\rangle\mathrm{e}^{\mathrm{i}\varphi}\langle\mathrm{C}_2| + |\mathrm{C}_2\rangle\mathrm{e}^{-\mathrm{i}\varphi}\langle\mathrm{C}_1|$, and when the eigenvalue $+1$ is found, I will have succeeded." should reread footnote 17 and then ask himself *Can I do it in practice?*

phases acquire different unknown values whenever the experiment is repeated. Averaging over the random phases removes the cross terms,

$$\{\text{cross terms}\} \to 0, \tag{72}$$

and we arrive at an effective statistical operator for the detectors,

$$\rho_{\mathrm{D}} \to \rho_{\mathrm{D,\,eff}} = |\mathrm{NC}\rangle p_0 \langle \mathrm{NC}| + \sum_{k=1}^{4} |\mathrm{C}_k\rangle p_k \langle \mathrm{C}_k|. \tag{73}$$

The adjective "effective" means that all observable properties calculated with $\rho_{\mathrm{D,eff}}$ are the same as those calculated with ρ_{D}. In other words, the cross terms of (70) are ineffective, they are of no consequence. This, of course, is not a recent insight. There are, for example, the accounts by Süssmann [14] and Peres [44][27] of 1958 and 1980, respectively.

So, ρ_{D} "has decohered" into $\rho_{\mathrm{D,eff}}$ – the transition (73) is now regarded as a physical process! – and we have a mixed state that is as if we had a Gibbs ensemble composed of a fraction p_0 of no-click cases, a fraction p_1 of clicks by detector D_1, ..., a fraction p_4 of clicks by detector D_4. One of the situations is the case, and we do not know which one until we take note of the outcome of the experiment. All is fine, or so it seems, because we just have this "classical ignorance" rather than the quantum-theoretical indefiniteness of ρ_{D}, and we no longer need to worry why we do not see the detectors with properties described by superpositions such as $\big(|\mathrm{C}_1\rangle + |\mathrm{C}_2\rangle\big)/\sqrt{2}$.

All is fine? No, we are fooling ourselves. Remembering Section 4.1, we note that the effective statistical operator does not have this one unique as-if reality associated with it, but very many of them. For example, we have

$$\begin{aligned}\rho_{\mathrm{D,eff}} &= |\mathrm{NC}\rangle p_0 \langle \mathrm{NC}| + |\mathrm{C}_1\rangle p_1 \langle \mathrm{C}_1| + |\mathrm{C}_2\rangle p_2 \langle \mathrm{C}_2| \\ &\quad + |\mathrm{C}_{34+}\rangle \frac{p_3+p_4}{2} \langle \mathrm{C}_{34+}| \\ &\quad + |\mathrm{C}_{34-}\rangle \frac{p_3+p_4}{2} \langle \mathrm{C}_{34-}| \end{aligned} \tag{74}$$

with

$$|\mathrm{C}_{34\pm}\rangle = |\mathrm{C}_3\rangle \sqrt{\frac{p_3}{p_3+p_4}} \pm |\mathrm{C}_4\rangle \sqrt{\frac{p_4}{p_3+p_4}} \tag{75}$$

and, clearly, the mixture described by $\rho_{\mathrm{D,eff}}$ can also be blended by p_0 parts of no-click cases, p_1 parts of clicks by D_1, p_2 parts of clicks by D_2, plus $\frac{1}{2}(p_3+p_4)$ parts each for the situations (of unknown phenomenology) that are described by the superpositions $|\mathrm{C}_{34+}\rangle$ and $|\mathrm{C}_{34-}\rangle$. As always, there is an infinity of blends for the one mixture and a corresponding infinity of as-if realities.

Another reason why we are fooling ourselves is that we still need the principle of random realization for selecting one of the five possibilities and Born's rule for calculating the probabilities from $\rho_{\mathrm{D,eff}}$. It is not the case, as self-suggesting this may appear, that we can just read off the probabilities by looking at the weights of our preferred as-if reality. Such a reading-off can sometimes give the right answer, but we know that it does only after an application of Born's rule has confirmed our educated guess.

Fourth, the transition (73) is an improvement of our description, it is not a physical process. The decoherence argument with the unknown phase factors $\mathrm{e}^{\mathrm{i}\phi_k}$ and so forth is a confession that we don't know the dynamics with sufficient precision, and our starting point (66) is wrong, too. We do not have enough information for describing the initial "ready" state of the detectors by a state ket $|\mathrm{R}\rangle$ or its corresponding rank-one statistical operator $|\mathrm{R}\rangle\langle\mathrm{R}|$. We can, at most, have limited knowledge about a few degrees of freedom of the detectors and account for that by a highly mixed state. The state kets of (66), (68), and (69) border on utter nonsense.

When preparing the experiment and getting the detectors into the "ready" state, we are not controlling all quantum degrees of freedom of the detectors perfectly but, at best, the few relevant ones with some limited precision. There are, therefore, very many pure quantum states that we could imagine for the "ready" detectors rather than a unique state associated with the ket $|\mathrm{R}\rangle$. Strictly speaking, there is not even the pure-polarization ket $|\wp\rangle$, but that is acceptable as a pretty good approximation in a well-controlled experiment.

Rather than the convex sum of projectors in (73) and (74), we have, at best, a convex sum of highly mixed statistical operators for the detectors,

$$\rho_{\mathrm{D}} = p_0 \rho_{\mathrm{D}}^{(\mathrm{NC})} + \sum_{k=1}^{4} p_k \rho_{\mathrm{D}}^{(k)}, \tag{76}$$

where $\rho_{\mathrm{D}}^{(k)}$ is for a click of the kth detector and $\rho_{\mathrm{D}}^{(\mathrm{NC})}$ for the no-click case. But this ρ_{D}, too, is more an illusion than a reality. Even the smallest detectors contain a gargantuan number of atoms, and we don't really know what exactly the detectors are composed of – removing a single molecule, say, from the surface of one of the wires in the apparatus has no bearing on the relevant aspects of the physical situation. Therefore, we will be hard-pressed to write down meaningful expressions for the $\rho_{\mathrm{D}}^{(k)}$s.[28]

This is not a problem, though, because there is no use of the quantum description of the detectors that have clicked or not. There are no probabilities of subsequent events to be calculated from ρ_{D}. The photon-absorption event has been amplified or not, one of the detectors has clicked or the photon has escaped detection: The measurement of this photon's polarization is over.

Fifth, since neither decoherence nor any other mechanism select one particular outcome (see Sect. 8), the whole "measurement problem" reduces to the question *Why is there one specific outcome?* which is asking *Why are there randomly realized events?* in the particular context considered. This harkens back to Section 1, where we noted that quantum theory cannot give an answer.

[27] Reprinted in [11].

[28] He who is still pondering the question in footnote 26, may find a useful hint here.

In summary, then, the alleged "measurement problem" does not exist as a problem of quantum theory. Those who want to pursue the question *Why are there events?* must seek the answer elsewhere.

11.2 Schrödinger's cat

In this context one can hardly avoid the mentioning of Schrödinger's infamous cat[29]. When commenting on the physical significance of the quantum-theoretical wave function – it is an *expectation catalog* (*Katalog der Erwartung*) in Schrödinger's own words – he gave, as a warning, the "ridiculous" (*burlesk*) example of a cat whose vital state serves as the indicator whether a radioactive decay has occurred or not. That passage reads

> "One can even set up quite ridiculous cases. A cat is penned up in a steel chamber, along with the following diabolical device (which must be secured against direct interference by the cat): in a Geiger counter there is a tiny bit of radioactive substance, *so* small, that *perhaps* in the course of one hour one of the atoms decays, but also, with equal probability, perhaps none; if it happens, the counter tube discharges and through a relay releases a hammer which shatters a small flask of hydrocyanic acid. If one has left this entire system to itself for an hour, one would say that the cat still lives *if* meanwhile no atom has decayed. The first atomic decay would have poisoned it. The ψ-function of the entire system would express this by having in it the living and the dead cat (pardon the expression) mixed or smeared out in equal parts." (77)

in Trimmer's translation [25–28]. In other words, the diabolic device establishes the strong correlations

$$\begin{aligned} \text{no decay} &\longleftrightarrow \text{live cat}, \\ \text{yes decay} &\longleftrightarrow \text{dead cat}, \end{aligned} \tag{78}$$

so that the cat is a *bona fide* detector for the radioactive decay event. It has the obvious advantage that the detector is easy to read.

Schrödinger is, of course, overshooting the mark. Just like there is no ket $|\mathrm{D}\rangle$ for the detectors in Section 11.1, there is no ψ-function that has "the living and dead cat smeared out". The imagined superpositions of the dead and the live cat, which have inspired so many authors, are fantastic phantoms of the same kind as the detector states associated with the kets $\bigl(|\mathrm{C}_1\rangle + |\mathrm{C}_2\rangle e^{i\varphi}\bigr)/\sqrt{2}$. We do not know how to recognize them and, therefore, we cannot see them. In all of that, the cat is really superfluous, it just adds drama to the plot; the position of the hammer, say, already tells the story.

It would be fitting if any such correlated state of a quantum degree of freedom and a macroscopic detector would be called a *Schrödinger-cat state* in today's jargon. But that is not the case. The term is routinely applied to superpositions of states of a few quantum degrees of freedom, or even of a single one, that have vastly different quantum numbers. In this sense, the silver atom emerging from a Stern-Gerlach apparatus is in a Schrödinger-cat state. This usage of the term does not do justice to what Schrödinger had in mind; it should be discouraged, the more so because such superpositions are nothing special, they are common in standard interferometric devices.

[29] Heisenberg's dog [45] does not concern us here.

12 Summary

The foundations of quantum theory have been laid, and laid well, by the founding fathers – Planck, Einstein, Bohr, Heisenberg, Schrödinger, Born, Dirac, and others. Much has happened since those early days of the 1920s. The phenomenology and the formalism are much richer today and, to supplement this enrichment, the interpretation has more links between the formalism and the phenomena. This is a natural consequence of our communal duty to explore the consequences, implications, and applications of quantum theory, and we are still in the process of doing just that.

It is true that, occasionally, the founding fathers and their peers found it difficult to reconcile the lessons of quantum theory with the world view they had acquired earlier. The younger players, notably Heisenberg and Dirac, overcame these difficulties more quickly and more easily than their seniors, Planck and Einstein among them. Einstein, for example, did not like the random nature of events. He told Franck the following[30]:

> "I can, if the worse comes to worst, still realize that God may have created a world in which there are no natural laws. In short, chaos. But that there should be statistical laws with definite solutions, i.e., laws that compel God to throw dice in each individual case, I find highly disagreeable." (79)

While Einstein's metaphor of the dice-throwing God is charming and memorable, and his personal difficulties are of great historical interest, it is ultimately irrelevant what he, or any other individual, finds disagreeable. The facts count – and Einstein, of course, understood this. In science, truth is the daughter of time, not of authority.

One must also acknowledge that, after the lapse of fourscore years, the terminology is more precise today than it was during the founding years. Concepts get clearer in time, and one learns to avoid sloppy terms and misleading phrases. The foundations, however, have not been touched by these refinements.

As way of summary, here are our answers to the questions asked at the beginning:

- Yes, quantum theory well defined.
- Yes, quantum theory has a clear interpretation.

[30] As quoted by Snow in [46].

- Yes, quantum theory is a local theory.
- No, quantum evolution is not reversible.
- No, wave functions do not collapse; you reduce your state.
- No, there is no instant action at a distance.
- Heisenberg's cut is where you put it.
- No, Schrödinger's cat is not half dead and half alive.
- No, there is no "measurement problem".

Tersely: Quantum theory is a well-defined local theory with a clear interpretation. No "measurement problem" or any other foundational matters are waiting to be settled.

What, then, about the steady stream of publications that offer solutions for alleged fundamental problems, each of them wrongly identified on the basis of one misunderstanding of quantum theory or another? Well, one could be annoyed by that and join van Kampen [47] in calling it a scandal when a respectable journal prints yet another such article. No-one, however, is advocating censorship, even of the mildest kind, because the scientific debate cannot tolerate it. Yet, is it not saddening that so much of diligent effort is wasted on studying pseudo-problems?

The Centre for Quantum Technologies is a Research Centre of Excellence funded by the Ministry of Education and the National Research Foundation of Singapore.

References

1. *Compendium of Quantum Physics: Concepts, Experiments, History and Philosophy*, edited by D. Greenberger, K. Hentschel, F. Weinert (Springer, Heidelberg, 2009)
2. I. Newton, *Philosophiae Naturalis Principia Mathematica* (Streator, London, 1687)
3. J.C. Maxwell, *A Treatise on Electricity and Magnetism* (Clarendon Press, Oxford, 1873)
4. M. Born, Z. Phys. **37**, 863 (1926)
5. W. Gerlach, O. Stern, Z. Phys. **9**, 353 (1922)
6. B. Friedrich, D. Herschbach, Phys. Today **56**, 53 (2003)
7. J. Bernstein, `arXiv:1007.2435[physics.hist-ph]`
8. R. Haag, Commun. Math. Phys. **132**, 245 (1990)
9. R. Haag, Found. Phys. **43**, 1295 (2013)
10. A. Einstein, B. Podolsky, N. Rosen, Phys. Rev. **47**, 777 (1935)
11. *Quantum Theory and Measurement*, edited by J.A. Wheeler, W.H. Zurek (Princeton University Press, Princeton, 1983)
12. N.H.D. Bohr, Nature **136**, 65 (1935)
13. N.H.D. Bohr, Phys. Rev. **48**, 696 (1935)
14. G. Süßmann, Abh. Bayer. Akad. Wiss. **88** (1958)
15. N. Gisin, Helv. Phys. Acta **62**, 363 (1989)
16. N. Gisin, Phys. Lett. A **143**, 1 (1990)
17. B.-G. Englert, Z. Naturforsch. **52a**, 13 (1997)
18. J. Schwinger, M.O. Scully, B.-G. Englert, Z. Phys. D **10**, 135 (1988)
19. B.-G. Englert, M.O. Scully, J. Schwinger, Found. Phys. **18**, 1045 (1988)
20. K. Kraus, *States, Effects, and Operations: Fundamental Notions of Quantum Theory*, volume 190 of Lecture Notes in Physics, edited by A. Böhm, J.D. Dollard, W.K. Wootters (Springer, Heidelberg, 1983)
21. D. Bohm, *Quantum Theory* (Prentice Hall, New York, 1951)
22. A. Buchleitner, K. Hornberger, *Coherent Evolution in Noisy Environments*, volume 611 of Lecture Notes in Physics (Springer, Heidelberg, 2002)
23. J.S. Bell, Physics **1**, 195 (1964)
24. J.F. Clauser, M.A. Horne, A. Shimony, R.A. Holt, Phys. Rev. Lett. **23**, 880 (1969)
25. E. Schrödinger, Naturwissenschaften **23**, 807 (1935)
26. E. Schrödinger, Naturwissenschaften **23**, 823 (1935)
27. E. Schrödinger, Naturwissenschaften **23**, 844 (1935)
28. J.D. Trimmer, Proc. Am. Philos. Soc. **124**, 323 (1980)
29. A.K. Ekert, Phys. Rev. Lett. **67**, 661 (1991)
30. A. Ling, M.P. Peloso, I. Marcikic, V. Scarani, A. Lamas-Linares, C. Kurtsiefer, Phys. Rev. A **78**, 020301 (2008)
31. R. Haag, *Local Quantum Physics* (Springer, Heidelberg, 1996)
32. S. Popescu, D. Rohrlich, Found. Phys. **24**, 379 (1994)
33. J. Bell, Phys. World **3**, 33 (1990)
34. N.G. van Kampen, Phys. World **3**, 20 (1990)
35. N.G. van Kampen, Phys. World **4**, 16 (1991)
36. R. Peierls, Phys. World **4**, 19 (1991)
37. K. Gottfried, Phys. World **4**, 34 (1991)
38. P.C.W. Davies, J.R. Brown, *The Ghost in the Atom: A discussion of the Mysteries of Quantum Physics* (Cambridge University Press, Cambridge, 1986)
39. M. Schlosshauer, J. Kofler, A. Zeilinger, `arXiv:1301.1069[quantph]`
40. C. Sommer, `arXiv:1303.2719[quantph]`
41. M. Tegmark, Fortschr. Phys. **46**, 855 (1998)
42. C.A. Fuchs, A. Peres, Phys. Today **53**, 70 (2000)
43. N.G. van Kampen, Physica A **153**, 97 (1988)
44. A. Peres, Phys. Rev. D **22**, 879 (1980)
45. J.A. Bergou, B.-G. Englert, J. Mod. Opt. **45**, 701 (1998)
46. A.P. French, *Einstein: A Centenary Volume* (Harvard University Press, Cambridge, 1979)
47. N.G. van Kampen, Am. J. Phys. **76**, 989 (2008)

Ulrich Mohrhoff

Sri Aurobindo International Centre of Education, Pondicherry, India

Paper Title: **My two fateful encounters with Berge (so far)**

Reference: Original article (2014)

My two fateful encounters with Berge (so far)

Feb 11, 2014

My first published paper [1], which appeared in the *American Journal of Physics* in 1996, was titled "Restoration of interference and the fallacy of delayed choice: Concerning an experiment proposed by Englert, Scully and Walther." Not long after its publication, while I was thinking about a response to David Mermin's article "What is quantum mechanics trying to tell us?" [2], which became my article "What quantum mechanics is trying to tell us" [3], I realized that I had made a mistake. I hadn't heeded Nico van Kampen's warning that "whoever endows the wave function with more meaning than is needed for computing observable phenomena is responsible for the consequences" [4].

In this case the first consequence was that I was asked by the then Editor of *American Journal of Physics* to review an article by Englert, Scully, and Walther that was intended "to set the record straight." Subsequently Berge, the corresponding author, and I were asked to prepare two papers to be published back to back, and to get back to the editorial offices of the journal only after we had seen our final versions and had no further comments to make. Thence ensued a very friendly email correspondence between Berge and myself, which resulted in ESW's "Quantum erasure in double-slit interferometers with which-way detectors" [5] and my "Objectivity, retrocausation, and the experiment of Englert, Scully, and Walther" [6].

We agreed more or less on what was wrong, but we couldn't quite agree on what was right. On one hand, Berge *et al.* would insist that when the experimenters find the photon in the first resonator, "they *know* that the atom went through the first slit.... The photon stored in the resonators supplies genuine WW [=which-way] knowledge." On the other hand, they would insist that saying "this atom went through the first slit" is "not a statement about the atom's past.... the properties 'went (or will go) through the first slit' or 'went (or will go) through the second slit' are not actual, but only potential, until a measurement result establishes which one is the case." They resolved this apparent contradiction by pointing out that "physics is always about as-if-realities. Physics deals with the phenomena that we perceive and which we can communicate about. An imagined reality beyond our perception... is the business of philosophers."

Being a philosopher at heart, I would insist that finding the photon in the first resonator does warrant the claim that the atom *really* went through the first slit. At the same time I fully agreed with Berge *et al.* that the following assumption "flies in the face of the fundamental quantum-mechanical indeterminacy": "'went (or will go) through the first slit' is an objective property possessed by the atom to begin with..., a property that we are merely recognizing when we find the photon in the first resonator." I resolved the apparent contradiction by pointing out that this is not what I mean by saying that the atom *really* went through the first slit. Finding the photon in the first resonator does not warrant the claim that the atom went through the first slit *and that it would have done so even if the photon had not been detected.* The atom went through the first slit *because* the photon was found in the first resonator. The way I see it, properties in the quantum domain are possessed if and only if their possession is indicated by an actual event or state of affairs. In the quantum domain, to *be* is to be *measured*. If what happens there depends on what happens or is the case in the macroworld, what happens in the quantum domain at an earlier time can depend on what happens in the macroworld at a later time. The implicit claim is not that we can change the past but that we can contribute to creating a past where there was none.

I think that both attitudes are legitimate, except that my use of the term "retrocausation" was unfortunate. If a property is not actual until a measurement result establishes that it is (or was) possessed, a more appropriate term would be "retroactualisation."

I also believe that humankind's quest for knowledge has two important aspects. One is lab-oriented. It has produced amazing feats of ingenuity and the most unexpected experimental data. The other is armchair-oriented. It seeks to situate the experimental data in a larger narrative, one that can deal with such humanly important issues as the reality of qualities and values, which cannot be addressed by the quantitative and value-free methods of empirical science.

This quest, I came to realize, can hardly succeed without abandoning that classical sleight of hand which consists in the reification of calculational tools. The fundamental theoretical framework of physics is a probability calculus, and the events to which (and on the basis of which) it serves to assign probabilities, are measurement outcomes. We now have enough no-go theorems to conclude that there is no physical mechanism or natural process by which the correlations between measurement outcomes can be explained. If there is something that is responsible for the existence of the universe, and if, as I believe, its creative power is unlimited, then this should not come as a surprise, for it would be self-contradictory to explain the working of an infinite power in terms of physical mechanisms or natural processes. If this works under constraints, we need to know to what end, and we need to know why the constraints have the particular form that they do. The answer to the second question will of course depend on the answer to the first.

In 2002, I published a paper [7] in which I argued that the Standard Model plus General Relativity are preconditions of the possibility of an "interesting" world, defined (by E. Squires [8]) as one that contains chemistry, and suggested that the validity of these theories may be required for the very consistency of quantum mechanics: Quantum mechanics presupposes the macroscopic events which it correlates; the existence of these events requires a sufficient variety of chemical elements; and this implies the validity of the Standard Model and General Relativity, at least as effective theories. In another paper [9], published in the *International Journal of Quantum Information*, I derived quantum mechanics from the existence of reasonably stable objects that "occupy space" while being composed of finite numbers of objects that do not "occupy space." I suspect it was mostly this paper, and the criticism it contained of the standard axiomatic approach, that led to my second encounter with Berge. (Axioms are supposed to be clear and compelling. The standard axioms of quantum mechanics are anything but.)

In November 2008, I received an email from a Senior Editor of World Scientific Publishing, informing me that they are

> "looking for suitable author(s) to write book(s) in the field of quantum mechanics. We were very impressed with your extensive publication and your strong background in this field. Prof. Berthold-Georg Englert, National University of Singapore, Managing Editor of the International Journal of Quantum Information, also recommended to us that you might be the right person for this book project."

Not bad for a university dropout who has not seen the inside of a physics laboratory since his student days at the University of Göttingen in the mid-Seventies! This is how, in 2011, I came to publish a textbook with the preposterous title *The World According to Quantum Mechanics: Why the Laws of Physics Make Perfect Sense After All.* It is based on a philosophically oriented optional course of quantum physics, which I have offered since year 2000 to higher secondary and undergraduate students at the Sri Aurobindo International Centre of Education in Pondicherry. So far the book has copped three reviews at Amazon.com, all unsolicited and five-star. It is probably the only book with a five-star review that says, "You will hate this book".[1] Thank you, Berge, and Happy Birthday to you!

Bibliography

[1] U. Mohrhoff, *Restoration of interference and the fallacy of delayed choice: Concerning an experiment proposed by Englert, Scully and Walther*, Am. J. Phys. **64**, 1468–1475 (1996).
[2] N. D. Mermin, *What is quantum mechanics trying to tell us?*, Am. J. Phys. **66**, 753–767 (1998).
[3] U. Mohrhoff, *What quantum mechanics is trying to tell us*, Am. J. Phys. **68**, 728–745 (2000).
[4] N. van Kampen, *Ten theorems about quantum mechanical measurements*, Physica A **153**, 97–113 (1988).
[5] B.-G. Englert, M. O. Scully, and H. Walther, *Quantum erasure in double-slit interferometers with which-way detectors*, Am. J. Phys. **67**, 325–329 (1999).
[6] U. Mohrhoff, *Objectivity, retrocausation, and the experiment of Englert, Scully, and Walther*, Am. J. Phys. **67**, 330–335 (1999).
[7] U. Mohrhoff, *Why the laws of physics are just so*, Found. Phys. **32**, 1313–1324 (2002).
[8] E. Squires, *Do we live in the simplest possible interesting world?*, Eur. J. Phys. **2**, 55–57 (1981).
[9] U. Mohrhoff, *Quantum mechanics explained*, Int. J. Quant. Info. **7**, 435–458 (2009).

[1] Excerpt: "The way this book covers the two-slit experiment, everything falls into place and makes perfect sense. There is no wave particle dualism, just the naked necessity of a probabilistic regime. It is so simple. Painfully obvious. Easy to grasp with just a minimum of mathematical rigor. It boggles the mind that QM has not been understood this way from the get go. This feels like 20/20 hindsight writ large. ... If you've been trying to make sense of QM you will hate this book. It'll make you feel stupid for not having been able to see this all along. Time to eat some humble pie."

Chapter 5

Cold Atoms

Kazimierz Rzążewski

Center for Theoretical Physics of the Polish Academy of Sciences, Warsaw, Poland
Physikalisches Institut, Universität Stuttgart, Stuttgart, Germany

Paper Title: **Two cold atoms in a harmonic trap**

Reference: Foundations of Physics **28**, 549–559 (1998)

Personal message:

It was 1998. I was enjoying my Humboldt Award in Konstanz. A young Masters degree student, Thomas Busch, was struggling with a problem of two atoms confined in a harmonic trap, a problem posed by his advisor Martin Wilkens. I explained to him the need to use the Fermi pseudopotential as a mathematically correct description of the contact interaction in three dimensions. What followed was my convoluted solution of the spectrum for this problem. I told Berge all the details and in a few days he came up with the most elegant and simple solution, correcting my mistake. The resulting paper has been cited more than 260 times. More importantly, a delight for every theoretician: Tilman Esslinger from ETH in Zurich has confirmed our result in an ingenious experiment.

Foundations of Physics, Vol. 28, No. 4, 1998

Two Cold Atoms in a Harmonic Trap[1]

Thomas Busch,[2] Berthold-Georg Englert,[3] Kazimierz Rzążewski,[4] and Martin Wilkens[2]

Received January 21, 1997

Two ultracold atoms moving in a trap interact weakly at a very short distance. This interaction can be modeled by a properly regularized contact potential. We solve the corresponding time-independent Schrödinger equation under the assumption of a parabolic, spherically symmetric trapping potential.

1. INTRODUCTION

The interaction of ultracold atoms in a trap is a topic of intense research with applications ranging from collective phenomena in optical lattices[(1)] to the recently observed Bose–Einstein condensation of alkaline vapors.[(2)] The distinct feature which makes the interaction in these systems differ from the related scenario of ultracold collisions in free space is the presence of the trapping potential. Whereas in free space the pair interaction is probed via differential cross sections and collisional phase shifts, for example, these quantities lose meaning in the presence of a trapping potential since the confinement does not admit the construction of asymptotic in and out states. Instead, in a trap the interacting many-atom system is characterized by a discrete energy spectrum and possibly also by transition matrix elements, similar to the situation in molecular physics.

The many-body problem of trapped atoms is rather complicated, in particular when using a realistic atom–atom interaction. To avoid this

[1] Dedicated to the memory of Professor Asim O. Barut.

[2] Institut für Physik, Universität Potsdam, Am Neuen Palais 10, 14469 Potsdam, Germany.

[3] Sektion Physik, University of Munich, Theresienstr. 37, 80333 München, Germany; presently at Atominstitut der Österreichischen Universitäten, Stadionallee 2, 1020 Wien, Austria.

[4] Center for Theoretical Physics and College of Science, Polish Academy of Sciences, Al. Lotników 32/46, 02-668 Warsaw, Poland.

difficulty one commonly employs the shape-independent approximation in which the physical interaction is replaced by a pointlike potential of zero range.[5] For ultracold atoms such a replacement is well justified since their de Broglie wavelength is so large that the finer details of the physical interaction potential do not matter.

Whereas in one spatial dimension a zero-range potential is easily modeled by a simple δ-function,[(4)] in three dimensions the frequently employed $\delta^{(3)}(\mathbf{r})$ function must be treated with great care because the corresponding Hamiltonian is not self-adjoint.[(5)] In particular, any attempt is void which would aim at an exact diagonalization of such a Hamiltonian, since it would only reproduce the trivialities of the noninteracting theory.[(6)]

This notwithstanding we shall here diagonalize exactly the problem of two particles in a three-dimensional trap which interact via a pointlike potential. However, in order to ensure the self-adjointness of the corresponding Hamiltonian we shall use the so-called regularized δ-function $\delta^{(3)}_{reg}(\mathbf{r}) \equiv \delta^{(3)}(\mathbf{r})(\partial/\partial r)\, r$ to model the atom–atom pair interaction.[6] Within the domain of square integrable functions that are regular at the origin, this distribution acts in the same manner as the pure δ-function. The domain of a Hamiltonian of the general structure $-\nabla^2 + \delta^{(3)}_{reg}(\mathbf{r})$ is slightly larger though, as it also includes functions which diverge $\sim r^{-1}$ for $r \to 0$.[7] It is worth remembering that the regularized δ-potential has also been used in other contexts; let us just mention two historic papers: the work of Breit and Zilsel in nuclear physics[(10)] and Bersons' study[(11)] of the ionization of negative ions.[8]

2. THREE-DIMENSIONAL TRAP

A. Schrödinger equation

Assuming now an external isotropic harmonic oscillator potential, we use the following Hamiltonian to describe the motion of two atoms that interact with each other via a pointlike force:

$$H = -\tfrac{1}{2}\nabla_1^2 - \tfrac{1}{2}\nabla_2^2 + \tfrac{1}{2}r_1^2 + \tfrac{1}{2}r_2^2 + 4\pi a_0\, \delta^{(3)}_{reg}(\mathbf{r}_1 - \mathbf{r}_2) \tag{1}$$

[5] For a mathematical theory of pointlike interactions see Ref. 3.

[6] This regularization has been introduced by Fermi.[(7)] In more general terms, regularizations of zero-range potentials are discussed in Refs. 3, 5, and 8. If one wishes to avoid mathematical subtleties associated with regularization, one can regard zero-range potentials as limiting cases of separable potentials; some aspects thereof are treated in Refs. 4 and 9.

[7] Quite remarkably, these functions are neither in the domain of the kinetic energy operator $-\frac{1}{2}\nabla^2$ nor in the one of the potential energy operator $\delta^{(3)}_{reg}(\mathbf{r})$ alone, but only in the domain of the whole Hamilton operator.

[8] The two-dimensional analog of $\delta^{(3)}_{reg}$ has proved extremely useful in mathematical physics, as exemplified by Šeba's solution of a chaotic quantum billiard.[(12)]

where $\mathbf{r}_1$ and $\mathbf{r}_2$ are the position vectors of the two atoms. We are using harmonic-oscillator units, in which all energies are in units of $\hbar\omega$ and all lengths are in units of the extension $\sqrt{\hbar/(m\omega)}$ of the ground state of the trap. The strength of the interaction is characterized by the scattering length a_0 (in oscillator units) and the interaction potential is chosen to be the regularized δ-function.

Because of the quadratic form of both the kinetic energy and the external potential energy in Eq. (1), this Hamiltonian can be separated into a center-of-mass part and a relative part, $H = H_{\mathrm{CM}} + H_{rel}$ with

$$\begin{aligned} H_{\mathrm{CM}} &= -\frac{1}{2}\nabla_R^2 + \frac{1}{2}\mathbf{R}^2 \\ H_{rel} &= -\frac{1}{2}\nabla_r^2 + \frac{1}{2}\mathbf{r}^2 + \sqrt{2}\,\pi a_0\,\delta^{(3)}(\mathbf{r})\frac{\partial}{\partial r}r \end{aligned} \tag{2}$$

which can then be diagonalized separately. Here $\mathbf{R} = \sqrt{1/2}\,(\mathbf{r}_1 + \mathbf{r}_2)$ is the center-of-mass coordinate and $\mathbf{r} = \sqrt{1/2}\,(\mathbf{r}_1 - \mathbf{r}_2)$ is the relative coordinate.[9] One can see immediately that the eigenfunctions of the center-of-mass motion are the well known states of the harmonic oscillator. So one is just left with the problem of solving the Schrödinger equation for the relative motion,

$$\left(H_{osc} + \sqrt{2}\,\pi a_0\,\delta^{(3)}(\mathbf{r})\frac{\partial}{\partial r}r\right)\Psi(\mathbf{r}) = E\Psi(\mathbf{r}) \tag{3}$$

in which H_{osc} stands for the Hamiltonian of the three-dimensional, isotropic, harmonic oscillator. The two-particle problem has been reduced to an effective one-particle problem; we emphasize that the harmonicity of the trapping potential is essential for this reduction.

To solve the Schrödinger equation (3) analytically, we expand the unknown wave function $\Psi(\mathbf{r})$ into the complete set of the familiar wave functions $\varphi_n(\mathbf{r})$ of the harmonic oscillator,

$$\Psi(\mathbf{r}) = \sum_{n=0}^{\infty} c_n\varphi_n(\mathbf{r}) \tag{4}$$

Only $l = 0$ states enter the sum in (4) since all other states vanish at $\mathbf{r} = 0$. These $l \neq 0$ states do not "feel" the pointlike interaction and stay unperturbed.

[9] The true center-of-mass position is, of course, given by $\frac{1}{2}(\mathbf{r}_1 + \mathbf{r}_2) = \mathbf{R}/\sqrt{2}$ and the true position of atom 1 relative to atom 2 by $\mathbf{r}_1 - \mathbf{r}_2 = \sqrt{2}\,\mathbf{r}$. The unconventional factors of $\sqrt{2}$ are included so that the effective masses for the center-of-mass motion and the relative motion are the same.

We insert the expansion (4) into Eq. (3) and obtain ($H_{osc}\varphi_n = E_n\varphi_n$)

$$\sum_{n=0}^{\infty} c_n(E_n - E)\,\varphi_n(\mathbf{r}) + \sqrt{2}\,\pi a_0\,\delta^{(3)}(\mathbf{r})\,\frac{\partial}{\partial r}\,r \sum_{m=0}^{\infty} c_m\varphi_m(\mathbf{r}) = 0 \tag{5}$$

The expansion coefficients c_n can be determined by projecting this expression onto $\varphi_n^*(\mathbf{r})$, which yields

$$c_n(E_n - E) + \sqrt{2}\,\pi a_0\,\varphi_n^*(0)\left[\frac{\partial}{\partial r}\left(r\sum_{m=0}^{\infty} c_m\varphi_m(\mathbf{r})\right)\right]_{r\to 0} = 0 \tag{6}$$

Consequently, the coefficients c_n are of the form

$$c_n = A\,\frac{\varphi_n^*(0)}{E_n - E} \tag{7}$$

where the value of the n-independent numerical constant A is fixed by the normalization of $\Psi(\mathbf{r})$.[10] Upon inserting the ansatz (7) into (6) we arrive at

$$\sqrt{2}\,\pi\left[\frac{\partial}{\partial r}\left(r\sum_{n=0}^{\infty}\frac{\varphi_n^*(0)\,\varphi_n(\mathbf{r})}{E_n - E}\right)\right]_{r\to 0} = -\frac{1}{a_0} \tag{8}$$

which determines the $l = 0$ eigenvalues of H_{rel} of (2). We note in passing that we are dealing here with the $r' \to 0$ limit of the Green's function solution discussed in Ref. 13.

B. Energy shifts

In order to simplify (8) the summation must be carried out first. This can be done by inserting the following representation of the harmonic oscillator s-wave functions

$$\varphi_n(\mathbf{r}) = \pi^{-3/4}[L_n^{(1/2)}(0)]^{-1/2}\,e^{-r^2/2}L_n^{(1/2)}(r^2) \qquad (\text{to } E_n = \tfrac{3}{2} + 2n) \tag{9}$$

which leads to

$$\frac{1}{\sqrt{2}\pi}\left[\frac{\partial}{\partial r}\left(re^{-r^2/2}\sum_{n=0}^{\infty}\frac{L_n^{(1/2)}(r^2)}{n - \nu}\right)\right]_{r\to 0} = -\frac{1}{a_0} \tag{10}$$

[10] Actually, the equality $|A|^2 = \sqrt{2}\,\pi a_0^2\,\partial E/\partial a_0$ holds with the functional relation (16) between E and a_0.

where $\nu \equiv \frac{1}{2}E - \frac{3}{4}$ is the noninteger analog of the principle quantum number n of the oscillator. The integral representation

$$\frac{1}{n-\nu} = \int_0^\infty \frac{dy}{(1+y)^2} \left(\frac{y}{1+y}\right)^{n-\nu-1} \tag{11}$$

which is valid for $n - \nu > 0$, shifts the n dependence of the sum in Eq. (10) from the denominator to an exponent,

$$\sum_{n=0}^{\infty} \frac{L_n^{(1/2)}(r^2)}{n-\nu} = \sum_{n=0}^{\infty} \int_0^\infty \frac{dy}{(1+y)^2} \left(\frac{y}{1+y}\right)^{n-\nu-1} L_n^{(1/2)}(r^2) \tag{12}$$

The summation variable n appears now only as the degree of the Laguerre polynomial and in the exponent so that we encounter one of the generating functions of the Laguerre polynomials, viz.

$$\sum_{n=0}^{\infty} L_n^{(1/2)}(x)\, z^n = (1-z)^{-3/2} \exp\left(\frac{xz}{z-1}\right) \tag{13}$$

Accordingly, we get

$$\begin{aligned} \sum_{n=0}^{\infty} \frac{L_n^{(1/2)}(r^2)}{n-\nu} &= \int_0^\infty \frac{dy}{\sqrt{1+y}}\, e^{-yr^2} \left(\frac{y}{1+y}\right)^{-\nu-1} \\ &= \Gamma(-\nu)\, U\left(-\nu, \frac{3}{2}, r^2\right) \end{aligned} \tag{14}$$

where the last step recognizes a standard integral representation of the confluent hypergeometric function U.[11] In view of the well-known behavior of $U(n, m, x)$ at $x = 0$, here:[12]

$$\Gamma(-\nu)\, U\left(-\nu, \frac{3}{2}, r^2\right) = -\sqrt{\pi}\left(\frac{2\Gamma(-\nu)}{\Gamma(-\nu - 1/2)} - \frac{1}{r} + O(r)\right) \tag{15}$$

Eq. (10) thus reduces to

$$\sqrt{2}\,\frac{\Gamma(-E/2 + 3/4)}{\Gamma(-E/2 + 1/4)} = \frac{1}{a_0} \tag{16}$$

[11] See, for example, entry 13.2.5 in Ref. 14.
[12] Entries 13.1.3 and 13.1.2 in Ref. 14.

To begin with, this equation is valid for $E < \frac{3}{2}$ only [recall the restriction of (11) to $\nu < n$] but its range of validity is immediately extended to all $E \neq \frac{3}{2} + 2n$ by analytical continuation.

Upon combining Eqs. (4), (7), (9), and (14) we observe that the $\ell = 0$ solutions of the Schrödinger equation (3) are given by

$$\Psi(\mathbf{r}) = \tfrac{1}{2}\pi^{-3/2} A e^{-r^2/2} \Gamma(-\nu)\, U(-\nu, \tfrac{3}{2}, r^2) \tag{17}$$

One verifies without effort that this obeys (3) for $r > 0$ independent of the value of $E = 2\nu + \frac{3}{2}$, but the $\delta^{(3)}$ singularities at $r = 0$ cancel each other only for those E values for which (16) holds.

The left-hand side of Eq. (16) has simple poles at the s-state energies $E = \frac{3}{2}, \frac{7}{2}, \frac{11}{2}, \ldots$ of the unperturbed oscillator, and zeros for $E = \frac{1}{2}, \frac{5}{2}, \frac{9}{2}, \ldots$. The left-hand side of (16) is therefore positive in the intervals

$$E < \tfrac{1}{2}, \qquad \tfrac{3}{2} < E < \tfrac{5}{2}, \qquad \tfrac{7}{2} < E < \tfrac{9}{2}, \ldots$$

and negative in the intervals

$$\tfrac{1}{2} < E < \tfrac{3}{2}, \qquad \tfrac{5}{2} < E < \tfrac{7}{2}, \qquad \tfrac{9}{2} < E < \tfrac{11}{2}, \ldots$$

These matters are illustrated in Fig. 1. For a positive scattering length, which is characteristic of the scattering by a repulsive interaction, the energy levels are shifted upward and for sufficiently large values of a_0 they saturate at a shift of $\Delta E = 1$. When the scattering length is negative the levels are shifted downward and saturate at a shift of $\Delta E = -1$.[13]

In the absence of the harmonic trapping potential, the three-dimensional regularized δ-potential gives rise to a single bound state if the scattering length a_0 is positive;[14] the corresponding binding energy $1/a_0^2$ tends to ∞ as $a_0 \to 0^+$.[15] Owing to the trapping potential this state is now shifted upward in energy. We note that both for $a_0 < 0$ and for $a_0 > 0$ the ground state of the system lies below the one of the unperturbed system.

The velocity dependence associated with the $\partial/\partial r$ differentiation thus leads to a counterintuitive feature of the regularized δ-potential: For $a_0 > 0$ it behaves like an attractive interaction as far as bound-state properties are concerned, but it scatters like a repulsive potential. And for $a_0 < 0$ matters are reversed—no bound state, but scattering as if an attractive interaction

[13] Incidentally, the equality reported in footnote 10 requires $\partial E/\partial a_0 > 0$.

[14] Since we are employing oscillator units this bound state is recovered from (16) in the limit $-E \gg 1$ when the left-hand side equals $\sqrt{-E}$.

[15] In the same limit the wave function of this bound state $\Psi \sim e^{-\sqrt{2}\, r/a_0}/r$ has a vanishing overlap with any (a_0-independent) square-integrable wave function.

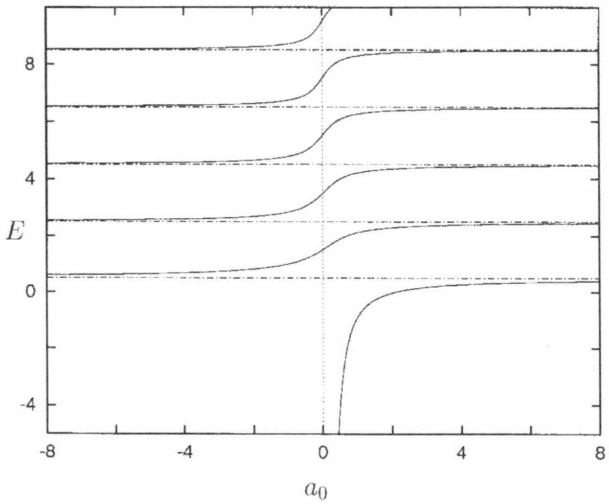

Fig. 1. Spectrum of the $l = 0$ states in the two-particle system with a regularized zero-range interaction in three dimensions. The solid lines (——) represent the solutions for E of (16) in their dependence on the strength of the interaction as measured by the scattering length a_0. The broken lines (—.—.—) indicate the asymptotic values that are reached for $a_0 \to \pm\infty$.

were at work. The limit $a_0 \to 0^+$ exhibits the anomaly *in nuce*: the binding energy grows without bound and the scattering cross section vanishes.

A further observation is that for growing energy the level shift is saturating for smaller values of the scattering length, that is, for weaker interaction strengths. A numerical criterion is the slope of the linear a_0-dependence near $a_0 = 0$, where the solutions of (16) are approximately given by[16]

$$E \simeq \frac{3}{2} + 2n + \sqrt{2/\pi}\binom{n+1/2}{n} a_0, \qquad (n = 0, 1, 2, \ldots) \tag{18}$$

Indeed, the slope grows proportional to $\sqrt{n}$. This is due to the fact that $\varphi_n(0)$, the s-state oscillator wave function at the center, increases with the principle quantum number n as $n^{1/4}$. So the zero-range interaction modeled by the regularized δ-function is getting more and more effective for higher

[16] In passing we note that a weak perturbation is characterized by $|a_0| \ll 1$, which indicates that the scattering properties of $\delta^{(3)}_{\text{reg}}$ are essential here, not the binding properties.

energy states.[17] Incidentally, this $n^{1/4}$ growth is the reason why one must evaluate the sum in (8) prior to the required differentiation.[18]

3. ONE-DIMENSIONAL AND TWO-DIMENSIONAL ANALOGS

These findings in three dimensions can be compared with the one-dimensional and two-dimensional analogs,[19] shown in Figs. 2 and 3. In all three cases the s-wave scattering amplitude is of the form $e^{i\theta}\sin\theta = (-i+\cot\theta)^{-1}$. For zero-range interactions the scattering phase θ has a particularly simple dependence on the wave number k, viz.

$$\cot\theta = \begin{cases} -(ka_0)^{-1} & \text{in three dimensions} \\ \dfrac{2}{\pi}\log(ka_0) & \text{in two dimensions} \\ ka_0 & \text{in one dimension} \end{cases} \tag{19}$$

The same relations hold for (almost) arbitrary finite-range interactions in the long-wavelength limit.

The scattering length a_0 is identified in this limit. The three-dimensional relation is quite familiar; the conventions we employ in one and two dimensions are exhibited in (19). They are such that matters are as similar as possible: (i) $\cot\theta$ increases with growing a_0; (ii) there is a bound state for $a_0 > 0$; (iii) its binding energy equals $1/a_0^2$.

An analysis that is closely analogous to the three-dimensional one can be performed in one and two dimensions as well. Remarkably, the shifted energies in the one-dimensional case are again determined by Eq. (16), except that now the unperturbed even-state energies $E = \frac{1}{2}, \frac{5}{2}, \frac{9}{2},\ldots$ are obtained for $1/a_0 = 0$, not for $a_0 = 0$. In marked contrast to the three-dimensional case the values of the (even) wave functions at the center decrease as $n^{-1/4}$ here. In the limit of weak interaction strength ($1/a_0 \ll 1$) the one-dimensional analog of (18) reads

$$E \cong \frac{1}{2} + 2n - \sqrt{2/\pi}\binom{n-1/2}{n}\frac{1}{a_0}, \qquad (n = 0, 1, 2,\ldots) \tag{20}$$

[17] The modeling of the physical interaction by a zero-range potential is an approximation which is valid as long as the interparticle de Broglie wavelength is larger than the range of the physical force. Consequently, this approximation is invalid if the relevant energies are too large.

[18] We also note this consequence: The coefficients c_n of (7) decrease like $c_n \sim n^{-3/4}$ for $n \gg 1$ and therefore the wave functions $\Psi(\mathbf{r})$ of (4) are not in the domain of H_{osc}. This remark is, of course, analogous to the one in footnote 7.

[19] In one dimension one has simply $V(x_1 - x_2) = -(2/a_0)\,\delta^{(1)}(x_1 - x_2)$; matters are more complicated in two dimensions because of logarithmic singularities.[3, 12]

Indeed, the slope decreases proportional to $\sqrt{n}$ with n. Hence, with growing energy the saturation takes place at larger values of the interaction strength $1/a_0$. These remarks are confirmed by Fig. 2.

In two dimensions the situation is somewhat more complicated. The scattering length is always positive, and the shifted energies are given by the solutions of

$$\psi(-E/2+1/2) = \log\left(\frac{1}{2a_0^2}\right) \tag{21}$$

where $\psi(\ )$ is the logarithmic derivative of Euler's Γ-function. In the vicinity of the unperturbed energies $E = 1, 3, 5, \ldots$, one has

$$E \cong 1 + 2n + 2\left[\log\left(\frac{1}{2a_0^2}\right)\right]^{-1}, \qquad (n = 0, 1, 2, \ldots) \tag{22}$$

so that the coupling strength is measured by $[\log(\frac{1}{2}/a_0^2)]^{-1}$. Here, the derivative of E with respect to the coupling strength does not depend on n in the weak-coupling limit. Figure 3 demonstrates the two-dimensional case.

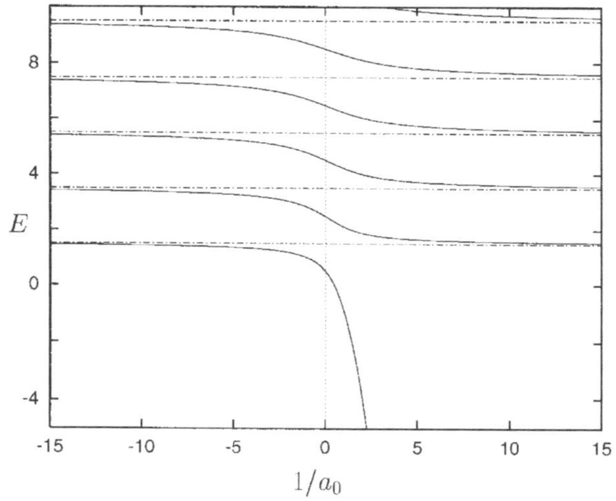

Fig. 2. Spectrum of the even states in the two-particle system with a zero-range interaction in one dimension. The solid lines (——) represent the solutions for E of (16) in their dependence on the strength of the interaction as measured by the inverse $1/a_0$ of the scattering length. The broken lines (—·—·—) indicate the asymptotic values that are reached for $1/a_0 \to \pm\infty$.

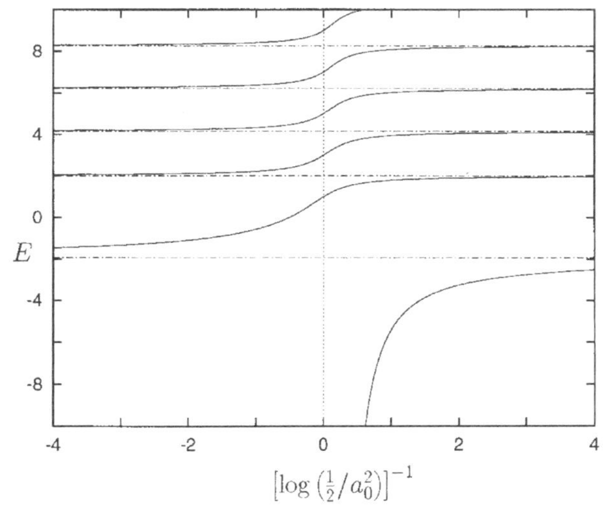

Fig. 3. Spectrum of the $m = 0$ states in the two-particle system with a zero-range interaction in two dimensions. The solid lines (——) represent the solutions for E of (21) in their dependence on the strength of the interaction which is given by $[\log(\frac{1}{2}/a_0^2)]^{-1}$ in terms of the scattering length a_0. The broken lines (—.—.—) indicate the asymptotic values $E = -1.923264$, 2.008166, 4.146997, 6.221442, 8.270587,... that are reached in the limit of very strong coupling.

4. SUMMARY

We have presented an analytical solution to the problem of two cold atoms with a zero-range interaction in a harmonic trap. We have thus added another example to the short list of solvable problems in quantum mechanics. In addition to its pedagogical value, the problem studied is of potential significance for the physics of ultracold vapors.

ACKNOWLEDGMENTS

We would like to thank C. Eberlein for bringing Ref. 4 to our attention. BGE expresses his gratitude for the hospitable environment provided by H. Paul and his group in Adlershof. KR and MW would like to thank J. Mlynek for the hospitality extended to them during their stay in Konstanz. KR gratefully acknowledges financial support by the Optik Zentrum Konstanz. TB and BGE each thank the Center of Theoretical Physics for supporting a visit to Warsaw.

REFERENCES

1. P. Verkerk, B. Lounis, C. Salomon, and C. Cohen-Tannoudji, *Phys. Rev. Lett.* **68**, 3861 (1992); P. S. Jessen, C. Gerz, P. D. Lett, W. D. Phillips, S. L. Rolston, R. J. C. Spreeuw, and C. I. Westbrook, *Phys. Rev. Lett.* **69**, 49 (1992); A. Hemmerich, C. Zimmermann, and T. W. Hänsch, *Europhys. Lett.* **22**, 89 (1993).
2. M. H. Anderson, J. R. Ensher, M. R. Matthews, C. E. Wieman, and E. A. Cornell, *Science* **269**, 198 (1995); C. C. Bradley, C. A. Sackett, J. J. Tollett, and R. G. Hulet, *Phys. Rev. Lett.* **75**, 1687 (1995); K. B. Davis, M.-O. Mewes, M. R. Andrews, N. J. van Druten, D. S. Durfee, D. M. Kurn, and W. Ketterle, *Phys. Rev. Lett.* **75**, 3969 (1995).
3. S. Albeverio, F. Gesztesy, R. Hoegh-Krohn, H. Holden, *Solvable Models in Quantum Mechanics* (Springer, Heidelberg, 1988).
4. G. Barton and D. Waxman, "Wave equations with point-support potentials having dimensionless strength parameters," University of Sussex, unpublished.
5. P. Šeba, *Czech. J. Phys. B* **36**, 559 (1986).
6. K. Huang, *Statistical Mechanics*, 2nd edn. (Wiley, New York, 1987).
7. E. Fermi, *Ricera Sci.* **7**, 12 (1936).
8. K. Wódkiewicz, *Phys. Rev. A* **43**, 68 (1991).
9. B.-G. Englert, *Lett. Math. Phys.* **34**, 239 (1995).
10. G. Breit and P. T. Zilsel, *Phys. Rev.* **71**, 232 (1947).
11. I. J. Bersons, *J. Phys. B* **8**, 3078 (1975).
12. P. Šeba, *Phys. Rev. Lett.* **64**, 1855 (1990).
13. R. K. Janev and Z. Maric, *Phys. Lett.* **46A**, 313 (1974).
14. M. Abramowitz and I. A. Stegun, eds., *Handbook of Mathematical Functions* (Dover, New York, 1972).

Stephen A. Fulling

Departments of Mathematics and Physics, Texas A&M University, College Station, Texas, USA

Paper Title: **Statistics of dressed modes in a thermal state**

Reference: Optics Communications **208**, 139–144 (2002)

Personal message:

Around the beginning of this century, Berge Englert spent several years at Texas A&M University because Marlan Scully thought he was a handy guy to have around. I quickly learned that he was right.

At some point I gave an informal talk about a problem I was working on, applying semiclassical approximation to the quantum propagator of a particle confined to a plane and also subjected to a constant force field. To do that you need to solve the classical equation of motion for trajectories that start and end at given space-time points and that either do or don't bounce off the "ceiling". That leads to a cubic equation. Who learns how to solve cubic equations nowadays? I was reading up on the classic Scipione-Tartaglia-Cardano formula and trying to get a useful answer out of it and was running into numerical difficulty. The next day Berge presented me with a much more practical formula in terms of trigonometric functions (basically found by Viète in the century after all those Italians). This whole project was one of the most interesting I have worked on, and you can see a presentation of it to undergraduates, including the history lesson, at http://www.math.tamu.edu/~fulling/amusecei.pdf .

Around that time Marlan and the Kocharovsky brothers derived a formula for the structure of a thermal equilibrium state of a weakly interacting Bose gas, but Marlan was dissatisfied with the length and clarity of the derivation and challenged everyone around him to find a better one. I recognized that the problem involved a product of two Bogolubov transformations, one that Bogolubov invented to connect bare and dressed bosons in the ground state, and one that can be used to construct a thermal state as a pure state of a system with twice as many degrees of freedom and hence comes up in the theories of Hawking and Unruh about the temperature of relativistic event horizons. Marlan was impressed by that idea and took me to a conference in Jackson Hole to talk about it, and that was how I found myself giving a lecture on black holes to a quantum optics audience. But I ran into trouble reducing the resulting formulas to anything simple. Marlan persuaded Berge and Mark Pilloff, a visiting graduate student from Berkeley, to help out. Berge solved the problem by a completely different method, using Wigner functions; he insisted that Mark and I should be coauthors. (Well, at least we did check the algebra, and Mark did a related calculation that had to be summarized in the paper by a two-sentence remark because of space limitations.)

Meanwhile we continued to wrestle with the Bogolubov transformations and wrote an exposition of that approach and its conceptual and aesthetic advantages. Berge produced yet another derivation and generalization of the final formulas, based on generating functions, so in the end we had four different calculations, counting Marlan's original one. We never managed to prove that the Bogolubov formulas are equivalent to the others, but we verified that numerically for a selection of cases.

In the course of this I came up with a slightly complicated function that needed to be expanded in a power series. Berge took one look at it and said, "That's the generating function for the Legendre polynomials," and wrote down the series. As I said, a handy guy to have around.

A further reference: S. A. Fulling, B.-G. Englert, and M. D. Pilloff, *Interacting bosons at finite temperature: How Bogolubov visited a black hole and came home again*, Found. Phys. **33**, 87–110 (2003).

1 July 2002

ELSEVIER

Optics Communications 208 (2002) 139–144

OPTICS COMMUNICATIONS

www.elsevier.com/locate/optcom

Statistics of dressed modes in a thermal state

Berthold-Georg Englert [a,b,*], Stephen A. Fulling [a,b], Mark D. Pilloff [b,c]

[a] Department of Mathematics, Texas A&M University, College Station, TX 77843-3368, USA
[b] Department of Physics, Institute for Quantum Studies, Texas A&M University, College Station, TX 77843-4242, USA
[c] Department of Physics, University of California, Berkeley, CA 94720, USA

Received 23 April 2002; accepted 25 April 2002

Abstract

By a Wigner-function calculation, we evaluate the trace of a certain Gaussian operator arising in the theory of a boson system subject to both finite temperature and (weak) interaction. Thereby we rederive (and generalize) a recent result by Kocharovsky, Kocharovsky, and Scully [Phys. Rev. A 61(2000) 053606] in a way that is technically much simpler. One step uses a special case of the response of Wigner functions to linear transformations, and we demonstrate the general case by simple means. As an application we extract the counting statistics for each mode of the Bose gas. © 2002 Elsevier Science B.V. All rights reserved.

Keywords: Gaussian operator; Wigner function; Cold Bose gas

In a recent paper on dilute boson gases [1], Kocharovsky, Kocharovsky, and Scully faced the problem of evaluating the trace of a product of two 2-mode Gaussian operators. They solved it by exploiting the algebra of 2-mode squeezing operators in a remarkable *tour de force* occupying the appendix of [1]. The main objective of the present Brief Report is to put on record an alternative derivation that is simpler and more direct.

In a few words, the essence of our calculation is this: We recall (i) that the trace of a product of two operators can be obtained as a phase space integral of the product of the respective Wigner functions [2]; (ii) that the Wigner functions of Gaussian operator functions are Gaussian functions of the phase space variables; and (iii) that integrals of Gaussians are immediate.

The trace of interest is

$$T(u,v;\epsilon) = \mathrm{tr}\{G(u,v)\rho_\epsilon\} = \langle G(u,v)\rangle_\epsilon \tag{1}$$

with

$$G(u,v) = \mathrm{e}^{\mathrm{i}ua^\dagger a + \mathrm{i}vb^\dagger b} \tag{2}$$

and

$$\rho_\epsilon = (1-\mathrm{e}^{-\epsilon})^2 \mathrm{e}^{-\epsilon(A^\dagger A + B^\dagger B)}, \tag{3}$$

where $a^\dagger$, a and $b^\dagger$, b are the ladder operators of the physical excitations (two independent harmonic oscillators), and the Bogolubov transformation

$$A = a\cosh\vartheta - b^\dagger \sinh\vartheta, \quad A^\dagger = a^\dagger\cosh\vartheta - b\sinh\vartheta,$$
$$B = b\cosh\vartheta - a^\dagger\sinh\vartheta, \quad B^\dagger = b^\dagger\cosh\vartheta - a\sinh\vartheta \tag{4}$$

* Corresponding author. Fax: +1-979-485-1235.
E-mail address: bge@jewel.tamu.edu (B.-G. Englert).

PII: S0030-4018(02)01557-2

(with real ϑ) relates them to the ladder operators of the "dressed modes." The statistical operator ρ_ϵ is a thermal state (with temperature parameter $\epsilon > 0$) that is diagonal in the dressed number operators $A^\dagger A$ and $B^\dagger B$. Table 1 states how our notation is related to the one used in [1]. The factor $G(u, v)$ is the generating function for the count of physical excitations inasmuch as

$$\delta_{a^\dagger a,m}\delta_{b^\dagger b,n} = \int_{(2\pi)} \frac{du}{2\pi} \int_{(2\pi)} \frac{dv}{2\pi} e^{-imu-inv} G(u, v), \tag{5}$$

the integrations covering any interval of length 2π. Accordingly, upon expanding $T(u, v; \epsilon)$ in powers of e^{iu} and e^{iv} we get the probabilities for having certain numbers of quanta in the physical excitations.

1. Phase–space evaluation

Our starting point is the 1-mode thermal state,

$$\rho_\epsilon^{(1)} = (1 - e^{-\epsilon})e^{-\epsilon a^\dagger a}, \tag{6}$$

and its well-known Wigner function,

$$[\rho_\epsilon^{(1)}]_W = 2\tanh\left(\frac{1}{2}\epsilon\right)e^{-2\alpha^*\alpha\tanh(\epsilon/2)}, \tag{7}$$

normalized in accordance with

$$\mathrm{tr}\{\rho_\epsilon^{(1)}\} = \int \frac{d^2\alpha}{\pi} [\rho_\epsilon^{(1)}]_W, \tag{8}$$

where α is the complex phase space variable associated with a, and the standard parameterization of the phase space integral – specified by $\alpha = x + iy$, $\alpha^* = x - iy$, and $d^2\alpha = dx dy$ – suffices for our purposes.

Table 1
Dictionary for switching from the present notation to the one in Eqs. (63), (68), (69) of [1]

Here	There	Here	There
$T(u, u; \epsilon)$	$\Theta_{\pm\mathbf{k}}(u)$	u, v	u, u
$a, a^\dagger$	$\hat\beta_{\mathbf{k}}, \hat\beta_{\mathbf{k}}^\dagger$	ϵ	$\varepsilon_{\mathbf{k}}/T$
$b, b^\dagger$	$\hat\beta_{-\mathbf{k}}, \hat\beta_{-\mathbf{k}}^\dagger$	$\tanh\vartheta$	$A_{\mathbf{k}}$
$A, A^\dagger$	$\hat b_{\mathbf{k}}, \hat b_{\mathbf{k}}^\dagger$	Y_+	$z(-A_{\mathbf{k}})$
$B, B^\dagger$	$\hat b_{-\mathbf{k}}, \hat b_{-\mathbf{k}}^\dagger$	Y_-	$z(A_{\mathbf{k}})$

First, we need the Wigner function (or Weyl transform) of $G(u, v)$ in (2), which (among other methods) can be found efficiently by analytic continuation from (6) and (7) [taking $\epsilon \to -iu$ and $\epsilon \to -iv$ in two copies of (7), discarding its prefactor]:

$$[G(u, v)]_W = e^{-i(u+v)/2}\sec\left(\frac{1}{2}u\right)\sec\left(\frac{1}{2}v\right) \times e^{2i\alpha^*\alpha\tan(u/2)+2i\beta^*\beta\tan(v/2)}, \tag{9}$$

where β is the phase space variable for b.

Second, to get the Wigner function of the statistical operator (3) we recall a fundamental property of Wigner functions, namely that linear similarity transformations of the ladder operators carry over to the Wigner function (see Section 4 below). For the Bogolubov transformation (4) this means that $[F(a^\dagger, a, b^\dagger, b)]_W = f(\alpha^*, \alpha, \beta^*, \beta)$ implies [3]

$$[F(A^\dagger, A, B^\dagger, B)]_W = f(\alpha^*\cosh\vartheta - \beta\sinh\vartheta, \ldots, \beta\cosh\vartheta - \alpha^*\sinh\vartheta), \tag{10}$$

and therefore we have

$$[\rho_\epsilon]_W = 4\tanh^2\left(\frac{1}{2}\epsilon\right) \times e^{-2[(\alpha^*\alpha+\beta^*\beta)\cosh(2\vartheta)-(\alpha^*\beta^*+\alpha\beta)\sinh(2\vartheta)]\tanh(\epsilon/2)}. \tag{11}$$

Given (9) and (11), the trace (1) is a 4-dimensional Gaussian integral that is routinely evaluated,

$$\begin{aligned} T(u, v; \epsilon) &= \int \frac{d^2\alpha}{\pi} \int \frac{d^2\beta}{\pi} [G(u, v)]_W [\rho_\epsilon]_W \\ &= 4(e^\epsilon - 1)^2\Big[(e^\epsilon - 1)^2(1 + e^{iu})(1 + e^{iv}) \\ &\quad + (e^\epsilon + 1)^2(1 - e^{iu})(1 - e^{iv}) \\ &\quad + 2(e^{2\epsilon} - 1)(1 - e^{iu+iv})\cosh(2\vartheta)\Big]^{-1}. \end{aligned} \tag{12}$$

While this formula is the most direct and natural result of the calculation, it may appear physically inscrutable. However, it can be worked into the form

$$T(u,v;\epsilon) = \frac{2(Y_+-1)(Y_--1)}{(Y_+-e^{iu})(Y_--e^{iv})+(Y_--e^{iu})(Y_+-e^{iv})} = T(v,u;\epsilon), \tag{13}$$

where

$$Y_+ = \frac{e^\epsilon + \tanh\vartheta}{1+e^\epsilon\tanh\vartheta}, \quad Y_- = \frac{e^\epsilon - \tanh\vartheta}{1-e^\epsilon\tanh\vartheta}. \tag{14}$$

This matches the result, for the special case $u = v$, reported in (69) of [1] (translated with Table 1):

$$T(u,u;\epsilon) = \frac{Y_+-1}{Y_+-e^{iu}}\frac{Y_--1}{Y_--e^{iu}} = \left\langle e^{iu(a^\dagger a+b^\dagger b)}\right\rangle_\epsilon. \tag{15}$$

The challenge of rederiving (15) by simpler means has thus been met; in fact, (13) is more general because u and v need not be the same.

Remark. One can also calculate the trace (1) by using the Q-function for ρ_ϵ and the P-function for $G(u,v)$, rather than the Wigner function for both. But that approach is more involved because the statement corresponding to (10) is more complicated.

2. Occupation probabilities

Let us now see what we can learn from this gain in flexibility. Special cases are those of $v = 0$ and $u = 0$, the generating functions for the count of excitations in either the a-mode or the b-mode,

$$T(u,0;\epsilon) = \langle e^{iua^\dagger a}\rangle_\epsilon = \frac{Y-1}{Y-e^{iu}}, \quad T(0,v;\epsilon) = \langle e^{ivb^\dagger b}\rangle_\epsilon = \frac{Y-1}{Y-e^{iv}}, \tag{16}$$

where

$$Y = \frac{2Y_+Y_- - Y_+ - Y_-}{Y_+ + Y_- - 2} = \frac{e^\epsilon + \tanh^2\vartheta}{1+e^\epsilon\tanh^2\vartheta} > 1. \tag{17}$$

The symmetry in (16) is, of course, just a particular example of $T(u,v;\epsilon) = T(v,u;\epsilon)$, which we noted at (13) in passing. The geometric series that result from expanding the right-hand sides of (16) in powers of e^{iu} or e^{iv}, respectively, tell us the individual counting statistics for each mode,

$$\langle\delta_{a^\dagger a,n}\rangle_\epsilon = \langle\delta_{b^\dagger b,n}\rangle_\epsilon = (1-Y^{-1})Y^{-n}. \tag{18}$$

So, if there were no statistical correlations between the two modes, the probability for having a total of N quanta in both modes would be given by

$$\sum_{n=0}^{N}\langle\delta_{a^\dagger a,n}\rangle_\epsilon\langle\delta_{b^\dagger b,N-n}\rangle_\epsilon = (N+1)(1-Y^{-1})^2Y^{-N}. \tag{19}$$

But there *are* correlations and, indeed, the correct value of $\langle\delta_{a^\dagger a+b^\dagger b,N}\rangle_\epsilon$, obtained by expanding $T(u,u;\epsilon)$ of (15) in powers of e^{iu},

$$\langle\delta_{a^\dagger a+b^\dagger b,N}\rangle_\epsilon = (1-Y_+^{-1})(1-Y_-^{-1})\frac{Y_+^{-N-1}-Y_-^{-N-1}}{Y_+^{-1}-Y_-^{-1}}, \tag{20}$$

differs noticeably from (19), unless we are in the limiting situation of $\vartheta = 0$ when $Y_+ = Y_- = Y$. In that limit we have

$$T(u,v;\epsilon)|_{\vartheta=0} = \frac{e^\epsilon-1}{e^\epsilon-e^{iu}}\frac{e^\epsilon-1}{e^\epsilon-e^{iv}}, \tag{21}$$

and this factorization states the obvious: For $\vartheta = 0$ the modes do not get dressed and so they maintain their statistical independence.

Rather more involved is the expression for the probability of m quanta in mode a and n quanta in mode b,

$$\langle\delta_{a^\dagger a,m}\delta_{b^\dagger b,n}\rangle_\epsilon = (1-Y_+^{-1})(1-Y_-^{-1})\left[\frac{1}{2}(Y_+^{-1}+Y_-^{-1})\right]^{m+n} \times\sum_{k=0}^{\infty}\frac{(m+n-k)!}{k!(m-k)!(n-k)!}\left[\frac{-4Y_+Y_-}{(Y_++Y_-)^2}\right]^k, \tag{22}$$

where the summation terminates when k equals the smaller one of m and n. Note that the right-hand side is invariant under the interchange $m \leftrightarrow n$, which is another manifestation of $T(u,v;\epsilon) = T(v,u;\epsilon)$. Incidentally, one could write (22) in terms of a Jacobi polynomial, but that does not seem to add transparency to the result.

3. Bogolubov transformation of a single mode

Also worthy of study is a Bogolubov transformation mixing one mode with itself by

$$A = a\cosh\vartheta - a^\dagger \sinh\vartheta \tag{23}$$

in place of (4). In fact, the dressing of a pair of modes as in (4) can be diagonalized into two decoupled transformations of the form (23); in the Bose condensation problem this corresponds to replacing traveling waves by standing waves.

In this case we should evaluate

$$T(u;\epsilon) = \int \frac{d^2\alpha}{\pi}[G(u)]_W[\rho_\epsilon]_W, \tag{24}$$

where $[G(u)]_W$ is the $v=0$ version of (9) and

$$[\rho_\epsilon]_W = 2\tanh\left(\frac{1}{2}\epsilon\right) \times e^{-[2\alpha^*\alpha\cosh(2\vartheta)-(\alpha^{*2}+\alpha^2)\sinh(2\vartheta)]\tanh(\epsilon/2)} \tag{25}$$

is obtained by $\alpha \to \alpha\cosh\vartheta - \alpha^*\sinh\vartheta$ in (7). [Note the subtle difference in the relative sizes of the terms in the exponent in (25) versus (11).] The Gaussian integration is more elementary than in the previous case and yields

$$\begin{aligned} T(u;\epsilon) = 2(e^\epsilon - 1)[&(e^\epsilon-1)^2(1+e^{iu})^2 \\ &+ (e^\epsilon+1)^2(1-e^{iu})^2 + 2(e^{2\epsilon}-1) \\ &\times (1-e^{2iu})\cosh(2\vartheta)]^{-\frac{1}{2}} \end{aligned} \tag{26}$$

as the analog of (12). Note that $T(u;\epsilon)^2 = T(u,u;\epsilon)$, as it should. On the other hand, $T(u;\epsilon)T(v;\epsilon)$ does not equal $T(u,v;\epsilon)$, as these functions are giving the statistics for two different bases of single-quantum states, which have different dynamics. If desired, the counterpart of (22) could be obtained by expanding (26) in powers of e^{iu}, which is easily done because we meet here a familiar generating function for Legendre's polynomials.

One motivation for investigating a single mode was the hope of elucidating the structure of equations (14), which cry out to be interpreted as addition formulas for the hyperbolic tangent function with $e^{-\epsilon} = \tanh(\ldots)$. There is, in fact, a way of constructing the thermal state of a boson mode by means of another Bogolubov transformation [4]; one would think that (14) then expresses the composition of the two transformations. However, the appropriate thermal parameter in that construction is $\frac{1}{2}\epsilon$ rather than ϵ. This discrepancy may be related to the fact, stressed in [1], that quanta in the interacting Bose gas are present in strongly correlated *pairs*. Calculations following [4] are in progress, but so far have not yielded expressions as simple as those provided by the method reported here.

4. Linear similarity transformations

It is striking that the fundamental transformation property of Wigner functions, of which (10) is the special case we make use of, is not mentioned in any of the reviews [5–7] or textbooks [8,9], although it has been much exploited for various applications (e.g., in [10,11]). Since it deserves to be known more widely, we present the following compact derivation, which is more general than all previous ones.

As a warm-up we will first consider the single-mode case of Section 3. The Wigner function (or Weyl transform) $F_W(\alpha^*,\alpha)$ of an operator function $F(a^\dagger,a)$ is given by

$$\begin{aligned} F_W(\alpha^*,\alpha) &= \int\frac{d^2\beta}{\pi}e^{\beta\alpha^*-\beta^*\alpha}\mathrm{tr}\left\{e^{\beta^*a-\beta a^\dagger}F(a^\dagger,a)\right\} \\ &= 2\mathrm{tr}\left\{e^{\alpha a^\dagger-\alpha^*a}Se^{\alpha^*a-\alpha a^\dagger}F(a^\dagger,a)\right\}, \end{aligned} \tag{27}$$

where we encounter the unitary displacement operator $e^{\alpha^*a-\alpha a^\dagger}$,

$$\begin{aligned} e^{\alpha a^\dagger-\alpha^*a}a^\dagger e^{\alpha^*a-\alpha a^\dagger} &= a^\dagger - \alpha^*, \\ e^{\alpha a^\dagger-\alpha^*a}a e^{\alpha^*a-\alpha a^\dagger} &= a - \alpha, \end{aligned} \tag{28}$$

and the reflection operator

$$S = \int\frac{d^2\beta}{2\pi}e^{\beta^*a-\beta a^\dagger}, \tag{29}$$

which is both unitary and Hermitian and therefore its own inverse, $S^\dagger = S = S^{-1}$ [12]. Its name derives from the unitary transformation that S effects,

$$S^{-1}a^\dagger S = -a^\dagger, \quad S^{-1}aS = -a. \tag{30}$$

A linear similarity transformation turns $F(a^\dagger,a)$ into $G(a^\dagger,a)$, another operator function,

$$G(a^\dagger, a) = V^{-1}F(a^\dagger, a)V = F(\mu a^\dagger + \nu a, \sigma a + \tau a^\dagger), \tag{31}$$

where

$$V^{-1}a^\dagger V = \mu a^\dagger + \nu a, \quad Va^\dagger V^{-1} = \sigma a^\dagger - \nu a,$$
$$V^{-1}aV = \sigma a + \tau a^\dagger, \quad VaV^{-1} = \mu a - \tau a^\dagger. \tag{32}$$

The commutation relation $[a, a^\dagger] = 1$ puts the restriction $\mu\sigma - \nu\tau = 1$ on the numerical coefficients μ, ν, σ, τ, and otherwise they can take on any complex values permitted by $|\mu|, |\sigma| \geqslant |\nu|, |\tau|$, which condition ensures that $V^{-1}a^\dagger V$ has eigenbras and $V^{-1}aV$ has eigenkets. It is not necessary that (32) be a unitary transformation, such as (23). In particular, the non-unitary transformation $a^\dagger \to a^\dagger - a$, $a \to \frac{1}{2}(a + a^\dagger)$ is permissible and useful [10].

The crucial observation is now that the reflection operator is invariant under this similarity transformation,

$$V^{-1}SV = S \quad \text{or} \quad S^{-1}VS = V. \tag{33}$$

This is so because V is the exponential of a linear combination of $a^{\dagger 2}$, $a^\dagger a$, and a^2 (or a product of such exponentials), as illustrated by

$$V = e^{(\tau/\mu)a^{\dagger 2}/2}\mu^{-a^\dagger a}e^{-1/2(\nu/\mu)a^2}. \tag{34}$$

Clearly, the reflection (30) has no effect on V, which is the second statement of (33).

Now, perform the replacements

$$F(a^\dagger, a) \to V^{-1}F(a^\dagger, a)V = G(a^\dagger, a)$$
$$S \to S = V^{-1}SV \tag{35}$$

in (27) to establish

$$G_W(\alpha^*, \alpha) = 2\mathrm{tr}\{Ve^{\alpha a^\dagger - \alpha^* a}V^{-1}SVe^{\alpha^* a - \alpha a^\dagger}V^{-1}F(a^\dagger, a)\}, \tag{36}$$

then note that

$$Ve^{\alpha^* a - \alpha a^\dagger}V^{-1} = e^{(\mu\alpha^* + \nu\alpha)a - (\sigma\alpha + \tau\alpha^*)a^\dagger}, \tag{37}$$

and arrive at

$$G_W(\alpha^*, \alpha) = F_W(\mu\alpha^* + \nu\alpha, \sigma\alpha + \tau\alpha^*). \tag{38}$$

Indeed, the linear similarity transformation (31) of the operator functions carries over to their Wigner functions.

The arguments in higher dimensions are quite analogous. There are now n pairs of ladder operators and n pairs of phase-space variables associated with them. We write them compactly as $2n$-component rows,

$$\vec{a} = (a_1^\dagger, a_1, a_2^\dagger, a_2, \ldots, a_n^\dagger, a_n),$$
$$\vec{\alpha} = (\alpha_1^*, \alpha_1, \alpha_2^*, \alpha_2, \ldots, \alpha_n^*, \alpha_n). \tag{39}$$

The n-dimensional displacement and reflection operators are then

$$\prod_{k=1}^{n} e^{\alpha_k^* a_k - \alpha_k a_k^\dagger} = e^{\vec{\alpha}\mathsf{K}\vec{a}^{\mathrm{T}}}, \quad S = \int \frac{d^{2n}\beta}{(2\pi)^n} e^{\vec{\beta}\mathsf{K}\vec{a}^{\mathrm{T}}} \tag{40}$$

with the $2n \times 2n$ block matrix

$$\mathsf{K} = \begin{pmatrix} 0 & -1 & & & \\ 1 & 0 & & & \\ & & \ddots & & \\ & & & 0 & -1 \\ & & & 1 & 0 \end{pmatrix} = -\mathsf{K}^{-1} = -\mathsf{K}^{\mathrm{T}}, \tag{41}$$

so that the analog of (27) reads

$$F_W(\vec{\alpha}) = 2^n \mathrm{tr}\{e^{-\vec{\alpha}\mathsf{K}\vec{a}^{\mathrm{T}}}Se^{\vec{\alpha}\mathsf{K}\vec{a}^{\mathrm{T}}}F(\vec{a})\}. \tag{42}$$

A linear similarity transformation in n dimensions is of the form

$$V^{-1}\vec{a}V = \vec{a}\mathsf{V}, \quad V\vec{a}V^{-1} = \vec{a}\mathsf{V}^{-1} \tag{43}$$

where V is a $2n \times 2n$ matrix restricted mainly by

$$\mathsf{V}\mathsf{K}\mathsf{V}^{\mathrm{T}} = \mathsf{K}, \tag{44}$$

which is the analog of $\mu\sigma - \nu\tau = 1$ for the coefficients in (32). Here, too, V is the exponential of a bilinear form of the ladder operators, $V = \exp(\vec{a}\mathscr{V}\vec{a}^{\mathrm{T}})$, with a (largely) arbitrary complex $2n \times 2n$ coefficient matrix $\mathscr{V}$, and so (33) is equally true in n dimensions. By arguments analogous to those of Eqs. (35)–(38) it then follows that the Wigner function of the transformed operator

$$G(\vec{a}) = V^{-1}F(\vec{a})V = F(\vec{a}\mathsf{V}) \tag{45}$$

is given by

$$G_W(\vec{\alpha}) = 2^n \mathrm{tr}\left\{Ve^{-\vec{\alpha}\mathsf{K}\vec{a}^{\mathrm{T}}}V^{-1}SVe^{\vec{\alpha}\mathsf{K}\vec{a}^{\mathrm{T}}}V^{-1}F(\vec{a})\right\}$$
$$= 2^n \mathrm{tr}\left\{e^{-\vec{\alpha}\mathsf{V}\mathsf{K}\vec{a}^{\mathrm{T}}}Se^{\vec{\alpha}\mathsf{V}\mathsf{K}\vec{a}^{\mathrm{T}}}F(\vec{a})\right\}$$
$$= F_W(\vec{\alpha}\mathsf{V}). \tag{46}$$

This closes the case.

Acknowledgements

We thank Marlan Scully for insisting that a simpler derivation of (15) was possible and worthwhile. The authors acknowledge the support from Air Force Research Laboratories (Rome, New York), DARPA-QuIST, TAMU Telecommunication and Informatics Task Force (TITF) initiative, the Office of Naval Research, and the Welch Foundation.

References

[1] V.V. Kocharovsky, Vl.V. Kocharovsky, M.O. Scully, Phys. Rev. A 61 (2000) 053606.
[2] See the seminal papers by E. Wigner, Phys. Rev. 40 (1932) 749, and J.E. Moyal, Proc. Camb. Phil. Soc. 45 (1949) 99, as well as the more recent reviews [5–7] and textbook expositions [8,9]. As in [10] we use the same normalization for Wigner functions (statistical operators) and Weyl transforms (observables).
[3] Eq. (10) is the slightest generalization of the fact [2] that the quantum dynamics of a harmonic oscillator in the Wigner representation is identical to the classical dynamics (Liouville flow) of a phase–space probability distribution.
[4] Y. Takahashi, H. Umezawa, Collect. Phenom. 2 (1975) 55.
[5] V.I. Tatarskii, Sov. Phys. Usp. 26 (1983) 311.
[6] N. Balazs, B.K. Jennings, Phys. Rep. 104 (1984) 347.
[7] M. Hillery, R.F. O'Connell, M.O. Scully, E.P. Wigner, Phys. Rep. 106 (1984) 121.
[8] M.O. Scully, M.S. Zubairy, Quantum Optics, Cambridge University Press, Cambridge, 1997.
[9] W.P. Schleich, Quantum Optics in Phase Space, Wiley-VCH, New York, 2001.
[10] B.-G. Englert, J. Phys. A 22 (1989) 625.
[11] G. García-Calderón, M. Moshinsky, J. Phys. A 13 (1980) L185;
A.K. Ekert, P.L. Knight, Phys. Rev. A 42 (1990) 487.
[12] The appearance of the reflection operator is central to experimental schemes for measuring Wigner functions. See e.g., B.-G. Englert, N. Sterpi, H. Walther, Opt. Commun. 100 (1993) 526;
K. Banaszek, K. Wódkiewicz, Phys. Rev. Lett. 76 (1996) 4344;
L.G. Lutterbach, L. Davidovich, Phys. Rev. Lett. 78 (1997) 2547.

Kean Loon Lee

Centre for Quantum Technologies, National University of Singapore, Singapore
Joint Quantum Centre Durham–Newcastle, School of Mathematics and Statistics, Newcastle University, Newcastle upon Tyne, United Kingdom

Paper Title: **Ultracold fermions in a graphene-type optical lattice**

Reference: Physical Review A **80**, 043411 (2009)

Personal message:

Berge has a broad interest in physics. In CQT, he started the FermiCold project group, which focuses on the theoretical study of cold fermionic atoms. This paper is a product of this theoretical effort. Berge played a key role in the proposal to simulate Dirac fermions with optical lattice, with the possibility of easily tuning the mass of the Dirac fermions. Such a setup has since been experimentally realized.

PHYSICAL REVIEW A **80**, 043411 (2009)

Ultracold fermions in a graphene-type optical lattice

Kean Loon Lee (李健伦),[1,2,3,*] Benoît Grémaud,[2,1,4] Rui Han (韩睿),[1] Berthold-Georg Englert,[1,4] and Christian Miniatura[5,1,4]

[1]*Centre for Quantum Technologies, National University of Singapore, 3 Science Drive 2, Singapore 117543, Singapore*
[2]*Laboratoire Kastler-Brossel, UPMC-Paris 6, ENS, CNRS; 4 Place Jussieu, F–75005 Paris, France*
[3]*NUS Graduate School for Integrative Sciences and Engineering, National University of Singapore, Singapore 117597, Singapore*
[4]*Department of Physics, National University of Singapore, 2 Science Drive 3, Singapore 117542, Singapore*
[5]*Institut Non Linéaire de Nice, UMR 6618, UNS, CNRS; 1361 route des Lucioles, 06560 Valbonne, France*

(Received 23 June 2009; published 19 October 2009)

Some important features of the graphene physics can be reproduced by loading ultracold fermionic atoms in a two-dimensional optical lattice with honeycomb symmetry and we address here its experimental feasibility. We analyze in great detail the optical lattice generated by the coherent superposition of three coplanar running laser waves with respective angles $2\pi/3$. The corresponding band structure displays Dirac cones located at the corners of the Brillouin zone and close to half-filling this system is well described by massless Dirac fermions. We characterize their properties by accurately deriving the nearest-neighbor hopping parameter t_0 as a function of the optical lattice parameters. Our semiclassical instanton method proves in excellent agreement with an exact numerical diagonalization of the full Hamilton operator in the tight-binding regime. We conclude that the temperature range needed to access the Dirac fermions regime is within experimental reach. We also analyze imperfections in the laser configuration as they lead to optical lattice distortions which affect the Dirac fermions. We show that the Dirac cones do survive up to some critical intensity or angle mismatches which are easily controlled in actual experiments. In the tight-binding regime, we predict, and numerically confirm, that these critical mismatches are inversely proportional to the square root of the optical potential strength. We also briefly discuss the interesting possibility of fine tuning the mass of the Dirac fermions by controlling the laser phase in an optical lattice generated by the incoherent superposition of three coplanar independent standing waves with respective angles $2\pi/3$.

DOI: 10.1103/PhysRevA.80.043411 PACS number(s): 37.10.Jk, 03.75.Lm, 03.75.Ss, 71.10.Fd

I. INTRODUCTION

In 2004, researchers in Manchester isolated one-atom thick sheets of carbon atoms, with the atoms organized in a planar honeycomb structure [1]. Such a material is referred to as graphene and is of utmost importance in condensed-matter physics since by stacking it one gets the graphite structure and by wrapping it one gets carbon nanotubes and fullerenes [2]. Graphene is also of great theoretical interest because it provides a physical realization of two-dimensional field theories with quantum anomalies [3]. Indeed, the effective theory that describes the low-energy electronic excitations in graphene is that of two-dimensional massless Weyl-Dirac fermions. In graphene these massless fermions propagate with about one 300th of the speed of light. Triggered by the Manchester discovery, an intense activity has flourished in the field, and continues to flourish, as witnessed by Refs. [4–9], for example. The reported and predicted phenomena include the Klein paradox (the perfect transmission of relativistic particles through high and wide potential barriers) [7], the anomalous quantum Hall effect induced by Berry phases [10,11], and its corresponding modified Landau levels [12].

It is now well established that some condensed-matter phenomena can be reproduced by loading ultracold atoms into optical lattices [13,14]. The great advantage is that the relevant parameters (shape and strength of the light potential, atom-atom interaction strength via Feshbach resonances [15], etc.) are accessible and can be accurately controlled while spurious effects that destroy the quantum coherence are absent such as the analog of the electron-phonon interaction. In [16], Zhu *et al.* proposed to observe Dirac fermions with cold atoms in a hexagonal optical lattice. Our present objective is to analyze in detail this scheme capable of reproducing in atomic physics the unique situation found in graphene. It consists of creating a two-dimensional honeycomb optical lattice and loading it with ultracold fermions such as the neutral ^{6}Li or ^{40}K atoms.

Parts of this paper recall known results. In addition to the need of setting the stage and introducing the notational conventions, there is also the intention to bridge the solid-state community and the atomic physics community on the particular subject of massless Dirac fermions as observed in graphene sheets and its counterpart in atomic physics. We also present extensions of previous solid-state works in the atomic physics context and report a number of results.

We analyze the various experimental parameters that need to be controlled in order to reproduce, with cold atoms trapped in an optical lattice, the physics at work in graphene. After briefly introducing optical lattices, we first explain how to create an optical lattice with the honeycomb symmetry and analyze its crystallographic features. We then calculate the band structure in the tight-binding approximation and by exact diagonalization, thereby providing evidence for the occurrence of the so-called Dirac points. Next, we evaluate the nearest-neighbor hopping amplitude by using a semiclassical instanton method. For the benefit of possible experiments we

*leekeanl@sps.nus.edu.sg

1050-2947/2009/80(4)/043411(18)

give the necessary requirements for reaching the massless Dirac fermion regime. Finally, we examine how massless Dirac fermions survive lattice distortions that could result from intensity-unbalanced or misaligned laser beams. These distortions open the way to new physics related to the quantum Hall effect [17]. We will close by briefly mentioning possible experiments to target for noninteracting and interacting ultracold fermions [18,19].

II. HONEYCOMB OPTICAL LATTICE

A. Radiative forces and optical lattices

A two-level atom (with angular frequency separation ω_{at} and excited-state angular frequency width Γ) that interacts with a monochromatic laser field with complex amplitude $\mathcal{E}(\boldsymbol{r},t)=\boldsymbol{E}(\boldsymbol{r})e^{-i\omega_L t}$ gets polarized and experiences radiative forces due to photon absorption and emission cycles [20,21]. When the light frequency is tuned far away from the atomic resonance, i.e., when the light detuning $\delta=\omega_L-\omega_{at}$ is much larger than Γ, the field-induced saturation effects are negligible and the atom essentially keeps staying in its ground state. In this situation, the atom-field interaction is dominated by stimulated emission processes where the atomic dipole absorbs a photon from one Fourier component of the field and radiates it back into the same or another one of these Fourier modes. In each such stimulated cycle, there is a momentum transfer to the atom and, as a net result, the atom experiences an average force in the course of time. This dipole force exerted by the field onto the atom in its ground state is conservative. It derives from the polarization energy shift of the atomic levels (ac Stark or light shifts) [22] and the dipole potential $V(\boldsymbol{r})$ is given by

$$V(\boldsymbol{r})=\frac{\hbar\Gamma}{8}\frac{\Gamma}{\delta}\frac{I(\boldsymbol{r})}{I_s}, \qquad (1)$$

where $I(\boldsymbol{r})=\epsilon_0 c|\boldsymbol{E}(\boldsymbol{r})|^2/2$ is the light field intensity (time-averaged energy current density) at the center-of-mass position $\boldsymbol{r}$ of the atom and I_s is the saturation intensity of the atom under consideration.

For multilevel atoms, the situation is more complicated as the dipole potential now depends on the particular atomic ground state sublevel under consideration. However, if the laser detuning δ is much larger than the fine and hyperfine structure splittings of the atomic electronic transition, then all ground-state atomic sublevels will essentially experience the same dipole potential. This common potential turns out to be given by Eq. (1) as well. Hence, by conveniently tailoring the space and time dependence of the laser field, one can produce a great variety of dipole potentials and thus manipulate the ground-state atomic motion.

Optical lattices are periodic intensity patterns of light obtained through the interference of several monochromatic laser beams [23]. By loading ultracold atoms into such artificial crystals of light one obtains periodic arrays of atoms. Indeed, as seen from Eq. (1), when the light field is blue-detuned from the atomic resonance ($\delta>0$), then the atoms can be trapped in the field-intensity minima whereas for red-tuned light ($\delta<0$) they can be trapped at the field-intensity maxima. Such arrays of ultracold atoms trapped in optical lattices have been used in a wide variety of experiments. As recently evidenced by the observation of the Mott-Hubbard transition with degenerate gases [24], they have proven to be a unique tool to mimic, test, and go beyond phenomena observed until now in the condensed-matter realm [14,25]. They also have a promising potential for the implementation of quantum simulators and for quantum information processing purposes [13,26,27].

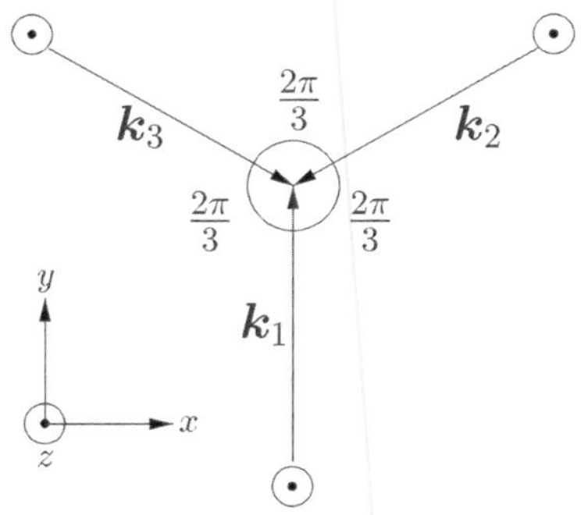

FIG. 1. The coplanar three-beam configuration used to generate the honeycomb lattice. All beams have the same frequency, strength, and linear polarization orthogonal to their common propagation plane. The honeycomb lattice under consideration is obtained for blue-detuned beams with respective angles $2\pi/3$. For these symmetric laser beams, the time-averaged radiation pressure—albeit small at large detuning—vanishes in this configuration. By reversing the propagation direction of one of the lasers, such that $\boldsymbol{k}_1=\boldsymbol{k}_2+\boldsymbol{k}_3$, say, a triangular lattice of a different geometry is formed. We will, however, exclusively deal with the $\boldsymbol{k}_1+\boldsymbol{k}_2+\boldsymbol{k}_3=0$ case.

B. Optical lattice with honeycomb structure

1. Field configuration and associated dipole potential

The simplest possible optical lattice with honeycomb structure is generated by superposing three coplanar traveling plane waves that have the same angular frequency $\omega_L=ck_L$, the same field strength $E_0>0$, the same polarization and the three wave vectors $\boldsymbol{k}_a$ form a trine: their sum vanishes and the angle between any two of them is $2\pi/3$,

$$\boldsymbol{k}_1+\boldsymbol{k}_2+\boldsymbol{k}_3=0, \quad \boldsymbol{k}_a\cdot\boldsymbol{k}_b=k_L^2\left(\frac{3}{2}\delta_{ab}-\frac{1}{2}\right) \qquad (2)$$

with $a,b=1,2,3$ and δ_{ab} is the Kronecker symbol [23]. As is illustrated in Fig. 1, we choose the x,y plane as the common plane of propagation and, to be specific, use

$$\boldsymbol{k}_1=k_L\boldsymbol{e}_y, \quad \left.\begin{matrix}\boldsymbol{k}_2\\ \boldsymbol{k}_3\end{matrix}\right\}=k_L\frac{\mp\sqrt{3}\boldsymbol{e}_x-\boldsymbol{e}_y}{2} \qquad (3)$$

for the parametrization of the wave vectors.

Further, we take all fields to be linearly polarized orthogonal to the plane, so that the three complex field amplitudes are given by

$$\mathcal{E}_a(\boldsymbol{r},t) = E_0 e^{i(\boldsymbol{k}_a\cdot\boldsymbol{r}-\phi_a)} e^{-i\omega_L t}\boldsymbol{e}_z, \tag{4}$$

where ϕ_a is the phase of the ath field for $t=0$ at $\boldsymbol{r}=0$. We note that a joint shift of the reference points in time and space,

$$t \to t - \frac{1}{3\omega_L}\sum_a \phi_a, \quad \boldsymbol{r} \to \boldsymbol{r} + \frac{2}{3k_L^2}\sum_a \phi_a \boldsymbol{k}_a, \tag{5}$$

removes the phases ϕ_a from Eq. (4), so that the simple choice $\phi_1=\phi_2=\phi_3=0$ is permissible, and we adopt this convention. In an experimental implementation, one would need to stabilize the phase differences $\phi_a-\phi_b$ to prevent a rapid jitter of the lattice that could perturb the atoms trapped in the potential minima.

Dipole potential (1) generated by the electric field $\boldsymbol{E}=\Sigma_a \mathcal{E}_a$ is of the form

$$V(\boldsymbol{r}) = V_0|f(\boldsymbol{r})|^2 = V_0 v(\boldsymbol{r}) \quad \text{with} \quad V_0 = \frac{\hbar\Gamma}{8}\frac{\Gamma}{\delta}\frac{I_0}{I_s}, \tag{6}$$

where I_0 is the intensity associated with the field strength E_0. The total dimensionless field amplitude $f(\boldsymbol{r})$ and the dimensionless optical potential $v(\boldsymbol{r})$ are given by

$$f(\boldsymbol{r}) = 1 + \exp(-i\boldsymbol{b}_1\cdot\boldsymbol{r}) + \exp(i\boldsymbol{b}_2\cdot\boldsymbol{r}) \tag{7}$$

and

$$v(\boldsymbol{r}) = 3 + 2\cos(\boldsymbol{b}_1\cdot\boldsymbol{r}) + 2\cos(\boldsymbol{b}_2\cdot\boldsymbol{r}) + 2\cos[(\boldsymbol{b}_1+\boldsymbol{b}_2)\cdot\boldsymbol{r}], \tag{8}$$

where $\boldsymbol{b}_1=\boldsymbol{k}_3-\boldsymbol{k}_1$ and $\boldsymbol{b}_2=\boldsymbol{k}_1-\boldsymbol{k}_2$ feature the reciprocal primitive vectors. For parametrization (3), we have

$$\left.\begin{matrix}\boldsymbol{b}_1\\ \boldsymbol{b}_2\end{matrix}\right\} = \kappa\frac{\boldsymbol{e}_x \mp \sqrt{3}\boldsymbol{e}_y}{2} \tag{9}$$

with $\kappa=|\boldsymbol{b}_a|=\sqrt{3}k_L$. One may further notice that the periodic patterns associated to each of the cosine terms in Eq. (8) have the same spatial period of $(2\pi/k_L)/\sqrt{3}$, about 58% of the laser wavelength $\lambda_L=2\pi/k_L$.

Linear combinations of the Brillouin vectors with integer coefficients define the reciprocal lattice $\widetilde{\mathcal{B}}$, a regular pattern in $\boldsymbol{k}$ space,

$$\widetilde{\mathcal{B}} = \{n_1\boldsymbol{b}_1 + n_2\boldsymbol{b}_2 | n_1,n_2 = 0, \pm 1, \pm 2, \ldots\}. \tag{10}$$

The reciprocal lattice is central to all studies of the dynamics of particles that move under the influence of the given periodic potential [28].

In particular, one domain in reciprocal space of utmost importance is the first Brillouin zone Ω defined as the so-called primitive Wigner-Seitz cell [28] of $\widetilde{\mathcal{B}}$; see Fig. 2. It is a *regular* hexagon but with the subtle feature that opposite edges are to be identified with each other since they can be related by a displacement vector in $\widetilde{\mathcal{B}}$. By the same token the three corners K_a (respectively, K_a') have to be identified with one another and we collectively denote them by K (respectively, K'). These two different corners K and K' are known in the graphene literature as the Dirac points for a reason that will become clear in the next section. Upon denoting $\boldsymbol{K}$ $\equiv\boldsymbol{K}_1$ and $\boldsymbol{K}'\equiv\boldsymbol{K}_1'$, their positions in Ω are given by the wave vector of the lasers that generate the optical honeycomb potential,

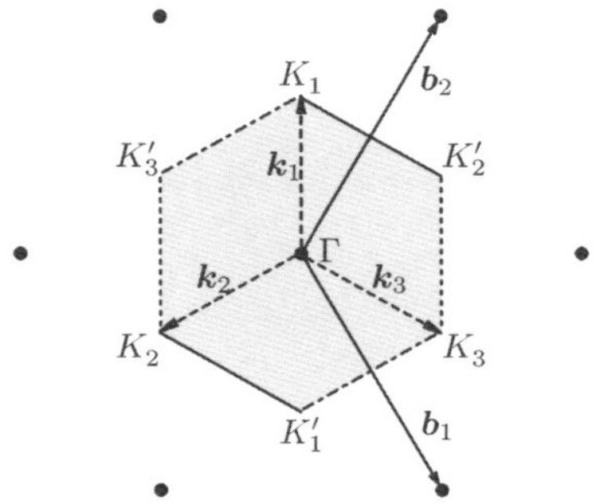

FIG. 2. The triangular reciprocal lattice $\widetilde{\mathcal{B}}$ associated with the triangular Bravais lattice of Fig. 4. It is spanned by the reciprocal primitive vectors $\boldsymbol{b}_1$ and $\boldsymbol{b}_2$ of Eq. (9) and is also a triangular lattice (as indicated by the full dots). The shaded region identifies the first Brillouin zone Ω which is here a regular hexagon. Its center is conventionally named Γ in the solid-state literature. Opposite edges are in fact identical as they only differ by a translation in the reciprocal lattice. This feature is emphasized by drawing the identical edges with the same (solid, dashed, or dash-dotted) line. For the same reason, the three corners K_a $(a=1,2,3)$ are to be identified with each other, and likewise the three corners K_a' are really only one point in Ω. Thus, only two of the six corners, collectively labeled as K and K' and known as the Dirac points, are different. Also shown are the wave vectors of the three coplanar plane waves (dashed arrows).

$$\boldsymbol{K} = -\boldsymbol{K}' = \frac{1}{3}(\boldsymbol{b}_2 - \boldsymbol{b}_1) = \boldsymbol{k}_1, \tag{11}$$

and $\boldsymbol{K}_2=\boldsymbol{k}_2=\boldsymbol{K}-\boldsymbol{b}_2$, $\boldsymbol{K}_3=\boldsymbol{k}_3=\boldsymbol{K}+\boldsymbol{b}_1$, as well as $\boldsymbol{K}_a=-\boldsymbol{K}_a'$.

2. Triangular Bravais lattice

The dimensionless potential (8) consists of a periodic two-dimensional array of maxima, minima, and saddle points generated by repeated translations of a primitive unit tile called the *basis*. The underlying lattice geometry itself is encapsulated in the associated Bravais lattice $\mathcal{B}$, that is

$$\mathcal{B} = \{m_1\boldsymbol{a}_1 + m_2\boldsymbol{a}_2 | m_1,m_2 = 0, \pm 1, \pm 2, \ldots\}, \tag{12}$$

such that the value of the potential is not affected by any displacement $\boldsymbol{R}\in\mathcal{B}$, $v(\boldsymbol{r}+\boldsymbol{R})=v(\boldsymbol{r})$.

The Bravais primitive vectors $\boldsymbol{a}_a$ are constructed based on the relation

$$\boldsymbol{a}_a\cdot\boldsymbol{b}_b = 2\pi\delta_{ab}. \tag{13}$$

In other words, the Bravais lattice $\mathcal{B}$ and the Brillouin lattice $\widetilde{\mathcal{B}}$ constitute dual spaces. Supplementing Eq. (9), we have the explicit parametrization

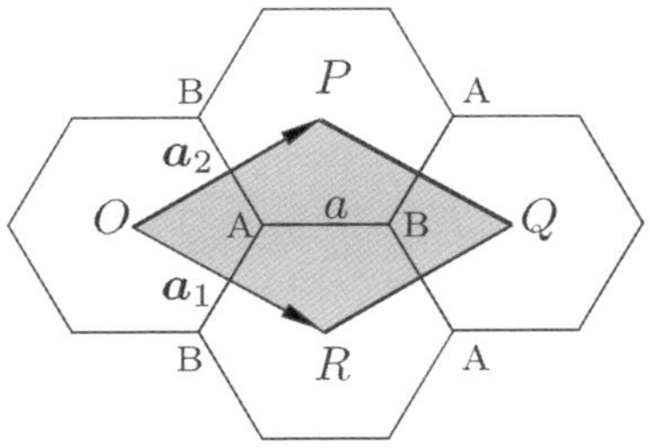

FIG. 3. The underlying Bravais lattice $\mathcal{B}$ of a two-dimensional honeycomb is the two-dimensional triangular Bravais lattice with a two-point basis A and B. The gray-shaded area is the primitive cell Σ. The honeycomb lattice constant a is defined as the distance between nearest-neighbor sites.

$$\left.\begin{matrix} \boldsymbol{a}_1 \\ \boldsymbol{a}_2 \end{matrix}\right\} = \Lambda \frac{\sqrt{3}\boldsymbol{e}_x \mp \boldsymbol{e}_y}{2}, \tag{14}$$

where $\Lambda = |\boldsymbol{a}_a| = 4\pi/(3k_L) = 2\lambda_L/3$ is the common length of the Bravais primitive vectors.

The Bravais lattice defined by Eq. (14) is a *triangular* one. We opt here for the diamond-shaped primitive cell Σ delineated by the two Bravais lattice vectors as a tiling for optical potential (8); see Fig. 3. Another possible choice would have been the hexagonal Wigner-Seitz cell [28]. This cell is useful when discussing the symmetry group of the lattice.

To proceed further one now needs to analyze the structure of optical potential (8) inside the primitive cell. In passing, we mention here that red-detuned ($\delta < 0$) lasers give $V_0 < 0$ and there is only one potential minimum in each primitive cell Σ. Upon trapping atoms in these potential minima, one gets a triangular lattice that is not of graphene type. This situation is interesting in view of quantum magnetism and frustration phenomena [14] but it is not the situation we want to study here.

3. Honeycomb structure

When the optical lattice is instead blue-detuned ($\delta > 0$), V_0 is positive and atoms are "weak-field seekers." The potential minima coincide with the minima of the electric field strength and the maxima coincide as well. By choice of coordinate system, the maxima locate at the Bravais sites and the dimensionless potential (8) has its maximal value of $v(\mathbf{0}) = 9$ at the corners O, P, Q, R of the diamond-shaped primitive cell Σ; see Fig. 4.

Two different potential minima, given by the zeros of the total dimensionless field amplitude $f(\boldsymbol{r})$, are found in Σ at

$$\boldsymbol{r}_{\mathrm{A}} = \frac{1}{3}(\boldsymbol{a}_1 + \boldsymbol{a}_2) = \frac{\Lambda}{\sqrt{3}}\boldsymbol{e}_x \text{ and } \boldsymbol{r}_{\mathrm{B}} = 2\boldsymbol{r}_{\mathrm{A}}, \tag{15}$$

respectively. From a crystallographic point of view, Σ is a primitive cell with a *two-point* basis. By applying repeated Bravais translations on Σ, one generates two different sublattices of potential minima, one made up of A-type sites and the other made of B-type sites; see Figs. 3 and 4. Altogether the potential minima are organized in a honeycomb structure reminiscent of the positions of the carbon atoms in graphene sheets.

The three displacements that move an A site to a neigh-

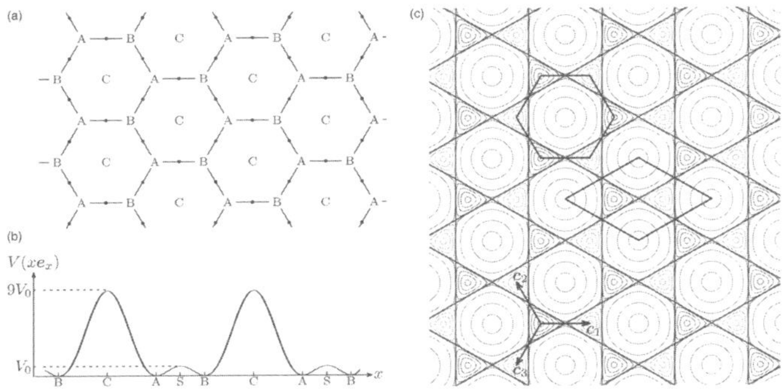

FIG. 4. (Color online) Left: the honeycomb pattern composed of the triangular lattices of minima at sites A and B, of maxima at sites C, as well as of the saddle points between neighboring A and B sites (marked by dots). The bottom plot shows the potential along the x axis which is one of the ...ABCABC... lines with $x=0$ at a C site. The saddle points S appear as local maxima here, with a height that is one ninth of the global maxima at sites C. Cold atoms trapped in this optical potential would be found at the A and B sites. Right: equipotential lines for the optical honeycomb potential (6). Along the straight black lines that connect the saddle points, we have $V(\boldsymbol{r}) = V_0$. The (red) closed circular curves filling out a hexagonal area are centered at the points of maximal potential; from inside out the respective values are $V(\boldsymbol{r}) = 8V_0$, $5V_0$, $2V_0$, and $1.05V_0$. The closed curves filling out areas of the shape of equilateral triangles are centered at the minima that constitute the A sublattice (blue) or the B sublattice (green); along the curves the potential has the values $V(\boldsymbol{r}) = 0.95V_0$, $0.6V_0$, $0.3V_0$, and $0.05V_0$. One primitive diamond-shaped unit tile Σ spanned by $\boldsymbol{a}_1$ and $\boldsymbol{a}_2$ is traced out. It contains two different minima, one of A type (in blue, on the left inside) and one of B type (in green, on the right inside). The trine of the A$\rightarrow$B displacement vectors (16) is indicated as well. Finally, for completeness, we also trace out the Bravais Wigner-Seitz unit tile. It is a hexagon centered at a potential maximum and with potential minima at its corners.

boring B site—they translate the A sublattice to the B sublattice—are parametrized by

$$c_1 = \frac{1}{3}(a_1 + a_2) = ae_x,$$

$$c_2 = \frac{1}{3}(a_2 - 2a_1) = a\frac{-e_x + \sqrt{3}e_y}{2},$$

$$c_3 = \frac{1}{3}(a_1 - 2a_2) = a\frac{-e_x - \sqrt{3}e_y}{2}, \qquad (16)$$

where $a=|c_j|=\Lambda/\sqrt{3}=4\pi/(3\kappa)=2\lambda_L/\sqrt{27}$ is the honeycomb lattice constant. It is the distance from an A site to a neighboring B site or the distance from the center of the hexagon of minima to one of its corners.

Halfway between two neighboring minima, the potential has saddle points where $v(r)=1$. They are located at the center and at the middle of the edges of Σ; see Fig. 4. As the saddle points on opposite sides of Σ are connected by Bravais displacements, there are, therefore, three nonequivalent triangular sublattices of saddle points and we thus count three saddle points per primitive cell.

We also note that the potential is invariant under 120° rotations around the locations of the potential minima and maxima and, therefore, that the potential is isotropic in the vicinity of these points. We anticipate that the local harmonic oscillator potential at a minimum will be isotropic; see Eq. (35) below. By contrast, the corresponding local potential at a saddle point is not isotropic.

All these matters are illustrated in Fig. 4, where we clearly identify the various triangular sublattices. Cold fermionic atoms trapped in this optical potential would be found at the A and B sites, similar to the binding of electrons in graphene to the carbon ions.

As a side remark, it may be worth mentioning that the saddle points affect the classical dynamics of a particle evolving in the honeycomb potential with a sufficiently large energy. Since the potential is nonseparable and angular momentum is not conserved here, the saddle points could be the seed for instabilities in which case the motion could turn out to be nonintegrable and chaotic. If so, this chaotic behavior should then be revealed, for example, in the statistical properties of the quantum spectra, whose level spacing fluctuations are expected to be described by the Gaussian orthogonal ensemble [29].

4. Optical honeycomb potential and graphene

In graphene sheets, the electrostatic potential that governs the dynamics of electrons, the sum of the Coulomb potentials of the carbon ions, exhibits the symmetries associated to a honeycomb pattern. Of course, in the finer details, the optical dipole potential of Eqs. (6) and (8) differs markedly from the graphene potential. In particular, the very strong forces that the electrons in graphene experience close to the ions have no counterpart in the optical lattice and the interaction between the atoms loaded into the optical potential is quite different from the electric repulsion between electrons. Nevertheless, the common symmetry group implies great similarities between the band structures of the two potentials, and in the respective parameter regimes where the tight-binding approximation is valid, the effective Hamilton operators are virtually identical. In particular, experiments made with atoms offer new knobs to play with and, with due attention to the difference between the two physical systems, these observations may deepen our understanding about phenomena observed with graphene samples.

In a very definite sense, honeycomb potential (6) is the simplest of all graphene-type potentials [30]. Their general form is a Fourier sum over the Brillouin vectors,

$$V(r) = \sum_{Q\in\tilde{B}} e^{iQ\cdot r} v_Q \text{ with } v_{-Q} = v_Q^*. \qquad (17)$$

The various symmetry properties of a honeycomb potential ensure that the v_Qs are grouped into sets of closely related coefficients. If one coefficient in Eq. (17) is nonzero, a whole set of closely related coefficients have corresponding nonzero values as well.

Other than the trivial constant solution $V(r)=v_0$, the simplest case is obtained when all coefficients vanish except for the set associated with $v_{b_1}=V_0$ and, by convention, $v_0=3V_0$. This yields honeycomb potential (6) with $v(r)$ of Eq. (8).

III. MASSLESS DIRAC FERMIONS

A. Band structure in the hopping picture

In the hopping picture, one envisions the particle as hopping from site to site with some quantum mechanical hopping (or tunneling) amplitude. In the simplest situation, all sites have the same energy, only hops between nearest-neighbor sites are considered, and all hopping amplitudes take on the same complex value t_0. The one-particle quantum dynamics is then conveniently described using second quantization. In the present situation, as we have two different sublattices, one has to introduce two sets of fermionic annihilation and creation operators, one for the A sites, $(a_{i\sigma}, a_{i\sigma}^\dagger)$, and one for the B sites $(b_{j\sigma}, b_{j\sigma}^\dagger)$, where i and j label the sites in the two-dimensional lattices while σ stands for the spin index or any other pertinent quantum number of the particle. The second-quantized Hamilton operator then reads

$$H = \sum_{\langle i,j\rangle,\sigma} (t_0 b_{i\sigma}^\dagger a_{j\sigma} + t_0^* a_{i\sigma}^\dagger b_{j\sigma}) + \epsilon \sum_{i\sigma} (a_{i\sigma}^\dagger a_{i\sigma} - b_{i\sigma}^\dagger b_{i\sigma}), \qquad (18)$$

where $\langle i,j\rangle$ means that only nearest neighbors are included in the sum. The model defined by this Hamilton operator accounts for hopping to neighboring lattice sites but does not permit a change in the internal quantum number σ during the hop. We have also included a possible energy mismatch ϵ between the A and B sites [3]. Using the Fourier transform in Ω of the fermionic operators, the right-hand side of Eq. (18) can be recast into the form

$$H = \sum_{k\in\Omega,\sigma} (a_{k\sigma}^\dagger, b_{k\sigma}^\dagger) \begin{pmatrix} \epsilon & z_k \\ z_k^* & -\epsilon \end{pmatrix} \begin{pmatrix} a_{k\sigma} \\ b_{k\sigma} \end{pmatrix} \qquad (19)$$

with

$$z_k = t_0 \sum_n e^{ik\cdot c_n}, \tag{20}$$

from which we get the band spectrum

$$\epsilon_\pm(k) = \pm\sqrt{\epsilon^2 + |z_k|^2}. \tag{21}$$

As expected from the fact that the honeycomb lattice consists of two distinct sublattices, we find two bands: a conduction band (+) and a valence band (−). These bands are here independent of the spin index σ meaning that each $k \in \Omega$ accommodates $2\sigma+1$ internal states per subband. Without any real loss of generality, we will stick to spin-$\frac{1}{2}$ fermions in the sequel. As readily checked, z_k vanishes when

$$1 + e^{ik\cdot a_1} + e^{ik\cdot a_2} = 0, \tag{22}$$

which is solved by the corners K and K' of Ω since $K\cdot a_2 = K'\cdot a_1 = 2\pi/3$. We thus see that the conduction and the valence bands are gapped by ϵ, a situation typical of a metal when the lattice is filled with particles. When there is exactly one particle per site (a situation known as half-filling), all levels in the valence band are filled at zero temperature and the Fermi energy E_F (the energy of the highest filled level) precisely cuts the energy surface at the K and K' points. In this case the low-energy excitations of the system can be described by linearizing the band spectrum in the neighborhood of K and K'. Denoting by $q=p/\hbar$ the small displacement from either K or K', the linearization of z_k gives

$$|z_k| \approx \frac{3a|t_0|}{2}|q| = \hbar v_0|q| = |p|v_0, \tag{23}$$

where the quantity $v_0 = 3a|t_0|/(2\hbar)$ is called the *Fermi velocity* in the solid-state community. We adopt this terminology although it is somewhat unfortunate, because it has nothing to do with the standard Fermi velocity $\sqrt{2E_F/m}$, which depends on the actual mass of the particle.

The dispersion relation now takes on the very suggestive form

$$\epsilon_\pm(p) \approx \pm\sqrt{m_*^2 v_0^4 + p^2 v_0^2} \tag{24}$$

that is typical of a relativistic dispersion relation with particle-hole symmetry. The effective mass m_*, defined through $\epsilon = m_* v_0^2$, appears thus as the rest mass of the excitations and relates to the energy imbalance of the two sublattices. The Fermi velocity v_0 is the analog of the velocity of light in relativity.

The effective Hamilton operator that is derived from these considerations and describes the dynamics of the excitations around K and K',

$$H = \int \frac{(dr)}{(2\pi)^2}\bar{\psi}(r)\begin{pmatrix} i\gamma\cdot\nabla + m_* & 0 \\ 0 & i\gamma\cdot\nabla - m_* \end{pmatrix}\psi(r), \tag{25}$$

where $\psi(r)$ is a four-component Dirac spinor encapsulating the excitations around K and K' while $\bar{\psi} = \psi^\dagger\begin{pmatrix}\gamma^0 & 0\\ 0 & \gamma^0\end{pmatrix}$ generates an equation of motion that resembles the Weyl-Dirac equation in two dimensions. This is why the name *Dirac points* is given to K and K' (see Refs. [3,9] for more details). In this two-dimensional context, the Dirac matrices are $\gamma^\mu = (\gamma^0, \gamma) = (\sigma_z, i\sigma_x, i\sigma_y)$ in terms of the standard Pauli matrices.

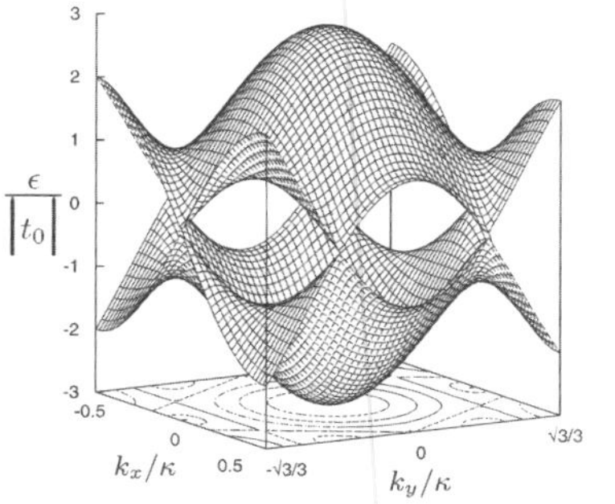

FIG. 5. The tight-binding band structure of graphene (in units of the tunneling strength $|t_0|$) as a function of $k \in \Omega$ in units of $\kappa = \sqrt{3}k_L$. The origin of energy has been chosen at the Dirac points and the axis ranges are $|k_x/\kappa| \leq 1/2$ and $|k_y/\kappa| \leq \sqrt{3}/3$. The bottom contour lines are lines of constant $|\epsilon|/|t_0|$.

When ϵ vanishes, as is the case in real graphene where all lattice sites have the same energy, then $\epsilon_\pm(k) = \pm|z_k|$ and the two bands are degenerate at the corners of Ω where they display circular conical intersections (see Fig. 5). In the literature, this situation is referred to as a semimetal or a zero-gap semiconductor and the corresponding low-energy excitations are known as massless Dirac fermions. The total band width is $W = 6|t_0|$ and, at half-filling, the Fermi energy $E_F = 3|t_0|$ (taking the energy origin at the lower band minimum) precisely slices the energy bands at the Dirac points. Hence, the Fermi surface reduces to these two points, so that the density of states vanishes there [9]; see Fig. 6.

B. Tight-binding approximation

Mindful of possible experiments, the hopping parameter t_0 appears to be an important amplitude to evaluate. We report three different methods for estimating its strength $|t_0|$. We will start with the familiar tight-binding approximation using localized Wannier functions [31,32] that are further approximated by Gaussians. We will then develop a more

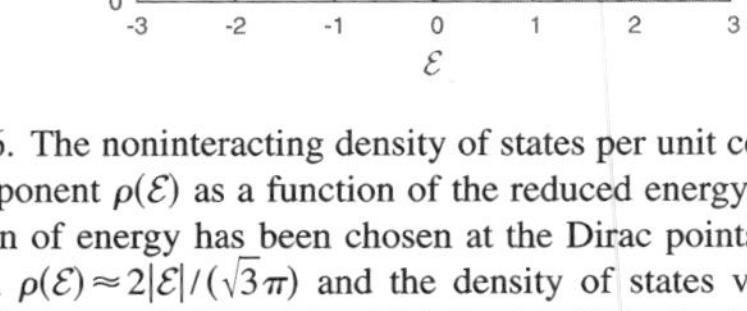

FIG. 6. The noninteracting density of states per unit cell and per spin component $\rho(\mathcal{E})$ as a function of the reduced energy $\mathcal{E} = E/|t_0|$. The origin of energy has been chosen at the Dirac points. When $\mathcal{E} \ll 1$, then $\rho(\mathcal{E}) \approx 2|\mathcal{E}|/(\sqrt{3}\pi)$ and the density of states vanishes at $\mathcal{E} = 0$, a signature of the semimetal behavior. Note the logarithmic Van Hove singularity at $|\mathcal{E}| = 1$.

accurate semiclassical calculation based on an instanton approach [33]. We will compare both results to a brute-force exact numerical computation.

As a consequence of Bloch's theorem [28,34], the energy spectrum of an atom of mass m moving in the honeycomb lattice potential is obtained from

$$\mathcal{H}\psi_{nk}(\boldsymbol{r}) = \left[-\frac{\hbar^2\boldsymbol{\nabla}^2}{2m} + V(\boldsymbol{r})\right]\psi_{nk}(\boldsymbol{r}) = \epsilon_n(\boldsymbol{k})\psi_{nk}(\boldsymbol{r}), \quad (26)$$

where we dropped the spin index σ which is not essential here. The Bloch waves ψ_{nk} are given by

$$\psi_{nk}(\boldsymbol{r}) = e^{i\boldsymbol{k}\cdot\boldsymbol{r}}u_{nk}(\boldsymbol{r}) \quad (27)$$

with $\boldsymbol{k}\in\Omega$, n the band index, and $u_{nk}(\boldsymbol{r})$ is a $\mathcal{B}$-periodic function. The latter can be conveniently expanded using Wannier functions [28,34,35] in accordance with

$$u_{nk}(\boldsymbol{r}) = \sum_{\boldsymbol{R}\in\mathcal{B}} e^{-i\boldsymbol{k}\cdot(\boldsymbol{r}-\boldsymbol{R})}w_n(\boldsymbol{r}-\boldsymbol{R}). \quad (28)$$

Wannier functions are very useful in describing models where particles are localized in space such as the Hubbard model [36]. They form an orthonormal basis set of functions centered at different Bravais lattice sites which are copies of the same "seed" functions defined in a given primitive cell. The localization properties of the Wannier functions crucially depend on the analyticity properties of u_{nk} as a function of $\boldsymbol{k}$ and decay exponentially in the simple cases [37–40].

In the tight-binding approximation, the atoms are assumed to be sufficiently deeply localized in the different potential wells where they only populate the lowest vibrational levels. Vibrational states in different wells are also assumed to have small overlap: the atomic motion is thus "frozen" except for the small tunneling amplitude between neighboring wells and are then effectively confined to move in the lowest bands of the lattice. Since the Wannier functions display the same symmetry as the local potential structure [41,42], the natural idea here is thus to construct tight-binding Wannier functions from linear combinations of wave functions deeply localized in the two potential wells of the primitive cell (the so-called atomic orbitals) [32,43]. This trial wave function exploits at best the sublattice structure of the honeycomb lattice and should give good results at least for the first two bands.

After dropping the band index n, this approach, reminiscent of the LCAO method (linear combination of atomic orbitals) [28,34], leads to the ansatz

$$\psi_k(\boldsymbol{r}) = \alpha_k\psi_k^{(\mathrm{A})}(\boldsymbol{r}) + \beta_k\psi_k^{(\mathrm{B})}(\boldsymbol{r}), \quad (29)$$

where the quasi-Bloch wave functions

$$\psi_k^{(\mathrm{A})}(\boldsymbol{r}) = \sum_{\mathrm{A}} e^{i\boldsymbol{k}\cdot\boldsymbol{r}_\mathrm{A}}w_\mathrm{A}(\boldsymbol{r}-\boldsymbol{r}_\mathrm{A}),$$

$$\psi_k^{(\mathrm{B})}(\boldsymbol{r}) = \sum_{\mathrm{B}} e^{i\boldsymbol{k}\cdot\boldsymbol{r}_\mathrm{B}}w_\mathrm{B}(\boldsymbol{r}-\boldsymbol{r}_\mathrm{B}) \quad (30)$$

essentially live on the type-A sublattice and the type-B sublattice, respectively. The sublattice Wannier functions $w_\mathrm{A}(\boldsymbol{r})$ and $w_\mathrm{B}(\boldsymbol{r})$ are normalized to unity. In the present case, we even have $w_\mathrm{B}(\boldsymbol{r})=w_\mathrm{A}(-\boldsymbol{r})$ due to the reflection symmetry of the potential, $V(-\boldsymbol{r})=V(\boldsymbol{r})$ [41]. We define the on-site energies as $E_a=\langle w_a|\mathcal{H}|w_a\rangle$ $(a=\mathrm{A},\mathrm{B})$ and use the parametrization $E_\mathrm{A}=E_0+\Delta$ and $E_\mathrm{B}=E_0-\Delta$ in the following, with E_0 the mean on-site energy and Δ half the on-site energy difference. Most importantly, the sublattice Wannier functions are orthogonal. However, obtaining their exact expressions is a difficult task and one often resorts to simple approximations that do not have this property. This is why, in view of this very common practical situation, we will consider in the following that the Wannier functions $w_\mathrm{A}(\boldsymbol{r})$ and $w_\mathrm{B}(\boldsymbol{r})$ can overlap.

Plugging now the ansatz [Eqs. (29) and (30)] into Eq. (26), and only considering coupling between nearest-neighbor lattice sites, we get the 2×2 homogeneous linear system

$$\begin{pmatrix} \Delta - E & Z_k - ER_k \\ Z_k^* - ER_k^* & -(\Delta+E)\end{pmatrix}\begin{pmatrix}\alpha_k \\ \beta_k\end{pmatrix} = 0, \quad (31)$$

where $E=\epsilon(\boldsymbol{k})-E_0$ and with the matrix entries

$$Z_k = \sum_a t_a e^{i\boldsymbol{k}\cdot\boldsymbol{c}_a},$$

$$t_a = \langle w_\mathrm{A}|(\mathcal{H}-E_0)|w_{\mathrm{B}_a}\rangle,$$

$$R_k = \sum_a \langle w_\mathrm{A}|w_{\mathrm{B}_a}\rangle e^{i\boldsymbol{k}\cdot\boldsymbol{c}_a}. \quad (32)$$

Here $\mathrm{B}_a=\mathrm{A}+\boldsymbol{c}_a$ is a short-hand notation for the three B sites next to the A site.

Several remarks are in order. First one notes that the off-diagonal matrix entries depend on the energy as soon as the sublattice Wannier functions overlap. Second, as readily checked, the hopping amplitudes t_a and E are independent of any energy shift in the Hamilton operator and are thus independent of any particular choice for the energy origin as one expects. Note also that the values of E_A and of E_B do not depend on the particular choice for point A or point B since $\mathcal{H}$ is $\mathcal{B}$-translation invariant. By the same token, the values of t_a and of $\langle w_\mathrm{A}|w_{\mathrm{B}_a}\rangle$ do not depend on the particular choice of A, but B must be one of its three nearest neighbors.

To have a nonzero solution, the 2×2 determinant associated to Eq. (31) has to vanish, from which we get the band structure. When the overlaps of the sublattice Wannier functions are small, $\langle w_\mathrm{A}|w_{\mathrm{B}_a}\rangle \ll 1$, the band structure is very well approximated by

$$\epsilon_\pm(\boldsymbol{k}) \approx E_0 \pm \sqrt{\Delta^2+|Z_k|^2}, \quad (33)$$

a form reminiscent of Eq. (21). For the honeycomb lattice, for which $\mathcal{H}$ is $\mathcal{B}$ periodic and invariant under reflections, we further have $E_\mathrm{A}=E_\mathrm{B}=E_0$ and $\Delta=0$, which implies that the effective mass m_* of the Dirac fermions is indeed zero. As a consequence, we get the two first bands as $\epsilon_\pm(\boldsymbol{k})=E_0\pm|Z_k|$. Furthermore, since $V(\boldsymbol{r})$ is also invariant under $2\pi/3$ rotations about any lattice site A, all three tunneling amplitudes t_a from A to B_a acquire the same value and Z_k of Eq. (32) turns into z_k of Eq. (20) with

$$t_0=\int(d\boldsymbol{r})w_A^*(\boldsymbol{r})(\mathcal{H}-E_0)w_A(\boldsymbol{r}-\boldsymbol{c}), \quad (34)$$

where $\mathcal{H}$ is the differential operator of Eq. (26) and $\boldsymbol{c}$ is either one of the three displacement vectors in Eq. (16).

C. Harmonic approximation

To proceed further one needs an approximation for the Wannier functions w_A and w_B. One possibility is to rely on the harmonic approximation of the potential wells around sites A and B, that is, to approximate w_A and w_B by the corresponding harmonic ground-state wave functions. We find

$$V(\boldsymbol{r}_a+\boldsymbol{r})\approx\frac{3}{4}V_0\kappa^2r^2=\frac{m\omega_0^2}{2}r^2 \text{ for } a=\text{A,B}$$
$$\text{with} \quad \hbar\omega_0=3\sqrt{V_0E_R}, \quad (35)$$

where $E_R=\hbar^2k_L^2/(2m)$ is the recoil energy of the atom. In terms of $\ell=\sqrt{\hbar/(m\omega_0)}$, the familiar length unit of the harmonic oscillator, the ground-state wave function is

$$w_A(\boldsymbol{r}_A+\boldsymbol{r})=w_B(\boldsymbol{r}_B+\boldsymbol{r})\approx\frac{1}{\sqrt{\pi}\ell}\exp\left(-\frac{1}{2}r^2/\ell^2\right). \quad (36)$$

From this we get $E_A=E_B=E_0\approx\hbar\omega_0$ and the overlap integrals are simply

$$\langle w_A|w_{B_a}\rangle=\exp\left(-\frac{2\pi^2}{9}\sqrt{\frac{V_0}{E_R}}\right). \quad (37)$$

Keeping in mind that $V_0\gg E_0\gg E_R$ in the tight-binding regime, $\langle w_A|w_{B_a}\rangle\ll1$ and we find from Eq. (34)

$$t_0\approx-\left(\frac{\pi^2}{3}-1\right)V_0\exp\left(-\frac{2\pi^2}{9}\sqrt{\frac{V_0}{E_R}}\right) \quad (38)$$

at leading order. However, since the hopping amplitude is given by the overlap integral of the localized wave functions w_A and w_B of two neighboring sites, we see that the value of t_0 crucially depends on the tails of these wave functions. Wannier functions often decay exponentially and, therefore, they cannot be realistically approximated by Gaussian wave functions. Hence, Eq. (38) can, at best, serve as a rough underestimate [44]. In the next section we will derive a reliable and accurate estimate of the tunneling amplitudes in the tight-binding regime by use of the instanton method.

D. Semiclassical estimate

Using k_L^{-1}, $\sqrt{V_0/m}$, V_0, and $\sqrt{m/(k_L^2V_0)}$ as length, velocity, energy, and time units, respectively, the Schrödinger equation can be conveniently recast into a dimensionless form that features an effective Planck's constant $\hbar_e$ (we keep the same symbols for the rescaled variables for simplicity),

$$i\hbar_e\partial_t\psi=-\frac{\hbar_e^2}{2}\nabla^2\psi+v(\boldsymbol{r})\psi, \quad \hbar_e=\sqrt{\frac{2E_R}{V_0}}, \quad (39)$$

with $v(\boldsymbol{r})$ given by Eq. (8), here expressed in rescaled units. In the tight-binding approximation it is assumed that $V_0\gg E_R$, and thus $\hbar_e\ll1$. In this situation, semiclassical methods provide very efficient and very accurate ways for evaluating dynamical and spectral quantities of interest. They generally amount to evaluating integrals with the aid of semiclassical expressions for the quantum propagator derived from its Feynman-path integral formulation through stationary-phase approximations around the classical trajectories [45].

For example, it is well known that the energy splitting between the two lowest energy levels of an atom moving in a one-dimensional symmetric double well can be accurately calculated using the WKB method [45]. This WKB method can be extended to several dimensions and in the sequel we will derive a semiclassical estimate of t_0 for the honeycomb lattice using the method proposed in [33]. It amounts to evaluating t_0 using the classical complex trajectory (in rescaled units) that connects A and B through the classically forbidden region—the so-called instanton trajectory.

Using $\hbar\omega_0$ as an order of magnitude for the vibrational level inside a potential well, we see that in the rescaled units, this energy is $\hbar\omega_0/V_0=3\hbar_e/\sqrt{2}\ll1$. So we can simply look for the instanton trajectory at zero energy. In rescaled units, the hopping amplitude is then expressed as

$$\frac{|t_0|}{V_0}=\alpha\sqrt{\hbar_e}e^{-S_0/\hbar_e}, \quad (40)$$

where S_0 is the (rescaled) classical action along the zero-energy instanton trajectory, and the numerical factor α is obtained from integrating out the fluctuations around the zero-energy instanton trajectory (see below).

As the zero-energy instanton fully runs in the classically forbidden region, the variables take on complex values. For our particular case, the good parametrization turns out to keep $\boldsymbol{r}$ real while taking $t=i\tau$ and $\boldsymbol{p}=-i\tilde{\boldsymbol{p}}$ purely imaginary with τ and $\tilde{\boldsymbol{p}}$ real. Hamilton's classical equations of motion in the new variables are just obtained from the original ones by flipping $V(\boldsymbol{r})$ to $-V(\boldsymbol{r})$. The symmetry of the potential dictates that the zero-energy instanton trajectory is simply the straight line connecting site A to B (see Fig. 4). In the following we calculate the instanton between A and A$+\boldsymbol{c}_1$. Integrating the equation of motions, one gets the instanton trajectory in the rescaled form $\boldsymbol{r}_0(\tau)=k_Lax_0(\tau)\boldsymbol{e}_x$ with

$$\tan[\pi x_0(\tau)/3]=-\sqrt{3}\coth[3\sqrt{2}\tau/4]. \quad (41)$$

The boundary conditions are $x_0=1$, $\dot{x}_0=0$ when $\tau\to-\infty$ and $x_0=2$, $\dot{x}_0=0$ when $\tau\to\infty$, meaning that the instanton starts at A with zero velocity and ends at B with zero velocity; the whole process requiring an infinite amount of time. This is indeed what is expected as both end points of the instanton are instable in the reversed potential picture. Since the energy associated with this instanton trajectory is zero, the classical action is simply

$$S_0=\int_{ka}^{2ka}dx|f(x,y=0)|=4\sqrt{2}\left(1-\frac{\pi}{3\sqrt{3}}\right)\approx2.237, \quad (42)$$

where $f(\boldsymbol{x},\boldsymbol{y})$ is given by Eq. (7).

The computation of α proves technically more demanding. Following [33], it is given by the product $\alpha_1\alpha_2$ with

$$\alpha_1 = \sqrt{\frac{S_0}{2\pi}}\sqrt{\frac{\det[-\partial_\tau^2+\omega_0^2]}{\det'[-\partial_\tau^2+\omega_x^2(\tau)]}},$$

$$\alpha_2 = \sqrt{\frac{\det[-\partial_\tau^2+\omega_0^2]}{\det[-\partial_\tau^2+\omega_y^2(\tau)]}}. \tag{43}$$

Here $\omega_a^2(\tau)=(\partial_a^2 v)(\boldsymbol{r}_0)$ $(a=x,y)$ is the curvature of the rescaled potential along the zero-energy instanton trajectory $\boldsymbol{r}_0(\tau)$ while ω_0 is the curvature of the rescaled harmonic potential approximation around A; see Eq. (35). In rescaled units, we have $\omega_0=3/\sqrt{2}$. The prime in the formula for α_1 means that the determinant is calculated by excluding the eigenspace of the operator $-\partial_\tau^2+\omega_x^2$ with the smallest eigenvalue.

The determinants of the differential operators involved in the computation of α stem from the linear stability analysis of the dynamical flow in the neighborhood of the zero-energy instanton trajectory as encapsulated in the monodromy matrix. They can be straightforwardly computed from solutions of the linear Jacobi-Hill equations of degree 2 associated with these differential operators [46]. For example, α_2 is solved as

$$\alpha_2 = \lim_{T\to\infty}\sqrt{\frac{J_0(T)}{J(T)}}, \tag{44}$$

where the Jacobi fields $J(\tau)$ and $J_0(\tau)$ satisfy the differential equations

$$\frac{d^2J(\tau)}{d\tau^2} - \omega_y^2(\tau)J(\tau) = 0,$$

$$\frac{d^2J_0(\tau)}{d\tau^2} - \omega_0^2 J_0(\tau) = 0, \tag{45}$$

with initial conditions

$$J_0(-T) = J(-T) = 0,$$

$$\dot{J}_0(-T) = \dot{J}(-T) = 1. \tag{46}$$

The interested reader is referred to [33,46] for details. We simply give here the final result for the honeycomb lattice:

$$\alpha_1 = \sqrt{\frac{27\sqrt{2}}{\pi}} \approx 3.486, \quad \alpha_2 \approx 0.449, \quad \alpha \approx 1.565. \tag{47}$$

Recasting the semiclassical calculation of the tunneling amplitude in units of the recoil energy finally yields

$$\frac{|t_0|}{E_R} \approx 1.861\left(\frac{V_0}{E_R}\right)^{3/4}\exp\left[-1.582\sqrt{\frac{V_0}{E_R}}\right]. \tag{48}$$

The same type of scaling laws has been obtained in the case of the two-dimensional square optical lattice [47,44]. In the square-lattice geometry, however, the potential is separable and the semiclassical calculation proves much simpler as it reduces to using the well-known Mathieu equation.

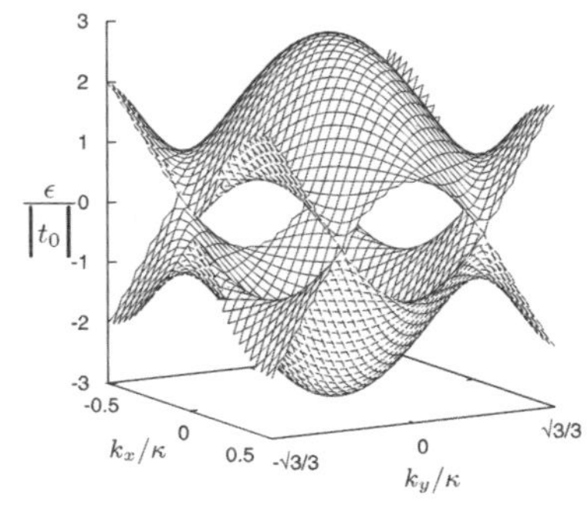

FIG. 7. Numerically calculated band structure of the two lowest energy bands for $\hbar_e$=0.25 at discrete points in the Brillouin zone Ω. The same conventions as in Fig. 5 are adopted. The value of $|t_0|$ is determined by requiring that $\epsilon_\pm = \pm 3|t_0|$ at the center Γ of the Brillouin zone. The similarity with Fig. 5 shows that at V_0=32E_R the tight-binding regime has already been reached.

E. Numerical computation of the band structure

Using Bloch's theorem and plugging Eq. (27) into Eq. (26), we get a family of partial differential equations for the $u_{n\boldsymbol{k}}$s labeled by the Bloch vector $\boldsymbol{k}\in\Omega$. After scaling variables with the same units as in the previous paragraph, the band structure is then extracted by numerically solving

$$\mathcal{H}_{\boldsymbol{k}}u_{n\boldsymbol{k}}(\boldsymbol{r}) = \epsilon_{n\boldsymbol{k}}u_{n\boldsymbol{k}}(\boldsymbol{r}),$$

$$\mathcal{H}_{\boldsymbol{k}} = \frac{\hbar_e^2}{2}(-i\boldsymbol{\nabla}+\boldsymbol{k})^2 + v(\boldsymbol{r}) \tag{49}$$

for each $\boldsymbol{k}\in\Omega$ (expressed now in units of k_L).

The $u_{n\boldsymbol{k}}$s being $\mathcal{B}$ periodic, they are conveniently Fourier expanded in the reciprocal lattice $\tilde{\mathcal{B}}$ according to

$$u_{n\boldsymbol{k}}(\boldsymbol{r}) = \sum_{\boldsymbol{Q}\in\tilde{\mathcal{B}}} C_{n\boldsymbol{Q}}e^{i\boldsymbol{Q}\cdot\boldsymbol{r}}. \tag{50}$$

The matrix representation of $\mathcal{H}_{\boldsymbol{k}}$ is sparse and banded. It is then appropriately truncated and diagonalized such that only a small number of coefficients $C_{n\boldsymbol{Q}}$ are actually significant for the corresponding energy bands. The energy bands obtained in this way are exact and one can investigate their dependence on $\hbar_e$ as done in Figs. 7 and 8.

The essential feature is to realize that the band degeneracies at points K and K' are generic and do not depend on the actual value of the effective Planck's constant. Indeed the existence of two degeneracy points in the first Brillouin zone for the honeycomb lattice is a general consequence of the lattice symmetries [48,49]. The lattice symmetries are encapsulated in the point group of the lattice which is the set of operations that leave fixed one particular point of the lattice. The corresponding elements are rotations, reflections, inversions, and their combinations. Combined with $\mathcal{B}$ translations, one gets the space group of the lattice. The graphene point

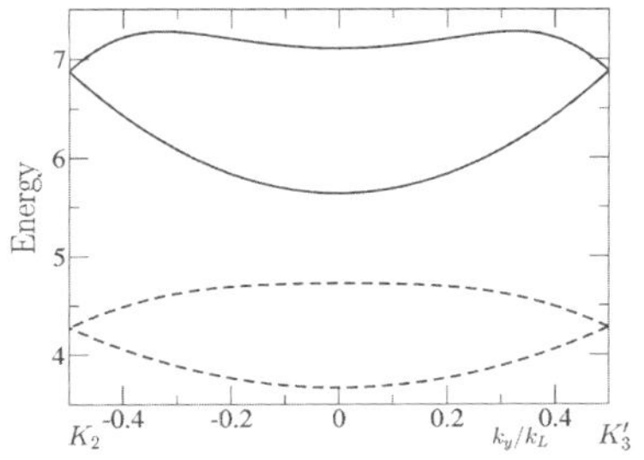

FIG. 8. Band structure for nearly free particles moving in a weak honeycomb optical potential in units of V_0. The first two levels are plotted as a function of k_y/k_L at $k_x/k_L=\sqrt{3}/2$, so along the vertical edge of Ω from K_2 to K_3'; see Fig. 2. The solid curves are obtained for $\hbar_e=\sqrt{2E_R/V_0}=\sqrt{10}$ and the dashed ones for $\hbar_e=\sqrt{5}$. As one can see the band structure is rather flat in the band center but the level curvature increases when $\hbar_e$ is increased. For small V_0 values, the energies of the bands are larger than V_0, that is, above the saddle point separating the A and B sites, emphasizing that the corresponding Bloch states are not anymore localized in these two sites. The Dirac degeneracies in the ground state obtained at the Brillouin zone corners are generic and can be inferred from group-theoretic considerations. Note, however, that the conical intersections do not extend much over the first Brillouin zone when the potential is weak but start to spread when $\hbar_e$ is decreased.

group has been analyzed by Lomer [48] and contains 12 elements. In terms of Bloch wave functions ψ_{nk}, the lattice space-group operations translate into point-group operations on $\boldsymbol{k}\in\Omega$, possibly followed by a reciprocal lattice translation to bring back the resulting new wave vector in Ω. The key point is that degeneracies can only occur at Bloch wave vectors which are invariant (up to reciprocal lattice translations) under the action of a non-Abelian subgroup G of the point group. For the graphene this happens at the Dirac points. For example, at corner K_1, beside unity, G is made of two rotations of angles $\pm 2\pi/3$ about the center Γ of the Brillouin zone and three reflexions about the lines connecting Γ to the three points labeled K. This group of order 6 admits an irreducible two-dimensional representation which explains the band spectrum degeneracy at the Dirac points.

This can be nicely illustrated in the weak V_0 limit (or equivalently when $\hbar_e$ is large). In this case, the particles are quasifree; i.e., the energies of the bands are larger than V_0, that is, above the saddle point separating the A and B sites (see Fig. 8). The band spectrum can be understood in two steps. First, one folds the parabolic dispersion relation of the free particle into the first Brillouin zone (repeated-zone scheme [28]) and then one couples crossing levels at Bragg planes by the weak potential. At K_1, three plane waves fold with the same kinetic energy, namely, $\boldsymbol{K}_1=\boldsymbol{k}_1$, $\boldsymbol{K}_2=\boldsymbol{k}_2$, and $\boldsymbol{K}_3=\boldsymbol{k}_3$ (see Fig. 2). The weak periodic potential then couples these three plane-wave states and the coupling matrix elements are all identical. The eigenstates of this 3×3 matrix split into a singlet and a doublet. When V_0 is negative, the singlet is the ground state which is consistent with the triangular Bravais lattice obtained in this case ($\delta<0$). When V_0 is positive ($\delta>0$), the doublet becomes the ground state and

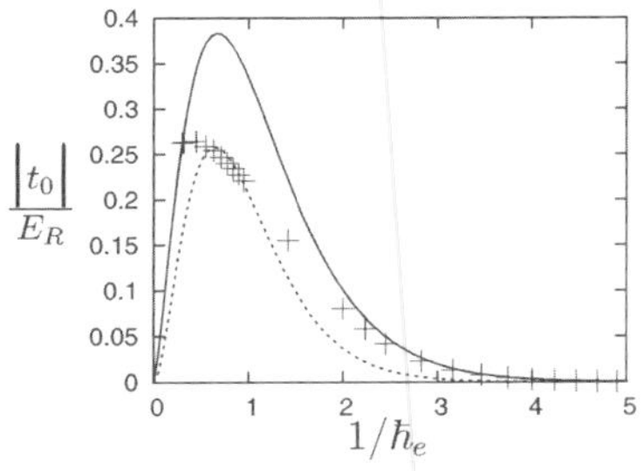

FIG. 9. The hopping parameter $|t_0|$ in units of the recoil energy E_R (crosses) as a function of the inverse of the effective Planck's constant $\hbar_e=\sqrt{2E_R/V_0}$ as obtained from the exact numerical computation. The harmonic approximation (dashed curve) and the semiclassical calculation (solid curve) of the hopping parameter have been added for comparison even if their range of validity is restricted to the tight-binding regime $\hbar_e \ll 1$.

features the tip of the conical intersection between the two subbands when the quasimomentum is moved away from K; see Fig. 8.

From the exact numerical calculation, one can extract the slope of the dispersion relation at the Dirac points and then the corresponding tunneling strength $|t_0|$ as a function of $\hbar_e^{-1}$; see Fig. 9. Figure 10 gives the comparison between the exact calculation, the harmonic, and the semiclassical calculations as a function of $\hbar_e^{-1}$ in the tight-binding regime where $\hbar_e \ll 1$. As one can see, the harmonic approximation is way off whereas the semiclassical estimate proves excellent.

F. Reaching the massless Dirac fermion regime

To access the massless Dirac fermion regime one first needs to completely fill the ground-state band alone, a situation known as half-filling. This is achieved by having exactly one fermion with spin state $\sigma=\pm 1/2$ per Bravais cell.

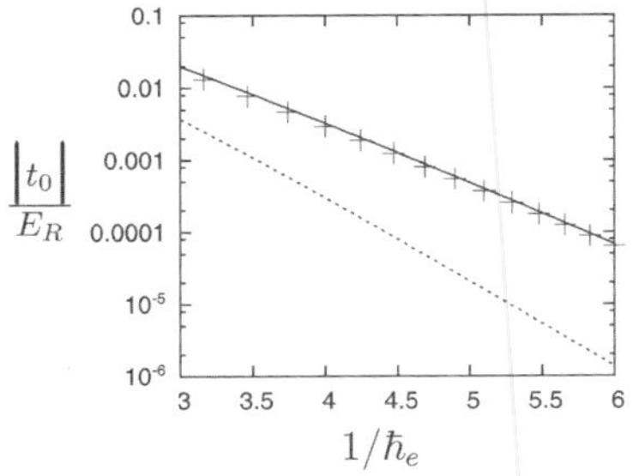

FIG. 10. The hopping parameter $|t_0|$ (in units of the recoil energy E_R) as a function of the inverse effective Planck's constant $\hbar_e=\sqrt{2E_R/V_0}$ in the tight-binding regime where $\hbar_e \ll 1$. As one can see, the harmonic approximation (dashed curve) is completely off. For example at $V_0=32E_R$ (or $\hbar_e=0.25$) $|t_0|$ is underestimated by a factor 10 and the discrepancy gets worse as V_0 increases. On the other hand, the agreement between the semiclassical calculation (solid curve) and the exact numerical computation (crosses) just proves excellent.

Starting from a spin-unpolarized cloud of fermions, half-filling is thus reached by loading the optical honeycomb lattice with exactly two fermions per Bravais cell, corresponding to an average number density $\rho=1$ in the tight-binding picture. When this is achieved, the Fermi energy slices the band structure at the Dirac points. For experiments that study transport phenomena, one would also need to subsequently dope the sample away from half-filling such that the Fermi energy of the system is varied in the linear part of the band structure.

In a usual experiment, atoms are generally held in an external harmonic potential and the optical lattice potential is superimposed. Reaching half-filling could then be done in two steps, first by significantly increasing repulsive interactions U between fermions through a Feshbach resonance and then by driving the system into the Mott-Hubbard phase with one fermion per site as done in [50,51]. Then setting U to zero again should maintain the system at number density $\rho=1$. Obvious candidates for such experiments are ^{40}K as well as ^{6}Li atoms [50–52]. In the external trap, the Mott insulator appears first where the local filling is approximately one atom per site and one needs to ensure that adding more atoms (or increasing the chemical potential μ) does not favor the appearance of the doubly occupied Mott phase. This will be the case for very strong repulsion $U \gg \mu, t_0, k_B T$ in which case one expects the entire center of the trap to contain a Mott insulating phase with single occupancy and negligible thermally activated doubly occupied sites. In the case of the honeycomb lattice, starting from a spin-unpolarized sample, it is known that half-filling is reached for $U_c \sim 5t_0$, the atoms displaying at the same time an antiferromagnetic order [18]. Note that $U_c \sim W$, where $W=2E_F=6|t_0|$ is the bandwidth.

Doping the system could be done in the following way. The external harmonic confinement (with angular trap frequency Ω_t) defines a characteristic length $\zeta=\sqrt{2|t_0|/(m\Omega_t)}$ over which the energy is shifted by precisely the tunneling energy t_0 [53–55]. This length defines the distance over which one given lattice site is coupled by tunneling to its neighbors. In turn, having loaded N_F fermions into the trap, one can define a characteristic filling factor through $\tilde{\rho}=N_F(a/\zeta)^2$. Varying $|t_0|$ by changing the lattice potential height V_0 or tightening or loosening the trap by changing Ω_t would thus allow to tune $\tilde{\rho}$ in a controlled way and hence to dope the system.

For the conical intersection at the Dirac points to significantly spread over the Brillouin zone Ω, one needs to reach the tight-binding regime where V_0 is large enough (typically $V_0 > 10E_R$ will do). Inspection of the Taylor expansion of Eq. (20) then shows that it is sufficient to have $|\boldsymbol{q}|a \ll 2$ ($\boldsymbol{q}$ being the small displacement from a Dirac point) for the band structure to be well approximated by a linear dispersion relation around the Dirac points. The available energy range ΔE is thus set by the bandwidth itself, namely, $\Delta E \ll W$. So tuning the filling factor away from half-filling and residual thermal fluctuations will keep the system in the massless Dirac fermion regime provided $\mu, k_B T \ll W$ (Fig. 11). For example, at $V_0=32E_R$, the temperature constraint, as derived from Eq. (48), is $T<T_R/50$ whereas it is $T<T_R/2$ at $V_0=10E_R$. There is thus room left for reaching the massless Dirac fermion regime within the current state-of-the-art cooling technology.

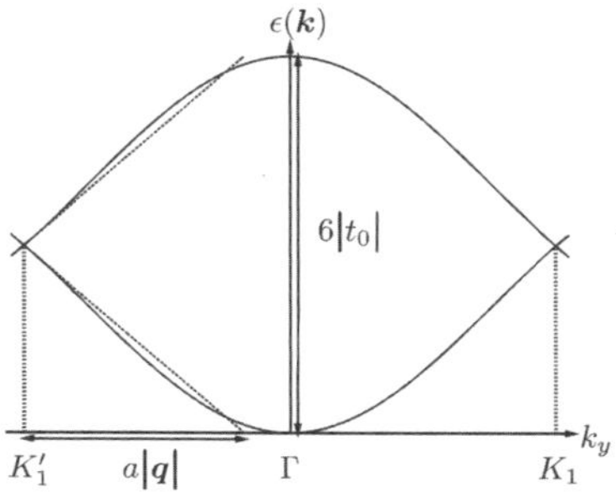

FIG. 11. Cut of the linear dispersion approximation along Oy at $k_x=0$ in the first Brillouin zone Ω as compared to the actual band spectrum in the tight-binding regime. At half-filling, the Fermi energy cuts the band spectrum at the Dirac points K and K'. Doping the system away from half-filling moves the Fermi energy up or down but the system can still be described in terms of massless Dirac fermions provided $a|\boldsymbol{q}| \ll 2$, i.e., provided the change in the Fermi energy is much less than the bandwidth $W=6|t_0|$ itself. By the same token, thermal excitations of the system can still be described as thermal massless Dirac fermions provided $k_B T \ll W$.

IV. ROBUSTNESS OF THE MASSLESS DIRAC FERMIONS

As the very existence of the massless Dirac fermion regime rests on the two conical degeneracies in the band structure, one may wonder if this regime would resist imperfections of the system. Indeed the argument we gave to explain the conical degeneracies relied on group-theoretic arguments which were specific to the hexagonal symmetry of the honeycomb structure. In practice, it is impossible to control the laser configuration to the point where all intensities and alignment angles would all be exactly equal. Such imperfections in the system would obviously break the hexagonal symmetry and one could think that the Dirac fermions would just be destroyed. In fact, as we will see shortly, massless Dirac fermions are quite robust and survive small imperfections that are easily within experimental reach.

A. Imbalanced hopping amplitudes

To understand why massless Dirac fermions are robust, we will start by analyzing the case of imbalanced hopping amplitudes as done in [56]. For real graphene, this would correspond to stretching the graphene sheet. In this case, the tight-binding band structure is given by $\epsilon_{\pm}(\boldsymbol{k}) = \pm|Z_{\boldsymbol{k}}|$, where $Z_{\boldsymbol{k}}$ is defined in Eq. (20). The degeneracies are found at points $\boldsymbol{k}_D \in \Omega$ canceling $Z_{\boldsymbol{k}}=0$. This condition boils down to sum up three vectors to zero in the two-dimensional plane; see Fig. 12. As such, a solution is only possible provided the hopping amplitudes satisfy one of the norm inequalities given by

$$||t_2| - |t_3|| \leq |t_1| \leq |t_2| + |t_3| \tag{51}$$

and cyclic permutations. If this is the case, defining the angles $\varphi_{1,2}=\arg t_{2,3}-\arg t_1$, the Dirac points solve

$$\cos(\boldsymbol{k}_D \cdot \boldsymbol{a}_1 - \varphi_1) = \frac{|t_3|^2 - |t_2|^2 - |t_1|^2}{2|t_1 t_2|},$$

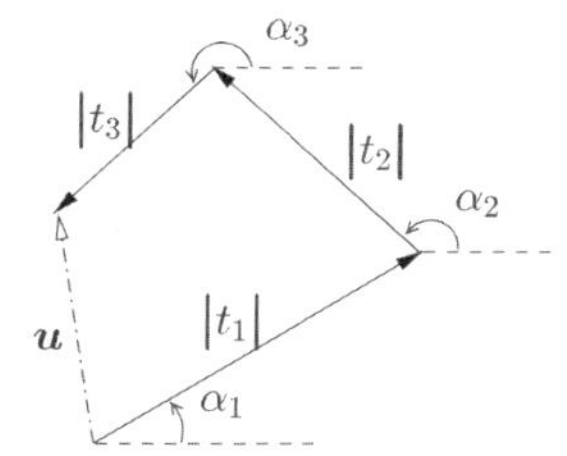

FIG. 12. The condition $Z_k=0$ is equivalent to cancel the resultant vector **u** of three vectors, each with length $|t_n|$ and polar angle $\alpha_n=\boldsymbol{k}\cdot\boldsymbol{c}_n+\arg(t_n)$. There will always be a solution provided one of the norm inequalities $||t_2|-|t_3||\leq|t_1|\leq|t_2|+|t_3|$ (and cyclic permutations) is satisfied.

$$\cos(\boldsymbol{k}_D\cdot\boldsymbol{a}_2-\varphi_2)=\frac{|t_2|^2-|t_3|^2-|t_1|^2}{2|t_1t_3|}, \tag{52}$$

subject to the condition

$$|t_2|\sin(\boldsymbol{k}_D\cdot\boldsymbol{a}_1-\varphi_1)+|t_3|\sin(\boldsymbol{k}_D\cdot\boldsymbol{a}_2-\varphi_2)=0. \tag{53}$$

We find the important result that the system self-adapts to changes in the hopping amplitudes by shifting the Dirac points away from the corners of the Brillouin zone until the norm inequalities (51) break and degeneracies disappear. Thus, provided the hopping imbalance is not too strong, the massless Dirac fermions do survive imperfections in the system and the hexagonal symmetry breaking.

We illustrate this important feature in the simple case of only one imbalanced hopping amplitude, namely, $t_1=\gamma t_0$ and $t_2=t_3=t_0$. We further choose γ real and $0<|\gamma|\leq 2$ for the Dirac points to exist. We then find two Dirac points D_γ and D'_γ given by $\boldsymbol{k}_D=-\boldsymbol{k}'_D=\varphi_0(\boldsymbol{b}_2-\boldsymbol{b}_1)$ where $\varphi_0\in[0,1/2]$ solves $\cos(2\pi\varphi_0)=-\gamma/2$. This means that the two Dirac points D_γ and D'_γ move along opposite paths in the Brillouin zone Ω. The fact that Dirac points always come in by pairs of opposite location in Ω is generic [57]. When γ is increased from 0 to 2, D_γ starts at $\boldsymbol{k}_0=(3k_L/4)\boldsymbol{e}_y$ for $\gamma=0$, then moves along axis Oy and reach corner K_1 at $\gamma=1$. Note that when $\gamma\to 0$, the physical situation is that of weakly coupled "zigzag" linear chains. For $\gamma>1$, D_γ leaves Ω but a translation in reciprocal lattice brings it back on the vertical edges of Ω (technically we get two copies of the same point). D_γ reaches the middle of the vertical edge at $\gamma=2$ where it merges with D'_γ into a *single* Dirac point; see Fig. 13. Interesting physics occurs at $\gamma=2$ in connection with the quantum Hall effect [17,58]. As soon as $\gamma>2$, the degeneracy is lifted and the massless Dirac fermions do not exist anymore. For negative γ, D_γ and D'_γ move back from $\pm(3k_L/4)\mathbf{e}_y$ to the center Γ of the Brillouin zone where they merge and disappear; see Fig. 13. The fact that Dirac points can only merge at the center and mid-edge points of Ω is also generic [57]. Hence, far from being a nuisance, we see that controlling the hopping amplitude imbalance proves an interesting way of exploring the massless Dirac fermions physics under different circumstances by moving around the Dirac points in the Brillouin zone.

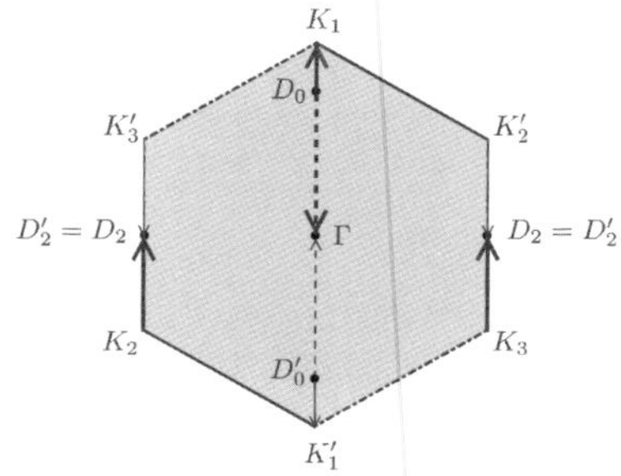

FIG. 13. When the three hopping amplitudes t_n are unbalanced, the Dirac points are shifted in the Brillouin zone Ω and disappear when the norm inequality $||t_2|-|t_3||\leq|t_1|\leq|t_2|+|t_3|$ is no longer satisfied. We depict here how the Dirac points D_γ and D'_γ move in Ω when only one hopping amplitude is imbalanced, namely, $t_1=\gamma t_0$ and $t_2=t_3=t_0$. Points D_γ (thick path) and D'_γ (thin path) move along opposite paths. Increasing γ from 0, point D_γ starts at D_0 and moves upward. It reaches point K_1 at $\gamma=1$ (balanced amplitudes case) then moves along the vertical edge of Ω where it reaches its middle point D_2 at $\gamma=2$. The Dirac points cease to exist when $\gamma>2$. For negative γ, D_γ moves downward from D_0 (dotted thick path), reaches the zone center Γ for $\gamma=-2$, and then ceases to exist for $\gamma<-2$.

B. Optical lattice distortions

The previous discussion concentrated on the impact of imbalanced hopping amplitudes irrespective of the change in symmetry of the lattice potential. We will now analyze these lattice distortions in more detail and give quantitative estimates about the experimental degree of control which is required to target the massless Dirac fermion regime. We will consider in-plane laser beams with different (positive) strengths $E_n=s_nE_0$ and with respective angles away from $2\pi/3$; see Fig. 14. It is important to note that we will always stick to imperfections which are compatible with a two-point

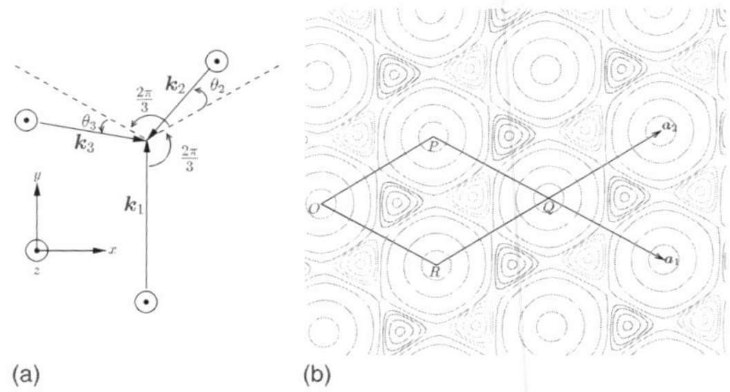

FIG. 14. (Color online) (a) The asymmetric in-plane three-beam configuration. Three monochromatic and linearly polarized laser beams with wave vectors $\boldsymbol{k}_n$ interfere with different strengths $E_n=s_nE_0$ $(n=1,2,3)$. The respective angles depart from $2\pi/3$. (b) Distorted optical lattice obtained with $\vartheta_3=\vartheta_2=5\times10^{-2}$ and $s_1=1$, $s_2=1.03$, and $s_3=0.97$. The color convention is the same as in Fig. 4. For weak enough distortions, the primitive diamond-shaped cell Σ still contains two field minima as evidenced in the plot.

Bravais cell. They will only induce distortions of the hexagonal spatial structure of field minima but without breaking this pattern.

The new optical lattice potential is now given by $V'(\boldsymbol{r}) = V_0 |f'(\boldsymbol{r})|^2$ with the new total dimensionless field amplitude

$$f'(\boldsymbol{r}) = s_1 + s_2 \exp(-i\boldsymbol{b}'_1 \cdot \boldsymbol{r}) + s_3 \exp(i\boldsymbol{b}'_2 \cdot \boldsymbol{r}). \quad (54)$$

Here the $\boldsymbol{b}'_n$ $(n=1,2)$ feature the new reciprocal lattice basis vectors. They define in turn a new set of Bravais lattice basis vectors $\boldsymbol{a}'_n$ giving rise to a new primitive diamond-shaped cell Σ'. Unless the angle mismatches vanish, the new Bravais and reciprocal lattices are no longer hexagonal but oblique with no special symmetry except for inversion. As a consequence, the new first Brillouin zone Ω' is still a hexagon but *no longer a regular one*.

Since we assume a two-point primitive cell, the minima of the new optical potential still identify with zeros of $f'(\boldsymbol{r})$. Similarly with the case of imbalanced hopping amplitudes, we find two solutions if and only if the field strengths s_n satisfy one of the norm inequalities $|s_2 - s_3| \leq s_1 \leq s_2 + s_3$ (and cyclic ones). In this case the minima are given by

$$\cos(\boldsymbol{b}'_1 \cdot \boldsymbol{r}) = \frac{s_3^2 - s_2^2 - s_1^2}{2 s_1 s_2},$$

$$\cos(\boldsymbol{b}'_2 \cdot \boldsymbol{r}) = \frac{s_2^2 - s_3^2 - s_1^2}{2 s_1 s_3}, \quad (55)$$

subject to the condition $s_2 \sin(\boldsymbol{b}'_1 \cdot \boldsymbol{r}) = s_3 \sin(\boldsymbol{b}'_2 \cdot \boldsymbol{r})$. In the following we will examine separately the effect of strength imbalance and angle mismatch.

1. Critical field strength imbalance

To give an estimate of the critical field strength imbalance beyond which the Dirac points cannot survive, we consider the simple case of only one imbalanced laser beam and no angle mismatch, namely, $\theta_2 = \theta_3 = 0$, $s_1 = 1 + \eta$, and $s_2 = s_3 = 1$. In this case the Bravais lattice, the reciprocal lattice, the primitive cell Σ, and the Brillouin zone Ω are *not* modified. The new optical potential $V'(\boldsymbol{r}) = V_0 v'(\boldsymbol{r})$ reads

$$v'(\boldsymbol{r}) = v(\boldsymbol{r}) + 2\eta \delta v(\boldsymbol{r}) + \eta(\eta + 2),$$

$$\delta v(\boldsymbol{r}) = \cos(\boldsymbol{b}_1 \cdot \boldsymbol{r}) + \cos(\boldsymbol{b}_2 \cdot \boldsymbol{r}), \quad (56)$$

where $v(\boldsymbol{r})$ is given by Eq. (8). Note that when only one field strength is imbalanced, the corresponding potential still displays a reflection symmetry. In the present case, it is the Ox-reflection symmetry because $V'(\boldsymbol{r})$ is invariant under the exchange $\boldsymbol{b}_1 \leftrightarrow \boldsymbol{b}_2$. Requiring now that the primitive cell Σ exhibits two field minima imposes $-1 \leq \eta \leq 1$. Their positions in Σ are given by $\boldsymbol{r}'_{\mathrm{A,B}} = \varphi_{\mathrm{A,B}}(\boldsymbol{a}_1 + \boldsymbol{a}_2)$ with $\cos(2\pi\varphi_{\mathrm{A,B}}) = -(1+\eta)/2$. Their midpoint $\boldsymbol{r}'_S = (\mathbf{r}'_{\mathrm{A}} + \mathbf{r}'_{\mathrm{B}})/2 = (\boldsymbol{a}_1 + \boldsymbol{a}_2)/2$ is a saddle point and defines the potential barrier height V'_S to cross to go from A and B in Σ. One finds $V'_S = (\eta - 1)^2 V_0$.

As a whole the field minima organize in a hexagon which is stretched (η negative) or compressed (η positive) along Ox; see Fig. 15. As a consequence two of the three new vectors $\boldsymbol{c}'_n$ joining one minimum to its three nearest neighbors will have equal length. In the present situation we get $|\boldsymbol{c}'_2| = |\boldsymbol{c}'_3| \neq |\boldsymbol{c}'_1|$. The potential barrier height V''_S to cross to go from A to B along $\boldsymbol{c}'_2$ and $\boldsymbol{c}'_3$ is given by the corresponding saddle points located at the middle of the edges of Σ. One finds $V''_S = (\eta + 1)^2 V_0$.

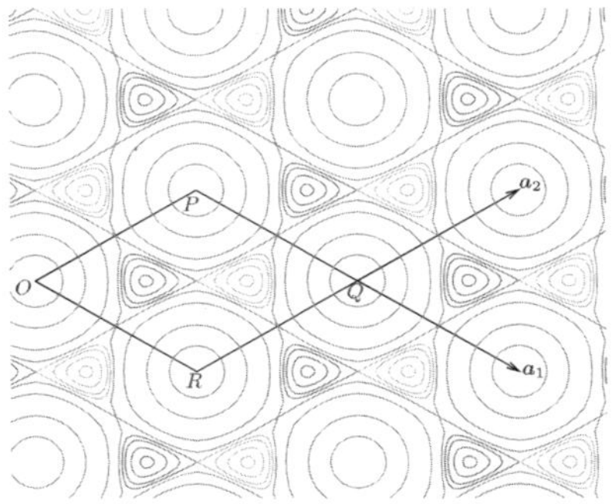

FIG. 15. (Color online) Slightly distorted lattice obtained with vanishing mismatch angles and one imbalanced field strength, namely, $s_1 = 10/9$ and $s_2 = s_3 = 1$. The color convention is the same as in Fig. 4. In this particular case the hexagon of field minima is slightly squeezed along the horizontal axis Ox and the vectors $\boldsymbol{c}'_n$ connecting a given minimum to its three nearest neighbors have now different lengths. In the situation depicted $|\boldsymbol{c}'_2| = |\boldsymbol{c}'_3| \neq |\boldsymbol{c}'_1|$. In turn, due to the reflection symmetry about Ox, the tight-binding hopping amplitudes satisfy $|t_2| = |t_3| \neq |t_1|$.

Now, when η is increased from 0, the minima move closer along $\boldsymbol{c}'_1$ and move away along $\boldsymbol{c}'_2$ and $\boldsymbol{c}'_3$. At the same time, the potential barrier V'_S along $\boldsymbol{c}'_1$ is lowered and the potential barrier V''_S along $\boldsymbol{c}'_2$ and $\boldsymbol{c}'_3$ is increased. As a net effect, in the tight-binding picture, we expect the tunneling amplitude $|t_1|$ to increase while $|t_2|$ and $|t_3|$ decrease. We get the opposite conclusion when η is lowered from 0. Since the potential is invariant through $\boldsymbol{b}_1 \leftrightarrow \boldsymbol{b}_2$, we further have $|t_2| = |t_3|$ and we recover the case of one imbalanced hopping amplitude analyzed in the previous section.

One could try to derive a semiclassical expression of the t_n as a function of η using the instanton method but, actually, such a tedious calculation proves unnecessary, at least when η is small. Indeed, by inspection of the semiclassical expression (40), we expect the ratio $|t_1/t_2|$ to scale as $\exp[\Delta S(\eta)/\hbar_e]$ at leading order, where $\Delta S(\eta)$ is the action difference between the two instanton trajectories linking sites A and B along $\boldsymbol{c}'_2$ and $\boldsymbol{c}'_1$, respectively. For small enough η we expect $\Delta S(\eta)$ to grow linearly with η, the slope being positive since the ratio $|t_1/t_2|$ should increase with η. The Dirac degeneracies disappear when this ratio is 2 (see previous section), thus we get the semiclassical prediction that this will happen when $\eta \propto \hbar_e$. This result can also be inferred by saying that the Dirac points will disappear as soon as the perturbing potential $2\eta\delta V(\boldsymbol{r})$, see Eq. (56), strongly mixes the unperturbed states. This will happen when the corresponding coupling energy equals the mean level spacing of the unperturbed system and we get back to the prediction $\eta \propto \hbar_e$.

To check our semiclassical prediction we have computed, for each value of the effective Planck's constant $\hbar_e$, the

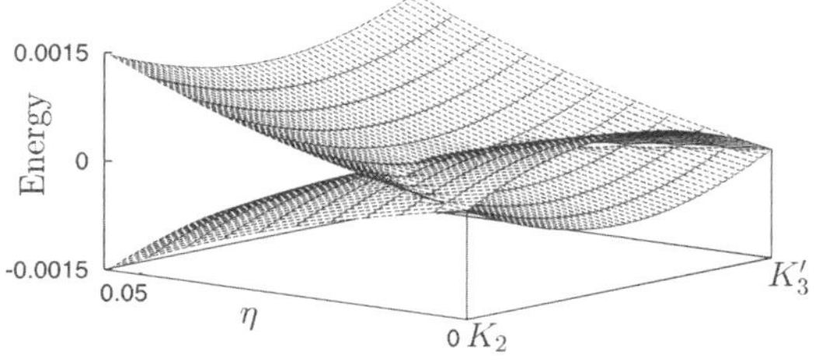

FIG. 16. The band diagram for the two lowest levels as a function of η for $V_0=80E_R$ ($\hbar_e \approx 0.158$). The bands are plotted along the vertical straight line joining the Dirac points K_2 and K_3' of the balanced situation; see Fig. 2. The origin of energy is fixed at the Fermi energy for a half-filled band and all bands have been shifted such that the upper and lower bands intersect at zero energy difference.

ground-state and first-excited-state levels for different values of η and we have extracted the corresponding critical value η_c for which the Dirac degeneracies are lifted. Figure 16 gives an example of the band structure obtained at $\hbar_e = 1/\sqrt{40} \approx 0.16$ for η ranging from 0 to 0.054. We have then plotted η_c as a function of $\hbar_e$; see Fig. 17. We have fitted the data with the quadratic fit function $\alpha\hbar_e + \beta\hbar_e^2$ and found $\alpha \approx 0.1074$ and $\beta \approx 0.0624$ enforcing the very good agreement obtained with our linear prediction in the semiclassical regime $\hbar_e \ll 1$. The quadratic correction could certainly be inferred from semiclassical higher-order corrections.

We would like to emphasize at this point that increasing or decreasing η from 0 is not symmetrical. When η is decreased from 0, the Dirac degeneracies are predicted to disappear when $|t_1/t_2| \to 0$. However, the best that we can do is to let $\eta \to -1$. This unfortunately means that one laser beam is almost extinguished and the situation is more that of very weakly coupled one-dimensional chains, a situation we postpone to future study as it proves interesting for high-T_c superconductivity [59]. We thus see that decreasing slightly η from 0 does not harm the Dirac degeneracies. They move inside Ω but do survive. By contrast, increasing slightly η from 0 does destroy the Dirac degeneracies as soon as $\eta \sim \hbar_e$.

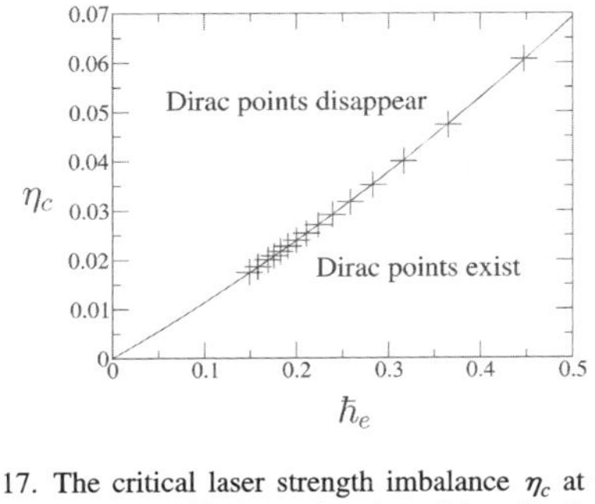

FIG. 17. The critical laser strength imbalance η_c at which the Dirac degeneracies are lifted as a function of the effective Planck's constant $\hbar_e = \sqrt{2E_R/V_0}$. The solid line corresponds to a quadratic fit of the numerical data. The linear coefficient is $\alpha \approx 0.1074$ while the quadratic one is $\beta \approx 0.0624$. As one can see our numerical results are in good agreement with our semiclassical prediction $\eta_c \propto \hbar_e$. The degree of control of the intensity imbalance of the laser fields gets more stringent as the optical lattice depth V_0 is increased. Nevertheless, at already $V_0=20E_R$ ($\hbar_e \approx 0.3$), the laser intensities should all equal within 8% which does not sound particularly demanding.

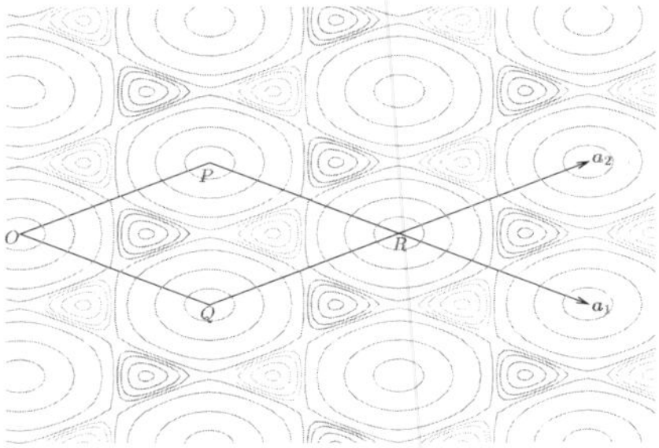

FIG. 18. (Color online) Distorted lattice obtained with balanced field strengths $s_n=1$ and angle mismatch $\theta_3=-\theta_2=-\pi/10$. The color convention is the same as in Fig. 4. In this particular case the hexagon of field minima is stretched along the horizontal axis Ox and the vectors $\boldsymbol{c}_n'$ connecting a given minimum to its three nearest neighbors have now different lengths. In the situation depicted $|\boldsymbol{c}_2'| = |\boldsymbol{c}_3'| \neq |\boldsymbol{c}_1'|$. In turn, due to the reflection symmetry about Ox, the tight-binding hopping amplitudes satisfy $|t_2| = |t_3| \neq |t_1|$.

As one can see from the plots, the tolerance about the intensity mismatch of the laser beams increases with $\hbar_e$ or equivalently when the optical lattice depth V_0 decreases. On the other hand, as we already saw, the Dirac cones do not extend much over the Brillouin zone if V_0 is too small. So there is a trade-off to make. The situation is, however, really favorable since the intensity mismatch tolerance is already in the 10% range for $V_0 \sim 10E_R$. This means that the massless Dirac fermions prove quite robust and should be easily accessed experimentally.

2. Critical in-plane angle mismatch

We now estimate the critical angle mismatch when all laser beams have the same intensities ($s_1=s_2=s_3=1$). We see from Eq. (54) that the new optical potential still displays the exchange symmetry $\boldsymbol{b}_1' \leftrightarrow \boldsymbol{b}_2'$ and thus a reflection invariance with respect to their bisectrix. In the following we stick to the simple case where $\theta_3=-\theta_2=\theta$ and θ is small. In this case both the Bravais lattice, the reciprocal lattice, the Brillouin zone Ω and the diamond-shaped primitive cell Σ get modified. The new reciprocal basis vectors turn out to be $\boldsymbol{b}_1' = \boldsymbol{b}_1 + \delta\boldsymbol{b}_1$ and $\boldsymbol{b}_2' = \boldsymbol{b}_2 + \delta\boldsymbol{b}_2$ where $\delta\boldsymbol{b}_1 = (\theta/\sqrt{3})\boldsymbol{b}_2$ and $\delta\boldsymbol{b}_2 = (\theta/\sqrt{3})\boldsymbol{b}_1$. Since the exchange symmetry $\boldsymbol{b}_1 \leftrightarrow \boldsymbol{b}_2$ is again preserved, the new potential continues to display the Ox-reflection invariance. Figure 18 gives a plot of the new potential structure for $\theta=-\pi/10$.

This situation boils down again to the case of one imbalanced tunneling amplitude. Indeed, the angle between the $\boldsymbol{b}_1'$ and $\boldsymbol{b}_2'$ decreases when θ is increased from 0. In turn the angle between the corresponding $\boldsymbol{a}_n'$ increases and the hexagon structure made by the A and B minima get compressed

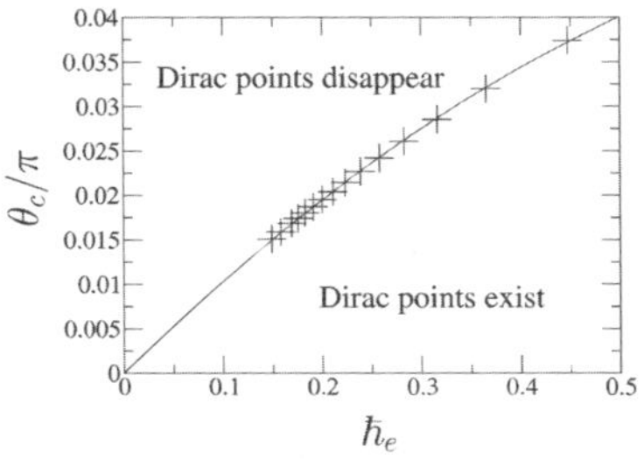

FIG. 19. The critical angle mismatch θ_c (in units of π) beyond which the Dirac degeneracies disappear as a function of the effective Planck's constant $\hbar_e=\sqrt{2E_R/V_0}$. The dashed line corresponds to a quadratic fit of the numerical data. The linear coefficient is 0.109 while the quadratic one is −0.0577. As one can see our numerical results are in good agreement with our semiclassical prediction $\theta_c \propto \hbar_e$. The degree of control of the angle mismatch gets more stringent as the optical lattice depth V_0 is increased. Nevertheless, at already $V_0=20E_R$ ($\hbar_e \approx 0.3$), the angle mismatch should be less than 5° which is not particularly demanding.

along Ox. The opposite conclusion holds when θ is decreased from 0. We get again the situation where $|t_2|=|t_3| \neq |t_1|$ and $|t_1/t_2| \gtrless 1$ when $\theta \gtrless 0$ and vice versa. Like for the field strength imbalance, the situations $\theta>0$ and $\theta<0$ are not symmetric. The massless Dirac fermions prove more sensitive to *closing* the angle between the $\boldsymbol{b}_n'$, so for $\theta_3=-\theta_2=\theta>0$ because $|t_1/t_2|$ then increases and the threshold $|t_1/t_2|=2$ is more rapidly hit. This is the situation we explore.

Applying the same reasoning as before, we thus predict the critical angle mismatch beyond which the massless Dirac fermions are destroyed to scale as $\theta_c \propto \hbar_e$. Again, to get θ_c as a function of $\hbar_e$, we numerically compute the band structure at a given $\hbar_e$ for different in-plane mismatch angles θ and then extract the value θ_c for which the Dirac degeneracy is lifted. We then repeat the procedure for different $\hbar_e$. As one can see, our prediction is in very good agreement with the numerical calculations, see Fig. 19, and well supported by a quadratic fit. As θ_c increases with $\hbar_e$, there is a trade-off to make between reaching the tight-binding regime where V_0 is large and achieving an experimentally reasonable angle mismatch tolerance which requires V_0 to be small. The trade-off turns out to be a favorable one since already for $V_0=20E_R$ ($\hbar_e \approx 0.3$), one gets a tolerance of about 5° on the laser beam alignment. We expect the same type of scaling for small out-of-plane angle mismatches. Furthermore, when several small distortions combine, their effects should add up and thus the critical imperfection threshold should still scale with $\hbar_e$.

As an overall conclusion we see that massless Dirac fermions are quite robust to moderate lattice distortions. Demonstrating them in an experiment should not be particularly demanding in terms of the control of the laser configuration.

C. Inequivalent potential wells

We finally briefly mention how to distort the optical lattice in a systematic manner as it allows for an experimental control of the mass of the Dirac fermions as well as for a continuous switch from a honeycomb lattice to a triangular one. In Sec. II B 4, we observed that honeycomb potential (6) is the simplest of all graphene-type potentials characterized by choosing v_0 and $v_{\boldsymbol{b}_1}$ real (in fact, positive) while putting all unrelated coefficients in Eq. (17) to zero. Now, letting $v_{\boldsymbol{b}_1}$ to acquire a phase φ, such that $e^{-i\varphi}v_{\boldsymbol{b}_1}$ is positive, will break the reflection symmetry of the honeycomb potential [30].

In the $\boldsymbol{r}$-dependent part of the dimensionless potential (8), this phase φ is introduced by the replacement

$$\sum_{a=1}^{3} \cos(\boldsymbol{b}_a \cdot \boldsymbol{r}) \to \sum_{a=1}^{3} \cos(\boldsymbol{b}_a \cdot \boldsymbol{r} + \varphi), \tag{57}$$

where $\boldsymbol{b}_3=-\boldsymbol{b}_1-\boldsymbol{b}_2$. This can be implemented by superimposing three *independent* standing waves, of the same wavelength and with equal intensity, whose wave vectors form the trine of Fig. 1 [60]. As a consequence of the incoherent superposition, the t replacement of Eq. (5) is not available and the $\boldsymbol{r}$ replacement alone cannot remove all three phases of the standing waves. One can, however, shift $\boldsymbol{r}$ such that the three phases are the same, and then one has an intensity pattern proportional to the right-hand side of Eq. (57).

Most of the hexagon structure of Fig. 4 remains unchanged by this modification: lattice sites A, B, and C continue to be the locations of local minima and maxima, whereas the saddle points S acquire new positions on the ...ABCABC... lines.

Figure 20 confirms that, for small φ values, the minima of the honeycomb dipole potential are still organized in a hexagonal pattern but we now have *different* potential depths at sites A and B. The potential energy mismatch is $2\epsilon \approx 8V_0|\varphi|/\sqrt{3}$. In view of Eqs. (21) and (24), this means that the Dirac fermions acquire a mass $m_* \propto |\varphi|$ or, in other words, that the Dirac degeneracies are lifted. The possibility of fine tuning the mass of the Dirac fermions through the parameter φ is an interesting experimental knob to play with.

Increasing $|\varphi|$ further, one can also see that, for the particular values $|\varphi|=\pi/6$ and $\pi/2$, the three sublattices of saddle points merge into a single triangular lattice, which coincides with the A, B, or C lattice, respectively; see Fig. 20. This merging of a potential minimum or maximum with three saddle points leads to a peculiar third-order saddle point. For $\varphi=\pi/6$, say, the S sites merge with the B sites and we have

$$\sum_a \cos(\boldsymbol{b}_a \cdot \boldsymbol{r} + \varphi)|_{\varphi=\pi/6} \approx -\frac{1}{6}\sum_a [\boldsymbol{b}_a \cdot (\boldsymbol{r}-\boldsymbol{r}_{\rm B})]^3 \tag{58}$$

for $|\boldsymbol{r}-\boldsymbol{r}_{\rm B}| \ll \kappa^{-1}$, hence a cubic saddle point rather than the usual quadratic saddle point.

An unpolarized ultracold gas of spin-$\frac{1}{2}$ fermions loaded into such a potential at half-filling would lead to two fermions per well. By driving the system through attractive interactions, one could even get a Mott insulator of fermion pairs. By switching off all interactions and setting $\varphi=0$, one should be able to study oscillations of atoms between the A and B

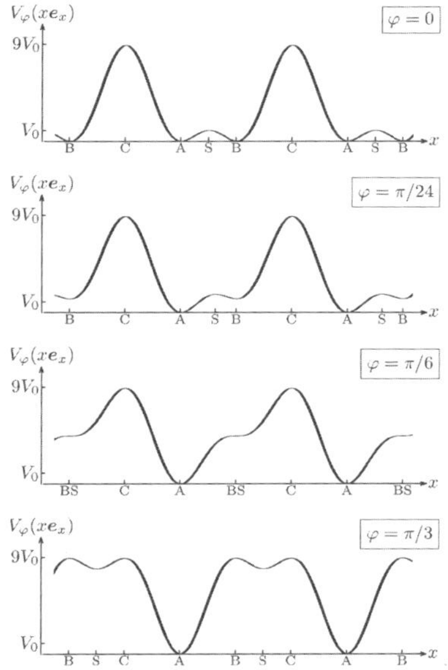

FIG. 20. For various values of the phase parameter φ of Eq. (57), the plot shows the potential energy along a ...ABCABC... line in Fig. 4. The top plot, for $\varphi=0$, repeats the bottom-left plot of Fig. 4 for reference. The degeneracy between sites A and B is lifted for the small φ value of $\varphi=\pi/24$; the saddle points have moved closer to the B sites, where we continue to have local minima. In this situation the Dirac fermions acquire a mass $m_* \propto |\varphi|$. When $\varphi = \pi/6$, the saddle points S coincide with the B sites and we have cubic saddle points there. Finally, in the bottom plot, we have $\varphi = \pi/3$ and the saddle points are halfway between adjacent B and C sites, with potential maxima at both of them. Except for a displacement, the potential in the bottom plot is the negative of the potential in the top plot and thus identical with honeycomb potential (6) for red rather than blue detuning of the three running wave lasers. For ease of comparison, the potential constants are adjusted such that the maxima and minima are at $V=0$ and $V=9V_0$, respectively, for all φ values.

sublattices. We will analyze this situation in a follow-up paper.

V. CONCLUSION

Motivated by the vivid field of graphene physics, we have explained and analyzed how to reproduce massless Dirac fermions by loading ultracold fermions in an optical lattice with honeycomb structure. We have described the two-dimensional laser configuration that gives rise to an optical potential where field minima are organized in a honeycomb structure (with lattice constant a) and we have thoroughly detailed the corresponding crystallographic features. The behavior of atoms propagating in such an optical potential in the tight-binding regime is in one-to-one correspondence with the behavior of electrons propagating in a graphene sheet. The ground-state and first-excited levels of the band structure exhibit two conical degeneracies located at the corners of the first Brillouin zone, as dictated by symmetry arguments. In the neighborhood of these degeneracies, the band spectrum is linear.

When the lattice is loaded with fermions at half-filling, the Fermi energy slices the band structure at these degeneracy points known as the Dirac points. Around half-filling, the tight-binding Hamilton operator can then be recast in a form reminiscent of the relativistic Weyl-Dirac Hamilton operator and featuring so-called massless Dirac fermions. The important parameter driving the dynamics turns out to be the hopping amplitude t_0 between nearest-neighbor sites as it gives the bandwidth $W=6|t_0|$ and the "Fermi velocity" $v_0 = 3a|t_0|/(2\hbar)$. We have derived a semiclassical expression for $|t_0|$ in terms of the effective Planck's constant of the problem, namely, $\hbar_e = \sqrt{2E_R/V_0}$ (with V_0 the optical potential strength and E_R the recoil energy) and have compared it to an exact numerical calculation of the band spectrum. From this we have derived quantitative experimental criteria (such as the required initial temperature of the atomic gas) to reach the massless Dirac fermion regime.

We have also examined the robustness of the massless Dirac fermions to imperfections of the laser configuration (field strengths imbalances and angle mismatches). Massless Dirac fermions turn out to be quite robust as the equality of the beam intensities should be controlled within the few percent range while the respective beam angles should equal $2\pi/3$ within the few degrees range. By appropriately controlling these lattice distortions, one can even control and move the Dirac points in the Brillouin zone. Lastly, we introduce an irremovable phase to the honeycomb potential, hence lifting the degeneracy between two sublattices. In turn, the Dirac fermions acquire a mass proportional to this phase. We also briefly mention the peculiar properties of saddle points and the possibility to study oscillations of atoms between two sublattices as a consequence of this irremovable phase.

As an overall conclusion, mimicking graphene physics with ultracold fermions is within experimental reach. For noninteracting fermions, one could think of implementing transport experiments (in the presence of disorder or not). For example, by rotating the whole honeycomb lattice around a given axis perpendicular to the lattice plane [61] or by implementing the scheme proposed in [62], one would mimic effective magnetic fields able to reproduce the quantum Hall effect situation. In the rest frame of the atoms, the centrifugal effects are described by a fictitious vector potential. This leads to Landau levels and paves the way to physical effects analogous to the quantum Hall effect. The possibility to move the Dirac points in the Brillouin zone even offers new physical effects to test [17].

Interacting systems on a lattice prove also particularly interesting as they can be mapped (at least for strong interactions) on Heisenberg models and thus offer ways of exploring quantum magnetism [63]. In the case of the honeycomb lattice, quantum phase transitions are predicted to occur when the interaction strength $|U|$ is strong enough. For repulsive interactions, quantum Monte Carlo calculations predict antiferromagnetic order to occur at half-filling [18]. For attractive interactions, mean-field calculations have started to analyze the BEC-BCS crossover and predict a semimetal/superconductor transition [19]. Recent Monte Carlo studies have even started to analyze this BEC-BCS crossover [64] and one can expect an increase in such studies in the near future. On the other hand, Wu and Sarma have begun studies on the higher-energy orbitals on a honeycomb lattice and they conclude that the band structure leads to enhanced interaction effects giving rise to various charge and bond ordered states at commensurate fillings [65]. Very recently, implementations of massless Dirac fermions in square lattices have been proposed [66,67]. The situation seems thus mature for an experimental effort toward loading ultracold fermions in a honeycomb optical lattice.

ACKNOWLEDGMENTS

B.G. and C.M. would like to thank Dominique Delande, Gilles Montambaux, Jean-Noël Fuchs, Mark Goerbig, and David Wilkowski for stimulating discussions. L.K.L. acknowledges support from the French Merlion-PhD program (Grant No. CNOUS 20074539). This work was also supported by the CNRS PICS Grant No. 4159 (France) and the France-Singapore Merlion program (SpinCold Grant No. 2.02.07). Centre for Quantum Technologies is a Research Centre of Excellence funded by Ministry of Education and National Research Foundation of Singapore.

---

[1] K. S. Novoselov, A. K. Geim, S. V. Morozov, D. Jiang, Y. Zhang, S. V. Dubonos, I. V. Grigorieva, and A. A. Firsov, Science **306**, 666 (2004).
[2] R. Saito, G. Dresselhaus, and M. S. Dresselhaus, *Physical Properties of Carbon Nanotubes* (Imperial College Press, London, 1998).
[3] G. W. Semenoff, Phys. Rev. Lett. **53**, 2449 (1984).
[4] M. Wilson, Phys. Today **59** (1), 21 (2006).
[5] A. C. Neto, F. Guinea, and N. M. Peres, Phys. World **19** (11), 33 (2006).
[6] S. Y. Zhou, G.-H. Gweon, J. Graf, A. V. Fedorov, C. D. Spataru, R. D. Diehl, Y. Kopelevich, D.-H. Lee, S. G. Louie, and A. Lanzara, Nat. Phys. **2**, 595 (2006).
[7] M. I. Katsnelson, K. S. Novoselov, and A. K. Geim, Nat. Phys. **2**, 620 (2006).
[8] M. I. Katsnelson and K. S. Novoselov, Solid State Commun. **143**, 3 (2007).
[9] A. H. Castro Neto, F. Guinea, N. M. R. Peres, K. S. Novoselov, and A. K. Geim, Rev. Mod. Phys. **81**, 109 (2009).
[10] K. Novoselov, A. Geim, S. Morozov, D. Jiang, M. Katsnelson, I. Grigorieva, S. Dubonos, and A. Firsov, Nature (London) **438**, 197 (2005).
[11] Y. Zhang, Y. Tan, H. Stormer, and P. Kim, Nature (London) **438**, 201 (2005).
[12] G. Li and E. Andrei, Nat. Phys. **3**, 623 (2007).
[13] I. Bloch, Nat. Phys. **1**, 23 (2005).
[14] M. Lewenstein, A. Sanpera, V. Ahufinger, B. Damski, A. Sen(de), and U. Sen, Adv. Phys. **56**, 243 (2007).
[15] E. Timmermans, P. Tommasini, M. Hussein, and A. Kerman, Phys. Rep. **315**, 199 (1999).
[16] S.-L. Zhu, B. Wang, and L.-M. Duan, Phys. Rev. Lett. **98**, 260402 (2007).
[17] P. Dietl, F. Piéchon, and G. Montambaux, Phys. Rev. Lett. **100**, 236405 (2008).
[18] T. Paiva, R. T. Scalettar, W. Zheng, R. R. P. Singh, and J. Oitmaa, Phys. Rev. B **72**, 085123 (2005).
[19] E. Zhao and A. Paramekanti, Phys. Rev. Lett. **97**, 230404 (2006).
[20] C. Cohen-Tannoudji, J. Dupont-Roc, and G. Grynberg, *Atom-Photon Interactions: Basic Processes and Applications* (Wiley-Interscience, New York, 1998).
[21] H. Wallis, Phys. Rep. **255**, 203 (1995).
[22] R. Grimm, M. Weidemüller, and Y. B. Ovchinnikov, Adv. At., Mol., Opt. Phys. **42**, 95 (2000).
[23] G. Grynberg and C. Robilliard, Phys. Rep. **355**, 335 (2001).
[24] M. Greiner, O. Mandel, T. Esslinger, T. W. Hänsch, and I. Bloch, Nature (London) **415**, 39 (2002).
[25] D. Jaksch, C. Bruder, J. I. Cirac, C. W. Gardiner, and P. Zoller, Phys. Rev. Lett. **81**, 3108 (1998).
[26] D. Jaksch, H.-J. Briegel, J. I. Cirac, C. W. Gardiner, and P. Zoller, Phys. Rev. Lett. **82**, 1975 (1999).
[27] X. Chen, B. Zheng, Z.-C. Gu, B. Yoshida, and I. L. Chuang, Phys. Rev. Lett. **102**, 220501 (2009).
[28] N. W. Ashcroft and N. D. Mermin, *Solid State Physics* (Thomson Learning, United Kingdom, 1976).
[29] O. Bohigas, Nucl. Phys. A. **751**, 343 (2005).
[30] The symmetry aspects of honeycomb potential will be discussed in more details in a follow-up paper.
[31] G. Wannier, Phys. Rev. **52**, 191 (1937).
[32] P. Wallace, Phys. Rev. **71**, 622 (1947).
[33] G. Mil'nikov and H. Nakamura, J. Chem. Phys. **115**, 6881 (2001).
[34] M. Marder, *Condensed Matter Physics* (Wiley, New York, 2000).
[35] P. B. Blakie and C. W. Clark, J. Phys. B **37**, 1391 (2004).
[36] J. Hubbard, Proc. R. Soc. London, Ser. A **276**, 238 (1963).
[37] W. Kohn, Phys. Rev. **115**, 809 (1959).
[38] J. des Cloizeaux, Phys. Rev. **135**, A685 (1964).
[39] G. Nenciu, Commun. Math. Phys. **91**, 81 (1983).
[40] C. Brouder, G. Panati, M. Calandra, C. Mourougane, and N. Marzari, Phys. Rev. Lett. **98**, 046402 (2007).
[41] J. des Cloizeaux, Phys. Rev. **129**, 554 (1963).
[42] J. des Cloizeaux, Phys. Rev. **135**, A698 (1964).
[43] S. Reich, J. Maultzsch, C. Thomsen, and P. Ordejón, Phys. Rev. B **66**, 035412 (2002).
[44] I. Bloch, J. Dalibard, and W. Zwerger, Rev. Mod. Phys. **80**, 885 (2008).
[45] M. Gutzwiller, *Chaos in Classical and Quantum Mechanics* (Springer, New York, 1990).
[46] L. Schulman, *Techniques and Applications of Path Integration* (Wiley, New York, 1981).
[47] W. Zwerger, J. Opt. B: Quantum Semiclassical Opt. **5**, S9 (2003).
[48] W. M. Lomer, Proc. R. Soc. London, Ser. A **227**, 330 (1955).

[49] J. Slonczewski and P. Weiss, Phys. Rev. **109**, 272 (1958).
[50] R. Jördens, N. Strohmaier, K. Günter, H. Moritz, and T. Esslinger, Nature (London) **455**, 204 (2008).
[51] U. Schneider, L. Hackermüller, S. Will, T. Best, I. Bloch, T. A. Costi, R. W. Hemes, D. Rasch, and A. Rosch, Science **322**, 1520 (2008).
[52] T. Bourdel, L. Khaykovich, J. Cubizolles, J. Zhang, F. Chevy, M. Teichmann, L. Tarruell, S. J. J. M. F. Kokkelmans, and C. Salomon, Phys. Rev. Lett. **93**, 050401 (2004).
[53] M. Rigol, A. Muramatsu, G. G. Batrouni, and R. T. Scalettar, Phys. Rev. Lett. **91**, 130403 (2003).
[54] M. Rigol and A. Muramatsu, Phys. Rev. A **70**, 043627 (2004).
[55] M. Köhl, H. Moritz, T. Stöferle, K. Günter, and T. Esslinger, Phys. Rev. Lett. **94**, 080403 (2005).
[56] Y. Hasegawa, R. Konno, H. Nakano, and M. Kohmoto, Phys. Rev. B **74**, 033413 (2006).
[57] G. Montambaux, F. Piechon, J.-N. Fuchs, and M. O. Goerbig, e-print arXiv:0904.2117.
[58] M. O. Goerbig, J.-N. Fuchs, G. Montambaux, and F. Piéchon, Phys. Rev. B **78**, 045415 (2008).
[59] P. Lee, N. Nagaosa, and X. Wen, Rev. Mod. Phys. **78**, 17 (2006).
[60] We thank David Wilkowski for telling us about this way of introducing the phase parameter φ into the optical potential.
[61] S. Tung, V. Schweikhard, and E. A. Cornell, Phys. Rev. Lett. **97**, 240402 (2006).
[62] A. S. Sørensen, E. Demler, and M. D. Lukin, Phys. Rev. Lett. **94**, 086803 (2005).
[63] E. Demler, W. Hanke, and S. Zhang, Rev. Mod. Phys. **76**, 909 (2004).
[64] S.-Q. Su, K.-M. Tam, and H.-Q. Lin, e-print arXiv:0902.3242.
[65] C. Wu and S. Das Sarma, Phys. Rev. B **77**, 235107 (2008).
[66] J.-M. Hou, W.-X. Yang, and X.-J. Liu, Phys. Rev. A **79**, 043621 (2009).
[67] N. Goldman, A. Kubasiak, A. Bermudez, P. Gaspard, M. Lewenstein, and M. A. Martin-Delgado, Phys. Rev. Lett. **103**, 035301 (2009).

Chapter 6

Information about Quantum Systems

Jaroslav Řeháček

Department of Optics, Palacký University, Olomouc, Czech Republic

Paper Title: **Minimal qubit tomography**

Reference: Physical Review A **70**, 052321 (2004)

Personal message:

My first visit to Singapore happened on Berge's kind invitation more than a decade ago. I really enjoyed the city and the friendly atmosphere about the quantum lab, which was still young and small at that time. On this occasion I had the opportunity to join Berge and his co-workers in analyzing a new highly symmetrical tomography protocol, which we know today is a pertinent example of SIC (Symmetrical Informationally Complete) measurements. Our paper on "Minimal qubit tomography" that appeared in 2004 in Physical Review A would become one of my most cited papers. I wish Berge all the best!

PHYSICAL REVIEW A **70**, 052321 (2004)

Minimal qubit tomography

Jaroslav Řeháček,[1] Berthold-Georg Englert,[2] and Dagomir Kaszlikowski[2]
[1]*Department of Optics, Palacky University, 17. listopadu 50, 772 00 Olomouc, Czech Republic*
[2]*Department of Physics, National University of Singapore, Singapore 117542, Singapore*
(Received 15 May 2004; published 22 November 2004)

We present, and analyze thoroughly, a highly symmetric and efficient scheme for the determination of a single-qubit state, such as the polarization properties of photons emitted by a single-photon source. In our scheme there are only four measured probabilities, just enough for the determination of the three parameters that specify the qubit state, whereas the standard procedure would measure six probabilities.

DOI: 10.1103/PhysRevA.70.052321 PACS number(s): 03.67.−a, 03.65.Wj, 07.60.Fs

I. INTRODUCTION

Experiments that exploit the polarization degree of freedom of single photons, detected one by one, have become an almost routine matter in recent years. In particular, a whole class of experiments that demonstrate the technical feasibility of quantum cryptography, or quantum key distribution, use the photon polarization as the carrier of the quantum bit, or *qubit*. Other experiments make use of a spatial degree of freedom, essentially the path qubit of a two-path interferometer, which is sometimes translated into the alternative of early or late arrival for the sake of easier transmission.

In applications like these, as well as many others, one must be able to characterize the qubit source and the transmission channel. For this purpose a complete determination of the state of the qubit is required, both as it is emitted from the source and as it arrives after transmission. To be able to perform the regular on-the-fly calibration of the setup, so as to compensate for the unavoidable drifts, one needs an efficient diagnostics that does not consume more qubits than really necessary.

The standard procedure measures three orthogonal components of the relevant qubit analog of Pauli's spin vector operator, so that *six* probabilities are estimated for the determination of the *three* real parameters that specify the qubit state. But clearly, *four* measured probabilities should suffice to establish the values of three parameters. Indeed, such minimal schemes for state determination are possible, and it is the objective of this paper to present and analyze one such scheme, a highly symmetric one.

In Sec. II we briefly review the standard six-output measurement scheme and then introduce the minimal four-output scheme, followed by remarks on state determination for qubit pairs. We then proceed to describe, in Sec. III, optical implementations for the measurement of a photon's polarization qubit—polarimeters or ellipsometers in the jargon of classical optics.

The question of how one infers a reliable estimate for the qubit state after the detection of a finite, possibly small, number of qubits is addressed in Sec. IV. After discussing the optimality of the highly symmetric four-output scheme in Sec. V and remarking on some peculiar aspects of measuring pure qubit states in Sec. VI, we analyze adaptive measurement strategies in Sec. VII and then close with a summary.

II. QUBIT TOMOGRAPHY

A. Standard six-state tomography

We describe, as usual, the binary quantum alternative of the qubit by a Pauli vector operator $\vec{\sigma}=(\sigma_x,\sigma_y,\sigma_z)$. The physical nature of the qubit is irrelevant for the present general discussion—it might just as well be the spin-$\frac{1}{2}$ degree of freedom of an electron or a pseudospin such as the path degree of freedom in a two-path interferometer or the internal degree of freedom of a two-level atom—but in the particular application that we have in mind it is the polarization degree of freedom of a photon. Then we use the convention specified by

$$\sigma_x = |\mathsf{h}\rangle\langle\mathsf{v}| + |\mathsf{v}\rangle\langle\mathsf{h}|,$$

$$\sigma_y = i(|\mathsf{h}\rangle\langle\mathsf{v}| - |\mathsf{v}\rangle\langle\mathsf{h}|),$$

$$\sigma_z = |\mathsf{v}\rangle\langle\mathsf{v}| - |\mathsf{h}\rangle\langle\mathsf{h}|, \tag{2.1}$$

where $|\mathsf{v}\rangle$ and $|\mathsf{h}\rangle$ are the ket vectors for vertical and horizontal polarization, respectively.

The statistical operator of the qubit emitted by a given source,

$$\rho = \frac{1}{2}(1 + \vec{s}\cdot\vec{\sigma}), \tag{2.2}$$

is parametrized by the Pauli vector $\langle\vec{s}\rangle=\langle\vec{\sigma}\rangle=\mathrm{tr}\{\vec{\sigma}\rho\}$, the expectation value of $\vec{\sigma}$. The positivity of ρ restricts the Pauli vectors to the Bloch ball, $s=|\vec{s}|\leq 1$. The experimental characterization of the source requires, therefore, a complete measurement of $\vec{s}$ with sufficient precision. Any procedure that can yield this information is an example of *qubit tomography*.

In the standard approach one measures σ_x for some qubits supplied by the source, σ_y for some others, and σ_z for yet others. Assuming an unbiased procedure—that is, for each qubit there is an equal chance for either one of the three measurements to happen—there are six possible outcomes that occur with the probabilities

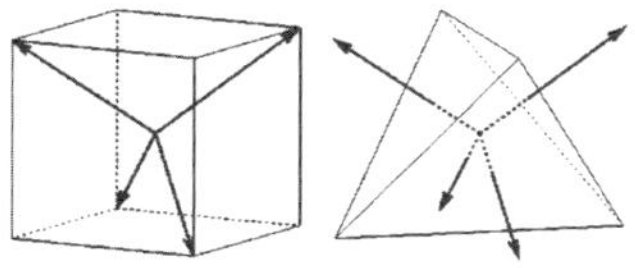

FIG. 1. Picturing the vector quartet of Eq. (2.5) as pointing from the center to four nonadjacent corners of a cube or as the vectors normal on the faces of a tetrahedron.

$$p_{\xi\pm} = \left\langle \frac{1}{6}(1 \pm \sigma_\xi) \right\rangle \equiv \langle P_{\xi\pm}\rangle \text{ for } \xi = x,y,z. \quad (2.3)$$

Each operator $P_{\xi\pm}$ is a third of a projector, and since these non-negative operators decompose the identity,

$$\sum_{\xi=x,y,z} (P_{\xi+} + P_{\xi-}) = 1, \quad (2.4)$$

they constitute the positive operator-valued measure (POVM) of this standard *six-state tomography*.

This POVM is an example of a tomographically complete set of measurements of pairwise complementary observables; namely, σ_x, σ_y, and σ_z, so that their eigenstates constitute sets of mutually unbiased bases. As Wootters and Fields have shown [1], such sets are particularly well suited for tomographic purposes, inasmuch as the statistical errors in the estimates based on a finite number of measurements are minimal. The sets themselves are not of minimal size, however, because one measures six probabilities to determine three parameters, the components of the Pauli vector $\vec{s}$. Indeed, the six probabilities of Eq. (2.3) are subject to the three constraints $p_{\xi+}+p_{\xi-}=\frac{1}{3}$, $\xi=x,y,z$, rather than to the single constraint of unit sum. A minimal POVM, by contrast, would refer to only four outcomes and their probabilities, with unit sum as the only constraint.

B. Minimal four-state tomography

We construct such a minimal POVM of high internal symmetry by first choosing four unit vectors $\vec{a}_1,\ldots,\vec{a}_4$ with equal angle between each pair:

$$\vec{a}_j \cdot \vec{a}_k = \frac{4}{3}\delta_{jk} - \frac{1}{3} = \begin{cases} 1 & \text{for } j=k, \\ -1/3 & \text{for } j \neq k. \end{cases} \quad (2.5)$$

Geometrically speaking, such a quartet consists of the vectors pointing from the center of a cube to nonadjacent corners, as illustrated in Fig. 1 and exemplified by

$$\vec{a}_1 = 3^{-1/2}(1,1,1),$$

$$\vec{a}_2 = 3^{-1/2}(1,-1,-1),$$

$$\vec{a}_3 = 3^{-1/2}(-1,1,-1),$$

$$\vec{a}_4 = 3^{-1/2}(-1,-1,1). \quad (2.6)$$

Alternatively, one may picture these vectors as the normal vectors for the faces of the tetrahedron that is defined by the other four corner of the cube.

The linear dependence of the $\vec{a}_j$'s is stated by their null sum

$$\sum_{j=1}^{4} \vec{a}_j = 0 \quad (2.7)$$

and their completeness by the decomposition of the unit dyadic:

$$\frac{3}{4}\sum_{j=1}^{4} \vec{a}_j\vec{a}_j = \overleftrightarrow{1}. \quad (2.8)$$

The perfect symmetry of the tetrahedron geometry manifests itself in the simplicity of this completeness relation and the inner products of Eq. (2.5). As discussed in Sec. V B below, the tetrahedron geometry is optimal in the sense that any other vector quartet would define a less efficient scheme for four-state tomography.

Each such quartet of $\vec{a}_j$'s defines a POVM for minimal four-state tomography in accordance with

$$\sum_{j=1}^{4} P_j = 1 \text{ with } P_j \equiv \frac{1}{4}(1 + \vec{a}_j \cdot \vec{\sigma}). \quad (2.9)$$

This POVM is an example of a "symmetric informationally complete POVM" [2]. Upon measuring it and so determining the probabilities [3]

$$p_j = \langle P_j \rangle = \frac{1}{4}(1 + \vec{a}_j \cdot \vec{s}), \quad (2.10)$$

the Pauli vector is readily available,

$$\vec{s} = 3\sum_j p_j \vec{a}_j, \quad (2.11)$$

and so are the statistical operator and its square,

$$\rho = 6\sum_j p_j P_j - 1 = \sum_j \langle P_j \rangle (6P_j - 1),$$

$$\rho^2 = \rho - 1 + 3\sum_j p_j^2. \quad (2.12)$$

It follows that, in addition to being restricted to the range $0 \leq p_j \leq \frac{1}{2}$, the probabilities p_j obey the inequalities

$$\frac{1}{4} \leq \sum_j p_j^2 = \frac{3+s^2}{12} \leq \frac{1}{3}. \quad (2.13)$$

The upper bound is reached by all pure states, $\rho=\rho^2$ and $s=1$, the lower bound for the completely mixed state, $\rho=\frac{1}{2}$ and $s=0$.

C. Qubit-pair tomography

The minimal property of the four-state POVM of Eq. (2.9) carries over to multiqubit states. In case of n qubits, one has 4^n joint probabilities for the 4^n-1 independent parameters of the $2^n \times 2^n$ matrix elements of the statistical operator, so that the count is just right. By contrast, if one were to measure n realizations of the six-state POVM of Eq. (2.4), one would have 6^n joint probabilities which contain quite a lot of redundant information.

More specifically, consider the $n=2$ situation of a source emitting qubit pairs. Using the P_j's from above for one qubit and corresponding operators Q_k for the other, we obtain the 16 joint probabilities $\langle P_j Q_k \rangle$ by measurning the two four-state POVMs. They are the numerical ingredients in

$$\rho = \sum_{j,k} (6P_j - 1)\langle P_j Q_k \rangle (6Q_k - 1), \qquad (2.14)$$

the explict construction of the two-qubit statistical operator. If there are no correlations in the joint probabilities, so that $\langle P_j Q_k \rangle = \langle P_j \rangle \langle Q_k \rangle$, then ρ is the product of two factors of the one-qubit form in Eq. (2.12), as it should be. The generalization of the $n=2$ expression (2.14) to $n>2$ is immediate.

Qubit-pair tomography of this kind requires that the vector quartet $\vec{b}_k$ associated with the Q_k's have a known orientation relative to the quartet $\vec{a}_j$ of the P_j's. One can determine this orientation by "quantum measurement tomography" [4]; that is, by measuring the joint probabilities $\langle P_j Q_k \rangle$ for a source with a known output [5]. In the simplest situation, for example, the source emits pairs that are perfectly anticorrelated:

$$\rho = \frac{1}{4}(1 - \vec{\sigma}^{(1)} \cdot \vec{\sigma}^{(2)}). \qquad (2.15)$$

The orthogonal dyadic $\overleftrightarrow{O}$ that turns the $\vec{a}_j$ quartet and the $\vec{b}_k$ quartet into each other,

$$\vec{b}_k = \overleftrightarrow{O} \cdot \vec{a}_k, \quad \vec{a}_j = \vec{b}_j \cdot \overleftrightarrow{O}, \qquad (2.16)$$

is then given by

$$\overleftrightarrow{O} = -9 \sum_{j,k} \vec{a}_j \langle P_j Q_k \rangle \vec{a}_k = -9 \sum_{j,k} \vec{b}_j \langle P_j Q_k \rangle \vec{b}_k, \qquad (2.17)$$

and can thus be determined experimentally.

III. ELLIPSOMETRY

A. Standard six-outcome ellipsometer

Devices for characterizing the polarization properties of a light source are called *ellipsometers*, a term that makes reference to elliptic polarization as the generic outcome of the measurement. Figure 2 shows the schematic setup of a standard six-state ellipsometer, which realizes the POVM of Eqs. (2.3) and (2.4) for the photonic polarization qubit. The input beam splitter (BS) reflects one-third of the light to a polarizing BS (PBS) that reflects vertically polarized photons and transmits horizontally polarized ones and so directs them to two photodetectors. This branch thus realizes a measurement of σ_z and accounts for P_{z+} and P_{z-} in the sum of Eq. (2.4).

Two-thirds of the light are transmitted at the input BS and are then equally split at a second BS. A photon transmitted there will be detected by either one of the two detectors behind another PBS. This PBS is rotated by 45°, so that a measurement of σ_x is realized in this branch and the term $P_{x+} + P_{x-}$ is accounted for in Eq. (2.4).

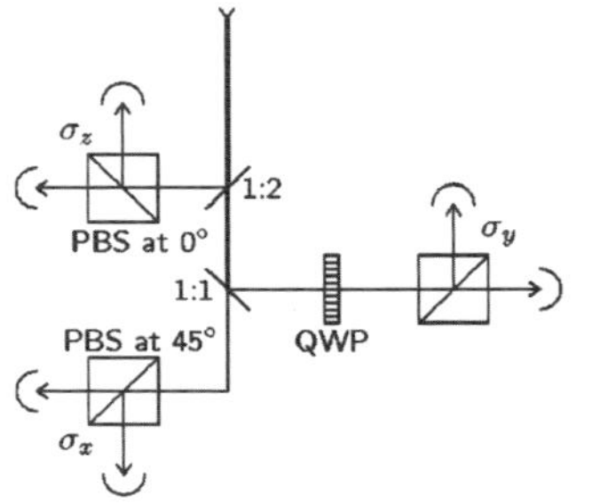

FIG. 2. Schematic setup of an ellipsometer that implements the standard six-state POVM; see text.

Finally, the photons that are reflected at the second BS pass through a quarter-wave plate (QWP) before a PBS directs them to a third pair of detectors. This branch implements the measurement of σ_y and accounts for the remaining terms in Eq. (2.4); namely, P_{y+} and P_{y-}.

B. Minimal four-outcome ellipsometer

There are a number of alternative schemes for an ellipsometer that realizes the minimal four-state POVM of Eq. (2.9). To demonstrate the case, we present one scheme here and discuss a few other schemes, which are much simpler and much more practical, elsewhere [6,7].

The principle of one minimal ellipsometer is illustrated by Fig. 3(a). It is an asymmetric four-path interferometer. At the input, one-half of the light intensity is directed into the uppermost path for reference, whereas each of the other three paths gets one-sixth of the intensity. Photons in the lower

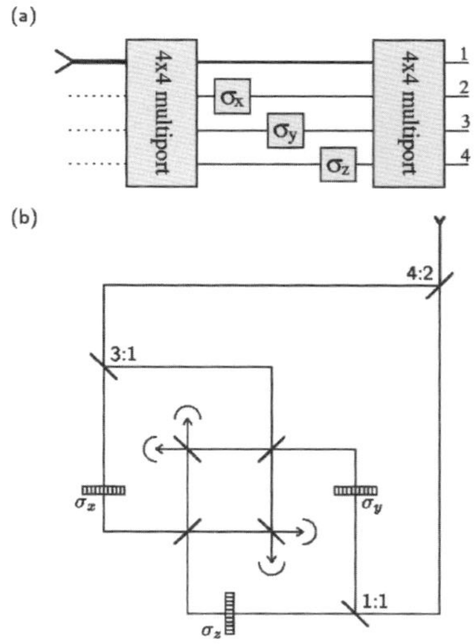

FIG. 3. (a) Principle of a four-path interferometer for minimal ellipsometry and (b) optical network for the implementation; see text.

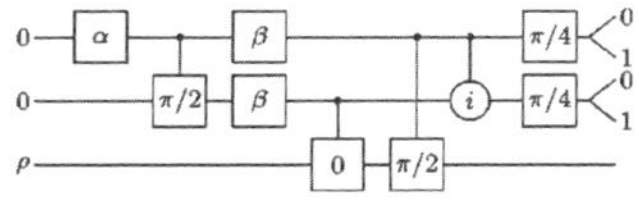

FIG. 4. Quantum computation network for minimal qubit tomography. The bottom qubit is the one of interest; it enters in the state ρ whose properties are to be determined. The two top qubits are auxiliary qubits that enter in state 0 of the computational basis and are eventually measured. The probabilities for the four different measurement results—00, 01, 10, and 11—are the p_j's of Eq. (2.10) with the $\vec{a}_j$'s of Eq. (2.6). The network uses eight gates of the generalized Hadamard type (3.1), three of them controlled by one of the auxiliary qubits, and a controlled phase gate.

paths pass through wave plates that realize the unitary polarization transformations $\rho \rightarrow \sigma_\xi \rho \sigma_\xi$ with $\xi = x$, or y, or z, respectively. Then all four paths are recombined by a balanced beam merger, which distributes the intensity of each input evenly among the four outputs. An unpolarized photon has, therefore, a probability of 25% for being detected by a particular one of the four detectors. The probabilities for a polarized photon are the p_j's of Eq. (2.10), so that the polarization POVM of Eq. (2.9) is measured indeed.

An optical network for a four-path interferometer of this kind is shown in Fig. 3(b). The BS at the input reflects 4/6 of the intensity and transmits 2/6, and the subsequent BSs either split the beam 3:1 or 1:1; together, these three BSs implement the initial stage of Fig. 3(a). At the central stage, there are wave plates in three partial beams (labeled by σ_x, σ_y, and σ_z, respectively). And the final beam merger consists of four 1:1 BSs which direct the photons to the four photodetectors [8].

The interferometer of Fig. 3(b) has several loops. Some alternative setups have two loops [9] or need a single loop only, among them the interferometer of the experiment by Clarke *et al.* [10] and the minimal one-loop scheme of Ref. [6], and yet another setup has no loop at all [7]. We note that Clarke *et al.* did not perform ellipsometry; their experiment served a different purpose, and although their setup could be used for ellipsometry, there is no mentioning of this possible application in Ref. [10].

It is worth mentioning that such a setup can also be viewed as a quantum computation network for three qubits, one being the polarization qubit of interest, the other two qubits representing the four paths of the interferometer [11]. The network is depicted in Fig. 4. It consists of a sequence of generalized Hadamard gates,

$$-\boxed{\phi}- \; : \; \begin{pmatrix} \langle 0| \\ \langle 1| \end{pmatrix} \rightarrow \begin{pmatrix} \cos\phi & \sin\phi \\ \sin\phi & -\cos\phi \end{pmatrix} \begin{pmatrix} \langle 0| \\ \langle 1| \end{pmatrix}. \tag{3.1}$$

At the first stage, the auxiliary qubits are prepared in a superposition state that has amplitude $1/\sqrt{2}$ for $|00\rangle \triangleq |\mathsf{vv}\rangle$ and amplitudes $1/\sqrt{6}$ for $|01\rangle, |10\rangle$, and $|11\rangle \triangleq |\mathsf{hh}\rangle$. We achieve this by a controlled gate with $\phi = \frac{1}{2}\pi$ (this is a controlled-not gate, in fact) that is sandwiched by a gate with $\phi = \alpha$ and two gates with $\phi = \beta$, where α and β are such that $\sin(2\alpha) = \frac{1}{3}(\sqrt{3}-1)$ and $\tan(2\beta) = \sqrt{3}+1$.

The central stage has controlled gates acting on the qubit of interest, a gate with $\phi = 0$ for σ_z, another one with $\phi = \frac{1}{2}\pi$ for σ_x. For their product to realize $\sigma_y = i\sigma_x\sigma_z$, we provide the factor of i by a subsequent controlled phase gate, which implements the phase change $|11\rangle \rightarrow i|11\rangle$, but has no effect on $|00\rangle$, $|01\rangle$, and $|10\rangle$.

At the final stage, the two auxiliary qubits are passed through standard Hadamard gates, for which $\phi = \frac{1}{4}\pi$ in Eq. (3.1), and are then measured in the computational basis. The probabilities for getting 00, 01, 10, or 11 are p_1, p_4, p_2, and p_3 of Eqs. (2.10) and (2.6), respectively.

Conditioned on the measurement outcome, the qubit of interest emerges in the corresponding reduced state $2P_j\rho P_j/p_j$. For the optical implementation of Fig. 3(b), these final states of the polarization qubit are fictitious, however, unless the photodetectors are of a fantastic nondemolition kind: sensitive to the passage of a photon without absorbing the photon or affecting its polarization [12].

IV. COUNTING QUBITS

A. Maximum-likelihood estimator

In an actual experiment, we do not measure the probabilities of Eqs. (2.3) or (2.10), but rather relative frequencies that are statistically determined by these probabilities. The available information consists of the counts of detector clicks, n_1, n_2, n_3, n_4 for the minimal four-state tomography of Sec. II B and $n_{x\pm}$, $n_{y\pm}$, $n_{z\pm}$ for the standard six-state scheme of Sec. II A. In what follows, we focus on the novel four-detector situation.

In view of the intrinsic probabilistic nature of quantum phenomena, a given total number of N qubits does not result in a definite, predictable number of clicks for each detector. It is, therefore, clear that an observed breakup

$$N = n_1 + n_2 + n_3 + n_4 \tag{4.1}$$

is consistent with not just one Pauli vector $\vec{s}$ in Eq. (2.2), but with many. One expects that, as a rule, the obvious guess that obtains from Eq. (2.11) upon the replacement $p_j \rightarrow n_j/N$, namely,

$$\vec{S} = 3\sum_j \nu_j \vec{a}_j \quad \text{with } \nu_j \equiv \frac{n_j}{N} \tag{4.2}$$

gives a reliable estimate of the true Pauli vector $\vec{s}$. But it may happen that the length of this inferred $\vec{S}$ exceeds unity, and this is in fact a typical situation if the true $\vec{s}$ is of unit length; that is, the source emits qubits in a pure state.

Let us thus proceed to show how one infers a plausible and physically correct answer on the basis of the registered data. The experimental data consist in fact not just of the total counts n_j in the breakup (4.1), but of a particular sequence of detector clicks. *Provided that* the source emits qubits in the state specified by the Pauli vector $\vec{s}$, with no statistical correlations between different qubits, the probability $\mathcal{L}(\vec{s}; n_1, \ldots, n_4)$ for getting the observed click sequence is, therefore, given by

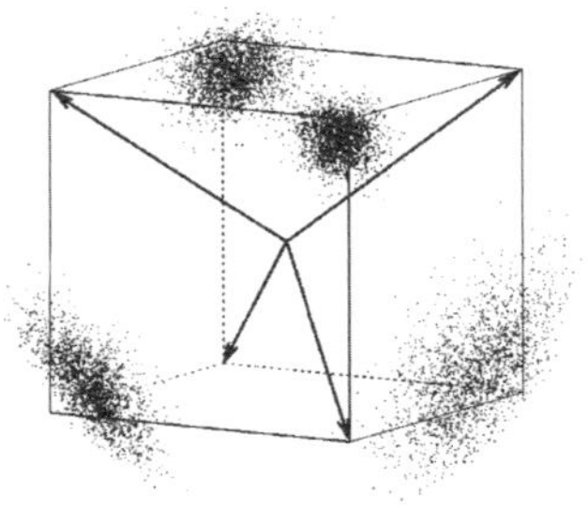

FIG. 5. The likelihood function as a probability density for the Pauli vector $\vec{s}$. The plot shows the "likelihood clouds" for simulated measurements for which the most likely Pauli vector has length $s=0.84$ and is pointing toward one of the corners of the cube of Fig. 1. A total of 100 qubits have been detected for the bottom right corner, 200 for the bottom left corner, 400 for the top rear corner, and 800 for the top front corner. The successive shrinking of the cloud is clearly visible.

$$\mathcal{L}=\prod_{j=1}^{4} p_j(\vec{s})^{n_j}, \tag{4.3}$$

where $p_j(\vec{s})$ is the probability (2.10) that a qubit is registered by the jth detector.

Conversely, in the spirit of the Bayesian principle of statistical inference, we can regard $\mathcal{L}(\vec{s};n_1,\ldots,n_4)$ as the likelihood that the source is characterized by $\vec{s}$, *given that* a click sequence with total detector counts $n_1,\ldots,n_4$ was observed. When many qubits have been counted, the likelihood function is sharply peaked and essentially vanishes outside the immediate vicinity of its maximum. These matters are illustrated by the four "likelihood clouds" in Fig. 5.

The maximum-likelihood (ML) estimator $\vec{S}$ picks out the most likely Pauli vector, the one for which $\mathcal{L}$ is largest:

$$\max_{\vec{s}} \mathcal{L}(\vec{s})=\mathcal{L}(\vec{S}). \tag{4.4}$$

For the purposes of this paper, we accept this $\vec{S}$ as our plausible guess for $\vec{s}$, while being fully aware of other strategies [13–15].

An important theorem by Fisher states that ML estimators become efficient in the large-N limit [16,17]. So the performance, for large N, of an experimental tomographic setup can be quantified by the accuracy of the ML estimator. Usually an analytical expression for the ML estimator is not available and one has to solve a nonlinear operator equation for $\rho=\frac{1}{2}(1+\vec{S}\cdot\vec{\sigma})$ to find it [18]. In the present context, however, the high symmetry of the vector quartet of Eq. (2.5) makes it possible to simplify this problem considerably.

There is a benefit in maximizing the right-hand side of Eq. (4.2) with respect to the probabilities p_j rather than with respect to $\vec{s}$. Lagrange multipliers are used to account for the two constraints. One is the unit sum of the p_j, $\Sigma_j p_j=1$, which is the unit trace of ρ; the other is the positivity of ρ, $\rho\geqslant 0$, which is the upper bound in Eq. (2.13).

Without this positivity constraint, the likelihood (4.3) would be maximized by $p_j=\nu_j$, which in turn would imply the Pauli vector of Eq. (4.2). But if the actual counts violate the inequality

$$\sum_j \nu_j^2 \leqslant \frac{1}{3}, \tag{4.5}$$

this simple estimation fails to provide a physically meaningful result. When this happens, both constraints must be taken into account and the likelihood $\mathcal{L}(\vec{s})$ attains its maximum on the boundary of the set of qubit states—that is, for a Pauli vector of unit length [19].

The variation of the likelihood vanishes at the extremal point:

$$\delta \ln \mathcal{L}=\sum_j \frac{n_j}{p_j}\delta p_j=0. \tag{4.6}$$

The variations δp_j are subject to the two constraints

$$\sum_j \delta p_j=0, \quad \sum_j p_j\delta p_j=0, \tag{4.7}$$

for which we use the Lagrange multipliers $N\lambda$ and $3N\mu$, respectively. Denoting the p_j values at the extremal point by $\tilde{p}_j$, Eq. (4.6) then implies

$$\frac{\nu_j}{\tilde{p}_j}=\lambda+3\mu\tilde{p}_j \quad \text{for } j=1,\ldots,4. \tag{4.8}$$

We exploit $\Sigma_j\nu_j=1$, $\Sigma_j\tilde{p}_j=1$, and $\Sigma_j\tilde{p}_j^2=\frac{1}{3}$ to establish

$$\sum_j \frac{\nu_j}{\tilde{p}_j}=4\lambda+3\mu \tag{4.9}$$

and

$$1=\lambda+\mu, \tag{4.10}$$

the first by summing the four equations in (4.8), the second by summing them after multiplication with $\tilde{p}_j$.

Taken together, Eqs. (4.8)–(4.10) make up a set of six equations for the six unknowns $\tilde{p}_1,\ldots,\tilde{p}_4$, λ, and μ. We solve the quadratic equations (4.8) for $\tilde{p}_j$, or rather $\nu_j/\tilde{p}_j$,

$$\frac{\nu_j}{\tilde{p}_j}=\frac{1}{2}(\lambda+\sqrt{\lambda^2+12\mu\nu_j}), \tag{4.11}$$

and Eq. (4.10) for λ, and substitute these into Eq. (4.9) to arrive at a single equation for μ:

$$\mu+2-\frac{1}{2}\sum_j\sqrt{(1-\mu)^2+12\mu\nu_j}=0. \tag{4.12}$$

There is always the solution $\mu=0$, and thus $\lambda=1$, but Eq. (4.11) amounts to Eq. (4.2) for these values, and therefore this solution is not acceptable, when the inequality (4.5) is not obeyed, as is the case in the present discussion. Upon recognizing that Eq. (4.12) can also be written as

$$6\mu^2\sum_j \frac{(3\nu_j - 1)\nu_j}{1-\mu+6\mu\nu_j+\sqrt{(1-\mu)^2+12\mu\nu_j}} = 0, \quad (4.13)$$

we can discard the unphysical solution $\mu=0$ and find the relevant solution in the range $0<\mu\leqslant 2$ as the root of this sum over j. The upper bound $\mu=2$ is reached when all qubits are detected by the same detector, so that one of the ν_j equals 1 and the others vanish.

In summary, we determine the ML estimator

$$\vec{S} = 3\sum_j \tilde{p}_j\vec{a}_j \quad (4.14)$$

as follows. If the inequality (4.5) is obeyed by $\nu_j=n_j/N$, we take $\tilde{p}_j=\nu_j$. Otherwise, we find μ as the positive root of Eqs. (4.12) and (4.13) and then get the four $\tilde{p}_j$'s from Eq. (4.11) with $\lambda=1-\mu$. In the extremal situation of $\nu_j=\delta_{jk}$, we have $\tilde{p}_j=\frac{1}{6}+\frac{1}{3}\delta_{jk}$ and $\vec{S}=\vec{a}_k$.

B. Many detector clicks

In this procedure for finding the ML estimator, there is a crucial difference between relative frequencies ν_j that obey the inequality (4.5), and can therefore serve as probabilities, and those that violate the inequality. When the total number N of detected qubits is small, statistical fluctuations are relatively large and a violation is hardly surprising. But what is the typical situation for a large number of detector clicks; should we expect Eq. (4.5) to be obeyed or violated?

For the likelihood (4.3), only the breakup (4.1) matters, not the particular sequence of detector clicks. There are $N!/(n_1!n_2!n_3!n_4!)$ different sequences for a given breakup, so that there is a multinomial statistics for the probability of getting a particular breakup for the given probabilities $p_j(\vec{s})$ of Eq. (2.10). We denote the corresponding averages over possible breakups by overbars, as illustrated by

$$\overline{n_j} = N\overline{\nu_j} = Np_j \quad (4.15)$$

and

$$\overline{\nu_j\nu_k} = p_jp_k + \frac{1}{N}(\delta_{jk}p_k - p_jp_k). \quad (4.16)$$

Upon recalling that $\Sigma_j p_j^2=(3+s^2)/12$ [cf. Eq. (2.13)], the latter averages imply

$$\sum_j \overline{\nu_j^2} = \frac{1}{N} + \frac{N-1}{N}\sum_j p_j^2 = \frac{1}{3} - \frac{1-s^2}{12} + \frac{9-s^2}{12N}, \quad (4.17)$$

so that, *on average*, the inequality (4.5) is violated for pure states (for which $s=1$) and is obeyed for mixed states (for which $s<1$). We thus expect that the detector counts for a pure qubit state will typically violate inequality (4.5), whereas the counts for a mixed state will tend to obey it.

As a more precise statement about this matter, we note that the fraction of detector click sequences that violate the inequality is

$$\text{prob(violation)} = \frac{1}{2} + \frac{1}{2}\text{erf}\left(\frac{\sqrt{N}}{2\kappa}\left(\sum_j \overline{\nu_j^2} - \frac{1}{3}\right)\right), \quad (4.18)$$

where erf() is the standard error function and κ is given by

$$\kappa^2 = \sum_{j,k} p_jp_k(p_j-p_k)^2 = 2\sum_j p_j^3 - 2\left(\sum_j p_j^2\right)^2. \quad (4.19)$$

Equation (4.18) applies for $N\gg 1$; in order to derive it, first observe that the central limit theorem ensures that

$$\overline{\exp\left(i\sum_j \alpha_j(\nu_j-p_j)\right)} = \exp\left(-\frac{1}{2N}\sum_{j,k}\alpha_j(\delta_{jk}p_k - p_jp_k)\alpha_k\right) \quad (4.20)$$

for large N [20]. Then use this generating function for the mean values of products of the ν_j's to calculate the probability that $\Sigma_j\nu_j^2>\frac{1}{3}$, with consistent approximations for $N\gg 1$.

For pure states, the argument of the error function in Eq. (4.18) is $1/(3\kappa\sqrt{N})$, which is always positive, but decreases with growing N, so that click sequences that violate the inequality are more frequent than the ones obeying it, but they are not much more frequent. A remarkable exception is the situation of the Pauli vector $\vec{s}$ being exactly opposite to one of the directions of the vector quartet of Eqs. (2.5) and (2.6). Then one of the p_j's vanishes and the other three are all equal to $\frac{1}{3}$, so that $\kappa=0$ and there is a unit probability for getting a violation.

For mixed states, the argument of the error function is negative and increases in magnitude $\propto\sqrt{N}$ (more about this in Sec. V C). Accordingly, a violation of the inequality is highly improbable. In the extreme situation of the completely mixed state, $\rho=\frac{1}{2}$, all p_j's are $\frac{1}{4}$, so that $\kappa=0$ here too, and the probability for violating the inequality vanishes.

C. Asymptotic efficiency

After detecting many qubits, we expect the ML estimator $\vec{S}$ to be quite close to the true Pauli vector $\vec{s}$. If inequality (4.5) is obeyed, the average squared distance

$$(\Delta\vec{s})^2 = \overline{(\vec{S}-\vec{s})^2} = \overline{\left(3\sum_j(\nu_j-p_j)\vec{a}_j\right)^2} = \frac{12}{N}\left(1-\sum_j p_j^2\right)$$
$$= \frac{9-s^2}{N} \quad (4.21)$$

is immediately available as a consequence of Eq. (4.16).

The analysis is more involved when Eq. (4.5) is not obeyed. Let us consider once more the extreme situation of $\vec{s}$ being exactly opposite to one of the $\vec{a}_j$, such as $\vec{s}=-\vec{a}_4$, say, so that $p_4=0$ and $\nu_4=0$, and ν_1, ν_2, ν_3 differ little from $p_1=p_2=p_3=\frac{1}{3}$. To leading order in $\nu_j-\frac{1}{3}$, the solution of Eq. (4.12) is then

$$\mu = 1 + \frac{9}{8}\sum_{j=1}^{3}\left(\nu_j - \frac{1}{3}\right)^2 \quad (4.22)$$

and the resulting ML estimator is

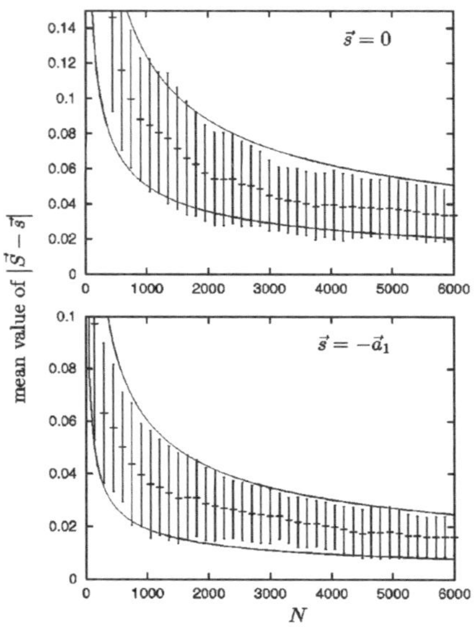

FIG. 6. Mean value of the distance from the ML estimator $\vec{S}$ to the true Pauli vector $\vec{s}$ for 40 simulated experiments with up to 6000 detected qubits per run. The top plot is for $\vec{s}=0$, the bottom plot for $\vec{s}=-\vec{a}_4$. The solid lines indicate one standard deviation to each side of the mean value, as obtained from the large-N approximations for $(\Delta\vec{s})^2$ in Eqs. (4.21) and (4.24), respectively.

$$\vec{S}=\frac{1}{2}\sum_{j=1}^{3}(1+3\nu_j)\vec{a}_j=\vec{s}+\frac{3}{2}\sum_{j=1}^{3}\left(\nu_j-\frac{1}{3}\right)\vec{a}_j. \qquad (4.23)$$

In conjunction with Eq. (4.16), we thus get the large-N approximation

$$(\Delta\vec{s})^2=3\overline{\nu_1^2+\nu_2^2+\nu_3^2}-1=\frac{2}{N} \qquad (4.24)$$

in this case of perfect antialignment between $\vec{s}$ and the quartet of $\vec{a}_j$'s. The sensitivity to small misalignments is discussed in Sec. VII C.

As demonstrated by the numerical results of Fig. 6, these asymptotic approximations are actually quite reliable for $N \gtrsim 1000$. The plots report the mean values, with statistical error bars of one standard deviation, of the distance $|\vec{S}-\vec{s}|$ as obtained from 40 simulated experiments, whereby up to 6000 qubits are detected in each run. The top plot refers to a true state that is completely mixed, so that $s=0$ in Eq. (4.21). In the bottom plot we have $\vec{s}=-\vec{a}_4$ and Eq. (4.24) applies.

By comparison, for the standard six-output ellipsometer with counts $n_{\xi\pm}$ of the respective detectors, the ML estimator is

$$\vec{S}=\left(\frac{n_{x+}-n_{x-}}{n_{x+}+n_{x-}},\frac{n_{y+}-n_{y-}}{n_{y+}+n_{y-}},\frac{n_{z+}-n_{z-}}{n_{z+}+n_{z-}}\right), \qquad (4.25)$$

provided that the length of this $\vec{S}$ does not exceed unity. If it does, the search has to be constrained to unit vectors, much like in the discussion of Eqs. (4.6)–(4.14). When Eq. (4.25) applies, the error corresponding to Eq. (4.21) is given by

$$(\Delta\vec{s})^2=\frac{9-3s^2}{N}. \qquad (4.26)$$

For $s>0$, this is a bit smaller than the value in Eq. (4.21), so that the six-output scheme gives a slightly more reliable estimate. But when s gets close to unity, the four-output scheme can provide better estimates because there are states for which the error is substantially smaller, as the extreme situation of Eq. (4.24) shows. For the six-output ellipsometer, the privileged pure states ρ project on an eigenstate of σ_x, or σ_y, or σ_z, so that one of the six detectors in Fig. 2 never registers a qubit. Then we have

$$(\Delta\vec{s})^2=\frac{8}{3N}, \qquad (4.27)$$

which is a bit larger than its analog (4.24). The main lesson is thus that the two schemes are of comparable asymptotic efficiency.

V. OPTIMALITY OF MINIMAL FOUR-STATE TOMOGRAPHY

A. Cramér-Rao bound

As a preparation of the discussion below on the optimality and efficiency of the minimal four-state tomography, we recall some well-known facts about state estimation. We regard the qubit state as parametrized by the three components of its Pauli vector $\vec{s}=(s_x,s_y,s_z)$, rather than by the four probabilities p_j, because the latter are constrained. Then, the error of the components of the ML estimator $\vec{S}$ can be estimated by the Cramér-Rao lower bound [17,21]

$$(\Delta s_\xi)^2 \geqslant (I_F^{-1})_{\xi\xi} \quad \text{with } \xi=x,y,z, \qquad (5.1)$$

where I_F is the Fisher information matrix,

$$(I_F)_{\xi\zeta}=\overline{\frac{\partial\log\mathcal{L}}{\partial s_\xi}\frac{\partial\log\mathcal{L}}{\partial s_\zeta}}, \qquad (5.2)$$

and the averaging is done over the data—that is, over the multinomial distribution that we exploited already in Sec. IV B.

There is no unique numerical measure for the comparison of the estimated state $\rho_{\text{est}}=\frac{1}{2}(1+\vec{S}\cdot\vec{\sigma})$ with the true state $\rho_{\text{true}}=\frac{1}{2}(1+\vec{s}\cdot\vec{\sigma})$. One can make a case for the trace-class distance

$$\mathcal{D}_{\text{tr}}=\frac{1}{2}\text{tr}\{|\rho_{\text{est}}-\rho_{\text{true}}|\}=\frac{1}{2}|\vec{S}-\vec{s}|, \qquad (5.3)$$

for the Hilbert-Schmidt distance

$$D_{\text{HS}}=\frac{1}{2}[\text{tr}\{(\rho_{\text{est}}-\rho_{\text{true}})^2\}]^{1/2}=\frac{1}{2}|\vec{S}-\vec{s}|, \qquad (5.4)$$

for the Uhlmann fidelity

$$\mathcal{U} = \mathrm{tr}\{|\sqrt{\rho_{\mathrm{est}}}\sqrt{\rho_{\mathrm{true}}}|\} = \frac{1}{2}(1+\vec{S}\cdot\vec{s}+u)^{1/2} + \frac{1}{2}(1+\vec{S}\cdot\vec{s}-u)^{1/2}, \quad (5.5)$$

with

$$u = \sqrt{(\vec{S}+\vec{s})^2 - (\vec{S}\times\vec{s})^2}, \quad (5.6)$$

and for some more. In the single-qubit situation of present interest, there is no difference between $\mathcal{D}_{\mathrm{tr}}$ and $\mathcal{D}_{\mathrm{HS}}$, but they are not the same for higher-dimensional systems, such as the qubit pairs of Sec. II C. We note in passing that the Uhlmann fidelity and the trace-class distance are natural pairs, inasmuch as they obey a fundamental inequality

$$\mathcal{D}_{\mathrm{tr}}^2 + \mathcal{U}^2 \leq 1, \quad (5.7)$$

irrespective of the dimension of the Hilbert space.

As long as one is comparing pure states with each other, the actual choice between the quantitative measures of Eqs. (5.3)–(5.5) does not matter much, because then all of them are monotonic functions of the length of the difference $\vec{S}-\vec{s}$ of the two Pauli vectors. If one state is mixed, however, this is not true for the Uhlmann fidelity. For the sake of computational simplicity, we opt for the (square of twice) the Hilbert-Schmidt distance and thus quantify the measure of the estimate by the size of $(\vec{S}-\vec{s})^2$, whose statistical average is already considered in Sec. IV C. A convenient conservative estimate is given by the Cramér-Rao bound of Eq. (5.1).

According to Fisher's theorem [16] the ML estimator of ρ attains the Cramér-Rao bound for large N. Strictly speaking, this statement applies only to mixed states, because only for them are the ML estimators unbiased. For pure states, the positivity constraint plays an important role, and the Cramér-Rao bound derived for unbiased estimators tends to overestimate the error, and therefore we shall consider pure states separately below. As a rule of thumb, the Cramér-Rao bound is reliable when the lion's share of the likelihood function is contained within the Bloch ball. This can, in fact, be used as an operational definition of "mixedness." Indeed, as the analysis in Sec. IV B shows, for mixed states there is no significant fraction of the likelihood function outside the Bloch ball if N is sufficiently large.

B. Optimality of the tetrahedron geometry

The four-state tomography with the tetrahedron geometry of Eqs. (2.5)–(2.9) is the best minimal qubit-state measurement, in the sense that, among the measurements with four output channels, it provides the greatest accuracy. This can be seen as follows.

For the multinomial statistics of Sec. IV B, the Fisher information simplifies to

$$(I_F)_{\xi\zeta} = N\sum_j \frac{1}{p_j}\frac{\partial p_j}{\partial s_\xi}\frac{\partial p_j}{\partial s_\zeta}, \quad (5.8)$$

where p_j is the probabilities of detecting a qubit in the jth output channel. Consider now this variational problem: For a given input state, the average distance D is to be minimized over all possible four-element POVMs. The functional in question is

$$D = \mathrm{Sp}(I_F^{-1}) - \lambda\sum_j \Pi_j, \quad (5.9)$$

where Sp() is the trace ("spur") of the 3×3 matrix, and a Lagrange operator Λ takes care of the constraint $\Sigma_j\Pi_j = 1$.

First, let us find the POVM that minimizes the functional (5.9) for the maximally mixed input state. In that case the extremal equations read

$$R_j\Pi_j = \Lambda\Pi_j \quad \text{for } j=1,\ldots,4,$$

$$\Lambda = \sum_j R_j\Pi_j, \quad (5.10)$$

where

$$R_j = \mathrm{Sp}(I_F^{-2}T_j) \quad (5.11)$$

involves the operator matrix T_j whose matrix elements are given by

$$T_{j,\xi\zeta} = \mathrm{tr}\{\Pi_j\sigma_\xi\}\sigma_\zeta + \sigma_\xi\mathrm{tr}\{\sigma_\zeta\Pi_j\}. \quad (5.12)$$

Note the joint appearance of 3×3 Sp() traces and quantum-mechanical tr{} traces. One verifies by inspection that any POVM of the tetrahedron geometry of Eqs. (2.5)–(2.9) satisfies these extremal equations and hence identifies the maximally mixed state with greatest accuracy.

Obviously, for biased states the optimal POVM itself becomes biased. The generalization of the extremal equations to this case is straightforward; the operators T_j will then contain one more term proportional to the true state ρ. Although an analytical solution may be difficult to obtain, one can always find the extremal measurement by an iterative procedure. As the input state is usually not known and selected in random, operators R_j should be averaged over the Bloch ball. In this way one obtains an algorithm providing the optimal sequential measurement, in the sense that the Hilbert-Schmidt distance is consistently reduced in each iteration step. Numerical results show that the tetrahedron POVM is also optimal for uniformly distributed input states.

Quite explicitly, the optimal distance for the tetrahedron POVM is

$$D_{\mathrm{opt}} = \frac{9-s^2}{N}, \quad (5.13)$$

in agreement with, or as a consequence of, the mean square distance of Eq. (4.21). As one would expect, the accuracy of the ellipsometer is somewhat better for pure input states than for mixed states. But, most importantly, the accuracy does not depend on the orientation of $\vec{s}$ relative to the measurement tetrahedron.

C. Orientation of the measurement tetrahedron

Their relative orientation is, however, not completely irrelevant. For example, the value of κ^2 in Eq. (4.19) clearly depends on it. More generally, we can obtain the large-N mean values of functions of $\vec{S}-\vec{s}$ from the asymptotic generating function

$$\overline{\exp[i\vec{r}\cdot(\vec{S}-\vec{s})]}=\exp\left(-\frac{1}{2N}\vec{r}\cdot\overleftrightarrow{K}\cdot\vec{r}\right), \tag{5.14}$$

where the dyadic

$$\overleftrightarrow{K}=9\sum_{j,k}\vec{a}_j(\delta_{jk}p_j-p_jp_k)\vec{a}_k \tag{5.15}$$

depends on the positioning of $\vec{s}$ relative to the vector quartet of the $\vec{a}_j$'s. This is, of course, an immediate consequence of Eq. (4.20).

As an application, let us consider the asymptotic mean value of the Uhlmann fidelity of Eqs. (5.5) and (5.6). We first note that

$$\mathcal{U}=1-\frac{1}{8}\left[(\vec{S}-\vec{s})^2+\frac{[\vec{s}\cdot(\vec{S}-\vec{s})]^2}{1-s^2}\right] \tag{5.16}$$

holds when $\vec{S}-\vec{s}$ is small. Then we recall Eq. (4.21) and extract

$$\overline{[\vec{s}\cdot(\vec{S}-\vec{s})]^2}=\frac{1}{N}\vec{s}\cdot K\cdot\vec{s}=\frac{72\kappa^2}{N} \tag{5.17}$$

from Eq. (4.25) to arrive at

$$\overline{\mathcal{U}}=1-\frac{9-s^2}{8N}-\frac{9\kappa^2}{N(1-s^2)}, \tag{5.18}$$

where we meet the orientation-dependent quantity κ^2 of Eq. (4.19).

There are extremal orientations of three kinds. The value of κ^2 is largest when $\vec{s}$ is parallel to one of the vectors of the tetrahedron quartet and smallest when $\vec{s}$ is antiparallel. In addition to these four maxima and four minima of κ^2, there are also six saddle points that have $\vec{s}$ parallel to the sum of two different vectors of the quartet. The maximal and minimal values of κ^2 are given by the upper and lower signs in

$$\kappa^2=\frac{(1\pm s)(3\mp s)s^2}{72}\quad\text{for } \vec{s}=\pm s\vec{a}_j \quad (\text{any } j), \tag{5.19}$$

respectively, and the value at the saddle points is

$$\kappa^2=\frac{(3-s^2)s^2}{72}\quad\text{for } \vec{s}=\frac{\sqrt{3}}{2}s(\vec{a}_j+\vec{a}_k)\ (\text{any } j\neq k). \tag{5.20}$$

Accordingly, the approach of $\overline{\mathcal{U}}$ to unity is fastest for the $\vec{s}=-s\vec{a}_j$ orientation, for which the large-N approximation

$$\overline{\mathcal{U}}=1-\frac{(3+s)(3+2s)}{8N(1+s)} \tag{5.21}$$

applies. A small value of κ is also advantageous in Eq. (4.18), as it ensures a large negative argument of the error function and thus a small probability for violating the inequality (4.5).

VI. MEASURING PURE QUBIT STATES

When the measured quantum system is known to be in a pure state—which is a bold overidealization of any realistic situation—this knowledge can be exploited systematically when estimating the otherwise unknown state. A somewhat more realistic situation arises when the input state is pure but we do not know this to begin with, although this case is quite a bit artificial as well, because one can hardly assume that real sources are not affected by classical noise or that the experimental setup is totally decoupled from the environment. Put differently, it is far-fetched to assume that the experimenter will ever have the perfect control that is necessary to ensure that a source emits a pure state. Nevertheless, there is an interest in such idealized scenarios because they come up in analyses of eavesdropping attacks on schemes for quantum cryptography, where—as a matter of principle—it is assumed that the eavesdropper is only limited by the laws of physics, not by practical limitations.

For $s=1$, the Uhlmann fidelity (5.5) simplifies to

$$\mathcal{U}=\sqrt{\frac{1+\vec{s}\cdot\vec{S}}{2}} \tag{6.1}$$

and is equivalent to

$$\mathcal{U}=\sqrt{1-\frac{1}{4}(\vec{S}-\vec{s})^2} \tag{6.2}$$

if the estimator is a pure state as well, $S=1$. When indeed estimating pure states with pure states, the situation to be considered now, it is customary to take the average of $\mathcal{U}^2$,

$$\overline{F}=\overline{\mathcal{U}^2}=1-\frac{1}{4}\overline{(\vec{s}-\vec{S})^2}=1-\frac{1}{4}(\Delta\vec{s})^2, \tag{6.3}$$

as the fidelity measure that judges the quality of the estimation procedure. One must keep in mind that both the true Pauli vector $\vec{s}$ and its estimator $\vec{S}$ are unit vectors here, as Eq. (6.3) applies only under this restriction. Knowing that the source generates pure states means having a lot of prior knowledge because the set of pure states is much smaller than the set of mixed states (the Bloch sphere rather than the Bloch ball), and so one can safely expect that a better, possibly much better, accuracy of the estimation can be achieved.

The average fidelity of the optimal *joint* measurement on N copies is known to obey the inequality [22,23]

$$\overline{F}\leq\frac{N+1}{N+2}, \tag{6.4}$$

so that the corresponding error $1-\overline{F}$ will decrease as $1/N$ in the large-N limit. Any reasonable estimator should show this dependence.

In our scheme the qubits are measured individually, not jointly. Nevertheless, as a consequence of the Fisher theorem in general and of the findings in Sec. IV C in particular, the variance $(\Delta\vec{s})^2$ in Eq. (6.3) is proportional to $1/N$ in the large-N limit, and so is then $1-\overline{F}$. But we need to reconsider the asymptotics of the maximum-likelihood estimation, now taking into account that both $\vec{s}$ and $\vec{S}$ are restricted to the Bloch sphere, so that we get the estimator $\vec{S}$ from Eqs. (4.10)–(4.13), whether inequality (4.5) is obeyed or not.

Here, too, the estimation is sensitive to the orientation of the Pauli vector $\vec{s}$ relative to the vector quartet of the measurement tetrahedron. We deal first with the case of "not antialigned"; that is, $\vec{s}\neq-\vec{a}_j$ for all $\vec{a}_j$'s. Then, ν_j-p_j and μ are of the order of $1/\sqrt{N}$ and μ is given by

$$\mu=\frac{2}{3\kappa^2}\sum_j p_j(\nu_j-p_j), \tag{6.5}$$

with $\kappa^2>0$ from Eq. (4.19). The resulting averages

$$\overline{(\nu_j-p_j)\mu}=\frac{2}{9N\kappa^2}(3p_j-1)p_j,$$

$$\overline{\mu^2}=\frac{2}{9N\kappa^2}, \tag{6.6}$$

available as a consequence of Eq. (4.16), are used in

$$\overline{F}=1-3\sum_j\overline{[(\nu_j-p_j)-(3p_j-1)p_j\mu]^2} \tag{6.7}$$

to arrive at

$$1-\overline{F}=\frac{2}{N}-\frac{2}{3N\kappa^2}\sum_j[(3p_j-1)p_j]^2$$
$$=\frac{4}{N}-\frac{2}{9N\kappa^2}\left(27\sum_j p_j^4-1\right). \tag{6.8}$$

In particular, when the tetrahedron is aligned with the Pauli vector, the upper-sign case of Eq. (5.19) with $s=1$, we have $\kappa^2=1/18$ and $27\Sigma_j p_j^4=7/4$, so that

$$1-\overline{F_{\uparrow\uparrow}}=\frac{1}{N} \tag{6.9}$$

for this parallel strategy ($\uparrow\uparrow$). This is only half as big as what one would get for the $s=1$ value of Eq. (4.21) and thus demonstrates the advantage of estimating the pure true state by pure-state estimators only.

There is no such advantage for the antiparallel strategy ($\uparrow\downarrow$) that has the tetrahedron antialigned with $\vec{s}$, because the argument in Sec. IV B establishes that the ML estimator is always a pure state then. Since this is now the lower-sign case of Eq. (5.19) for $s=1$, we have $\kappa^2=0$ and Eqs. (6.5)–(6.8) are not valid. Instead, we recall that Eqs. (4.22)–(4.24) apply and conclude that

$$1-\overline{F_{\uparrow\downarrow}}=\frac{1}{2N} \tag{6.10}$$

holds here. The antiparallel strategy has thus half the average error of the aligned strategy.

On the other hand, for a generic orientation of the input state there will always be an uncertainty in the length of the input state vector and therefore the fidelity will approach unity at a much slower rate proportional to $1/\sqrt{N}$. Only if one knows beforehand that the input state is pure can the $1/N$ rate be achieved even for general orientations of the tetrahedron. Let us illustrate this last remark by the example of the parallel strategy. For the parallel orientation, only about half of the measurement outcomes will violate inequality (4.5); the others would not (see Sec. IV B). Knowing that the input state is pure one can, however, ignore this inequality and always solve Eqs. (4.8)–(4.10) for the input-state unit Pauli vector.

VII. ADAPTIVE STRATEGIES

A. Premeasurement strategy

The good performance of the antiparallel strategy hints at a very simple adaptive procedure that provides the fast $1/N$ asymptotic behavior without any prior knowledge about the purity of the input state: Let us split the input ensemble into two halves. After the first $N/2$ particles have been registered and the direction of the input Pauli vector estimated, the experimenter adopts the antiparallel strategy for measuring the rest of the ensemble. Notice that in this simple adaptive scenario, the first half of the particles are used for a premeasurement and serve only for adjusting the measurement apparatus, while it is the second half which provide the actual estimate of the input state.

Let θ denote the angle between the Pauli vector $\vec{s}$ of the input state and the Pauli vector $\vec{s}_1$ estimated from the first half of the ensemble. This angle θ can be estimated with an accuracy of $\overline{1-\cos\theta}\propto 1/N$ in the first stage. This means that, in the second stage, the mean probability of detecting a particle in the channel antiparallel to $\vec{s}_1$ will be proportional to $1/N$. No matter how large is N, only a few particles will be detected in this channel.

The maximal uncertainty in the length of the Pauli vector is then easily calculated with the aid of

$$S^2=\frac{12}{N^2}\sum_j n_j^2-3 \tag{7.1}$$

for the estimator of Eq. (4.2). Let us set $n_1=\delta$ with δ a small number independent of N and look for the minimal length of the estimated Pauli vector compatible with the given δ. Since the right-hand side of Eq. (7.1) is a concave function of n_2, n_3, and n_4, it is minimal when all of them are equal to each other, $n_2=n_3=n_4=(N-\delta)/3$, with the consequence

$$S_{\min}=1-\frac{4\delta}{N} \tag{7.2}$$

for large N. This guarantees that in the second stage both the orientation and length of the Pauli vector, and so the fidelity, will be determined with an error proportional to $1/N$. Numerical simulations show that for large N (up to $N=10^5$) the mean error $1-\overline{F}$ of this simple protocol is about twice as large as, and thus worse than, that of the optimal joint POVM measurement, in which all N qubits are measured together. Let us emphasize that while our protocol would provide this performance for any input pure state, it also provides a meaningful estimate for any input mixed state, of course, with a larger uncertainty proportional to $1/\sqrt{N}$. In contrast to that, the optimal joint measurement that attains the ultimate limit of Eq. (6.4) would not work for mixed input states.

B. Self-learning strategies

It is known that sequential measurements when combined with self-learning adaptive strategies can come close to the quantum estimation limit [24–26]. Their improvement on the conventional sequential measurement depends on the purity of the input state [24]. Adaptive techniques are more sensitive to pure states than to states with a lot of classical noise.

The optical network of Fig. 3(b) can easily be adapted to a self-learning procedure and so can other optical implementations. After each detection the current information about the input state can be evaluated, and the operations σ_x, σ_y, and σ_z acting on the next particle inside the interferometer can be modified by a common unitary transformation. This is economically achieved by performing the required unitary transformation on the approaching photon before it enters the interferometer proper.

From the discussion in Sec. VI one might get the impression that a particularly good adaptive strategy would be to always keep one of the measured half-projectors antiparallel to the current estimate. Matters are, however, not so simple.

Let us illustrate the difference between the adaptive and nonadaptive procedures at the extreme example of measuring only two qubits. In the *nonadaptive* case, everything is as discussed above. In particular, the probability that the first qubit is detected by the jth detector and the second by the kth detector is $p_j p_k$ with the p_j's of Eq. (2.10) and, in accordance with Sec. IV, the ML estimator is $\vec{S}=\vec{a}_j$ if $j=k$ and $\vec{S}=\sqrt{3/4}(\vec{a}_j+\vec{a}_k)$ if $j\neq k$. Upon averaging $(\vec{S}-\vec{s})^2$ first over all measurement results for a given input Pauli vector $\vec{s}$ and then over all possible inputs, we thus get

$$\overline{(\vec{S}-\vec{s})^2}=\frac{5-\sqrt{3}}{3}=1.089 \tag{7.3}$$

as the figure of merit.

In the *adaptive case*, the tetrahedron is realigned for the second qubit after the first qubit has been detected. Since the ML estimator $\vec{S}$ obtained after the detection of the first qubit will coincide with one of the tetrahedron vectors, the antialigning of the tetrahedron for the second qubit amounts to the replacement $\vec{a}_j\to-\vec{a}_j$. Now the probability that the first qubit is detected by the jth detector and the second by the kth detector is $\mathcal{L}=p_j(\frac{1}{2}-p_k)$. The resulting ML estimator is then given by $\vec{S}=\sqrt{3/8}(\vec{a}_j-\vec{a}_k)$, and upon averaging over all measurement results and all input Pauli vectors we get

$$\overline{(\vec{S}-\vec{s})^2}=\frac{11-\sqrt{24}}{6}=1.017, \tag{7.4}$$

which is markedly smaller than the nonadaptive value in Eq. (7.3). It is also smaller than the value for the adaptive strategy with parallel alignment because that is identical with the nonadaptive procedure when only two qubits are detected.

We note that the averages in Eqs. (7.3) and (7.4) are taken over pure input states. If one averages over all input states,

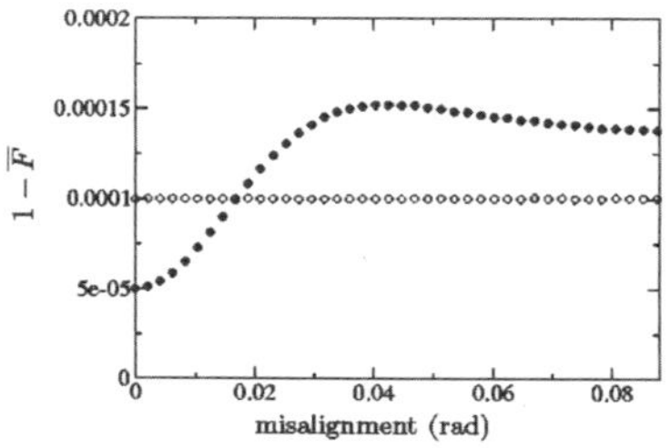

FIG. 7. Mean estimation errors of the POVM tetrahedrons that, on the Bloch sphere, differ from the exact parallel (squares) and antiparallel (circles) orientations by the known angle given on the abscissa.

pure and mixed, one obtains the respective numbers $(7-\sqrt{3})/5=1.054$ and $(7-\sqrt{6})/5=0.910$, again with a clear advantage for the antialigning adaptive strategy.

C. Numerical simulations

Although these numbers speak clearly in favor of the antialigning adaptive scheme, one should, however, keep in mind that they apply only for the exactly aligned or antialigned settings of the apparatus. But in such an adaptive scheme, after the first particle is observed, the uncertainty of the input state Pauli vector orientation is still quite large, which may result in a significant misalignment in the second adaptive step. In fact, the antialigning strategy has a greater sensitivity to such misalignments. This is illustrated by the simulation data shown in Fig. 7, where the estimation errors of both strategies are shown in dependence on the misalignments of the apparatuses for the chosen input intensity of 10^4 particles. Each point has been obtained by averaging over 5×10^5 ML estimates.

As expected, the antialigning adaptive strategy performs better if no misalignment is present. However, even a small misalignment (of the order of 1° in this case) is enough to wash out this advantage. For even larger misalignments the aligned setting provides much better performance [27]. Different sensitivities of both measurement strategies to this kind of error might be of quite some importance for the potential applications of the minimal ellipsometer in quantum communication protocols and quantum cryptography.

One can understand this extreme sensitivity of the antialigned setting and why it becomes immediately worse than the aligned setting by taking a second look at Eqs. (6.8) and (6.10). The $\kappa^2=0$ result (6.10) is *not* the $\kappa^2\to0$ limit of the $\kappa^2>0$ result (6.8). In fact we have (see Fig. 8)

$$1-\overline{F}\to\frac{4}{3N}\quad\text{as }\kappa^2\to0\quad\text{in Eq. (6.8)}, \tag{7.5}$$

which is larger than the error of $\overline{F_{\uparrow\uparrow}}$ in Eq. (6.9). Therefore, the slightest misalignment takes us from the $1/(2N)$ error of Eq. (6.10) to this $4/(3N)$ because Eq. (7.5) applies to tiny nonzero values of κ^2, while Eq. (6.10) holds only if $\kappa^2=0$ exactly.

This observation also resolves the apparent contradiction between the general upper bound of Eq. (6.4) and the large-

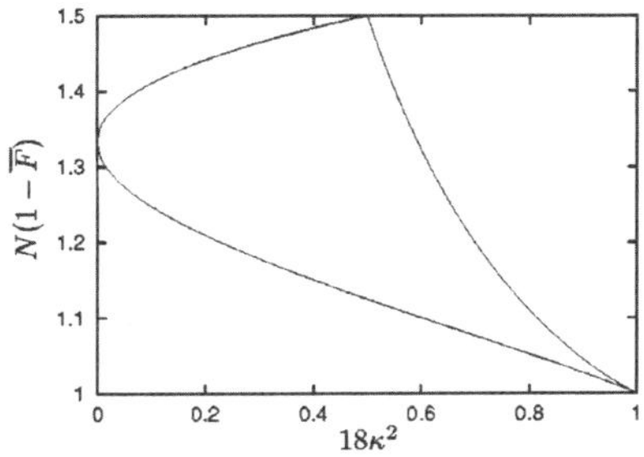

FIG. 8. For each value of κ^2 between $\kappa^2=0$ (perfect antialignment) and $\kappa^2=1/18$ (perfect alignment), the possible values of the coefficient of the $1/N$ term in Eq. (6.8) are in the area bounded by the two curves. The smallest coefficient obtains for the case of perfect alignment, when Eq. (6.9) applies.

N error for perfect antialignment in Eq. (6.10), which does not respect that upper bound. Nevertheless, this example is not a valid counterexample, because it refers to an absurdly artificial situation: The experimenter has perfect *a priori* knowledge of the state to be measured and has perfect control over his measurement apparatus, such as to ensure the perfect antialignment to which Eq. (6.10) applies. In other words, *when* it applies, there is no need for a state estimation to begin with.

Having thus compared the performances of the two extreme strategies, we now calculate the mean fidelity of the following adaptive measurement: After detecting each new qubit the information about the input state is updated and a new ML estimate is calculated. Then one of the measured half-projectors is aligned along this current estimate. These two steps are repeated until all input particles are used up. Figure 9 shows mean fidelities and errors that were obtained by averaging over 200 000 randomly selected pure input states. The quantum limit, Eq. (6.4), is shown for comparison. It is evident that this bound can be attained only for large N, while the most pronounced difference is seen for moderately sized ensembles. Such a behavior is typical for all sequential self-learning estimation strategies.

Finally, let us compare the efficiency of the parallel adaptive strategy with a very simple sequential measurement

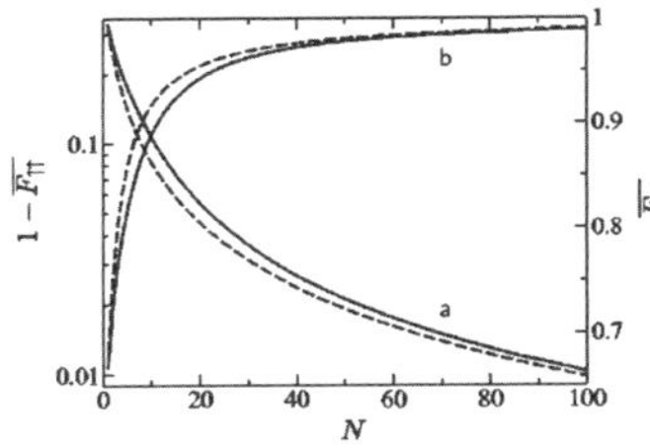

FIG. 9. Mean error (curves a, decreasing) and mean fidelity (curves b, increasing) of the minimal qubit tomography as a function of N, the size of the measured ensemble. Solid lines: the proposed network of Figs. 3 and 4 with the adaptive parallel strategy described in the text. Dashed lines: the quantum limit of Eq. (6.4). Notice that the proposed optical network attains the quantum limit asymptotically for large N.

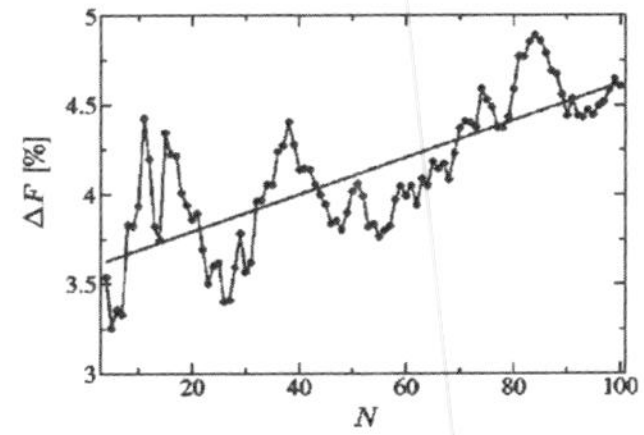

FIG. 10. Relative difference ΔF, in percent, between the errors of the random and parallel adaptive strategies (black dots) and least-squares linear fit (straight line). The lines connecting the dots guide the eye but have no further significance. The irregularities stem from the limited number of input states (200 000) used for the averaging.

where the orientation of the measured half-projectors is chosen at random in each step (see Fig. 10). As expected, the adaptive strategy is better and its benefit grows with increasing size of the measured ensemble.

VIII. SUMMARY

We have presented a minimal measurement scheme for single-qubit tomography that has no more than the necessary number of four outputs. The scheme is conceptually simple, highly symmetric and optimal among all four-output schemes, and can be realized with the present technology for the polarization qubit of photons emitted by a single-photon source. As a demonstration, we designed a simple, but not simplest, optical network.

Our thorough analysis showed that the scheme is efficient in the sense that it enables one to estimate the qubit state reliably without first detecting an enormous number of qubits—a few thousand are sufficient for most practical applications; a few hundred may be enough if extreme precision is not required. The efficiency can be increased by adaptive procedures in which the apparatus is adjusted in accordance with the current estimate for the qubit state.

Since the four-output setup provides optimal complete tomography with the minimal number of output channels, it is particularly well suited as a detection device for certain quantum communication protocols such as tomographic quantum cryptography [28]. Indeed, there are protocols for quantum key distribution that exploit the tetrahedron quartet of states [29], among them a highly efficient tomographic protocol [30].

ACKNOWLEDGMENTS

We are very grateful for the valuable discussions with Artur Ekert, Christian Kurtsiefer, Antia Lamas-Linares, Ng Hui Khoon, Tin Kah Ming, and Goh Choon Guan. J.Ř. is grateful for the kind hospitality received during his visits to Singapore. We gratefully acknowledge financial support from Temasek Grant No. WBS: R-144-000-071-305, from NUS Grant No. WBS: R-144-000-089-112, and from Grant No. LN00A015 of the Czech Ministry of Education.

[1] W. K. Wootters and B. D. Fields Ann. Phys. (N.Y.) **191**, 363 (1989).

[2] J. M. Renes, R. Blume-Kohout, A. J. Scott, and C. M. Caves, J. Math. Phys. **45**, 2171 (2004).

[3] Incidentally, these probabilities are closely related to Feynman's "negative probabilities" [R. P. Feynman, in *Quantum Implications: Essays in Honour of David Bohm*, edited by B. Hiley and D. Peat (Routledge London, 1987)], which are the expectation values of $\frac{1}{2}(1+\sqrt{3}\vec{a}_j\cdot\vec{s})$. In fact, they are not probabilities, but rather the natural, discrete qubit analog of Wigner's phase-space function for continuous degrees of freedom; see W. K. Wootters, Ann. Phys. (N.Y.) **176**, 1 (1987) and D. Galetti, and A. F. R. De Toledo Piza, Physica A **149**, 267 (1988), for further details about these matters in general and, e.g., A. Luis, and J. Peřina, J. Phys. A **31**, 1423 (1998), for some particular aspects.

[4] In *state tomography* one must know the relevant properties of the measurement device; in *measurement tomography* one must know the state that is measured.

[5] We thank Ch. Kurtsiefer, for asking whether this can be done.

[6] B.-G. Englert, K. M. Tin, C. G. Goh, and H. K. Ng, Laser Phys. (to be published), e-print physics/0409015.

[7] B.-G. Englert, C. G. Goh, Ch. Kurtsiefer, A. Lamas Linares, H. K. Ng, and K. M. Tin (unpublished).

[8] It may be worth mentioning that, in principle if not in practice, any single-photon POVM can be implemented with an optical multiport in a rather transparent fashion. The setup of Fig. 3(b) is a four-port: the photodetectors register the photons exiting at the four output ports; one of the two entry ports at the 4:2 beam splitter is used, whereas no use is made of the two other entry ports at the 3:1 and 1:1 beam splitters. The setup of Ref. [9] is also a four-port, whereas the setups in Refs. [6,7] are two ports.

[9] See, for example, J. M. Renes, Ph. D. thesis, University of New Mexico, 2004, Fig. 6.5.

[10] R. B. M. Clarke, V. M. Kendon, A. Chefles, S. M. Barnett, E. Riis, and M. Sasaki, Phys. Rev. A **64**, 012303 (2001); see also A. Chefles, in *Quantum State Estimation* [13].

[11] Networks with only one auxiliary qubit are also possible. An example is given by T. Decker, D. Janzing, and T. Beth, Int. J. Quantum Inf. **2**, 353 (2004).

[12] The qubit of interest is always measured itself in the simpler networks of Ref. [11] and also in the practical schemes of Clarke *et al.* [10], the minimal single-loop setup of Ref. [6], and the no-loop setup of Ref. [7].

[13] *Quantum State Estimation*, edited by M. Paris and J. Řeháček, Lecture Notes in Physics Vol. 649 (Springer, Berlin, 2004).

[14] For example, one could take the likelihood of Eq. (4.3) as the weight function for the calculation of the mean of all Pauli vectors $\vec{s}$, thus averaging over the part of the cloud in Fig. 5 that is inside the Bloch sphere, and then accept this average as the best guess. Consult the classic text by Helstrom [15] for a general introduction to quantum-state estimation, or turn to Ref. [13] for a very recent account.

[15] C. W. Helstrom, *Quantum Detection and Estimation Theory* (Academic Press, New York, 1976).

[16] R. A. Fisher, Proc. Cambridge Philos. Soc. **22**, 700 (1925).

[17] H. Cramér, *Mathematical Methods of Statistics* (Princeton University Press, Princeton, 1946).

[18] Z. Hradil, Phys. Rev. A **55**, R1561 (1997).

[19] By construction, the procedure of [14] would always give a physically acceptable estimate.

[20] A detailed analysis of how large N must be for such asymptotic large-N approximations to be valid has been performed by S. L. Braunstein, J. Phys. A **25**, 3813 (1992).

[21] C. R. Rao, Bull. Calcutta Math. Soc. **37**, 81 (1945).

[22] S. Massar and S. Popescu, Phys. Rev. Lett. **74**, 1259 (1995).

[23] R. Derka, V. Bužek, and A. K. Ekert, Phys. Rev. Lett. **80**, 1571 (1998).

[24] D. G. Fischer and M. Freyberger, Phys. Lett. A **273**, 293 (2000).

[25] D. G. Fischer, S. H. Kienle, and M. Freyberger, Phys. Rev. A **61**, 032306 (2000).

[26] T. Hannemann, D. Reiß, C. Balzer, W. Neuhauser, P. E. Toschek, and C. Wunderlich, Phys. Rev. A **65**, 050303 (2002).

[27] Further support in favor of the aligned adaptive scheme is given by the figures of merit obtained for the two-qubits-only situation of Sec. VII B by the procedure of [14]. The numbers corresponding to those in Eqs. (7.3) and (7.4) are 11/21 and 21/40, when all input states are averaged over, as well as 19/24 and 4/5, when the average is over pure states only. In both cases, the aligned scheme does slightly better than the antialigned scheme.

[28] Y. C. Liang, D. Kaszlikowski, B.-G. Englert, L. C. Kwek, and C. H. Oh, Phys. Rev. A **68**, 022324 (2003).

[29] J. M. Renes, e-print quant-ph/0402135.

[30] B.-G. Englert, D. Kaszlikowski, H. K. Ng, W. K. Chua, and J. Anders (unpublished).

Zdeněk Hradil

Department of Optics, Palacký University, Olomouc, Czech Republic

Yong Siah Teo

Department of Optics, Palacký University, Olomouc, Czech Republic
Centre for Quantum Technologies, National University of Singapore, Singapore

Paper Title: **Quantum-State Reconstruction by Maximizing Likelihood and Entropy**

Reference: Physical Review Letters **107**, 020404 (2011)

Personal messages:

Zdeněk Hradil March 10, 2014

I had been meeting Berge frequently at quantum optics conferences, but I had the pleasure to get to know him better at Atominstitut, Vienna in the middle of the '90s. One particular occasion was the panel discussion on statistical methods in quantum optics, where we argued about Bayesian, maximum entropy and maximum likelihood methods, their advantages and disadvantages. I still remember the peaceful end of our discussions in the Schweizerhaus at Prater, a place where they serve very good Czech bier from Budweis. This is why I am selecting our common paper with Berge, published almost 20 years later, where some of those concepts are unified in a single maximum-likelihood maximum-entropy (MLME) reconstruction protocol.

Teo Yong Siah February 24, 2014

This work discusses a method to maximize the likelihood and entropy for general informationally incomplete quantum measurements in order to infer the unknown quantum state of a given source. A crucial discussion with Prof. Berge Englert that took place in the Centre for Quantum Technologies, Singapore, compared the proposal of Robin Blume-Kohout [Phys. Rev. Lett. **105**, 200504 (2010)] with the idea of entropy maximization. The fruit of the discussion was an important identification that the simultaneous maximization of the likelihood and entropy may be understood as a prior-modified method for the usual maximum-likelihood principle. This pivotal insight was further expanded on in Olomouc, the Czech Republic, to yield an efficient optimization procedure by considering a specific linear combination of the two functions, which is the main result of this paper. This, in the author's opinion, clearly demonstrates Berge's astuteness in making accurate and relevant connections between different aspects of physics.

PRL 107, 020404 (2011) PHYSICAL REVIEW LETTERS week ending 8 JULY 2011

Quantum-State Reconstruction by Maximizing Likelihood and Entropy

Yong Siah Teo,[1,2] Huangjun Zhu,[1,2] Berthold-Georg Englert,[1,3] Jaroslav Řeháček,[4] and Zdeněk Hradil[4]

[1]*Centre for Quantum Technologies, National University of Singapore, Singapore 117543, Singapore*
[2]*NUS Graduate School for Integrative Sciences and Engineering, Singapore 117597, Singapore*
[3]*Department of Physics, National University of Singapore, Singapore 117542, Singapore*
[4]*Department of Optics, Palacky University, 17. listopadu 12, 77146 Olomouc, Czech Republic*

(Received 13 February 2011; published 8 July 2011)

Quantum-state reconstruction on a finite number of copies of a quantum system with informationally incomplete measurements, as a rule, does not yield a unique result. We derive a reconstruction scheme where both the likelihood and the von Neumann entropy functionals are maximized in order to systematically select the most-likely estimator with the largest entropy, that is, the least-bias estimator, consistent with a given set of measurement data. This is equivalent to the joint consideration of our partial knowledge and ignorance about the ensemble to reconstruct its identity. An interesting structure of such estimators will also be explored.

DOI: 10.1103/PhysRevLett.107.020404 PACS numbers: 03.65.Wj, 03.65.Ud, 03.67.−a

The acquisition of information about a given ensemble composed of many copies of a quantum system, which is used to identify its quantum state, for instance, always depends on the way the measurements are performed. Such a protocol of probing the ensemble is known as quantum tomography. Since our measurement resources are always limited, the information is never complete, and the inference of the nature of this ensemble from the measurement data needs to account for the fact that some aspects of the ensemble are fully characterized while others are not.

This is especially the case if the ensemble to be characterized is complex and no feasible set of *informationally complete* measurements is available. In this case, some aspects of the quantum ensemble are not measured, and reconstructions of its properties are not unique. For example, the true quantum state of a mode of light made up of an ensemble of photons can be described by a statistical operator $\rho_{\rm true}$ residing in an infinite-dimensional Hilbert space, and, no matter how ingeniously a measurement scheme is designed to probe this ensemble, an infinite amount of information about it will always remain unknown.

The standard approach to this problem is to apply an *ad hoc* truncation on the Hilbert space and perform the state reconstruction in a particular subspace. This results in a smaller number of unknown parameters that can then be *uniquely* determined by the measurement scheme. Since the truncation is largely based on the experimentalist's intuition about the expected result, that is, the true state that describes an ensemble of infinitely many copies of such quantum systems, this cannot be an objective method [1]. A more objective alternative is to consider the largest possible reconstruction subspace which is compatible with any existing *prior* knowledge about the ensemble. For example, if an experimentalist has prior knowledge of the range of the energy spectrum of a particular quantum ensemble, he should consider the largest possible reconstruction subspace that contains quantum states describing the ensemble in this range of energies. This inevitably introduces more unknown parameters which cannot be uniquely determined by the measurements, and we should select the state estimator in this subspace that is least biased. In this Letter, we show how to carry out this procedure by using two celebrated principles—maximum likelihood (ML) [2,3] and maximum entropy (ME) [4]. They will be utilized concurrently to yield a unique and objective state reconstruction scheme, the *maximum likelihood–maximum entropy* reconstruction scheme (MLME), a synthesis of knowledge and ignorance. This scheme will be applicable for any general set of quantum measurements, in particular, for those which are noncommuting.

By maximizing the likelihood, information is extracted from the measured data (*knowledge*) in an optimal way [3], and an estimator for the true state describing a given ensemble of quantum systems can be reconstructed. However, due to the informational incompleteness of the set of measurements, there exists in general a convex set of estimators which are all maximally likely. We shall select from this set the estimator with the largest entropy, which represents the least-bias guess of the true state consistent with the measurement data. As both the entropy and likelihood functionals are convex, our proposed reconstruction scheme will guarantee a unique solution even for informationally incomplete measurement schemes. This reconstruction procedure minimizes the spurious details coming from the parameters which are not uniquely determined by the measurements.

Given a source that produces identical copies of quantum systems, each in the state described by the statistical operator $\rho_{\rm true}$, one can perform measurements on N such copies. These measurement outcomes are described by a set of positive operators Π_j that compose a probability operator measurement (POM), with j running over all outcomes, and a list of outcome occurrences n_j, $\sum_j n_j = N$, or correspondingly the measured frequencies $f_j = \frac{n_j}{N}$, is the raw data obtained from these measurements. Throughout this analysis, we shall assume that the measurements are perfect in the sense of $\sum_j \Pi_j = 1$. The next step is to infer the unknown state $\rho_{\rm true}$ from the measurement data. For this, we look for estimators—the ML estimators—which maximize the *likelihood functional* defined as

$$\mathcal{L}(\rho) = \Big(\prod_j p_j^{f_j}\Big)^N, \tag{1}$$

with $p_j = \mathrm{tr}\{\rho\Pi_j\}$. Using this method, one can always obtain positive estimators suitable for statistical predictions [5].

A POM is *informationally complete* if the mapping of statistical operators ρ on the probabilities p_j is injective. This results in a unique operator $\hat{\rho}_{\rm ML} \geq 0$ that maximizes $\mathcal{L}(\rho)$, where a hat is used to denote an estimator. We shall focus on the case in which the POM is not informationally complete. This means that there is a convex set of estimators that maximize $\mathcal{L}(\rho)$ for a given set of f_j's. The next task is now to choose the one (the MLME estimator) which maximizes the von Neumann entropy [6] $S(\rho) = -\mathrm{tr}\{\rho\log\rho\}$ over the set of ML estimators. It is well known that the ME estimator is of the form

$$\hat{\rho}_{\rm ME} = \frac{e^{\sum_j \mu_j \Pi_j}}{\mathrm{tr}\{e^{\sum_j \mu_j \Pi_j}\}} \tag{2}$$

with real μ_j's. Note that our proposed strategy is fundamentally different from the standard ME scheme. The latter involves searching for the ME estimator $\hat{\rho}_{\rm ME}$ which produces the measured frequencies of POM outcomes or moments of a particular observable and maximizing its entropy subjected to these linear constraints [7,8]. Since these constraints can become incompatible with one another due to the presence of statistical noise, this procedure may fail because there simply is no ME estimator.

In our current strategy, instead of taking the data as strict *linear constraints*, information is extracted from the data via the *nonlinear* ML technique and mapped onto the convex subset of quantum states constituting the plateau of the likelihood functional. Thereafter, the entropy is maximized to yield a unique unbiased estimator. Conceptually, this can be turned into a convex optimization problem of maximizing the objective functional

$$I(\lambda;\rho) = \lambda S(\rho) + \frac{1}{N}\log\mathcal{L}(\rho), \qquad \lambda \geq 0, \tag{3}$$

which involves a single parameter λ as a weight on $S(\rho)$. This $I(\lambda;\rho)$ is the amalgam of two separate measures of information and so reflects the joint consideration of knowledge and ignorance. In fact, we note that, up to an irrelevant additive constant, $I(\lambda;\rho)$ is a weighted sum of two entropies: $S(\rho)$, which quantifies the "lack of information," and the negative of $S(\{f_j\}\,|\,\{p_j\}) = \sum_j f_j \log(f_j/p_j)$ (relative entropy), which quantifies the "gain of information" from the measurements. Hence, Eq. (3) is indeed a natural combination of two complementary aspects of information.

Since $\delta\mathcal{L}(\rho)/\delta\rho = 0$ due to the constraint of maximal likelihood, varying $I(\lambda;\rho)$ with respect to ρ for *fixed* λ gives

$$\frac{\delta I(\lambda;\rho)}{\delta\rho} = \lambda\frac{\delta S(\rho)}{\delta\rho}. \tag{4}$$

In order to maximize $I(\lambda;\rho)$, we need the derivative to be zero, and this is obtained only when $\lambda \to 0$. In other words, in order to search for the MLME estimator via Eq. (3), it is necessary to take *both* our knowledge and ignorance of the unknown true state into consideration in such a way that our ignorance takes an infinitesimal weight. Geometrically, the log-likelihood functional is much larger than the entropy functional in this limit, and, since the log-likelihood functional has a plateau corresponding to the convex set of most-likely estimators, a tiny admixture of the entropy functional introduces a gentle convex hill top *within* the plateau to select the maximum entropy estimator.

By parameterizing a given statistical operator ρ as $\rho = A^\dagger A/\mathrm{tr}\{A^\dagger A\}$ to ensure positivity, one can invoke the method of steepest ascent and derive an iterative routine for MLME. Here, we simply mention that one can start from the maximally mixed state and iterate the equations

$$\rho_{k+1} = \frac{(1+\epsilon T_k)\rho_k(1+\epsilon T_k)}{\mathrm{tr}\{(1+\epsilon T_k)\rho_k(1+\epsilon T_k)\}}, \tag{5}$$

$$T_k = R_k - 1 - \lambda(\log\rho_k - \mathrm{tr}\{\rho_k\log\rho_k\}) \tag{6}$$

with a step size ϵ and $R_k = \sum_j (f_j/p_j^{(k)})\Pi_j$, until the extremal equations $T_{k'}\rho_{k'} = \rho_{k'}T_{k'} = 0$ are satisfied for a particular $k = k'$ with some numerical precision. We denote this operator as the MLME estimator $\hat{\rho}_{\rm MLME} \equiv \rho_{k'}$. If $\lambda = 0$, we recover the ML iterative scheme [5].

We compare the MLME scheme with the standard ME scheme by using the simple example of a trine POM defined by the three outcomes $\Pi_0 = (1+\sigma_z)/3$ and $\Pi_\pm = (1 \pm \sqrt{3}\sigma_x/2 - \sigma_z/2)/3$, where σ_x and σ_z are standard Pauli operators. A straightforward calculation shows that when $n_0 = 6$, $n_+ = 2$, and $n_- = 1$ after measuring $N = 9$ copies, for instance, the standard ME scheme fails, as no quantum state has the frequencies $f_0 = 2/3$, $f_+ = 2/9$, and $f_- = 1/9$ as probabilities. On the other hand, the MLME scheme still gives a positive estimator described by the Bloch vector $(0.194, 0, 0.981)$ for those frequencies, thus showing its versatility. Only when the frequencies are probabilities giving positive estimators may we use the ME scheme, and, in this case, the MLME scheme naturally incorporates these constraints.

To discuss the method of choosing λ, we shall apply the MLME scheme to homodyne tomography, a technique

PRL **107,** 020404 (2011) PHYSICAL REVIEW LETTERS week ending 8 JULY 2011

which is used to reconstruct quantum states of light [9]. This is typically done by measuring a POM which resembles a set of eigenstate projectors $|x_\vartheta\rangle\langle x_\vartheta|$ of quadrature operators $X\cos\vartheta + P\sin\vartheta$ for various ϑ values, where X and P are, respectively, the position and momentum quadrature operators and x and ϑ are parameters specifying these projectors. It is clear that a finite set of such measurements is never informationally complete in the infinite-dimensional Hilbert space, and thus the MLME scheme is necessary to obtain a unique estimator. Figure 1 shows the dependence of $\log(\mathcal{L}(\hat\rho))/N$ and $S(\hat\rho)$ on λ such that $\delta I(\lambda \to 0; \hat\rho) = 0$. In practice, λ can be chosen from a range near zero, within which $\log(\mathcal{L}(\hat\rho))/N$ and $S(\hat\rho)$ remain almost constant.

Homodyne tomography is commonly used not only in quantum tomography on the true state but also in quantum diagnostics where a given true state is to be classified as being classical-nonclassical or separable-entangled. A typical quantity which is often investigated as an indication of whether an unknown true state is nonclassical is the value of the Wigner functional at the phase space origin evaluated with a reconstructed estimator $\hat\rho$ for the unknown true state. This is defined as $W_{00} = 2\,\mathrm{tr}\{\hat\rho\mathcal{P}\}$, with $\mathcal{P} = \int dx|x_\vartheta\rangle\langle -x_\vartheta|$ for any ϑ. In particular, a negative value for W_{00} implies that $\hat\rho$ is a nonclassical state. To obtain an estimator $\hat\rho$, one would need to choose a subspace from the infinite-dimensional Hilbert space in which the reconstruction procedure is tractable. This means that the value of W_{00} will depend on this truncation, which in turn relies on the prior knowledge one has about the true state. Using our scheme, we perform a simulation, shown in Fig. 2, to illustrate this dependence.

If the true state lies outside the subspace of interest, then the estimated value of W_{00} can drastically deviate from the true value. It is clear that a truncation of the Hilbert space into a smaller reconstruction subspace can lead to diagnostics which are highly incompatible with the true result. So, if one is interested in performing an objective quantum tomography experiment on a given ensemble of quantum systems with some prior knowledge regarding its true state, an option would be to reconstruct the MLME estimator in the largest possible subspace based on this prior knowledge. By enlarging the reconstruction subspace, many more admissible states are taken into consideration, and more reliable state estimations and quantum diagnostics can thus be performed. We now have an operational reconstruction scheme that combines our knowledge and ignorance about the unknown true state to give us a unique state estimator in an objective way.

There exists an interesting structure in these MLME estimators, and, to explore it, one needs some knowledge on the structure of the POM used and its influence on the D-dimensional Hilbert space. Suppose the set of K POM elements Π_j are informationally incomplete. A consequence of this is that the number of linearly independent Π_j's is less than D^2. To determine their linear

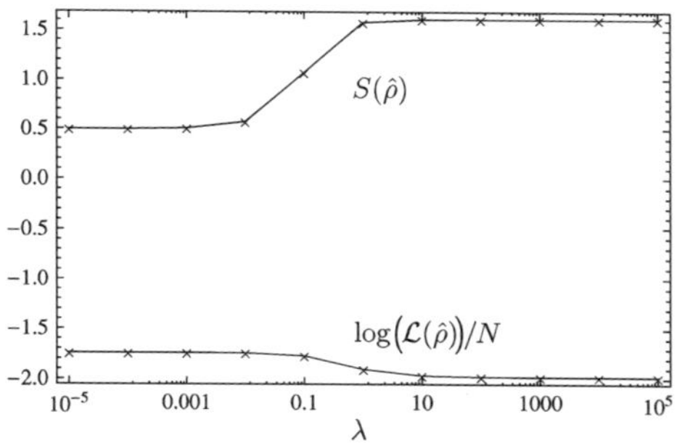

FIG. 1. A simulation on quantum tomography on a randomly generated mixed state of light in the five-dimensional Fock space. In this plot, the number of copies of quantum systems measured is fixed at $N = 10^4$. A choice of 20 quadrature eigenstates made up of four different ϑ settings, with five x values corresponding to each setting, which are projected onto this space was used, and state estimators are constructed for different values of λ. As λ decreases, both the entropy and likelihood functionals approach their respective optimal values obtained from MLME (i.e., when $\lambda \to 0$). When λ is zero, there is a convex set of estimators giving the optimal likelihood value. For very large λ values, the estimators approach the maximally mixed state, and hence $S(\rho)$ approaches the maximal value log 5.

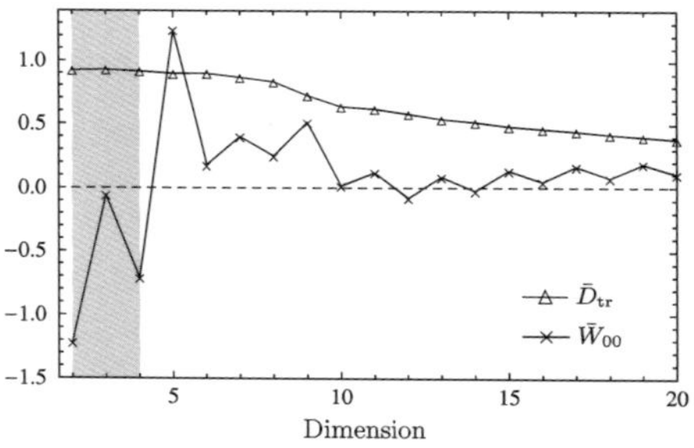

FIG. 2. A simulation on quantum tomography on a randomly generated mixed state ρ_{true} of light in the 20-dimensional Fock space with a slightly positive $W_{00} = 0.141$. $\bar D_{\mathrm{tr}}$ and $\bar W_{00}$, respectively, denote the trace-class distance between the reconstructed estimator and the true state and the Wigner function at the phase space origin, both averaged over 50 experiments with $N = 10^4$. The same set of 20 quadrature eigenstates as in Fig. 1, projected onto this space, was used, and this set of measurements is informationally complete in the two-, three-, and four-dimensional Fock subspaces (shaded region). The values $\bar W_{00}$ and $\bar D_{\mathrm{tr}}$ were obtained by ML [9] in subspaces of dimensions two to four and by the MLME scheme in dimensions greater than four. The plot shows a strong dependence of $\bar W_{00}$ and $\bar D_{\mathrm{tr}}$ on the subspace dimension. In this case, it is obvious that the negativity of $\bar W_{00}$ inferred by a reconstruction in a subspace too small is just an artifact of the truncations. Also, $\bar D_{\mathrm{tr}}$ decreases as the reconstruction subspace increases in dimension. This demonstrates the advantages of the MLME scheme over the ML method.

PRL 107, 020404 (2011) PHYSICAL REVIEW LETTERS week ending 8 JULY 2011

independence, we can look for the eigenvalues of the $K \times K$ Gram matrix M with the matrix elements defined as

$$M_{jk} = \mathrm{tr}\{\Pi_j \Pi_k\}. \tag{7}$$

The number of positive eigenvalues $n_{>0} \leq D^2$ of M determines the number of linearly independent measurement outcomes. The largest value of $n_{>0}$ is D^2, since this is the maximum number of linearly independent operators spanning the space of Hermitian operators as a basis. Hence a set of informationally incomplete Π_j's acting on the D-dimensional Hilbert space is such that $n_{>0} < D^2$.

Any D-dimensional positive operator can be represented by a set of D^2 Hermitian basis operators $\Gamma_j = \Gamma_j^\dagger$ satisfying the trace-orthonormality condition $\mathrm{tr}\{\Gamma_j \Gamma_k\} = \delta_{jk}$. For dimension two, an example of such a basis is the familiar set of four operators $1/\sqrt{2}$, $\sigma_x/\sqrt{2}$, $\sigma_y/\sqrt{2}$, and $\sigma_z/\sqrt{2}$. Once the number of independent measurement outcomes $n_{>0}$ is known, one can construct a set $\{\Gamma_j\}_{j=1}^{n_{>0}}$ of $n_{>0}$ trace-orthonormal Hermitian basis operators directly from the K POM elements. In other words, each of the K POM elements can be expressed as a linear combination of the $n_{>0}$ basis operators

$$\Pi_j = \sum_{k=1}^{n_{>0}} a_{jk} \Gamma_k, \tag{8}$$

where all coefficients a_{jk} are real. This implies that the $n_{>0}$-dimensional subspace is spanned by the basis operators that *uniquely* specify the POM outcomes. We will coin this subspace the *measurement subspace*. The rest of the $D^2 - n_{>0}$ Hermitian basis operators, which are trace-orthonormal to the previous set and span the subspace that is the complement to the measurement subspace, can also be constructed.

Suppose a state estimator $\hat{\rho}_{\mathrm{ML}}$ is generated by using the ML procedure on a set of measurement data obtained from the POM outcomes Π_j. We can represent this estimator by a set of Hermitian trace-orthonormal basis operators inasmuch as

$$\hat{\rho}_{\mathrm{ML}} = \underbrace{\sum_{k=1}^{n_{>0}} c_k^{\mathrm{ML}} \Gamma_k}_{\equiv \tilde{\rho}_{\mathrm{ML}}} + \underbrace{\sum_{k=n_{>0}+1}^{D^2} c_k^{\mathrm{ME}} \Gamma_k}_{\equiv \tilde{\rho}_{\mathrm{ME}}}. \tag{9}$$

The part $\tilde{\rho}_{\mathrm{ML}}$ resides in the measurement subspace, which is spanned by the measurement outcomes Π_j giving the measurement data, and is uniquely fixed for all ML estimators by the ML procedure for the same set of measurement data. The part $\tilde{\rho}_{\mathrm{ME}}$ resides in the complementary subspace, which is orthogonal to the measurement subspace, and thus does not contribute to the p_j's. In other words, $\mathrm{tr}\{\tilde{\rho}_{\mathrm{ME}} \Pi_j\} = 0$, and this can imply the existence of a family of $\tilde{\rho}_{\mathrm{ME}}$'s that gives the same set of ML probabilities as long as the $\hat{\rho}_{\mathrm{ML}}$'s are positive.

Therefore, the MLME scheme can be understood as an optimization over the complementary subspace to maximize $S(\rho)$ under the constraint $\hat{\rho}_{\mathrm{MLME}} \geq 0$. However, one notes that only certain sets of c_j^{ME}'s are allowed during the optimization due to this positivity constraint. This is especially important when $\hat{\rho}_{\mathrm{MLME}}$ is rank deficient and lies on the boundary of the state space. Geometrically, the plateau of most-likely states is generally a much smaller subspace contained in the complementary subspace. In some cases, this plateau contains a *single* ML estimator due to the positivity constraint even when the measurements are informationally incomplete. In general, the boundary of the plateau is complicated and deserves further study.

In summary, we have developed a state reconstruction scheme which is applicable to any set of measurements, particularly those which are informationally incomplete, as in homodyne tomography, for instance. We emphasize that, in order to carry out least-bias state reconstructions and quantum diagnostics on an ensemble of quantum systems, a good way is to do this over a large subspace of states compatible with some prior knowledge on the ensemble in order to avoid inaccurate results, which can have detrimental effects on statistical predictions. We then explored the geometrical structure of the state estimators obtained from this scheme and gave an alternative understanding to the state reconstruction procedure.

This work is supported by NUS Graduate School (NGS) and the Centre for Quantum Technologies which is a Research Centre of Excellence funded by Ministry of Education and National Research Foundation of Singapore, as well as the Czech Ministry of Education, Project No. MSM6198959213, and the Czech Ministry of Industry and Trade, Project No. FR-TI1/364.

[1] J. Řeháček, D. Mogilevtsev, and Z. Hradil, New J. Phys. **10**, 043022 (2008).
[2] R. A. Fisher, Phil. Trans. R. Soc. A **222**, 309 (1922).
[3] C. W. Helstrøm, *Quantum Detection and Estimation Theory* (Academic, New York, 1976).
[4] E. T. Jaynes, Phys. Rev. **106**, 620 (1957); **108**, 171 (1957).
[5] J. Řeháček, Z. Hradil, E. Knill, and A. I. Lvovsky, Phys. Rev. A **75**, 042108 (2007).
[6] J. von Neumann, *Mathematical Foundations of Quantum Mechanics* (Princeton University, Princeton, NJ, 1955).
[7] V. Bužek, G. Adam, and G. Drobný, Ann. Phys. (N.Y.) **245**, 37 (1996).
[8] A. R. Rossi, and M. G. A. Paris, Eur. Phys. J. D **32**, 223 (2005).
[9] D. T. Smithey, M. Beck, M. G. Raymer, and A. Faridani, Phys. Rev. Lett. **70**, 1244 (1993); A. Ourjoumtsev, R. Tualle-Brouri, and P. Grangier, Phys. Rev. Lett. **96**, 213601 (2006); J. S. Neergaard-Nielsen, B. M. Nielsen, C. Hettich, K. Mølmer, E. S. Polzik, Phys. Rev. Lett. **97**, 083604 (2006).

Jun Suzuki

Graduate School of Information Systems,
The University of Electro-Communications
Tokyo, Japan

Paper Title: **Accessible information about quantum state: An open optimization problem**

Reference: Chapter 11, pp. 309–348 of the book *"Mathematics of Quantum Computation and Quantum Technology"*, edited by Goong Chen, Louis Kauffman, and Samuel J. Lomonaco, Chapman & Hall/CRC (2007).

Editors' note: Due to copyright restrictions, only the first two pages of the book are reprinted here.

My encounter with Berge

Jun Suzuki

Graduate School of Information Systems, The University of Electro-Communications,
1-5-1 Chofugaoka, Chofu-shi, Tokyo, 182-8585 Japan
junsuzuki@is.uec.ac.jp

Abstract

Here is a short story of how I met Berge and how he taught me physics and other matters.

1. My first encounter with Berge

I met Berge on 14th of May in 2005 just after I finished my Ph.D. in the USA. I arrived at Changi airport around midnight of that day and I was picked up by a Ph.D. student from NUS, Andreas Keil, who is now back in Germany with his family. After he took me to his condominium, we went to a pub and enjoyed beer till past 3 am. Andreas told me during a beer that my new boss (Berge) wanted to have lunch together the next day since he was leaving for Europe. So we woke up at 7 am and had a short tour of Singapore in the morning before lunch. (Harbourfront $\xrightarrow{\text{bus}}$ Holland village $\xrightarrow{\text{bus}}$ Botanic garden $\xrightarrow{\text{walk}}$ Russian and other embassies[1] $\overset{\text{walk}}{\rightsquigarrow}$ Orchard $\overset{\text{bus}}{\rightsquigarrow}$ NUS in four hours! No Circle Line existed at that time.)

When I arrived at NUS after that exhausting long walk, I met Berge and Ola at one of the NUS Staff Clubs. My first impression was that he looked like a very serious professor, that is, no smile, nice beard. Berge explained to me that he was leaving for Europe the next day for about two months. I expected that he would say something like, "adjust yourself for tropical climate and prepare for new projects when I come back from Europe." In fact, he welcomed me warmly and gave me a stack of papers (about 50 pages) with just a single staple on the corner. He told me I should study quantum cryptography (quantum key distribution, QKD) which he was currently working on with his group. This stack of paper contained about 10 articles of his recent publications on QKD. I thought this was a (German) practical way of training.

Since Berge knew that I had never worked on this subject before, he thought this was the shortest path to work as his postdoc. Indeed, all papers written by him and his group were self-contained and well-written even for a non-expert on the subject. The only non-practical part of Berge's gift was that this stack was too heavy to carry around and they were so firmly stapled...

When I was communicating with Berge before joining in his group, I always started with, "Dear Prof. Englert," to show my respect for him. He never asked me to just call him Berge during email communication. When I started to work with him after his summer trip, I continued addressing him as "Dr. Englert." This was because he never offered me to call him by his rather long first name, "Berthold-Georg," which I had no idea how to pronounce. After a few days, during lunch time at Science canteen, he suddenly commented to me with a serious face: "Jun, nobody calls me Dr. or Prof. Englert here," rather than simply saying "Please call me Berge."

I worked on the problem of security analysis for QKD protocols with Berge and his students. As a part of this problem, we had to carry out some non-trivial optimization problem, known as accessible information. I became interested in this fundamental problem in quantum information theory. We spent some time to work out the details and our efforts were written up as a review article [3], which I selected as my contribution to this Festschrift. I briefly summarize how we came across this problem and why I think this is important in the next section.

2. Accessible information about quantum states

I think one of Berge's basic interests is the problem of quantifying the amount of information about quantum states by performing a good measurement. This is of fundamental interest in understanding the nature of quantum mechanics, which is inherently probabilistic. It also has many practical applications, such as wave-particle duality, quantum tomography, security analysis of QKD protocols, quantum state estimation, and so on. The major difference among these problems is that one needs to deal with different figures of merit depending upon how one quantifies one's knowledge.

In 2005, I was mainly working on the problem of finding accessible information for non-tomographic

[1] By accident. We were walking from the north side of the Botanic garden to Orchard. You can see their locations on a map.

QKD protocols with Berge and his Ph.D. student, Syed Muhamad Assad, who is now working at ANU in Australia. The main part of the accessible information problem is to calculate the accessible information of given quantum states by optimizing the classical mutual information over all possible measurements. This problem is rather old and was studied from the information-theoretic perspective in 1970s. Revival of interests on this subject began when the study of various QKD protocols became a hot subject in the physics community and security analyses were required. The problem is well-stated as an optimization over all possible measurements and yet the solution is rather hard to obtain analytically except for a few simple cases. An apparent reason for this difficulty is that the cost function is non-linear. The other technical aspect is that one has to consider general measurements rather than simple projective measurements in general.

I found this problem more and more interesting as I spent more time on it, but I wanted to understand why this optimization was so hard that only very simple cases could be solved by hand. Based on various examples that he had worked out, Berge had a strategy to find the optimal solution to a given problem. A simple yet very powerful numerical algorithm was developed by Berge and his colleagues to find the optimal solution for a given problem [1, 2]. So, one can produce many nice plots to show that a given QKD protocol is secure under certain attacks by an eavesdropper. (Later, we moved to analyze the Holevo bound which gives more precise security bounds under all kinds of attacks.) For ordinary researchers, this line of analysis would be sufficient, at least from a physicist's point of view on QKD security, for publication.

Berge obviously likes to tackle the hard(er) problems rather than the easy ones. He also insists on getting satisfactory and complete solutions before turning our results into a publication. Numerical search for this kind of research should be used to get analytical insights into the problem as he emphasized. He also said we would otherwise only produce nice looking figures without understanding what was going on behind the problem. I also like to understand problems by finding the hidden order. We realized that there were some hidden symmetries when dealing with ensembles of quantum states used in QKD protocols. This symmetry — group covariance — is not obvious in the beginning when states are subjected to eavesdropper's attack, but it shows up when working with a proper basis. It is natural that any attack which breaks the symmetry disturbs more than one that does not break it, and hence the eavesdropper should respect the same symmetry. This was indeed the case for our problem, and we could find several non-trivial examples where analytical solutions were written down which perfectly matched numerical findings. These matters were summarized in my joint paper with Assad and Berge.

3. What I've learnt from Berge

There are so many things that I learned from Berge. Some major lessons are listed below.

- Physics is more important than proving mathematical theorems. (Be a good physicist)
- Do not only work on problems which others believe are important.[2] (Taste of physics)
- Do not publish your results before you complete the problem satisfactorily. (No compromising)
- Properly assigned symbols and good notations are very important to simplify the problem. (Elegance)
- You should correct others when they make wrong statements, regardless of who they are.[3] (Honesty)
- You can close your eyes during seminars, but you should correct the presenter when he/she makes wrong statements. (Berge's magic)

A few more extras are:

- When you are with your family, you should be away from thinking about physics.
- A house man should wash dishes for his wife.
- To impress the wife that physics is not all we think about all day, when wife is away, cook dinner for the kids, for instance, cheese pasta.[4]

4 Acknowledgement

I would like to thank Berge for bringing me into the field of quantum information theory. This was indeed the best start of my career as you taught me how to be a good physicist. My experiences with you have been one of the exciting and valuable treasures in my life. It was great honor and pleasure to work with you during my postdoc time. I would also like to thank you for never forcing me to work on any projects before I agreed to them. Whenever we have common interest, I look forward to enjoyable collaboration with you in the near future.

[2]Typical examples are entanglement, non-locality, man-made artificial solvable models.

[3]This applies when you go to a museum even though you may embarrass a guide [4].

[4]The original sentence was "Man has to be able to cook by himself to survive when his wife is away. This includes cheese pasta." My wife, Clarice, recalled that Berge made a stronger statement than what I decried here.

A Notes on inequality for average quantities for sorted ensemble

In this appendix, we give a simple observation about Berge's work on duality relationship given in Ref. [5].

A.1 Setup

Let $\mathcal{H}$ be a finite dimensional Hilbert space and $\mathcal{S}(\mathcal{H})$ be the set of all density operators on $\mathcal{H}$. We consider two real-valued functionals A_j $(j = 1, 2)$:

$$A_j : \rho \longmapsto A_j(\rho) =: a_j \in \mathbb{R} \quad (j = 1, 2).$$

In general, the A_js are not necessarily linear functionals of states ρ. If so, Riesz-Frechet's lemma simplifies the problem, i.e., there exists an operator A_j on $\mathcal{H}$ such $A_j(\rho) = \langle A_j, \rho\rangle$. Physically speaking, the values a_j correspond to some experimentally measurable quantities which are not necessarily expressed as expectation values of some hermitian operators. Examples of this kind are $A(\rho) = |\mathrm{Tr}\{A\rho\}|$ with A an operator (not necessarily hermitian), $A(\rho) = [\mathrm{Tr}\{A^2\rho\} - (\mathrm{Tr}\{A\rho\})^2]^{1/2}$ (variance of an observable A), information-theoretic quantities such as the von Neumann entropy, q-covariance functions, operator monotone metrics, etc.

Let $I_j \subset \mathbb{R}$ be the range of A_j, and we assume that the A_js are continuous and bounded so that the I_js are closed sets. In this case, we can always redefine A_j in such a way that I_j is non-negative[5].

Suppose there exists a function $f : I_1 \times I_2 \to \mathbb{R}$ which represents some amount of joint information on two quantities defined by functionals A_js. By the same logic as before, we assume that f is a non-negative function. An interesting situation is when $f(a_1, a_2)$ is bounded by a positive number $f^* \in \mathbb{R}$ from above (or below by f_*):

$$f(a_1, a_2) \le f^* \quad (\text{or} \ge f_*). \tag{1}$$

Here the upper bound can depend on the state $\rho \in \mathcal{S}(\mathcal{H})$ in general. In such a case, we simply set f^* to be the maximum of this upper bound over all possible states, i.e., $\max_{\rho\in\mathcal{S}(\mathcal{H})} f^*(\rho)$. Thus, we only consider the case where f^* is a constant and is independent of states. A familiar example of this kind is[6]

$$f(x_1, x_2) = x_1^2 + x_2^2,$$

for wave-particle duality, which is of the additive type and bounded by 1. Another one is of the product type (e.g., Heisenberg-Robertson relationship)[7]:

$$f(x_1, x_2) = x_1 x_2,$$

which is bounded from below. In many interesting cases studied in literature, the inequality (1) can be interpreted as a trade-off relation between a_1 and a_2. Namely, if the inequality is saturated when a_1 reaches the maximum value, then a_2 is enforced to be the minimum, and vice versa.

A.2 Sorted ensemble via measurement

Given a quantum state ρ_Q on $\mathcal{H}_Q$, a (indirect) measurement is described by two steps: First is to couple ρ_Q with the external degrees of freedom which represents a detector (also called, apparatus, ancilla, environment, etc.) and the second is to perform a measurement on the detector.

We consider a quantum map, (completely-positive and trace-preserving map or quantum channel) which couples a state on $\mathcal{H}_Q$ with the detector $\mathcal{H}_D$ [8]:

$$\mathcal{E} : \mathcal{S}(\mathcal{H}_Q) \longrightarrow \mathcal{S}(\mathcal{H}_Q \otimes \mathcal{H}_D).$$

We then perform a measurement $\Pi = \{\Pi_x \,|\, \Pi_x \ge 0, \sum_x \Pi_x = I_{\mathcal{H}_D}\}_{x\in\mathcal{X}}$ on the detector system, i.e., Π on $\mathcal{S}(\mathcal{H}_D)$ [9]. The resulting (conditional) states on $\mathcal{S}(\mathcal{H}_Q)$ will be given according to the measurement outcomes which are labeled by an index $x \in \mathcal{X}$:

$$\rho_x = \frac{1}{p_x}\mathrm{Tr}_D\left\{I_{\mathcal{H}_Q}\otimes\sqrt{\Pi_x}\,\mathcal{E}(\rho)\,I_{\mathcal{H}_Q}\otimes\sqrt{\Pi_x}\right\},$$
$$p_x = \mathrm{Tr}_{QD}\left\{\mathcal{E}(\rho)\,I_{\mathcal{H}_Q}\otimes\Pi_x\right\}.$$

The set $\{(p_x, \rho_x)\}_{x\in\mathcal{X}}$ is called a sorted ensemble [5].

Since measurement outcomes are random according to the probability distribution p_x, a particular measurement outcome provides a large value for $a_1(x) = A_j(\rho_x)$ and/or a rather small value for $a_2(x) = A_2(\rho_x)$, and so on. It is then useful to define the following average quantities:

$$\overline{a_j(\Pi|\mathcal{E})} := \sum_{x\in\mathcal{X}} p_x A_j(\rho_x) \quad (j = 1, 2). \tag{2}$$

These numbers quantify the average quantities A_j extracted from the measurement Π. We note that important and relevant quantities are $\overline{a_j(\Pi|\mathcal{E})}$ when measurement events are random[10].

[5] We can also rescale A_js so that $\forall\rho\in\mathcal{S}(\mathcal{H})$, $0 \le A_j(\rho) \le 1$. But we do not use this convention here.

[6] More generally, $f(x_1, x_2) = x_1^\alpha + x_2^\alpha$ with $\alpha > 1$ or $\alpha < 0$.

[7] This function is neither jointly convex nor concave. By taking the inverse, $f(x_1, x_2) = x_1^{-1}x_2^{-1}$ is jointly convex.

[8] Usually, one writes $\mathcal{E}$: $\rho_Q \otimes \rho_D \mapsto \mathcal{E}(\rho_Q \otimes \rho_D)$, but the initial state of the detector can be considered as a part of this map.

[9] By taking a large Hilbert space, one can restrict to the class of projective measurements, but this doesn't matter here.

[10] Of course, one can also calculate other quantities by a post-selection (choosing some outcomes only), but that is a completely different story.

A.3 Inequality for average quantities

A function $f(x_1, x_2)$ is said to be jointly convex on $I_1 \times I_2$ if the following inequality holds:

$$\forall \lambda \in [0,1],\ \forall x_j, x'_j \in I_1 \times I_2\ \ (j=1,2)$$
$$f(\lambda x_1 + (1-\lambda)x'_1, \lambda x_2 + (1-\lambda)x'_2) \leq \lambda f(x_1, x_2) + (1-\lambda) f(x'_1, x'_2).$$

Here we show a simple observation based on convexity property of the function $f(x_1, x_2)$.

> Observation: For a given state $\rho \in \mathcal{S}(\mathcal{H})$, functionals A_j $(j=1,2)$ satisfying the inequality (1), if f is jointly convex, the averaged quantities (2) also obey the same inequality for any quantum map and measurement:
> $$f(\overline{a_1}, \overline{a_2}) \leq f^*.$$

The reasoning for this observation is:

$$f(\overline{a_1}, \overline{a_2}) = f\Big(\sum_{x\in\mathcal{X}} p_x A_1(\rho_x), \sum_{x\in\mathcal{X}} p_x A_2(\rho_x),\Big) \quad (3)$$
$$\leq \sum_{x\in\mathcal{X}} p_x f\big(A_1(\rho_x), A_2(\rho_x)\big) \quad (4)$$
$$\leq \sum_{x\in\mathcal{X}} p_x\, f^* \quad (5)$$
$$= f^*. \quad (6)$$

The first inequality (4) follows from the joint convexity of f (Jensen's inequality) and the second inequality (5) is due to the definition of f^*.

A few remarks are in order before closing this appendix. First, the above argument also applies to the case of more than two real functionals, i.e., A_j $(j=1,2,\ldots,n)$ obeying $f(a_1, a_2, \ldots, a_n) \leq f^*$. Second, a similar result holds if the function is jointly concave and is bounded from below. Third, a state-dependent upper bound $f(a_1, a_2) \leq f^*(\rho)$ rather than the state-independent one treated here should give a tighter bound than Eq. (5). Fourth, we can apply the above argument to the results reported in Ref. [6] (general path knowledge and visibility functions in higher dimension) and we can show that average quantities must obey the same tradeoff relations. The only non-trivial part would be to obtain the maximum (supremum) and the minimum (infimum) of average quantities over all possible measurements.

Last, a particularly interesting situation is when this tradeoff function is monotonically increasing with respect to each variable. Assume that one knows the upper and lower bounds for the average quantities over all possible measurements:

$$a_j^* := \sup_{\Pi} \overline{a_j(\Pi)},$$
$$a_{j*} := \inf_{\Pi} \overline{a_j(\Pi)}.$$

In this case, we obtain two extreme statements as

$$f(a_1^*, a_{2*}) \leq f^* \quad \text{and} \quad f(a_{1*}, a_2^*) \leq f^*.$$

This was shown in Ref. [5] in the context of wave-particle duality and the erasure relation on general grounds, where they proved that for a functional $A(\rho) = \big|\mathrm{Tr}\{A\rho\}\big|$, the solution is

$$a^* = \sup_{\Pi} \overline{a(\Pi)} = \mathrm{Tr}_D\big\{\big|\mathrm{Tr}_Q\{\mathcal{E}(\rho) A \otimes I_{\mathcal{H}_D}\}\big|\big\},$$
$$a_* = \min_{\Pi} \overline{a(\Pi)} = \big|\mathrm{Tr}_Q\{A\,\mathrm{Tr}_D\{\mathcal{E}(\rho)\}\}\big|,$$

with $\mathrm{Tr}\{|A|\} = \mathrm{Tr}\{\sqrt{A^\dagger A}\}$ the trace norm of A. The minimum follows from the convexity of $A(\rho)$ with respect to states, whereas a non-trivial upper bound is obtained by proving the following lemma which holds for any linear operator A on $\mathcal{H}$:

$$\sup_{\Pi} \sum_{x\in\mathcal{X}} \big|\mathrm{Tr}\{A\Pi_x\}\big| = \mathrm{Tr}\{|A|\}.$$

Another challenging question inspired by this problem is to obtain the maximum and the minimum for given general functionals with some reasonable assumptions such as monotonicity, convexity, additivity, symmetry, and so on.

References

[1] J. Řeháček, B.-G. Englert, and D. Kaszlikowski, "*Iterative procedure for computing accessible information in quantum communication*," Phys. Rev. A **71**, 054303 (2005).

[2] K. L. Lee, W. K. Chua, S. Y. Looi, and B.-G. Englert, "*SOMIM: An open-source program code for the numerical Search for Optimal Measurements by an Iterative Method*" eprint arXiv:0805.2847 [quant-ph]. URL: http://theory.quantumlah.org/project/SOMIM/

[3] J. Suzuki, S. M. Assad, and B.-G. Englert, "*Accessible information about quantum states: An open optimization problem*," in Mathematics of Quantum Computation and Quantum Technology, ed. by G. Chen, S. J. Lomonaco, and L. Kauffman (Chapman & Hall/CRC, Boca Raton, 2007), pages 309–348.

[4] Witnessed by J. Suzuki at one of the museums in Innsbruck, 2006.

[5] B.-G. Englert and J. A. Bergou, "*Quantitative quantum erasure*," Opt. Comm. **179**, 337–355 (2000).

[6] B.-G. Englert, D. Kaszlikowski, L. C. Kwek, and W. H. Chee, "*Wave-particle duality in multi-path interferometers: General concepts and three-path interferometers*," Int. J. Quant. Inf. **6**, 129–157 (2008).

Chapter 11

Accessible information about quantum states: An open optimization problem

Jun Suzuki, Syed M. Assad, and Berthold-Georg Englert

Abstract We give a brief summary of the current status of the problem of extracting the accessible information when a quantum system is received in one of a finite number of pre-known quantum states. We review analytical methods as well as a numerical strategy. In particular, the group-covariant positive-operator-valued measures are discussed, and several explicit examples are worked out in detail. These examples include some that occur in the security analysis of schemes for quantum cryptography.

11.1 Introduction

A sender, traditionally called Alice, sends quantum states, one by one, to a receiver, Bob. Bob then wishes to perform measurements on the quantum states he receives to find out, the best he can, what Alice has sent. Generally speaking, owing to the nature of quantum mechanics, it is impossible for Bob to obtain full knowledge about the states which he is receiving. Instead, he has to choose his measurements judiciously from all measurements permitted by quantum mechanics. A natural question one might ask is then:

What is the best strategy for Bob to maximize his knowledge about the states he is receiving from Alice? (11.1)

The answer to this question is not only of importance for our understanding of the implications of quantum mechanics, it also has great practical significance for most areas in quantum information, in particular for the capacity of quantum channels and the security analysis of schemes for quantum cryptography under powerful eavesdropping attacks. Indeed, our own interest in the matter originates in its relevance to the security of "tomographic quantum cryptography," a class of protocols for quantum key distribution developed in Singapore [1, 2, 3, 4].

The main objective of this chapter is to provide a concise introduction to this problem with a summary of ongoing research in this field. For this purpose we will not give a rigorous mathematical exposition, and we will be content with stating most of theorems without proof. We suggest that readers who are interested in the technical mathematical details consult the pertinent literature referred to in the text.

Here is a brief preview of coming attractions. In Section 11.2 we remind the reader of a few basic concepts and, at the same time, establish the terminology and the notational conventions we are using. Then, in Section 11.3, we state question (11.1) as an optimization problem, for which the mutual information between Alice and Bob is the figure of merit. Section 11.4 reports essential properties of this mutual information and important theorems about known properties of the solution. A numerical procedure for searching the optimum by a steepest-ascent method is described in Section 11.5. Examples are presented in Section 11.6, where we limit the choice to cases with a structure as one meets it in the security analysis of quantum cryptography schemes. We close with a summary and outlook.

11.2 Preliminaries

11.2.1 States and measurements

We set the stage by first providing a brief mathematical description of the physical situation that (11.1) refers to, that is: Alice sends certain physical states to Bob who measures them to find out which state she sent. For simplicity and for concreteness, we consider only finite-dimensional systems.

Andreas Buchleitner

Physikalisches Institut, Albert-Ludwigs-Universität Freiburg, Freiburg, Germany

Paper Title: **Entanglement of identical particles and the detection process**

Reference: Fortschritte der Physik **61**, 225–237 (2013)

Personal message:

Happy 60th, Berge!

Fortschr. Phys. **61**, No. 2 – 3, 225 – 237 (2013) / **DOI** 10.1002/prop.201200079

Entanglement of identical particles and the detection process

Malte C. Tichy[1,2,*], **Fernando de Melo**[1,3,4], **Marek Kuś**[5], **Florian Mintert**[1,6], and **Andreas Buchleitner**[1]

[1] Physikalisches Institut, Albert-Ludwigs-Universität Freiburg, Hermann-Herder-Str. 3, 79104 Freiburg, Germany

[2] Lundbeck Foundation Theoretical Center for Quantum System Research, Department of Physics and Astronomy, University of Aarhus, 8000 Aarhus C, Denmark

[3] Instituut voor Theoretische Fysica, Katholieke Universiteit Leuven, Celestijnenlaan 200D, 3001 Heverlee, Belgium

[4] Centrum Wiskunde & Informatica, Science Park 123, 1098 XG Amsterdam, Netherlands

[5] Center for Theoretical Physics PAS, Al. Lotników 32/46, 02-668 Warszawa, Poland

[6] Freiburg Institute for Advanced Studies, Albert-Ludwigs-Universität, Albertstrasse 19, 79104 Freiburg, Germany

Received 13 April 2012, revised 13 April 2012, accepted 23 May 2012
Published online 18 June 2012

Key words Entanglement, indistinguishability, detection process, quantum and classical correlations, symmetrization postulate.

We introduce *detector-level entanglement*, a unified entanglement concept for identical particles that takes into account the possible deletion of many-particle which-way information through the detection process. The concept implies a measure for the *effective indistinguishability* of the particles, which is controlled by the measurement setup and which quantifies the extent to which the (anti-)symmetrization of the wave-function impacts on physical observables. Initially indistinguishable particles can gain or loose entanglement on their transition to distinguishability, and their quantum statistical behavior depends on their initial entanglement. Our results show that entanglement cannot be attributed to a state of identical particles alone, but that the detection process has to be incorporated in the analysis.

1 Introduction

Entanglement and the indistinguishability of particles lead to a variety of counter-intuitive quantum many-body phenomena. Entanglement manifests itself most prominently in measurements at remote detectors that exhibit stronger correlations than allowed by local and realistic theories [1]. Such *quantum correlations* imply the fundamental impossibility to assign a *physical reality* [2], i.e. a complete set of physical properties, to an entangled particle individually – a virtually unsettling consequence [3]. For example, the polarization of a photon that is entangled to the electronic state of an atom does not possess a value before its measurement.

Similarly, it is impossible to assign an identifying label to any particle in a system of two or more identical particles, and the many-body wave-function has to be (anti)symmetrized for (fermions) bosons [4]. Again, a postulate of quantum mechanics dictates our very description of reality and sets strict bounds on the information that we can retrieve from a quantum system. In this article, we deal with the *physical* consequences of the symmetrization postulate for entanglement, which are insinuated by the aforementioned analogy.

* Corresponding author E-mail: tichy@phys.au.dk

For different types of particles, like electrons and protons, i.e. non-identical particles, entanglement can be defined rigorously [5]. Since the particles that carry the entangled degrees of freedom are intrinsically distinct, they have an undetachable, definite *identity* and can thus be *labeled*: they possess well-defined properties such as rest mass or charge by which they can be discriminated unambiguously. Consequently, any measurement on one particle can be assigned to the density matrix that acts on the Hilbert space of this particle. The abstract quantum-information paradigm of parties that control quantum systems [6] – here: particle properties measured by detectors – can be applied immediately for an analysis of the physical situation.

The identification of particles and Hilbert spaces, however, breaks down as soon as identical particles are involved. Other distinctive properties are thus necessary to discriminate the particles. If they are localized far from each other, the particles define two distinguishable entities, related unambiguously to one local detector each. Between these entities, entanglement can then be defined. The (anti)symmetrization procedure does not affect any observable defined on these entities, and can thus be neglected [7]. This is already realized in quantum mechanics textbooks, e.g. [8] states that

> *"No quantum prediction, referring to an atom located in our laboratory, is affected by the mere presence of similar atoms in remote parts of the universe."*

For example, in a scenario with an atom on the moon that is possibly entangled to an atom on the earth, one can safely neglect the (anti)symmetrization procedure, since the atom on the moon never triggers the detector on the earth, and vice-versa.

Dealing with the (anti)symmetry of the many-particle wave-function is therefore a rather formal problem when particles are spatially well-separated and clearly distinguishable by this preparation. The apparent correlations in the particle labels must not be accounted for as *physical*, while correlations between remote detectors should. This is achieved by entanglement definitions for identical particles provided in [9–13], which we will refer to as *a priori* entanglement.

Yet, in other settings, such clear differentiation is sometimes ambiguous or even impossible. Electrons that leave an ionizing helium atom, photons passing simultaneously through the input arms of a beam splitter in a Hong-Ou-Mandel setting (HOM) [14, 15], or atoms in a Bose-Einstein condensate BEC [16] are strongly overlapping in space, such that their external states, i.e. their spatial preparation, do not allow to address the particles individually. Their indistinguishability then becomes relevant and affects the results of correlation measurements. In such situations, the impossibility to assign a label to a particle may indeed imply the impossibility to assign a physical reality to a particle that is measured in a detector. The (anti)symmetrization of the wave-function then results in quantum correlations, and cannot be neglected by any means. Quite in general, the dynamical behavior of entanglement during transitions from overlapping, indistinguishable, to well localized, distinguishable particles is heretofore barely understood, while the interest in its role in atomic and molecular [17] as well as in biological systems [18] is growing. Since such natural systems do not conform to a clear paradigmatic setting of remote parties, as in quantum information, it is necessary to include the measurement prescription in the description of entanglement.

In the present article, we introduce a such unified view on entanglement, which allows the consistent quantification of the entanglement between identical particles in arbitrary settings: Our description comprises the aforementioned situation with separately prepared, thus *distinguishable* particles, but also any possible scenario with particles that strongly overlap in space and that do not suggest any natural subsystem structure, i.e. when they indeed constitute *indistinguishable* particles. In order to quantify the impact of the indistinguishability on the measured quantum correlations, appropriate physical quantities are identified. We show that the indistinguishability of the particles depends on the experimental context and that it has a manifold impact: Particles can gain or loose quantum correlations through the measurement process.

We first review a priori entanglement in Sect. 2 and discuss an experimental situation in which this concept fails. Section 3 then introduces detector-level entanglement and the quantities that govern its behavior. By means of a model-system, we discuss the implications of the indistinguishability of particles for measured correlations in Sect. 4, and conclude in Sect. 5.

2 A priori entanglement

In previous approaches to the entanglement of identical particles [9–13], which we refer to as *a priori* entanglement, the problem of the (anti)symmetrization of the many-particle wave-function is tackled by effectively labeling the particles by some distinctive properties, e.g. their external state. In analogy to the Schmidt coefficients, [9, 10] use the Slater coefficients to determine the entanglement of identical particles. For example, if the minimal number of Slater determinants needed to describe a many-fermion state is not unity, the state is said entangled [13].

Much like distinguishable particles, two a priori non-entangled, *identical* particles will remain always unentangled unless an interaction between them takes place. The interaction between the particles may be direct, or mediated through ancilla particles. While the states of the particles may change in time, their identity, as given by their initially chosen distinctive properties, will be preserved through unitary time evolution.

The absence of a priori entanglement implies that it is, *in principle*, possible to perform measurements on a particle in the system by which one obtains a certain known result *deterministically*, i.e. one merely *reads off* a pre-existing value of a physical property – a situation well in accordance with local realism [2]. A strong and crucial assumption is, however, made here: it is taken for granted that experimentalists are indeed able and willing to always choose those detectors that address the well-defined properties of the particles.

Particles can be distinguished by their external preparation when they are spatially well separated, such that the assumption is then well justified, since each detector is then assigned exactly one particle. Such situation is also depicted in Fig. 1(a), where the spatial wave-functions of two particles have a finite overlap with exactly one detector each. The experimentalists then simply read off the preparation of the spin state of the particle in one of the detectors. We will refer to such a situation as an *unambiguous* setup.

In many situations, like, e.g., the one depicted in Fig. 1(b,c), the external degrees of freedom and the detectors do, however, not coincide anymore, and the a priori assumption is no longer justified – the experimentalists then choose to not merely read off pre-existing, known values, but to use an *ambiguous* detector setting in which the particles appear as indistinguishable. In an HOM-experiment, illustrated in Fig. 1(d), entanglement in the output arms of a beam splitter is created by the passage of two unentangled photons of opposite polarization through both input arms, and further post-selection on the coincident events [14, 19]. The initial state of the system reads

$$|\Psi_{\mathrm{HOM,ini}}\rangle = \frac{1}{\sqrt{2}}\left(|A,H;B,V\rangle + |B,V;A,H\rangle\right), \tag{1}$$

where $|A,H\rangle$ ($|B,V\rangle$) denotes the quantum state of a photon in the input mode A (B) with horizontal (vertical) polarization, and $|\Psi;\Phi\rangle$ is the short-hand notation for $|\Psi\rangle_1 \otimes |\Phi\rangle_2$. At first sight, $|\Psi_{\mathrm{HOM,ini}}\rangle$ might seem to be entangled since it cannot be written as a single factor. It is, however, correctly recognized as non-entangled by the a priori definitions [9–11]: a measurement of the polarization of the photons at the input arms indeed does not exhibit any correlations, the photon in mode $A(B)$ is always horizontally (vertically) polarized. The correlations in the particle labels in (1) – particle "1" seems to be entangled with particle "2" – are not physical, since no physical operator can be implemented to measure the property of a particle that is specified by the mere label.

The time evolution of the external quantum states induced by the scattering on the beam splitter reads, by virtue of Fresnel's equations [20],

$$|A\rangle \to \frac{1}{\sqrt{2}}\left(|L\rangle + |R\rangle\right) =: |\tilde{A}\rangle, \tag{2}$$

$$|B\rangle \to \frac{1}{\sqrt{2}}\left(|L\rangle - |R\rangle\right) =: |\tilde{B}\rangle, \tag{3}$$

where $|L\rangle\,, |R\rangle$ denote the output modes of the beam splitter. With the above definition of $|\tilde{A}\rangle$ and $|\tilde{B}\rangle$, the final state of the photons in the output arms becomes

$$|\Psi_{\mathrm{HOM,fin}}\rangle = \frac{1}{\sqrt{2}}\left(|\tilde{A}, H; \tilde{B}, V\rangle + |\tilde{B}, V; \tilde{A}, H\rangle\right). \tag{4}$$

The functional form of (1) and (4) is apparently unchanged during the scattering, and we can understand (4) with a mere re-labeling of the modes applied on (1). This is also immediate from the fact that only one-particle unitary evolutions occur, and no interaction between the photons takes place: initially orthogonal quantum states of two particles will necessarily remain orthogonal in the absence of mutual interaction. Entanglement measures such as in [9–11], which only account for the form of the quantum state, vanish, both for the initial state (1) and for the final state (4). The physical situation before and after the scattering on the beam splitter is, however, totally distinct: Both photons have a finite probability to be detected by the *same* local detector *after* the scattering process, whereas *before*, a local detector placed at one local optical mode can only detect one of the particles.

Since the two two-particle paths ($A \to R, B \to L$) and ($A \to L, B \to R$) are indistinguishable, they are *coherently* superimposed and entanglement between the particles is recorded for those detection events where both detectors click *coincidentally* [14,21]. The emerging quantum correlations are also immediate when we consider the state obtained upon post-selection of the coincident events, i.e. when one particle is found in each output arm:

$$|\Psi_{\mathrm{HOM,fin,proj}}\rangle = \frac{1}{2}\left(-|L, H; R, V\rangle + |L, V; R, H\rangle - |R, V; L, H\rangle + |R, H; L, V\rangle\right). \tag{5}$$

This form shows that the polarization of the photon found in the state $|L\rangle$ is correlated with the polarization of the one detected in $|R\rangle$. The quantum correlations recorded by the detectors can be explained by the *ambiguity* of the detector setting: Particles are not distinguished by their spatial states, $|\tilde{A}\rangle\,, |\tilde{B}\rangle$, but they are measured by the detectors, $|L\rangle\langle L|$ and $|R\rangle\langle R|$, with which each photon has a finite overlap.

It is important to retain that the entanglement in the state (5) crucially relies on the symmetrization of the two-photon wave-function, which results in the indistinguishability of the two two-particle paths shown in Fig. 1(b) and in a *coherent* superposition as in (5). If non-identical particles are initially prepared in both input arms of the beam splitter, one still looses the information on the internal state of the measured particles and finds anti-correlations between the local measurement outcomes. The *nature* of these anti-correlations, however, is then purely classical: The setup corresponds to a classical machine that randomly distributes pairs of particles to the two detectors. An every-day analogy with colored marbles or socks then captures the essence of the process [22].

In view of this experiment and of similar situations in which the spatial states of particles do not have an unambiguous, one-to-one relationship with the detectors, a characterization of entanglement that incorporates the measurement process as described above is necessary to overcome the discrepancies between hitherto available entanglement measures and entanglement that is actually detected in a specific experimental situation.

3 Detector-level entanglement

For this very purpose, we introduce *detector-level entanglement*, which incorporates the effects of the measurement process itself and therefore does not suffer from the problems mentioned above. In contrast to earlier treatments [2], we do not assign an absolute value of entanglement to a state on its own. Instead, we identify the particles as those entities which are eventually measured by some specified, distinct detectors, which we assume to have 100% detection efficiency and to be represented by mutually orthogonal projection operators. For an *unambiguous* detector scheme, each detector is associated with exactly one of the particles, which can then be treated as distinguishable such that *detector-level entanglement* reduces to the one determined by the previously introduced criteria [11]. In contrast, an *ambiguous* detector setting,

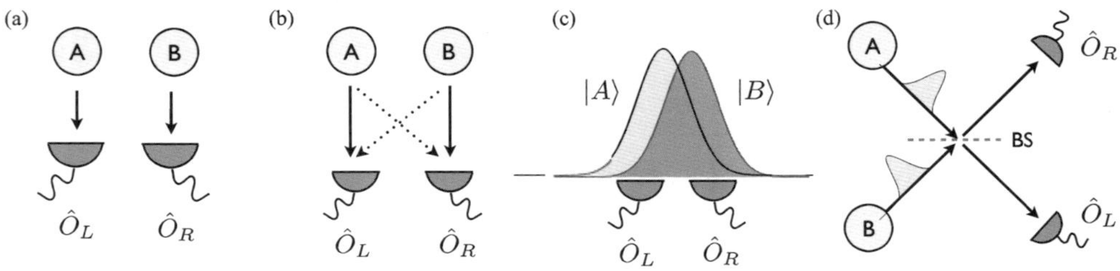

Fig. 1 Setups for the measurement of identical particles. (a) Unambiguous scheme, each particle is assigned to exactly one detector. The A (B)-particle will always be detected by the left (right) detector, denoted by $\hat{O}_L$ ($\hat{O}_R$). (b) Ambiguous scheme, both particles can trigger either detector. The two two-particle paths are depicted by the solid and the dashed arrows. (c) The generic situation of (b) is realized by two wave-functions that overlap in space, such that either detector can be triggered by either particle. (d) Hong-Ou-Mandel setup, two photons prepared in A and B scatter off the beam splitter (BS) and fall into each output mode with equal amplitude. This setup also realizes the ambiguous scheme shown in (b).

in which each particle can trigger each detector with a certain probability, can erase information about the provenience of the particles: the identification of the particles measured by the detectors with the particles prepared in the external states breaks down, since two *distinct* and possibly indistinguishable two-particle paths contribute to the event characterized by the detection of one particle in each detector, as illustrated in Fig. 1(b).

Our subsequent considerations apply for a general setting of arbitrarily many detectors, which we assume to detect particles in states of their external, i.e. spatial, degrees of freedom. The particles additionally carry an internal degree of freedom such as spin, in which they may be entangled. For the sake of clarity, we restrict ourselves hereafter to the case of two particles and two detectors. While the first- and second-quantized formulations of the problem lead to the same physical picture [23], it is here most convenient to use the suggestive language of first quantization in which most mathematical expressions possess a simple, intuitive form.

3.1 Detector-level density matrix

Experimentally observed expectation values are associated with the observable

$$\hat{O}_d(\hat{\alpha}, \hat{\beta}) = \underbrace{\hat{O}_L \otimes \hat{\alpha}}_{1} \otimes \underbrace{\hat{O}_R \otimes \hat{\beta}}_{2} + \underbrace{\hat{O}_R \otimes \hat{\beta}}_{1} \otimes \underbrace{\hat{O}_L \otimes \hat{\alpha}}_{2} , \tag{6}$$

where $\hat{O}_L$ and $\hat{O}_R$ are two external, mutually orthogonal projectors and describe spatial detectors we call *left* and *right*, respectively. $\hat{\alpha}$ and $\hat{\beta}$ are (not necessarily distinct) observables on the internal spaces of the particles. The terms labeled by 1 (2) act on the external and internal spaces of the first (second) particle.

The two detectors assign an identity (left and right) to each of the detected particles, and thereby specify the entities which may or may not carry entanglement. The *detector-level entanglement* of a state $\hat{\rho}_a$, with respect to a given set of detectors, can thus be inferred by application of any entanglement measure on the detector-level density matrix $\hat{\rho}_d$, *reconstructed by quantum state tomography* [24] *in the local basis defined by the detectors*. This corresponds to the measurement of a complete orthonormal set of observables, $\hat{\chi}_i, \hat{\chi}_j$, on the internal degrees of freedom of the particles, with the two external detectors given by (6):

$$\hat{\rho}_d = N \sum_{i,j} \hat{\chi}_i \otimes \hat{\chi}_j \mathrm{Tr}\left(\hat{O}_d(\hat{\chi}_i, \hat{\chi}_j)\hat{\rho}_a\right) , \tag{7}$$

where the normalization N ensures that $\mathrm{Tr}(\hat{\rho}_d) = 1$. Thus, (7) describes the density matrix of the internal degrees of freedom as reconstructed by the detection procedure. In contrast to previous approaches, our formulation takes into account that the spatial overlap of identical particles, and hence their distinguishability, might change with time, even if their quantum state may not. The density matrix ρ_d obtained in (7) possesses a subsystem structure which always reflects the actual experimental setting. It is thus sensitive to the dynamics of the structure of the state – and thus to the a priori entanglement of the system – and to the particle-detector relationship.

Thereby the problem of the indistinguishability of particles is overcome: Entanglement measures applied on $\hat{\rho}_d$ determine the *detector-level* entanglement of $\hat{\rho}_a$ and reproduce the entanglement measured between the two detectors. The encountered quantum correlations may already be present as a priori entanglement in the state or they may also be induced by the measurement process itself. As we will see below, also an interplay of a priori and measurement-induced entanglement may take place.

3.2 Path weights

In the actual reconstruction (7) of the detector-level density matrix ρ_d, several terms appear, which can be given physical interpretation.

In the first place, the probability for two particles injected into the (not necessarily orthogonal) external quantum states $|A\rangle$ and $|B\rangle$ to trigger a coincident event is governed by the following quantities:

$$D_{LR} \equiv \langle A| \hat{O}_L |A\rangle \langle B| \hat{O}_R |B\rangle \,, \tag{8}$$

$$D_{RL} \equiv \langle A| \hat{O}_R |A\rangle \langle B| \hat{O}_L |B\rangle \,. \tag{9}$$

D_{LR} (D_{RL}) is the probability for the particle prepared in $|A\rangle$ ($|B\rangle$) to be detected in the left detector, while the particle prepared in $|B\rangle$ ($|A\rangle$) is detected in the right detector. We will refer to these quantities as *path weights*. The two distinct two-particle paths shown in Fig. 1(b) correspond to these two events. If one path weight vanishes, a coincident event in the detectors reveals full information about which external state the particles were initially prepared in. The scheme is then *unambiguous*, and the situation boils down to the scheme depicted in Fig. 1(a). If both path weights, D_{LR} and D_{RL}, are non-zero, a click in one of the detectors does not completely reveal the provenience of the particles anymore, which corresponds to an ambiguous setting.

3.3 Effective indistinguishability

Even in the ambiguous case, however, particles can still be effectively distinguishable, since one may discriminate the initial preparations by a further measurement: If $\langle B| \hat{O}^\dagger_{L/R}\hat{O}_{L/R} |A\rangle = 0$, the scalar product of the projected states $\hat{O}_{L/R} |A\rangle$ and $\hat{O}_{L/R} |B\rangle$ vanishes, the states are orthogonal after detection. They can thus be distinguished in the sense that their previous preparation could be inferred by another measurement, i.e. the which-way information has not been deleted. At the same time, expectation values for measurements on the internal degrees of freedom at the detectors are not affected by the (anti)symmetrization of the wave-function since all cross-terms involving expressions with factors like $\langle A| \hat{O}_L |B\rangle \langle B| \hat{O}_R |A\rangle$ necessarily vanish. An analogous scenario with distinguishable particles and symmetrized observables would hence yield the very same expectation values. The (anti)symmetrization of the state only plays a role if the overlap of the projected states $\hat{O}_{R/L} |A\rangle$ and $\hat{O}_{R/L} |B\rangle$ does not vanish at both detectors. A quantitative description for the impact of indistinguishability is therefore given by the product of the overlaps at both detectors:

$$\gamma \equiv \langle A| \hat{O}_L |B\rangle \langle B| \hat{O}_R |A\rangle \,, \tag{10}$$

where we used that $\hat{O}^\dagger_{L/R}\hat{O}_{L/R} = \hat{O}_{L/R}$, for projector-valued $\hat{O}_{L/R}$. We baptize γ the *effective indistinguishability*. It quantifies how strongly the measurement process erases information about the previous preparation of the particles, and thus provides a measure for the indistinguishability of the particles. In other words, the larger γ is, the more coherently the two two-particle paths $(A \to L, B \to R)$ and

$(A \to R, B \to L)$ – see Fig. 1(b) – are super-imposed. The effective indistinguishability includes, in contrast to previous approaches [25] to distinguishability, the dependence on the detector setting. The case $\gamma = 0$ corresponds to the situation above in which particles are effectively distinguishable, and the information on the provenience of the particles is preserved during the detection process. For $\gamma \neq 0$, the (anti)symmetrization of the wave-function does, however, affect expectation values of physical observables.

We stress that the path weights D_{LR}, D_{RL} and the effective indistinguishability γ constitute independent parameters. Only the maximal modulus of γ depends on the path weights, which follows from the Cauchy-Schwarz inequality:

$$\begin{aligned} |\gamma| &= |\langle A|\, \hat{O}_L\, |B\rangle \langle B|\, \hat{O}_R\, |A\rangle\, | \\ &= |\langle A|\, \hat{O}_L \hat{O}_L\, |B\rangle \langle B|\, \hat{O}_R \hat{O}_R\, |A\rangle\, | \\ &\leq |\hat{O}_L\, |A\rangle\, | \cdot |\hat{O}_L\, |B\rangle\, | \cdot |\hat{O}_R\, |B\rangle\, | \cdot |\hat{O}_R\, |A\rangle\, | \\ &= \sqrt{D_{LR} D_{RL}} =: \gamma_{\text{max}} \end{aligned} \tag{11}$$

The absolute value of γ measures how strongly which-way information is erased by a measurement: For $|\gamma| = \gamma_{\text{max}}$, the two states $\hat{O}_{R/L}\, |A\rangle\, , \hat{O}_{R/L}\, |B\rangle$ are linearly dependent. Hence, one cannot design any discriminating measurement in this case. For particles which are prepared initially in linearly independent external quantum states, such detection erases the information about their provenience. They are hence completely indistinguishable for such a detector setting.

For particles with some spatial overlap at the detectors, situations occur in which $\gamma \neq 0$, even if $\langle A|B\rangle = 0$. In other words, the detection setup can effectively inhibit the differentiation of the particles that was possible before the measurement process. In turn, for any non-orthogonal states with $0 < |\langle A|B\rangle| < 1$, there are projectors which *do* differentiate the two states. Physical distinguishability hence strongly depends on the settings of the detectors. For our purpose of exploring the consequences of indistinguishability on entanglement, it is sufficient to consider real and negative γ, which corresponds to the case of HOM interferometry.

Physically speaking, the balance of the path weights D_{LR} and D_{RL} governs the ambiguity of the setup, whereas the effective indistinguishability γ quantifies the coherence of the two two-particle paths. The (anti)symmetrization of the wave-function has an impact on observables only if the path weights do both not vanish *and* $\gamma \neq 0$.

4 Applications

Let us now discuss the implications of the above for two identical particles prepared in external quantum states $|A\rangle$, $|B\rangle$. We assume that $\langle A|B\rangle = 0$ and take the internal degree of freedom of the particles to be equivalent to a spin-1/2-system. The following paradigmatic state will turn out to be strongly affected by ambiguous detector settings:

$$|\Psi\rangle = \frac{1}{\sqrt{2}} \{\cos\epsilon\, |A, \uparrow; B, \downarrow\rangle + \sin\epsilon\, |A, \downarrow; B, \uparrow\rangle + \delta\, (\cos\epsilon\, |B, \downarrow; A, \uparrow\rangle + \sin\epsilon\, |B, \uparrow; A, \downarrow\rangle)\} \tag{12}$$

The parameter $\delta = 1(-1)$ refers to bosons (fermions), ϵ is real and continuous and controls the a priori entanglement between the particles. The entanglement measure concurrence [26] of (12) reads

$$C_a(\epsilon) = 2|\cos\epsilon \sin\epsilon|. \tag{13}$$

For $\epsilon = 0, \pi/2$, the state is a priori non-entangled, and the particles in $|A\rangle$ and $|B\rangle$ can each be assigned a certain definite spin. For $\epsilon = \pi/4$, the state is maximally entangled, no information at all is available on the results of a spin-measurement on either particle.

4.1 Exchange interaction

Let us first investigate the interplay of a priori entanglement and the bosonic/fermionic nature of the particles. For this purpose, we transfer a text-book example for the exchange interaction [27] to the realm of a priori entangled identical particles in the state (12). The expectation value of the squared distance of the particles, $(\hat{x}_1 \otimes \mathbb{1}_2 - \mathbb{1}_1 \otimes \hat{x}_2)^2$, reads

$$\langle \Psi | (\hat{x}_1 \otimes \mathbb{1}_2 - \mathbb{1}_1 \otimes \hat{x}_2)^2 | \Psi \rangle = \overbrace{\langle A | \hat{x}^2 | A \rangle + \langle B | \hat{x}^2 | B \rangle - 2 \langle A | \hat{x} | A \rangle \langle B | \hat{x} | B \rangle}^{(\Delta x_{\text{class}})^2} \\ -\delta 4 \cos \epsilon \sin \epsilon | \langle A | \hat{x} | B \rangle |^2, \tag{14}$$

where the identity $\mathbb{1}_s$ that acts on the spin degree of freedom is omitted, for readability.

The first line equates to the intuitive expectation value for the squared distance $(\Delta x_{\text{class}})^2$ for a state $|A; B\rangle$ of distinguishable particles. In addition, we find a correction that makes the particles appear closer together or further apart, depending on the sign of $\delta \cos \epsilon \sin \epsilon$. This term is similar to the contribution that is found for non-entangled identical bosons and fermions [27], and it appears only when there is a finite spatial overlap of the particles, such that $\langle A | \hat{x} | B \rangle$ does not vanish, hence it matches the common sense interpretation of "exchange interaction".

In contrast to non-entangled bosons or fermions, the exchange term does not depend only on the species δ (bosonic or fermionic) of the particles, but also on the parameter ϵ that governs the a priori entanglement. By suitably manipulating their a priori entanglement properties, we can thus make bosons behave like fermions, and vice-versa. It is the *interplay* of the (anti-)symmetrization of the wave-function with the a priori entanglement between the particles that leads to this impact on the expectation value of the inter-particle distance. For $\epsilon = 0$, the state describes two particles of definite and orthogonal spin, and no exchange interaction takes place – indeed, this latter situation corresponds to two particles that can be distinguished by their spin.

In our detection setting, we find an analogous effect, and the specific choice of the a priori entangled state affects the coincident detection rate T. The latter is given by the expectation value of $O_d(\mathbb{1}, \mathbb{1})$ and reads

$$T = D_{LR} + D_{RL} + 4\delta\gamma \cos \epsilon \sin \epsilon. \tag{15}$$

The last term in this sum is proportional both to the *a priori* concurrence $C_a(\epsilon)$, Eq. (13), and to the *effective indistinguishability* γ, Eq. (10). Through the latter it includes the overlap of the single-particle states at the detectors, and the resulting (anti)bunching effects can be interpreted as a further signature of exchange interaction, similar to the contribution in (14). The case $D_{LR} = D_{RL} = 1/4, \gamma = -\gamma_{\text{max}}$ results in complete indistinguishability, corresponding to the situation in a HOM interferometer: Two photons in the $|\Psi^+\rangle$ Bell-state ($\epsilon = \pi/4$) always bunch ($T = 0$), while in the $|\Psi^-\rangle$ Bell-state ($\epsilon = 3\pi/4$) they antibunch ($T = 1$) [28].

4.2 Detector-level state

By virtue of (7), the state (12) yields the detector-level density matrix,

$$\rho_d = \frac{1}{2T} \begin{pmatrix} 0 & 0 & 0 & 0 \\ 0 & T+a & b & 0 \\ 0 & b & T-a & 0 \\ 0 & 0 & 0 & 0 \end{pmatrix}, \tag{16}$$

in the detector level basis

$$\{ |L \uparrow, R \uparrow\rangle, |L \uparrow, R \downarrow\rangle, |L \downarrow, R \uparrow\rangle, |L \downarrow, R \downarrow\rangle \}, \tag{17}$$

with

$$a = (D_{LR} - D_{RL}) \cos 2\epsilon, \tag{18}$$

$$b = 2\delta\gamma + 2(D_{LR} + D_{RL}) \sin\epsilon \cos\epsilon. \tag{19}$$

For unambiguous detector settings, $D_{LR} = 1, D_{RL} = 0$, we simply recover the spin properties of the a priori state, detecting a particle in the left (right) detector amounts to detecting the particle that was prepared in the external state $|A\rangle$ ($|B\rangle$). In general, however, the state at detector level, ρ_d, is possibly mixed. We stress that this loss of purity is not rooted in the detector inefficiency or any decoherence mechanism, but it is born out by ambiguous detector settings with non-maximal γ, i.e. for which the two two-particle paths are not super-imposed in a fully coherent way.

The quantitative and qualitative consequences of the loss of which-way information in the setup are captured by the following quantities:

- The *single-particle predictability* $P_d^{(R/L)}$ [29] of the particle detected by the right/left detector,

$$\left(P_d^{(R/L)}\right)^2 = \left|\mathrm{Tr}\left(\sigma_z \varrho_{R/L}\right)\right|^2, \tag{20}$$

where $\varrho_{R/L}$ is the reduced density matrix that describes the particle in the right/left detector, quantifies the observer's ability to predict the outcome of a measurement in the σ_z-basis. For our setup, the coherences of the reduced density matrices $\varrho_{R/L}$ always vanish by construction, since each single particle is never prepared in superpositions of eigenstates of σ_z, according to (12). The product of the coherences can be interpreted as the *visibility* [30] or interference capability, and this quantity thus also vanishes for the state under consideration. Consequently, the predictability quantifies all single-particle information in our case, and its value is equal for the two detected particles,

$$P_d^2 := \left(P_d^{(R)}\right)^2 = \left(P_d^{(L)}\right)^2 = \frac{(D_{LR} - D_{RL})^2 \cos^2 2\epsilon}{T^2}. \tag{21}$$

When the setup is fully ambiguous, $D_{LR} = D_{RL}$, and no information on the provenience of a detected particle is available, such that the predictability vanishes. Also when a maximally entangled state is prepared ($\cos 2\epsilon = 0$), no information on the preparation of any particle is available, with the analogous consequence, $P_d^2 = 0$.

- The *two-particle linear entropy*,

$$S_d = 1 - \mathrm{Tr}(\rho_d^2) = \frac{(D_{LR}D_{RL} - \gamma^2)(1 + \cos 4\epsilon)}{T^2} \tag{22}$$

quantifies the impurity of the two-particle state. In our case, the pair of particles (12) is always fully anti-correlated (there is always exactly one particle in each internal state, $|\uparrow\rangle, |\downarrow\rangle$), such that measurements exhibit anti-correlations in the σ_z-basis. The two-particle linear entropy S_d directly constitutes a measure for the *classical correlations* that are induced by the ambiguous detection scheme. Note that only for $|\gamma| \neq \gamma_{\max}$ can S_d assume a non-vanishing value, i.e. the only source of classical correlations is the incoherent distribution of the particles among the detectors.

- The *detector-level concurrence* C_d [26] measures the strength of the quantum correlations between the two detected particles. For our specific detector-level state (16), it is proportional to the product of the non-diagonal elements of ρ_d,

$$C_d^2 = \left(\frac{b}{T}\right)^2 = \frac{4\left(\delta\gamma + (D_{LR} + D_{RL}) \sin\epsilon \cos\epsilon\right)^2}{T^2}. \tag{23}$$

For the state under consideration, the three quantities are complementary [30], by virtue of

$$1 = C_d^2 + P_d^2 + 2S_d. \tag{24}$$

Note that this relation does not hold for general density matrices [30], but only for the detector-level state ρ_d of the specific form (16).

We can thus understand a loss of predictability P_d as an increase of correlations, whose (quantum or classical) nature is defined by the balance of C_d^2 and S_d.

If $\gamma = 0$, the detector-level concurrence equals the a priori value ($C_a = C_d$), and ambiguous detector settings with $D_{LR} \neq 0 \neq D_{RL}$ can still manifest themselves, but only as a higher two-particle entropy and lower single-particle predictability with respect to the unambiguous case. The identical particles behave distinguishably, hence just like particles of a different kind, and *classical correlations* are induced by the ambiguous detector setting. In contrast, non-vanishing values of γ change the situation dramatically, and the concurrence C_d will be affected as well. For illustration, let us discuss two exemplary transitions and their experimental implementations in HOM-type setups.

4.3 Vanishing predictability

First, we consider a scenario with vanishing predictability, $P_d = 0$, by setting $D_{LR} = D_{RL} = 1/4$, see (21). The setup is thus maximally ambiguous, which corresponds to the balanced HOM-type beam-splitter setup shown in Fig. 1(d). Each particle has the same probability to trigger either detector. Neither party can predict at all any single spin measurement outcome, while the measurement outcomes at the two detectors are perfectly anti-correlated.

The nature of these correlations depends on the effective indistinguishability γ and on the a priori entanglement defined by ϵ. For a priori non-entangled ($\epsilon = 0$) and effectively distinguishable ($\gamma = 0$) particles, the two-particle state (16) is maximally mixed, since $b = 0$ by virtue of (19), and the uncertainty is purely classical, $C_d = 0$, $2S_d = 1$, Eqs. (22) and (23). This situation corresponds to a setup in which distinguishable particles are randomly distributed among the two detectors. The two two-particle paths are not super-imposed coherently, and the induced correlations are purely classical.

For effectively indistinguishable particles ($\gamma = -\gamma_{\text{max}}$), the two two-particle paths are super-imposed in a fully coherent way. Consequently, the state at detector level ρ_d is pure, the single-particle predictability vanishes due to *detector-level* entanglement, and the detector-level concurrence (23) attains its maximal value.

The case $\gamma = -\gamma_{\text{max}}$ is analogous to the situation in ideal HOM interferometry, in which pairs of maximally entangled particles are created [14, 28]. If, on the other hand, the temporal delay between the photons is too large such that the provenience of the particles could be in principle inferred by a time-resolved measurement (and $\gamma = 0$), the particles result not to be entangled. That is, by variation of γ, the correlations are converted from merely classical correlations into *detector-level* entanglement!

For an a priori non-entangled state with $\epsilon = 0$ or $\epsilon = \frac{\pi}{2}$, the detector-level concurrence increases monotonically with $|\gamma|$. The less the particles can effectively be distinguished, the more they are entangled. In contrast, the dependence is non-monotonic and therefore more intricate for a priori entangled states, as illustrated in Fig. 2. This behavior is due to the competition between the (measurement-induced) term proportional to γ, and that proportional to $\cos\epsilon \sin\epsilon$, i.e. the a priori concurrence, in the parameter b (19).

4.4 Maximal indistinguishability

As a second example, we now fix the *effective indistinguishability* to $\gamma = -\gamma_{\text{max}}$, and tune the detector bias from unambiguous to completely ambiguous by variation of D_{LR} between 0 and $1/4$, while $D_{RL} = (1 - \sqrt{D_{LR}})^2$. This transition can be implemented experimentally with several beam splitters that assume different transmission probabilities. Increasing D_{LR} corresponds to smoothly breaking the unambiguous bond between particle and detector, while always maintaining maximal indistinguishability. The induced uncertainty for single spin-measurements is then maximal for $D_{LR} = D_{RL} = 1/4$.

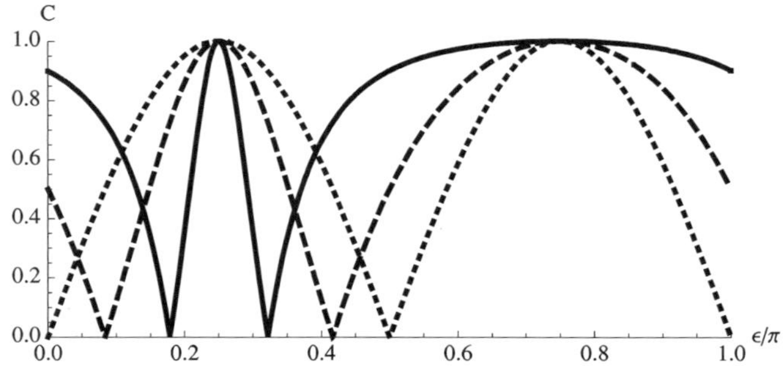

Fig. 2 Detector-level concurrence of two bosons in the state (12), as a function of ϵ (which controls the a priori entanglement (13) of the state), for $D_{LR} = D_{RL} = 1/4$, $\gamma = 0$ (dotted line), $\gamma = -0.5\gamma_{\rm max}$ (dashed line), $\gamma = -0.9\gamma_{\rm max}$ (solid line). For $\gamma = 0$, the detector-level concurrence C_d, Eq. (23) equates the a priori concurrence (13), whereas for $\gamma = -\gamma_{\rm max}$, $C_d = 1$ for all values of ϵ. The transition between distinguishable ($\gamma = 0$) and indistinguishable ($\gamma \neq 0$) particles is, however, not always monotonic.

No classical uncertainty is created by the detection, since $\gamma = -\gamma_{\rm max}$, and ρ_d remains pure. In other words, the two two-particle paths are always coherently super-imposed, whereas their amplitudes are varied by D_{LR} and D_{RL}. If $D_{LR} \neq D_{RL}$, the occurrence of a coincident event itself contains some statistical information on the possible provenience of the particles, the setup is then not maximally ambiguous and the *detector-level* entanglement is not maximal. While the measurement destroys the information about the preparation of the detected particles totally ($\gamma = -\gamma_{max}$), the probability for a certain type ($|A\rangle$ measured in the left or right detector) of coincident event to occur is biased, i.e. the two possible two-particle paths shown in Fig. 1(b) have different amplitudes. Again, predictability (and, due to the above-mentioned complementarity, concurrence) does not depend on D_{LR} monotonically, with remarkable implications: when bunching occurs, i.e. $\epsilon < \pi$, the detector-level concurrence C_d can be smaller than the a priori concurrence C_a, and the predictability (21) consequently grows. In other words, an ambiguous spatial measurement that does not discriminate spins can actually *enhance* the probability of a certain spin-measurement outcome with respect to the unambiguous case. We illustrate this behavior in Fig. 3, where the squared detector-level predictability is plotted for different values of ϵ, as a function of D_{LR}.

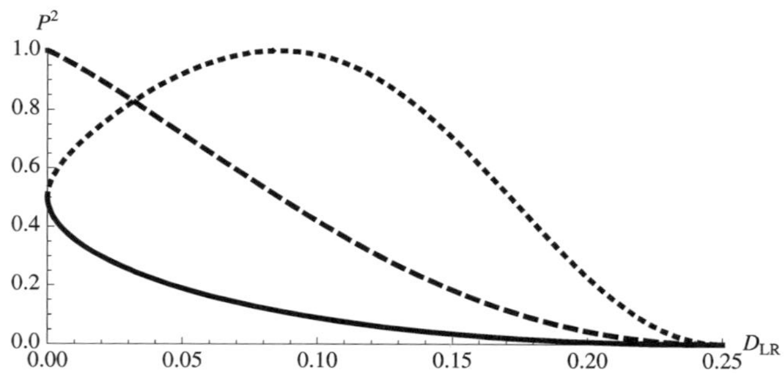

Fig. 3 Squared detector-level single-particle predictability P_d^2 as a function of D_{LR}, for $\gamma = -\gamma_{\rm max}$, $\epsilon = 3\pi/8$ (dotted line), $\epsilon = 0$ (dashed line) and $\epsilon = 5\pi/8$ (solid line). The predictability is not always a monotonically decreasing function of D_{LR} ($D_{LR} = 1/4$ corresponds to a completely ambiguous detector setting).

Although we focused on bosons throughout this section, the same situation can be recovered for fermions, merely exchanging the sign of the effective indistinguishability, $\gamma \rightarrow -\gamma$. In other words, the *species* of the particles (bosonic or fermions) merely boils down to the sign of a parameter, and it is the indistinguishability which is mainly responsible for the encountered coherent many-particle effects [23].

5 Conclusions

Entanglement and the indistinguishability of particles *interfere* on several levels – in the literal and in the figurative sense of the word: The very notion of entanglement needs to be adjusted in order to allow for the impossibility to label identical particles and for the consequent loss of subsystem structure. Thus, discriminating degrees of freedom need to be incorporated explicitly. This leads to the concept of a priori entanglement, which deals with the assignment of the notion of entanglement to a state on its own. On the other hand, the indistinguishability of identical particles also comes with a *many-body coherence property* which drastically influences the observable entanglement in many situations. These physical consequences of the symmetrization postulate were so far not captured by the available entanglement concepts in the literature.

Our here introduced notion of *detector-level entanglement* takes into account both aspects, such that it is suitable for the characterization of entanglement in systems in which particles may overlap in space (and where this overlap may be time-dependent under some nontrivial dynamics) and the assumptions for a priori entanglement are not met, since each detector can be triggered by both particles. We introduced with (7) the density matrix of a quantum state of two identical particles at the level of the detectors, such that in all situations, including the HOM setup described at the outset, the *physical* impact of indistinguishability is taken into account. Our examples show the important discrepancy between a priori entanglement and *detector-level* entanglement, and illustrate how the parameter γ provides a measure for the erasure of information on the particle preparation through the detection process. For vanishing γ, particles can be treated as distinguishable, since the (anti)symmetrization does not affect non-classical correlations. However, for $\gamma \neq 0$ the (anti)symmetrization is crucial for physical observables in general, and, in particular, it affects the entanglement between the measured particles. Our quantitative definition of *detector-level* entanglement incorporates these effects and thus matches the experimental reality.

Whereas the indistinguishability of particles is routinely exploited in linear-optics experiments with photons [31], it is usually neglected in experiments with interacting constituents, in which the mutual interaction constitutes the source of quantum correlations [32]. In the future, the combined action of these effects – interaction and indistinguishability – may be probed, e.g. in experiments with cold atoms in optical lattices, whose indistinguishability *and* interaction are under control [33]: The internal degrees of freedom provide tunable distinguishability, the interaction can be tuned via Feshbach resonances [34].

The understanding and the differentiation of the aforementioned effects is especially important for the characterization of entanglement in systems that do not follow the paradigm of quantum information, such as atoms and molecules [17, 35, 36]. In these systems, the encountered correlations are typically rooted in, both, interaction-induced a priori entanglement, and measurement-induced entanglement.

Acknowledgements M. C. T. acknowledges financial support by the German National Academic Foundation. F. de M. acknowledges support from the Alexander von Humboldt-Stiftung, the Belgian Interuniversity Attraction Poles Programme P6/02, and Vidi grant 639.072.803 from the Netherlands Organization for Scientific Research (NWO). Partial support of M. K. by Polish NCN grant DEC-2011/01/M/ST2/0379 is acknowledged, as well as support of F. M. by DFG grant MI 1345/2-1. A.B. and M.K. acknowledge partial support through the EU-COST Action MP1006 "Fundamental Problems in Quantum Physics".

References

[1] J. Bell, Physics **1**, 195 (1964).
[2] G. Ghirardi and L. Marinatto, Fortschr. Phys. **51**, 379 (2003).
[3] A. Einstein, B. Podolsky, and N. Rosen, Phys. Rev. **47**, 777 (1935).
[4] A. M. L. Messiah and O. W. Greenberg, Phys. Rev. **136**, B248 (1964).
[5] R. F. Werner, Phys. Rev. A **40**, 4277 (1989).
[6] M. A. Nielsen and I. L. Chuang, Quantum Computation and Quantum Information (Cambridge University Press, Cambridge, 2000).
[7] F. Herbut and M. Vujičić, J. Phys. A **20**, 5555 (1987).
[8] A. Peres, Quantum Theory: Concepts and Methods (Kluwer Academic Publishers, New York, 1993).
[9] J. Schliemann, J. Cirac, M. Kuś, M. Lewenstein, and D. Loss, Phys. Rev. A **64**, 022303 (2001).
[10] K. Eckert, J. Schliemann, D. Bruß, and M. Lewenstein, Ann. Phys. (New York) **299**, 88 (2002).
[11] G. Ghirardi, L. Marinatto, and T. Weber, J. Stat. Phys. **108**, 49 (2002).
[12] G. Ghirardi and L. Marinatto, Phys. Rev. A **70**, 012109 (2004).
[13] A. R. Plastino, D. Manzano, and J. S. Dehesa, Europhys. Lett. **86**, 20005 (2009).
[14] Z. Y. Ou and L. Mandel, Phys. Rev. Lett. **61**, 50 (1988).
[15] C. K. Hong, Z. Y. Ou, and L. Mandel, Phys. Rev. Lett. **59**, 2044 (1987).
[16] C. Simon, Phys. Rev. A **66**, 052323 (2002).
[17] D. Akoury et al., Science **318**, 949 (2007).
[18] J. Cai, S. Popescu, and H. J. Briegel, Phys. Rev. E **82**, 021921 (2010).
[19] S. Bose and D. Home, Phys. Rev. Lett. **88**, 050401 (2002).
[20] E. Hecht and A. Zajac, Optics (Addison-Wesley, Reading, 1987).
[21] Y. H. Shih and C. O. Alley, Phys. Rev. Lett. **61**, 2921 (1988).
[22] J. S. Bell, Speakable and Unspeakable in Quantum Mechanics (Cambridge University Press, Cambridge, 1987).
[23] M. C. Tichy, Entanglement and Interference of Identical Particles, PhD thesis, Universität Freiburg (2011), http://www.freidok.uni-freiburg.de/volltexte/8233/.
[24] D. F. James, P. G. Kwiat, W. J. Munro, and Phys. Rev. A **64**, 052312 (2001).
[25] F. W. Sun and C. W. Wong, Phys. Rev. A **79**, 013824 (2009).
[26] W. K. Wootters, Phys. Rev. Lett. **80**, 2245 (1998).
[27] D. J. Griffiths, Introduction to Quantum Mechanics (Prentice Hall, Upper Saddle River, 1995).
[28] S. Bose et al., Phys. Rev. A **68**, 052309 (2003).
[29] D. M. Greenberger and A. Yasin, Phys. Lett. A **128**, 391 (1988).
[30] M. Jakob and J. Bergou, Phys. Rev. A **76**, 52107 (2007).
[31] J. W. Pan et al., Rev. Mod. Phys. **84**, 777 (2012).
[32] J. I. Cirac and P. Zoller, Phys. Rev. Lett. **74**(05), 4091 (1995).
[33] J. F. Sherson et al., Nature **467**, 68 (2010).
[34] A. J. Leggett, Rev. Mod. Phys. **73**, 307 (2001).
[35] M. S. Schöffler et al., Science **320**, 920 (2008).
[36] M. C. Tichy, F. Mintert, and A. Buchleitner, J. Phys. B **44**, 192001 (2011).

Alexander Ling

Centre for Quantum Technologies, National University of Singapore, Singapore

Paper Title: **Experimental polarization state tomography using optimal polarimeters**

Reference: Physical Review A **74**, 022309 (2006)

Personal message:

Berge's independent re-discovery of efficient polarimetry using only four measurement settings was extended to quantum systems (Phys. Rev. A **70**, 052321 (2004)). We thought it was an interesting idea and put it into an experiment and even used it for tomography of two qubits. It was definitely an improvement over previous methods of two-qubit tomography (36 settings) and also clearer than a "minimal" tomography proposed by others that had little theoretical basis. So Berge's insight definitely helped to push quantum technology forward.

PHYSICAL REVIEW A **74**, 022309 (2006)

Experimental polarization state tomography using optimal polarimeters

Alexander Ling,* Kee Pang Soh, Antía Lamas-Linares, and Christian Kurtsiefer
Department of Physics, National University of Singapore, 117542 Singapore, Singapore
(Received 14 March 2006; published 8 August 2006)

We report on the experimental implementation of a polarimeter based on a scheme known to be optimal for obtaining the polarization vector of ensembles of spin-$\frac{1}{2}$ quantum systems and the alignment procedure for this polarimeter. We also show how to use this polarimeter to estimate the polarization state for identically prepared ensembles of single photons and photon pairs and extend the method to obtain the density matrix for generic multiphoton states. State reconstruction and performance of the polarimeter is illustrated by actual measurements on identically prepared ensembles of single photons and polarization entangled photon pairs.

DOI: 10.1103/PhysRevA.74.022309 PACS number(s): 03.67.−a, 03.65.Wj, 07.60.Fs, 42.25.Ja

I. INTRODUCTION

Many promising applications in quantum information (e.g., quantum computation and quantum communication) demand accurate state estimation. For many of them it is compelling to implement state estimation techniques that are both fast and consume as few copies of the state as possible. Research in improving the efficiency of quantum state estimation techniques is an area of active theoretical study [1–7] with much focus on spin-$\frac{1}{2}$ systems (qubits). Experimental reports on state estimation are fewer [8], partly because many schemes call for a joint measurement of an ensemble of qubits which is not always possible to implement.

One way to realize qubits experimentally is to use single photons and consider their polarization degree of freedom, as this can be described by a two-dimensional Hilbert space. Estimating the polarization state of an ensemble of single photons (called polarimetry in classical optics) is equivalent to estimating the state of the qubit ensemble. This makes efficient polarimetry interesting to quantum information.

Polarimetry that uses the least number of measurement outcomes is said to be minimal. Minimal polarimetry techniques in classical optics have been known for a long time, and a lot of work in their optimization has been done [9–13]. While these classical methods perform well in estimating the polarization state for single-photon ensembles in the limit of large numbers, their performance in the regime of extremely low light intensity (single-photon level) was uncertain and it was not obvious how to use them in estimating nonclassical states of light. For this reason, progress in polarimetry at the single-photon limit will assist in many areas, including characterization of faint sources of light, classical ellipsometry [9], advanced quantum key distribution protocols [14–16], and studies of the fundamental aspects of quantum theory [17].

In discriminating between different estimation techniques we distinguish between methods that are minimal and those that are minimal *and* optimal [3,7]: optimal methods have the best asymptotic efficiency in determining an unknown state when averaged over all possible input states. This can be used for an operational definition of minimal and optimal state estimation for ensembles of prepared quantum systems. It is the technique that provides the best improvement to our estimated state for each additional copy taken from the ensemble. Recently Řeháček *et al.* proposed such a method for state estimation of polarization-based single qubits [7], which can be viewed as an extension of classical techniques [10,11,18] to quantum systems.

In this paper we address the experimental problem of implementing the optimal state estimation method described in [7] by using a complete four-output polarimeter with no moving parts. We describe that polarimeter in Sec. II by reviewing the theory of optimal polarization state estimation and explain our implementation. In Sec. III polarization state estimation of multiphoton states is addressed. In Sec. IV we elaborate on the alignment procedure to make the polarimeter perform optimally. Experimental state reconstruction on ensembles of single photons and photon pairs with high fidelity will be illustrated in Secs. V and VI.

II. STATE ESTIMATION USING THE OPTIMAL POLARIMETER

The polarization state of light can be characterized using the three Stokes parameters S_1, S_2, and S_3, possibly augmented by an intensity S_m. Together they form a Stokes vector $\vec{S}$ and, when normalized, it is written as $\vec{S}=(1,S_1/S_m,S_2/S_m,S_3/S_m)$. A *reduced* Stokes vector $\vec{S}_r=(S_1,S_2,S_3)/S_m$ identifies a point in the Poincaré sphere (in this paper reduced Stokes vectors are denoted by an r subscript) [19].

A minimal scheme of estimating the Stokes vector requires exactly four detector readings, which corresponds to finding the overlap of the unknown Stokes vector with four noncoplanar vectors that define a tetrahedron in the Poincaré sphere (Fig. 1). These four noncoplanar vectors define four measurement operators B_j that govern the detector readings and form a set of complete positive operator value measurements (POVM's) [20]. The tetrahedron geometry defines the largest volume that can be enclosed by a vector quartet in the Poincaré sphere, making it the optimal estimation technique when using four POVM's [11,21]. Such a state estimation technique is also unbiased in the asymptotic limit because

*Electronic address: phylej@nus.edu.sg; http://www.quantumlah.org

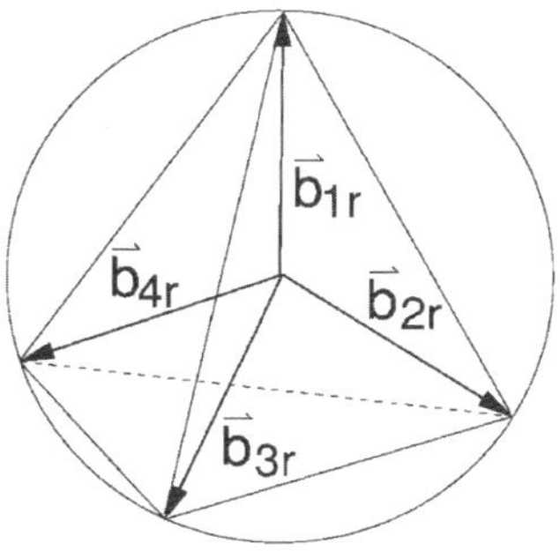

FIG. 1. Four (reduced) Stokes vectors in the Poincaré sphere that form a tetrahedron define the optimal POVM operators used for polarization state estimation. The tetrahedron gives the largest volume encompassable by a vector quartet in the sphere, making it the optimal measurement when using four POVM's.

the total distance of any vector in the Poincaré sphere to all four POVM vectors depends only on the vector's magnitude. In other words, the orientation of the unknown vector does not affect the accuracy with which it is estimated [7].

We shall denote the tetrahedron's reduced Stokes vectors by $\vec{b}_{1r}$, $\vec{b}_{2r}$, $\vec{b}_{3r}$, and $\vec{b}_{4r}$ as shown in Fig. 1 and write their corresponding normalized vectors as $\vec{b}_1$, $\vec{b}_2$, $\vec{b}_3$, and $\vec{b}_4$. Each measurement operator B_j may be expressed as

$$B_j = \frac{1}{4}(\vec{b}_j \cdot \vec{\sigma}), \tag{1}$$

where $\vec{\sigma}=(\sigma_0,\sigma_1,\sigma_2,\sigma_3)$, σ_0 being the unit matrix and $\sigma_{1,2,3}$ the Pauli matrices.

In an experiment we associate each operator B_j with a detector b_j. The average intensity falling on detector b_j is denoted as I_j. Thus expectation values of the tetrahedron operators are related to detected intensities as follows:

$$\frac{I_j}{I_t} = \langle B_j \rangle = \frac{1}{4}(\vec{b}_j \cdot \vec{S}), \tag{2}$$

with

$$I_t = \sum_{j=1}^{4} I_j.$$

Writing the intensities as a vector $\vec{I}=(I_1,I_2,I_3,I_4)/I_t$ gives us the Stokes vector

$$\vec{I} = \Pi \cdot \vec{S} \Leftrightarrow \vec{S} = \Pi^{-1} \cdot \vec{I}, \tag{3}$$

where Π is referred to as the instrument matrix. Each row of this matrix is composed from a vector $\vec{b}_j$. One possible instrument matrix of the ideal polarimeter is

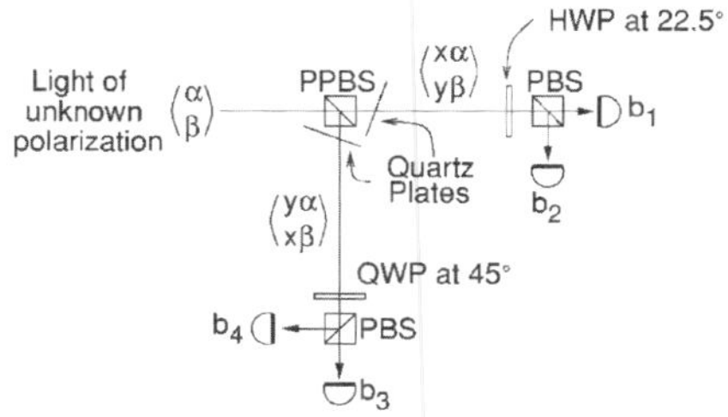

FIG. 2. Practical implementation of the tetrahedron polarimeter that achieves the ideal instrument matrix. Each detector b_j is associated to the tetrahedron vector $\vec{b}_j$. The partially polarizing beam splitter (PPBS) separates incoming light according to polarization, and quartz plates remove unwanted phase shifts. Light leaving the PPBS is passed through wave plates and polarizing beam splitters (PBS) to be projected on two different bases (±45° basis for transmitted light and the circular basis for reflected light).

$$\Pi = \frac{1}{4}\begin{pmatrix} 1 & \sqrt{\frac{1}{3}} & \sqrt{\frac{2}{3}} & 0 \\ 1 & \sqrt{\frac{1}{3}} & -\sqrt{\frac{2}{3}} & 0 \\ 1 & -\sqrt{\frac{1}{3}} & 0 & -\sqrt{\frac{2}{3}} \\ 1 & -\sqrt{\frac{1}{3}} & 0 & \sqrt{\frac{2}{3}} \end{pmatrix}. \tag{4}$$

Experimental realization of this instrument matrix is achieved by the polarimeter shown in Fig. 2. The first component of the polarimeter is a partially polarizing beam splitter (PPBS) that has a particular amplitude splitting ratio for incoming light, most easily determined using Jones vector notation for polarization. The amplitude division coefficients of the PPBS, x and y, obey energy conservation $|x^2|+|y^2|=1$. The PPBS takes horizontally polarized (H) light $\binom{1}{0}$ to the polarizations $\binom{x}{0}$ and $\binom{y}{0}$ in the transmitted and reflected arms, respectively, and vertically (V) polarized light $\binom{0}{1}$ to $\binom{0}{y}$ in transmission and $\binom{0}{x}$ in reflection.

If we project light in the transmitted arm of the PPBS on the ±45° polarization basis and light in the reflected arm onto the left and right circular polarization bases, the tetrahedral arrangement of the vectors $\vec{b}_j$ is ensured with the following relations:

$$x^2 = \frac{1}{2} + \frac{1}{2\sqrt{3}}, \quad y^2 = \frac{1}{2} - \frac{1}{2\sqrt{3}}. \tag{5}$$

Detailed steps are given in the Appendix .

Partially polarized light can be described using a density matrix (or coherency matrix). If we write the entries of the density matrix as a column vector $\vec{\rho}$, we can determine them from the Stokes vector [22] using the following transformation:

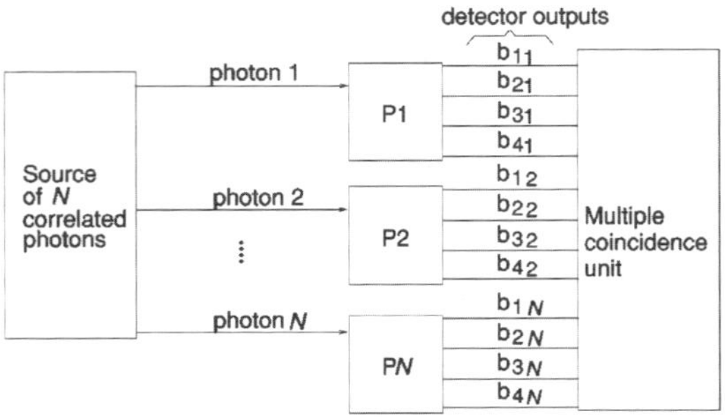

FIG. 3. Scheme for estimating the polarization state of an ensemble of N-correlated photons using N polarimeters $(P1, P2, \ldots, PN)$. A multiple-coincidence circuit identifies the 4^N possible coincidence combinations. For photon pairs (N=2), two polarimeters are used, giving 16 possible coincidence combinations. Several copies of the state are processed, giving a coincidence pattern used in estimating the polarization state of the ensemble.

$$\vec{\rho} = \frac{1}{2}\Gamma_1 \cdot \vec{S} = \frac{1}{2}\begin{pmatrix} 1 & 1 & 0 & 0 \\ 0 & 0 & 1 & i \\ 0 & 0 & 1 & -i \\ 1 & -1 & 0 & 0 \end{pmatrix} \cdot \vec{S}. \tag{6}$$

The columns of the matrix Γ_1 are the Pauli operators written as column vectors $\Gamma_1 = (\vec{\sigma}_0, \vec{\sigma}_1, \vec{\sigma}_2, \vec{\sigma}_3)$. The matrices B^{-1} and Γ_1 can be combined into a single matrix

$$T := \frac{1}{2}\Gamma_1 \Pi^{-1} \Rightarrow \vec{\rho} = T \cdot \vec{I}, \tag{7}$$

which might be referred to as a tomography matrix as it directly relates the detected intensities to the density matrix of the state.

III. POLARIZATION STATE TOMOGRAPHY FOR ENSEMBLES WITH MULTIPHOTONS

The instrument matrix scheme above can be extended to perform polarization state tomography on ensembles of multiphoton states. James *et al.* [8] have described a similar state estimation method. We follow their approach but use our optimal and instrumentally motivated measurement operators, thereby reducing any ambiguity over the choice of operators.

The simplest multiphoton system is a photon pair detected by testing for coincidence in the detection time of their component photons. In our measurement process each member of the photon pair is passed through a polarimeter. Given two polarimeters 1 and 2, each with four detectors b_{i_1} and b_{i_2}, respectively $(i_1, i_2 = 0, 1, 2, 3)$, we will have 16 possible coincidence combinations between the detectors (see Fig. 3). Each coincidence rate is governed by an operator composed from the individual detectors' measurement operators. If we denote again the measurement operator of detectors b_{i_1} and b_{i_2} as B_{i_1} and B_{i_2} and the coincidence count between them as c_{i_1,i_2}, we can express the coincidence rates as a linear function of a two-photon polarization state vector S_2:

$$\frac{c_{i_1,i_2}}{c_t} = \langle B_{i_1} \otimes B_{i_2} \rangle = \left(\frac{1}{4}\vec{b}_{i_1} \otimes \frac{1}{4}\vec{b}_{i_2}\right) \cdot \vec{S}_2, \tag{8}$$

with

$$c_t = \sum_{i_1,i_2=1}^{4} c_{i_1,i_2}.$$

Here, $\vec{S}_2$ is the Stokes vector equivalent for a two-photon system [8] and c_t is the total number of observed coincidences. We now have the set of measurement operators governing the coincidence pattern. The 16 coincidences c_{i_1,i_2} can be written in column vector format $\vec{C}_2 = (c_{1,1}, c_{1,2}, \ldots, c_{4,4})$. If we define the two-polarimeter instrument matrix as Π_2, we obtain an instrument response analogous to (3):

$$\vec{C}_2 = \Pi_2 \cdot \vec{S}_2 \Leftrightarrow \vec{S}_2 = \Pi_2^{-1} \cdot \vec{C}_2. \tag{9}$$

Thus we obtain the density matrix of the two-photon state by constructing the analogous two-photon expression for Eq. (6):

$$\vec{\rho}_2 = \frac{1}{2^2}\Gamma_2 \cdot \vec{S}_2 = T_2 \cdot \vec{C}_2. \tag{10}$$

Each column of Γ_2 is the product of two Pauli operators $\sigma_{i_1} \otimes \sigma_{i_2}$ $(i_1, i_2 = 0, 1, 2, 3)$ written in column vector format, and T_2 is the tomography matrix for the two-photon state.

It is now straightforward to generalize this concept to obtain the density matrix for states of N correlated photons. Using N polarimeters, we obtain the pattern of N-fold coincidences to build up the coincidence vector $\vec{C}_N$ which is used to find the N-photon Stokes vector and density matrix:

$$\vec{S}_N = \Pi_N^{-1} \cdot \vec{C}_N, \tag{11}$$

$$\vec{\rho}_N = \frac{1}{2^N}\Gamma_N \cdot \vec{S}_N = T_N \cdot \vec{C}_N. \tag{12}$$

Each row of the instrument matrix Π_N is given by $\left(\frac{1}{4}\vec{b}_{i_1} \otimes \frac{1}{4}\vec{b}_{i_2} \cdots \otimes \frac{1}{4}\vec{b}_{i_N}\right)$ and each column of Γ_N is the product of N Pauli matrices $\sigma_{i1} \otimes \sigma_{i_2} \cdots \otimes \sigma_{i_n}$ $(i_n = 0, 1, 2, 3$ and $n = 1, 2, \ldots, N)$. This generalized approach will work for all four-detector polarimeters in multiphoton analysis schemes (Fig. 3). We note that—although the polarimeters are optimal for estimating single qubit states—it is an open question if the scheme above is optimal in estimating multiphoton systems.

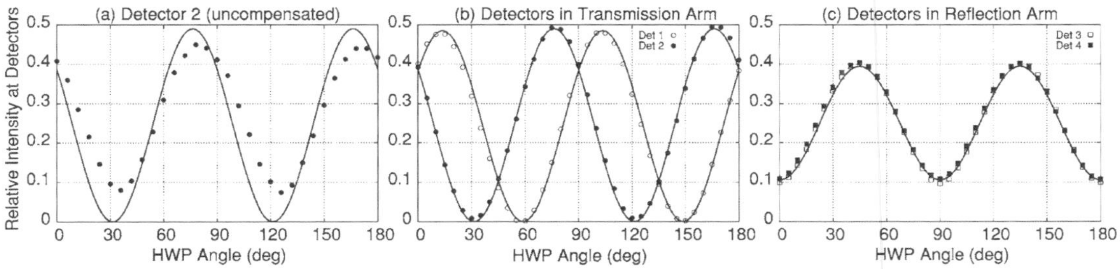

FIG. 4. Instrument response of the polarimeter to linearly polarized light. The data points show the variation in relative intensity at each detector with respect to the angle of the half-wave plate (HWP) in the polarization state preparation. The solid lines show the expected intensity modulation for an ideal device for each HWP setting [(13)], scaled for appropriate detector efficiencies. Error bars are smaller than the point markers. Panel (a) shows the relative intensity at detector 2 without compensation plates. Panels (b) and (c) are taken with compensation for phase shifts. The oscillation in (a) is out of phase and also of lower amplitude compared to the phase-corrected behavior of detector 2 in panel (b). Plots in (c) shows a lower amplitude because light in the reflected arm is not projected on a linear polarization basis.

IV. PHASE CORRECTION AND POLARIMETER CALIBRATION

A. Removing unwanted phase shifts

In the presented polarimeter, an ideal PPBS has the nominal beam splitting ratio (5) and also rotates the polarization state of light leaving the beam splitter into the correct polarization basis [12]. Such beam splitters, however, are not easily available and their design is the focus of active research [23]. We therefore use beam splitters with only the nominal intensity splitting ratio.

A PPBS without a phase shift diverts light in state $-\vec{b}_{jr}$ (which is conjugate to a tetrahedron vector $\vec{b}_{jr}$) from detector b_j. General beam splitters, however, lack this phase-preserving property. The result is that input of conjugate states $-\vec{b}_{jr}$ does not stop light from reaching the associated detectors. This suggests an easy alignment method for correcting any unwanted phase shifts with birefringent compensation plates.

For phase correction we first prepare high-quality H-polarized light using polarizers of extinction ratio 10^5. With one subsequent half-wave plate (HWP) and one quarter-wave plate (QWP) we can then prepare any polarization state on the surface of the Poincaré sphere. Compensator plates (0.5-mm-thick quartz) mounted on rotating stages were placed at each output arm of the PPBS, and light with a conjugate polarization state was sent to the polarimeter. For each polarization state $-\vec{b}_{jr}$ the compensator in the relevant output arm was rotated until the detector b_j received no light. Two input states (one for each output arm) were sufficient to compensate for the unwanted phase shifts.

We verified the compensated polarimeter behavior with linearly polarized light prepared using only the polarizer and HWP (this reduces preparation errors due to residual errors in the QWP). The prepared states have a Stokes vector of the form $(1, \cos 4\psi, \sin 4\psi, 0)$, where ψ is the angle of the HWP, so the normalized response of detector 1, for example, will be

$$I_1 = 1 + \sqrt{\frac{1}{3}} \cos 4\psi + \sqrt{\frac{2}{3}} \sin 4\psi. \qquad (13)$$

Passively quenched silicon avalanche photodiodes were used as detectors allowing us to perform photon counting. The number of photons accumulated at each detector output was noted for each angle of the HWP. The results are shown in Fig. 4.

The results show that the response of the compensated polarimeter is very close to ideal. The extrema of our measured intensities are less than 1° (of HWP angle) away from their nominal positions. This means that our actual measurement vectors are pointing in the same direction as the ideal tetrahedron vectors, although their magnitudes will be different due to imbalanced detection efficiencies. While this renders the asymptotic efficiency of our polarimeter less than ideal, it still represents the optimal setup for the collection efficiencies we can achieve. In other words, we are maximizing the volume defined by our experimental POVM vectors [21].

This measurement result is limited by the accuracy of our rotation controllers. Our wave plates are mounted on rotary motors with an accuracy of 0.3°. The polarizing beam splitters in the output arms have an extinction ratio of 10^4 and the wave plates' optical path lengths differ from their nominal values by less than 2%.

B. Calibrating the polarimeter

We calibrate the instrument matrix of this polarimeter to account for all residual phase shifts and coupling inefficiencies. A general calibration technique for four detector polarimeters ("equator-pole method") was described by Azzam *et al.* [18]. Incidentally, the phase dependence measurement shown in Fig. 4 was an essential part of this calibration.

Using this technique we are able to find the correction terms needed to be made to our ideal instrument matrix. A typical corrected instrument matrix is

$$\Pi_c = \frac{1}{4}\begin{pmatrix} 0.962 & 1.051\sqrt{\frac{1}{3}} & 0.920\sqrt{\frac{2}{3}} & 0.005 \\ 0.991 & 1.031\sqrt{\frac{1}{3}} & -0.956\sqrt{\frac{2}{3}} & -0.005 \\ 1.010 & -1.045\sqrt{\frac{1}{3}} & 0.005 & -0.945\sqrt{\frac{2}{3}} \\ 1.032 & -1.009\sqrt{\frac{1}{3}} & 0.029 & 1.011\sqrt{\frac{2}{3}} \end{pmatrix}.$$

The uncertainty for each of the correction terms above is on the order of 0.002. We see that the deviation from entries in the ideal instrument matrix (4) is on the order of a few percent.

The phase correction and calibration steps presented above must take into account the wavelength of the input light because optical elements are specified to perform only within a certain bandwidth. The polarimeter was built to study the polarization state of light coming from a spontaneous parametric down-conversion (SPDC) source [24] with a spectral bandwidth of 4.740 ± 0.014 nm centered around 702 nm. The same light source was used for phase correction and polarimeter calibration and the experiments described in the remaining sections.

V. EXPERIMENTAL STATE TOMOGRAPHY FOR ENSEMBLES OF SINGLE PHOTONS

The ability of the tetrahedron polarimeter to estimate polarization states without bias was tested by preparing a set of pure polarization states equally distributed over the Poincaré sphere. In this way we could better identify regions that suffer poor state estimation (if any).

Computer-controlled motors were used to rotate wave plates (after a H filter) in preparing the set of polarization states. The Stokes vector of a pure polarization state can be expressed as $\vec{S}=(1, \cos 2\delta \cos 2[\psi+\delta], -\cos 2[\psi+\delta]\sin 2\delta, -\sin 2[\psi+\delta])$, where δ and ψ are the QWP and HWP angles, respectively. Thus any set of coordinates (characterized by the polar and azimuthal angles) on the Poincaré sphere can be expressed in terms of the wave plate angles.

For each set of angles, the detectors accumulated photon detection events for 1 s giving a particular vector $\vec{I}$ from which an estimated Stokes vector $\vec{S}_e$ and probability density matrix ρ_e can be obtained via Eqs. (3) and (6). To calculate the distance of the estimated state from the (ideal) prepared state ρ_i $(\vec{S}_i)$, we use the Uhlmann fidelity, defined as $(\mathrm{tr}[\sqrt{\sqrt{\rho_i}\rho_e\sqrt{\rho_i}}])^2$ [25,26]. For pure states this quantity reduces to the overlap of their Stokes vectors $\frac{1}{2}(\vec{S}_i \cdot \vec{S}_e)$.

The fidelity was mapped to the appropriate polar and azimuthal coordinates on the Poincaré sphere (Fig. 5); linear polarization states correspond to a polar angle of 0°. The average fidelity for the whole map is 99.8% with a minimum fidelity of $98.4\% \pm 0.9\%$ (the cumulative photon count per point is approximately 2000). There are no systematic areas of low fidelity even when wedge errors in the state preparation wave plates cause count rates to drop. This indicates that the polarimeter estimates all pure polarization states equally well.

Fidelity does not distinguish between errors introduced in state preparation from errors in the state estimation process. Therefore we have characterized our state preparation apparatus independently and are confident that their contribution to the error in calculated fidelity above is on the order of $\pm 0.01\%$. Thus we assign the residual difference in fidelity to imperfections in the detection apparatus.

VI. EXPERIMENTAL STATE TOMOGRAPHY FOR A TWO-PHOTON ENSEMBLE

We will now illustrate how to use two polarimeters to perform polarization state tomography on a two-photon state generated from an SPDC source. First, two polarimeters were correctly aligned and after calibration their instrument matrices were found to be

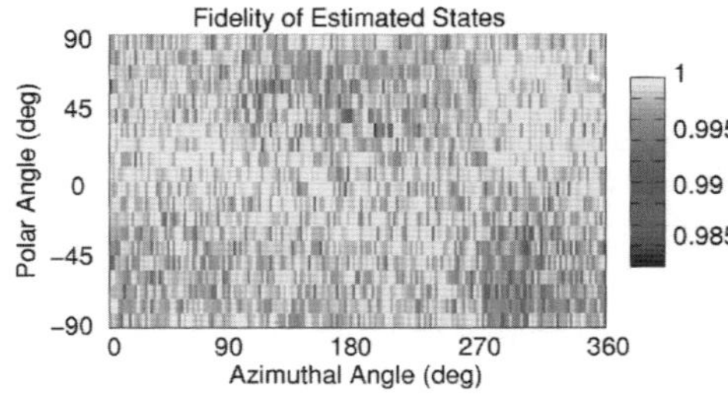

FIG. 5. (Color online) A set of polarization states $(\vec{S}_i)$ equally distributed over the Poincaré sphere surface was generated; photons from each of these states were sent to the polarimeter, from which an estimated state $(\vec{S}_e)$ is obtained. The figure shows the fidelity of the estimated state to the prepared state, $\frac{1}{2}(\vec{S}_e \cdot \vec{S}_i)$. It is roughly constant over the Poincaré sphere, showing that the polarimeter is an unbiased polarization state estimator.

$$\frac{1}{4}\begin{pmatrix} 0.903 & 0.927\sqrt{\frac{1}{3}} & 0.9997\sqrt{\frac{2}{3}} & -0.041 \\ 1.124 & 1.135\sqrt{\frac{1}{3}} & -1.014\sqrt{\frac{2}{3}} & 0.0602 \\ 0.995 & -1.079\sqrt{\frac{1}{3}} & 0.001 & 0.913\sqrt{\frac{2}{3}} \\ 0.978 & -0.983\sqrt{\frac{1}{3}} & 0.003 & -0.936\sqrt{\frac{2}{3}} \end{pmatrix}$$

and

$$\frac{1}{4}\begin{pmatrix} 1.074 & 1.171\sqrt{\frac{1}{3}} & 0.913\sqrt{\frac{2}{3}} & -0.082 \\ 0.983 & 0.8804\sqrt{\frac{1}{3}} & -1.044\sqrt{\frac{2}{3}} & 0.004 \\ 1.082 & -1.172\sqrt{\frac{1}{3}} & 0.001 & -0.9625\sqrt{\frac{2}{3}} \\ 0.862 & -0.88\sqrt{\frac{1}{3}} & -0.002 & 0.867\sqrt{\frac{2}{3}} \end{pmatrix}.$$

We then arranged for the SPDC source to generate photon pairs that are detected as a maximally entangled Bell state $|\Psi^+\rangle$. Bell states created via SPDC are typically characterized by a polarization correlation experiment, from which a visibility value can be obtained [24]. The visibility measured in the HV and ±45° basis was above 97.7%±2%; such a high value is usually taken as evidence of a high degree of entanglement.

The photon pairs were passed through the polarimeters and the pattern of coincidences between them was observed. The 16 observed coincidence rates (collected using the scheme similar to [27]) make up the coincidence vector $\vec{C}$=(21444, 1505, 24104, 26002, 979, 24716, 23210, 22447, 21661, 30752, 24061, 268, 19010, 23692, 339, 17695).

Using this vector with Eqs. (9) and (10) we obtain the density matrix whose real components are

$$\mathrm{Re}[\rho]=\begin{pmatrix} -0.002 & -0.01 & -0.03 & -0.024 \\ -0.01 & 0.506 & 0.485 & 0.025 \\ -0.03 & 0.485 & 0.498 & 0.009 \\ -0.024 & -0.024 & 0.009 & -0.003 \end{pmatrix},$$

while the magnitude of the imaginary components are below a value of 0.04 (see Fig. 6).

The uncertainty in each of the above terms is on the order of 0.011. The Uhlmann fidelity of this state to the ideal $|\Psi^+\rangle$ state was found to be 0.990±0.014. Error bars in all cases were computed by numerical derivation and propagated Poissonian counting noise. The propagated error bars result in an estimated density matrix compatible with the ideal $|\Psi^+\rangle\langle\Psi^+|$ state.

VII. CONCLUSION

In this paper we have illustrated a simple alignment procedure for optimizing the tetrahedron polarimeter. Phase shifts introduced by a commercially available PPBS were easily corrected by phase compensation plates in the output arms. The response of the compensated polarimeter was measured over a dense sampling of states on the Poincaré sphere and found to be similar to that of an ideal device. This shows that beam splitters need only have the nominal intensity splitting ratio, making optimal polarimeters more accessible.

We also described an instrumentally motivated method for constructing the measurement operators governing light distribution to each output of the polarimeter. This instrument-based approach also allows a convenient generalization to obtain measurement operators governing multiphoton coincidences. These operators can then be applied to the linear reconstruction of multiphoton Stokes vectors and their density matrices.

Optimal polarimeters were then used for estimating the polarization state of experimentally prepared ensembles of single photons and photon pairs in a Bell state. The estimated states were evaluated by computing their fidelity to the (ideal) prepared states. We found an average fidelity above 99.8% in all our experiments. Thus we have built and demonstrated the use of optimal four-output polarimeters in multiphoton polarization state tomography.

While preparing this document, it came to our attention that a similar four output polarimeter was suggested independently in a recent paper [17].

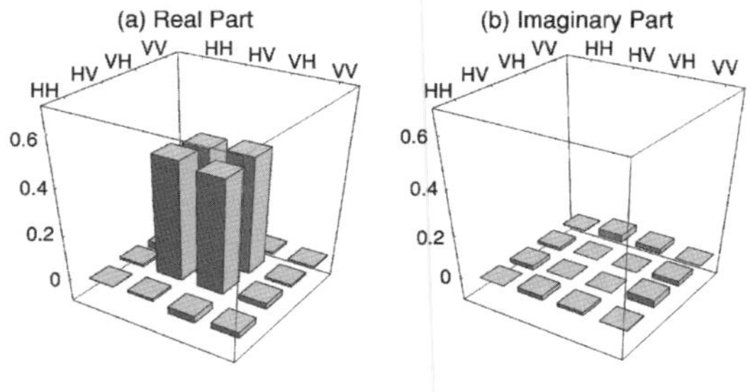

FIG. 6. (Color online) The density matrix of a Bell state $|\Psi^+\rangle\langle\Psi^+|$ obtained by linear reconstruction from photon pairs.

ACKNOWLEDGMENTS

We would like to thank Ivan Marcikic, Berge Englert, and Janet Anders for helpful discussions. This work was supported by DSTA Grant No. R-394-000-019-422 and A*Star Grant No. R-144-000-071-305.

APPENDIX

To derive the intensity splitting ratio of the PPBS, we first express the polarization states are expressed using Jones vectors unless we are describing the tetrahedron vectors $\vec{b}_j$. The tetrahedron (Stokes) vectors $\vec{b}_{j,k}$ have the scalar product property

$$\vec{b}_j\cdot\vec{b}_k=\frac{2}{3}+\frac{4}{3}\delta_{jk}. \tag{A1}$$

Recalling the parameters of the intensity splitting ratio of the PPBS, x and y, we see that a general input polarization state $\binom{\alpha}{\beta}$ leads to the polarizations $\binom{x\alpha}{y\beta}$ and $\binom{y\alpha}{x\beta}$ in the transmitted and reflected arms of the PPBS, respectively. In our polarimeter, light leaving the arms of the PPBS must be analyzed in two different polarization bases. Two orthogonal vectors that form a basis may be expressed as $\binom{\cos\theta}{e^{i\phi}\sin\theta}$ and $\binom{-e^{-i\phi}\sin\theta}{\cos\theta}$. This leads, for example, to the normalized light intensity falling on detector b_1:

$$I_1/I_t = |\alpha x \cos\theta + \beta y e^{-i\phi}\sin\theta|^2. \tag{A2}$$

We choose a different measurement basis for detectors 3 and 4: for example, light reaching detector 3 is

$$I_3/I_t = |\alpha y \cos\theta' + \beta x e^{-i\phi'}\sin\theta'|^2. \tag{A3}$$

Using the vector $\vec{b}_1$ as an example, Eq. (2) allows us to express the operator B_1 in terms of the measurement basis to fulfill Eq. (A2):

$$\langle B_1\rangle = |\alpha x \cos\theta + \beta y e^{-i\phi}\sin\theta|^2 \tag{A4}$$

$$= \left|\begin{pmatrix} x\cos\theta & y e^{-i\phi}\sin\theta\end{pmatrix}\begin{pmatrix}\alpha\\ \beta\end{pmatrix}\right|^2. \tag{A5}$$

The following choice of B_1 fulfills this condition:

$$B_1 = \begin{pmatrix} x\cos\theta \\ y\sin\theta e^{i\phi}\end{pmatrix}\begin{pmatrix} x\cos\theta & y\sin\theta e^{-i\phi}\end{pmatrix}. \tag{A6}$$

Since the tetrahedron can be oriented arbitrarily we choose for convenience to measure the 45° linear polarization basis ($\theta=\pi/4$, $\phi=0$) in the transmitted arm and the circular polarization basis ($\theta'=\pi/4$ $\phi'=\pi/2$) in the reflected arm. This reduces the measurement operators to only the beam splitting parameters x and y,

$$\left.\begin{matrix}B_1\\B_2\end{matrix}\right\} = \frac{1}{2}\begin{pmatrix} x^2 & \pm xy \\ \pm xy & y^2\end{pmatrix}, \quad \left.\begin{matrix}B_3\\B_4\end{matrix}\right\} = \frac{1}{2}\begin{pmatrix} y^2 & \mp ixy \\ \pm ixy & x^2\end{pmatrix},$$

which together with Eq. (2) allows us to express all tetrahedron vectors in terms of x and y:

$$\left.\begin{matrix}\vec{b}_1\\ \vec{b}_2\end{matrix}\right\} = \begin{pmatrix} 1 \\ x^2-y^2 \\ \pm 2xy \\ 0\end{pmatrix}, \quad \left.\begin{matrix}\vec{b}_3\\ \vec{b}_4\end{matrix}\right\} = \begin{pmatrix} 1 \\ y^2-x^2 \\ 0 \\ \mp 2xy\end{pmatrix}. \tag{A7}$$

From Eq. (A1), we can write

$$\vec{b}_1\cdot\vec{b}_2 = \frac{2}{3}, \quad \vec{b}_1\cdot\vec{b}_3 = \frac{2}{3}. \tag{A8}$$

This allows us to obtain an equation in x alone:

$$36x^8 - 24x^4 + 1 = 0. \tag{A9}$$

The last equation gives two solution sets; we choose the set where $x^2=\frac{1}{2}+\frac{1}{2\sqrt{3}} \Rightarrow y^2=\frac{1}{2}-\frac{1}{2\sqrt{3}}$.

[1] S. Massar and S. Popescu, Phys. Rev. Lett. **74**, 1259 (1995).
[2] R. Derka, V. Bužek, and A. K. Ekert, Phys. Rev. Lett. **80**, 1571 (1998).
[3] J. I. Latorre, P. Pascual, and R. Tarrach, Phys. Rev. Lett. **81**, 1351 (1998).
[4] R. D. Gill and S. Massar, Phys. Rev. A **61**, 042312 (2000).
[5] R. Schack, T. A. Brun, and C. M. Caves, Phys. Rev. A **64**, 014305 (2000).
[6] E. Bagan, M. Baig, R. Muñoz-Tapia, and A. Rodriguez, Phys. Rev. A **69**, 010304(R) (2004).
[7] J. Řeháček, B.-G. Englert, and D. Kaszlikowski, Phys. Rev. A **70**, 052321 (2004).
[8] D. F. V. James, P. G. Kwiat, W. J. Munro, and A. G. White, Phys. Rev. A **64**, 052312 (2001).
[9] R. M. A. Azzam, in *Handbook of Optics*, 2nd ed., edited by M. Bass, E. W. van Stryland, D. R. Williams, and W. L. Wolfe (McGraw-Hill, New York, 1995), Vol. II, Chap. 27.
[10] A. Ambirajan and D. C. Look, Jr., Opt. Eng. (Bellingham) **34**, 1651 (1995).
[11] A. Ambirajan and D. C. Look Jr., Opt. Eng. (Bellingham) **34**, 1656 (1995).
[12] R. Azzam and A. De, J. Opt. Soc. Am. A **20**, 955 (2003).
[13] D. S. Sabatke, M. R. Descour, E. L. Dereniak, W. C. Sweatt, S. A. Kemme, and G. S. Phipps, Opt. Lett. **25**, 802 (2000).
[14] D. Bruß, M. Christandl, A. Ekert, B.-G. Englert, D. Kaszlikowski, and C. Macchiavello, Phys. Rev. Lett. **91**, 097901 (2003).
[15] Y. C. Liang, D. Kaszlikowski, B.-G. Englert, L. C. Kwek, and C. H. Oh, Phys. Rev. A **68**, 022324 (2003).
[16] B.-G. Englert, D. Kaszlikowski, H. K. Ng, W. K. Chua, J. Řeháček, and J. Anders, e-print quant-ph/0412075.
[17] E. Andersson, S. M. Barnett, and A. Aspect, Phys. Rev. A **72**, 042104 (2005).
[18] R. Azzam, I. Elminyawi, and A. El-Saba, J. Opt. Soc. Am. A **5**, 681 (1988).
[19] In the language of spin-$\frac{1}{2}$ systems the reduced Stokes vector is the Pauli vector and the Poincare sphere is called the Bloch sphere.
[20] K. Kraus, *States, Effects and Operations. Fundamental Notions of Quantum Theory, Lecture Notes in Physics*, Vol. 190 (Springer-Verlag, Berlin, 1983).
[21] B.-G. Englert, K. M. Tin, C. G. Goh, and H. K. Ng, Laser Phys. **15**, 7 (2005).
[22] E. L. O'Neill, in *Introduction to Statistical Optics* (Dover, New York, 1991), Chap. 9.
[23] R. M. A. Azzam and F. F. Sudrajat, Appl. Opt. **44**, 190 (2005).
[24] C. Kurtsiefer, M. Oberparleiter, and H. Weinfurter, Phys. Rev. A **64**, 023802 (2001).
[25] A. Uhlmann, Rep. Math. Phys. **9**, 273 (1976).
[26] R. Jozsa, J. Mod. Opt. **41**, 2315 (1994).
[27] S. Gaertner, H. Weinfurter, and C. Kurtsiefer, Rev. Sci. Instrum. **76**, 123108 (2005).

Huangjun Zhu

Perimeter Institute for Theoretical Physics, Waterloo, Ontario, Canada

Paper Title: **Minimal tomography with entanglement witnesses**

Reference: Physical Review A **81**, 052339 (2010)

Personal message:

I would like to express my sincere gratitude to Berge for guidance and help during my PhD and thereafter. Happy 60th birthday, Berge! Wish you a healthy and prosperous life in the coming years!

Entanglement is a valuable resource for many quantum protocols such as quantum communication, quantum key distribution, and quantum computation. The detection and characterization of entanglement in experiments are crucial for the implementation of all these nonclassical tasks. Entanglement can be detected by measuring entanglement witnesses. How many witnesses do we have to measure to determine whether a generic unknown quantum state is entangled or separable? If we were to rely on each witness separately, we might have to measure an infinite number of them even in the case of two qubits.

In this work, guided by Berge's insight, we introduce informationally complete measurements whose outcomes are entanglement witnesses. The witnesses can be measured successively; if all of them give an inconclusive result, one exploits their tomographic completeness for a reconstruction of the quantum state and can then determine its entanglement properties by data processing. Therefore, the number of witnesses needed is no more than the dimension of the state space. The optimized witness-based measurement can provide exponential improvement with respect to witness efficiency in high-dimensional Hilbert spaces, at the price of a reduction in the tomographic efficiency. For the case of two polarization qubits of photons, we show how existing technology can be used to implement the optimized witnesses in a very efficient way. Owing to the details of the implementation, which actually measures the eigenstate basis of the witness rather than solely determining the expectation value of the witness, one does not need to measure more than six witnesses in this example of a 16-dimensional state space. Recently, the idea proposed in this work has been realized in an experiment, see Phys. Rev. Lett. **113**, 170402 (2014) (Editors' note: this is the article contributed by Jibo Dai at the end of this section).

PHYSICAL REVIEW A **81**, 052339 (2010)

Minimal tomography with entanglement witnesses

Huangjun Zhu,[1,2] Yong Siah Teo,[1,2] and Berthold-Georg Englert[1,3]
[1]*Center for Quantum Technologies, National University of Singapore, Singapore 117543, Singapore*
[2]*NUS Graduate School for Integrative Sciences and Engineering, Singapore 117597, Singapore*
[3]*Department of Physics, National University of Singapore, Singapore 117542, Singapore*
(Received 19 March 2010; published 26 May 2010)

We introduce informationally complete measurements whose outcomes are entanglement witnesses and so answer the question of how many witnesses need to be measured to decide whether an arbitrary state is entangled or not: as many as the dimension of the state space. The witnesses can be measured successively; if all of them give an inconclusive result, one exploits their tomographic completeness for a reconstruction of the quantum state and can then determine its entanglement properties by data processing. There are witnesses that are optimal for this purpose. The optimized witness-based measurement can provide exponential improvement with respect to witness efficiency in high-dimensional Hilbert spaces, at the price of a reduction in the tomographic efficiency. We describe a systematic construction and illustrate the matter with the example of two qubits. For the case of two polarization qubits of photons, we show how existing technology can be used to implement the optimized witnesses in a very efficient way. Owing to the details of the implementation, which actually measures the eigenstate basis of the witness rather than solely determining the expectation value of the witness, one does not need to measure more than six witnesses in this example of a 16-dimensional state space.

DOI: 10.1103/PhysRevA.81.052339 PACS number(s): 03.67.Mn, 03.65.Ud, 03.65.Wj

I. INTRODUCTION

Quantum entanglement is a useful resource with a wide range of applications such as quantum teleportation [1–3], quantum key distribution [4], and quantum computation [5]. The detection and characterization of entanglement in experiments are crucial for the implementation of all these nonclassical tasks. Entanglement can be detected by measuring entanglement *witnesses* [6,7]. How many witnesses do we have to measure to determine whether a generic unknown quantum state is entangled or separable? [8] If we were to rely on each witness separately, we would have to measure an infinite number of them even in the case of two qubits—this is well known [9,10] but overlooks a crucial point: Each measurement of a witness provides information about the quantum state and eventually there will be enough data for a complete state reconstruction.

State tomography with an informationally complete (IC) probability operator measurement (POM) [11–14], which may consist of no more than d^2 outcomes (d is the dimension of the Hilbert space, a *minimal* IC POM has d^2 outcomes) can provide accurate information about an unknown input state, if sufficient copies of the input state are measured. With this tomographic information at hand, we can reconstruct the unknown input state and then determine its separability, possibly exploiting well-known criteria, such as the positive partial transpose (PPT) criterion [6,15], the range criterion [16], the matrix realignment criterion [17,18], the covariance matrix criterion [19], or perhaps others, or making use of available algorithms [20,21]. If we can design a minimal IC POM such that every outcome is a witness, d^2 witnesses will suffice to determine the separability of any unknown input state. It should be emphasized that after the reconstruction of the input state, it may still not be easy to determine the separability of the input state, but no additional witnesses need to be tested in experiments, and no more quantum resources will be consumed.

We establish that such witness operator measurements (WOMs) do exist and show how to construct WOMs with optimal witnesses [22] as outcomes from rank-one POMs—POMs in which each outcome is a subnormalized projector onto a pure state. These WOMs can provide exponential improvement of the witness efficiency at the price of a much lower tomographic efficiency. The trade-off between witness efficiency and tomographic efficiency turns out to be quite common, and it is impossible to construct a minimal IC POM which achieves both the best witness efficiency and optimal tomographic efficiency.

We illustrate the matter for the case of two qubits and propose an experimental setup for measuring optimal witnesses for two photonic polarization qubits. This setup is feasible with current technology and, in fact, provides more information than just the expectation values of the witness observables. In the course of measuring the witnesses one learns the probabilities with which their eigenstates occur, and we show how one can take good advantage of this more specific information.

Here is a brief outline of the article. We begin, in Sec. II, with a discussion of the general features of WOMs, for witnesses derived from pure states, for optimal witnesses based on mixed states, and for IC POMs of optimal witnesses. In Sec. III, we deal with the case of two qubits and discuss the proposal for a witness basis measurement for polarization qubits. We offer an outlook on future work that explores such measurements of witness bases systematically (Sec. IV) and we close with a summary.

II. WITNESS OPERATOR MEASUREMENTS: GENERAL FEATURES

We shall be concerned with a bipartite system whose parts have subsystem dimensions d_1 and d_2, with $d_1 \leqslant d_2$, and the total dimension is $d = d_1 d_2$. Usually, a witness is a Hermitian operator with nonnegative mean values for all separable states and negative mean values for some entangled states [6], but this particular convention about the threshold value is somewhat inconvenient in the current context. We will use the following more general definition: A witness is a positive operator with mean values less than or equal to some threshold for all separable states and mean values above this threshold for some entangled states.

A simple example of such a witness in the two-qubit case is the projector onto a Bell state with the threshold at $\frac{1}{2}$. Geometrically speaking and irrespective of the threshold conventions, a witness defines a hyperplane in the state space which separates a partial set of entangled states from all other states. A witness is called *optimal* if no other witnesses can detect all entangled states detected by this witness and also some other ones [22]. An optimal witness defines a hyperplane in the state space which touches the convex set of separable states.

A. Pure-state witnesses

All unit-trace positive operators can serve as statistical operators. The largest mean value of a statistical operator ρ_w is its largest eigenvalue $\lambda_{\max}$, and if the largest mean value μ of ρ_w with all separable states satisfies $\mu < \lambda_{\max}$, then ρ_w is a witness with the threshold at μ. Accordingly, ρ_w can serve as a witness if its eigenspace to the largest eigenvalue does not contain any product state, and only then. As a consequence, every entangled pure state is a witness, and because almost all pure states are entangled, it is easy to construct an IC POM with rank-one witnesses as outcomes.

To improve the witness efficiency (the probability of detecting a random entangled state), it would be favorable to have optimal witnesses as outcomes, with the optimality criterion as specified in Ref. [22]. Since rank-one POMs are the best choice for tomographic purposes, it is of much concern whether a pure-state witness can be an optimal witness. As we shall see shortly, the efficiency of pure-state witnesses is poor, and there are more efficient witnesses for the construction of WOMs.

First, we note that, in the case of $d_2 > d_1$, there are no pure-state witnesses that are optimal witnesses, and in the case of $d_1 = d_2$, the only candidates for optimal pure-state witnesses are maximally entangled states. The optimality of a pure-state witness is determined by its Schmidt coefficients, and without loss of generality we can assume

$$\rho_w = |\Psi\rangle\langle\Psi|, \quad |\Psi\rangle = \sum_{i=0}^{d_1-1} |ii\rangle\sqrt{\lambda_i}, \tag{1}$$

with the Schmidt coefficients $\sqrt{\lambda_i}$ in nonincreasing order,

$$1 > \lambda_0 \geqslant \lambda_1 \geqslant \cdots \geqslant \lambda_{d_1-1} > 0, \quad \sum_{i=0}^{d_1-1} \lambda_i = 1. \tag{2}$$

The eigenvalues of the partial transpose $\rho_w^{T_2}$ are

$$\begin{aligned} &\lambda_i \quad \text{for} \quad i = 0,\ldots,d_1-1 \\ \text{and } &\pm\sqrt{\lambda_i\lambda_j} \quad \text{for} \quad i,j = 0,\ldots,d_1-1, \quad \text{with} \quad i<j. \end{aligned} \tag{3}$$

It follows that

$$\mathrm{Tr}\{\rho_{\mathrm{sep}}\rho_w\} = \mathrm{Tr}\{\rho_{\mathrm{sep}}^{T_2}\rho_w^{T_2}\} \leqslant \lambda_0 \tag{4}$$

for all separable states ρ_{sep}, and since $\rho_{\mathrm{sep}} = |00\rangle\langle 00|$ achieves this bound, the threshold of the pure-state witness ρ_w can be set at λ_0. To determine the optimality of such a witness, we construct the corresponding witness in the usual sense; that is: with the threshold at zero, $W = \lambda_0\mathbf{1} - |\Psi\rangle\langle\Psi|$, where $\mathbf{1}$ denotes the identity operator. Its partial transpose $Q = W^{T_2}$, which is a positive operator, is

$$\begin{aligned} Q = {}& \lambda_0 P^{(\mathrm{r})} + \sum_{i=0}^{d_1-1} |ii\rangle(\lambda_0 - \lambda_i)\langle ii| \\ &+ \sum_{\substack{i,j=0 \\ (i<j)}}^{d_1-1} [(\lambda_0 - \sqrt{\lambda_i\lambda_j})P_{ij}^{(+)} + (\lambda_0 + \sqrt{\lambda_i\lambda_j})P_{ij}^{(-)}], \end{aligned} \tag{5}$$

with

$$\begin{aligned} P_{ij}^{(\pm)} &= (|ij\rangle \pm |ji\rangle)\frac{1}{2}(\langle ij| \pm \langle ji|), \\ P^{(\mathrm{r})} &= \sum_{i=0}^{d_1-1}\sum_{j=d_1}^{d_2-1} |ij\rangle\langle ij|. \end{aligned} \tag{6}$$

If $d_2 > d_1$ or $|\Psi\rangle\langle\Psi|$ is not maximally entangled, the range of Q always contains some product ket and, according to Theorem 2 in Ref. [22], W is not optimal. If $d_1 = d_2$, and $|\Psi\rangle\langle\Psi|$ is a maximally entangled state, then Q is proportional to the projector onto the antisymmetric subspace, which contains no product ket in its range, and W could be an optimal witness.

The probability that a random pure state is detected by a pure-state witness decreases exponentially with the dimension of the Hilbert space. To demonstrate this point we note that pure states $|\Phi\rangle\langle\Phi|$ can be represented as points on a $2d_1d_2 - 1$ dimensional unit sphere (many points may correspond to the same state, but this does not matter, as we are only concerned with the ratio) and that the set of pure product states is of measure zero. The states detected by ρ_w form a spherical cap determined by $\langle\Phi|\rho_w|\Phi\rangle > \lambda_0$. The detection probability is equal to the ratio of the hyper-area of the spherical cap to that of the sphere, and this ratio is given by

$$(1-\lambda_0)^{d_1d_2-1}, \quad \text{with} \quad \lambda_0 \geqslant 1/d_1, \tag{7}$$

where λ_0 is the square of the largest Schmidt coefficient of $|\Psi\rangle$ in $\rho_w = |\Psi\rangle\langle\Psi|$. The maximum detection probability of $(1 - 1/d_1)^{d_1d_2-1}$ is achieved when $|\Psi\rangle\langle\Psi|$ is maximally entangled. This maximum detection probability approaches e^{-d_2} for $d_1 \gg 1$. So, even for the best pure-state witnesses, the detection probability decreases exponentially with the dimension and, therefore, it is desirable to find more efficient witnesses.

B. Optimal witnesses

There are two kinds of witnesses, namely, decomposable witnesses and nondecomposable witnesses. Decomposable witnesses can be written in the form

$$W = P + Q^{T_2}, \tag{8}$$

where both P and Q are positive (semidefinite) operators. Witnesses that cannot be written in this form are nondecomposable. Decomposable witnesses can only detect negative partial transpose (NPT) entangled states, while nondecomposable witnesses can also detect PPT entangled states [22]. To simplify the following discussion, we restrict our attention to decomposable witnesses; nondecomposable witnesses can be considered as well, but there is no convenient parametrization at hand.

An optimal decomposable witness W can be written as the partial transpose of some positive operator Q whose range contains no product ket [22]. We take a probabilistic approach in the study of detection efficiency of a generic witness constructed from the partial transpose of a statistical operator Q, see also [23], because the method used in the calculation of detection efficiency of a pure-state witnesses is generally difficult to carry out and may not give an intuitive result. When $|\Phi\rangle\langle\Phi|$ is distributed according to the normalized unitarily invariant Haar measure $d\mu(\Phi)$, the expectation value and variance of the random variable $\langle\Phi|Q^{T_2}|\Phi\rangle$ are

$$E[\langle\Phi|Q^{T_2}|\Phi\rangle] = \frac{1}{d},\qquad \mathrm{Var}\,[\langle\Phi|Q^{T_2}|\Phi\rangle] = \frac{\mathrm{Tr}\{(Q-\mathbf{1}/d)^2\}}{d(d+1)}. \tag{9}$$

We note that the variance is proportional to the squared Hilbert-Schmidt (HS) distance between Q and the maximally mixed state.

In a high-dimensional Hilbert space ($d \gg 1$), there are generally many positive and negative eigenvalues of Q^{T_2} distributed randomly for a generic Q, and as a consequence of the central-limit theorem, the distribution of the random variable $\langle\Phi|Q^{T_2}|\Phi\rangle$ will approximate a Gaussian distribution, so that the probability of obtaining a negative value (the signature that $|\Phi\rangle\langle\Phi|$ is nonseparable) is mainly determined by the ratio of the expectation value to the standard deviation.

Accordingly, it would be favorable to maximize the standard deviation in order to increase the detection efficiency. Recall that the expectation values of the purity $\mathrm{Tr}\{Q^2\}$ with respect to the HS measure and Bures measure are [24,25]

$$E[\mathrm{Tr}\{Q^2\}]_{\mathrm{HS}} = \frac{2d}{d^2+1},\qquad E[\mathrm{Tr}\{Q^2\}]_{\mathrm{B}} = \frac{5d^2+1}{2d(d^2+2)}, \tag{10}$$

respectively, which both scale $\propto 1/d$ for $d \gg 1$. So, if we choose Q randomly according to either of these two measures, or other measures commonly used, the ratio of $(E[\langle\Phi|Q^{T_2}|\Phi\rangle])^2$ to $\mathrm{Var}[\langle\Phi|Q^{T_2}|\Phi\rangle]$ is on the order of d with high probability, and the detection probability decreases exponentially with growing d. It is now clear why a pure-state witness can only detect a very small fraction of entangled states in a high-dimensional Hilbert space: When turned into a witness W in the usual sense, the positive operator $Q = W^{T_2}$ is highly mixed [see Eq. (5)].

However, if Q is a pure entangled state, the witness W is optimal [26], and the ratio is of order 1. Also, in a high-dimensional Hilbert space, a randomly chosen pure state is approximately a maximally entangled state with high probability. For a maximally entangled state Q, the detection probability of W can be calculated similarly to the case of pure-state witnesses, with the result

$$\frac{\int_0^{\pi/4} d\alpha\,(\cos\alpha)^{d_1(d_1-1)-1}(\sin\alpha)^{d_1(d_1+1)-1}}{\int_0^{\pi/2} d\alpha\,(\cos\alpha)^{d_1(d_1-1)-1}(\sin\alpha)^{d_1(d_1+1)-1}} = \frac{\Gamma(d_1^2)}{2^{d_1^2}\Gamma\left(\frac{d_1(d_1+1)}{2}\right)}\sum_{k=0}^{d_1}\frac{\binom{d_1}{k}(-1)^k\Gamma\left(\frac{k+1}{2}\right)}{\Gamma\left(\frac{d_1(d_1-1)+k+1}{2}\right)}. \tag{11}$$

This ratio becomes

$$\frac{1}{\sqrt{2\pi}}\int_1^{\infty} dx\, e^{-\frac{1}{2}x^2} \approx 0.1573 \tag{12}$$

in the limit $d_1 \to \infty$; see also [23]. A numerical calculation shows that, when $d_1 \geqslant 7$, the deviation of the ratio from the limit is within 1%, and the largest deviation occurs for $d_1 = 2$, where the ratio is $\frac{1}{8}$. In conclusion, in a high-dimensional Hilbert space, a witness constructed from the partial transpose of a randomly chosen pure state will, with high probability, detect about 15.7% of pure entangled states, thus achieving exponential improvement over a pure-state witness.

The lesson to be drawn from these considerations is not about detecting entangled pure states $|\Phi\rangle\langle\Phi|$, which are themselves of little interest because the unknown input state is almost never a pure state. Rather, we learn here that the witness $W = Q^{T_2}$ is very efficient if Q projects onto a pure entangled state.

C. Construction of informationally complete witness operator measurements from optimal witnesses

We can now construct WOMs with optimal witnesses as outcomes. Given a rank-one POM $\sum_i w_i\rho_i = \mathbf{1}$ with outcomes $w_i\rho_i$, where $w_i > 0$ and the ρ_is are projectors onto pure entangled states, we can construct a WOM with outcomes $\propto w_i\rho_{iw}$, where $\rho_{iw} = \lambda_{\max}\mathbf{1} - \rho_i^{T_2}$ with $\lambda_{\max}$ being the maximum of the largest eigenvalues of all $\rho_i^{T_2}$s. This WOM can achieve exponential improvement with respect to witness efficiency, and if the POM is IC, so is the WOM.

We have restricted our discussion to decomposable witnesses in Sec. II B. But this restriction is not essential for the construction of witness operator measurements. Our construction can be generalized to incorporate nondecomposable witnesses straightforwardly once they are found, then both NPT and PPT entangled states can be detected directly. The difficulty with nondecomposable witnesses is their characterization and construction [22].

The outcomes $\propto w_i\rho_{iw}$ of the WOM constructed previously are highly mixed, and this decreases the tomographic efficiency. To compare the tomographic efficiencies of the WOM and the POM, we consider an example in which the POM is a symmetric IC POM (SIC POM), which is conjectured to exist in all finite dimensions [27–29]; a comprehensive account is given in Ref. [30]. The mean square errors (according to the HS distance) achieved by the SIC POM and the optimal WOM are, respectively, given by [31,32]

$$E\left(\|\hat{\rho}-\rho\|_{\mathrm{HS}}^2\right)_{\mathrm{POM}} = \frac{1}{N}(d^2+d-1-\mathrm{Tr}\{\rho^2\}) \sim \frac{d^2}{N} \tag{13}$$

and

$$\begin{aligned} E(\|\hat{\rho}-\rho\|_{\mathrm{HS}}^2)_{\mathrm{WOM}} \\ &= \frac{1}{N}\left(\frac{1}{d}[(d+1)^2(d-1)(d\lambda_{\max}-1)^2+1]-\mathrm{Tr}\{\rho^2\}\right) \\ &\sim \frac{\lambda_{\max}^2 d^4}{N}, \end{aligned} \tag{14}$$

where $\hat{\rho}$ is an estimate of the input state ρ in accordance with the measurement results and N is the number of copies of the input state available for state tomography. Note that $\lambda_{\max} \geqslant 1/d_1$ (if every outcome of the SIC POM is approximately maximally entangled, we have $\lambda_{\max} \sim 1/d_1$); so in a high-dimensional Hilbert space, to achieve the same mean square error, the number of copies of an input state required in the WOM is at least d_2^2 times larger. For different rank-one POMs, the specific reduction of the tomographic efficiency may be different, but the order of magnitude should be similar.

The trade-off between the tomographic efficiency and the witness efficiency is not restricted to this specific example, as we have seen that rank-one outcomes, which are best for tomographic purposes, are generally very poor as witnesses, the more so in a high-dimensional Hilbert space. Generally speaking, positive operators with high purity tend to be poor witnesses, and those with low purity are not desirable for tomography.

The detection of entanglement of mixed states turns out to be much more difficult [23], even if Q is a pure state. For $W = (|\Psi\rangle\langle\Psi|)^{T_2}$, the distribution of the random variable $\mathrm{Tr}\{\rho W\} = \langle\Psi|\rho^{T_2}|\Psi\rangle$ is the same in form as that of the random variable $\langle\Phi|Q^{T_2}|\Phi\rangle$ discussed previously, provided ρ varies among all unitarily equivalent statistical operators according to the Haar measure. In particular, the expectation value and variance of the random variable are the same as those in Eq. (9) after replacing Q by ρ. Since the fraction of separable states approaches zero in a high-dimensional Hilbert space, by similar arguments as before, we can conclude that the probability of detecting entanglement of a random mixed state will generally decrease exponentially as the purity of the input state decreases.

Despite the exponential improvement with respect to witness efficiency, the WOM with optimal witnesses as outcomes is still not efficient enough to detect highly mixed entangled states. In that case, entanglement detection through efficient state tomography may be a better choice. To estimate the number of copies required in state tomography to reach sufficient accuracy for determining the separability of an input state, we recall that the radius of the largest separable ball in the space of statistical operators is $\sqrt{1/(d-1)d} \approx 1/d$ [33], the dimension of the state space is $d^2-1 \approx d^2$, and a reasonable estimate of accuracy requirement is

$$\begin{aligned} E(\|\hat{\rho}-\rho\|_{\mathrm{HS}}) &\sim 1/d^3, \\ E(\|\hat{\rho}-\rho\|_{\mathrm{HS}}^2) &\sim 1/d^6; \end{aligned} \tag{15}$$

for state tomography with a SIC POM this means that one needs to measure about $N \sim d^8$ copies. Clearly, it is difficult to detect an entangled state that has separable states in its vicinity.

III. TWO-QUBIT WITNESS OPERATOR MEASUREMENTS

A. Generic situation

For illustration, we now turn to the situation of two qubits ($d_1 = d_2 = 2, d = 4$) and analyze the difference between rank-one POMs and WOMs with optimal witnesses. In the case of two qubits, there are only decomposable witnesses [22]. Since every rank-two subspace contains at least one product ket [34,35], the only candidates of optimal witnesses (in the usual sense) are the partial transposes of some entangled pure states, and they are indeed optimal [22]. We first compare pure-state witnesses and witnesses constructed from their partial transposes.

By a suitable local unitary transformation, any pure entangled two-qubit state can be turned into the form $\rho(\alpha) = |\Psi\rangle\langle\Psi|$ with

$$\begin{aligned} |\Psi\rangle &= |00\rangle\cos\alpha + |11\rangle\sin\alpha, \\ 0 < \alpha &= \frac{1}{2}\arcsin(q) \leqslant \frac{\pi}{4}, \quad p = \sqrt{1-q^2}, \end{aligned} \tag{16}$$

where q is the concurrence of the pure state, p is the length of the Bloch vector of each reduced single-qubit state, and α is introduced for convenience. Because any one of the three parameters q, p, or α specifies $|\Psi\rangle$, we will use them interchangeably to denote a pure state with given concurrence. Two pure states are *aligned* if there exists a local unitary transformation which turns both of them into the form of Eq. (16), possibly with different parameters α_1 and α_2 or, equivalently, if the Bloch vectors of their reduced statistical operators are parallel and the fidelity between them is equal to that between their reduced statistical operators. The fidelity between two pure states with given concurrences obtains the maximum value of $\cos(\alpha_1-\alpha_2)^2$ when they are aligned.

When $\alpha < \frac{\pi}{4}$, the witness constructed from the pure state is not optimal, and Lemma 2 in Ref. [22] implies that the witnesses constructed from its aligned pure states with larger concurrence are finer. The probability that a random pure entangled state is detected by a pure-state witness with concurrence q is $[\frac{1}{2}(1-p)]^3$ as follows from Eq. (7), with the maximum probability of $\frac{1}{8}$ achieved for a Bell state witness. For two aligned pure states $\rho(\alpha_1)$ and $\rho(\alpha_2)$, state $\rho(\alpha_2)$ can detect state $\rho(\alpha_1)$ if

$$0 < \mathrm{Tr}\{\rho(\alpha_1)\rho(\alpha_2)\} - \cos(\alpha_2)^2 = \sin(\alpha_1)\sin(2\alpha_2-\alpha_1), \tag{17}$$

which means $2\alpha_2 > \alpha_1$. A counterintuitive consequence of this observation is that, when $\alpha \leqslant \frac{\pi}{8}$ or equivalently $q \leqslant \frac{1}{2}\sqrt{2}$, $\rho(\alpha)$ cannot detect any Bell states—although one might naturally expect that these maximally entangled states are the ones that are most easily detected. An analogous phenomenon also exists for $d > 4$: If one Schmidt coefficient of an entangled pure state is particularly large, then it cannot detect any maximally entangled states.

The four eigenvalues of

$$\begin{aligned} \rho(\alpha)^{T_2} &= |00\rangle\frac{1+p}{2}\langle 00| + |11\rangle\frac{1-p}{2}\langle 11| \\ &\quad + (|01\rangle+|10\rangle)\frac{q}{4}(\langle 01|+\langle 10|) \\ &\quad - (|01\rangle-|10\rangle)\frac{q}{4}(\langle 01|-\langle 10|) \end{aligned} \tag{18}$$

are $\frac{1}{2}(1 \pm p), \pm\frac{1}{2}q$, so that statistical operators that are optimal witnesses related to $\rho(\alpha)$ can be constructed as

$$\rho_w(\alpha) = \frac{\frac{1+p}{2}\mathbf{1} - \rho(\alpha)^{T_2}}{1+2p}. \tag{19}$$

Such a $\rho_w(\alpha)$ has concurrence $q/(1+2p)$ and negativity $(1-p)/(1+2p)$, and $\mu = (1+p)/(2+4p)$ is its witness threshold. These states are quite special, they are Bell states if $\alpha = \frac{\pi}{4}$ and, for other values of α, they are maximally entangled mixed states in the sense of Ref. [36]; that is, their concurrence cannot be increased by any global unitary transformation.

When $\rho_w(\alpha_1)$ and $\rho_w(\alpha_2)$ are aligned—we call $\rho_w(\alpha_1)$ and $\rho_w(\alpha_2)$ aligned if $\rho(\alpha_1)$ and $\rho(\alpha_2)$ are aligned—$\rho_w(\alpha_2)$ can detect $\rho_w(\alpha_1)$ if

$$0 > \mathrm{Tr}\{\rho_w(\alpha_1)\rho(\alpha_2)^{T_2}\} = \frac{\sin(\alpha_2)\sin(\alpha_2 - 2\alpha_1)}{1+2p_1}, \tag{20}$$

which means $2\alpha_1 > \alpha_2$, a condition that is dual to the condition of Eq. (17). If $\alpha_1 > \frac{\pi}{8}$, so that $\rho_w(\alpha_1)$ could be any Bell state, then $\rho_w(\alpha_1)$ can be detected by any aligned optimal witness $\rho_w(\alpha_2)$. On the other hand, an optimal witness $\rho_w(\alpha_2)$ cannot detect any state $\rho_w(\alpha_1)$ with $\alpha_1 \leqslant \alpha_2/2$; in particular, Bell states cannot detect any states $\rho_w(\alpha_1)$ with $\alpha_1 \leqslant \frac{\pi}{8}$. As $\alpha, q \to 0$, the $\rho_w(\alpha)$s approach separable states, and they can detect fewer and fewer entangled states but, surprisingly, they are still the best witnesses for detecting even more weakly entangled states $\rho_w(\alpha')$ with $\alpha' \leqslant \alpha$. As a consequence, an infinite number of witnesses are needed to detect all entangled states, if we rely on each witness separately. This illustrates the remark at the end of the first paragraph in Sec. I.

Taking a SIC POM [27–29] as an example, we shall now show the sharp difference between a rank-one POM and a WOM consisting of optimal witnesses. Most known examples of SIC POMs are constructed from fiducial states under the action of the generalized Pauli group, or Heisenberg-Weyl group. In dimension four, one such fiducial state is given in Eq. (147) of [29]. In the case of two qubits, if we choose the standard product basis $|00\rangle, |01\rangle, |10\rangle, |11\rangle$ for Appleby's $|e_0\rangle, |e_1\rangle, |e_2\rangle, |e_3\rangle$, then the SIC POM thus constructed has the very peculiar property that the concurrence of all fiducial states are the same, namely, $\sqrt{2/5}$ [37]. We note that these fiducial states are typical in the sense that their squared concurrence equals the average squared concurrence of all pure two-qubit states [24].

With the aid of the general procedure presented in Sec. II C, we can construct a WOM with optimal witnesses as outcomes. Moreover, the WOM is literally also SIC, except that the outcomes are not pure. Curves "a" and "b" in Fig. 2 show the entanglement detection ratio for the SIC POM and the WOM for both pure and mixed states, with each witness acting separately. There is a huge improvement of the WOM over the SIC POM with respect to witness efficiency, especially for the detection of mixed states. The SIC POM cannot detect any Bell states, because the concurrence of each fiducial state is less than $\sqrt{1/2}$; yet, the detection ratio of the WOM is virtually 100% when the concurrence of the input state exceeds 0.75. If mixed states are distributed according to the HS measure, the overall detection probability of the WOM is about 13%; for states with concurrence larger than $\frac{1}{2}$, the probability is about 84%.

The improvement in the witness efficiency comes at the price of a much lower tomographic efficiency, the mean square error achieved by the SIC POM and the WOM given in Eqs. (13) and (14) now reads

$$\begin{aligned} E(\|\hat{\rho} - \rho\|^2_{\mathrm{HS}})_{\mathrm{POM}} &= \frac{1}{N}(19 - \mathrm{Tr}\{\rho^2\}), \\ E(\|\hat{\rho} - \rho\|^2_{\mathrm{HS}})_{\mathrm{WOM}} &= \frac{1}{N}(64 + 15\sqrt{15} - \mathrm{Tr}\{\rho^2\}), \end{aligned} \tag{21}$$

where we have inserted the value $\lambda_{\max} = \frac{1}{2}(1+\sqrt{3/5})$. The reduction in the tomographic efficiency of the WOM compared to the SIC POM is roughly by a factor of $\frac{2}{13}$.

B. Optimal witnesses for qubits carried by photons

1. Two photons carry one polarization qubit each

For polarization qubits of photons, witnesses of the kind (19) can be measured with existing linear-optics devices. We identify the labels "0" and "1" with vertical and horizontal polarization, respectively, $|01\rangle \,\hat{=}\, |\mathrm{VH}\rangle$, for instance. In particular, the setup that was used for an incomplete Bell-basis measurement in the teleportation experiment of Bouwmeester *et al.* [2] is suitable after a slight modification that enables one to identify states that have both photons in the same polarization state; see Fig. 1.

Owing to their specific signatures (see Table I), this device can then discriminate between the two Bell states with kets $(|\mathrm{VH}\rangle \pm |\mathrm{HV}\rangle)/\sqrt{2}$ and the two product states with kets $|\mathrm{VV}\rangle$ and $|\mathrm{HH}\rangle$, which constitute the four eigenstates of the witness operator of Eq. (19). After measuring sufficient copies of the two-qubit state of interest and thereby determining the probabilities (actually, the relative frequencies) of occurrence for these four two-qubit states, we calculate the expectation

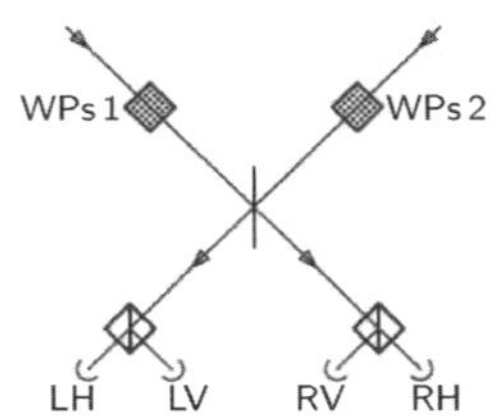

FIG. 1. A linear-optics setup that offers an experimental implementation of the optimal witness of Eq. (19) for polarization qubits. Two photons that are indistinguishable by their spatial-spectral properties are simultaneously incident on a half-transparent mirror, photon 1 from the left and photon 2 from the right. They carry one polarization qubit each, with their unknown two-qubit state to be analyzed. After being transmitted through, or reflected off, the half-transparent mirror, the photons are detected behind polarizing beam splitters that reflect vertically polarized photons and transmit horizontally polarized ones. The four detectors LH, LV, RV, and RH must be able to discriminate between one-photon and two-photon events. The four eigenstates of the witness operator are distinguished by different signatures; see Table I. By letting the photons pass through polarization changing wave plates in the input ports, labeled by WPs1 and WPs2, one realizes other witnesses that differ from the witness of Eq. (19) by local unitary transformations.

TABLE I. Signatures of the eigenvalues of the optimal witness $\rho_w(\alpha)$ of Eq. (19) for polarization qubits ($0 \mathrel{\hat{=}} \mathrm{V}$, $1 \mathrel{\hat{=}} \mathrm{H}$), detected by the setup in Fig. 1, with no wave plates in the input ports. As a consequence of the Hong-Ou-Mandel effect [38], the cases (1,0,0,1) and (0,1,1,0) do not occur.

Eigenvalue	Eigenket	Counts (LH,LV,RV,RH)
0	$\|\mathrm{VV}\rangle$	(0,2,0,0) or (0,0,2,0)
$\frac{p}{1+2p}$	$\|\mathrm{HH}\rangle$	(2,0,0,0) or (0,0,0,2)
$\frac{1+p-q}{2+4p}$	$(\|\mathrm{VH}\rangle + \|\mathrm{HV}\rangle)/\sqrt{2}$	(1,1,0,0) or (0,0,1,1)
$\frac{1+p+q}{2+4p}$	$(\|\mathrm{VH}\rangle - \|\mathrm{HV}\rangle)/\sqrt{2}$	(1,0,1,0) or (0,1,0,1)

value of $\rho_w(\alpha)$ as a weighted sum of its eigenvalues and then learn whether this witness conclusively identifies the two-qubit state as a nonseparable state. If the outcome is inconclusive, we measure a second witness that is defined by suitably chosen wave plates WPs1 and WPs2 in the input ports of the measuring device of Fig. 1. If the second witness gives an inconclusive result as well, we measure a third witness, and so forth.

But, contrary to the general situation discussed in Sec. II, we do not need to continue this process until as many as 16 witnesses are measured because each witness measurement gives us four relative frequencies and not only a single expectation value—we are not just measuring the witness, we are measuring the *witness basis*. Put differently, when measuring one witness, we get estimates of three two-qubit observables. For the witness of Eq. (19), identified as witness 1 in Table II, these observables are $Z\mathbf{1} + \mathbf{1}Z$, ZZ, and $XX + YY$. Here, X, Y, and Z are the standard Pauli operators for a polarization qubit,

$$\begin{aligned} X &= |\mathrm{H}\rangle\langle\mathrm{V}| + |\mathrm{V}\rangle\langle\mathrm{H}|, \\ Y &= |\mathrm{H}\rangle\mathrm{i}\langle\mathrm{V}| - |\mathrm{V}\rangle\mathrm{i}\langle\mathrm{H}|, \\ Z &= |\mathrm{V}\rangle\langle\mathrm{V}| - |\mathrm{H}\rangle\langle\mathrm{H}|, \end{aligned} \tag{22}$$

and a product such as $Z\mathbf{1}$, for instance, means that Z is applied on the first qubit and the identity $\mathbf{1}$ on the second qubit.

Indeed, the six witnesses listed in Table II make up an IC POM—together they provide estimates for all 15 expectation values that uniquely specify the two-qubit state. Clearly, the reduction from 16 to 6 witnesses is of great importance for practical implementations.

TABLE II. The six witnesses of the kind depicted in Fig. 1 that enable full tomography of the two-qubit state. The second and third columns list the unitary operators U_1 and U_2 that describe the effect of the wave plates WPs1 and WPs2, respectively, on the polarization of the incoming photons. The fourth column states the three two-qubit operators whose expectation values are determined when the eigenstate basis of the corresponding witness is measured.

Witness	U_1	U_2	Observables
1	$\mathbf{1}$	$\mathbf{1}$	$Z\mathbf{1}+\mathbf{1}Z,\ ZZ,\ XX+YY$
2	$\mathbf{1}$	X	$Z\mathbf{1}-\mathbf{1}Z,\ ZZ,\ XX-YY$
3	$C^\dagger$	C	$X\mathbf{1}+\mathbf{1}Y,\ XY,\ YZ+ZX$
4	$C^\dagger$	XC	$X\mathbf{1}-\mathbf{1}Y,\ XY,\ YZ-ZX$
5	C	$C^\dagger$	$Y\mathbf{1}+\mathbf{1}X,\ YX,\ ZY+XZ$
6	C	$XC^\dagger$	$Y\mathbf{1}-\mathbf{1}X,\ YX,\ ZY-XZ$

The wave plates in the input ports implement the unitary transformations U_1 and U_2 on the polarization of photons 1 and 2, respectively, and so the incoming two-photon state ρ_{2ph} is transformed in accordance with

$$\rho_{\mathrm{2ph}} \rightarrow U_1 \otimes U_2 \rho_{\mathrm{2ph}} (U_1 \otimes U_2)^\dagger \tag{23}$$

before the photons arrive at the half-transparent mirror. In effect, then, the eigenstate basis of the transformed witness

$$(U_1 \otimes U_2)^\dagger \rho_w(\alpha) U_1 \otimes U_2 \tag{24}$$

is measured rather than that of the original witness $\rho_w(\alpha)$. The unitary operator C that appears in Table II, permutes the Pauli operators cyclically,

$$CX = YC, \quad CY = ZC, \quad CZ = XC; \tag{25}$$

it is implemented by wave plates that yield the polarization changes

$$\begin{aligned} |\mathrm{V}\rangle &\rightarrow C|\mathrm{V}\rangle = (|\mathrm{V}\rangle + |\mathrm{H}\rangle)/\sqrt{2}, \\ |\mathrm{H}\rangle &\rightarrow C|\mathrm{H}\rangle = \mathrm{i}(|\mathrm{H}\rangle - |\mathrm{V}\rangle)/\sqrt{2}, \end{aligned} \tag{26}$$

possibly accompanied by an irrelevant overall phase factor.

We note that there are in fact three pairs of witnesses: After measuring witnesses 1 and 2, we know the expectation values of the five observables $Z\mathbf{1}$, $\mathbf{1}Z$, XX, YY, and ZZ; and each of the witness pairs 3,4 and 5,6 gives another five expectation values. Tomography of this kind requires the measurement of six orthogonal bases, one more than the five bases that would be required in a minimal scheme, because each witness pair has redundant information about one observable (namely, ZZ for witnesses 1 and 2). This feature is a consequence of the structure of the witness bases: two Bell states plus two product states.

Another twist is this: Since the device of Fig. 1 is the same for all values of the witness parameter α, we are in fact measuring a whole one-parameter family of witnesses in one go. It follows that, rather than comparing the expectation value of one $\rho_w(\alpha)$ with its threshold value, we can apply the more stringent criterion

$$\max_\alpha \left\{ \langle \rho_w(\alpha) \rangle - \frac{1+p}{2+4p} \right\} \leqslant 0 \tag{27}$$

or

$$\min_\alpha \langle \rho(\alpha)^{T_2} \rangle \geqslant 0, \tag{28}$$

which holds for all separable states and is violated by the nonseparable states that are detected by this family of witnesses. In Eq. (28), the values of α are not restricted to the range specified in Eq. (16), and the minimization over α implies that

$$4\nu_1\nu_2 \geqslant (\nu_3 - \nu_4)^2 \tag{29}$$

is true for all separable states. Here, ν_1 and ν_2 are the relative frequencies with which the two product states are observed, while ν_3 and ν_4 are the relative frequencies of the two Bell states. In summary, then, the family of linear witnesses $\{\rho_w(\alpha)\}$

is equivalent to the witness basis measurement (WBM) specified by the device of Fig. 1 in conjunction with the quadratic WBM criterion (29).

Figure 2 shows the entanglement detection ratio for the SIC POM, the WOM (with two different measurement schemes), and the six witnesses in Table II for both pure and mixed states. The witness efficiency of the six witnesses and the WBM criterion is better than that of the WOM with the conventional measurement scheme. Moreover, the efficiency of tomography with the six witness bases is comparable with that of the SIC POM. If the six witness bases are measured with equal probability, and the standard reconstruction scheme is adopted, the mean square can be computed with the methods described in Refs. [31,32]. For the completely mixed state, the result is

$$E\left(\|\hat{\rho}-\rho\|_{\mathrm{HS}}^2\right)_{\mathrm{WBM}}=\frac{81}{4N}, \tag{30}$$

and for a generic input state the mean square error is also close to this value and it can be reduced with the maximum likelihood reconstruction scheme (see Refs. [39–41]). Compared with Eq. (21), the tomographic efficiency of the WBM composed of the six witness bases is close to that of the SIC POM and much better than that of the WOM. This is a consequence of using expectation values of rank-one operators, that is, probabilities, rather than expectation values of the rank-four witness operators for the state reconstruction. In addition, there is a certain amount of redundancy ($6\times 4=24$ probabilities rather than 16 expectation values) in the WBM scheme, and this also improves the accuracy of the tomography.

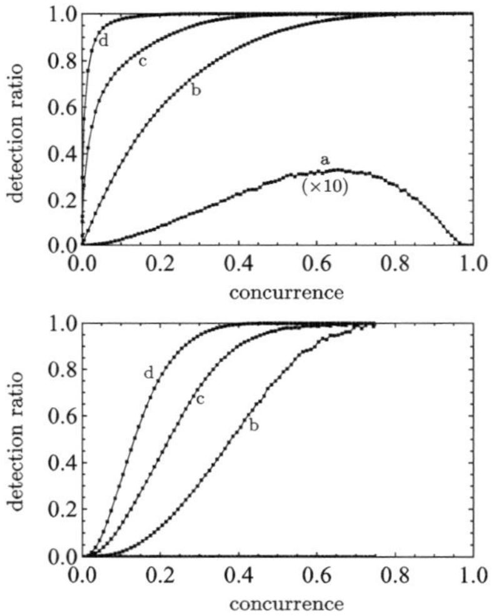

FIG. 2. Simulation results for the entanglement detection for two-qubit states. The various curves show the detection ratio for the SIC POM (curve a) and the optimal WOM (curves b), with the outcomes used as standard linear witnesses, as well as for the witness-basis measurements for the six witnesses in Table II (curves c) and the optimal WOM (curves d), measured with the setup in Fig. 1 with the quadratic WBM criterion (29). The top plot is for randomly chosen pure states, the bottom plot is for randomly chosen mixed states. The detection ratio of the SIC POM (curve a in the top plot) has been multiplied by 10 for ease of viewing; note that the SIC POM does not detect states with a concurrence close to 1. In the bottom plot, which displays the most relevant data, 10^7 mixed states were generated randomly according to the HS measure, applying a procedure described in Ref. [24]; the SIC POM gave virtually no counts for mixed states. Curves b, c, and d stop at concurrence 0.75 because there are too few counts for reliable statistics for larger concurrence values.

2. *One photon carries a polarization qubit and a path qubit*

Another situation is that of a single photon carrying both a polarization qubit and a path qubit. All unitary transformations on this two-qubit system can be realized with linear optical elements [42]. For each of the six witnesses in Table II, there is such a transformation that maps the four eigenstates of the witness onto the pure product states with kets $|00\rangle$, $|01\rangle$, $|10\rangle$, and $|11\rangle$, which are then easily distinguished. Standard photodetectors will serve the purpose here because there is no need to discriminate between one-photon and two-photon events. And for the same reasons as in Sec. III B1, these six witnesses are sufficient for full state tomography.

IV. WITNESS BASIS MEASUREMENTS

This strategy of measuring witness bases, rather than determining solely the expectation values of the witness observables, can be applied as well to entangled states of more complicated quantum degrees of freedom. This is unexplored territory that deserves further study, and we will here be content with offering a few remarks.

For $d_1=d_2=\sqrt{d}$, the two-qudit basis composed of the d_1 product kets $|ii\rangle$ and the $(d_1-1)d_1$ entangled kets $(|ij\rangle \pm |ji\rangle)/\sqrt{2}$ with $i<j$ is a witness basis; see (1)–(6). Upon denoting by ν_{ii} and $\nu_{ij}^{(\pm)}$ the relative frequencies with which the respective detectors fire, the qudit version of the qubit criterion (29) reads

$$\mathcal{W}\geqslant 0, \tag{31}$$

obeyed by all separable states and violated by the detected entangled states, where $\mathcal{W}$ is the real symmetric $d_1\times d_1$ matrix defined by

$$\mathcal{W}_{ij}=\mathcal{W}_{ji}=\begin{cases}\nu_{ii} & \text{if } i=j,\\ \frac{1}{2}\left(\nu_{ij}^{(+)}-\nu_{ij}^{(-)}\right) & \text{if } i<j.\end{cases} \tag{32}$$

The minimal number of bases of this kind that one would need to measure for tomographic completeness is currently unknown when $d>4$. Preliminary studies have established that 11 witness bases are surely enough for two qutrits ($d=3^2$), rather than 81 witnesses. We leave it at that and will return to this matter on another occasion.

V. SUMMARY

In summary, we have introduced WOMs and developed a systematic method of constructing a WOM with optimal witnesses as outcomes from a rank-one POM. This WOM can provide exponential improvement with respect to the witness efficiency in a high-dimensional Hilbert space, at the price of a reduction in the tomographic efficiency by a factor of order d_2^{-2}. The trade-off between the witness efficiency and the tomographic efficiency is quite common—this reflects the intrinsic difficulty of entanglement detection.

For the situation of two entangled polarization qubits carried by photons ($d^2 = 16$), we have given a detailed description of how such optimized witnesses can be implemented in laboratory experiments with existing technology. In fact, the experimental device is such that one does not need to measure 16 witnesses before full state tomography is possible, six settings of the apparatus are enough to collect sufficient data for the reconstruction of the two-qubit state under investigation. This is the benefit of actually measuring the eigenstate basis of the witness rather than determining solely the expectation value of the witness observable. Another bonus is the much increased tomographic efficiency.

This strategy of measuring witness bases can also be applied to entangled states of more complicated quantum systems. The signature of entanglement for a witness basis is known, but many details require further study.

ACKNOWLEDGMENTS

We are grateful for valuable discussions with Artur Ekert and Karol Życzkowski. H.Z. thanks Maciej Lewenstein and Lin Chen for assistance and advice. This work is supported by the National Research Foundation and the Ministry of Education, Singapore.

[1] C. H. Bennett, G. Brassard, C. Crepeau, R. Jozsa, A. Peres, and W. K. Wootters, Phys. Rev. Lett. **70**, 1895 (1993).
[2] D. Bouwmeester, J.-W. Pan, K. Mattle, M. Eible, H. Weinfurter, and A. Zeilinger, Nature (London) **390**, 575 (1997).
[3] D. Boschi, S. Branca, F. De Martini, L. Hardy, and S. Popescu, Phys. Rev. Lett. **80**, 1121 (1998).
[4] A. K. Ekert, Phys. Rev. Lett. **67**, 661 (1991).
[5] R. Jozsa, in *The Geometric Universe: Science, Geometry, and the Work of Roger Penrose*, edited by S. A. Huggett, L. J. Mason, K. P. Tod, S. Tsou, and N, M. J. Woodhouse (Oxford University Press, Oxford, UK, 1998); also available as e-print arXiv:quant-ph/9707034.
[6] M. Horodecki, P. Horodecki, and R. Horodecki, Phys. Lett. A **223**, 1 (1996).
[7] M. Bourennane, M. Eibl, C. Kurtsiefer, S. Gaertner, H. Weinfurter, O. Gühne, P. Hyllus, D. Bruß, M. Lewenstein, and A. Sanpera, Phys. Rev. Lett. **92**, 087902 (2004).
[8] B.G.E. thanks Artur Ekert for asking this question in Vienna in 2000.
[9] O. Gühne and G. Tóth, Phys. Rep. **474**, 1 (2009).
[10] R. Horodecki, P. Horodecki, M. Horodecki, and K. Horodecki, Rev. Mod. Phys. **81**, 865 (2009).
[11] E. Prugovečki, Int. J. Theor. Phys. **16**, 321 (1977).
[12] P. Busch, Int. J. Theor. Phys. **30**, 1217 (1991).
[13] G. M. D'Ariano, P. Perinotti, and M. F. Sacchi, J. Opt. B: Quantum Semiclass. Opt. **6**, S487 (2004).
[14] POM, with its emphasis on "probability" and "measurement" is preferable quantum physics jargon. The corresponding popular mathematical term POVM (positive-operator-valued measure) makes reference to measure theory.
[15] A. Peres, Phys. Rev. Lett. **77**, 1413 (1996).
[16] P. Horodecki, Phys. Lett. A **232**, 333 (1997).
[17] O. Rudolph, J. Phys. A: Math. Gen. **33**, 3951 (2000).
[18] K. Chen, L.-A. Wu, Quantum Inf. Comput. **3**, 193 (2003).
[19] O. Gühne, P. Hyllus, O. Gittsovich, and J. Eisert, Phys. Rev. Lett. **99**, 130504 (2007).
[20] M. Lewenstein and A. Sanpera, Phys. Rev. Lett. **80**, 2261 (1998).
[21] F. Hulpke and D. Bruß, J. Phys. A: Math. Gen. **38**, 5573 (2005).
[22] M. Lewenstein, B. Kraus, J. I. Cirac, and P. Horodecki, Phys. Rev. A **62**, 052310 (2000).
[23] M. Žnidarič, T. Prosen, G. Benenti, and G. Casati, J. Phys. A: Math. Theor. **40**, 13787 (2007).
[24] K. Życzkowski and H. J. Sommers, J. Phys. A: Math. Gen. **34**, 7111 (2001).
[25] I. Bengtsson and K. Życzkowski, *Geometry of Quantum States, An Introduction to Quantum Entanglement* (Cambridge University Press, Cambridge, UK, 2006).
[26] M. Lewenstein (private communication).
[27] G. Zauner, Ph.D. thesis, University of Vienna, 1999. Available online at [http://www.mat.univie.ac.at/~neum/papers/physpapers.html].
[28] J. M. Renes, R. Blume-Kohout, A. J. Scott, and C. M. Caves, J. Math. Phys. **45**, 2171 (2004).
[29] D. M. Appleby, J. Math. Phys. **46**, 052107 (2005).
[30] A. J. Scott and M. Grassl, J. Math. Phys. **51**, 042203 (2010).
[31] A. J. Scott, J. Phys. A: Math. Gen. **39**, 13507 (2006).
[32] H. Zhu, Y. S. Teo, and B.-G. Englert (unpublished).
[33] L. Gurvits and H. Barnum, Phys. Rev. A **66**, 062311 (2002).
[34] S. Hill and W. K. Wootters, Phys. Rev. Lett. **78**, 5022 (1997).
[35] B.-G. Englert and N. Metwally, in *Mathematics of Quantum Computation*, edited by R. K. Brylinski and G. Chen (Chapman & Hall/CRC Press, Boca Raton, 2002).
[36] F. Verstraete, K. Audenaert, and B. De Moor, Phys. Rev. A **64**, 012316 (2001).
[37] H. Zhu, Y. S. Teo, and B.-G. Englert (unpublished).
[38] C. K. Hong, Z. Y. Ou, and L. Mandel, Phys. Rev. Lett. **59**, 2044 (1987).
[39] Z. Hradil, Phys. Rev. A **55**, R1561 (1997).
[40] Z. Hradil, J. Řeháček, J. Fiurášek, and M. Ježek, in *Quantum State Estimation*, edited by M. G. A. Paris and J. Řeháček (Springer, Berlin/New York, 2004), *Lecture Notes in Physics*, Vol. 649.
[41] Y. S. Teo, H. Zhu, and B.-G. Englert, Opt. Commun. **283**, 724 (2010).
[42] B.-G. Englert, C. Kurtsiefer, and H. Weinfurter, Phys. Rev. A **63**, 032303 (2001).

Arun Sehrawat

Departamento de Física, Universidad de Guadalajara, Jalisco, Mexico
Centre for Quantum Technologies, National University of Singapore, Singapore

Paper Title: **Optimal error regions for quantum state estimation**

Reference: New Journal of Physics **15**, 123026 (2013)

Personal message:

I would like to contribute our paper with Berge, *"Optimal error regions for quantum state estimation"*, New J. Phys. 15, 123026 (2013) to this Festschrift. I choose this article because the main idea that comes from Berge is very simple, and it still fascinates me. During the progress of this article, I learned many fine details of quantum state estimation. I cannot forgot all those discussions with the whole group which settled all the concepts used and introduced in this article.

Greetings and lots of well wishes to our dear Berge!

New Journal of Physics
The open access journal for physics

Optimal error regions for quantum state estimation

Jiangwei Shang[1,6], Hui Khoon Ng[1,2,4], Arun Sehrawat[1,5], Xikun Li[1] and Berthold-Georg Englert[1,3]

[1] Centre for Quantum Technologies, National University of Singapore, 3 Science Drive 2, Singapore 117543, Singapore

[2] Applied Physics Lab, DSO National Laboratories, 20 Science Park Drive, Singapore 118230, Singapore

[3] Department of Physics, National University of Singapore, 2 Science Drive 3, Singapore 117542, Singapore

E-mail: jiangwei.shang@quantumlah.org

New Journal of Physics **15** (2013) 123026 (26pp)
Received 20 July 2013
Published 13 December 2013
Online at http://www.njp.org/
doi:10.1088/1367-2630/15/12/123026

Abstract. An estimator is a state that represents one's best guess of the actual state of the quantum system for the given data. Such estimators are points in the state space. To be statistically meaningful, they have to be endowed with error regions, the generalization of error bars beyond one dimension. As opposed to standard ad hoc constructions of error regions, we introduce the maximum-likelihood region—the region of largest likelihood among all regions of the same size—as the natural counterpart of the popular maximum-likelihood estimator. Here, the size of a region is its prior probability. A related concept is the smallest credible region—the smallest region with pre-chosen posterior probability. In both cases, the optimal error region has constant likelihood on its boundary. This surprisingly simple characterization permits concise reporting of the error regions, even in high-dimensional problems. For illustration, we identify optimal error regions for single-qubit and two-qubit states from computer-generated data that simulate incomplete tomography with few measured copies.

[4] Current address: Yale-NUS College, 6 College Avenue East, Singapore 138614, Singapore.

[5] Current address: Departamento de Física, Universidad de Guadalajara, 44420 Guadalajara, Jalisco, Mexico.

[6] Author to whom any correspondence should be addressed.

Contents

1. Introduction

Quantum state estimation (see e.g. [1]) is central to many, if not all, tasks that process quantum information. The characterization of a source of quantum carriers, the verification of the properties of a quantum channel and the monitoring of a transmission line used for quantum key distribution—all three require reliable quantum state estimation, to name just the most familiar examples.

In the typical situation that we are considering, several independently and identically prepared quantum-information carriers are measured one-by-one by an apparatus that realizes a probability-operator measurement (POM), suitably designed to extract the wanted information. The POM has a number of outcomes, with detectors that register individual information carriers (photons in the majority of current experiments), and the data consist of the observed sequence of detection events ('clicks')[7].

The quantum state to be estimated is described by a statistical operator, the *state*, and the data can be used to determine an *estimator* for the state—another state that, so one hopes, approximates the actual state well. There are various strategies for finding such an estimator. Thanks to the efficient methods that Hradil, Řeháček, and their collaborators developed for calculating maximum-likelihood estimators (MLEs, reviewed in [2]; see also [3]), MLEs have

[7] It is advisable to verify that the observed sequence does not have systematic correlations that speak against the assumption of independently and identically prepared quantum-information carriers.

become the estimators of choice. For the given data, the MLE is the state for which the data are more likely than for any other state.

Whether one prefers the MLE or a point estimator found by another method, the data have statistical noise and, therefore, one needs to supplement the point estimator with error bars of some sort—*error regions*, more generally, for higher-dimensional problems. Many recipes, which are often ad hoc in nature, have been proposed for attaching a vicinity of states to an estimator. These usually rely on having a lot of data [4, 5], involve data resampling [6] or consider all data that one might have observed [7, 8]. By contrast, we systematically construct error regions from the data *actually* observed.

For this purpose, we propose *maximum-likelihood regions* (MLRs) and *smallest credible regions* (SCRs). These are regions in the space of quantum states (more precisely: in the reconstruction space; see section 2). The MLR is that region of pre-chosen size for which the given data are more likely than for any other region of the same size. The SCR is the smallest region with pre-chosen credibility—the credibility of a region being the probability of finding the actual state in the region, conditioned on the data (see e.g. [9]). Whether one chooses the MLR or the SCR as the optimal error region depends on the situation at hand.

Central to both concepts is the notion of the *size* of a region. In fact, some notion of size must underlie *any* useful definition of error regions, since one usually aims at reporting an error region that is not unnecessarily large—a judgement that can only be made with a suitable concept of size. We agree with Evans *et al* [10] that, in the context of state estimation, it is most natural to measure the size of a region by its prior—before any data are at hand—probability of finding the actual state in the region: regions with the same prior probability are considered as having the same size. Hence, the size of a region expresses the relative importance of that region of states.

The identification 'size ≡ prior probability' is technically possible because both quantities simply add when disjointed regions are combined into a single region. While for some tasks one prefers not to assign a prior[8], since state estimation expresses our best attempt at guessing the state, any prior information we possess should be taken into account in the estimation process alongside the data. Much guidance on choosing priors can be found in the standard statistics literature; in appendix A, we provide a summary that focuses on points relevant in quantum contexts. Ultimately, the choice of prior is up to the user, but it should be *consistent*: the estimation results should be dominated by the data, not the prior, if many copies of the state are measured.

As we show below, the problems of finding the MLR and the SCR are duals of each other. In both cases, the optimal regions contain all states for which the likelihood of the data exceeds a threshold value. This provides a concise way of communicating one's uncertainty of the estimate. That the optimal error regions possess such a simple description is surprising, since our construction imposes no restriction on the shape of the regions to be considered. The shape of the optimal regions are uniquely determined by the likelihood function, in sharp contrast to the arbitrariness in the shape of a confidence region (see appendix B), a concept that is the subject of recent discussion [7, 8]. Yet the two are not unrelated: our SCRs provide natural starting points for the construction of the confidence regions considered in [7].

[8] For tasks like quantum key distribution, one may want to adopt a different attitude and assume the worst possible scenario rather than relying on one's information to assign a prior. Then, the confidence regions of [7, 8] are appropriate as error regions.

While the chosen MLR or SCR depends on the prior, the set of candidate regions is prior-independent: it depends only on the likelihood function for the given data. Also reassuring is the fact that every MLR or SCR is a small vicinity of the MLE, in the respective limits of small size or small credibility. This is reminiscent of standard ellipsoidal error regions constructed around the MLE, which are, however, applicable only in the limit of a large amount of data when the central limit theorem can be invoked and the uncertainty can be characterized by the Fisher information (see, for instance, [4]).

Here is a brief outline of the paper. We set the stage in section 2 where we introduce the reconstruction space, discuss the size of a region, and define the various joint and conditional probabilities. Equipped with these tools, we then formulate in section 3 the optimization problems that identify the MLRs and SCRs and find their solutions in section 4. We illustrate the matter by simulated single-qubit and two-qubit experiments in sections 5 and 6, and close with an outlook. Two appendices supply additional material: a guide for choosing priors, and a comparison with confidence regions.

2. Setting the stage

2.1. Reconstruction space

The outcomes Π_k, $k = 1, 2, \ldots, K$, of the POM, with which the data are acquired, are positive ($\Pi_k \geqslant 0$) Hilbert-space operators with $\sum_{k=1}^{K} \Pi_k = 1$. If the state ρ describes the system, the probability p_k that the kth detector clicks is

$$p_k = \mathrm{tr}\{\Pi_k \rho\} = \langle \Pi_k \rangle \quad \text{(Born rule).} \tag{1}$$

The positivity of ρ and its normalization to unit trace ensure $p_k \geqslant 0$ for all k and $\sum_{k=1}^{K} p_k = 1$. Probabilities $p = (p_1, p_2, \ldots, p_K)$ for which there is a state ρ such that (1) holds are *permissible* probabilities. They make up the *probability space*. The probability space for a K-outcome POM is usually smaller than that of a tossed K-sided die because not all positive p_k with unit sum are permissible. The quantum nature of the state estimation problem enters *only* in these additional restrictions on p: quantum state estimation is standard statistical state estimation with constraints of quantum-mechanical origin. The rich methods of statistical inference immediately apply, modified where necessary to account for the restricted probability space.

Whereas p is uniquely determined by ρ through (1), the converse is true only if the POM is informationally complete. In any case, there is always a *reconstruction space* $\mathcal{R}_0$, a set of ρ that contains exactly one ρ for each permissible p, consistent with the Born rule. If there is more than one reconstruction space, it does not matter which one we choose. As an example, consider a harmonic oscillator with its infinite-dimensional state space. If the POM has two outcomes, with p_1 equal to the probability of finding the oscillator in its ground state and $p_2 = 1 - p_1$, the reconstruction space is the set of convex combinations of the projector to the ground state and another state with no ground-state component. In this situation, there are very many reconstruction spaces to choose from because *any* other state serves the purpose, and all one can infer from the data is an estimate of the ground-state probability.

Since the probability space is unique, while there can be many different reconstruction spaces, it is often more convenient to work in the probability space. In particular, the probability space has the desirable property that it is always convex; it is, however, not always possible to find a convex reconstruction space. The primary objective of state estimation is then to find an estimator, or a region of estimators, for the probabilities p. The conversion of p into a state ρ can be performed later, if at all. At this stage, if the POM is not informationally complete, one must invoke additional criteria—beyond what the data tells us—for a unique mapping $p \to \rho$. For example, one could follow Jaynes's guidance [11, 12] and maximize the entropy [13] (see also [14]). However, following the tradition in this topic, we will formally work in a reconstruction space $\mathcal{R}_0$, although all actual calculations are performed in the probability space. Estimators are states in $\mathcal{R}_0$, and regions are sets of states there.

2.2. *Size and prior content of a region*

The reconstruction space is an abstract construct that is often not endowed with a self-suggesting unique metric. Instead, a region's prior probability—the quantity that matters most in the present context of statistical inference—offers a natural notion of size. This relieves us of the need to invoke additional, possibly artificial, criteria for the assignment of size, for instance, one that has more to do with a simple parameterization of the state space than the relative importance of different regions in terms of our prior expectations.

We denote by $(\mathrm{d}\rho)$ the size ($\equiv$ prior probability) of the infinitesimal vicinity of state ρ in $\mathcal{R}_0$. The size $S_{\mathcal{R}}$ of a region $\mathcal{R}$ is then

$$S_{\mathcal{R}} = \int_{\mathcal{R}} (\mathrm{d}\rho) \quad \text{with} \int_{\mathcal{R}_0} (\mathrm{d}\rho) = 1. \tag{2}$$

A pertinent remark: we exclude pathological cases of improper priors where the prior density $(\mathrm{d}\rho)$ cannot be normalized; should an improper prior be useful in a particular context, it should come about as the limit of a well-defined sequence of proper priors.

By construction, $S_{\mathcal{R}}$ does not depend on the parameterization used for the numerical representation of $(\mathrm{d}\rho)$. The primary parameterization is in terms of the probabilities,

$$(\mathrm{d}\rho) = (\mathrm{d}p)\, w(p) \quad \text{with } (\mathrm{d}p) = \mathrm{d}p_1\, \mathrm{d}p_2 \cdots \mathrm{d}p_K, \tag{3}$$

where the prior density $w(p)$ is positive for all permissible probabilities and vanishes for non-permissible probabilities. To enforce positivity and normalization of p, $w(p)$ always contains

$$w_0(p) = \eta(p_1)\, \eta(p_2) \cdots \eta(p_K)\, \delta\left(\sum_k p_k - 1\right) \tag{4}$$

as a factor, where $\eta(\,)$ denotes Heaviside's unit step function and $\delta(\,)$ is Dirac's delta function. If there are no other constraints, we have the probability space of a K-sided die. For genuine quantum measurements, however, there are additional constraints, some accounted for by more delta-function factors, others by step functions. The delta-function constraints reduce the dimension of the reconstruction space from $K-1$ to the number of independent probabilities. Accordingly, there is a factor of constraint $w_{\mathrm{cstr}}(p)$ (containing $w_0(p)$) that specifies the probability space and appears in all possible priors. In particular, there are two specific priors we will employ as examples below: the *primitive prior*

$$(\mathrm{d}\rho) \propto (\mathrm{d}p)\, w_{\mathrm{cstr}}(p) \tag{5}$$

and the *Jeffreys prior* [15]

$$(\mathrm{d}\rho) \propto (\mathrm{d}p)\, w_{\mathrm{cstr}}(p)\, \frac{1}{\sqrt{p_1 p_2 \cdots p_K}}, \tag{6}$$

which is a popular choice of an unprejudiced prior [16].

For the harmonic-oscillator example in section 2.1, which has the same probability space as a tossed coin, the factor $w_0(p)$ selects the line segment with $0 \leqslant p_1 = 1 - p_2 \leqslant 1$ in the $p_1 p_2$ plane. If we choose the primitive prior $(\mathrm{d}\rho) = (\mathrm{d}p)\, w_0(p)$, the subsegment with $a \leqslant p_1 \leqslant b$ has size $b - a$. For the Jeffreys prior

$$(\mathrm{d}\rho) = (\mathrm{d}p)\, w_0(p) \frac{1}{\pi \sqrt{p_1 p_2}}, \tag{7}$$

the same subsegment has size $\frac{2}{\pi}[\sin^{-1}(\sqrt{b}) - \sin^{-1}(\sqrt{a})]$.

In this example, and also in those we use for illustration in section 5 below, it is easy to state quite explicitly the restrictions on the set of permissible probabilities that follow from the Born rule. In other situations, including the examples of section 6, this is more difficult. In yet more complicated situations it could be impossible. It is, however, possible to check numerically if a certain $\tilde{p} = (\tilde{p}_1, \ldots, \tilde{p}_K)$ is permissible. For example, one calculates a MLE (which can be done efficiently) for relative frequencies $n_k/N = \tilde{p}_k$ (see below), and if the resulting probabilities p are such that $p = \tilde{p}$, then $\tilde{p}$ is permissible; otherwise it is not. For practical reasons, it may be necessary to truncate the full state space—which can be, and often is, infinite-dimensional—to a test space of manageable size. With such a truncation one accepts that not all permissible probabilities are investigated. Therefore, a criterion for judging if the test space is large enough is to verify that the estimated probabilities do not change significantly when the space is enlarged. Examples for the artifacts that result from test spaces that are too small can be found in [17].

2.3. Point likelihood, region likelihood and credibility

The data D consist of a sequence of detector clicks, with n_k clicks in total of the kth detector after measuring $N = n_1 + n_2 + \cdots + n_K$ copies of the state[9]. The probability of obtaining D, *if* ρ is the state, is the *point likelihood*

$$L(D|\rho) = p_1^{n_1} p_2^{n_2} \cdots p_K^{n_K}, \tag{8}$$

which attains its maximum value when ρ is the MLE $\widehat{\rho}_{\mathrm{ML}} \in \mathcal{R}_0$; the MLE is fully determined by the relative frequencies n_k/N. The joint probability of finding the state ρ in the region $\mathcal{R}$ and obtaining the data D is

$$\mathrm{prob}(D \wedge \mathcal{R}) = \int_{\mathcal{R}} (\mathrm{d}\rho)\, L(D|\rho)\,. \tag{9}$$

For $\mathcal{R} = \mathcal{R}_0$, $\mathrm{prob}(D \wedge \mathcal{R}_0) = L(D)$ is the *prior likelihood*. Since one of the click sequences is surely observed, we have

$$\sum_D L(D|\rho) = 1, \quad \sum_D L(D) = \int_{\mathcal{R}_0} (\mathrm{d}\rho) = 1, \tag{10}$$

where the sum is taken over all possible data for N clicks.

[9] One can account for detector inefficiencies and dark counts, but such technical details, which are important for practical applications, are immaterial to the current discussion.

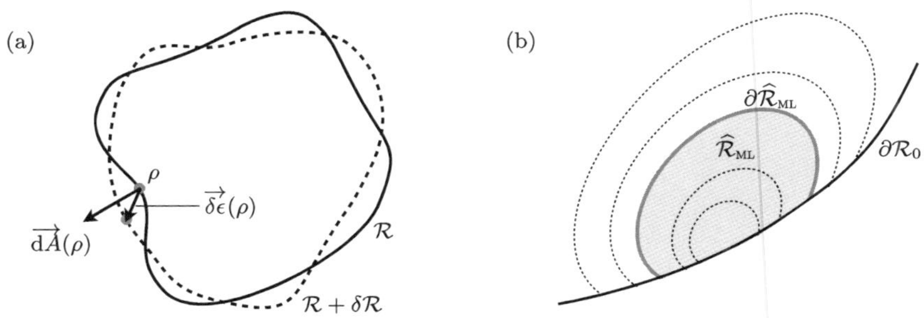

Figure 1. (a) Infinitesimal variation of region $\mathcal{R}$. The boundary $\partial\mathcal{R}$ of region $\mathcal{R}$ (solid line) is deformed to the boundary of region $\mathcal{R}+\delta\mathcal{R}$ (dashed line). $\overrightarrow{\mathrm{d}A}(\rho)$ is the vectorial surface element of $\partial\mathcal{R}$ at ρ, and $\overrightarrow{\delta\epsilon}(\rho)$ is the infinitesimal displacement of ρ. (b) Dotted lines indicate ILSs. The boundary $\partial\widehat{\mathcal{R}}_{\mathrm{ML}}$ of $\widehat{\mathcal{R}}_{\mathrm{ML}}$ can contain part of the surface $\partial\mathcal{R}_0$ of the reconstruction space.

We factor the joint probability in two ways

$$\mathrm{prob}(D\wedge\mathcal{R}) = L(D|\mathcal{R})S_{\mathcal{R}} = C_{\mathcal{R}}(D)L(D) \tag{11}$$

and so identify the *region likelihood* $L(D|\mathcal{R})$ and the *credibility* $C_{\mathcal{R}}(D)$. These are conditional probabilities: $L(D|\mathcal{R})$ is the probability of obtaining the data D if the actual state is in the region $\mathcal{R}$; $C_{\mathcal{R}}(D)$ is the probability that the actual state is in $\mathcal{R}$ if D were obtained—the posterior probability of $\mathcal{R}$.

3. Optimal error regions

For the given data D, we desire a region with the largest likelihood—the MLR. For this purpose, we maximize the region likelihood $L(D|\mathcal{R})$ under the constraint that only regions with a pre-chosen size s participate in the competition, with $0 < s < 1$,

$$\max_{\mathcal{R}\subseteq\mathcal{R}_0} L(D|\mathcal{R}) = L(D|\widehat{\mathcal{R}}_{\mathrm{ML}}) \quad \text{with } S_{\mathcal{R}} = s; \tag{12}$$

an unconstrained maximization of $L(D|\mathcal{R})$ is not meaningful as it gives the limiting region comprising nothing but the point $\widehat{\rho}_{\mathrm{ML}}$. The resulting MLR $\widehat{\mathcal{R}}_{\mathrm{ML}}$ depends on D and s. Since the size is fixed, we can equivalently maximize the joint probability $\mathrm{prob}(D\wedge\mathcal{R})$ under the size constraint.

The answer to this maximization problem is given in [10, corollary 4], which we translate into our present context as follows:

The MLRs of various sizes s consist of all states ρ for which the point likelihood exceeds a threshold value, with higher thresholds for smaller sizes. (13)

This is justified by a proof of considerable mathematical sophistication in [10]. We offer an alternative argument that is more accessible to the working physicist. Consider infinitesimal variations of a region $\mathcal{R}$ by deforming its boundary. The maximum property of the MLR and its

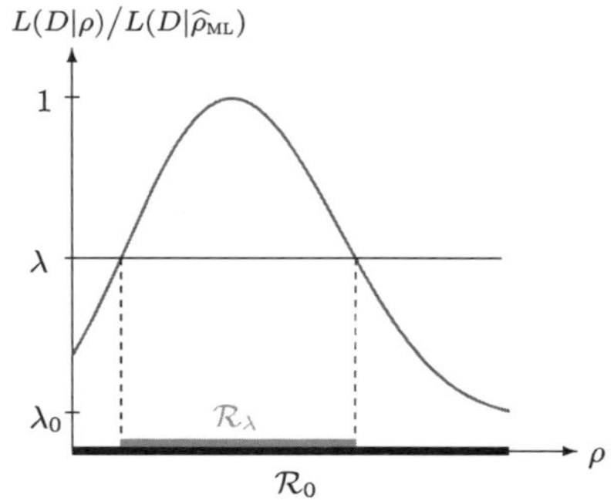

Figure 2. Illustration of a BLR: $\mathcal{R}_0$ is the reconstruction space; the region $\mathcal{R}_\lambda$ is a BLR, delineated by the threshold value $\lambda L(D|\widehat{\rho}_{\mathrm{ML}})$; λ_0 marks the minimum ratio $L(D|\rho)/L(D|\widehat{\rho}_{\mathrm{ML}})$ over $\mathcal{R}_0$.

fixed size require prob$(D \wedge \mathcal{R})$ and $S_{\mathcal{R}}$ to be stationary under all variations about $\mathcal{R} = \widehat{\mathcal{R}}_{\mathrm{ML}}$ (see figure 1(a)),

$$\begin{aligned} \delta S_{\mathcal{R}} &= \int_{\partial\mathcal{R}} \overrightarrow{\mathrm{d}A}(\rho) \cdot \overrightarrow{\delta\epsilon}(\rho) = 0, \\ \delta \mathrm{prob}(D \wedge \mathcal{R}) &= \int_{\partial\mathcal{R}} \overrightarrow{\mathrm{d}A}(\rho) \cdot \overrightarrow{\delta\epsilon}(\rho)\, L(D|\rho) = 0. \end{aligned} \tag{14}$$

Both hold only if $L(D|\rho)$ is constant on the boundary $\partial\widehat{\mathcal{R}}_{\mathrm{ML}}$ of $\widehat{\mathcal{R}}_{\mathrm{ML}}$, for an $\widehat{\mathcal{R}}_{\mathrm{ML}}$ entirely contained inside $\mathcal{R}_0$ (so that $\overrightarrow{\delta\epsilon}(\rho)$ can have any direction): $\partial\widehat{\mathcal{R}}_{\mathrm{ML}}$ is an *iso-likelihood surface* (ILS). Concavity of the log-likelihood further requires $\widehat{\mathcal{R}}_{\mathrm{ML}}$ to be the interior of this ILS. $\partial\widehat{\mathcal{R}}_{\mathrm{ML}}$ can also contain part of the boundary of $\mathcal{R}_0$ (see figure 1(b)), in which case only the part of $\partial\widehat{\mathcal{R}}_{\mathrm{ML}}$ inside $\mathcal{R}_0$ is an ILS. In either case, $\widehat{\mathcal{R}}_{\mathrm{ML}}$ is a *bounded-likelihood region* (BLR), comprising all states in $\mathcal{R}_0$ with point likelihood exceeding a certain threshold. BLRs have appeared previously in standard statistical analysis (see [18] and references therein).

We specify the threshold value as a fraction λ of the maximum value $L(D|\widehat{\rho}_{\mathrm{ML}})$ of the point likelihood; see figure 2. The BLR $\mathcal{R}_\lambda$ has the characteristic function

$$\chi_\lambda(\rho) = \eta\Big(L(D|\rho) - \lambda L(D|\widehat{\rho}_{\mathrm{ML}})\Big), \tag{15}$$

and

$$s_\lambda = \int_{\mathcal{R}_0} (\mathrm{d}\rho)\, \chi_\lambda(\rho) \tag{16}$$

is the size of $\mathcal{R}_\lambda$. We have $\mathcal{R}_\lambda = \mathcal{R}_0$ and $s_\lambda = s_0 = 1$ for $\lambda \leqslant \lambda_0$ with $\lambda_0 \geqslant 0$ given by $\min_\rho L(D|\rho) = \lambda_0 L(D|\widehat{\rho}_{\mathrm{ML}})$. As λ increases from λ_0 to 1, s_λ decreases monotonically from 1 to 0. The size s specified in (12) is obtained for an intermediate λ value and the corresponding BLR is the looked-for MLR. Note that the MLE is contained in all MLRs: as $s \to 0$, the MLR becomes an infinitesimal vicinity of the MLE, and $L(D|\widehat{\mathcal{R}}_{\mathrm{ML}}) \to L(D|\widehat{\rho}_{\mathrm{ML}})$.

The MLR is the region for which the observed data are particularly likely. With a reversal of emphasis, we now look for a region that contains the actual state with high probability. Ultimately, this is the SCR $\widehat{\mathcal{R}}_{\mathrm{SC}}$—the smallest region with the pre-chosen credibility value c. For the given D, the optimization problem

$$\min_{\mathcal{R} \subseteq \mathcal{R}_0} S_{\mathcal{R}} = S_{\widehat{\mathcal{R}}_{\mathrm{SC}}} \quad \text{with } C_{\mathcal{R}}(D) = c \tag{17}$$

is dual to that of (12). Here, we minimize the size for given joint probability; there we maximize the joint probability for a given size. It follows that the BLRs are not only the MLRs but they are also the SCRs: each MLR is a SCR, each SCR is an MLR.

The BLR $\mathcal{R}_\lambda$ has credibility

$$c_\lambda = \frac{1}{L(D)} \int_{\mathcal{R}_0} (\mathrm{d}\rho)\, \chi_\lambda(\rho) L(D|\rho), \tag{18}$$

which, like s_λ, decreases monotonically from 1 to 0 as λ increases from λ_0 to 1. The credibility c specified in (17) is obtained for an intermediate value, and the corresponding BLR is the looked-for SCR.

That the general definitions of the MLR and the SCR, which allow for regions of arbitrary shapes, permit such a simple characterization in terms of BLRs is remarkable. BLRs are reminiscent of standard ellipsoidal error regions constructed by analyzing the neighborhood of the peak of the likelihood function—a procedure justified only for large enough N for the central limit theorem to apply (see, for instance, [4]); yet, our result employs no such assumption. Also surprising is that, while λ depends on the choice of prior, the set of regions that enter the competition is independent of that choice; the prior enters only in the size, region likelihood, and credibility of the MLR/SCR.

Once the data are obtained, there is *the* MLR and *the* SCR for these data, and other MLRs or SCRs associated with unobserved data play no role. This is in sharp contrast to confidence regions, whose construction requires consideration of all data that could have been obtained, since the confidence level is a property of the entire set of confidence regions, one for each possible data (see appendix B). Nevertheless, they are not unrelated. Christandl and Renner [7] showed that high-credibility regions offer starting points for constructing confidence regions—a set of SCRs with high credibility immediately suggests itself—and Blume–Kohout [8] argued that BLRs can be good confidence regions.

4. Reporting error regions

For a BLR, s_λ and c_λ are linked by

$$L(D)\frac{\partial}{\partial\lambda}c_\lambda = L(D|\widehat{\rho}_{\mathrm{ML}})\,\lambda\frac{\partial}{\partial\lambda}s_\lambda. \tag{19}$$

Therefore, once s_λ is known as a function of λ, we obtain c_λ from

$$c_\lambda = \frac{\lambda s_\lambda + \int_\lambda^1 \mathrm{d}\lambda'\, s_{\lambda'}}{\int_0^1 \mathrm{d}\lambda'\, s_{\lambda'}}. \tag{20}$$

This is, of course, consistent with the limiting values for $\lambda \leqslant \lambda_0$ and $\lambda = 1$, and also establishes that $c_\lambda > s_\lambda$ for $\lambda_0 < \lambda < 1$. Furthermore, (19) and (20) tell us that in the $\lambda \to 1$ limit, when both s_λ and c_λ vanish, their ratio is finite and exceeds unity,

$$\frac{c_\lambda}{s_\lambda} \to \frac{1}{\int_0^1 \mathrm{d}\lambda'\, s_{\lambda'}} \to \frac{L(D|\widehat{\rho}_{\mathrm{ML}})}{L(D)} > 1 \quad \text{for } \lambda \to 1. \tag{21}$$

We note that this provides the value of $L(D)$, since the maximal value $L(D|\widehat{\rho}_{\mathrm{ML}})$ of the point likelihood is computed earlier as it is needed when identifying the BLRs.

Inasmuch as the value of s_λ quantifies our prior belief that the actual state is in $\mathcal{R}_\lambda$, we are surprised when the data tell us that the probability for finding the state in $\mathcal{R}_\lambda$ is larger.

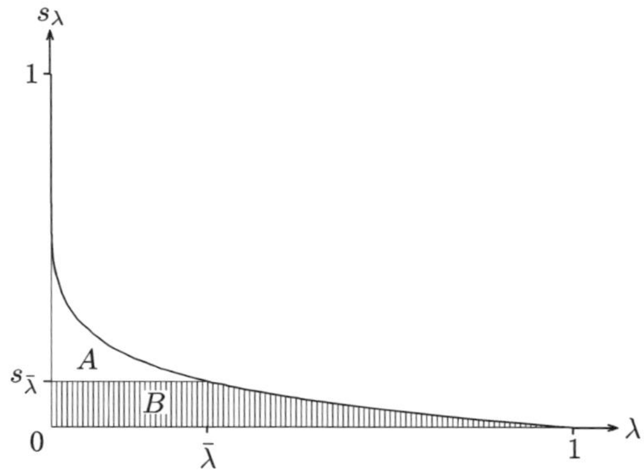

Figure 3. Geometrical meaning of the relation (20) between the size s_λ and the credibility c_λ. For the chosen value of λ, say $\bar{\lambda}$, the horizontal line from $(0, s_{\bar{\lambda}})$ to $(\bar{\lambda}, s_{\bar{\lambda}})$ divides the area under the graph of s_λ into the two pieces A and B indicated in the plot. The credibility is the fractional size of area B, that is: $c_{\bar{\lambda}} = B/(A + B)$.

Accordingly, the SCR is the region for which we are most surprised for the given prior belief. This matter and other aspects of Bayesian inference based on the concept of relative surprise are discussed in [10].

Relation (20) has a simple geometrical meaning in terms of areas under the graph of s_λ, as explained in figure 3. This relation is of considerable practical importance because we only need to evaluate the integrals of (16), but not those of (18). Since the latter integrals require well-tailored Monte Carlo methods to handle the typically sharply peaked point likelihood, the numerical effort is substantially reduced if we only need to evaluate (16). The error regions for the observed data are then concisely communicated by reporting s_λ and c_λ as functions of λ. With these, the end user interested in the MLR with the size s of his liking or the SCR of his wanted credibility c can determine the required value of λ. It is then easy to check if a state is inside the specified error region. The example of section 6 illustrates the matter for an eight-dimensional reconstruction space, for which the error regions will be impossible to visualize, but can still be easily specified through reporting the s_λ and c_λ values.

5. Example: incomplete single-qubit tomography

For a first illustration, we consider the simplest situation that exhibits the typical features: the quantum-information carriers have a qubit degree of freedom, which is measured by one of two standard POMs that are not informationally complete.

5.1. Probability-operator measurements and priors

For both POMs, the unit disk in the xy plane suggests itself for the reconstruction space $\mathcal{R}_0$. The first POM is the crosshair measurement that combines projective measurements of σ_x and

σ_y into a four-outcome POM ($K = 4$) with probabilities

$$\left.\begin{matrix} p_1 \\ p_2 \end{matrix}\right\} = \frac{1}{4}(1 \pm x), \quad \left.\begin{matrix} p_3 \\ p_4 \end{matrix}\right\} = \frac{1}{4}(1 \pm y) \quad \text{with } x = \langle\sigma_x\rangle, \ y = \langle\sigma_y\rangle\,. \tag{22}$$

The permissible probabilities are identified by

$$w_{\text{cstr}}(p) \doteq \eta(p)\,\delta(p_1 + p_2 - \tfrac{1}{2})\,\delta(p_3 + p_4 - \tfrac{1}{2})\,\eta(3 - 8p^2), \tag{23}$$

where

$$\eta(p) = \prod_{k=1}^{K} \eta(p_k) \quad \text{and} \quad p^2 = \sum_{k=1}^{K} p_k^2\,. \tag{24}$$

The dotted equals sign in (23) stands for 'equal up to a multiplicative constant', namely the factor that ensures the unit size of the reconstruction space.

The second POM is the three-outcome trine measurement ($K = 3$), whose outcomes are subnormalized projectors on the eigenstates of σ_x and $(-\sigma_x \pm \sqrt{3}\,\sigma_y)/2$ with eigenvalue +1. It has the probabilities

$$p_1 = \frac{1}{3}(1 + x), \quad \left.\begin{matrix} p_2 \\ p_3 \end{matrix}\right\} = \frac{1}{6}(2 - x \pm \sqrt{3}\,y), \tag{25}$$

for which

$$w_{\text{cstr}}(p) \doteq \eta(p)\,\delta(p_1 + p_2 + p_3 - 1)\,\eta(1 - 2p^2) \tag{26}$$

summarizes the constraints that the permissible values of p_1, p_2, p_3 obey.

Both POMs have the same primitive prior

$$(\mathrm{d}\rho) = \mathrm{d}x\,\mathrm{d}y\,\frac{1}{\pi}\,\eta(1 - x^2 - y^2) = \mathrm{d}r^2\,\frac{\mathrm{d}\varphi}{2\pi}, \tag{27}$$

where $0 \leqslant r \leqslant 1$ and φ covers any convenient range of 2π. This prior is uniform in x and y, and in r^2 and φ. The polar-coordinate version, that is $x + \mathrm{i}y = r\,\mathrm{e}^{\mathrm{i}\varphi}$, is the more natural parameterization of the unit disk. The Jeffreys prior for the four-outcome POM is

$$(\mathrm{d}\rho) = \frac{2}{\pi^2}\frac{\mathrm{d}r\,r\,\mathrm{d}\varphi}{\sqrt{1 - r^2 + \frac{1}{4}r^4\sin(2\varphi)^2}}. \tag{28}$$

For the three-outcome POM, we have the Jeffreys prior

$$(\mathrm{d}\rho) = \frac{1}{4\pi - 24\sin^{-1}\left(\frac{1}{3}\right)}\frac{\mathrm{d}r\,r\,\mathrm{d}\varphi}{\sqrt{1 - \frac{3}{4}r^2 + \frac{1}{4}r^3\cos(3\varphi)}}. \tag{29}$$

5.2. Computer-generated data

Figures 4(a) and (b) show SCRs obtained for simulated experiments in which $N = 24$ copies of a qubit state are measured. The actual state used for the simulation has $(x, y) = (0.6, 0.2)$. Its position in the reconstruction space is indicated by the red star ($\star$).

In figure 4(a), we see the SCRs for the four-outcome POM. Two experiments were simulated, with $(n_1, n_2, n_3, n_4) = (8, 5, 10, 1)$ and $(6, 3, 10, 5)$ clicks of the detectors, respectively, and the triangles ($\triangle$) show the positions of the corresponding MLEs. For each data, the plot reports the SCRs with credibility $c = 0.5$ and 0.9, both for the primitive prior of (27)

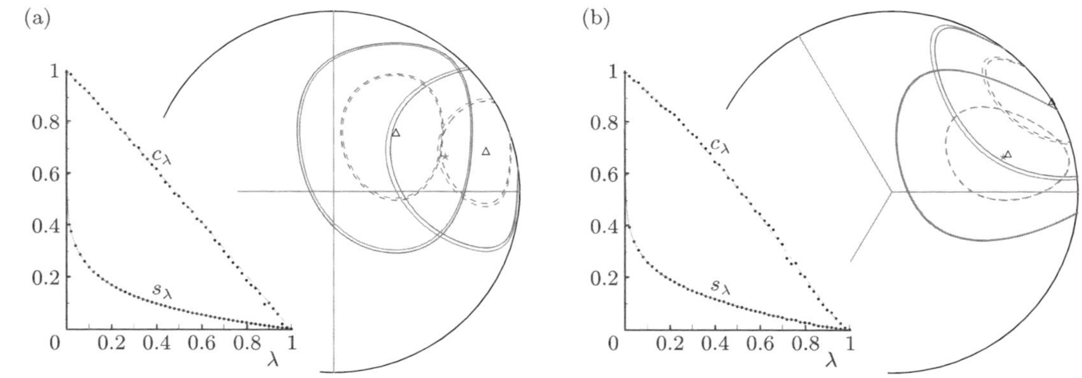

Figure 4. Smallest credible regions for simulated experiments. Twenty-four copies are measured by the POMs of section 5.1, which have the unit disk of figure A.1 as the reconstruction space. Plot (a) is for the four-outcome POM with the crosshair indicating the orientations of the two projective measurements. Plot (b) is for the three-outcome measurement with the orientation of the trine indicated. The red star ($\star$) at $(x, y) = (0.6, 0.2)$ marks the actual state that was used for the simulation. For each POM, there are SCRs for the data of two simulated experiments, with black triangles ($\triangle$) indicating the respective MLEs. The boundaries of the SCRs with credibility $c = 0.9$ are traced by the continuous lines; all of these SCRs contain the actual state. The dashed lines are the boundaries of the SCRs with credibility $c = 0.5$; the actual state is inside half of these SCRs. The red lines are for the primitive prior of (27), the blue lines are for the Jeffreys priors of (28) and (29), respectively.—the insets in the lower left corners show the size s_λ and the credibility c_λ for the BLRs of two different sets of data. Inset (a) is for (6, 3, 10, 5) counts for the four-outcome POM and the Jeffreys prior. Inset (b) is for (13, 7, 4) counts for the three-outcome POM and the primitive prior. The dots show the values computed with a Monte Carlo algorithm. There is much more scatter in the c_λ values than the s_λ values. The red lines are fits to the s_λ values, with the fits using twice as many values as there are dots in the insets. The green lines that approximate the c_λ values are obtained from the red lines with the aid of (20).

and for the Jeffreys prior of (28). The actual state is inside two of the four SCRs with credibility $c = 0.5$ and is contained in all four SCRs with credibility $c = 0.9$.

Not unexpectedly, we get quite different regions for the two rather different sets of detector click counts. Yet, we observe that the choice of prior has little effect on the SCRs, although the total number of measured copies is too small for relying on the consistency of the priors. The same remarks apply to the SCRs for the three-outcome POM in figure 4(b); here we counted $(n_1, n_2, n_3) = (15, 8, 1)$ and $(13, 7, 4)$ detector clicks in the simulated experiments.

In section 4 we remarked that the estimator regions are properly communicated by reporting s_λ and c_λ as functions of λ. This is accomplished by the insets in figure 4 for two of the four simulated experiments. The dots give the values obtained by numerical integration that uses a Monte Carlo algorithm. The scatter of these numerical values confirms the expected: the computation of s_λ only requires sampling the probability space in accordance with the prior and determining the fraction of the sample that is in $\mathcal{R}_\lambda$; for the computation of c_λ we need to add the values of $L(D|\rho)$ for the sample points inside $\mathcal{R}_\lambda$; and since $L(D|\rho)$ is a sharply peaked function of the probabilities, the s_λ values are more trustworthy than the c_λ values for the same computational effort. The line fitted to the s_λ values is a Padé approximant (see e.g. [19, section 5.12]) that takes the analytic forms near $\lambda = \lambda_0 = 0$ and $\lambda = 1$ into account. The line approximating the c_λ values is then computed in accordance with (20).

6. Example: incomplete two-qubit tomography

For a second illustration, we consider the situations that arise in the quantum-key-distribution schemes by Bennett and Brassard (BB84 [20]) and the trine–antitrine (TAT) scheme of reference [21]. Both schemes can be implemented by having a source of entangled qubit pairs distribute one qubit each to the two communicating parties. Prior to any key generation, the two-qubit state emitted by the source needs to be characterized. It is desirable to achieve quantum state estimation with reliable error regions without sacrificing many data that are then not available for the key generation.

6.1. Probability-operator measurements and computer-generated data

In the BB84 scheme, each qubit is measured by the crosshair POM of (22); the resulting two-qubit POM has sixteen outcomes that obey eight constraints that give delta-function factors in $w_{\text{cstr}}(p)$. In the TAT scheme, one qubit is measured by the trine POM of (25) and the other qubit by the antitrine POM that has the signs of x and y reversed in (25); the resulting two-qubit POM has nine outcomes subject to the single delta-function constraint of unit sum. Accordingly, the probability space is eight-dimensional for both schemes[10], and we cannot report the SCRs by showing the optimal error regions in the reconstruction space, as was possible for the two-dimensional probability space in figure 4. Therefore, we employ the strategy of section 4 and report the size s_λ and the credibility c_λ of the respective BLRs as functions of λ.

For the generation of the simulated data, we first add noise to the singlet state by putting it through a random Pauli channel[11]. The resulting true state has the probabilities for the two-qubit POMs given in the top row of table 1. For example, the '12' entry in the 4×4 table for the double-crosshair POM is the probability for outcome $\Pi_1 \otimes \Pi_2 = \frac{1}{4}(1+\sigma_x) \otimes \frac{1}{4}(1-\sigma_x)$. The '11' entry of the 3×3 table for the TAT POM is $16/9$ times that number. Note that all marginal probabilities (sums of rows and sums of columns) are equal; this is so because the

[10] In actual experiments, the probability space is nine-dimensional because one must account for the no-click probability of the qubit pairs that do not give rise to coincidence clicks. Furthermore, the state estimation could also exploit the data collected for single-qubit detection without the coincidental detection of the partner qubit. Consistent with the footnote in section 2.3, we are here content with the idealized situation of perfect detection devices because our objective is to give an example for a higher-dimensional space rather than evaluating real experimental data.

[11] A random Pauli channel is used as a simple model for noise in a communication protocol. The channel acts on an input state as $\rho \to \sum_{jk} r_{jk}\, \sigma_j \otimes \sigma_k\, \rho\, \sigma_j \otimes \sigma_k$, where $j, k = 0, x, y, z$ (σ_0 denotes the single-qubit identity operator), and the r_{jk} are sixteen randomly chosen probabilities. The 60 copies of the true state come from passing 60 copies of the singlet state through one instance of the random Pauli channel, i.e., the r_{jk} are randomly picked once, with r_{00} given a higher weight of 0.7 to simulate weak noise.

Table 1. Computer-generated data for the estimation of a two-qubit state from measuring 60 identically prepared copies. The first row gives the joint probabilities of the true state. The broken second row shows the number of detector-click pairs obtained in the simulated experiment (and their expected values) together with the single-qubit marginals. The third row reports the joint probabilities of the MLEs for the data in the second row. In each row, we have a 4×4 table on the left for the double-crosshair POM of the BB84 scenario and a 3×3 table on the right for the nine-outcome POM of the TAT scheme. The rows of a 4×4 table for the double-crosshair POM refer to the four Π_j of the first qubit in the pair and the columns refer to the Π_k of the second qubit; entry 'jk' is the probability for outcome $\Pi_j \otimes \Pi_k$. Analogously, entry 'jk' in a 3×3 table for the TAT scheme is the expectation value of $\Pi_j \otimes \Pi_k$ with trine outcome Π_j and antitrine outcome Π_k.

True-state probabilities

Double-crosshair POM:

	1	2	3	4
1	0.0206	0.1044	0.0625	0.0625
2	0.1044	0.0206	0.0625	0.0625
3	0.0625	0.0625	0.0212	0.1038
4	0.0625	0.0625	0.1038	0.0212

TAT POM:

	1	2	3
1	0.1856	0.0739	0.0739
2	0.0739	0.1848	0.0747
3	0.0739	0.0747	0.1848

Computer-generated data (expected number of clicks)

Double-crosshair POM:

	1	2	3	4	
1	0 (1.24)	4 (6.26)	6 (3.75)	4 (3.75)	14 (15)
2	6 (6.26)	3 (1.24)	8 (3.75)	4 (3.75)	23 (15)
3	3 (3.75)	1 (3.75)	0 (1.27)	8 (6.23)	12 (15)
4	1 (3.75)	7 (3.75)	4 (6.23)	1 (1.27)	13 (15)
	10 (15)	15 (15)	18 (15)	17 (15)	

TAT POM:

	1	2	3	
1	11 (11.14)	4 (4.43)	5 (4.43)	20 (20)
2	2 (4.43)	10 (11.09)	5 (4.48)	17 (20)
3	4 (4.43)	6 (4.48)	13 (11.09)	23 (20)
	17 (20)	20 (20)	23 (20)	

MLE probabilities

Double-crosshair POM:

	1	2	3	4
1	0.0056	0.1012	0.0497	0.0571
2	0.0939	0.0493	0.0821	0.0611
3	0.0630	0.0344	0.0025	0.0949
4	0.0365	0.1160	0.1293	0.0232

TAT POM:

	1	2	3
1	0.1833	0.0667	0.0833
2	0.0333	0.1667	0.0833
3	0.0667	0.1000	0.2167

reduced single-qubit states of the true state are completely mixed. For the same reason, both tables are symmetric and the lower-left and upper-right 2×2 subtables of the 4×4 table have entries of $1/16 = 0.0625$. More generally, there is a one-to-one correspondence between the 16 permissible probabilities in the 4×4 table and the nine permissible probabilities in the 3×3 table, because all table entries are determined by the expectation values of $A \otimes B$ with $A, B = 1, \sigma_x, \sigma_y$.

Simulated measurements of 60 qubit pairs in the true state for each POM produced the counts of detector-click pairs in the second row of table 1; expected values are given in parentheses. Owing to the statistical fluctuations, the tables of counts are not symmetric[12] and the marginal counts are not equal.

The third row of table 1 shows the corresponding MLE probabilities. These probabilities are equal to the relative frequencies of the counts for the 9-outcome POM but are different from the relative frequencies for the 16-outcome POM. This tells us that the computer-generated data are not typical for the double-crosshair POM, whereas we have typical data for the TAT POM.

6.2. Size and credibility of the bounded-likelihood regions

As noted in section 4, the primary task of the data evaluation is the computation of the multi-dimensional integrals that give the size s_λ of the BLRs for the whole range of $0 < \lambda < 1$. For the data in table 1, these are integrals over eight-dimensional regions. We used a random-sampling technique for this purpose.

As a preparation, we generated a random sample of 648 785 permissible sets of probabilities, uniformly distributed in accordance with the primitive prior (see section 6.3 below). In view of the one-to-one correspondence between the permissible probabilities of the 16-outcome POM and the 9-outcome POM, the same random sample can be, and was, used for both POMs.

The actual data processing consists of two steps. In the first step, we determine the size s_λ for the 160 values of λ with $-\log_{10}\lambda = 0.1(0.1)16.0$. This requires a simple counting of how many samples are inside the BLR $\mathcal{R}_\lambda$ if the primitive prior is used. In the case of the Jeffreys prior, one adds the weights $(p_1 p_2 \cdots)^{-1/2}$ of the samples inside the BLR. The correct normalization follows from $s_{\lambda=0} = 1$.

In the second step, the integrals needed in (20) are evaluated, for which a simple linear interpolation between adjacent (λ, s_λ) pairs is sufficiently accurate. Then, c_λ is known as a function of λ and the λ values for which we have 99% or 95% credibility are determined.

We show s_λ and c_λ as functions of λ in figure 5. Table 2 reports the λ values of the 99% and 95% credibility thresholds. We observe that for the 16-outcome POM, the true state is inside the SCRs with 99% credibility for both the primitive prior and the Jeffreys prior, whereas it is inside the 95% SCR only for the primitive prior but not for the Jeffreys prior. This is more evidence that these data are untypical. By contrast, for the 9-outcome POM, the true state is inside all SCRs for both priors and both values of the credibility.

Typicality, or lack thereof, can also be noticed in figure 5. Since the Jeffreys prior gives more weight to the regions near the boundary of the probability space than the primitive prior, and less weight to regions deep inside, one expects that the values of s_λ for the primitive prior are larger than those for the Jeffreys prior if the data are typical and, accordingly, the MLE is not close to the boundary. This is indeed the case for the TAT data, but not for the double-crosshair data.

[12] One usually restores the symmetry by so-called 'twirling' before the key generation protocol is executed. The characterization of the source, however, should be done without the twirling.

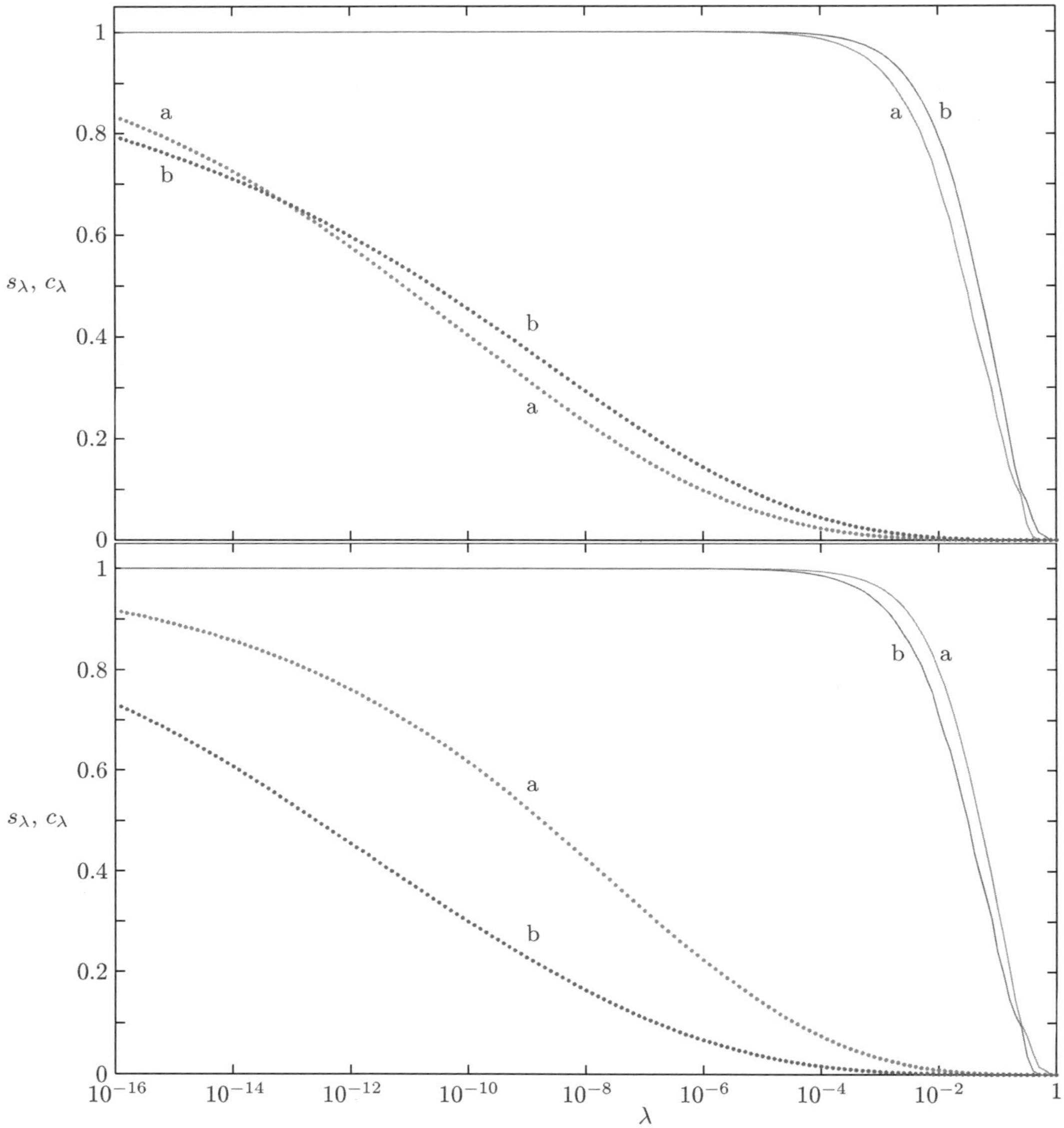

Figure 5. The size s_λ (dotted lines) and the credibility c_λ (solid lines) as functions of λ for the data of table 1. The top plot is for the double-crosshair POM, the bottom plot is for the TAT POM. Curves 'a' are for the primitive prior, curves 'b' are for the Jeffreys prior. The abscissa is linear in $\log\lambda$. For $\lambda \lesssim 1$, the BLRs are so small that only very few sample points are inside and the sizes s_λ have comparatively large fluctuation errors. This statistical noise is visible in the bottom-right corners of the plots. It has no bearing, however, on the accuracy of the credibility c_λ in the important range of smaller λ values, as one notes upon recalling figure 3.

Table 2. Threshold λ values for 99% and 95% credibility for the data of table 1 and figure 5, and the sizes of the respective BLRs. The true state is inside the $\mathcal{R}_\lambda$s with $\lambda < 3.368 \times 10^{-3}$ for the 16-outcome POM (with its untypical data), and inside the BLRs with $\lambda < 0.2486$ for the 9-outcome POM.

	16-outcome POM			9-outcome POM		
	λ	s_λ	c_λ	λ	s_λ	c_λ
Primitive prior	6.70×10^{-5}	0.0279	0.99	1.92×10^{-4}	0.0601	0.99
	6.03×10^{-4}	0.0106	0.95	1.44×10^{-3}	0.0268	0.95
	λ	s_λ	c_λ	λ	s_λ	c_λ
Jeffreys prior	1.73×10^{-4}	0.0374	0.99	6.74×10^{-5}	0.0186	0.99
	1.35×10^{-3}	0.0161	0.95	6.20×10^{-4}	0.0070	0.95

6.3. Numerical effort

The two steps of data evaluation, the computation of the size s_λ and then the credibility c_λ, take a few seconds of central processing unit (CPU) time. The preparation of the random sample of permissible probabilities, which could be done ahead of the data taking, lasts much longer. Our sample of 648 785 probabilities took almost 100 h of CPU time on a standard desktop (Intel i7-870 CPU, using one of the four cores and 8 GB RAM). The procedure we employed was simple and reliable but not optimized for speed. There is clearly much room for improvement (see section 7).

For each potential sample of probabilities we first generate nine random numbers $x_1, x_2, \ldots, x_9$ uniformly and independently between 0 and 1. Then, the nine probabilities $p_k = \log(x_k)/\log(x_1 x_2 \cdots x_9)$ constitute a sample in the eight-dimensional simplex of the classical nine-sided die, and the samples are distributed in accordance with the primitive prior. The sample $p = (p_1, p_2, \ldots, p_9)$ is accepted if it is a permissible set of probabilities for the TAT POM with its nine outcomes. Only 9.27% of the 7×10^6 candidate probabilities generated were accepted.

Whereas the generation of another sample p is fast, the test of permissibility is the part that consumes most of the CPU time. After identifying the candidate p with the relative frequencies of a measurement with the 9-outcome POM, we calculate a MLE for these frequencies. If the probabilities of the MLE are equal to p, this sample probability is accepted, otherwise it is rejected.

7. Outlook

For the given data and chosen size or credibility, the MLR or the SCR is a neighborhood of the MLE. In this sense, one can regard them as systematically constructed error regions for the MLE. Although there are efficient methods for computing the MLE [2, 3], equally efficient algorithms for finding the MLR and the SCR need to be developed. In particular, there remains the challenge of evaluating the multi-dimensional integrals that give s_λ.

Random-sampling techniques are our methods of choice. In section 6, a simple strategy required 100 h of CPU time for creating a sample of permissible probabilities. However, this can be substantially reduced by optimizing the sampling process. Parallelizing the sampling over many different computers and, later, combining into a single dataset was suggested. The chance that a candidate probability is permissible (only 9.27% in the sampling of the example) can be much increased by employing cleverer Monte Carlo methods, where one makes use of information at the current physical point to stay within the physical state space. We have just begun to explore this and will report on our progress in due course. It is also worth noting that this computational time is an overhead that is incurred only once and the sampling can be done ahead of any actual data-taking in the laboratory.

Often, only a few parameters computed from the state are of interest. It is, therefore, possible to reduce the dimensionality of the problem by discarding nuisance parameters. A variant of the methodology described here can be used to determine small regions of high credibility in the few-parameter space of interest, without first determining SCRs in the reconstruction space. We will return to this matter on another occasion.

Acknowledgments

We have benefited greatly from our discussions with David Nott and thank him in particular for bringing [10, 16] to our attention. The Centre for Quantum Technologies is a Research Centre of Excellence funded by the Ministry of Education and the National Research Foundation of Singapore.

Note added in proof. Since the completion of this paper, we have improved the sampling algorithm and thereby reduced the CPU time consumed to less than 10% of what it was before. This speed-up results from a faster test of permissibility that is still easy to implement. Further improvement is likely.

Appendix A. Choosing the prior

The assignment of prior probabilities to regions in the reconstruction space should be done in an unprejudiced manner while taking into account all prior information that might be available. We cannot do justice to the rich literature on this subject and are content with noting that [16] reviews various approaches to constructing unprejudiced priors. Here, we discuss some criteria that are useful when choosing a prior, illustrating with examples familiar in quantum contexts.

A general remark: the chosen prior should give some weight to (almost) all states. It should not give extremely high weight to states in some part of the state space and extremely low weight to other states. This is to say, the prior should be *consistent* in the sense that the credibility of a region—its posterior content—is dominated by the data rather than by the prior if a reasonably large number N of copies is measured. In the examples of figure 4, $N = 24$ is close to being 'reasonably large', while $N = 2$ in figure B.1 is clearly not. Also, $N = 60$ in section 6 is not large enough to ensure data dominance because the λ values in table 2 for the primitive prior and for the Jeffreys prior are quite different and, hence, correspond to quite different BLRs.

Below, we describe a few criteria for choosing priors. We begin in section A.1 with the common choice of a uniform prior; section A.2 discusses priors motivated by the utility of the estimated state; section A.3 invokes symmetry arguments to restrict considerations to priors that possess some symmetry properties; section A.4 presents form-invariant prior constructions; section A.5 deals with the situation where one has a target state in mind; and section A.6 is about priors induced by marginalization of full-state-space priors according to what the data can tell us.

A.1. Uniformity

The time-honored strategy of choosing a uniform prior on $\mathcal{R}_0$ in which all states are treated equally gets us into a circular argument. Our identification of the size of a region with its prior content amounts to assigning equal probabilities to regions of equal sizes, prior to acquiring any data. But that just means that we now have to declare how we measure the size of a region without prejudice, and we are again faced with the original question about a uniform prior.

In fact, there is no unique meaning of the uniformity of a prior. In the sense that each prior tells us how to quantify the size of a region, each prior is uniform with respect to its induced size measure. To illustrate, reconsider the harmonic-oscillator example of section 2.1. For the primitive prior of (5), the parameterization

$$\begin{aligned} &p_1 = \tfrac{1}{2}(v+u), \quad p_2 = \tfrac{1}{2}(v-u), \\ &\mathrm{d}p_1\,\mathrm{d}p_2 = \mathrm{d}u\,\mathrm{d}v\,\tfrac{1}{2} \end{aligned} \tag{A.1}$$

gives

$$\begin{aligned} (\mathrm{d}\rho) = {} & \mathrm{d}u\,\mathrm{d}v\,\tfrac{1}{2}\eta(v+u)\eta(v-u)\delta(v-1) \\ & \to \mathrm{d}u\,\tfrac{1}{2} \quad \text{with} \quad -1 \leqslant u \leqslant 1, \end{aligned} \tag{A.2}$$

where we integrate over v in the last step and so observe that the primitive prior is uniform in u, that is, the size of the region $u_1 < u < u_2$ is proportional to $u_2 - u_1$. Likewise, the parameterization

$$\begin{aligned} &p_1 = v(\sin\alpha)^2, \quad p_2 = v(\cos\alpha)^2, \\ &\mathrm{d}p_1\,\mathrm{d}p_2 = \mathrm{d}\alpha\,\mathrm{d}v\; v\sin(2\alpha) \end{aligned} \tag{A.3}$$

gives

$$(\mathrm{d}\rho) \to \mathrm{d}\alpha\,\frac{2}{\pi} \quad \text{with } 0 \leqslant \alpha \leqslant \frac{\pi}{2} \tag{A.4}$$

for the Jeffreys prior of (6), which is uniform in α. Other priors can be treated analogously, each of them yielding a uniform prior in an appropriate single parameter.

Visualization of the uniformity for qubit priors can be found in figure A.1. Plot (b) shows uniform tiling of the unit disk by tiles of equal size. Here size is measured by the primitive prior of (27), which is uniform in x and y, and in r^2 and φ (the latter is used for the plot). Plots (c1) and (c2) show uniform tilings of the unit disk for the Jeffreys prior for the four-outcome POM of (28), while plots (d1) and (d2) show those for the three-outcome POM of (29). The crosshair symmetry of the four-outcome POM and the trine symmetry of the three-outcome POM are manifest in their respective uniform tilings.

The parameterizations in (A.1) and (A.3), and the tilings of figure A.1 exhibit in which explicit sense the primitive prior and the Jeffreys prior are uniform. However, the priors are what they are, irrespective of how they are parameterized. They are explicitly uniform in a particular parameterization and implicitly uniform in all others. Uniformity, it follows, cannot serve as a principle that distinguishes one prior from another.

This ubiquity of uniform priors for a continuous set of infinitesimal probabilities is in marked contrast to situations in which prior probabilities are assigned to a finite number of discrete possibilities, such as the 38 pockets of a double-zero roulette wheel. Uniform

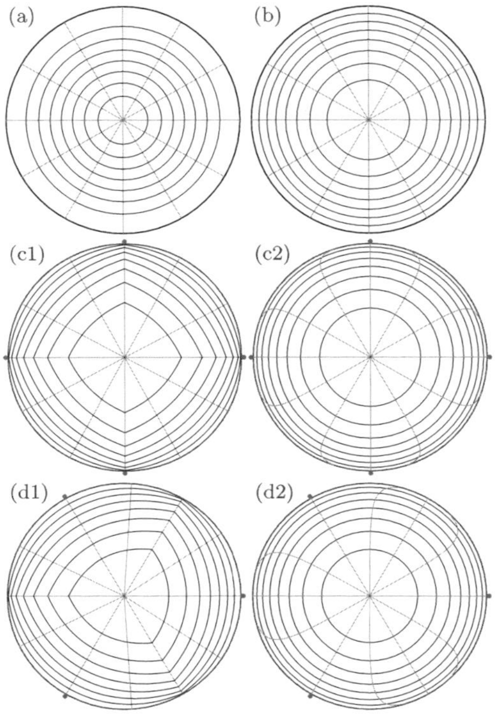

Figure A.1. Uniform tilings of the unit disk for four different priors. The disk is in the xy plane, with the x-axis horizontal, the y-axis vertical, and the disk center at $x = y = 0$. Tiling (a) depicts the marginal prior of (A.14); tiling (b) is for the primitive prior of (27); tilings (c1) and (c2) illustrate the Jeffreys prior of (28) with the blue dots (•) just outside the unit circle indicating the four directions onto which the POM outcomes project; and, tilings (d1) and (d2) are for the Jeffreys prior of (29), the blue dots marking the three directions of the trine projectors. In each tiling, we identify 96 regions of equal size by dividing the disk into eight 'tree rings' of equal size and twelve 'pie slices' of equal size. In the tilings (a), (b), (c1) and (d1), the boundaries of the pie slices are (red) rays and an arc of the unit circle. In the tilings (a), (b), (c2) and (d2), the tree rings have concentric circles as their boundaries.

probabilities of $1/38$ suggest themselves, are meaningful, and clearly distinguished from other priors, all of which have a bias.

Uniformity in a particularly natural parameterization of the probability space might also be meaningful. This, however, invokes a notion of 'natural' that others may not share.

A.2. Utility

In many applications, estimating the state is not a purpose in itself, but only an intermediate step on the way to determining some particular property of the physical system. The objective is to find the value of a parameter that quantifies the *utility* of the state.

For example, one could be interested in the fidelity of the actual state with a target state, or in an entanglement measure of a two-partite state, or in another quantity that tells us how useful are the quantum-information carriers for their intended task. In a situation of this kind, one should, if possible, use a prior that is uniform in the utility parameter of interest. In contrast to the situation of the previous section, where requiring uniformity in $\mathcal{R}_0$ may be ill-advised because uniformity is a parameterization-dependent notion, here we specify uniformity for the parameter we are interested in.

To illustrate, consider a single qubit. Suppose the utility parameter is the purity $\xi(\rho) = \mathrm{tr}\{\rho^2\}$ of the state ρ. With the Bloch-ball representation of a qubit state, $\rho = \frac{1}{2}(1 + \boldsymbol{\varrho}\cdot\boldsymbol{\sigma})$, where $\boldsymbol{\varrho} = \mathrm{tr}\{\boldsymbol{\sigma}\rho\} = \langle\boldsymbol{\sigma}\rangle$ is the Bloch vector and $\boldsymbol{\sigma}$ is the vector of Pauli matrices, the purity is

$$\xi(\rho) = \tfrac{1}{2}(1+\varrho^2) \quad \text{with } \varrho = |\boldsymbol{\varrho}|\,. \tag{A.5}$$

A prior uniform in purity induces a prior on the state space according to

$$(\mathrm{d}\rho) \propto \mathrm{d}\xi\,\mathrm{d}\Omega \propto \varrho\,\mathrm{d}\varrho\,\mathrm{d}\Omega\,, \tag{A.6}$$

where we parameterize the Bloch ball by spherical coordinates (ϱ, θ, ϕ). Here, $\mathrm{d}\Omega$ is the prior for the angular coordinates; the prior for the radial coordinate ϱ is fixed by our choice of uniformity in ξ. Irrespective of what we choose for $\mathrm{d}\Omega$, the marginal prior for ϱ is uniform in ξ.

If one can quantify the utility of an estimator by a cost function, an optimal prior can be selected by a minimax strategy. For each prior in the competition one determines the maximum of the cost function over the states in the reconstruction space, and then chooses the prior for which the maximum cost is minimal. In classical statistics, such minimax strategies are common (see, for instance, [22, chapter 5]); for an example in the context of quantum state estimation, see [23].

A.3. Symmetry

Symmetry considerations are often helpful in narrowing the search for the appropriate prior. For a particularly instructive example, see section 12.4.4 in Jaynes's posthumous book [24].

Returning to the uniform-in-purity prior of (A.6), one can invoke rotational symmetry in favor of the usual solid-angle element, $\mathrm{d}\Omega = \sin\theta\mathrm{d}\theta\,\mathrm{d}\phi$, as the choice of angular prior. The reasoning is as follows: the purity of a qubit state does not change under unitary transformations; unitarily equivalent states have the same purity. Now, regions that are turned into each other by a unitary transformation have identical radial content whereas the angular dependences are related by a rotation. Invariance under rotations, in turn, requires that the prior is proportional to the solid angle, hence the identification of $\mathrm{d}\Omega$ with the differential of the solid angle. Note that the resulting prior element $(\mathrm{d}\rho)$ is different from the usual Euclidean volume element, $\varrho^2\mathrm{d}\varrho\,\sin\theta\mathrm{d}\theta\,\mathrm{d}\phi$, which would be natural if the Bloch ball were an object in the physical three-dimensional space, but it is not.

Symmetry arguments should be used carefully and not blindly. For a fairly tossed coin, the prior should not be affected if the probabilities for heads and tails are interchanged, $w(p_1, p_2) = w(p_2, p_1)$. However, for the harmonic-oscillator example of section 2.1, which has the same reconstruction space as the coin, there is poor justification for requiring this symmetry because the two probabilities—of finding the oscillator in its ground state, or not—are not on equal footing.

A.4. Invariance

When one speaks of an *invariant prior*, one does not mean the invariance under a change of parameterization—all priors are invariant in this respect—but rather a *form-invariant* construction in terms of a quantity that, preferably, has an invariant significance. We consider two particular constructions that make use of the metric induced by the response of the selected function to infinitesimal changes of its variables.

The first construction begins with a quantity $F(p)$ that is a function of all probabilities $p=(p_1,\ldots,p_K)$. We include the square root of the determinant of the dyadic second derivative in the prior density as a factor,

$$(\mathrm{d}\rho)=(\mathrm{d}p)\left|\det\left\{\left(\frac{\partial^2 F}{\partial p_j\,\partial p_k}\right)_{jk}\right\}\right|^{1/2} w_{\mathrm{cstr}}(p), \tag{A.7}$$

where $w_{\mathrm{cstr}}(p)$ contains all the delta-function and step-function factors of constraint as well as the normalization factor that ensures the unit size of the reconstruction space. The prior defined by (A.7) is invariant in the sense that a change of parameterization, from p to α, say, does not affect its structure

$$(\mathrm{d}\rho)=(\mathrm{d}\alpha)\left|\det\left\{\left(\frac{\partial^2 F}{\partial \alpha_j\,\partial \alpha_k}\right)_{jk}\right\}\right|^{1/2} w_{\mathrm{cstr}}(p(\alpha)), \tag{A.8}$$

because the various Jacobian determinants take care of each other. Since $w_{\mathrm{cstr}}(p)$ enforces all constraints, the p_k are independent variables when $F(p)$ and $G(p,\nu)$ are differentiated in (A.7) and (A.9), respectively.

For the second construction, we use a data-dependent function $G(p,\nu)$ of the probabilities p and the frequencies $\nu=(\nu_1,\nu_2,\ldots,\nu_K)$ with $\nu_j=n_j/N$. Here, the square root of the determinant of the expected value of the dyadic square of the p-gradient of G is a factor in the prior density

$$(\mathrm{d}\rho)=(\mathrm{d}p)\left|\det\left\{\overline{\left(\frac{\partial G}{\partial p_j}\frac{\partial G}{\partial p_k}\right)_{jk}}\right\}\right|^{1/2} w_{\mathrm{cstr}}(p), \tag{A.9}$$

where $\overline{f(\nu)}$ denotes the expected value of $f(\nu)$,

$$\overline{f(\nu)}=\sum_D L(D|\rho)f(\nu)\,. \tag{A.10}$$

We have, in particular, the generating function

$$\overline{\exp\left(\sum_{k=1}^{K}a_k\nu_k\right)}=\left(\sum_{k=1}^{K}\mathrm{e}^{a_k/N}p_k\right)^N \tag{A.11}$$

for the expected values of products of the ν_k. The prior defined by (A.9) is form-invariant in the same sense, and for the same reason, as the prior of (A.7).

Table A.1. Form-invariant priors constructed by one of the two methods described in the text. The '$\sqrt{\det}$' column gives the p-dependent factors only and omits all p-independent constants. The first method of (A.7) proceeds from functions of the probabilities that have extremal values when all probabilities are equal or all vanish save one. The second method of (A.9) uses functions that quantify how similar are the probabilities and the frequencies. The 'hedged prior' is named in analogy to the 'hedged likelihood' [25].

Method	Primary function	$\sqrt{\det}$
1st	$-\sum_k p_k \log p_k$ (Shannon entropy)	$\frac{1}{\sqrt{p_1 p_2 \cdots p_K}}$ (Jeffreys prior)
1st	$\sum_k p_k^2$ (purity)	1 (primitive prior)
2nd	$\sum_k \nu_k p_k$ (inner product)	$\sqrt{p_1 p_2 \cdots p_K}$ (hedged prior)
2nd	$\sum_k \nu_k \log(\nu_k/p_k)$ (relative entropy)	$\frac{1}{\sqrt{p_1 p_2 \cdots p_K}}$ (Jeffreys prior)

Table A.1 reports a few examples of '$\sqrt{\det}$' factors constructed by one of these two methods. It is worth noting that the Jeffreys prior can be obtained from the entropy of the probabilities by the first method as well as from the relative entropy between the probabilities and the frequencies by the second method. The latter is a variant of Jeffreys's original derivation [15] in terms of the Fisher information.

A.5. Conjugation

Sometimes there are reasons to expect that the actual state is close to a certain target state with probabilities $t = (t_1, t_2, \ldots, t_K)$. This is the situation, for example, when a source is designed to emit the quantum-information carriers in a particular state. A *conjugate prior*

$$(\mathrm{d}\rho) = (\mathrm{d}p)\left(p_1^{t_1} p_2^{t_2} \cdots p_K^{t_K}\right)^{\beta} w_{\mathrm{cstr}}(p) \quad \text{with } \beta > 0 \tag{A.12}$$

could then be a natural choice. Such priors are called 'conjugate' in standard statistics literature because the $(\cdots)^{\beta}$ factor has the same structure as the point likelihood: a product of powers of the detection probabilities. The $(\cdots)^{\beta}$ factor is maximal for $p = t$ and the peak is narrower when β is larger.

The conjugate prior can be understood as the 'mock posterior' for the primitive prior that results from pretending that β copies have been measured in the past and data obtained that are most typical for the target state. Therefore, a conjugate prior is quite a natural way of expressing the expectation that the apparatus is functioning well. The posterior content of a region will be data-dominated only if N is much larger than β.

In this context, it may be worth noting that the Bayesian mean state,

$$\widehat{\rho}_{\text{BM}} = \int_{\mathcal{R}_0} (\mathrm{d}\rho)\,\rho, \tag{A.13}$$

computed with the conjugate prior above, is usually not the target state unless β is large. One could construct priors for which $\widehat{\rho}_{\text{BM}}$ is the target state, but the presence of the $w_{\text{cstr}}(p)$ factor requires a case-by-case construction.

A.6. Marginalization

All priors used as examples—the ones in (A.2), (A.4) and (A.12), and in table A.1—have in common that they are defined in terms of the probabilities and, therefore, they refer to the particular POM with which the data are collected. While this takes the significance of the data duly into account, it does not seem to square with the point of view that prior probabilities are solely a property of the physical processes that put the quantum-information carriers into the state that is then diagnosed by the POM.

When adopting this viewpoint, one begins with a prior density defined on the entire state space. In addition to the parameters that specify the reconstruction space (essentially the probabilities p), this full-space prior will depend on parameters whose values are not determined by the data. There could be very many nuisance parameters of this kind. In the harmonic-oscillator example of section 2.1, the data tell us only about the ground-state population but nothing about the population in any specific excited state. For a prior assigned on the formally infinite-dimensional state space, all but the ground-state population are nuisance parameters. Upon integrating the full-space prior over the nuisance parameters, one obtains a *marginal prior* on the reconstruction space. As a function on the reconstruction space, the marginal prior is naturally parameterized in terms of the probabilities and so fits into the formalism we are using throughout.

We note that the invoking of 'additional criteria' for a unique mapping from p to ρ, as mentioned at the end of section 2.1, is exactly what would be required if one wishes to report estimated values of the nuisance parameters. That, however, goes beyond making statements that are solidly supported by the data and is, therefore, outside the scope of our present discussion.

The symmetric uniform-in-purity prior of sections A.2 and A.3 provides an example for marginalization if the POM only gives information about $x = \langle\sigma_x\rangle$ and $y = \langle\sigma_y\rangle$ but not about $z = \langle\sigma_z\rangle$. We express the full-space prior in Cartesian coordinates, integrate over z, and arrive at

$$\begin{aligned}(\mathrm{d}\rho) &= \mathrm{d}x\,\mathrm{d}y\,\frac{1}{2\pi}\int_{-\infty}^{\infty}\mathrm{d}z\,\frac{\eta(1-x^2-y^2-z^2)}{\sqrt{x^2+y^2+z^2}}\\ &= \mathrm{d}x\,\mathrm{d}y\,\frac{1}{\pi}\,\eta(1-x^2-y^2)\cosh^{-1}\frac{1}{\sqrt{x^2+y^2}}\,.\end{aligned} \tag{A.14}$$

This marginal prior is a function on the unit disk in the xy plane, which is the natural choice of reconstruction space here. When one expresses $(\mathrm{d}\rho)$ in polar coordinates, one sees that $(\mathrm{d}\rho)$ is uniform in φ and in $r^2\cosh^{-1}(1/r) - \sqrt{1-r^2}$, which increases monotonically from -1 to 0 on the way from the center of the disk at $r = 0$ to the unit circle where $r = 1$. Plot (a) in figure A.1 illustrates the matter.

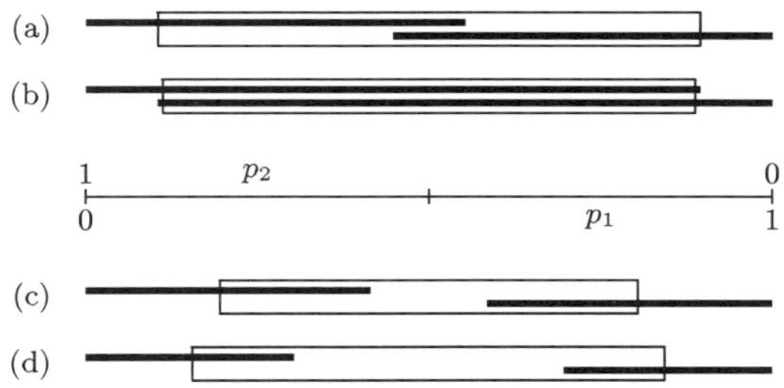

Figure B.1. Confidence regions and smallest credible regions. The bars indicate intervals of $p_1 = 1 - p_2$ for the harmonic-oscillator example of section 2.1, which has the reconstruction space of a tossed coin. Two copies are measured. The left solid bars indicate the regions for $(n_1, n_2) = (0, 2)$ counts; the right solid bars are for $(n_1, n_2) = (2, 0)$; and, the central open bars are for $(n_1, n_2) = (1, 1)$. Cases (a) and (b) show two sets of confidence regions for confidence level $\gamma = 0.8$. Regions (c) and (d) are the SCRs for the primitive prior and the Jeffreys prior, respectively, both for credibility $c = 0.8$.

Appendix B. Confidence regions

The confidence regions that were recently studied by Christandl and Renner [7], and independently by Blume-Kohout [8], are markedly different from the MLRs and the SCRs. The MLR and the SCR represent inferences drawn about the unknown state ρ from the data D that have actually been observed. By contrast, confidence regions are a set of regions, one region for each data, whether observed or not, from the measurement of N copies. The confidence regions would contain *any* state in, at least, a certain fraction of many N-copy measurements, if many measurements were performed. This fraction is the confidence level.

When denoting by $\mathcal{C}_D$ the confidence region for data D, the confidence level γ of the set $\mathbf{C}$ of $\mathcal{C}_D$s for all conceivable data (for fixed N) is

$$\gamma(\mathbf{C}) = \min_{\rho} \sum_{D} L(D|\rho)\, \eta_{\mathcal{C}_D}(\rho), \tag{B.1}$$

where the minimum is reached in the 'worst case'. For example, in the security analysis of a protocol for quantum key distribution, one wishes a large value of γ to protect against an adversary who controls the source and prepares the quantum-information carriers in the state that is best for her.

Any set $\mathbf{C}$, for which γ has the desired value, serves the purpose. A smaller set $\mathbf{C}'$, in the sense that $\mathcal{C}'_D$ is contained in $\mathcal{C}_D$ for all D, is preferable, but usually there is no smallest set of confidence regions. Here, 'smaller' is solely in this inclusion sense, with no reference to a quantification of the size of a region and, therefore, there is no necessity for specifying the prior probability of any region. Since the transition from set $\mathbf{C}$ to the smaller set $\mathbf{C}'$ requires the shrinking of some of the $\mathcal{C}_D$s without enlarging even a single one, it is easily possible to have two sets of confidence regions with the same confidence level and neither set smaller than the other.

For illustration, we consider the harmonic-oscillator example of section 2.1 yet another time. Figure B.1 shows two sets of confidence regions ($\gamma = 0.8$) and the corresponding three SCRs ($c = 0.8$) for the primitive prior and the Jeffreys prior. Both sets of confidence regions are optimal in the sense that one cannot shrink even one of the regions without decreasing the confidence level, but neither set is smaller than the other. In the absence of additional criteria that specify a preference, both work equally well as sets of confidence regions. This generic non-uniqueness of confidence regions, and the arbitrariness associated with it, are in marked contrast to the SCRs, which are always unique.

We also observe in this example that confidence regions tend to overlap a lot, which is indeed unavoidable if a large confidence level is desired; whereas, the SCRs for different data usually do not overlap unless the data are quite similar. In figure B.1, there is no overlap of the SCRs for $(n_1, n_2) = (0, 2)$ and $(2, 0)$.

Another important difference of considerable concern in all practical applications is the following. Once the data are obtained, there is *the* MLR and *the* SCR for these data, and it plays no role what other MLRs or SCRs are associated with different data that have not been observed. To find a confidence region for the actual data, however, one must first specify the whole set **C** of confidence regions because the confidence level of (B.1) is a property of the whole set.

References

[1] Paris M and Řeháček J (ed) 2004 *Quantum State Estimation* (*Lecture Notes in Physics* vol 649) (Heidelberg: Springer)

[2] Hradil Z, Řeháček J, Fiurášek J and Ježek M 2004 Maximum-likelihood methods in quantum mechanics *Quantum State Estimation* (*Lecture Notes in Physics* vol 649) ed M Paris and J Řeháček (Heidelberg: Springer) chapter 3

[3] Teo Y S 2012 Numerical estimation schemes for quantum tomography *PhD Thesis* Singapore (arXiv:1302:3399 [quant-ph])

[4] Řeháček J, Mogilevtsev D and Hradil Z 2008 *New J. Phys.* **10** 043022

[5] Audenaert K M R and Scheel S 2009 *New J. Phys.* **11** 023028

[6] Efron B and Tibshirani R J 1993 *An Introduction to the Bootstrap* (New York: Chapman and Hall)

[7] Christandl M and Renner R 2012 *Phys. Rev. Lett.* **109** 120403

[8] Blume-Kohout R 2012 Robust error bars for quantum tomography arXiv:1202.5270 [quant-ph]

[9] Berger J O 1985 *Statistical Decision Theory and Bayesian Analysis* 2nd edn (New York: Springer) chapter 4

[10] Evans M J, Guttman I and Swartz T 2006 *Can. J. Stat.* **34** 113

[11] Jaynes E T 1957 *Phys. Rev.* **106** 620

[12] Jaynes E T 1957 *Phys. Rev.* **108** 171

[13] Teo Y S, Zhu H, Englert B-G, Řeháček J and Hradil Z 2011 *Phys. Rev. Lett.* **107** 020404

[14] Bužek V 2004 Quantum tomography from incomplete data via MaxEnt principle *Quantum State Estimation* (*Lecture Notes in Physics* vol 649) ed M Paris and J Řeháček (Heidelberg: Springer) chapter 6

[15] Jeffreys H 1946 *Proc. R. Soc. Lond.* A **186** 453

[16] Kass R E and Wasserman L 1996 *J. Am. Stat. Assoc.* **91** 1343

[17] Teo Y S, Stoklasa B, Englert B-G, Řeháček J and Hradil Z 2012 *Phys. Rev.* A **85** 042317

[18] Wasserman L A 1989 *Ann. Stat.* **17** 1387

[19] Press W H, Teukolsky S A, Vetterling W T and Flannery B P 2007 *Numerical Recipes: The Art of Scientific Computing* 3rd edn (Cambridge: Cambridge University Press)

[20] Bennett C and Brassard G 1984 *IEEE Conf. Computers, Systems and Signal Processing, Bangalore, India* (New York: IEEE) p 175

[21] Tabia G and Englert B-G 2011 *Phys. Lett.* A **375** 817

[22] Lehmann E L and Casella G 1998 *Theory of Point Estimation* 2nd edn (Berlin: Springer)

[23] Ng H K, Phuah K T B and Englert B-G 2012 *New J. Phys.* **14** 085007

[24] Jaynes E T 2003 *Probability Theory—The Logic of Science* (Cambridge: Cambridge University Press)

[25] Blume-Kohout R 2010 *Phys. Rev. Lett.* **105** 200504

Jiangwei Shang

Centre for Quantum Technologies, National University of Singapore, Singapore

Paper Title: **Symmetric minimal quantum tomography by successive measurements**

Reference: Physical Review A **85**, 052116 (2012)

Personal message:

First of all, I want to highlight that it is my greatest honour to be Berge's first PhD student to graduate shortly after his 60th birthday. There is a Chinese saying that "Once my teacher, always my teacher"; it has been almost five years since I started my undergraduate honours year project with Berge. Besides being a great physicist, Berge also excels as a cook, an international chess player, and so on and so forth. The Christmas turkey that Berge makes every year is always the best; the snack that Berge serves during our mahjong sessions with Ola is probably the reason that I can win, haha! I am also impressed by Berge's broad knowledge about politics and history; I mean almost every civilization's history. For some time, I was even ashamed of knowing so much less about my own country's history than Berge did.

The paper I selected to include here is the very first published paper in a journal that bears my name. But, the importance of this paper to me is not just the academic achievement, but also the experience that I've gained from doing this project. In quantum information theory, the problem of constructing symmetric, informationally complete probability-operator measurements (SIC POMs) in a finite dimension is considered to be a hard one, and researchers have devoted a lot of energy and time to it. Another equally hard problem is the construction of a maximal set of mutually unbiased bases (MUB) in finite dimensions. In this paper, using a two-step measurement scheme, we found an unexpected operational relation between SIC POMs and MUB (in certain dimensions): The latter are used to construct the former. Berge's deep insights into SIC POMs and MUB made this paper possible and I've also benefitted a lot from his guidance about academic thinking and writing.

Happy birthday, Berge!!!

PHYSICAL REVIEW A **85**, 052116 (2012)

Symmetric minimal quantum tomography by successive measurements

Amir Kalev,[1] Jiangwei Shang,[1] and Berthold-Georg Englert[1,2]

[1]*Centre for Quantum Technologies, National University of Singapore, 3 Science Drive 2, 117543, Singapore*

[2]*Department of Physics, National University of Singapore, 2 Science Drive 3, 117542, Singapore*

(Received 1 March 2012; published 17 May 2012)

We consider the implementation of a symmetric informationally complete probability-operator measurement (SIC POM) in the Hilbert space of a d-level system by a two-step measurement process: a diagonal-operator measurement with high-rank outcomes, followed by a rank-1 measurement in a basis chosen in accordance with the result of the first measurement. We find that any Heisenberg-Weyl group-covariant SIC POM can be realized by such a sequence where the second measurement is simply a measurement in the Fourier basis, independent of the result of the first measurement. Furthermore, at least for the particular cases studied, of dimensions 2, 3, 4, and 8, this scheme reveals an unexpected operational relation between mutually unbiased bases and SIC POMs; the former are used to construct the latter. As a laboratory application of the two-step measurement process, we propose feasible optical experiments that would realize SIC POMs in various dimensions.

DOI: 10.1103/PhysRevA.85.052116 PACS number(s): 03.65.Wj, 03.67.−a, 42.50.Ex

I. INTRODUCTION

Quantum-state tomography is a measurement procedure designed to acquire complete information about the state of a given quantum system. It is an important component in most, if not all, quantum-computation and quantum-communication tasks. The successful execution of such tasks hinges in part on the ability to assess the state of the system at various stages.

A general measurement in quantum mechanics is a probability-operator measurement (POM). A POM is informationally complete (IC) if any state of the system is determined completely by the measurement statistics [1–3]. State tomography infers these probabilities from the data acquired with the aid of the POM.

A symmetric IC POM (SIC POM) is an IC POM of a particular kind. In a d-dimensional Hilbert space (of kets) it is composed of d^2 subnormalized rank-1 projectors with equal pairwise fidelity [4]. Their high symmetry and high tomographic efficiency have attracted the attention of many researchers, and a lot of work, both analytical and numerical, has been devoted to the construction of SIC POMs in various dimensions; see, e.g., Refs. [5–10]. While a rigorous proof for the existence of SIC POMs in any finite dimension is still missing, a great deal of numerical evidence suggests strongly that group-covariant SIC POMs indeed exist in all finite-dimensional Hilbert spaces.

A group-covariant SIC POM is a measurement which can be generated from a single projector (onto the so-called fiducial state) under the action of a group consisting of unitary operations. Almost all known SIC POMs are covariant with respect to the Heisenberg-Weyl (HW) group (also known as the generalized Pauli group). Two SIC POMs are said to be equivalent if there is a unitary operator that maps one SIC POM to the other.

In contrast to the major theoretical progress all experiments and even proposals for experiments implementing SIC POMs have been limited to the very basic quantum system, the two-level system (qubit) [11,12], with the exception of the recent experiment by Medendorp *et al.* [13], where a SIC POM for a three-level system was approximated. This is, in part, due to the fact that there is no systematic procedure for implementing SIC POMs in higher dimensions, in a simple experimental setup.

Recently, however, an experiment that realizes a SIC POM in the four-dimensional Hilbert space of a qubit pair was proposed [14]. The experimental scheme exploited a new approach to SIC POMs that uses a two-step process: a measurement with full-rank outcomes, followed by a projective measurement on a basis that is chosen in accordance with the result of the first measurement. In this work, following the ideas presented in Ref. [14], we explore the possibilities of implementing SIC POMs using a successive-measurement scheme. We start by "breaking" a given SIC POM into two successive measurements, each with d outcomes, with the intention that each measurement will be relatively easy to implement. Unexpectedly, we find that this approach provides a simple, systematic procedure to implement all HW group-covariant SIC POMs (HW SIC POMs). The latter could be realized by first implementing a POM with high-rank outcomes diagonal in a given basis followed by a rank-1 projective measurement, where the basis of the first measurement and the basis of the second measurement are related by the Fourier transform (FT).

Based on this approach, we propose a general experimental scheme implementing HW SIC POMs in the Hilbert space of a d-dimensional quantum system (a qudit). In this scheme, the qudit is carried by a single photon as a path qudit, and the implementation is accomplished by means of linear optics. In particular, we show that the one-parameter family of nonequivalent HW SIC POMs in dimension 3 could be implemented using the successive-measurement approach in a single experimental setup. Furthermore, we study the construction of the known SIC POMs in dimensions 2, 4, and 8 from two successive measurements. We find that the concept of mutually unbiased bases (MUB) plays a central role in the construction of SIC POMs in these dimensions—a hint at a possibly profound link between SIC POMs and MUB; Ref. [15] is a recent review on MUB.

The paper is organized as follows. Section II is concerned with finite-dimensional Hilbert spaces. There we discuss the formulation of SIC POMs in general, and the HW SIC POMs in particular, in terms of two successive measurements. Then we study the construction of known SIC POMs in particular dimensions. In Sec. III, we reformulate the SIC POM in dimension 2 (known as the tetrahedron measurement) in terms of successive measurements, and show that the actual implementation of it by Ling *et al.* [11] was indeed carried out using a sequential-measurement scheme. We also show how a relation between the SIC POM and MUB in dimension 2 is revealed through this formulation. In Sec. IV we study the decomposition of all known nonequivalent SIC POMs in dimension 3 into two successive measurements. We show that this decomposition allows the implementation of all (known) nonequivalent SIC POMs in dimension 3 with a single experimental setup. In Sec. V, we study the realization of the (known) SIC POMs in dimension 4 by successive measurements. Here we also find an interesting structural and operational relation between MUB and SIC POMs. We briefly describe a proposal for their implementation, using single-photon sources together with passive linear optical elements [14]. In Sec. VI, we discuss the construction of the three known, nonequivalent, group-covariant SIC POMs in dimension 8 in terms of successive measurements. We show that the one that is covariant with respect to the three-qubit Pauli group has the same structure as the SIC POMs in the other studied dimensions. Finally, we offer conclusions in Sec. VII.

II. THE GENERAL CASE

A general measurement on a quantum system is composed of a set of outcomes. These outcomes are mathematically represented by positive operators $\mathcal{P}_j$ that sum up to the identity, $\mathcal{P}_j \geqslant 0$ with $\sum_j \mathcal{P}_j = 1$. The probability of obtaining the outcome $\mathcal{P}_j$ is given by the Born rule: $p_j = \mathrm{tr}\{\mathcal{P}_j\rho\}$, where ρ is the premeasurement statistical operator of the system. If the jth outcome is found, the postmeasurement statistical operator of the system is given by $\rho_j = \frac{1}{p_j}P_j\rho P_j^\dagger$, where P_j is the relevant Kraus operator for the jth outcome, $\mathcal{P}_j = P_j^\dagger P_j$. Note that the decomposition of the $\mathcal{P}$s into the corresponding Kraus operators is not unique; for example, $P_j^\dagger P_j$ is invariant under the unitary transformation $P_j \to U_j P_j$, with different U_js corresponding to different implementations of the POM.

Suppose that a given system is subjected to a sequence of two POMs, each with d outcomes $\{\mathcal{A}_k = A_k^\dagger A_k\}_{k=1}^d$, followed by $\{\mathcal{B}_j^{(k)}\}_{j=1}^d$, where the superscript k indicates that in general the second measurement depends on the actual outcome of the first measurement. Following Born's rule, the probability of obtaining the nth and mth outcomes for the first and second measurements is given by $\mathrm{tr}\{\rho A_n^\dagger \mathcal{B}_m^{(n)} A_n\}$. Accordingly, the two successive measurements are equivalent to a single POM with d^2 outcomes that are labeled by a pair of indices $\mathcal{P}_{n,m} = A_n^\dagger \mathcal{B}_m^{(n)} A_n$, with $n,m = 1,\ldots,d$. Indeed, summing $\mathcal{P}_{n,m}$ over the outcomes labeled by m yields the outcome $\mathcal{A}_n$. Therefore, upon finding the overall outcome $\mathcal{P}_{n,m}$, we know that the nth outcome of the first POM and the mth outcome of the second POM are the case. In what follows we will identify the As and the $\mathcal{B}$s such that the $\mathcal{P}$s make up a SIC POM in the d-dimensional Hilbert space of a qudit.

A. HW SIC POMs

Let us begin by showing that *all* HW SIC POMs could be realized by a two-step measurement scheme with a rather simple structure—a high-rank, diagonal-operator measurement, followed by a measurement in the Fourier basis. In a d-dimensional Hilbert space, the HW group is composed of d^2 unitary operators $\{X^k Z^j\}$, $j,k = 1,\ldots,d$, with Z and X, the generators of the group, defined as

$$Z = \sum_{n=0}^{d-1} |n\rangle\omega^n\langle n|, \quad X = \sum_{n=0}^{d-1} |n\oplus 1\rangle\langle n|, \tag{1}$$

where $\omega = e^{2\pi i/d}$ is the fundamental dth root of unity and $\oplus$ stands for the sum modulo d. A HW SIC POM in a d-dimensional Hilbert space has d^2 outcomes $\mathcal{P}_{k,j}$ with $k,j = 1,\ldots,d$, which are generated from a single projector onto a fiducial state $|\psi_{\mathrm{fid}}\rangle$, under the action of the group elements,

$$\mathcal{P}_{k,j} = X^k Z^j |\psi_{\mathrm{fid}}\rangle\frac{1}{d}\langle\psi_{\mathrm{fid}}|Z^{j\dagger}X^{k\dagger}. \tag{2}$$

The fiducial state is chosen such that the $\mathcal{P}$s satisfy the defining property of a SIC POM [16],

$$\mathrm{tr}\{\mathcal{P}_{k,j}\mathcal{P}_{m,n}\} = \frac{1}{d^2}\left(\delta_{k,m}\delta_{j,n} + (1-\delta_{k,m}\delta_{j,n})\frac{1}{d+1}\right). \tag{3}$$

With the (normalized) fiducial state

$$|\psi_{\mathrm{fid}}\rangle = \sum_{n=0}^{d-1} |n\rangle\alpha_n \tag{4}$$

in Eq. (2), we obtain

$$\mathcal{P}_{k,j} = \frac{1}{d}\sum_{n,m=0}^{d-1} |m\oplus k\rangle\alpha_m\omega^{(m-n)j}\alpha_n^*\langle n\oplus k|. \tag{5}$$

At this point we note that the right-hand side of Eq. (5) has a two-step measurement structure. The Kraus operators corresponding to the outcomes of the first measurement are

$$A_k = \sum_{m=0}^{d-1} |m\oplus k\rangle\alpha_m\langle m\oplus k| \tag{6}$$

with $k = 1,\ldots,d$, while the outcomes of the second measurement are projections onto the eigenstates of the Fourier basis,

$$\mathcal{B}_j = \frac{1}{d}\sum_{m,n=0}^{d-1} |m\rangle\omega^{(m-n)j}\langle n|, \tag{7}$$

with $j = 1,\ldots,d$. Indeed, for Eqs. (5)–(7) we have

$$\mathcal{P}_{k,j} = A_k^\dagger\mathcal{B}_j A_k, \tag{8}$$

so that the HW SIC POM for the fiducial state of Eq. (4), when it exists, is realized as a two-step measurement. This demonstrates the case. If we relax the requirement that the sequential measurements in Eq. (8) compose a *symmetric* IC POM, one can show [17] that in any finite-dimensional Hilbert space there exist αs such that these measurements are IC.

The mathematical formulation of SIC POMs as a two-step measurement process hints at the possibility for their implementation. Here, we propose a general experimental

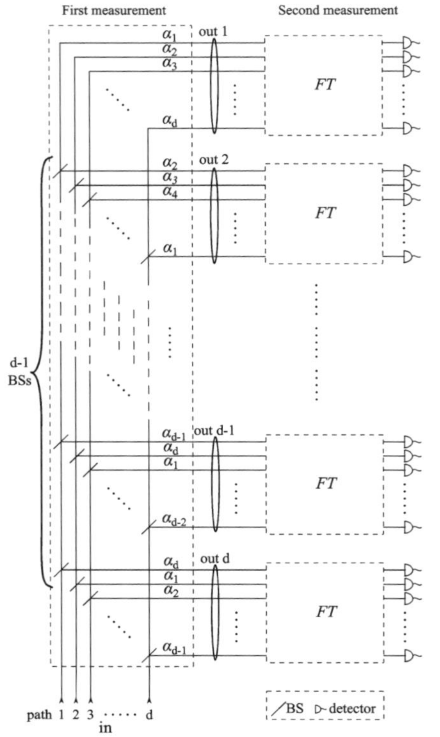

FIG. 1. An optical implementation of a HW SIC POM using a two-step measurement process.

scheme with which any HW SIC POM in a d-dimensional Hilbert space of a qudit could be realized. The qudit is carried by a single photon and is encoded in d spatial alternatives of the photon ("path qudit"). A unitary transformation on the qudit state amounts to sending the photon through a set of beam splitters (BSs) and phase shifters (PSs), similar to the methods presented in Ref. [18].

In this optical setting the HW SIC POMs are implemented as follows (see Fig. 1 as a reference): The first measurement setup is designed to implement the Kraus operators of Eq. (6) by appropriately choosing the reflection amplitudes of the $d-1$ BSs at each path. The choice $t_{n-1,k}r_{n,k} = \alpha_{k\ominus n}$, where $r_{n,k}$ and $t_{n,k}$ are the reflection and the transmission amplitudes of the nth BS at the kth path (here $n = 1,\ldots,d-1$, $k = 1,\ldots,d$, the BSs are counted from the entrance port, and the paths are numbered from left to right, as indicated in the figure; we define $t_{0,k} = 1$ for all k), ensures that a photon which enters the apparatus with a path statistical operator ρ exits at port k with the statistical operator $A_k\rho A_k^\dagger/\mathrm{tr}\{A_k\rho A_k^\dagger\}$. Upon exiting the first measurement apparatus, the photon is measured in the Fourier basis (indicated in the figure as a black box labeled by FT). This measurement could be realized by a collection of BSs and appropriate PSs [19].

B. Fuzzy measurements

So far, we considered the decomposition of a given HW SIC POM for a d-level system into a succession of two POMs, each with d outcomes. Now, we would like to follow the reverse path, namely, to start with a given structure for the two POMs, and study under what conditions they compose a SIC POM when measured in succession. In particular, we consider the situation where the first measurement is a "fuzzy measurement," where "fuzzy" means that each of the measurement outcomes corresponds to a projector onto the computational basis $Z_k = |k\rangle\langle k|, k = 1,\ldots,d$, mixed with the identity operator,

$$\mathcal{A}_k = \frac{1}{d}(1-\lambda) + \lambda Z_k, \tag{9}$$

whose positivity requires that $\frac{-1}{d-1} \leqslant \lambda \leqslant 1$. Up to a unitary transformation, the corresponding Kraus operators are

$$A_k = \sqrt{\frac{1-\lambda}{d}}\sum_{j(\neq k)} Z_j + \sqrt{\frac{1+(d-1)\lambda}{d}}Z_k. \tag{10}$$

For the second measurement, we consider a projective measurement on a basis that is chosen in accordance with the result of the first measurement,

$$\mathcal{B}_j^{(k)} = U_k^\dagger Z_j U_k \equiv |j^{(k)}\rangle\langle j^{(k)}|, \tag{11}$$

where $|j^{(k)}\rangle$ is the jth state of the kth basis. The unitary operator U_k specifies the basis for the second measurement. It is worth recalling that the outcomes of the first measurement are invariant under unitary transformations, $A_k \to U_k A_k$. Therefore we can write the overall outcomes of the two-step measurement process as

$$\mathcal{M}_{k,j} = A_k^\dagger \mathcal{B}_j^{(k)} A_k. \tag{12}$$

We now write the necessary and sufficient conditions that the $\mathcal{M}$s represent a SIC POM, that is, that this ansatz works. Equations (3) and (10)–(12) jointly require that

$$\lambda = \pm\frac{1}{\sqrt{1+d}} \quad \text{and} \quad |\langle m|n^{(m)}\rangle|^2 = \frac{1}{d}, \tag{13}$$

as well as for $k \neq m$,

$$|\langle n^{(m)}|[\alpha + (\beta-\alpha)(|m\rangle\langle m| + |k\rangle\langle k|)]|j^{(k)}\rangle|^2 = \frac{1}{d}, \tag{14}$$

with $\alpha = \sqrt{1-\lambda}$, $\beta = \sqrt{1+\lambda(d-1)}$, and all indices take on the values $1,\ldots,d$. From the condition on λ given right after Eq. (9), we get that $\lambda = 1/\sqrt{1+d}$ for $d \geqslant 4$. Note that the indices k and m label the first measurement while j and n label the second one. Recalling the definition of unbiased bases [15], Eq. (13) implies that different bases of the second measurement are unbiased to one of the states from the computational basis.

We are able to solve Eqs. (13) and (14) for dimensions 2 and 3 and can also show that the known SIC POM in dimension 4 is a solution for these equations. Unfortunately we did not manage to solve or prove the existence of a solution for these equations in higher dimensions. In the following sections we discuss the solutions for these equations.

III. DIMENSION 2: A QUBIT

The solution for Eqs. (13) and (14) in dimension 2 is fairly straightforward since $|m\rangle\langle m| + |k\rangle\langle k| = 1$ for $k \neq m$. Accordingly, for $\lambda = \pm 1/\sqrt{3}$, Eq. (14) reads $|\langle n^{(m)}|j^{(k)}\rangle|^2 = 1/2$. This means that one can realize a SIC POM in dimension 2 by a fuzzy measurement, with the corresponding diagonal Kraus operators

$$A_1 = \mathrm{diag}\left(\sqrt{\frac{1}{2} - \frac{1}{\sqrt{12}}}, \sqrt{\frac{1}{2} + \frac{1}{\sqrt{12}}}\right),$$
$$A_2 = \mathrm{diag}\left(\sqrt{\frac{1}{2} + \frac{1}{\sqrt{12}}}, \sqrt{\frac{1}{2} - \frac{1}{\sqrt{12}}}\right), \tag{15}$$

followed by a measurement in a basis which is chosen in accordance with the result of the first measurement. The solution indicates that the two bases of the second measurements must be unbiased to each other and to the computational basis. For example, if the As of Eq. (15) are diagonal in the σ_3 basis (where the σs are the Pauli operators), then the two MUB of the second measurements could be the σ_1 basis and the σ_2 basis. Actually all SIC POMs for a qubit are unitarily equivalent to the "tetrahedron measurement" (TM), whose outcomes correspond to four vectors that define a tetrahedron in the Bloch sphere [5,6]. As we just showed, the TM could be realized in a two-step measurement process, for example, by using a setup similar to the one presented in Fig. 1.

Actually, the TM was successfully implemented in an optical system [11], where the qubit was encoded in a photon's polarization ("polarization qubit") rather than in spatial alternatives (and therefore there was no need to stabilize interferometric loops in the setup). The setup of [11] also consisted of a sequence of two measurements, quite analogous to what is described above. In that setup, a partially polarizing beam splitter was used to implement the Kraus operators A_k and then, depending on whether the photon was transmitted or reflected, a measurement of σ_1 or σ_2 followed.

Evidently, there is an operational relation between the TM and the three MUB in dimension 2. The latter are used to construct the former by means of successive measurements. This relation actually stems from the common mathematical structure of the four kets (in the Hilbert space of a qubit) corresponding to the TM and the four kets composing the two bases unbiased to the computational basis and to each other. To see this more clearly, consider the columns of the following matrices:

$$\frac{1}{N}\begin{pmatrix} \chi & \chi \\ 1 & -1 \end{pmatrix}, \quad \frac{1}{N}\begin{pmatrix} 1 & 1 \\ i\chi & -i\chi \end{pmatrix}. \tag{16}$$

For $N = \sqrt{3+\sqrt{3}}$ and $\chi = \sqrt{2+\sqrt{3}}$, the four columns represent the kets corresponding to the TM. While for $N = \sqrt{2}$ and $\chi = 1$ the two columns of each matrix form a basis. The two bases are unbiased to each other and to the computational basis. We will see below that similar relations appear in dimensions 3, 4, and 8.

Finally, since the TM is equivalent to a HW SIC POM, it could also be implemented by a two-step process: a measurement with the corresponding Kraus operators of Eq. (6) followed by a measurement in the Fourier basis, Eq. (7) (call it the σ_1 basis). We note that while the outcomes of the first measurement for this process and the outcomes for the fuzzy measurement in the process discussed in this section are the same, the Kraus operators, and therefore the implementations, are different.

IV. DIMENSION 3: A QUTRIT

According to the conditions listed in Eqs. (13) and (14), the ansatz of Eqs. (10)–(12) yields a SIC POM in dimension 3, if and only if $|\langle m|n^{(m)}\rangle|^2 = 1/3$ and either $\lambda = 1/2$ and

$$|\langle n^{(m)}|(1 + |m\rangle\langle m| + |k\rangle\langle k|)|j^{(k)}\rangle|^2 = 1, \tag{17}$$

with $k \neq m$, or $\lambda = -1/2$ and

$$|\langle n^{(m)}|l\rangle\langle l|j^{(k)}\rangle|^2 = \tfrac{1}{9}, \tag{18}$$

with $k \neq m \neq l$. Whereas Eq. (17) does not have a solution, Eq. (18) can be solved. One possible solution is $|\langle n^{(m)}|l\rangle|^2 = |\langle l|j^{(k)}\rangle|^2 = 1/3$. This implies that a SIC POM in dimension 3 could be broken into a sequence of a fuzzy measurement with $\lambda = -1/2$,

$$A_k^\dagger = \frac{1}{\sqrt{2}}(|k\oplus 1\rangle\langle k\oplus 1| + |k\oplus 2\rangle\langle k\oplus 2|), \tag{19}$$

with $k = 1,2,3$, followed by a projective measurement onto a basis unbiased to the computational basis (in which the As are diagonal).

Actually, in dimension 3, there exists a one-parameter family of non-equivalent HW SIC POMs [6,7,10],

$$\Pi_{k,j} = \frac{1}{9}|\phi_{k,j}(\gamma)\rangle\langle\phi_{k,j}(\gamma)|,$$
$$|\phi_{k,j}(\gamma)\rangle = \frac{1}{\sqrt{2}}(|k\rangle - e^{i2\gamma}\omega^j|k\oplus 1\rangle), \tag{20}$$

where $k,j = 1,2,3$, $0 \leqslant \gamma \leqslant \pi/6$, and $\omega = \exp(2i\pi/3)$. This continuum of SIC POMs could be realized using our ansatz in the following way: First, a fuzzy measurement, with the corresponding Kraus operators of Eq. (19), is carried out. Then, if the kth outcome is found the system goes through the diagonal unitary transformation

$$U_k^\dagger = |k\rangle\langle k| - |k\oplus 1\rangle e^{i2\gamma}\langle k\oplus 1| + |k\oplus 2\rangle\langle k\oplus 2|. \tag{21}$$

And last, the system is subjected to a projective measurement onto a basis unbiased to the computational basis, say the Fourier basis; cf. Eq. (7) with $d = 3$.

This procedure implements the SIC POMs in Eq. (20) for all γ. From an operational point of view, this result shows that the entire family of nonequivalent SIC POMs could, in principle, be realized in a single setup. In Fig. 2 we present such an implementation in an optical setting for a path qutrit. The balanced (1:1) BSs in the first part are used to implement the fuzzy measurement, then the γ-dependent unitary transformations of Eq. (21) are implemented by path-dependent PSs placed in the appropriate paths, and finally the Fourier transformation is applied to the state of the qutrit after which the path of the photon is detected. The Fourier transformation is realized using three BSs: BS1, BS2, and BS3, which implement the transformations $(\sigma_3 + \sigma_1)/\sqrt{2}$, $(\sigma_3 + \sqrt{2}\sigma_1)/\sqrt{3}$, and $(\sigma_3 + \sigma_2)/\sqrt{2}$, respectively.

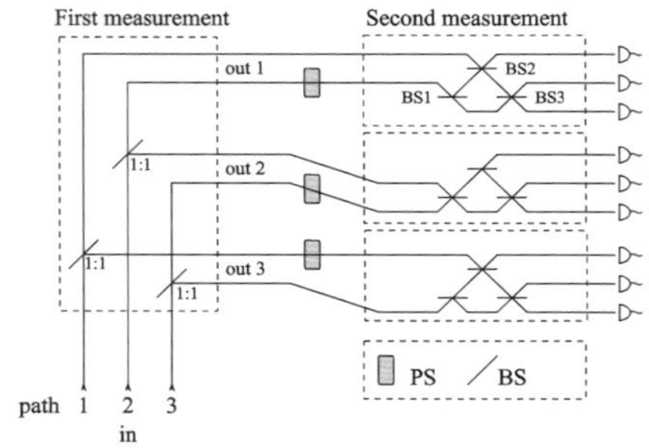

FIG. 2. An optical implementation of the one-parameter family of nonequivalent SIC POMs for a path qutrit.

In the above construction we chose the Fourier basis for the second measurement independent of the outcome of the first measurement. However, the same family of SIC POMs (or its unitary equivalent) could be realized with a different choice for the basis for the second measurement, as long as Eq. (18) is obeyed. For the bases of the second measurement, one may use three MUB which are also unbiased to the computational basis, for example the set

$$\mathfrak{B}_1 = \left\{ \begin{array}{l} \frac{1}{\sqrt{3}}(|0\rangle + |1\rangle + |2\rangle) \\ \frac{1}{\sqrt{3}}(|0\rangle + \omega\,|1\rangle + \omega^2\,|2\rangle) \\ \frac{1}{\sqrt{3}}(|0\rangle + \omega^2\,|1\rangle + \omega\,|2\rangle) \end{array} \right\},$$
$$\mathfrak{B}_2 = \left\{ \begin{array}{l} \frac{1}{\sqrt{3}}(|0\rangle + \omega^2\,|1\rangle + \omega^2\,|2\rangle) \\ \frac{1}{\sqrt{3}}(\omega^2\,|0\rangle + |1\rangle + \omega^2\,|2\rangle) \\ \frac{1}{\sqrt{3}}(\omega^2\,|0\rangle + \omega^2\,|1\rangle + |2\rangle) \end{array} \right\}, \qquad (22)$$
$$\mathfrak{B}_3 = \left\{ \begin{array}{l} \frac{1}{\sqrt{3}}(|0\rangle + \omega\,|1\rangle + \omega\,|2\rangle) \\ \frac{1}{\sqrt{3}}(\omega\,|0\rangle + |1\rangle + \omega\,|2\rangle) \\ \frac{1}{\sqrt{3}}(\omega\,|0\rangle + \omega\,|1\rangle + |2\rangle) \end{array} \right\}.$$

These bases can be used for all values of the parameter γ.

V. DIMENSION 4: TWO QUBITS

In dimension 4, there is only one known HW SIC POM, and all the other known SIC POMs are equivalent to it [7]. This SIC POM is composed of 16 subnormalized projectors onto 16 (fiducial) kets. The latter are represented in the following matrices as columns with $N = \sqrt{5+\sqrt{5}}$ and $\chi = \sqrt{2+\sqrt{5}}$ [20],

$$\frac{1}{N}\begin{pmatrix} \chi & \chi & \chi & \chi \\ 1 & -1 & 1 & -1 \\ 1 & 1 & -1 & -1 \\ 1 & -1 & -1 & 1 \end{pmatrix}, \quad \frac{1}{N}\begin{pmatrix} 1 & 1 & 1 & 1 \\ 1 & -1 & 1 & -1 \\ i\chi & i\chi & -i\chi & -i\chi \\ -i & i & i & -i \end{pmatrix},$$
$$\frac{1}{N}\begin{pmatrix} 1 & 1 & 1 & 1 \\ i\chi & -i\chi & i\chi & -i\chi \\ i & i & -i & -i \\ -1 & 1 & 1 & -1 \end{pmatrix}, \quad \frac{1}{N}\begin{pmatrix} 1 & 1 & 1 & 1 \\ i & -i & i & -i \\ 1 & 1 & -1 & -1 \\ -i\chi & i\chi & i\chi & -i\chi \end{pmatrix}. \qquad (23)$$

Each of these matrices could be written as a diagonal matrix times a unitary matrix. The set of bases corresponding to each unitary matrix together with the computational basis form the complete set of MUB in dimension 4. To be more specific, the diagonal matrices are

$$A_1 = \frac{1}{N}\mathrm{diag}(\chi,1,1,1), \quad A_3 = \frac{1}{N}\mathrm{diag}(1,1,\chi,1),$$
$$A_2 = \frac{1}{N}\mathrm{diag}(1,\chi,1,1), \quad A_4 = \frac{1}{N}\mathrm{diag}(1,1,1,\chi), \qquad (24)$$

and the unitary matrices are

$$\mathcal{U}_1 = \frac{1}{2}\begin{pmatrix} 1 & 1 & 1 & 1 \\ 1 & -1 & 1 & -1 \\ 1 & 1 & -1 & -1 \\ 1 & -1 & -1 & 1 \end{pmatrix},$$
$$\mathcal{U}_3 = \frac{1}{2}\begin{pmatrix} 1 & 1 & 1 & 1 \\ 1 & -1 & 1 & -1 \\ i & i & -i & -i \\ -i & i & i & -i \end{pmatrix},$$
$$\mathcal{U}_2 = \frac{1}{2}\begin{pmatrix} 1 & 1 & 1 & 1 \\ i & -i & i & -i \\ i & i & -i & -i \\ -1 & 1 & 1 & -1 \end{pmatrix}, \qquad (25)$$
$$\mathcal{U}_4 = \frac{1}{2}\begin{pmatrix} 1 & 1 & 1 & 1 \\ i & -i & i & -i \\ 1 & 1 & -1 & -1 \\ -i & i & i & -i \end{pmatrix}.$$

Noting that $\sum_j A_j^\dagger A_j = 1$, we identify the As with the Kraus operators of a measurement.

Actually, the operations of Eq. (25) transform the computational basis into the MUB,

$$\mathfrak{B}_1 = \left\{ \begin{array}{l} \frac{1}{\sqrt{2}}(|0\rangle + |1\rangle) \\ \frac{1}{\sqrt{2}}(|0\rangle - |1\rangle) \end{array} \right\} \otimes \left\{ \begin{array}{l} \frac{1}{\sqrt{2}}(|0\rangle + |1\rangle) \\ \frac{1}{\sqrt{2}}(|0\rangle - |1\rangle) \end{array} \right\},$$
$$\mathfrak{B}_2 = \left\{ \begin{array}{l} \frac{1}{\sqrt{2}}(|0\rangle + i\,|1\rangle) \\ \frac{1}{\sqrt{2}}(|0\rangle - i\,|1\rangle) \end{array} \right\} \otimes \left\{ \begin{array}{l} \frac{1}{\sqrt{2}}(|0\rangle + i\,|1\rangle) \\ \frac{1}{\sqrt{2}}(|0\rangle - i\,|1\rangle) \end{array} \right\},$$
$$\mathfrak{B}_3 = \mathrm{CZ}\left\{ \begin{array}{l} \frac{1}{\sqrt{2}}(|0\rangle + i\,|1\rangle) \\ \frac{1}{\sqrt{2}}(|0\rangle - i\,|1\rangle) \end{array} \right\} \otimes \left\{ \begin{array}{l} \frac{1}{\sqrt{2}}(|0\rangle + |1\rangle) \\ \frac{1}{\sqrt{2}}(|0\rangle - |1\rangle) \end{array} \right\}, \qquad (26)$$
$$\mathfrak{B}_4 = \mathrm{CZ}\left\{ \begin{array}{l} \frac{1}{\sqrt{2}}(|0\rangle + |1\rangle) \\ \frac{1}{\sqrt{2}}(|0\rangle - |1\rangle) \end{array} \right\} \otimes \left\{ \begin{array}{l} \frac{1}{\sqrt{2}}(|0\rangle + i\,|1\rangle) \\ \frac{1}{\sqrt{2}}(|0\rangle - i\,|1\rangle) \end{array} \right\},$$

where CZ stands for the CONTROLLED-Z (phase flip) operation, CZ$=\mathrm{diag}(1,1,1,-1)$. The bases $\mathfrak{B}_1$ and $\mathfrak{B}_2$ are composed of product states, while the bases $\mathfrak{B}_3$ and $\mathfrak{B}_4$ consist of maximally entangled states.

The structural relation between the fiducial kets, Eq. (23), and the kets that compose the four MUB in dimension 4, Eq. (25), is now clear. For $N = \sqrt{5+\sqrt{5}}$ and $\chi = \sqrt{2+\sqrt{5}}$, the columns of each matrix in Eq. (23) form the 16 fiducial kets, while for $N = 1/2$ and $\chi = 1$, the columns of each matrix form a basis. These bases are mutually unbiased to each other and to the computational basis; cf. Eq. (25).

The structure of the fiducial kets of Eq. (23) (which form the SIC POM in dimension 4) allows us to implement the SIC POM by two successive measurements: a measurement whose Kraus operators are listed in Eq. (24) and, depending on the measurement outcome, a measurement in one of the MUB of Eq. (26). We should not fail to mention that the Kraus operators of Eq. (24) correspond to a fuzzy measurement with $\lambda = 1/\sqrt{5}$; cf. Eq. (9).

A proposal for implementing this SIC POM for a two-qubit system by successive measurements was recently given in Ref. [14]. This proposal is based on the ideas presented in Ref. [18] where two-qubit states are encoded in a single-photon state. In this scheme the photon's polarization is one qubit, the polarization qubit, and the orbital alternative of the photon is the second qubit, the path qubit. Performing a unitary transformation on the two-qubit state amounts to sending the photon through a set of passive linear optical elements (optical plates) that unitarily change the state of the path and polarization qubits [18]. In particular, in order to realize the fuzzy measurement, two more path qubits were used as ancillas. In each path, the parameters of the different optical elements were set such that a photon which enters the apparatus with a polarization-path statistical operator ρ exits at path k with the two-qubit statistical operator $A_k\rho A_k^\dagger/\mathrm{tr}\{A_k\rho A_k^\dagger\}$. Thus, a projective measurement (with four possible outcomes) on the ancillary qubits effectively produces the desired POM on the two-qubit system $\{A_k\}$. At the second step, measurements in the MUB were implemented as described in Ref. [14].

VI. DIMENSION 8: THREE QUBITS

In dimension 8 there are three known nonequivalent SIC POMs. One of them is covariant with respect to the three-qubit Pauli group [10,21], while the other two are covariant with respect to the HW group [9]. According to the result presented in Sec. II, the latter two could be realized by a diagonal-operator measurement followed by a measurement in the Fourier basis of the three qubits. Interestingly, as we see in what follows, the former SIC POM (also known as Hoggar's SIC POM [21]) could be broken into a diagonal-operator measurement followed by projective measurements in eight MUB, similarly to what happens in dimensions 2, 3, and 4.

Hoggar's SIC POM is composed of (subnormalized) projectors onto 64 kets. The latter are constructed from the action of the three-qubit Pauli group elements on a fiducial ket $|\phi\rangle$,

$$\left|{}^{(k,l,m)}_{(n,r,s)}\right\rangle = Z_1^n X_1^k \otimes Z_2^r X_2^l \otimes Z_3^s X_3^m |\phi\rangle, \tag{27}$$

where all indices take on the values 1, 2. Here $Z = \sigma_3$ and $X = \sigma_1$ are the generators of the Pauli group in dimension 2, and their subscripts in Eq. (27) label the degree of freedom on which they act. In what follows we omit this subscript when no ambiguity arises. We refer to the basis in which σ_3 is diagonal as the computational basis. In this basis, the fiducial ket $|\phi\rangle$ is represented by

$$\phi = \frac{1}{\sqrt{6}}(\sqrt{2},0,-\omega^*,\omega^*,-\omega,-\omega^*,0,0)^\mathsf{T}, \tag{28}$$

where $\omega = \sqrt{i}$ is the fundamental eighth root of unity.

This SIC POM can be broken into two successive measurements: The Kraus operators corresponding to the first measurement are

$$\begin{aligned}
A_1 &= \frac{2}{\sqrt{3}}\mathrm{diag}(0,-i\sqrt{2},\omega^*,i\omega^*,-\omega^*,-i\omega,0,0),\\
A_2 &= \frac{2}{\sqrt{3}}\mathrm{diag}(-\omega^*,\omega^*,-i\sqrt{2},0,0,0,i\omega,i\omega^*),\\
A_3 &= \frac{2}{\sqrt{3}}\mathrm{diag}(\omega^*,i\omega^*,0,\sqrt{2},0,0,-i\omega^*,\omega),\\
A_4 &= \frac{2}{\sqrt{3}}\mathrm{diag}(-\omega,-\omega^*,0,0,-i\sqrt{2},0,-i\omega^*,-i\omega^*),\\
A_5 &= \frac{2}{\sqrt{3}}\mathrm{diag}(-\omega^*,i\omega,0,0,0,\sqrt{2},-i\omega^*,\omega^*),\\
A_6 &= \frac{2}{\sqrt{3}}\mathrm{diag}(0,0,i\omega,i\omega^*,i\omega^*,i\omega^*,\sqrt{2},0),\\
A_7 &= \frac{2}{\sqrt{3}}\mathrm{diag}(0,0,i\omega^*,\omega,-i\omega^*,\omega^*,0,i\sqrt{2}),\\
A_8 &= \frac{2}{\sqrt{3}}\mathrm{diag}(\sqrt{2},0,-\omega^*,\omega^*,-\omega,-\omega^*,0,0).
\end{aligned} \tag{29}$$

The basis for the second measurement is chosen in accordance with the result of the first measurement. The eight different bases for the second measurement, together with the computational basis, form a complete set of MUB in dimension 8. This set is constructed as follows. Consider the two unbiased bases in dimension 2—the ones that correspond to the Pauli matrices σ_1 and σ_2. We label these bases by $\{|e_v^b\rangle\}$ where $v = 1,2$ labels the vector in the basis $b = 1,2$. Furthermore, consider the operator

$$\begin{aligned}
\mathcal{G}(k,l,m) = \tfrac{1}{2}\big(&1 + \sigma_3^k \otimes \sigma_3^l \otimes \sigma_3^m + \sigma_3^{1-k} \otimes \sigma_3^{1-l} \otimes \sigma_3^{1-m}\\
&- \sigma_3 \otimes \sigma_3 \otimes \sigma_3\big),
\end{aligned} \tag{30}$$

with $k,l,m = 1,2$. The states defined by a fixed triplet (k,l,m),

$$\left|e_{(n,r,s)}^{(k,l,m)}\right\rangle = \mathcal{G}(k,l,m)\left|e_n^m\right\rangle \otimes \left|e_r^k\right\rangle \otimes \left|e_s^l\right\rangle, \tag{31}$$

form a basis, while bases with different (k,l,m) triplets are mutually unbiased. These bases together with the computational basis form a complete set of MUB for three qubits. The 64 fiducial kets in Eq. (27) are

$$\left|{}^{(k,l,m)}_{(n,r,s)}\right\rangle = A_{(k,l,m)}\left|e_{(n,r,s)}^{(k,l,m)}\right\rangle, \tag{32}$$

where the index of the Kraus operator is written in binary representation, with m the least significant bit. Indeed, the last equation implies that Hoggar's SIC POM can be realized by a measurement with the corresponding Kraus operators $A_{(k,l,m)}$ and, depending on the result, followed by a measurement in one of the MUB.

Finally, we note that the above construction of the SIC POM in dimension 8 is different in two points from the constructions given for the SIC POMs in dimensions 2, 3, and 4. First, the SIC POMs in dimensions 2, 3, and 4 are covariant with respect to the HW group while Hoggar's SIC POM is covariant with respect to the three-qubit Pauli group. And second, in dimensions 2, 3, and 4 the first measurement is a fuzzy measurement while in dimension 8 this is not the case.

VII. CONCLUSIONS

SIC POMs are considered to be hard to implement. Here, we are proposing to implement them by breaking the measurement process into two steps, having in mind that each step should be rather easy to implement. Based on this idea, we presented a systematic procedure that implements HW SIC POMs in finite-dimensional systems. The implementation is accomplished by a diagonal-operator measurement with high-rank outcomes followed by a rank-1 measurement in the Fourier basis. As an example we have considered the realization of HW SIC POMs for a path qudit encoded in a single photon. Moreover, we found that if we take the first measurement to be a fuzzy measurement and let the bases for the second measurement be chosen in accordance with the result of the first measurement, then in the particular studied cases (dimensions 2, 3, and 4) an operational link between SIC POMs and MUB appears, that is, the MUB are used to implement the SIC POMs in the successive-measurement scheme. A similar link was found in dimension 8 as well, but here the first measurement was not of the fuzzy kind.

There is still an open question as to the generality of such a relation and its origin. Currently it is unclear whether the successive-measurement approach will provide a reasonable scheme for implementing SIC POMs in arbitrary dimensions and thus reveal their structure in high-dimensional Hilbert spaces.

ACKNOWLEDGMENTS

We would like to thank Huangjun Zhu for valuable and stimulating discussions. The Centre for Quantum Technologies is a Research Centre of Excellence funded by the Ministry of Education and the National Research Foundation of Singapore.

[1] E. Prugovečki, Int. J. Theor. Phys. **16**, 321 (1977).
[2] P. Busch, Int. J. Theor. Phys. **30**, 1217 (1991).
[3] G. M. D'Ariano, P. Perinotti, and M. F. Sacchi, J. Opt. B: Quantum Semiclass. Opt. **6**, S487 (2004).
[4] One can lift the restriction that the POM outcomes are rank 1 while maintaining the SIC property, but we are not considering this more general situation here.
[5] J. Řeháček, B.-G. Englert, and D. Kaszlikowski, Phys. Rev. A **70**, 052321 (2004).
[6] J. M. Renes, R. Blume-Kohout, A. J. Scott, and C. M. Caves, J. Math. Phys. **45**, 2171 (2004).
[7] D. M. Appleby, J. Math. Phys. **46**, 052107 (2005).
[8] A. J. Scott, J. Phys. A **39**, 13507 (2006).
[9] A. J. Scott and M. Grassl, J. Math. Phys. **51**, 042203 (2010).
[10] G. Zauner, Int. J. Quantum Inf. **9**, 445 (2011).
[11] A. Ling, K. P. Soh, A. Lamas-Linares, and C. Kurtsiefer, Phys. Rev. A **74**, 022309 (2006).
[12] W. M. Pimenta, B. Marques, M. A. Carvalho, M. R. Barros, J. G. Fonseca, J. Ferraz, M. Terra Cunha, and S. Pádua, Opt. Express **18**, 24423 (2010).
[13] Z. E. D. Medendorp, F. A. Torres-Ruiz, L. K. Shalm, G. N. M. Tabia, C. A. Fuchs, and A. M. Steinberg, Phys. Rev. A **83**, 051801(R) (2011).
[14] A. Kalev, J. Shang, and B.-G. Englert, Phys. Rev. A **85**, 052115 (2012).
[15] T. Durt, B.-G. Englert, I. Bengtsson, and K. Ż yczkowski, Int. J. Quantum Inf. **8**, 535 (2010).
[16] The geometrical significance of Eq. (3) is discussed in H. K. Ng and B.-G. Englert, Int. J. Quantum Inf. **11**, 1250038 (2012), Appendix A.
[17] C. Carmeli, T. Heinosaari, and A. Toigo, Phys. Rev. A **85**, 012109 (2012).
[18] B.-G. Englert, C. Kurtsiefer, and H. Weinfurter, Phys. Rev. A **63**, 032303 (2001).
[19] M. Reck, A. Zeilinger, H. J. Bernstein, and P. Bertani, Phys. Rev. Lett. **73**, 58 (1994).
[20] I. Bengtsson, J. Phys.: Conf. Ser., **254**, 012007 (2010).
[21] S. G. Hoggar, Geom. Det. **69**, 287 (1998).

Jibo Dai

Centre for Quantum Technologies, National University of Singapore, Singapore

Paper Title: **Experimental detection of entanglement with optimal-witness families**

Reference: Physical Review Letters **113**, 170402 (2014)

Personal message:

Before I ever met Prof. Englert, I had actually heard many things about him. Some say that he is a very stern person. Some say that he is teaching the most difficult module and sets the most difficult tests. As a year-two undergraduate student at that time, the closest I had ever come to him is in admiring one of the three books that he wrote — *Lectures in quantum mechanics: Basic matters*. His image was kind of a mystery to me then. Hence when I knew that he was actually my academic mentor and eventually I had a chance to meet him, I was filled with anticipation, excitement, and yet a little bit of anxiety. But when I actually met him in the science canteen and he treated me to a drink from the cafe, my worry slowly faded away as we started to chat. He appeared not so unapproachable as many had described and actually had a very caring personality. He asked me how I was and where I came from etc. I told him I was from Chongqing, and got totally amazed when he said that he knew it and that he had read some books about it. Then he started to describe to me those books. Even I had never heard of those books. He also asked me how my studies were and what my plans were in the future. I told him that I really hoped to be a professor in the future but also had a difficult situation as I had a bond and could not go overseas for PhD. He tried to console me that things would be able to work out fine in the future and all I had to do first was to study and work hard. Prof. Englert left with me an impression of being a caring, knowledgeable, well-read and humorous person after that day. I had actually never imagined that I could meet him over a cup of coffee one day (and I continue to be amazed by how well and widely read he is, and laugh at the humorous jokes that he tells). Neither had I imagined that I could have an opportunity to be his PhD student one day.

That opportunity came in the year of 2011 when I got to know that the Centre for Quantum Technologies (CQT) was offering scholarship for PhD students and Berge was (and still is) one of the principal investigators. I gathered all the courage I had and asked him whether I could join his group as his PhD student. When he said yes, I couldn't describe how happy and excited I was at that time. Since then, I have truly learnt a lot from him.

Now, two and half years have passed and I have published two papers under his guidance. I would like to highlight the recent one. In this paper, we report an experiment done with a collaborator at DSI (Leonid Krivitsky) on how to detect entanglement in a two-photon system in the most efficient way by measuring optimal witness families. We also introduced adaptive schemes to greatly speed up the entanglement detection based on the partial information known about the state. The idea of measuring witness family was thought and introduced by Prof. Englert earlier and I find it really a nice idea in entanglement detection. Both the simulation and the experiment performed by us show that there is a great speed-up in entanglement detection.

Being a PhD student and doing research is not the only thing that I had benefitted from through Berge. As his PhD student, I also had the opportunity to be his teaching assistant (TA) for the undergraduate module, Classical Mechanics, that he teaches. I have been a TA for Classical Mechanics for three semesters and I really enjoy it a lot, both his teaching and jokes, as well as my tutorial times with the students. I finally have a chance to make up for the regret that I had when I was a undergraduate—not having taken any module taught by Berge. I have to confess that I was taught classic mechanics in a totally different way and in order to catch up with the way and level of difficulty that he teaches it, I have had to brush up my undergraduate course again and do a lot of work on my own. But I truly enjoyed solving the problems that he designed for this course.

Dear Berge, on this auspicious occasion of your 60th birthday, I would like to take this opportunity to thank you, for the continuous teaching, guidance and care that you have given me over the past years. I would like to wish you and Ola good health and happiness in all the years to come. Wish you even greater success in whatever you plan to achieve. I wish every day of yours is filled with joy and laughter.

Wish you a happy and blessed life!

PRL 113, 170402 (2014) PHYSICAL REVIEW LETTERS week ending 24 OCTOBER 2014

Experimental Detection of Entanglement with Optimal-Witness Families

Jibo Dai,[1,2] Yink Loong Len,[1,2,3] Yong Siah Teo,[2,4] Berthold-Georg Englert,[2,3] and Leonid A. Krivitsky[1]

[1]*Data Storage Institute, Agency for Science, Technology and Research, 5 Engineering Drive 1, Singapore 117608, Singapore*
[2]*Centre for Quantum Technologies, National University of Singapore, 3 Science Drive 2, Singapore 117543, Singapore*
[3]*Department of Physics, National University of Singapore, 2 Science Drive 3, Singapore 117542, Singapore*
[4]*Department of Optics, Palacký University, 17. listopadu 12, 77146 Olomouc, Czech Republic*

(Received 3 April 2014; published 21 October 2014)

We report an experiment in which one determines, with least tomographic effort, whether an unknown two-photon polarization state is entangled or separable. The method measures whole families of optimal entanglement witnesses. We introduce adaptive measurement schemes that greatly speed up the entanglement detection. The experiments are performed on states of different ranks, and we find good agreement with results from computer simulations.

DOI: 10.1103/PhysRevLett.113.170402 PACS numbers: 03.67.Mn, 03.65.Ud, 03.65.Wj

Introduction.—Entangled states play an important role in the manipulation of quantum information, be it for present-day quantum key distribution or future quantum computation. One may need to verify if a certain quantum state—perhaps emitted by a source of quantum-information carriers or obtained as the output of a quantum computation—is entangled or not. For this purpose, the expectation value of an entanglement witness is telling: The state is surely entangled if a negative value is obtained. A positive value, however, is inconclusive—the unknown state could be entangled or separable, the witness cannot tell, but other witnesses might be able to. The concept of a witness was first used by Horodecki, Horodecki, and Horodecki [1], and the term "witness" was introduced by Terhal [2]; for reviews that cover all important aspects of entanglement witnesses; see Refs. [3,4].

How many witnesses, then, does one need to measure until a conclusion is reached? The answer to this question is given in Ref. [5]: If one solely relies on the expectation values of the witnesses one by one, one may never get a conclusive answer; if, however, the expectation values of suitably chosen witnesses are jointly used for an estimation of the quantum state, D^2-1 witnesses suffice for a D-dimensional quantum system.

The number can be further reduced by exploiting all the information gathered when determining the expectation values of the witnesses measured in succession. In the case of a two-qubit state ($D=4$), for instance, one never has to measure more than six witnesses, rather than $15=4^2-1$. This can be demonstrated by an experiment such as the one proposed in Ref. [5]; we are here reporting its laboratory realization.

Witnesses and witness families.—A Hermitian observable W is an entanglement witness if $\mathrm{tr}\{\rho_{\mathrm{sep}}W\}\geq 0$ for *all* separable states ρ_{sep} and $\mathrm{tr}\{\rho_{\mathrm{ent}}W\}<0$ for at least one entangled state ρ_{ent}. For each entangled state, there are some witnesses that detect it (<0), but many other witnesses will give an inconclusive result (≥ 0).

We shall concern ourselves with two-qubit systems—in the experiment, they are polarization qubits of a down-converted photon pair—and focus on optimal decomposable witnesses [6] of the form $W=(|w\rangle\langle w|)^{T_2}$, where $|w\rangle$ is the ket of an entangled pure two-qubit state, and ^{T_2} denotes the partial transposition on the second qubit. This witness is optimal in the sense that no other witness can detect some entangled states in addition to the states already detected by W. The generic example is $|w\rangle = |00\rangle\cos(\frac{1}{2}\alpha)+|11\rangle\sin(\frac{1}{2}\alpha)$ with $\sin\alpha\neq 0$, on which all other $|w\rangle$s can be mapped by local unitary transformations.

For any α, the eigenkets of the resulting witness,

$$W^{(\alpha)} = |00\rangle\frac{1+\cos\alpha}{2}\langle 00| + |11\rangle\frac{1-\cos\alpha}{2}\langle 11| + |\Psi_+\rangle\frac{\sin\alpha}{2}\langle\Psi_+| - |\Psi_-\rangle\frac{\sin\alpha}{2}\langle\Psi_-|, \qquad (1)$$

are the same: the two product kets $|00\rangle$ and $|11\rangle$ as well as the two Bell kets $|\Psi_\pm\rangle = (|01\rangle \pm |10\rangle)/\sqrt{2}$. This whole family of entanglement witnesses can, therefore, be measured by the projective measurement of their common eigenstates—the witness basis of the family. For a separable state, $\mathrm{tr}\{\rho_{\mathrm{sep}}W^{(\alpha)}\}\geq 0$ for all α, and this requirement implies the witness-family criterion [5]

$$\mathcal{S} \equiv 4f_1f_2 - (f_3-f_4)^2 \geq 0. \qquad (2)$$

Here, f_1 and f_2 are the probabilities for the two product states, and f_3 and f_4 are those for the Bell states; we estimate these probabilities from the observed frequencies. Consequently, once the frequency data are obtained from the measurement, a negative value of $\mathcal{S}$ reveals that the unknown state ρ_{true} is entangled.

Three remarks are in order. (i) The expectation value of one witness $W^{(\alpha)}$ is a linear function of the f_js, whereas $\mathcal{S}$ is a quadratic function. This is reminiscent of, yet different from, the "nonlinear entanglement witness" of Ref. [7], which requires a joint measurement on two copies of the unknown state. Our witness-family measurement uses only one copy at a time. (ii) Also, the nonlinear witnesses of Ref. [8] are quite different; their evaluation requires complete or almost complete knowledge of the state. (iii) The witnesses $W^{(\alpha)}$ can also be measured by other schemes, such as that of Barbieri *et al.* [9], who extracted the expectation value of $W^{(\pi/2)}$ from local measurements that examine the two qubits individually. By contrast, we perform a joint measurement on both qubits, which implements the projective measurement in the eigenbasis of $W^{(\alpha)}$ and so realizes the most direct measurement of the witness. Additional information about (ii) and (iii) appears in the Supplemental Material [10].

Our witness-family measurement provides estimates for *three* two-qubit observables (four f_js with unit sum), whereas the expectation value of a single witness is only one number. As discussed in Ref. [5], this can be exploited for quantum-state reconstruction after measuring six witness families, related to each other by the six local unitary transformations of Table I: Each witness family provides the expectation values of one (of six) single-qubit observables and of two (of nine) two-qubit observables. Therefore, a measurement of all six witness families constitutes an informationally complete (IC) measurement for full tomography of an unknown two-qubit state; thereby, all six single-qubit parameters and six two-qubit parameters are obtained once, while three two-qubit parameters are determined twice. This offers the possibility of measuring an IC set of witness families such that, if all families give an inconclusive result ($\mathcal{S} \geq 0$), a full state estimation can be performed for identifying ρ_{true}. With ρ_{true} then at hand, its separability can be determined straightforwardly by, for example, checking the Peres-Horodecki criterion [1,15].

Scheme A: Random sequence.—Since ρ_{true} is unknown, there is no preference for a particular one of the six families to start with. Hence, one starts with a randomly chosen family and checks the inequality (2). If the result is inconclusive, one then chooses the next family at random from the remaining five families, and so forth, until a conclusive result is obtained. If all six families give inconclusive results, ρ_{true} is estimated from the data to establish if it is entangled or separable.

Scheme B: Adaptive measurements.—Alternatively, we can perform the witness-family measurements in an adaptive manner: We choose the next family to be measured in accordance with the data obtained from previous measurements. Each time a witness family is measured, a set of four frequencies is obtained and these informationally incomplete data are used to partially estimate ρ_{true} by, for example, jointly maximizing the likelihood and the entropy—the MLME strategy of Refs. [16,17]. The MLME estimators ρ_{MLME} tend to be highly mixed states and are thus hard to detect by entanglement witnesses; therefore, if the measurement of a witness family detects the entanglement of the MLME estimator by the criterion (2), measuring that family has a good chance of detecting the entanglement of ρ_{true}. The value of $\mathcal{S}$ is used for comparing the unmeasured witness families with the MLME estimator, with f_j replaced by the jth MLME probability $\text{tr}\{\rho_{\text{MLME}} U^\dagger \Pi_j U\}$, where $U = U_1 \otimes U_2$ is one of the six unitary operators of Table I, and Π_j projects to the jth ket in the witness bases. The family that gives the smallest value of $\mathcal{S}$ is measured next; this judicious choice of family reduces the average number of witness families that need to be measured before the entanglement is detected. Instead of fixing the above six families, based on the MLME estimator, one can also choose from all thinkable families in each step. However, it is not worth the trouble as such optimization hardly improves the entanglement detection [10].

Scheme C: Maximum-likelihood set.—All state estimators, including ρ_{MLME}, that maximize the likelihood compose a convex set, the maximum-likelihood (ML) set. They all give the same estimated probabilities for the witness families already measured. When the number of qubit pairs measured per witness family is large (10^4 pairs suffice in practice), ρ_{true} is very likely contained in the ML set. Then, if there is no separable state in the ML set, we can conclude that ρ_{true} is entangled. For finite data, this conclusion is correct within a certain error margin, as is the case for all conclusions drawn from entanglement witness measurements. For this nonseparability check, we compute the maximal values of the likelihood for both the entire state space ($\mathcal{L}_{\text{max}}$) and the entire space of separable states ($\mathcal{L}_{\text{max}}^{\text{sep}}$) [18]. If we find that $\mathcal{L}_{\text{max}} > \mathcal{L}_{\text{max}}^{\text{sep}}$, we infer that ρ_{true} is entangled. To further economize the adaptive scheme, this check is performed before looking for the unmeasured witness family with the smallest value of $\mathcal{S}$.

Simulations.—To investigate the efficiencies of the three schemes, we perform computer simulations of witness-family measurements with both pure and full-rank mixed two-qubit entangled states [10]. Figure 1 shows the histograms that summarize the cumulative distribution in the percentage of entangled states detected against the number of witness families needed using schemes A, B, and C. We observe that the average number of families is largest for scheme A and smallest for scheme C. For instance, with

TABLE I. The six witness families that enable full tomography of the two-qubit state. The single-qubit unitary operators U_1 and U_2 transform the first family into the other five families. The Pauli operator X permutes $|0\rangle$ and $|1\rangle$; the Clifford operator C permutes the three Pauli operators cyclically.

Family	U_1	U_2
1	1	1
2	1	X
3	$C^\dagger$	C
4	$C^\dagger$	XC
5	C	$C^\dagger$
6	C	$XC^\dagger$

PRL **113,** 170402 (2014) PHYSICAL REVIEW LETTERS week ending 24 OCTOBER 2014

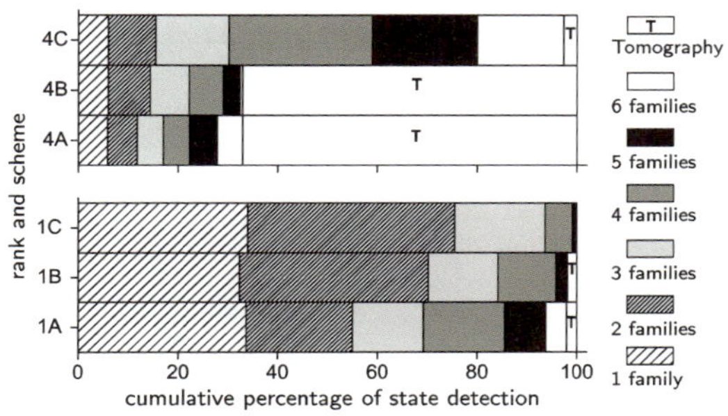

FIG. 1. Simulation results on the measurement of the set of six informationally complete entanglement witness families for 10^4 randomly chosen two-qubit entangled states: pure states (bottom), and full-rank mixed states (top). The cumulative histograms compare between measurements performed with scheme *A*, scheme *B*, and scheme *C*.

only three witness families measured, scheme *C* detects about 95% of the rank-1 entangled states whereas scheme *A* will need the measurement of five families to reach the same detection rate. When using either scheme *A* or scheme *B*, about 2% of the random pure states and about 67% of the full-rank mixed states are undetected by the six families without performing full tomography. The additional separability check in scheme *C* reduces the percentage of undetected pure states to virtually zero, and one needs no more than five witness families to detect entanglement for the rest of the pure states. The improvement is even more dramatic for the mixed states, with a reduction from about 67% to about 2.7%. We also observe that the mean number of witness families needed to detect entanglement for mixed states is larger than that for pure states. This is as expected, since mixed states generally have weaker entanglement and are, therefore, harder to detect.

Experiment.—We experimentally test the entanglement detection and tomographic scheme with three classes of states of different ranks. The first class of states are the pure states $\rho_{\text{true}}^{(1)} = |\vartheta\rangle\langle\vartheta|$, with $|\vartheta\rangle = |00\rangle \sin\vartheta + |11\rangle\cos\vartheta$ for $0 < \vartheta < \pi$, $\vartheta \neq \pi/2$. The second class of states are rank-two states of the form $\rho_{\text{true}}^{(2)} = |\Phi_+\rangle\mu\langle\Phi_+| + |\Phi_-\rangle(1-\mu)\langle\Phi_-|$ for $0 \leq \mu \leq 1$ and $\mu \neq 1/2$, where $|\Phi_\pm\rangle = (|00\rangle \pm |11\rangle)/\sqrt{2}$. The third class of states are the Werner states $\rho_{\text{true}}^{(3)} = |\Psi_-\rangle\lambda\langle\Psi_-| + (1-\lambda)/4$ for $1/3 < \lambda \leq 1$. The experiment uses the polarization qubits of a down-converted photon pair at 810 nm with, for example, ket $|10\rangle$ standing for the photon in mode 1 horizontally polarized and the photon in mode 2 vertically polarized.

The experimental setup is shown in Fig. 2. The rank-one states and rank-two states are produced as described in Refs. [19,20]. The rank-four states are produced by adding a controlled admixture of white noise to the singlet state by varying the coincidence time window of the detection electronics [21].

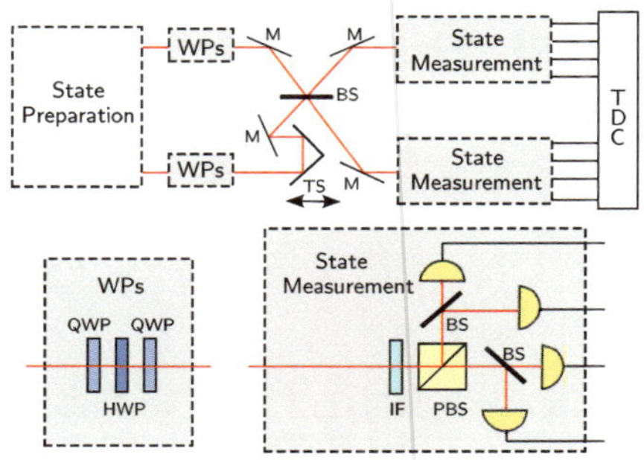

FIG. 2 (color online). Experimental setup. The polarization-entangled two-photon states are prepared by the method described in Ref. [20]. Upon emerging from the source, the two photons are guided with mirrors (M) to interfere at a 50 : 50 beam splitter, with the temporal overlap controlled by a translation stage (TS). After passing through interference filters (IF), the photons are sorted by polarizing beam splitters and registered by one of the photodetectors, four on each side. The detector outputs are addressed to a time-to-digital converter, and coincidences between counts of any two detectors are recorded. Two sets of wave plates, each composed of a half-wave plate (HWP) and two quarter-wave plates (QWP), implement the polarization changes that correspond to the unitary operators of Table I.

Owing to unavoidable imperfections, the actual states emitted by the source are not the ideal $\rho_{\text{true}}^{(1,2,3)}$ stated above, but full-rank approximations of them. For example, as reported in Ref. [20], fidelities above 97% are consistently achieved for the rank-two states, and the experimental rank-one and rank-four states are of similar quality. The simulations for Fig. 3 use the ideal states.

The signal and idler photons are directed to a Hong-Ou-Mandel (HOM) [22] interferometer with a 50 : 50 beam splitter (BS). On the way from the source to the BS, the photons pass through sets of wave plates (WPs) that change the polarization in accordance with one of the six local unitary transformations of Table I [10]. In each output port of the interferometer, the photons pass through an interference filter with a central wavelength at 810 nm and a full width at half maximum of 10 nm and are then sorted by a polarizing beam splitter (PBS). To discriminate between one-photon and two-photon events, another 50 : 50 BS is installed into each output port of the PBS, and eight single-photon avalanche photodiodes detect the photons. A time-to-digital converter (TDC) records the arrival times of the photons, and coincidences between any two of the eight detectors are obtained from the analysis of the time stamp record of the TDC. For each rank, we studied 21 different states.

Results.—We set the source to a particular state and performed witness-family measurements for 1 min per family, and so measured about 10^4 photon pairs for each state and family. The data were analyzed for all three schemes.

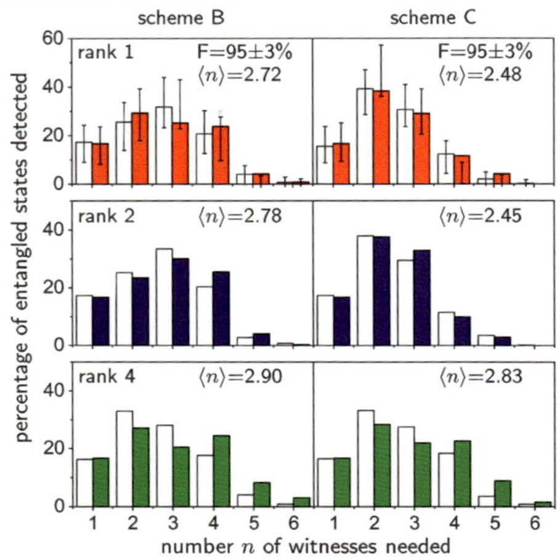

FIG. 3 (color online). A comparison of schemes B (left column) and C (right column) for rank-one states (top row), rank-two states (middle row), and rank-four states (bottom row). The histograms report the percentage of entangled states detected against the number n of witness families needed *without* performing state estimation; $\langle n\rangle$ is the average value. Both the simulation data (left empty bars) and the experimental data (right full bars) show that, for the three kinds of quantum states considered, scheme C provides further improvement over scheme B: It requires fewer families on average and the distributions are narrower. The similarity of the two histograms for the rank-one states is confirmed by their large fidelity F; similar values are obtained for the other histograms. Here, the simulation uses only states of the kind generated by the state preparation in the setup of Fig. 2, whereas no such restriction applies to the randomly chosen states for Fig. 1.

We also performed simulations for the three different classes of two-qubit states. For each class, data from 10^4 photon pairs (per witness family) for 10^3 states were simulated with Monte Carlo techniques, for the schemes A, B, and C. As expected, the simulation showed that for measurements done with the random order of scheme A, the number of families needed for entanglement detection is distributed almost evenly from one to six, with a mean of about 3.5, and fewer families need to be measured in the adaptive schemes.

This is confirmed by the experiment. For the three classes of states, only one of the witness families gives a conclusive result. Hence, for scheme A, the number of families needed for entanglement detection is equally likely to be one to six. The data for schemes B and C were analyzed as explained in the Supplemental Material [10] and the results are shown in Fig. 3, where one observes a significant improvement over the nonadaptive scheme A, which needs about 3.5 families on average. One rarely needs more than four witness families to detect the entanglement in these states; full tomography is never necessary since all rank-one, rank-two, and rank-four states are detected by one of the witness families.

The striking similarity between the results of the simulation and the experiment indicates that there are no significant systematic errors in the experimental data. For the rank-one states, the fidelity $F = \sum_j (p_j p'_j)^{1/2}$ between two probability distributions $\{p_j\}$ and $\{p'_j\}$ compares the histograms from simulation and experiment, and the large value of F is reassuring. The error bars for the rank-one states are obtained by bootstrapping the actual data 100 times; these error bars show the variation in the histograms that repeated measurements of this kind would display. The fidelity values and the error bars for the other histograms are of similar sizes and not displayed. We also collected IC data even if the state is known to be entangled before all six witness-family measurements are done so as to check the reliability of the tomography. The witness-family measurements indeed enable reliable quantum state tomography [10].

Conclusions.—We performed an experiment to verify the witness-family-based entanglement detection scheme introduced in Ref. [5]. In going beyond that proposal, we also introduced adaptive schemes that use the information acquired in previous measurements to reduce the average number of witness families that need to be measured. A few-witness way of detecting entanglement for photon-polarization qubits was thus demonstrated. Further, we showed that the witness-family measurements enable reliable quantum state tomography. With the necessary changes and within the limits set by what is experimentally feasible, the witness-family approach is also applicable to qubits of other physical kinds than photon polarization and to higher-dimensional or multipartite systems.

We are grateful for the insightful discussions with Jaroslav Řeháček and Huangjun Zhu. We thank Norbert Lütkenhaus for bringing Ref. [8] to our attention. This work is supported by the National Research Foundation and the Ministry of Education, Singapore, and is co-financed by the European Social Fund and the state budget of the Czech Republic, Project No. CZ.1.07/2.3.00/30.0004 (POST-UP).

[1] M. Horodecki, P. Horodecki, and R. Horodecki, Phys. Lett. A **223**, 1 (1996).

[2] B. Terhal, Phys. Lett. A **271**, 319 (2000).

[3] O. Gühne and G. Tóth, Phys. Rep. **474**, 1 (2009).

[4] R. Horodecki, P. Horodecki, M. Horodecki, and K. Horodecki, Rev. Mod. Phys. **81**, 865 (2009).

[5] H. Zhu, Y. S. Teo, and B.-G. Englert, Phys. Rev. A **81**, 052339 (2010).

[6] M. Lewenstein, B. Kraus, J. I. Cirac, and P. Horodecki, Phys. Rev. A **62**, 052310 (2000).

[7] M. Kotowski, M. Kotowski, and M. Kuś, Phys. Rev. A **81**, 062318 (2010).

[8] O. Gühne and N. Lütkenhaus, Phys. Rev. Lett. **96**, 170502 (2006).

[9] M. Barbieri, F. De Martini, G. Di Nepi, P. Mataloni, G. M. DAriano, and C. Macchiavello, Phys. Rev. Lett. **91**, 227901 (2003).
[10] See Supplemental Material at http://link.aps.org/supplemental/10.1103/PhysRevLett.113.170402, which includes Refs. [11–14], for comments on the experiment by Barbieri *et al.* [9] and the nonlinear witnesses of O. Gühne and N. Lütkenhaus [8]; discussions on general adaptive schemes; technical details on the generation of entangled states; details of the HOM calibration and settings of the WPs, data analysis, and quantum state tomography.
[11] K. Życzkowski and H.-J. Sommers, J. Phys. A **34**, 7111 (2001).
[12] B.-G. Englert, C. Kurtsiefer, and H. Weinfurter, Phys. Rev. A **63**, 032303 (2001).
[13] K. Banaszek, G. M. D'Ariano, M. G. A. Paris, and M. F. Sacchi, Phys. Rev. A **61**, 010304(R) (1999).
[14] J. Řeháček, Z. Hradil, E. Knill, and A. I. Lvovsky, Phys. Rev. A **75**, 042108 (2007).
[15] A. Peres, Phys. Rev. Lett. **77**, 1413 (1996).
[16] Y. S. Teo, H. Zhu, B.-G. Englert, J. Řeháček, and Z. Hradil, Phys. Rev. Lett. **107**, 020404 (2011).
[17] Y. S. Teo, B. Stoklasa, B.-G. Englert, J. Řeháček, and Z. Hradil, Phys. Rev. A **85**, 042317 (2012).
[18] J. Řeháček and Z. Hradil, Phys. Rev. Lett. **90**, 127904 (2003).
[19] P. G. Kwiat, E. Waks, A. G. White, I. Appelbaum, and P. H. Eberhard, Phys. Rev. A **60**, R773 (1999).
[20] J. Dai, Y. L. Len, Y. S. Teo, L. A. Krivitsky, and B.-G. Englert, New J. Phys. **15**, 063011 (2013).
[21] A. Ling, P. Han, A. Lamas-Linares, and C. Kurtsiefer, Laser Phys. **16**, 1140 (2006).
[22] C. K. Hong, Z. Y. Ou, and L. Mandel, Phys. Rev. Lett. **59**, 2044 (1987).

Chapter 7

Personal Anecdotes

Wilhelm Becker

Max Born Institute for Nonlinear Optics and Short-Pulse Spectroscopy, Berlin, Germany

Personal message:

True story!

In a little known brief episode of his career, Berge set up a small charter company in the American Southwest, called *Quantum Transportation.* However, customers opposed to being charged for paths they were sure they had not actually taken (but could have, as Berge insisted), and the company went out of business after a time interval too short to have been recorded. The picture shows, next to Berge, two satisfied clients, Seymour Alpert and Manfred Kleber.

Dagmar Bruß

Institut für Theoretische Physik III, Heinrich-Heine-Universität Düsseldorf, Düsseldorf, Germany

Kim-Suan Chiu

Centre for Quantum Technologies, National University of Singapore, Singapore
Kang Ching Road, Somewhere in Singapore

Leong Chuan Kwek

Centre for Quantum Technologies, National University of Singapore, Singapore

Chiara Macchiavello

Dipartimento di Fisica and INFN-Sezione di Pavia, Pavia, Italy

Article Title: **Berge's GUT**

Reference: Original article (2014)

Personal message:

Dedicated to Berge's 60th birthday!

(Editors' note: In a classic "Berge moment", Berge talked about the theory described in this contribution on April 1st, 2014, in a CQT seminar; his skill of delivering a talk of such profound content with a perfectly straight face is unparalleled and much admired.)

Berge's GUT

Dagmar Bruß[1], Kim-Suan Chiu[3,4], Leong Chuan Kwek[3] and Chiara Macchiavello[2]
[1]*Institut für Theoretische Physik III, Heinrich-Heine-Universität Düsseldorf, D-40225 Düsseldorf, Germany*
[2]*Dipartimento di Fisica and INFN-Sezione di Pavia, via Bassi 6, 27100 Pavia, Italy*
[3]*Centre for Quantum Technologies, National University of Singapore, 3 Science Drive 2, Singapore 117543, Singapore and*
[4] *Kang Ching Road, Somewhere in Singapore*

A short story about how Berge conquered GUT and produced alternative theories.

Evon is one of the chief architects and representatives of the Centre for Quantum Technologies (CQT) for this Festchrift. She (and all her colleagues at the main office at CQT) handles all the complex administrative arrangements for all the visitors. We told Evon that we will be writing Berge's GUT over the weekend. She was amused by the title — probably she was half expecting us to dissect Berge's body and exhibit gory details of his gut. Unfortunately, it is nothing of this sort, perhaps rather disappointingly to some people.

Actually, GUT refers to the Grand Unified Theory. This is a model in which particle physicists merge the three forces — the electromagnetic, the weak and the strong forces — into a single theory. The electromagnetic force describes the force between charges around us, the weak force describes the force governing a special type of radioactive decay, namely beta decay, and the strong force describes the force deep inside the nucleus of an atom. The attempt to unify these forces with gravity would then provide an ultimate theory often referred to as the theory of everything (TOE). Of course, from the human anatomy perspective, it is probably easier to study the toes rather than the gut, the latter begin hidden by skin and flesh and would need dissection to study carefully.

Anyhow, the crux of the ludicrous situation arose from a Youtube video [1] by Tony Padilla and Ed Copeland from the "Numberphile" [2] team, in which they showed that the sum of all positive integers, i.e., $1+2+3+4+\cdots$ is not an incredibly large number as you would expect, but a surprisingly small negative number, $-\frac{1}{12}$. Moreover, they showed that this result can be found in Joe Polchinski's wonderful book [3] on string theory (the best model for TOE) and the result finds application in "perturbative gauge theory". Naturally, Berge was extremely amused. He is always the champion of the weaks and the underdogs. The string theorists' claim that they know everything, possibly including the meaning of life, irritates him a lot. Berge knew about the analytic continuation of the Riemann zeta function $\zeta(z)$ at $z = -1$, but being Berge, he happily poked fun at the subject by repeating the argument several times.

For the benefit of readers who have not been exposed to the Youtube video, the argument [4] goes as follows: First we note that $S_1 = 1-1+1-1+1-1+\cdots$ is simply 1/2. Actually, according to Berge, you can prove this result rigorously! Why? Since $\frac{1}{1+x} = 1 - x + x^2 - x^3 + \cdots$ and the sum S_1 is simply the summation evaluated for $x = 1$. Next one needs to note that

$$\begin{aligned} S_2 &= 1-2+3-4+5-6+\cdots \\ \text{and} \quad S_2 &= \quad\;\; 1-2+3-4+5+\cdots \\ \Rightarrow 2S_2 &= 1-1+1-1+1-1+\cdots = \frac{1}{2} \\ \text{or} \quad S_2 &= \frac{1}{4}. \end{aligned}$$

If we now denote our required sum $1+2+3+4+\cdots$ as S, then it is not hard to see that

$$\begin{aligned} S - S_2 = &\ (1+2+3+4+\cdots) \\ &-(1-2+3-4+\cdots) \\ = &\ 0+4+0+8+0+12+\cdots \\ = &\ 4S. \end{aligned}$$

Simple algebra leads to $3S = -S_2$ or $S = -\frac{S_2}{3} = -\frac{1}{12}$.

The evening after he showed the details to one of us (Dagmar), Dagmar, Kim (who just insisted on being a co-author) and Kwek were supposed to go for an evening walk through the Southern Ridges in Singapore. Kim met us at CQT, and Berge was absolutely thrilled to confound Kim with the problem. Kim WAS supposed to be math trained but she was completely dumbfounded by Berge's argument. As we were rushing off to the Southern Ridges for our memorable walk, Kim even noticed a smug grin on Berge's face.

Naturally, Berge just could not allow the matter to rest here. He quickly came back to the whiteboard with a new "argument" and showed that he could actually prove that S is zero and not $-1/12$! The answer is simple: If we rename S_1 as A (after all, "a rose by any other name would smell as sweet" [Act II, Scene 2, Romeo and Juliet]) and let

$$\begin{aligned} A &= 1-1+1-1+1-1+\cdots \\ B &= 1+1+1+1+1+1+\cdots \end{aligned}$$

then,

$$\begin{aligned} A+B &= 2+0+2+0+2+0+\cdots = 2B \\ A-B &= 0-2+0-2+0-2+\cdots = -2B. \end{aligned}$$

It is not hard to conclude that $A = B$ and $A = -B$ from these equations, giving $A = B = 0$. See Fig. 1 for Berge's whiteboard scribbling.

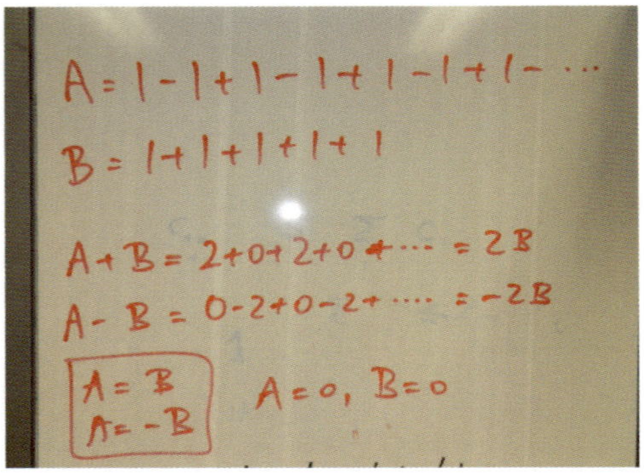

FIG. 1. Berge's corollary to his "proof".

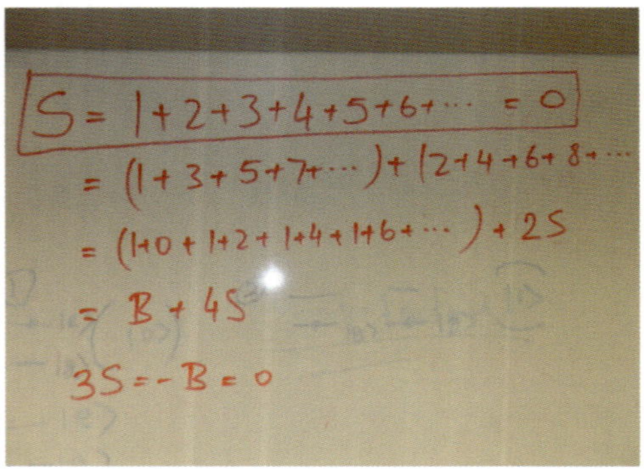

FIG. 2. The final argument. Both figures were recorded by Dagmar with her mobile phone.

This important Corollary leads us to the final Proof: As before, the symbol S denotes the sum of all natural numbers $1+2+3+\cdots$. One can bring some order into this sum by first writing the odd numbers, and then writing the even numbers; see the second line in the following set of equations. The sum of the even numbers is, obviously, just equal to $2S$. The sum of the odd numbers can be written as the sum of even numbers plus our previous sum B (it is clear that one gets all odd numbers by adding 1 to all even numbers, isn't it?); see the third line below. This leads to Berge's ingenious result, which in full beauty is displayed here:

$$\begin{aligned} S &= 1+2+3+4+5+6+\cdots \\ &= (1+3+5+7+\cdots)+(2+4+6+8+\cdots) \\ &= (1+0+1+2+1+4+1+6+\cdots)+2S \\ &= B+4S \\ \Rightarrow 3S &= -B = 0 \end{aligned}$$

Thus, Berge proved a new and fundamental cornerstone in string theory, namely the fact that $S=0$, or in words: all natural numbers sum up to zero! This final argument of Berge is shown in Fig. 2.

Let us come back to the main theme, Berge's GUT. As explained in the introduction, GUT refers to Grand Unified Theory. While still trying to come to terms with the depth of Berge's new insight, it occurred to us that he actually made an analogous step in the field of quantum information, some years before.

When we (Dagmar and Chiara) visited Singapore sometime in 2003, aside from the enjoyable time at the wonderful swimming pool, sipping one delicious Durian shake after another, we worked with Berge on an interesting problem concerning a comparison of quantum and classical key distillation for tomographic quantum cryptography. Our collaboration also involved Artur Ekert, Matthias Christandl and Dagomir Kaszlikowski. Together we showed that by exploiting tomographically complete measurements on entangled qudits (higher-dimensional quantum bits), the noise threshold for classical advantage distillation was the same as the threshold for the quantum version. The Quantum Lah group had a lot of visitors and we also benefited from discussions with many visiting experts, including Antonio Acin, Thomas Durt and Nobert Lütkenhaus (not to mention Kwek), for the work. Our paper published on this matter carries the subtitle: "Equivalence of Quantum and Classical Key Distillation" [5].

Time went on; work went on... Most interestingly, only a couple of years later, Dagomir, Kwek and Berge published another paper, also on a comparison of quantum and classical key distillation for tomographic quantum cryptography, with the subtitle "Nonequivalence of Quantum and Classical Key Distillation". This appeared ostensibly to contradict our first paper. Thus, this "anti-paper" can be seen as an annihilation of the previous work: The sum of the content of the two papers is zero. Using induction, we can easily and audaciously extend this argument to n papers: The sum of the content of all papers is again zero!

A proper GUT is supposed to unify three fundamental fields. It is straightforward to identify the third one. Even if it may not be obvious to the curious reader, both previous findings propose clearly a bridge to life sciences and philosophy, in particular to the insights of Nietzsche: *"every belief, every considering something-true, is necessarily false because there is simply no true world"* (Will to Power [notes from 1883-1888]). Nietzsche further argues that *"the highest values devaluate themselves. The aim is lacking; 'Why?' finds no answer."*

Consequently, as Berge masters the devaluation of values, his research can only be considered to be of the highest value: Via Berge's illuminating results, philosophy can now be unified with the other two cornerstones of modern Science, namely string theory and quantum information theory.

Happy birthday, Berge, and stay as GUT as you are!

3

---

[1] http://www.youtube.com/watch?v=w-I6XTVZXww.
[2] See the website for the "Numberphile" team: http://www.numberphile.com/team/.
[3] J. Polchinski, String Theory (Cambridge University Press, Cambridge, 2003).
[4] Evon always complained that she never quite understood why physicists speak in such strange language. However, the argument here is purely mathematical and fringes on primary school logic. I think she will have no excuse this time to not understand some of the intricacies of the situation.
[5] Tomographic Quantum Cryptography: Equivalence of Quantum and Classical Key Distillation, Dagmar Bruß, Matthias Christandl, Artur Ekert, Berthold-Georg Englert, Dagomir Kaszlikowski, and Chiara Macchiavello, Physical Review Letters 91, 097901 (2003).
[6] Coherent Eavesdropping Attacks in Quantum Cryptography: Nonequivalence of Quantum and Classical Key Distillation, Dagomir Kaszlikowski, Jenn Yang Lim, Leong Chuang Kwek, Berthold-Georg Englert, Physical Review A 72, 042315 (2005).
[7] https://archive.org/details/TheWillToPower-Nietzsche.

Christian Miniatura

INLN, Université de Nice-Sophia Antipolis, CNRS, Valbonne, France
Centre for Quantum Technologies, National University of Singapore, Singapore

Paper Title: **Through a Glass, Darkly: Scrambling and Painting Eggs with Berge**

Reference: Original article (2014)

Personal message:

"What were Mayakovsky's last words before he committed suicide in 1930?" "Don't shoot, Comrades, don't shoot!" From time to time, Berge likes to recall this Radio Yerevan black-humor communist joke about the Russian, and Soviet, poet Vladimir Mayakovsky. In 1915, Mayakovsky wrote a wonderful poem, The Backbone Flute, and I quote here its Prologue, with its prophetical (?) verse, in my own poor translation:

To you all
We used to love and love.
Icon sheltered in the cave of the soul
Like a glass of wine
At a banquet table
I raise my head filled with poems.
Often I say to myself, what if I put
The period of a bullet to my own demise?
Today just in case, I give
My farewell concert
Memory!
Gather within my brain
The countless ranks of my beloved ones,
Pour laughter from eyes to eyes
Light up the night with by-gone weddings,
From bodies and bodies, pour out the joy,
May no one forget this night.
Today I shall play the flute on my own backbone.

This is what my tribute to Berge's Sixtieth Anniversary intends to bring: memories from the past, memories from my encounters with him, raised like a glass of wine at a banquet table.

Happy Birthday, Berge! In every desert an oasis is tucked.

Through a Glass, Darkly: Scrambling and Painting Eggs with Berge

Christian Miniatura
INLN, Université de Nice-Sophia Antipolis, CNRS; 1361 route des Lucioles, 06560 Valbonne, France
Centre for Quantum Technologies and Physics Department,
National University of Singapore, 3 Science Drive 2, Singapore 117543, Singapore and
Institute of Advanced Studies, Nanyang Technological University, 60 Nanyang View, Singapore 639673, Singapore

Slice of Life. I have known Berge for twenty-two years now. Twenty years old: is it not the best time of life? I first met Berge during the workshop Optics and Interferometry with Atoms that was organized by Jürgen Mlynek in Konstanz in 1992. I still remember the first time I saw him. He was wearing a bolo tie, a cowboy outfit unusual to me except in American movies. This workshop was also when I first met his charming wife Ola. I was myself in the early stages of my scientific career (I had been hired at CNRS in 1989 and had defended my experimental thesis—yes indeed, I was an experimentalist at the time!—in 1990). I was working in the group of Jacques Baudon at Laboratoire de Physique des Lasers in Villetaneuse and we had developed a Stern-Gerlach atomic interferometer operating with metastable hydrogen atoms; see Figs. 1 and 2, which I include here as a wink from the past to Berge. At sufficiently low magnetic fields, these atoms behave as spin-1 particles and we were "splitting and recombining" them in various ways, looking at fringe patterns and their contrast, tracking topological effects like Berry phases, etc. It was the exciting early stages of atomic interferometry where results were shooting from all sides. Berge was already an accomplished and known physicist, and we came across his work with Julian Schwinger and Marlan Scully on the Humpty-Dumpty[1] effect, a case study of relevance to our experiments. We took the occasion of the Konstanz workshop to approach him and to invite him to Villetaneuse. He accepted and that was the start of a long, and still lively, friendship, often visiting each other at our homes. Life is sparked with blessings and important moments, and this was one. Of course I could not anticipate it at the time, but Berge was to profoundly and positively influence my life. As a matter of fact, Berge and I have only four joint publications (but about three still waiting on dusty shelves, in case he finally decides to work on them...). It is not a lot but, looking back into the past, we, together, have at least **constructed** a lot. As such, he is arguably the most important pillar in my scientific life. In the rest of this humble Essay, I will give some personal recollections[2] and anecdotes around the circumstances, context and scenery behind my interactions with this big-hearted man during these past twenty-two years[3].

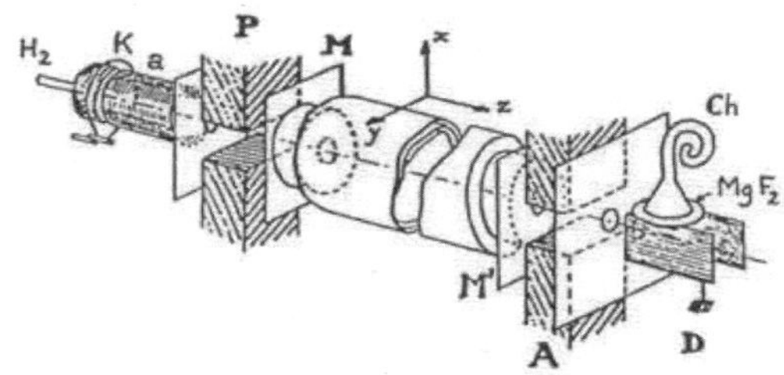

FIGURE 1. Villetaneuse longitudinal Stern-Gerlach interferometer: electron gun K, polarizer and analyzer P and A, mixers M and M', detector D and magnetic phase region R.

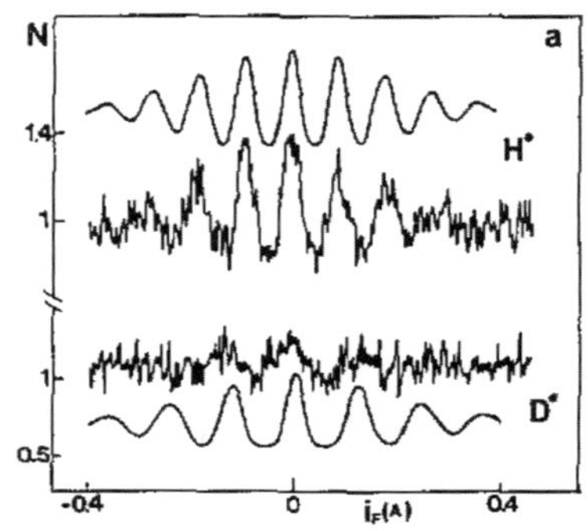

FIGURE 2. Example of atomic fringes obtained with the Villetaneuse device in the period 1992-1993 with metastable hydrogen and deuterium atoms (experiment and theory).

A. Lab on the Moon and One End of the World

Space oddity. I did my experimental PhD with Jacques Baudon from September 1987 to June 1990, at the Laboratoire de Physique des Lasers at Villetaneuse (University Paris XIII)[4]—a place more commonly known among us as the "Lab on the Moon", to highlight its seemingly improbable location in the north of Paris.

1. The Egg character in an English nursery rhyme and riddle, *Humpty Dumpty sat on a wall...*
2. Of course, prone to inaccuracies and mix-ups.
3. Apart from the Acknowledgements, parts in italics are winks to songs, movies, citations, etc.
4. Is 13 a lucky or unlucky number? I let you decide.

2

The joke was that some decision-maker, most probably under the influence of some illegal substance, may well have decided to implant a university there by just throwing a dart at a Paris map...and missing it. Let's just say that the university was embedded in a difficult neighborhood, the 9-3 (pronounced nine-three) suburb and its problem housing estates plagued by unemployment, drug trafficking and car burning. For security reasons, we were not allowed to stay in the lab at night and we had to vacate the place before 9pm. The most direct route to the Villetaneuse train station was along an almost abandoned rail track with no street lights, a nightmare at night after downpours. I still laugh when I recall Jacques swearing loudly, "C'est pas vrai, bordel de merde!" after setting foot in a pothole full of watery mud. Then, once at the station, Jacques, a smoker, would invariably say, "I'm sure that if I light up my cigarette, the stupid train[5] will hurry up". He would do so and soon yellow eyes would start to glow over the horizon, bringing the promise of landing soon in a glamorous Paris. Often, we would end up in the restaurant Le Vieux Paris near the Panthéon where we would enjoy a Château Tour de Marbuzet bottle (for the connoisseur, a Saint-Estèphe) over French south-west cuisine.

La Dolce Vita. My time at Villetaneuse was a happy one[6]! I was doing my PhD on atom-diatom collisions at thermal energies in a wonderful group led by a wonderful and very generous character. The group was composed of other wonderful characters: Jean Reinhardt (aka the Admiral[7]), Jacques Robert (aka Killing Finger[8]) and Georges Vassilev (aka the Third Man[9]) as well as Francisco Perales, another PhD student who started the same year as me, and, later, Sìle Nic Chormaic a red-blond fairy from Maynooth, Ireland. The Lucy[10] Fellowship. It is with them, and under Jean Bellissard's inspirational drive, but also, I confess, with the warm help of evening whisky drinks[11], that we realized that our set-up could be quickly and easily transformed into an atomic interferometer, which we did after I harvested my core PhD results. Our ideas granted me a staff position at CNRS in 1989 to start a new research programme on atomic interferometry which I did until 1996. The stage is now set for the main character of this Essay to come on stage.

But I have promises to keep, And miles to go before I sleep. I think it was Jacques Robert who first came to us with the Humpty-Dumpty set of papers and its motto "perfect recombination of wave-packets in the Stern-Gerlach apparatus requires cancellation of accumulated phases which in turn demands a microscopic precision never achieved in practice". Since we wanted to quantify our experimental results in that spirit, having in mind some (improbable and never implemented) metrological applications, then why not approach the (European) author? Who turned out to be invited to the Konstanz workshop. This is how it all started. Berge went to Villetaneuse in October 1993, enjoying an apartment on Île Saint-Louis[12]—but at a frivolous rent—and Berthillon ice creams. This was also the time that I realized that Berge, soon to turn forty and despite his numerous publications, hard and good work and growing visibility, had simply no permanent position in Germany and that his life, well, was not easy. Can you have peace of mind when you know that your current job contract expires soon, that you swiftly need to find and secure another contract while still doing visible and impactful science to stand a chance? Yet Berge was drawn and made for Science and Academia. Showing resilience, he never ever flinched.

Close Encounters of the Third Kind. During his stay, we (I should better say *he*) worked on a least-bias description of atomic beams, our first joint theoretical paper. I was impressed by his way of doing science: clear, deep and analytic thinking, dissecting things to demystify them. I have never worked myself on quantum information or on the foundations of quantum mechanics, but I must say that I like and share his down-to-earth views on the subject. It is always a delight for me to listen to him express his (strong) opinions on the matter. Berge does not like fake science, nor does he fancy "bubble science" and the buzz trend—and sometimes scams—around the "mysteries of quantum mechanics". If you have the chance, let him entertain you with his hilarious personal account of the quantum pendant case or of the intelligent design talk by a German priest. Let it be said straight: Berge speaks his mind and does not clap shoulders when the integrity of people and of science is at stake[13]. He is uncompromising on these matters. The most important thing for me is that, in October 1993, in the Lab on the Moon, in the 9-3, I happened to know a big-hearted and charming guy, a French cuisine, wine and cheese lover. We (and later our families) clicked and we became faithful friends.

---

5. Should I say space shuttle?
6. Heart beats sent to my dear friends Jessica, Marc, Bernard and Pascale wherever they are now: plastic spoons were taking funny curved shapes in our cups filled with hot tea or coffee.
7. *The black flag floats in the pot* and *Vermot Almanac not dead!* An incredible character, and passionate lover of classical music who knew La Callas, living in a barge house on the Seine river.
8. He was a master of Wushu, doing push-ups on fingers. Scary, isn't it?
9. He fled his communist home country, Bulgaria, and got political asylum and French citizenship. Looks like a KGB mole, doesn't he?
10. Name of my experimental set-up. With diamonds but no sky.
11. Clearly a shameful case of doping. Luckily a Scientific Olympic Committee running urine and blood tests does not yet exist.
12. The happy few who enjoyed such good fortune, raise your hands!
13. I guess he is a fearsome Editor, ☺.

3

Mandela Day. The following year, Berge came back to France as an invited lecturer for the follow-up workshop on optics and interferometry with atoms that I had organized for the end of June 1994, in the lost and picturesque, mountain-tucked, village of Peyresq in the south of France. Three future Nobel awardees were present, Claude Cohen-Tannoudji, Roy Glauber and Serge Haroche and other prominent physicists like Alain Aspect or Vladilen Letokhov were also present. The Lucy Fellowship as well. Julian Schwinger had also accepted our invitation to join, but he fell sick and he sadly passed away in July that year. Berge had been his post-doc, but I never had the occasion to meet this great man. One day, Berge opened a bottle of red wine from Sattui Winery that we drank together[14]. In Peyresq, I remember we were taken aback by the cold and cloudy weather during the first two days: We had to travel to the closest civilized place and hurriedly buy pullovers in the local shops. I remember Berge sitting next to me in the bus, whispering to me mischievously that, would the bus end up in the ravine bordering the mountain road we were traveling, the face of atomic physics would dramatically change. Indeed! But the weather changed, summer and heat finally arrived, and the atmosphere turned joyous. Unfortunately, I spent my time there as a petty officer, running from one place to another, settling bugs and details[15]. Keep it under your hat: Apart from some hilarious gossips[16] and embarrassing situations that I'll keep secret to avoid exposing the reader who was at Peyresq and involved, my memories dry up.

El Cóndor Pasa[17]. In that same year, out of the blue, Berge received an offer for a professorship position in Santiago de Chile and he went there at the end of 1994 to assess the situation. He brought me back a bottle of good Chilean wine, my first try at it. Serious or not, believe it or not, we chatted a few times about settling together at this "End of the World" where he would start a theory group while I would start an experiment group and we would work together. Unfortunately, the conditions attached to the position would not have been sufficient to secure his family's future and Berge had no choice but to decline this offer. I always fantasize about what our lives would have been like there...

Aux Sombres Héros de l'Amer. Knowing Berge better and better, I was genuinely ashamed and really concerned by the precarious and little-hope situations that his numerous temporary and short academic contracts were putting him and his family through, and I was trying my best to help him in any possible, even modest, ways. Chance was that, in 1994, our work on atomic interferometry got some national recognition, which I was able to use to invite Berge again in Villetaneuse from July to December 1995. No bourgeois place for him this time however. Berge spent the first two months at a place on the far side of Montmartre, the later four months near Arc de Triomphe. There is an unfortunate and infamous story attached to Montmartre: Berge opened a bank account in a branch at Barbès-Rochechouart. It turned out that the bank was involved in the greatest politico-financial scandal France had known since the Panama Canal Scandal in 1892, one century before! After his stay in Villetaneuse, he kept open this bank account in order to have some French money[18] available for any of his subsequent visits to France. It turned out that the bank siphoned off all his money through preposterous penalty fees, claiming his account was not used often enough. Gain, ill-gotten or not, is never enough when you need to cover as much as possible the colossal losses of the bank. There was nothing he nor I could do. We tried, we wrote, we phoned, we complained, nada! The amount was too small to be worth hiring a lawyer and filing a lawsuit: a clear case of a small and unpunished white collar crime[19]. From that year, I also remember an enjoyable and leisurely moment on Jean Reinhardt's barge house for the start of the French national day celebrations...and Berge mischievously making me realize my shameful oblivion of my wedding anniversary. During his 1995 stay, Berge worked on fringe visibility and which-way information[20], one of his most famous and cited papers, released in 1996.

Wild World. The year 1995 was also dreadful. On the light side, during the days of the presidential election of Jacques Chirac in May, my apartment was by coincidence flooded by wastewater[21] and for the next two months, my wife, our toddlers and I had to experience the Bohemian life, waiting for insurance companies to do their job and for sanitization and repair works to be conducted. On the darker side, for a decade we had been continuously losing our innocence, illusions, and idealistic dreams[22]. We had been shattered by AIDS[23], by the mad cow crisis, by the financial world and its firmer and firmer grip on the society, by the Bosnian War and the Rwanda genocide...I had better stop there...It is still vivid and depressing...Summer 1995 was the time of the Paris bombings, in particular the RER B bombing at Saint-Michel station on July 25[24], the Arc de Triomphe bombing on August 17, the Metro bombing at station Maison Blanche

14. I'll let Berge tell you about the Schwinger-Sattui connection.
15. Don't you know? Diva behavior does exist in our community.
16. *Five Million Vodka Bottles to the Moon.*
17. For picky characters: I know, it is in fact a Peru tune.
18. Euro coins and banknotes entered circulation only on 1st January 2002.
19. Do you really have any kind of illusion about bankers?
20. Read the Acknowledgements of that paper, ☺.
21. France was indeed in deep shit that day.
22. Rimbaud, where are you?
23. *Les Nuits Fauves.*
24. I still shudder when I come to think that Jacques Baudon was lucky enough to be on board the train scheduled just before.

4

later, on October 6. Many people were killed, many were wounded, many families suffered and mourned. Another RER B bombing occurred at Port Royal on December 3rd 1996, killing two Villetaneuse Maths PhD fellows. Awful days. But I also keep from these days a present Berge gave to me and which has never left me since then: a pocket, paperback version of the Abramowitz and Stegun handbook of mathematical functions, the holy bible of theorists.

B. La French Riviera

Born to Be Alive. The years 1995 and 1996 were kind of a sad admission of failure for me: It was too difficult to settle Berge for good in France. Positions were scarce, sure, but despite some local support, it was simply very unlikely that someone who was supposed to teach at the university but who did not speak French would be hired. Berge started to learn some French but the odds to convince the members of the hiring committee were too slim. Times are finally (slowly) changing in France now. At the same time, the Chilean romantic dream, even unrealistic as it was, was in fact hitting a nerve in me. In the fall of 1994 and in the beginning of 1995, I was mulling over the idea of giving my scientific life a new twist. I wanted to start a scientific activity revolving around disorder effects using light, cold atoms and quantum gases. During summer 1995, Robin Kaiser and I decided to write a thorough and detailed scientific proposal that we sent to Catherine Bréchignac, the head of physics at CNRS at the time. Much to our surprise, our project was very well received and, after a few twists and turns, got officially approved during the autumn 1996.

A Rock and a Hard Place. After a few months spent at Laboratoire Kastler-Brossel with Dominique Delande, Robin and I settled at the Institut Non Linéaire de Nice (INLN) in March 1997 where we started and built a new group. This was when and where I morphed into a theoretician, where I got my first official (and most brilliant to date) PhD student, Cord Müller. There we successfully developed a group which became a world leader in mesoscopic effects using light and cold atoms. We notably obtained the first-ever observation of coherent back-scattering (CBS) of light with cold atoms and Cord developed successfully the corresponding multiple-scattering theory. But, in these years, INLN did not prove, well let's just say, the welcoming place I hoped for and life there quickly became sourer and sourer. Yet, at the same time, and despite disgusting betrayals and irreversible internal tensions, our scientific activity was very productive and successful.

Gimme Shelter. But what about Berge? This was a time when we were seeing less of each other. Setting up the group at INLN was taking all of my time, energy and efforts, and Berge was meanwhile getting more and more involved in the quantum information business. So our scientific lives were kind of drifting apart. But, as friends, we always kept in contact and the emergence of the world wide web was very helpful for this! Thinking of Berge, I felt the years 1997-2003 were rueful missed opportunities. First there was the Berlin case in 1997-1999. Berge had applied for a professorship position in Berlin and was ranked first. We all thought "Finally, it's done!". Sadly only virtually, because fate had its malicious twist. The final hiring process lagged...and in the end, the position was simply canceled! Berge did not apply anywhere else in the meantime, so two years were lost for nothing and it was an awful blow to the morale. Then there was the Milano case in 2000-2002, actually the only one which came up with a real offer in the end, thanks to Rodolfo Bonifacio's workmanship. If the Berlin case boiled down to the German "efficiency" to save money, the Milano case is emblematic of the nasty tricks of Italian politics. The Italian government had decided to help talented Italians staying abroad, those not enjoying the relevant connections to secure a job in the Italian academic system. The plan was to bring them back home by opening academic positions. The programme also targeted talented foreign scientists, and this was how Berge was the first applicant drawn into the programme by Bonifacio's initiative and clairvoyance. Alas, Silvio Berlusconi came back into power and the whole programme lagged. It was difficult, even for him, to simply rule out a programme that had been officially voted for by the Italian parliament a few months earlier. So the Milano position finally materialized...but only for a duration of three years! This was how the Berlusconi administration finally sabotaged the programme and closed it. In hindsight, it is fortunate that Berge did not take the position.

All Quiet on the Western Front. Indeed, while the Milano case was still in the pipes, fate was at last smiling brightly on Berge. He was invited by Artur Ekert to come to Singapore and join the QuantumLah, the seed for a new, world-class and rather unique Centre on quantum information. After a 2-week visit, Berge set foot at the National University of Singapore in 2002. Singapore made him an offer he could not refuse, which he accepted in summer 2003. I always felt that Europe miserably and shamefully failed by letting Berge go. But to tell the truth, no single position Berge could have held in Europe could have put on the table working conditions competing with the ones offered by Singapore. A shift of paradigm had happened with globalization. Today, Berge is managing a group of twenty-five people, including me, and I know how deeply grateful to Singapore he is.

C. The other End of the World

Le Soleil se lève à l'Est. Berge's move did not go unnoticed to me. Indeed, my childhood friend Bernard Luciani was also in Singapore. He was the counsellor for culture, science and education at the French Embassy. The years 1998-2003 had been tough to me, and I needed a break, fresh air and inspiration. I had not seen my two friends for a long time, and South-East Asia was unknown but appealing to me. It was thus the perfect opportunity to kill two birds with one stone! Berge invited me for a ten-day visit to Singapore over the end of October—beginning of November 2003, and Bernard invited me to stay at his place. The truth is that, within a few days, I fell in love with Singapore. I could witness and appreciate how vibrant the Lion City was. It was definitely a good-spirited place where one could start something. And I was lucky: the French Embassy was running a double degree programme involving NUS and the best French Grandes Ecoles. My friend Bernard proposed that I join and teach in the NUS French Double Degree Programme and Berge proposed that I join and do some research with him at the QuantumLah. Some light at the end of the tunnel.

FIGURE 3. Time flies (both younger & fitter): at the QuantumLah, the seed for CQT, 31st October 2003. Pouring and drinking champagne in mugs (!) during my first stint in Singapore. In the picture, besides Berge, one can recognize Antia Lamas-Linares, Christian Kurtsiefer, Ajay Gopinathan, Hui Khoon Ng and me. Out of the picture, except for his arm, is Darwin Gosal.

Lost in Translation. Settling things both in France and in Singapore took some time and I finally set foot with my family in Singapore in July 2005, freshly promoted to Research Director at CNRS. *Monsieur le Directeur de Recherches* as Berge would delightedly say in French...but with a heavy German accent. CNRS had seconded me to NUS. The future looked soothing. I

FIGURE 4. Under the spell of Berge's intense and penetrating gaze: making the decision to move to S'pore, a crossroad in my life but not yet aware of its full implications. It will turn out to be a happy and lucky life-changing decision!

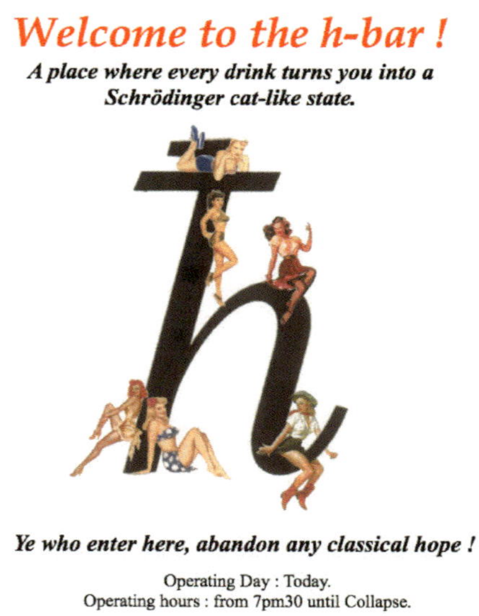

FIGURE 5. The h-bar flyer. The h-bar opened sporadically in Block F. Get lucky, prepare yourself to turn quantum!

spent eighteen wonderful months there, reuniting with long-time and faithful friends, enjoying very much my teaching, the exposure to a new world, food and culture, not to mention Singlish[25], meeting new dear friends

25. Can lah. Cannot also can.

like Lu Yin, Han Rui, Bess Fang[26], the beaming Kwek Leong Chuan, *Boss* Evon Tan, and Stéphane Bressan[27], visiting white-sand beaches, cities which had fired my imagination when I was in Europe, Bangkok, Phuket, Jakarta, Kuala Lumpur, Bali, Beijing, etc. From time to time, the h-bar was opening in block F[28]. *Tout paranormal, et pourtant*...A metamorphosis was under way.

Petit pays, je t'aime beaucoup. What I also remember from these eighteen months was how easy-going life in Singapore was (and still is)[29]. Singapore, the gigantic mall, the shopping paradise for my daughters[30]. It was during this first stay in Singapore that I did my work closest to quantum information. With Berge we showed that a certain loss of interference contrast observed in the CBS effect with cold atoms, and due to the Zeeman structure of the atoms, could be rephrased in terms of which-path information: the scattering of light storing information in the atomic scatterers, which acted as path-markers—a cute application of his 1996 duality paper. An embarrassing though hilarious anecdote of this period comes to my mind. In 2005-2006, I worked as a Panel Member of the University Research Council which evaluated and ranked the project proposals issued from the different Departments and Faculties. When the selection was being made, we were locked in a room that we could not leave until the session was over. This was a time when there were riots in France. During lunch, one of my Singaporean colleagues approached me, half-smiling, "Hey, Christian, let me ask you a question." "Yeah, sure, please shoot." "I saw the news about France on Channel5. Tell me, is it legal to burn cars in France?" "Well, no, of course not! Come on!" "Then, I don't understand. I saw the media coverage and there were plenty of policemen standing still and watching the people burn cars." "Well, hum, you know, things are complicated..." "So then, why have laws that you cannot enforce?" *Bang bang, he shot me down. Bang, bang, I hit the ground. Bang, bang*[31].

Random Access Memories. I also remember our (good!) work for the International Physics Olympiads held in Singapore in July 2006, as well as the 2006 FIFA World Cup Final between France and Italy on July 9 at the Olympiastadion, Berlin. Berge and I saw it together, late at night—2am—fueled with beers, at the Paulaner Brauhaus at Millennia Walk. The Australian customers around us were not particularly friendly to French people...That very same night my family was on board a Lufthansa plane, going back to France. I was crushed—Italy won on penalties after the match finished a 1-1 draw. Zidane was sent off in his last-ever match for head-butting Materazzi's chest in retaliation to verbal insults. *Zidane, il l'a frappé. Zidane, il l'a tapé. Coup de boule, coup de boule! La coupe on l'a ratée...*[32]

FIGURE 6. The Mentor: checking *Mademoiselle* Han's art work (or calculations?) during an Egg Painting session at Berge and Ola's apartment in Kent Vale, Easter 2008.

D. PICS, MoU, LIA, UMI and one hell of all these acronyms

Call Me. So, here we were, summer 2006 fading, my family back in France. I was staying at NUS until January 2007, and then ciao bye-bye S'pore, back to INLN where, with George Batrouni, we had created a new theory-experiment group on quantum gases and strongly correlated systems. A lot on our plates. But, at the same time, I *did* want to stay longer in Singapore. The only option then was to come back for a few months per year; better than nothing. Then, in September, came *the* phone call in my NUS office that changed it all. "Allo, this is Bernard Hébral from Physics Institute at CNRS. Singapore is under the spotlight in France and I'd like to come and visit the place. Could you please organize an official visit for me?" A higher-up calling

26. A cello virtuoso.
27. *Le Mangeur d'Archipels.*
28. F, like French. What else?
29. You find supermarkets open at night! People are nice!
30. Remember, Berge? Tigers glow in the dark.
31. The recent incidents in Little India showed that the Singapore Police can also stand still, ✌.

32. http://www.youtube.com/watch?v=XTFUhdd0ztA

me in my office? It turned out that, according to the CNRS Human Resource listings, I was the only CNRS physicist in Singapore.

Something in the Air. And indeed yes, Singapore was the hype in French media: My wife Paula was telling me that newspapers, radio, TV, all were having papers or news focused on S'pore. First there had been the International Olympic Committee Session, held from 2 to 9 July 2005 at the Raffles City Convention Centre. This was the first session ever organized in Singapore, so it got a lot of media coverage all over the world. London was chosen as the hosting city for the 2012 Summer Olympics, beating Paris. Tony Blair and Jacques Chirac came here. Lobbying efforts and diplomatic ballet, behind-the-curtain deals, rumors about bribing the IOC members were crawling, hush-hush! In the end the British were the best and won it all. Later that same year, in November 2005, Laurent Lafforgue, French Fields medalist in 2002, was forced to resign from the French High Education Council. The story stirred some turmoil in France. The polemic was raised by a private letter he sent to the president of the Council where he was criticizing the French education system and pointing at its debacle. Noticeably, in his letter, it was mentioned that Singapore had the best existing mathematics textbooks in the world. In 2010, Lafforgue even contributed to a math textbook designed for primary school and dubbed it the Singapore Method. One year later, on 17 September 2006, the annual meeting of the World Bank and of the International Monetary Fund took place in Singapore. Another important media focus, in particular with people protesting against the role of these world financial institutions, with Amnesty International condemning Singapore policy on freedom of expression, etc. But Singapore was also in the news for the sovereign wealth funds topping at 9 billion dollars or so, the newly-created National Research Foundation, etc. In short, Singapore was everywhere...and got the scientific attention of Bernard Hébral. I did organize his visit, he did come and meet official people at NUS—Berge included. He witnessed the ambition of Singapore in science, got the picture of the re-organization of research that NRF funding was implying, the opening of Singapore to high-standard international research and to foreign talent, got the flavor of the CQT proposal under preparation, saw that I was standing in the right effervescent spot...and left telling me "If you want to develop a long-term collaboration with Singapore, you'll get full support from us." "Roger, Sir!"

Matrix Reloaded. During the few months before my departure, Berge and I talked to Artur, and we started to mull over a scientific proposal around quantum gases experiments that could be developed at CQT. We had in mind mimicking graphene with cold atoms, and studying exotic phases when atoms are interacting, etc. This was the FermiCold proposal. Graphene was not yet widely known at the time, I came across it accidentally. In December 2005, I called my friend Jean Bellissard in Atlanta where he had taken an academic position and we were exchanging some news. Talking about this and that, I asked him "Jean, suppose you're a young man, right now, what kind of PhD subject in physics would you pick up?" He instantly replied "Graphene!". I did not know about graphene and he told me about the physical system, the band structure, the massless Dirac fermions and all the stuff. Looked cool indeed! After we hung up, I did the bibliography search and started to read. I immediately realized how easy it would be to mimic graphene with an optical lattice loaded with fermionic atoms. However, at the time I was involved in the supervision of a joint PhD student with Cord on Anderson localization and I shelved the idea. I revived it with Berge for the FermiCold project[33]. Back in France, INLN was to be evaluated by CNRS. It turned out that the Physics Deputy Director was Christian Chardonnet...whom I knew[34] from my moon raker years in the Lab on the Moon. Jorge Tredicce, the INLN Director and a good friend, was supportive of the FermiCold project. He arranged for me to sit next to Christian during an evening dinner. Christian asked me to write a proposal and to submit it to him within a week, which I did.

Ready, Steady, Go! Then everything ignited. In May 2007, Berge and Artur were in Paris to meet the CNRS officials. The project was discussed and the collaboration green-lighted in principle. With Berge, we put together a team: Benoît Grémaud (theory) and David Wilkowski (experiment). We all wrote a thorough proposal, together with Miklos Santha (who was developing a Quantum Computing activity at CQT) and submitted it both to CQT and CNRS for scientific evaluation. The project was finally approved by both sides in Spring 2008 and Benoît and I, with our families, we set foot in S'pore in July 2008, where we have been ever since[35]. In September 2008, the inaugural Singapore Formula One Grand Prix (done at night) took place. Since 2008, the official level of the collaboration has increased step by step: PICS→MoU→LIA→UMI. With the UMI we have now reached the top level: it is a legal joint lab between CNRS, NTU and NUS. Its name is Merlion MajuLab[36], Berge is the Deputy Director, I am the Director. Hum, wait! The UMI has been green-lighted, yes, but it is not yet signed. Hopefully it will happen before this summer...[37]

33. In 2009, it led to a joint publication with Berge that was highlighted by APS.
34. A top-spinner. I never could defeat him at table tennis.
35. David joined at the end of January 2009.
36. *Majulah Singapura!*
37. Note added: Yes indeed. The UMI has been officially signed on Friday 30th of May 2014 but Berge was out of town.

8

FIGURE 7. A milestone in the CQT-CNRS FermiCold collaboration: LIA signing ceremony at CQT on 21st November 2011. To find Berge, look for the beard! One can also recognize my collaborators and friends Benoît Grémaud and David Wilkowski as well as Jorge Tredicce. UMI is now in the stove!

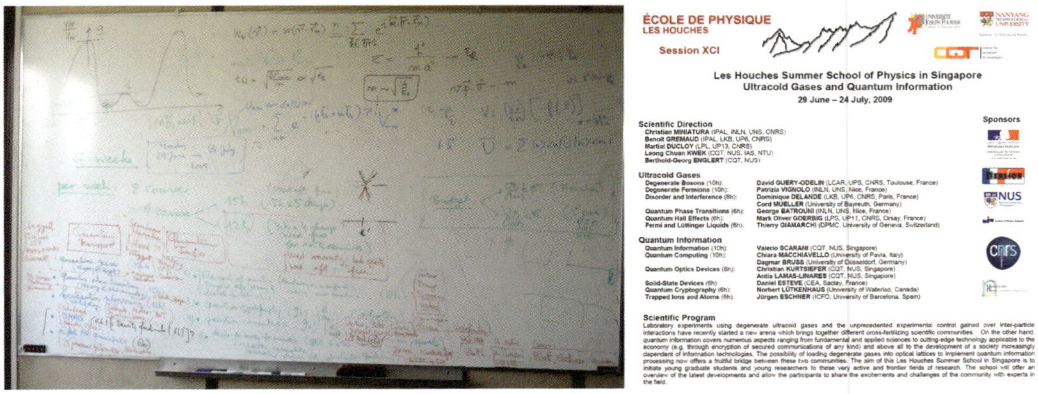

FIGURE 8. Whiteboard in my office at INLN, May-June 2007: Berge and I have filled up the details of our project to organize a Les Houches Summer School in Singapore. For real: a detail of the poster of the Les Houches in Singapore session organized from 29 June to 24 July 2009 at NTU—to date, the only Les Houches session having been organized abroad!

FIGURE 9. Left: Ransom was never paid, poor bear was never found...Right: Grumpy Old Man: Berge and I during a group BBQ event at Kent Vale, October 2009.

Jumpin' Jack Flash. I have of course plenty of other recollections. So many things have happened since July 2008, for example, Berge, Bess and I abducting a certain bear. I know that I'm not doing justice to all the other people who were instrumental in the scientific success of FermiCold, supporting officials, collaborators, students[38] and friends, but I think it is time for me to drop my pen. But before I do so, I would like to conclude by highlighting a project we organized and that I am very proud of. In April 2007, around a coffee at the Spinelli outlet at NUS University Hall, Berge, Kwek, Martial Ducloy[39] and I came up with the idea of a Les Houches summer school session that we would organize in Singapore. In May-June 2007, Berge stayed for one month at INLN, bringing along his students Bess, Han Rui, and Lu Yin. He also came with his family, settling in Antibes. It was in my office at INLN that Berge and I cooked up the program and budget of this Les Houches project. We were fortunate enough to convince the Les Houches committee to give us its green light and the session took place in summer 2009 at NTU. *Joli coup!* To date, it is the first and only session of the Les Houches School to have been organized abroad. A wonderful experience, beautiful memories!

E. The End?

Walking on the Moon. And now Berge, what's next my dear old chap? Game's *not* over. Why not the *outer* end of the World? Like Buzz Lightyear: To infinity...and beyond! ☺

ChM
Singapore, 11th of March 2014.

All these years here would not have been possible without the loving support of my wife Paula and my sweet tigers Agathe, Aurore and Apolline. Love!

38. Special thoughts to Lee Kean Loon, Wang Guangquan and Marta Wolak.
39. The director of the Lab on the Moon during my PhD and my early CNRS career. Whisky drinker, cigar smoker and friend. Always losing my bets with him.

Bess Fang

Institut d'Optique, Palaiseau, France

Personal message:

A physicist, a father, and a friend

I first met Berge in the common room of the former Quantum Information Technology Lab when I arrived as an intern in 2003. It wasn't obvious how I managed to see through that stern outlook of his, but I'm glad that I did, and found a great mind and a kind person behind. Through the times we worked together, he taught me many things, about physics, about life, about everything under the sun, and guided me in every way a caring father and loyal friend would. I run short of words to express the respect and affection I have for him, but perhaps the drawing here could speak for me.

Happy 60th, Berge!

Mark Hillery

Department of Physics and Astronomy, Hunter College of the City University of New York, New York, USA

Paper Title: **Berge and the Girl Named Florida**

Reference: Original article (2014)

Berge and the Girl Named Florida

March 6, 2014

It is a little known fact (probably because it isn't true, but let's ignore that detail for a moment) that during the period Berge spent in Los Angeles working with Julian Schwinger, he also worked as a private detective to make a bit of extra money. Here is a excerpt from his diary from that period.

It was a slow day; actually it had been a slow month. If people were stealing from each other, cheating each other, blackmailing each other, they weren't coming to see me about it. Maybe it was the heat. It was July, and five minutes outside brought on thoughts of heat stroke and ten made it a certainty. Today was cooler, and I actually had a client. A gentle breeze was coming in the window, but what came in through the door was something else altogether. She was six-feet tall with legs that didn't stop, a face a movie star would envy, and the rest of her said trouble. Her clothes were expensive and her manner was scared. She took a seat, started talking, and it all came out at once: a fat man, a thin man, a statue of a bird called the Corsican Penguin, someone named John Bell, cyclic groups, the Knights Templar, the Teutonic Knights, Peter Knight, ...

"Whoooaa honey, slow down," I said. "Slow down and let's take this from the beginning. First, why don't you tell me your name. Here, have something to calm you down." I poured us each two fingers of Scotch. She took it eagerly, and downed all of it in one gulp.

"I'm Florida," she said, "Florida Tallahassee."

Well, all that could turn into an interesting story, but it is probably time to start on the real one. I had gone to Innsbruck to visit Hans Briegel, and Berge was there too. He came to meet me at the train station, and we went to the hotel. After I got settled, we met for dinner, and Berge posed an interesting puzzle to me. It goes as follows. A family has two children, one of whom is a girl who has a rare name, say Florida. That wasn't the name Berge used, but it is the one that was used later in a popular science book, so we will go with it here. What is the probability that the other child is a girl? I, of course, said 1/3, which is the standard answer if you have no information about the girl's name. I mean, how can knowing that she has a rare name make any difference?

Berge said, "No, that's not right," and went on to explain over beer why not. Here is my version of Berge's explanation.

Let the probability of a girl being named Florida be ϵ. Now list the possibilities and their probabilities.

event	probability
2 boys	$p = 1/4$
one boy and a girl named Florida	$p = 2(1/4)\epsilon$
one boy and a girl not named Florida	$p = 2(1/4)(1-\epsilon)$
one girl named Florida and one girl not named Florida	$p = 2(1/4)\epsilon(1-\epsilon)$
two girls named Florida	$p = (1/4)\epsilon^2$
two girls not named Florida	$p = (1/4)(1-\epsilon)^2$

A quick check shows that the probabilities add up to one. Now we know that one of the children is a girl named Florida, and from the above list, that event has a probability of

$$\frac{\epsilon}{2} + \frac{1}{2}\epsilon(1-\epsilon) + \frac{1}{4}\epsilon^2 = \epsilon\left(1 - \frac{\epsilon}{4}\right).$$

The probability that both children are girls and at least one of them is named Florida is

$$\frac{1}{2}\epsilon(1-\epsilon) + \frac{1}{4}\epsilon^2 = \frac{\epsilon}{2}\left(1 - \frac{\epsilon}{2}\right).$$

Finally, the probability that given that one of the children is a girl named Florida the other child is a girl is

$$\frac{1}{2}\left(\frac{4-2\epsilon}{4-\epsilon}\right).$$

If ϵ is small, which it will be if the name is rare, this probability becomes $1/2$ instead of $1/3$. Note that as ϵ goes to 1 the above expression does go to $1/3$. Thus, what seems to be a completely extraneous bit of information makes a significant difference in the final probability. This is yet another illustration that probability is a very tricky subject.

One of the reasons Berge mentioned this cute problem to me was that he knew that a friend of mine, Leonard Mlodinow, is a science writer. Leonard and I have known each other since graduate school, we both had the same Ph.D. advisor, and he has gone on to an interesting and varied career. This includes a postdoc at Caltech, being a writer for the television series, *Star Trek, the Next Generation*, managing the development of educational software at both the Disney and Scholastic corporations, and now science writer. Berge thought that this example might prove useful in one of Leonard's books. Berge's timing was perfect, because Leonard was, at that time, writing a book on the role of chance in everyday events. It is called *The Drunkard's Walk*, and it was a New York Times bestseller. A month or two after that dinner in Innsbruck, I visited Leonard in Los Angeles, and passed along Berge's problem to him. Leonard loved it and incorporated it into the book. He was the one who named the girl Florida.

So, Berge, besides being an excellent physicist, is a collector of intellectual gems. Leonard and I are very happy that he shared this one with us, and I suspect a lot of readers of Leonard's book are too.

Happy 60th birthday, Berge, and remember that 60 is the new 40, whatever that is supposed to mean. All of us are looking forward to what interesting things, in physics and elsewhere, you will continue to show us in the coming years.

Marek Kuś

Center for Theoretical Physics, Polish Academy of Sciences, Warsaw, Poland

Personal message:

I have known Berge Englert for (as it looks now) many, many years. In fact we met towards the end of the previous century during one of the conferences organized by the Institute of Complex Systems in Dresden. We have never published a common paper; instead I owe to him two important things – switching my scientific interest to geometry of quantum mechanics and, what is even more important, the way of understanding quantum mechanics itself.

Fifteen years ago, still in the heroic times of quantum information theory, when two-qubit systems were the main objects of analysis, together with Karol Życzkowski, we published a paper about the differential geometry of such systems [1]. It was my first paper concerning quantum correlations and quantum information theory. At that time it was one of the first attempts to apply differential-geometric rather than linear-algebraic methods to describe quantum correlations and, as it could have been expected, it did not attract too much attention. Thus, it was to my astonishment, that during our next encounter, Berge praised the idea of such a geometric approach and encouraged me to proceed in this direction. In fact all my further (modest) contributions to the theory of entanglement have roots in this first paper. The paper itself was for the first time cited in Berge's work of Ref. [2]. A few months later, I published my second paper on entanglement, coauthored by Thomas Wellens, about the so-called best separable approximation for entangled mixed states. The problem had also a geometric flavor. We claimed that our solution of the problem was fairly general [3], but in fact the ultimate one was given by Berge and his coauthors only several years later [4]. As in the case of my first entanglement paper, this second one was a topic of our numerous discussions, again encouraging my further interest in the geometry of quantum systems.

Everybody who knows Berge knows his firm attitude towards fundamentals of quantum theory. Those who do not know should simply read his recent paper [5] (Editors' note: see also Hui Khoon Ng's contribution for this paper), probably the best short exposition I have ever read of what quantum mechanics really is and how it should (not) be (mis)interpreted. I had the privilege to learn it from the source itself, namely through numerous discussions with Berge.

Congratulating Berge on the occasion of his anniversary, I wish myself more and more such inspiration from encounters with him. All the best, dear Berge!

[1] M. Kuś and K. Życzkowski, *Geometry of entangled states*, Phys. Rev. A **63**, 032307 (2001).

[2] B.-G. Englert, N. Metwally, *Remarks on 2q–bit states*, Appl. Phys. B **72**, 35–42 (2001).

[3] Th. Wellens and M. Kuś, *Separable approximation for mixed states of composite quantum systems*, Phys. Rev. A **64**, 052302 (2001).

[4] G. C. Thiang, Ph. Raynal, and B.-G. Englert, *Optimal Lewenstein-Sanpera decomposition of two-qubit states using semidefinite programming*, Phys. Rev. A **80**, 052313 (2009).

[5] B.-G. Englert, *On Quantum Theory*, Eur. Phys. J. D **67**, 238 (2013).

Saverio Pascazio

Istituto Nazionale di Fisica Nucleare, Sezione di Bari, Bari, Italy
Dipartimento di Fisica and MECENAS, Università di Bari, Bari, Italy

Personal message:

I still remember when I first met Berge, many years ago. He gave a talk on the double-slit experiment, and discussed ideas that would forerun important concepts in the foundations of quantum mechanics, such as the quantum eraser, and which-path information.

But what struck me most during the talk was Berge's attitude towards physics and probably science in general: polite but firm; ready to listen, but also profoundly convinced of his own ideas. Above all, there was no room for compromise, in scientific matters. I believe that such an attitude has never changed, and in fact has characterised Berge's life and career.

I met Berge many more times, in many different places, and at many conferences. I listened to a number of his talks, read his papers as well as his recent books. They portray the same rigorous attitude towards physics. I sometimes wonder where such an attitude originates in, whether it is innate, or rather the result of interactions, collaborations and education. Sooner or later I will ask him.
Happy birthday, Berge!

Dario Poletti

Singapore University of Technology and Design, Singapore

Personal message:

Dear Berge,

It was an honor and a pleasure for me to be part of your group.

When I joined your group I knew that I would be impressed by your depth of knowledge, precise and detailed thinking, and your crystal-clear teaching skills. Working in your group and being assistant to one of your courses gave me a chance to experience that you are even better than I could possibly imagine. What I did not know about you, and I had the fortune to experience after I met you, is how generous you are.

I thank you for having accepted me in your group and for all you have taught me, not only in physics. I know I still have a lot to learn from you.

All the best in your future endeavours,
Dario